ANNALS OF THE NEW YORK ACADEMY OF SCIENCES

Volume 603

The New York Academy of Sciences
2 East 63rd Street
New York, New York 10021

BIOLOGICAL ACTIONS OF EXTRACELLULAR ATP

ANNALS OF THE NEW YORK ACADEMY OF SCIENCES
Volume 603

BIOLOGICAL ACTIONS OF EXTRACELLULAR ATP

Edited by George R. Dubyak and Jeffrey S. Fedan

The New York Academy of Sciences
New York, New York
1990

Library of Congress Cataloging-in-Publication Data

Biological actions of extracellular ATP/edited by George R. Dubyak and Jeffrey S. Fedan.
p. cm.—(Annals of the New York Academy of Sciences, ISSN 0077-8923; v. 603)
Based on a conference held in Philadelphia, Pa. on Nov. 27–29, 1990 by the New York Academy of Sciences.
Includes bibliographical references.
Includes index.
ISBN 0-89766-611-9 (cloth: alk. paper).—ISBN 0-89766-612-7 (paper: alk. paper)
1. Adenosine triphosphate—Physiological effect—Congresses. 2. Adenosine triphosphate—Receptors—Congresses. 3. Extracellular enzymes—Congresses. 4. Adenine—Derivatives—Physiological effect—Congresses. 5. Second messengers (Biochemistry)—Congresses. 6. Cellular signal transduction—Congresses. I. Dubyak, George R. II. Fedan, Jeffrey S. III. New York Academy of Sciences. IV. Series.
[DNLM: 1. Adenosine Triphosphate—physiology—congresses. 2. Receptors, Purinergic—drug effects—congresses. W1 AN626YL v. 603/QU 58 B61526 1990]
Q11.N5 vol. 603
[QP625.A3]
500 s—dc20
[574.19′25]
DNLM/DLC
for Library of Congress 90-13646
CIP

PCP
Printed in the United States of America
ISBN 0-89766-611-9 (cloth)
ISBN 0-89766-612-7 (paper)
ISSN 0077-8923

ANNALS OF THE NEW YORK ACADEMY OF SCIENCES

Volume 603
December 16, 1990

BIOLOGICAL ACTIONS OF EXTRACELLULAR ATP [a]

Editors and Conference Organizers
GEORGE R. DUBYAK and JEFFREY S. FEDAN

CONTENTS

[a] This volume is the result of a conference entitled Biological Actions of Extracellular ATP, which was held in Philadelphia, Pennsylvania on November 27-29, 1990 by the New York Academy of Sciences.

Part II. Adenine Nucleotide and Nucleoside Receptors: Chemistry and Classification

Part III. Adenine Nucleotide Receptors: Transmembrane Signaling Mechanisms

Part IV. Sources of Extracellular ATP

Part V. Metabolism and Utilization of Extracellular ATP by Ectoenzymes

Poster Papers

Financial assistance was received from:

- BURROUGHS WELLCOME CO.
- GLAXO RESEARCH LABORATORIES
- MERCK SHARP & DOHME RESEARCH LABORATORIES (MSDRL)
- MILES INC./BAYER AG
- NATIONAL SCIENCE FOUNDATION
- THE NELSON RESEARCH CENTER
- NORWICH EATON PHARMACEUTICALS, INC.

Preface

Surprise is the reaction often encountered when a biological researcher first learns that ATP, the "energy currency of the cell," triggers responses after the nucleotide is added to a cell's exterior. Upon entry into this research area, it is then, perhaps, difficult to become comfortable with other concepts that emerge, because it is typical of extracellular ATP that the unexpected becomes expected, and the unconventional becomes new dogma. For two decades Professor Geoffrey Burnstock, the acknowledged doyen of the field, has labored cleverly and intensely to understand various roles of extracellular ATP in the body, and to gain acceptance of these roles. It is because of his insights that we met as a scientific body to exchange information in this now-recognized, fast-growing adventure. This conference is a tribute to Professor Burnstock's perseverance, and to his resilience to the regular criticism of skeptics.

The conference was held because the subject of the biological actions of extracellular ATP had never before been treated as such. At many purine nucleoside/nucleotide meetings over the years, ATP was overshadowed by adenosine, primarily because adenosine research is a vast area in and of itself. This conference was held to bring together investigators who have been studying various aspects of extracellular ATP, and doing so in very diverse biological systems. The general plan of the program was to stimulate discussion and to disseminate information about ATP structure and chemistry, storage and release, metabolism, and transductive effects involving specific receptors in a wide range of cell types. The Table of Contents reflects a deliberate attempt at diversity. The conference was three days long and contained two poster sessions.

The assistance provided by the staff of The New York Academy of Sciences was remarkable, and, perhaps, unequaled in our experience. This allowed us to focus our attention on the pleasant aspects of planning, such as program content and inviting investigators to participate. For their great organizational skills, efficiency and encouragement, we are grateful to Ellen Marks, Renée Wilkerson, and Geri Busacco. The slides were never out of focus, and we thank Cyril Lichtensteiger for his keen vision. The expert guidance of our Conference Subcommittee, consisting of Drs. Anne Briscoe, Walter Scott, Fleur Strand, and Philip Siekevitz, was useful every step of the way. Without the financial support from the organizations listed on a previous page, the conference would have been critically compromised. On behalf of ourselves and all of the participants, we are especially grateful to the Conference Committee, and to The New York Academy of Sciences, for being willing to sponsor the conference.

Lastly, Samuel C. Silverstein, during the question period of the session he co-chaired, was inclined to reveal his ability to grapple with the age-old poetry-in-science chimera in penning:

Oh tell me Lord how it could be,
That though our cells make ATP,
It's not all used for energy,
But sometimes is secreted free.
It puzzles you, it puzzles me,
While Geoffrey Burnstock smiles with glee
At the many roles of ATP.

GEORGE R. DUBYAK
JEFFREY S. FEDAN

Overview

Purinergic Mechanisms

G. BURNSTOCK

Department of Anatomy and Developmental Biology
University College London
London WC1E 6BT, England

HISTORICAL BACKGROUND

The potent extracellular actions of purine nucleosides and nucleotides were first recognized by Drury and Szent-Györgyi.[1] Following this report, there was considerable activity within the field, with particular emphasis being placed on the actions of adenosine and ATP on the cardiovascular system and their resultant shock-inducing properties; the basic aspects of this work were summarized by Green and Stoner.[2] During the same period, the clinical effects of ATP administration in man were being widely explored, especially in geriatric patients with cardiovascular disorders. An extensive review of the medical literature was published by Boettge *et al.*,[3] who drew special attention to the physiological significance, pharmacological action, and therapeutic use of the adenyl compounds in man.

The potent vasodilatory actions of adenyl compounds led Holton and Holton[4] to suggest that ATP might be the vasodilatory substance that was released on antidromic stimulation of sensory nerves. Subsequently, it was shown that antidromic stimulation of the great auricular nerve, which results in vasodilatation of the rabbit ear vessels, was accompanied by ATP release.[5]

The sensitivity of the coronary vasculature to these same compounds prompted Berne[6] to propose that adenosine was the physiological mediator of the coronary vasodilatation associated with myocardial hypoxia. This hypothesis was based on the observation that inosine and hypoxanthine, the degradative products of adenosine, were found in the effluents of isolated perfused cat heart and in the coronary sinus blood of dog heart that had been subjected to severe hypoxia.[6] It was postulated that, during myocardial hypoxia, intracellular ATP was degraded to adenosine, which crossed the sarcolemma and induced relaxation of the vascular smooth muscle of the resistance vessels. Because ATP is usually more potent than adenosine in producing vasodilatation of coronary vessels, and because increased levels of ATP have now been found in the perfusates from hypoxic hearts, the possibility that ATP makes a more significant contribution than adenosine to the physiological regulation of coronary blood vessels was considered.[7]

A component that was neither adrenergic nor cholinergic was recognized in the autonomic nervous system in the early 1960s.[8,9] Nonadrenergic, noncholinergic nerves are strongly represented in the gastrointestinal tract of a wide range of vertebrate species and have also been identified in a variety of other organs, including lung, bladder, seminal vesicles, esophagus, uterus, eye, trachea, and parts of the cardiovascular system. Using the criteria summarized by Eccles[10] for the acceptance of putative neurotransmitters, Burnstock[11] proposed that the principal active substance released from at least some of these nerves was the purine nucleotide, ATP. Early evidence supporting the hypothesis included the observations that ATP was synthesized, stored, and released from nonadrenergic, noncholinergic nerves supplying the smooth muscle of the intestine and that it mimicked the effects of nerve stimulation on these muscles.[12] As a consequence, these nonadrenergic, noncholinergic nerves were tentatively termed "purinergic."[13]

Early experiments on the CNS led to the observation that adenosine and AMP were able to stimulate the accumulation of cAMP,[14] an effect that was theophylline sensitive.[15,16]

Early evidence for prejunctional purinergic receptors came from experiments carried out with the skeletal neuromuscular junction. In the isolated rat phrenic nerve-diaphragm preparation, adenosine and the adenine nucleotides reduced both the spontaneous and evoked release of acetylcholine (ACh) from motor nerve terminals.[17–19] Purine compounds were also shown to inhibit noradrenaline (NA) release from sympathetic nerves in a wide variety of tissues including rabbit kidney, canine adipose tissue, guinea pig vas deferens,[20] and rabbit central ear artery, saphenous vein, portal vein, and pulmonary artery.[21] Prejunctional modulation of ACh release from peripheral cholinergic nerves by the adenine nucleotides has been observed in the isolated guinea pig ileum and the myenteric plexus longitudinal muscle preparation.[22,23] Purine-mediated prejunctional modulation was also demonstrated at central synapses.[24] It was not until 1979, however, that prejunctional neuromodulation of transmitter release was shown to be mediated by receptors for adenosine, but not for ATP.[25]

The first indication that ATP might be released as a cotransmitter with NA was the demonstration that release of both [^{3}H]ATP and NA was blocked by guanethidine during stimulation of sympathetic nerves supplying the guinea pig taenia coli.[26] This concept was extended later to other tissues.[27,28]

A basis for distinguishing two main types of purinoceptors for adenosine and ATP/ADP was proposed in 1978.[29] The potent role of ATP as a vasodilator acting via receptors on vascular endothelial cells[30] is discussed by me elsewhere in this volume.[31]

SUBCLASSIFICATION OF PURINOCEPTORS

P_1- and P_2-Purinoceptors

In 1978 I proposed a basis for distinguishing two types of purinergic receptor, which was based largely on an analysis of the voluminous literature about the actions of purine nucleotides and nucleosides on a wide variety of tissues.[29] Since that time, many experiments have been carried out that support and extend this proposal.[32–34]

The original classification into P_1- and P_2-purinoceptors was based on four criteria: 1) the relative potencies of ATP, ADP, AMP, and adenosine; 2) the selective actions of antagonists, particularly methylxanthines; 3) the activation of adenylate cyclase by adenosine, but not by ATP; 4) the induction of prostaglandin synthesis by ATP, but not by adenosine. Thus the following classification was proposed: The P_1-purinoceptors are more responsive to adenosine and AMP than to ATP and ADP; methylxanthines such as theophylline and caffeine are selective competitive antagonists with respect to these receptors; and occupation of these receptors leads to inhibition or activation of an adenylate cyclase system with resultant changes in levels of intracellular cAMP. The P_2-purinoceptors are more responsive to ATP and ADP than to AMP and adenosine; these receptors are not antagonized by methylxanthines and do not act via an adenylate cyclase system; and their occupation may lead to prostaglandin synthesis.

Evaluation and expansion of purinoceptor classification has taken several directions, including studies of the stereoselectivity of P_1- and P_2-purinoceptors, the structural requirements for the actions of purines, and the chemistry of P_1- and P_2-purinoceptors; analysis of the influence of ectoenzymatic breakdown of nucleotides and the uptake of adenosine on measurements of relative agonist potencies; and development of more potent and selective P_1- and P_2-purinoceptor antagonists.[33–38]

Extracellular breakdown of ATP is rapid and involves a number of different enzymes.[39] This finding means that some of the actions of ATP and ADP might be mediated via P_1-purinoceptors following breakdown to AMP and adenosine.[40]

A_1, A_2, and A_3 Subclasses of the P_1-Purinoceptor

The P_1-purinoceptor was subdivided into A_1/R_i and A_2/R_a subtypes according to the relative potencies of a series of adenine analogues (and according to whether adenylate cyclase activity is increased or decreased in the presence of adenosine).[41,42] In general, A_1 receptors are preferentially activated by N^6-substituted adenosine analogues, whereas A_2 receptors are preferentially activated by 5′-substituted compounds. For A_1 receptors, adenine analogues may be listed in order of potency as follows: L-N^6-phenylisopropyladenosine (L-PIA) and N^6-cyclohexyladenosine (CHA) > 2-chloroadenosine (CADO) > 5′-*N*-ethyl-carboxamidoadenosine (NECA) and D-PIA (adenylate cyclase activity is decreased). For A_2 receptors: NECA > CADO > L-PIA and CHA (adenylate cyclase activity is increased).

There have been some problems with this subclassification, largely on the basis of inconsistent potency series in different tissues, particularly between central and peripheral tissues. The recent efforts to develop selective antagonists for A_1 and A_2 subclasses, however, are giving more credibility to this classification.[43–46]

An A_3 subclass of the P_1-purinoceptor has been claimed for an adenosine receptor, present in the heart and nerve endings, that is not coupled to adenylate cyclase.[47]

Another type of adenosine recognition site modulating the activity of adenylate cyclase, the intracellular P site, has also been described.[42,48] This P site is not susceptible to blockade by xanthines.

P_{2X}, P_{2Y}, and Other Subclasses of the P_2-Purinoceptor

The P_{2X}- and P_{2Y}-purinoceptor subclasses have been postulated on the basis of relative potencies of ATP analogues (and on the basis of selective antagonism).[49] For

P_{2X}-purinoceptors, ATP analogues may be listed in order of potency as follows: α,β-methylene ATP (α,β-meATP) and β,γ-meATP > ATP and 2-methylthio-ATP (2-Me-S-ATP) (arylazido aminoproprionyl ATP ($ANAPP_3$) is a selective antagonist, and prolonged exposure to α,β-meATP selectively desensitizes this receptor).[50] For P_{2Y}-purinoceptors, 2-Me-S-ATP > > ATP > α,β-meATP and β,γ-meATP (reactive blue 2, an anthraquinone sulfonic acid derivative, has been claimed to be a selective antagonist, at least over a limited concentration range).[51–54] Studies of the pharmacological actions of isopolar phosphonate analogues of ATP on guinea pig taenia coli and bladder have supported the P_{2X}/P_{2Y} subdivision of P_2-purinoceptors in smooth muscle and have also shown that L-adenosine 5′-(β,γ-methylene)triphosphonate and its analogues are selective agonists of the P_{2X}-purinoceptor,[55] and adenosine 5′-(2-fluorodiphosphate) has been shown to be a specific agonist for the P_{2Y}-purinoceptor, mediating relaxation of smooth muscle.[56] It has been claimed recently that the trypanoside, suramin, is a competitive antagonist of the P_2-purinoceptor,[38,57] although it does not appear to distinguish between P_{2X}- and P_{2Y}-purinoceptors.[58]

Because the receptors for ATP on platelets and mast cells (and lymphocytes) do not seem to fit this subclassification, they have been tentatively termed P_{2T}- and P_{2Z}-purinoceptors, respectively.[39]

TABLE 1 summarizes the purinoceptor classification currently in use.

DISTRIBUTION OF PURINOCEPTORS

Purinoceptors of various kinds have been identified on a wide variety of cell types. In general, adenosine is inhibitory in its actions, whereas ATP is either excitatory or inhibitory.

Nerves and Astrocytes

Adenosine, acting via prejunctional P_1-purinoceptors (usually of the A_1 subtype), is a potent modulator of transmitter release from terminal varicosities of peripheral adrenergic and cholinergic nerves.[59,60] The P_1-purinoceptors are particularly prominent in the brain, where their main role appears to be neuromodulatory.[34,61] The P_2-purinoceptors have been described on cell bodies of sensory neurons in nodose ganglion, spinal cord, and brain,[61–65] and also on intrinsic ganglionic neurons in heart and bladder.[66] There is recent evidence for P_2- as well as P_1-purinoceptors on astrocytes.[67,68]

Muscle

ATP has been proposed as a transmitter or cotransmitter in autonomic nerves supplying visceral and vascular organs.[28,69,70] Postjunctional receptors for ATP are implicit in the purinergic transmission mechanism: thus it is not surprising that P_2-purinoceptors are present in many smooth muscles. In some muscles, for example, those in the intestine and rabbit portal vein, ATP acting via P_{2Y}-purinoceptors is a

TABLE 1. Subtypes of Purinoceptors[a]

Purinoceptor	Subclass	Rank Order of Agonist Potency	Selective Antagonists[b]	Adenylate Cyclase Activity	Prostaglandin Synthesis
P_1	A_1 (R_i)	L-PIA, CHA > CADO > NECA, D-PIA > adenosine	CGS15943A, DPCPX	↓	—
	A_2 (R_a)	NECA > CADO > L-PIA, CHA > adenosine	PD116,948	↑	—
	A_3?	L-PIA, CHA, NECA > CADO		—	—
P_2	P_{2X}	α,β-meATP, β,γ-meATP > ATP = 2-Me-S-ATP	$ANAPP_3$, desensitization by α,β-meATP	—	↑
	P_{2Y}	2-Me-S-ATP > > ATP > α,β-meATP, β,γ-meATP	Reactive blue 2 (an anthraquinone sulfonic acid derivative)	—	↑
	P_{2Z}	ATP^{4-} > ATP		—	?
	P_{2T}	2-Me-S-ADP > ADP > α,β-meADP	ATP, AMP, adenosine (noncompetitively)	↓	?

[a] Adapted from reference 145.
[b] Nonselective antagonists for P_1-purinoceptors: caffeine, theophylline, 8-phenyltheophylline (8-PT), 1,3-dipropyl-8-(2-amino-4-chlorophenyl)xanthine (PACPX), 8-(*p*-sulfo)phenyltheophylline (8-SPT), 9-methyladenosine (9-MeA), and 1,3-dipropyl-8-cyclopentylxanthine (DPCPX).

potent relaxant, whereas in other muscles, for example, those in the urinary bladder and vas deferens (and most vascular smooth muscles), ATP acting via P_{2X}-purinoceptors has a potent contractile action.[49] The P_1-purinoceptors (usually of the A_2 subtype) are widespread in both vascular and visceral smooth muscle.

Both P_1- and P_2-purinoceptors have been identified in the vertebrate heart.[71,72] It is proposed that the P_1-purinoceptor present in heart is of the A_3 subtype.[47] When studying bullfrog atrial muscle cells, Friel and Bean[73] observed a biphasic electrical response to ATP. Such an observation is consistent with the evidence for release of ATP as a cotransmitter with adrenaline from sympathetic nerves supplying the frog heart.[74] From a study of the effects of ATP on the papillary and right ventricular muscles of the rat, it was suggested that P_2-purinoceptor activation induces both a positive inotropy and an increase in inositol-lipid metabolism.[75] The P_2-purinoceptors have been identified in the developing myotube.[76–78]

Endothelial and Epithelial Cells, Hepatocytes, and Pancreatic Secretory Cells

Potent actions of ATP on vascular endothelial cells leading to release of endothelium-derived relaxing factor (EDRF) and vasodilatation were first described in 1981 by De Mey and Vanhoutte, and have since been described in many vessels.[79–81] The endothelial ATP receptors have been shown to be of the P_{2Y} subclass.[82,83]

The P_2-purinoceptors have been shown to regulate ion transport in epithelial cells from a variety of different sources, including intestinal epithelial cells and kidney epithelium, where ATP stimulates Cl^- transport and alters Ca^{2+} distribution. ATP also regulates gastric acid secretion, and evidence has been presented for involvement of P_{2Y}-purinoceptors in ATP regulation of surfactant secretion from type II alveolar epithelial cells.[84,85]

It has been known for some time that ATP has glycogenolytic and hyperpolarizing actions on hepatocytes that are mediated by P_2-purinoceptors,[86] and it has been suggested that the receptor is of the P_{2Y}-purinoceptor subclass.[87] Adenosine, acting via an A_2 receptor, has also been claimed to stimulate hepatic glycogenolysis, but possibly by an indirect mechanism.[88]

Pancreatic B cells respond to ATP via P_{2Y}-purinoceptors to increase insulin secretion, whereas adenosine acts via the A_2 subtype of a P_1-purinoceptor in pancreatic A cells to increase glucagon secretion.[89]

Mast Cells and Cells of the Immune System

ATP induces Ca^{2+}-dependent histamine secretion from mast cells.[90] The agonist form is the tetrabasic acid ATP^{4-},[91,92] and this receptor has therefore been given the separate subclassification of P_{2Z}.[39]

The P_1-purinoceptors of the A_2 subtype have been described on various cells of the immune system, including lymphocytes and granulocytes.[93] ATP modifies cation fluxes and could thereby deliver the Ca^{2+} signal for lymphoctye activation.[94] Extracellular ATP has also been shown to stimulate transmembrane ion fluxes in macrophages, possibly via a P_{2Z}-purinoceptor.[95]

Platelets and Erythrocytes

ADP causes platelets to change shape rapidly, which leads to platelet aggregation, whereas P_1-purinoceptors mediate inhibition of ADP-induced platelet aggregation.[96] Because the platelet receptor is unique in being activated by ADP rather than ATP, it has been tentatively classified as a P_{2T}-purinoceptor.[39] The P_{2Y}-purinoceptors have been demonstrated in turkey erythrocytes.[97–99]

Fibroblasts and Other Cell Types

ATP receptors mediating membrane potential changes in fibroblasts have been described,[100] and the possibility has been raised that ATP released as a cotransmitter with noradrenaline from sympathetic nerves exerts some control of fibroblast function.[101] Purinoceptors have also been identified on spermatozoa, chemoreceptor cells in the carotid body, and neuroblastoma, adipose, thyroid, salivary acinar, and tumor cells.

PURINERGIC TRANSMITTERS

ATP as a Principal Neurotransmitter

Motor Nerves

Several pieces of evidence have been put forward since 1972 to strengthen and extend the knowledge of purinergic innervation of the smooth muscle of the intestine and bladder.[69] For example, release of ATP from nonadrenergic, noncholinergic nerves rather than from muscle has been clearly shown using the firefly ATP assay method.[102] Whereas there is a 2- to 6-fold increase in ATP release from the guinea pig taenia coli or bladder during isometric responses to purinergic nerve stimulation, there is no significant release of ATP during comparable responses elicited by direct muscle stimulation. Further, the nerve-mediated release of ATP is Ca^{2+} dependent; a 10-fold reduction in Ca^{2+} concentration resulted in an 80 to 90% reduction in both the mechanical response and the ATP release. Several studies have shown release of ATP from synaptosome or vesicle preparations of brain and gut.[103,104] Quinacrine, which binds to ATP, has been demonstrated with fluorescence histochemistry in nerve cell bodies and varicose nerve fibers in gut, bladder, and rabbit portal vein.[105,106]

The absence of a specific competitive antagonist to ATP receptors has clearly been an important gap in the supporting evidence for the purinergic hypothesis. Although apamin is not a specific competitive antagonist, low concentrations that act as a potent K^+ channel blocker also act so as to provide a parallel blocking of the responses of the taenia coli to nonadrenergic inhibitory nerves and to ATP. More recently, the photolyzable compound arylazido aminoproprionyl ATP ($ANAPP_3$) has been claimed to be a specific competitive blocker to ATP. It produces a competitive block of ATP and of noncholinergic excitatory responses of the bladder[107] and the vas deferens,[108] although it appears to be less effective in antagonizing the inhibitory responses of smooth muscle to ATP. Low concentrations of the stable analogue of ATP, α,β-methylene ATP, have been shown to desensitize the ATP receptor and to abolish noncholinergic excitatory nerve-mediated responses of the bladder.[50]

Evidence for purinergic innervation of the rabbit portal vein[109] and rat pulmonary mesenteric vessels[110] has also been presented.

Sensory Nerves

Recent experiments have done much to support the original proposal of Holton that ATP is released from primary afferent sensory fibers. Sensory neurons in the dorsal root ganglion that release ATP and stain with a nucleotide hydrolyzing acid phosphatase have been demonstrated, and a distinct population of neurons in lamina II of the dorsal horn grown in tissue culture has been shown to be highly sensitive to ATP.[62] In an *in vivo* study in the cat, Fyffe and Perl[63] provided evidence for a selective excitatory action of ATP on spinal neurons in the inner substantia gelatinosa or in the nucleus proprius of the dorsal horn receiving central terminals from a class of cutaneous mechanoreceptors with very fine afferent fibers. Finally, exogenously applied ATP produced a rapid and transient inwardly rectifying sodium current in single sensory neurons, enzymatically isolated from nodose, vestibular, trigemonal, and spinal ganglia of rats.[111]

ATP as a Cotransmitter

The possibility that some nerve fibers store and release more than one transmitter was raised in the mid-1970s,[28] and there is now considerable evidence in favor of the coexistence of established transmitters with various peptides and with purine nucleotides in nerve terminals in both central and autonomic systems. Detailed reviews of the evidence for coexistence and release of ATP with noradrenaline and acetylcholine are available.[59]

Noradrenaline and ATP

It has been known for a number of years that ATP is stored and released together with catecholamine from adrenal chromaffin cells. Coexistence and release of noradrenaline (NA) and ATP have also been claimed for sympathetic nerves supplying the aorta, mesenteric, ear, basilar, and pulmonary arteries.[31,70] ATP as well as NA release from guinea pig portal vein have been shown to be abolished following sympathectomy; fluorescence in nerves of the rat portal vein following incubation in quinacrine, which binds to ATP, is also abolished by sympathectomy.[112] In the rat tail artery the slow (but not the fast) depolarizations in response to sympathetic nerve stimulation are blocked by phentolamine; on the other hand, the fast depolarizations have recently been shown to be abolished by P_2-purinoceptor desensitization with α,β-methylene ATP.[113]

Evidence has been presented that ATP is stored and released as a cotransmitter together with NA from sympathetic nerves supplying the guinea pig vas deferens.[114,115] Westfall and his colleagues[108,114] showed that the initial phasic component of the excitatory response to sympathetic nerve stimulation is selectively antagonized by $ANAPP_3$, whereas the secondary, more tonic component of the response is antagonized by prazosin or reserpine. The Ca^{2+} channel blocker, nifedipine, has also been shown to block the initial (but not the secondary) responses of this preparation to nerve stimulation and contractions in response to ATP, but not to those in response to NA. Meldrum and Burnstock[116] provided further support for cotransmission by blocking the initial phase of the response of the vas deferens to sympathetic nerve stimulation following P_2-purinoceptor desensitization with α,β-methylene ATP. The phentolamine-resistant excitatory junction potential recorded in the vas deferens is also abolished by α,β-methylene ATP.[115] ATP and adenosine have been shown to inhibit NA release from adrenergic nerves supplying the vas deferens. The prejunctional receptor that mediates these actions is believed to be the P_1-purinoceptor because the inhibitory actions of ATP as well as adenosine are blocked by methylxanthines and because the slowly degradable methylene analogues of ATP are ineffective. Holck and Marks[117] showed that purine nucleotides or nucleosides also act as postjunctional neuromodulators in the vas deferens enhancing the contractile actions of NA, whereas NA can potentiate the responses of the vas deferens and seminal vesicle to ATP.

Thus NA and ATP are coexisting substances that act as synergistic neurotransmitters via postjunctional receptors, as well as substances that exert modulatory effects on each other via both pre- and postjunctional mechanisms. Although it has been assumed that ATP and NA are stored in the same vesicles,[118] recent evidence suggests that this may not be the case. Differential release of NA and ATP occurs during the prejunctional actions of a number of different agents, including prostaglandin E_2,[119] angiotensin III,[120,121] and calcitonin gene-related peptide.[122] Three possible explanations for these findings are illustrated in FIGURE 1.[123]

It is well known that neuropeptide Y coexists with ATP and NA in most sympathetic nerves and that it can act both as a prejunctional inhibitor of transmitter release and as a postjunctional potentiator of the actions of NA.[124] In recent studies with Jim Ellis in my laboratory, we have shown that neuropeptide Y has similar actions on ATP release and activity. It is interesting that at wide neuromuscular junctions (up to 2 μm), such as those in large elastic arteries, the dominant action of neuropeptide Y is to potentiate the postjunctional actions of ATP and NA (perhaps because they break down rapidly), whereas at close autonomic neuromuscular junctions (20-50 nm), such as those in vas deferens, the dominant action of neuropeptide Y is as a prejunctional modulator of transmitter release. In most blood vessels, neuropeptide Y has an initial transmitter potentiating action at low concentrations, but as the levels of neuropeptide Y build up in the cleft, it has prejunctional inhibitory actions (FIG. 2).

Acetylcholine and ATP

There are reports of coexistence and release of ATP with acetylcholine (ACh) at both somatic and autonomic motor nerve endings.[59,125] Some evidence suggests that

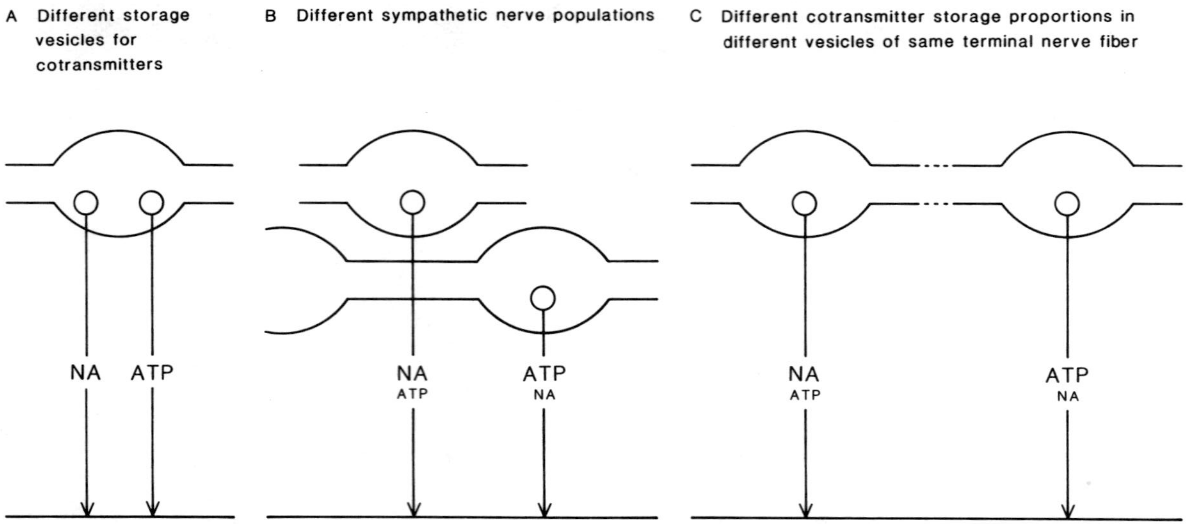

FIGURE 1. A diagram illustrating three possible explanations (**A**, **B**, and **C**) for differential prejunctional modulation of NA and ATP release from sympathetic nerve varicosities.

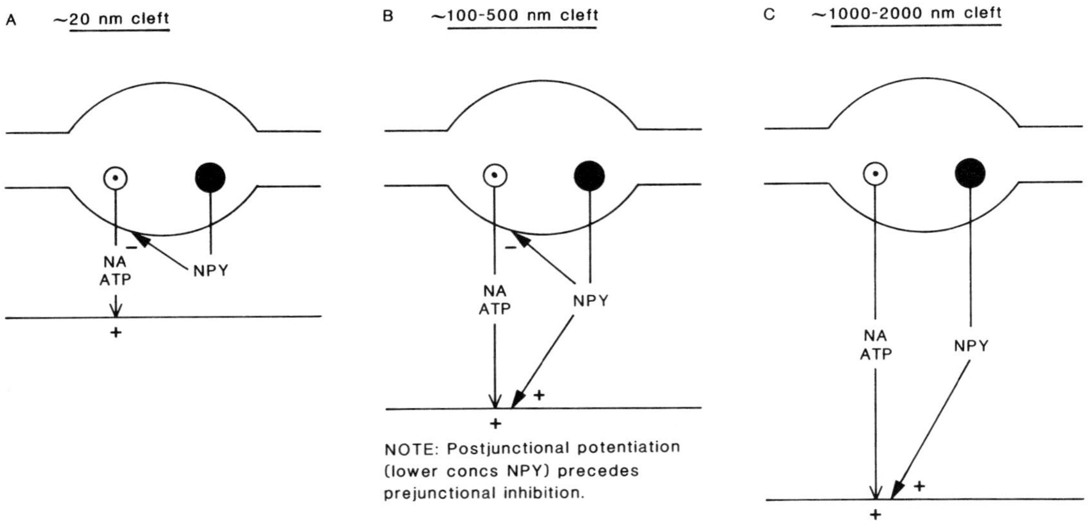

FIGURE 2. A diagram illustrating that pre- and/or postjunctional modulation of sympathetic cotransmission by neuropeptide Y (NPY) depends on the junctional cleft width. (**A**) A close (~20 mm) anterior neuromuscular cleft as seen, for example, in the vas deferens, where prejunctional modulation by NPY is dominant. (**B**) A medium size (~100-500 nm) junctional cleft characteristic of many blood vessels, where postjunctional modulation occurs with low concentrations of NPY and later with prejunctional modulation as the concentration of NPY in the cleft increases during transmission. (**C**) A wide (~1000-2000 nm) cleft typical of large elastic arteries, where postjunctional modulation by NPY is dominant.

ATP acts (via adenosine) as a neuromodulator of both the release and action of ACh; however, there is little information available to date to support a cotransmitter role for ATP at these junctions, apart from patch-clamp studies showing that micromolar concentrations of ATP activate certain channels in the membrane of cultured myoblasts and myotubes, as well as ACh.[76–78]

Cholinergic vesicles isolated from the electric organ of various elasmobranch fish contain ATP in addition to the principal transmitter, ACh. The ACh:ATP molar ratio in the three species studied is 4-10:1. The major nucleotide in these vesicles is ATP (83% of the total), with ADP (15%) and traces of AMP also being present. Studies on the turnover of adenine nucleotides in cholinergic synaptic vesicles have shown that ATP and ACh are depleted to the same extent (about 50%) during nerve stimulation, that adenosine is an effective precursor of vesicular adenine nucleotides, and that the new population of vesicles that appears following nerve stimulation has a high turnover rate for both ATP and ACh. Furthermore, in studies of a saturable uptake system for adenosine into nerve terminals isolated from the *Torpedo* electric organ, a K_m value of 1 μM has been reported, which is comparable to that of the high-affinity choline uptake system.[126] Evidence for axonal flow of ATP, as for ACh, in organelles other than mitochondria has been reported. ATP has also been shown to be released from the endings of phrenic nerves in the rat diaphragm during stimulation. The amounts of ATP released compare well with the amounts of ATP released on stimulation of some regions of the cortex or cortical synaptosomes. Release of [^{3}H]adenine derivatives has been shown to occur in the cholinergic septal system, where they were considered as possible cotransmitters with ACh.

ATP and adenosine have been shown to act on presynaptic purinergic receptors leading to modulation of the release of ACh from cholinergic motor nerves in other preparations, including brain and intestine.[59] These responses are blocked by methylxanthines, indicating that they are mediated by P_1-purinoceptors. An exception, where inhibition of ACh release from preganglionic frog neurons was shown to be mediated by ATP rather than adenosine, was reported recently by Silinsky and Ginsborg.[127]

CURRENT AND FUTURE DEVELOPMENTS

Transduction Mechanisms

cAMP and P_1-Purinoceptors

Cyclic AMP has been claimed to be the primary second messenger associated with A_1 and A_2 (but not A_3) purinoceptor subclasses.[47,128,129]

The original classification of adenosine receptors into A_1/R_i and A_2/R_a subtypes was based largely on the ability of adenosine and its analogues to stimulate or inhibit the production of cAMP. Few actions of adenosine, however, have been shown unequivocally to be mediated via changes in the level of cAMP. Although in many cases adenosine receptor agonists have been shown to alter the levels of cAMP, the involvement of such changes in the production of the final response is unclear. This underlines the view that a receptor is best conceived as being constructed of two units: a recognition component and a catalytic component. It is entirely possible that the same recognition component (such as A_1 or A_2) could be linked to a variety of catalytic components (such as stimulatory or inhibitory regulatory units of adenylate cyclase or Ca^{2+} channels) in the same or different cell types. Hence, it is preferable not to classify adenosine receptors according to their effect on adenylate cyclase, at least until the linkage between receptor occupation and cAMP levels is understood more thoroughly.

Ion Channels and P_2-Purinoceptors

The inhibitory actions of ATP acting on P_{2Y}-purinoceptors that lead to hyperpolarization of smooth muscle cells of the intestine appear to be associated with the selective opening of K^+ channels.[130]

The excitatory actions of ATP with respect to P_{2X}-purinoceptors on vascular and visceral smooth muscle cells appear to be associated with the opening of nonselective cation channels, resulting in depolarization and subsequent opening of voltage-dependent Ca^{2+} channels.[131,132] In addition, in some arterial smooth muscles, it has been claimed that increased calcium influx is also the result of direct activation of ATP-gated cation channels without any requirement for depolarization.[133,134]

In patch-clamp studies of developing chick skeletal muscle, external ATP has also been shown to activate cation-selective channels.[76] The effects of ATP in neuronal cells are complex, but one direct effect is a rapid depolarization caused by increased cation conductance.[65]

Phosphoinositol Transduction Mechanisms and P_2-Purinoceptors

Extracellular ATP at low (micromolar) concentration stimulates inositol 1,4,5-trisphosphate production and intracellular Ca^{2+} mobilization in hepatocytes,[87,135] adrenal medullary cells and other vascular endothelial cells,[136] aortic and ventricular myocytes,[137] erythrocytes,[97] Ehrlich ascites tumor cells,[138] and chick myotubes.[78] The P_{2Y}-purinoceptors coupled to phospholipase C activation and intracellular Ca^{2+} mobilization have also been demonstrated in primary cultures of sheep anterior pituitary cells[139] and turkey erythrocyte membranes.[98,99,140]

Molecular Biology of Receptors

The general direction that most receptor studies are taking at present is to clone the receptor following strong ligand binding and then to inject the appropriate mRNA into the *Xenopus* oocyte to express the receptor.[141] Lotan and colleagues, who have demonstrated a hyperpolarizing response in the oocyte to adenosine, have concluded that the adenosine-evoked outward current was carried by K^+. Responses to ATP, in contrast, appear to be mediated by Cl^- currents.[142]

Adenosine 5′-*0*-2-thio[^{35}S]diphosphate has been proposed as a radioligand for the P_{2Y}-purinoceptor in purified turkey erythrocyte membranes.[99,140] My own laboratory has recently identified [^{3}H]α,β-methylene ATP as a strongly binding ligand for the

P_2-purinoceptor,[143] and we are currently collaborating with molecular biologists to clone this receptor. We hope to use the *Xenopus* oocyte to examine the expression of its nucleic acid.

Cloning and expression of a cDNA coding for rat liver ecto-ATPase has recently been achieved.[144]

SUMMARY

An overview of the history of studies of the biological actions of extracellular ATP is presented. The basis of the subdivision of receptors for purines into P_1-purinoceptors for adenosine and P_2-purinoceptors for ATP and ADP are considered, as well as the recent proposal for subdivision of the ATP receptors into P_{2X}-, P_{2Y}-, P_{2Z}-, and P_{2T}-purinoceptor subtypes. These purinoceptor subtypes are discussed with respect to their transduction mechanisms, their distribution, and their physiological roles, including their roles in cotransmission and neuromodulation.

REFERENCES

1. Drury, A. N. & A. Szent-Györgyi. 1929. J. Physiol. **68:** 213-237.
2. Green, H. N. & H. B. Stoner. 1950. Biological Actions of the Adenine Nucleotides. Lewis. London.
3. Boettge, K., K. H. Jaeger & H. Mittenzwei. 1957. Arzneim.-Forsch. **7:** 24-59.
4. Holton, F. A. & P. Holton. 1954. J. Physiol. (London) **126:** 124-140.
5. Holton, P. 1959. J. Physiol. (London) **145:** 494-504.
6. Berne, R. M. 1963. Am. J. Physiol. **204:** 317-322.
7. Paddle, B. M. & G. Burnstock. 1974. Blood Vessels **11:** 110-119.
8. Burnstock, G., G. Campbell, M. Bennett & M. E. Holman. 1964. Int. J. Neuropharmacol. **3:** 163-166.
9. Burnstock, G. 1969. Pharmacol. Rev. **21:** 247-324.
10. Eccles, J. C. 1964. The Physiology of Synapses. Springer-Verlag. Berlin.
11. Burnstock, G. 1972. Pharmacol. Rev. **24:** 509-581.
12. Burnstock, G., G. Campbell, D. G. Satchell & A. Smythe. 1970. Br. J. Pharmacol. **40:** 668-688.
13. Burnstock, G. 1971. Nature **229:** 282-283.
14. Shimizu, H., J. W. Daly & C. R. Creveling. 1969. J. Neurochem. **16:** 1609-1619.
15. Sattin, A. & T. W. Rall. 1970. Mol. Pharmacol. **6:** 13-23.
16. Pull, I. & H. McIlwain. 1972. Biochem. J. **130:** 975-981.
17. Ginsborg, B. L. & G. D. S. Hirst. 1972. J. Physiol. (London) **224:** 629-645.
18. Ginsborg, B. L., G. D. S. Hirst & J. V. Maizels. 1973. Br. J. Pharmacol. **47:** 637P.
19. Ribeiro, J. A. & J. Walker. 1973. Br. J. Pharmacol. **49:** 724-725.
20. Hedqvist, P. & B. B. Fredholm. 1976. Naunyn-Schmiedeberg's Arch. Pharmacol. **293:** 217-223.
21. Su, C. 1978. J. Pharmacol. Exp. Ther. **204:** 351-361.
22. Sawynok, J. & K. H. Jhamadas. 1976. J. Pharmacol. Exp. Ther. **197:** 379-390.
23. Moritoki, H., M. Morita & T. Kanbe. 1976. Eur. J. Pharmacol. **35:** 185-198.
24. Phillis, J. W. & J. P. Edstom. 1976. Life Sci. **19:** 1041-1053.
25. De Mey, J., G. Burnstock & P. M. Vanhoutte. 1979. Eur. J. Pharmacol. **55:** 401-405.
26. Su, C., J. A. Bevan & G. Burnstock. 1971. Science **173:** 337-339.
27. Langer, S. Z. & J. E. B. Pinto. 1976. J. Pharmacol. Exp. Ther. **196:** 697-713.

28. BURNSTOCK, G. 1976. Neuroscience **1:** 239-248.
29. BURNSTOCK G. 1978. *In* Cell Membrane Receptors for Drugs and Hormones: A Multidisciplinary Approach. L. Bolis & R. W. Straub, Eds.: 107-118. Raven Press. New York, NY.
30. DE MEY, J. G. & P. M. VANHOUTTE. 1981. J. Physiol. (London) **316:** 347-355.
31. BURNSTOCK, G. 1990. Dual control of local blood flow by purines. Ann. N. Y. Acad. Sci. This volume.
32. BURNSTOCK, G., ED. 1981. Purinergic Receptors. Receptors and Recognition, Series B, Vol. 12. Chapman & Hall. London.
33. BURNSTOCK, G. & N. BUCKLEY. 1985. *In* Methods Used in Adenosine Research (Methods in Pharmacology). D. M. Paton, Ed. Vol. 6: 193-212. Plenum. New York, NY.
34. WILLIAMS, M. 1987. *In* Psychopharmacology: The Third Generation of Progress. H. Y. Meltzer, Ed.: 289-301. Raven Press. New York, NY.
35. HOGABOOM, G. K., J. P. O'DONNELL, & J. S. FEDAN. 1980. Science **208:** 1273-1275.
36. SATCHELL, D. 1984. Trends Pharmacol. Sci. **5:** 340-344.
37. DALY, J. W., O. HONG, W. L. PADGETT, M. T. SHAMIM, K. A. JACOBSON & D. UKENA. 1988. Biochem. Pharmacol. **37:** 655-664.
38. DUNN, P. M. & A. G. H. BLAKELEY. 1988. Br. J. Pharmacol. **93:** 243-245.
39. GORDON, J. L. 1986. Biochem. J. **233:** 309-319.
40. MOODY, C. J., P. MEGHJI & G. BURNSTOCK. 1984. Eur. J. Pharmacol. **97:** 47-54.
41. VAN CALKER, D., M. MÜLLER & B. HAMPRECHT. 1979. J. Neurochem. **33:** 999-1005.
42. LONDOS, C., J. WOLFF & D. M. F. COOPER. 1983 *In* Regulatory Function of Adenosine. R. M. Berne, T. W. Rall & R. Rubio, Eds.: 17-32. Martinus Nijhoff. Boston.
43. SCHWABE, U., P. UKENA & M. J. LOHSE. 1985. Naunyn-Schmiedeberg's Arch. Pharmacol. **330:** 212-221.
44. UKENEA, D., J. W. DALY, K. L. KIRK & K. A. JACOBSON. 1986. Life Sci. **38:** 797-807.
45. UKENA, D., M. T. SHAMIM, W. PADGETT & J. W. DALY. 1986. Life Sci. **39:** 743-750.
46. TRIVEDI, B. K. & R. F. BRUNS. 1988. J. Med. Chem. **31:** 1011-1014.
47. RIBEIRO, J. A. & A. M. SEBASTIÃO. 1986. Prog. Neurobiol. **26:** 179-209.
48. HASLAM, R. J., M. M. L. DAVIDSON & J. V. DESJARDINS. 1978. Biochem. J. **176:** 83-95.
49. BURNSTOCK, G. & C. KENNEDY. 1985. Gen. Pharmacol. **16:** 433-440.
50. KASAKOV, L. & G. BURNSTOCK. 1983. Eur. J. Pharmacol. **86:** 291-294.
51. KERR, D. I. B. & A. KRANTIS. 1979. Proc. Aust. Physiol. Pharmacol. Soc. **10:** 156P.
52. MANZINI, S., C. H. V. HOYLE & G. BURNSTOCK. 1986. Eur. J. Pharmacol. **127:** 197-204.
53. BURNSTOCK, G. & J. J. I. WARLAND. 1987. Br. J. Pharmacol. **90:** 383-391.
54. HOUSTON, D. A., G. BURNSTOCK & P. M. VANHOUTTE. 1987. J. Pharmacol. Exp. Ther. **241:** 501-506.
55. CUSACK, N. J., S. M. O. HOURANI, G. D. LOIZOU & L. A. WELFORD. 1987. Br. J. Pharmacol. **90:** 791-795.
56. HOURANI, S. M. O., L. A. WELFORD, G. D. LOIZOU & N. J. CUSACK. 1988. Eur. J. Pharmacol. **147:** 131-136.
57. SCHLICKER, E., E. URBANEK & M. GÖTHERT. 1989. J. Auton. Pharmacol. **9:** 371-380.
58. HOYLE, C. H. V., G. E. KNIGHT & G. BURNSTOCK. 1990. Br. J. Pharmacol. **99:** 617-621.
59. BURNSTOCK, G. 1986. *In* Coexistance of Neuronal Messengers: A New Principle in Chemical Transmission. Progress in Brain Research. T. Hökfelt, K. Fuxe & B. Pernow, Eds. Vol. 68: 193-203. Elsevier. Amsterdam.
60. PATON, D. M. 1987. *In* Pharmacology: Proceedings of the 10th International Congress of Pharmacology. M. J. Rand & C. Raper, Eds.: 267-270. Excerpta Medica. Amsterdam.
61. PHILLIS, J. W. & P. H. WU. 1981. Prog. Neurobiol. **16:** 187-239.
62. JAHR, C. E. & T. M. JESSEL. 1983. Nature **304:** 730-733.
63. FYFFE, R. E. W. & E. R. PERL. 1984. Proc. Natl. Acad. Sci. USA **81:** 6890-6893.
64. SALTER, M. W. & J. L. HENRY. 1985. Neuroscience **15:** 815-825.
65. KRISHTAL, O. A., S. M. MARCHENKO & A. G. OBUKHOV. 1988. Neuroscience **27:** 995-1000.
66. BURNSTOCK, G., T. G. J. ALLEN, C. J. S. HASSALL & B. S. PITTAM. 1987. *In* Histochemistry and Cell Biology of Autonomic Neurons and Paraganglia. C. Heym, Ed.: 323-328. Exp. Brain Res. Ser. 16. Springer-Verlag.

67. Gebicke-Haerter, P. J., S. Wurster, A. Schobert & G. Hertting. 1988. Naunyn-Schmiedeberg's Arch. Pharmacol. **338:** 704-707.
68. Pearce, B., S. Murphy, J. Jeremy, C. Morrow & P. Dandona. 1989. J. Neurochem. **52:** 971-977.
69. Burnstock, G. 1986. Acta Physiol. Scand. **126:** 67-91.
70. Burnstock, G. 1988. TIPS **9:** 116-117.
71. Burnstock, G. 1980. *In* Cardiovascular Receptors: Molecular, Pharmacological and Therapeutic Aspects. P. I. Korner & J. A. Angus, Eds. Circulation Res. **46**(Suppl. 1): 175-182.
72. Kakei, M. & A. Noma. 1984. J. Physiol. (London) **352:** 265-284.
73. Friel, D. D. & B. P. Bean. 1988. J. Gen. Physiol. **91:** 1-27.
74. Hoyle, C. H. V. & G. Burnstock. 1986. Eur. J. Pharmacol. **124:** 285-289.
75. Legssyer, A., J. Poggioli, D. Renard & G. Vassort. 1988. J. Physiol. (London) **401:** 185-199.
76. Kolb, H.-A. & M. J. O. Wakelam. 1983. Nature **303:** 621-623.
77. Hume, R. I. & M. G. Honig. 1986. J. Neurosci. **6:** 681-690.
78. Häggblad, J. & E. Heilbronn. 1988. FEBS Lett. **235:** 133-136.
79. Burnstock, G. & C. Kennedy. 1986. Circ. Res. **58:** 319-330.
80. Needham, L., N. J. Cusack, J. D. Pearson & J. L. Gordon. 1987. Eur. J. Pharmacol. **134:** 199-209.
81. Sauve, R., L. Parent, C. Simoneau & G. Roy. 1988. Pfluegers Arch. **412:** 469-481.
82. Burnstock, G. 1988. *In* Vasodilatation: Vascular Smooth Muscle, Peptides, Autonomic Nerves and Endothelium. P. M. Vanhoutte, Ed.: 1-14. Raven Press. New York, NY.
83. Pearson, J. D. & T. D. Carter. 1990. Ann. N. Y. Acad. Sci. This volume.
84. Rice, W. R. & F. M. Singleton. 1987. Br. J. Pharmacol. **91:** 833-838.
85. Gilfillan, A. M. & S. A. Rooney. 1988. Biochim. Biophys. Acta **959:** 31-37.
86. Häussinger, D., T. Stehle, W. Gerok, T.-A. Tran-Thi & K. Decker. 1987. Eur. J. Biochem. **169:** 645-650.
87. Keppens, S. & H. De Wulf. 1986. Biochem. J. **240:** 367-371.
88. Buxton, D. B., S. M. Robertson & M. S. Olson. 1986. Biochem. J. **237:** 773-780.
89. Loubatieres-Mariani, M. M. & J. Chapal. 1988. Diabete Metab. **14:** 119-126.
90. Dahlquist, R. & B. Diamant. 1974. Acta Pharmacol. Toxicol. **34:** 368-384.
91. Cockcroft, S. & B. D. Gomperts. 1979. J. Physiol. (London) **296:** 229-243.
92. Tatham, P. E. R., N. J. Cusack & B. D. Gomperts. 1988. Eur. J. Pharmacol. **147:** 13-21.
93. Bonnafous, J. C., J. Dornand, J. Favero & J. C. Mani. 1982. J. Receptor Res. **2:** 347-366.
94. Cameron, D. J. 1984. J. Clin. Lab. Immunol. **15:** 215-218.
95. Steinberg, T. H. & S. C. Silverstein. 1987. J. Biol. Chem. **262:** 3118-3122.
96. Haslam, R. J. & N. J. Cusack. 1981. Receptors Recog. Ser. B **12:** 221-285.
97. Berrie, C. P., P. T. Hawkins, L. R. Stephens, T. K. Harden & C. P. Downes. 1989. Mol. Pharmacol. **35:** 526-532.
98. Boyer, J. L., C. P. Downes & T. K. Harden. 1989. J. Biol. Chem. **264:** 884-890.
99. Cooper, C. L., A. J. Morris & T. K. Harden. 1989. J. Biol. Chem. **264:** 6202-6206.
100. Okada, Y., T. Yada, T. Ohno-Shosaku, S. Oiki, S. Ueda & K. Machida. 1984. Exp. Cell Res. **152:** 552-557.
101. Soares-da-Silva, P. & I. Azevedo. 1985. Blood Vessels **22:** 278-285.
102. Burnstock, G., T. Cocks, L. Kasakov & H. Wong. 1978. Eur. J. Pharmacol. **49:** 145-149.
103. Hammond, J. R., W. F. MacDonald & T. N. White. 1988. Can. J. Physiol. Pharmacol. **66:** 369-375.
104. White, T. D. 1988. Pharmacol. Ther. **38:** 129-168.
105. Burnstock, G., R. Crowe & H. K. Wong. 1979. Br. J. Pharmacol. **65:** 377-388.
106. Crowe, R. & G. Burnstock. 1981. Cell Tissue Res. **221:** 93-107.
107. Theobald, R. J., Jr. 1982. J. Auton. Pharmacol. **3:** 175-179.
108. Fedan, J. S., G. K. Hogaboom, J. P. O'Donnell, J. Colby & D. P. Westfall. 1981. Eur. J. Pharmacol. **69:** 41-53.

109. REILLY, W. M., V. L. SAVILLE & G. BURNSTOCK. 1987. Eur. J. Pharmacol. **140:** 47-53.
110. INOUE, T. & M. S. KANNAN. 1988. Am. J. Physiol. **254:** H1142-H1148.
111. KRISHTAL, O. A., S. M. MARCHENKO & V. I. PIDOPLICHKO. 1983. Neurosci. Lett. **35:** 41-45.
112. BURNSTOCK, G., R. CROWE, C. KENNEDY & J. TÖRÖK. 1984. Br. J. Pharmacol. **82:** 359-368.
113. SNEDDON, P. & G. BURNSTOCK. 1984. Eur. J. Pharmacol. **106:** 149-152.
114. SNEDDON, P. & D. P. WESTFALL. 1984. J. Physiol. (London) **347:** 561-580.
115. SNEDDON, P. & G. BURNSTOCK. 1984. Eur. J. Pharmacol. **100:** 85-90.
116. MELDRUM, L. A. & G. BURNSTOCK. 1983. Eur. J. Pharmacol. **92:** 161-163.
117. HOLCK, M. I. & B. H. MARKS. 1978. J. Pharmacol. Exp. Ther. **205:** 104-117.
118. LAGERCRANTZ, H. & G. FRIED. 1982. *In* Neurotransmitter Vesicles. R. L. Klein, H. Lagercrantz & H. Zimmermann, Eds.: 174-188. Academic Press. London.
119. TRACHTE, G. T. 1985. Prostaglandins **29:** 47-59.
120. TRACHTE, G. T. 1988. Eur. J. Pharmacol. **146:** 261-269.
121. ELLIS, J. L. & G. BURNSTOCK. 1989. Br. J. Pharmacol. **97:** 1157-1164.
122. ELLIS, J. L. & G. BURNSTOCK. 1989. Br. J. Pharmacol. **98:** 707-713.
123. KATSURAGI, T., T. TOKUNAGA, K. MIYAMOTO, L. KURATOMI & T. FURUKAWA. 1988. J. Pharmacol. Exp. Ther. **247:** 302-308.
124. POTTER, E. K. 1988. Pharmacol. Ther. **37:** 251-273.
125. RICHARDSON, P. J. & S. J. BROWN. 1987. J. Neurochem. **48:** 622-630.
126. ZIMMERMANN, H. 1982. *In* Co-transmission. A. C. Cuello, Ed.: 243-259. Macmillan. London.
127. SILINSKY, E. M. & B. L. GINSBORG. 1983. Nature. **305:** 327-328.
128. DALY, J. W. 1985. Adv. Cyclic Nucleotide and Protein Phosphorylation Res. **19:** 29-46.
129. BRUNS, R. F. 1990. Ann. N. Y. Acad. Sci. This volume.
130. HOYLE, C. H. V. & G. BURNSTOCK. 1989. *In* Handbook of Physiology. Section 6: The Gastrointestinal System, Motility and Circulation. J. D. Wood, Ed. Vol. 1: 435-464. American Physiological Society. Bethesda, MD.
131. NAKAZAWA, K. & N. MATSUKI. 1987. Pfluegers Arch. **409:** 644-646.
132. FRIEL, D. D. 1988. J. Physiol. (London) **401:** 361-380.
133. BENHAM, C. D. & R. W. TSIEN. 1987. Nature **328:** 275-278.
134. BENHAM, C. D. 1990. Ann N. Y. Acad. Sci. This volume.
135. CHAREST, R., V. PRPIĆ, J. H. EXTON & P. F. BLACKMORE. 1985. Biochem. J. **227:** 79-90.
136. BOEYNAEMS, J. M., S. PIROTTON, A. VAN COEVORDEN, E. RASPE, D. DEMOLLE & C. ERNEUX. 1988. J. Receptor Res. **8:** 121-132.
137. DANZIGER, R. S., S. RAFFAELI, R. MORENO-SANCHEZ, M. SAKAI, M. C. CAPOGROSSI, H. A. SPURGEON & R. G. HANSFORD. 1988. Cell Calcium **9:** 193-199.
138. DUBYAK, G. R. 1986. Arch. Biochem. Biophys. **245:** 84-95.
139. VAN DER MERWE, P. A., I. K. WAKEFIELD, J. FINE, R. P. MILLAR & J. S. DAVIDSON. 1989. FEBS Lett. **243:** 333-336.
140. HARDEN, T. K., J. L. BOYER, H. A. BROWN, C. L. COOPER, R. A. JEFFS & M. W. MARTIN. 1990. Ann. N. Y. Acad. Sci. This volume.
141. DASCAL, N. 1987. CRC Crit. Rev. Biochem. **22:** 317-388.
142. LOTAN, I., N. DASCAL, S. COHEN & Y. LASS. 1986. Pfluegers Arch. **406:** 158-162.
143. BO, X. & G. BURNSTOCK. 1989. J. Auton. Nerv. Syst. **28:** 85-88.
144. LIN, S.-H. 1990. Ann. N. Y. Acad. Sci. This volume.
145. BURNSTOCK, G. 1989. *In* Adenosine Receptors in the Nervous System. J. A. Ribeiro, Ed.: 1-14. Taylor & Francis. London.

DISCUSSION OF THE PAPER

S. C. SILVERSTEIN (*Columbia University, New York, NY*): What is the role of ATP at the neuromuscular junction? Is it possible that ATP is secreted through the plasma membrane, and not only via vesicular exocytosis? Formally, sequestration in vesicles is similar to transport through the plasma membrane.

BURNSTOCK: ATP is known to have three main roles at the skeletal neuromuscular junction. In the developing myotube, ATP has been shown to have direct actions on the postjunctional membrane, opening channels comparable to those opened by acetylcholine; in the adult, ATP acts both as a postjunctional modulator potentiating the actions of acetylcholine and, after ectoenzymatic breakdown to adenosine, as a prejunctional modulator of acetylcholine release via P_1-purinoceptors.

It is usually assumed that ATP, in common with other neurotransmitters, is stored and released largely from vesicles, but nonvesicular release has not been excluded.

B. B. FREDHOLM (*Karolinska Institutet, Stockholm, Sweden*): P_1 receptors are commonly subdivided into A_1 and A_2 subtypes, that is, not into P_{1X} and P_{1Y} subtypes, and no longer into R_i or R_s subtypes, as originally proposed by Londos. Would it not be possible to subdivide P_2 receptors into subtypes labeled, say, N1, N2, and so on? Doing so would emphasize that the nucleotides are chemically very different, are physiologically very different, and are associated with very different signal transduction mechanisms.

BURNSTOCK: P_{2X} and P_{2Y} subclasses of the P_2-purinoceptors are now widely used in the literature and have been officially used by the recent survey of receptor subclassification carried out by the British Pharmacological Society. I think that it would be confusing to introduce a change in terminology at this time, even if there are some logical advantages to your suggestion.

Cardiac Effects of Adenosine and ATP[a]

AMIR PELLEG, CARL M. HURT, AND ERIC L. MICHELSON

Likoff Cardiovascular Institute
Department of Medicine
Hahnemann University
Philadelphia, Pennsylvania 19102

INTRODUCTION

Since the appearance of the classic work of Drury and Szent-Györgyi,[1] it has become clear that adenosine and related compounds play an important physiologic role in the cardiovascular system which is independent of the biologic role that these compounds play in cellular metabolism. Adenosine and ATP are released in the heart under physiologic and pathophysiologic conditions[2–4] that are characterized by an altered oxygen supply to oxygen demand ratio. Furthermore, once in the extracellular space, adenosine and ATP can bind to specific cell surface receptors that mediate their effects on the coronary blood vessels, the specialized pacemaker and conducting tissues of the heart, and the contracting myocardial cells.

Although ATP has been shown to exert pronounced effects in the mammalian heart, the mechanism of its action in general and the specific receptors mediating its actions in particular are not fully characterized. The cardiac effects of adenosine and ATP in the mammalian heart have been extensively discussed in recent reviews,[5–10] and, consequently, emphasis is given in the present text to some provocative aspects of these effects and the directions for future research.

TABLE 1 summarizes the known cardiac effects of adenosine and ATP. The metabolic effect of adenosine, characterized by a shift of myocardial substrate utilization from fatty acids to carbohydrates, has been shown in a limited number of studies. Similarly, only a few reports have dealt with the ability of adenosine to reduce cardiac vascular and myocardial damage associated with reperfusion. Further studies are required to determine the ability of endogenous adenosine to exert these effects, the physiologic importance of these effects, and their implications for developing novel pharmacologic, therapeutic, and diagnostic agents.

[a] The original work was supported by grants-in-aid from the American Heart Association, Southeastern Pennsylvania Chapter, and by a grant from the National Institutes of Health (HL 43006-01A1).

VASOACTIVITY

Both adenosine and ATP exert pronounced effects on the coronary arteries and thereby affect coronary blood flow. There has been a suggestion that adenosine, a potent vasodilator of coronary arteries,[1,11] may also be a major physiologic regulator of coronary blood flow. This suggestion, known as the Adenosine Hypothesis,[12,13] has been given substantial support by numerous studies; however, other data have raised several important questions that are still unanswered.[14,15]

In contrast, ATP can either vasodilate or vasoconstrict arterial blood vessels, depending on the experimental conditions.[16–18] Evidence has been found in a feline model for the involvement of adenosine in a vasodilatory action of ATP *in vivo.*[19]

TABLE 1. Cardiac Effects of Adenosine and ATP

	Adenosine	ATP	Reference
Coronary vasodilation	+	+	1, 11-19
Coronary vasoconstriction	−	+	16, 18
Negative chronotropic effect (cardiac pacemakers)	+	+	5, 10
Negative dromotropic effect (atrioventricular nodal conduction)	+	+	5, 10
Direct electrophysiologic effects			
Atrial myocytes	+	+	10
Ventricular myocytes	−	+	20
Negative inotropic effect	+	+	10, 21
Triggered synthesis and release of prostaglandins	+	+	22, 23
Antiadrenergic effect	+	?	6, 24
Metabolic effect	+	?	25-28
Reduced reperfusion damage	+	?	29, 30

The mechanism of the vasodilating action of adenosine is not fully known. Studies with canine and human coronary arteries have indicated that the A_2 (or R_a) adenosine receptor subtype is mechanistically involved in this action of the nucleoside.[31,32] Studies in isolated guinea pig heart, however, have shown that the endothelium-mediated coronary vasodilation by adenosine is independent of endothelial adenylate cyclase activation, as expected from involvement of the A_2 receptor subtype.[33]

With regard to the vascular action of ATP, the involvement of two P_2 receptor subtypes, P_{2X} and P_{2Y}, has been suggested. These subtypes are believed to mediate the vasoconstricting and vasodilating actions of ATP, respectively.[16,18] It has been hypothesized that the vasodilating action is the result of ATP's effect on endothelial cells while the vasoconstricting action is the result of direct effect of ATP on vascular smooth muscle cells.[34] In addition, a role of adenosine and ATP has been suggested in acetylcholine-induced coronary vasodilation.[35]

ELECTROPHYSIOLOGIC EFFECTS

Negative Chronotropic and Dromotropic Effects

Both adenosine and ATP exert pronounced electrophysiologic effects in the mammalian heart. These include a negative chronotropic effect on cardiac pacemakers and a negative dromotropic effect on atrioventricular nodal conduction, in particular.[5,10] Studies *in vitro* as well as *in vivo* have indicated that adenosine depresses sinus node,[36–39] junctional,[40,41] and ventricular pacemakers.[42,43] Similarly, ATP depresses pacemaker activity and slows the rate of ventricular escape rhythm.[43]

It is well established that the action of adenosine on supraventricular pacemakers is mediated by A_1 cell surface receptors.[44,45] Binding of adenosine to these receptors results in activation of potassium ion channels, causing increased K^+-outward current ($i_K{}^+$) and hyperpolarization of cell membranes.[46,47] In this regard, the action of adenosine is similar to that of acetylcholine. Indeed, it has been shown that activation of adenosine receptors or muscarinic cholinergic receptors affects the same K^+ channel population.[46–48] This could explain the interaction between adenosine and acetylcholine observed *in vitro*[39] and the interaction between adenosine and vagal input to the heart observed *in vivo*[49] with regard to their depressant effects on sinus node pacemaker activity.

Less is known about the mechanism of the negative chronotropic action of ATP. It has been shown that the nucleotide is rapidly degraded to ADP, AMP, and adenosine,[50] which can mediate the electrophysiologic action of ATP. Indeed, evidence for this mechanism has been found both *in vitro*[51] and *in vivo.*[37,52] In addition, it has been suggested that ATP's action can be mediated by P_1-purinoceptors before ATP breaks down to adenosine.[53,54] In the cat, dog, and human, however, ATP triggers a vagal reflex that plays a major role in its electrophysiologic actions.[55]

Both adenosine and ATP inhibit ventricular automaticity.[42,43] The mechanism of this action is not fully known (see below); however, at least with regard to ATP, it seems that this action is mediated by the degradation of ATP to adenosine and is independent of a triggered vagal reflex.[43] Previous studies *in vitro* have indicated that adenosine can suppress pacemaker activity of canine Purkinje fibers.[56,57]

In recent years, numerous studies have confirmed the early observation of Drury and Szent-Györgyi[1] regarding the action of adenosine and related compounds on the atrioventricular node. This action is characterized by slowing of the conduction in the atrioventricular node resulting in prolongation of the time interval between right atrial and His bundle electrograms, culminating in complete atrioventricular block.[5–10] It is now well established that the negative dromotropic action of adenosine is mediated by a cell surface receptor, that is, the P_1-purinoceptor (of the A_1 subtype).[58,59] Thus, it is conceivable that atrioventricular nodal block observed in the clinical setting under specific conditions, such as inferior wall myocardial infarction, is mediated, at least in part, by endogenous adenosine released from ischemic tissues, and, consequently, may respond in part to aminophylline administration.[60–62]

The action of ATP on the atrioventricular node is mediated in part by the degradation of ATP to adenosine.[50,51] In the dog, the action of ATP is mediated in part also by a triggered vagal reflex[37] as well as by a direct action independent of the degradation of ATP to adenosine.[63] The lack of a specific antagonist of ATP at P_2-purinoceptor sites prevents better insight into the mechanism of action of the nucleotide.

Differential Sensitivity of Cardiac Pacemakers to Adenosine

One interesting aspect of the negative chronotropic action of adenosine is the differential sensitivity of cardiac pacemakers to the nucleoside. Observations were first made *in vitro,*[64] and recent studies *in vivo* have confirmed these observations.[65] Moreover, it was shown that ventricular pacemakers are more sensitive to adenosine than the sinus node.[65] In addition, catecholamine challenge increased the effect of adenosine in the sinus node, but attenuated it in the ventricle; this action was reversed by propranolol.[65] Based on these and previous findings, we have hypothesized that the hyperpolarization-activated current (i_f) plays a major role in the differential sensitivity to adenosine of sinus node and ventricular pacemakers.[65] The hypothesis outlined in FIGURE 1 assumes the following: 1) Under basal conditions, i_f plays a minor role in sinus node pacemaker activity, but it plays a major role in ventricular pacemaker activity. 2) Isoproterenol enhances i_f in both sinus node and ventricular pacemakers. 3) Adenosine suppresses isoproterenol-dependent i_f in the sinus node. 4) Quinidine suppresses adenosine-induced $i_K{}^+$ but not i_f. 5) Adenosine suppresses i_f in ventricular pacemakers even without its enhancement by isoproterenol.

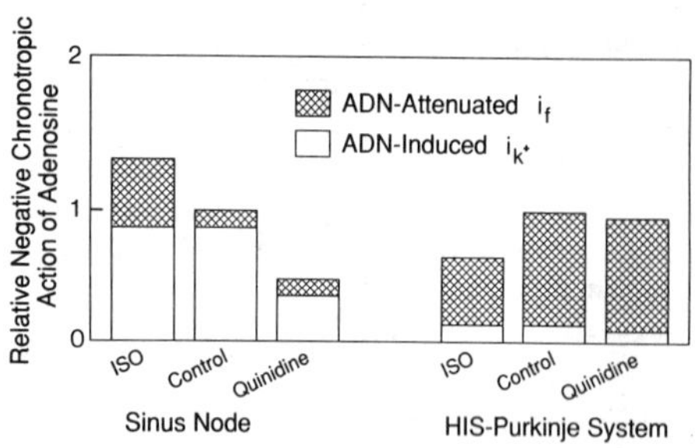

FIGURE 1. Schematic outline of the working hypothesis. The relative negative chronotropic action of adenosine (ADN) is represented by the columns. Each column contains clear and shaded areas representing the adenosine-induced potassium outward current ($i_k{}^+$) and adenosine-attenuated hyperpolarization activated inward current (i_f). It is assumed that i_f plays a minor role in sinus node pacemaker activity and a major role in ventricular pacemaker activity under baseline conditions (Control), and that adenosine can attenuate this current under these conditions as well as following its enhancement by isoproterenol (ISO). See text for further details.

Assuming that ventricular automaticity originates in His-Purkinje fibers, then assumptions 1 through 4 are supported by data previously obtained in several models *in vitro* and *in vivo.* Specifically, i_f is found in the sinus node and Purkinje fibers and its activation in the latter is a steep sigmoidal curve with a maximum at membrane potentials ranging from -90 to -110 mV and a minimum at membrane potentials between -50 and -60 mV.[66] Thus, it is expected that i_f plays a different role in the sinus node, where the maximal diastolic potential (MDP) is between -60 and -70 mV, from that played in Purkinje fibers, where MDP is near -90 mV. Indeed, data obtained in voltage-clamp experiments in sinus node cells and Purkinje fibers support this interpretation.[67–71]

The enhancement of i_f during catecholamine challenge of sinus node cells and Purkinje fibers has also been well documented.[67,68,71] Studies of sinus node cells, for example, have shown that adenosine attenuated i_f previously enhanced by isoproterenol.[67] Moreover, studies *in vitro* and *in vivo* have shown that the action of adenosine was accentuated during catecholamine challenge.[67,72,73] It seems, therefore, that the latter is due to the suppression of catecholamine-enhanced i_f, which is added to the direct action mediated by $i_K{}^+$.

Quinidine, a Class 1A antiarrhythmic agent, attenuated the electrophysiologic effects of adenosine *in vitro*,[74] probably by suppressing adenosine-induced $i_K{}^+$.[75] Therefore, it is expected that in the sinus node, where the action of adenosine is mediated mainly by $i_K{}^+$, quinidine should attenuate adenosine's action. In contrast, in His-Purkinje fibers, where adenosine's action is assumed to be mediated by its action on i_f, quinidine should have little or no effect on the negative chronotropic effect of adenosine. Indeed, data obtained in recent studies *in vivo* agree with these predictions.[65]

Studies of adenosine's effects on Purkinje fibers *in vitro* give support to assumption 5. For example, Bertolino and Di Francesco reported that adenosine (100 μM) shifted the i_f current activation curve to the left.[76] Furthermore, Rosen *et al.*[57] have reported that adenosine did not affect the spontaneous pacemaker activity of depolarized canine Purkinje fibers. Because the MDP of these fibers resembled that of intact sinus node cells, and because the adenosine effect in sinus node cells is mediated mainly by $i_K{}^+$ (see above), the lack of effect of adenosine under these conditions argues against a major role of adenosine-induced $i_K{}^+$ in Purkinje fibers. In addition, the inverse relationship between i_f and cell membrane potential makes it tempting to speculate that the association of depolarized cells with a lack of effect of adenosine is related to inactivation of i_f. This interpretation is supported by the effects of cesium chloride on ventricular escape rhythm and overdrive suppression in the isolated guinea pig heart with complete atrioventricular conduction block.[77] In this preparation, cesium, an effective blocker of i_f,[78–80] significantly prolonged ventricular cycle length and the pause following cessation of overdrive.[77] Furthermore, it has been recently shown that acetylcholine inhibits i_f in sinus node cells.[81,82] Thus, a direct action of adenosine on i_f could be another manifestation of the well-documented similarity between the electrophysiologic actions of adenosine and acetylcholine. On the basis of these considerations, it is tempting to speculate that the differential sensitivity of sinus node and ventricular pacemakers to adenosine is mainly due to the different pacemaker currents at these two sites and their different responses to adenosine.

INTERACTION WITH OTHER DRUGS

The interaction of adenosine with other drugs has been recently reviewed,[9] and there is no need to discuss it here. We would, however, like to emphasize one aspect of this interaction. Drugs that inhibit the transport of adenosine across cell membrane (for example, dipyridamole) and drugs that prevent the metabolism of adenosine (for example, adenosine deaminase inhibitors) tend to potentiate adenosine's actions because they increase the level of adenosine in the extracellular space.

In contrast, the presence of drugs that competitively inhibit adenosine's action at adenosine's receptor sites (for example, methylxanthines) will tend to attenuate adenosine's effects. Further studies are required to determine the clinical implications of these interactions.

ANTIADRENERGIC EFFECTS

The ability of adenosine to antagonize the positive inotropic effect of catecholamines was first demonstrated in the isolated perfused guinea pig heart by Schrader *et al.*[24]

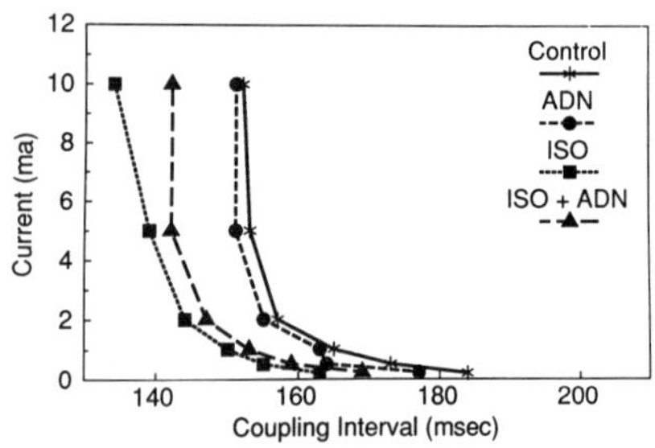

FIGURE 2. A typical example of the effect of adenosine on the strength-interval curve in the myocardium of an anesthetized open-chest dog with chronic myocardial infarction. The strength-interval curves were constructed from measurements made at an intramyocardial site within the area perfused by the left anterior descending coronary artery during infusion of either saline (Control) or adenosine (ADN, 5 μmol/min) into the artery in the absence and presence of isoproterenol (ISO, 0.02 μg/kg/min, i.v.). The strength-interval curve was shifted to the left during isoproterenol infusion. This effect was attenuated by localized administration of adenosine.

Since then, numerous studies *in vitro* have confirmed this observation and have widened the spectrum of effects attributable to this action of adenosine.[6,10] The antiadrenergic actions of adenosine are distinctively different from the direct actions that do not depend on prior stimulation by catecholamines.

The mechanism behind this phenomenon is not fully known; however, recent data have suggested the involvement of presynaptic action, characterized by reduced neurotransmitter release from nerve terminals,[83,84] and/or postsynaptic inhibition of the adenylate cyclase-cAMP system.[6,10] The question of the physiologic role of the antiadrenergic action is still open. Two studies *in vivo* failed to observe a significant antagonism between adenosine and catecholamine.[85,86] More recent work, however, has indicated that adenosine can modulate, at least with regard to some electrophysiologic parameters, the effects of catecholamine stimulation. For example, adenosine reversed the tachycardic effect of isoproterenol on ventricular escape rhythm in patients with complete atrioventricular block.[87] In addition, as can be seen in FIGURE 2, adenosine caused a shift to the right of the ventricular excitability-refractoriness strength-interval curve previously shifted to the left by intracoronary isoproterenol infusion in a canine model. These studies are by no means conclusive. Rather, they indicate that further studies are required before the physiologic importance of the antiadrenergic action of adenosine is fully known.

No data is available on antiadrenergic action of ATP in the mammalian heart. In view of the rapid degradation of the nucleotide to adenosine, however, a putative action mediated by ATP's degradation by ectoenzyme cannot be excluded.

TRIGGERED SYNTHESIS AND RELEASE OF PROSTAGLANDINS

ATP can trigger the synthesis and release of prostaglandins from the perfused heart and other organs[22] as well as from isolated blood vessels.[88] A similar effect was observed in cultured vascular endothelial cells.[89] Recent data indicate that the receptor that mediates this action of ATP is P_{2Y}.[89,90] Because prostaglandins can exert pronounced chronotropic, inotropic, and vasodilatory effects in the cardiovascular system,[91] it has been speculated that some of ATP's effects are mediated by prostaglandins. Indeed, preliminary studies have shown that ATP-induced sinus tachycardia in the isolated perfused rabbit heart is mediated by prostaglandin synthesis triggered by

ATP's action at P_2 receptor sites.[92] This could explain, at least in part, the potentiation of ATP-induced bradycardia in the presence of indomethacin.[93]

Less is known about adenosine and prostaglandins. In one study using isolated perfused rabbit heart, prostaglandins caused increased release of adenosine and, conversely, exogenous adenosine caused the release of prostaglandin-like substances.[23] More recently, it was shown that exogenous adenosine stimulates the synthesis of prostacyclin in isolated perfused rabbit heart.[94,95] The physiologic importance of these observations remains to be determined.

PHYSIOLOGIC ROLE OF ADENOSINE AND ATP

The overall effects of adenosine in the mammalian heart are schematically summarized in FIGURE 3. The release of adenosine from intracellular compartments is triggered by an altered oxygen supply to oxygen demand ratio, which may occur under either physiologic or pathophysiologic conditions. On the one hand, adenosine depresses cardiac pacemakers and atrioventricular nodal conduction resulting in the reduction of cardiac work and decreased oxygen demand. On the other hand, adenosine vasodilates coronary arteries, and thereby increases coronary blood flow and the supply of oxygen to the myocardium.

In addition, adenosine's indirect, antiadrenergic action has a similar effect: reduced oxygen demand and increased oxygen supply. Specifically, our preliminary data have indicated that adenosine can also antagonize the effect of catecholamines on intramyocardial pressure (IMP). Because the latter is now recognized as a major determinant of coronary blood flow, it is tempting to speculate that the reduction of IMP by adenosine, mediated by its antiadrenergic action, acts synergistically with the direct vasodilatory action to increase coronary blood flow, while simultaneously further decreasing myocardial oxygen demand.

The ability of adenosine to reduce vascular and myocardial reperfusion damage (associated with reflow following relatively short periods of complete coronary artery occlusion) fits well into this scheme, that is, a scheme in which adenosine may be viewed as an endogenous cardioprotective agent acting via several mechanisms to preserve the ischemic myocardium.

Less is known about the physiologic role of ATP. The specific actions of the nucleotide that are independent of its degradation to adenosine have only recently been defined. These include direct actions on vascular smooth muscle and myocardial

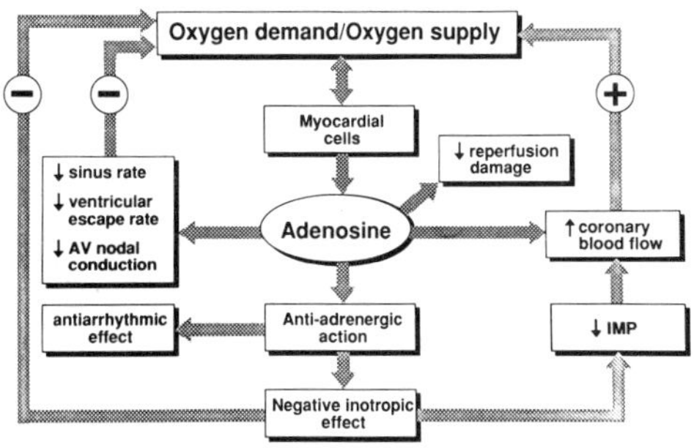

FIGURE 3. Schematic outline of the cardiac effects of adenosine. Adenosine, an endogenous metabolite, is released from myocardial cells under physiologic and pathophysiologic conditions characterized by reduced oxygen supply and/or increased oxygen demand. The effects of this nucleoside tend to reduce oxygen demand and increase oxygen supply. See text for details. AV: atrioventricular; IMP: intramyocardial pressure.

cells. The former have been suggested to be involved in the response of coronary arteries to localized endothelial cell damage and localized release of ATP associated with platelet activation. Further studies, however, are required before the physiologic role of extracellular ATP in the heart is fully understood.

ACKNOWLEDGMENT

We thank Ms. Karen Rose for her assistance in the preparation of this manuscript.

REFERENCES

1. DRURY, A. N. & A. SZENT-GYÖRGYI. 1929. The physiological activity of adenine compounds with especial reference to their action upon the mammalian heart. J. Physiol. (London) **68:** 213-237.
2. RUBIO, R. & R. M. BERNE. 1985. Release of adenosine by the normal myocardium in dogs and its relationship to the regulation of coronary resistance. Circ. Res. **25:** 460-464.
3. IMAI, S., A. L. RILEY & R. M. BERNE. 1964. Effect of ischemia on adenine nucleotides in cardiac and skeletal muscle. Circ. Res. **15:** 443-450.
4. FORRESTER, T. & C. A. WILLIAMS. 1977. Release of adenosine triphosphate from isolated adult heart cells in response to hypoxia. J. Physiol. **268:** 371-390.
5. BELHASSEN, B. & A. PELLEG. 1984. Electrophysiologic effects of adenosine triphosphate and adenosine in the mammalian heart: Clinical and experimental aspects. J. Am. Coll. Cardiol. **4:** 414-424.
6. PELLEG, A. 1987. Cardiac electrophysiology and pharmacology of adenosine and ATP: Modulation by the autonomic nervous system. J. Clin. Pharamacol. **27:** 366-372.
7. PELLEG, A. 1985. Cardiac cellular electrophysiologic effects of adenosine and adenosine triphosphate. Am. Heart J. **110:** 688-693.
8. PELLEG, A., E. L. MICHELSON & L. S. DREIFUS, EDS. 1987. Cardiac Electrophysiology and Pharmacology of Adenosine and ATP: Basic and Clinical Aspects. Alan R. Liss. New York, NY.
9. PELLEG, A. & R. S. PORTER. 1990. Pharmacology of adenosine. Pharmacotherapy: in press.
10. BELARDINELLI, L. 1989. The cardiac effects of adenosine. Prog. Cardiovasc. Dis. **32:** 73-97.
11. WEDD, A. M. 1931. The action of adenosine and certain related compounds on the coronary flow of the perfused heart of the rabbit. J. Pharmacol. Exp. Ther. **9:** 355-366.
12. BERNE, R. M. 1963. Cardiac nucleotides in hypoxia: Possible role in regulation of coronary blood flow. Am. J. Physiol. **204:** 317-322.
13. GERLACH, E., B. DEUTICKE & R. H. DREISBACH. 1963. Der nucleotid-abbau im herzmuskel bei sauerstoffmangel und seine mogliche bedeutung fur die coronardurchblutung. Naturwissenschaften **50:** 228-229.
14. BERNE, R. M., J. M. GIDDAY, H. E. HILL, R. R. CURNISH & R. RUBIO. 1987. Adenosine in the local regulation of blood flow: Some controversies. *In* Topics and Perspectives in Adenosine Research. Proceedings of the Third International Symposium on Adenosine (Munich, June 1986). E. Gerlach & B. F. Becker, Eds.: 395-405. Springer-Verlag. New York, NY.
15. SPARKS, H. V. & M. W. GORMAN. 1987. Adenosine in the local regulation of blood flow: Current controversies. *In* Topics and Perspectives in Adenosine Research. Proceedings of the Third International Symposium on Adenosine (Munich, June 1986). E. Gerlach & B. F. Becker, Eds.: 406-415. Springer-Verlag. New York, NY.
16. BURNSTOCK, G. & C. KENNEDY. 1985. A dual function for adenosine 5'-triphosphate in the regulation of vascular tone. Circ. Res. **58:** 319-330.
17. WHITE, T. D. & J. A. ANGUS. 1987. Relaxant effects of ATP and adenosine on canine large and small coronary arteries *in vitro.* Eur. J. Pharmacol. **143:** 119-126.

18. HOPWOOD, A. M. & G. BURNSTOCK. 1987. ATP mediates coronary vasoconstriction via P_{2X}-purinoceptors and coronary vasodilation via P_{2Y}-purinoceptors in the isolated perfused rat heart. Eur. J. Pharmacol. **136:** 49-54.
19. RIBEIRO, J. A. & M. S. LIMA. 1985. The hypotensive effect of intracarotid injections of adenosine triphosphate depends on its hydrolysis to adenosine. Pharmacol. Res. Commun. **17:** 255-260.
20. CHRISTIE, A. & S.-S. SHEU. 1990. Electrophysiologic effects of ATP on rat ventricular myocytes. Ann. N.Y. Acad. Sci. This volume.
21. BURNSTOCK, G. & P. MEGHJI. 1983. The effect of adenyl compounds on the rat heart. Br. J. Pharmacol. **79:** 211-218.
22. NEEDLEMAN, P., M. S. MINKES & J. R. DOUGLAS, JR. 1974. Stimulation of prostaglandin biosynthesis by adenine nucleotides: Profile of prostaglandin release by perfused organs. Circ. Res. **34:** 455-460.
23. ZEHL, U., C. RITTER & W. FORSTER. 1976. Influence of prostaglandins upon adenosine and of adenosine upon prostaglandin release in the isolated rabbit heart. Acta Biol. Med. Ger. **35:** K77-K82.
24. SCHRADER, J., G. BAUMANN & E. GERLACH. 1977. Adenosine as inhibitor of myocardial effects of catecholamines. Pfluegers Arch. **372:** 29-35.
25. RABERGER, G., O. KRAUPP, W. STUHLINGER, G. NELL & J. J. CHIRKDAJIAN. 1970. The effects of intracoronary infusion of adenosine on cardiac performance, blood supply and myocardial metabolism in dogs. Pfluegers Arch. **317:** 20-34.
26. LAW, W. R. & R. M. RAYMOND. 1987. Adenosine modulated myocardial insulin stimulated glucose uptake (abstr.). Fed. Proc. **46:** 402.
27. KAHLES, H., J. JUNGGEBURTH, T. LICK, W. SCHAFER & K. KOCHSIEK. 1987. Effects of a N(6)-disubstituted adenosine derivative on myocardial metabolism and ischemic stress following coronary occlusion. Drug Res. **37:** 1137-1140.
28. LASLEY, R. D., D. G. L. VAN WYLEN & R. M. MENTZER. 1989. Adenosine enhances glucose metabolism during low flow myocardial ischemia (abstr.). Circulation **80**(Suppl. II): 612.
29. OLAFSSON, B., M. B. FORMAN, D. W. PUETT, A. POU, C. C. CATES, G. C. FRIESINGER & R. VIRMANI. 1987. Reduction of reperfusion injury in the canine preparation by intracoronary adenosine: Importance of the endothelium and no-reflow phenomenon. Circulation **76:** 1135-1143.
30. BABBITT, D. G., R. VIRMANI & M. B. FORMAN. 1989. Intracoronary adenosine administered after reperfusion limits vascular injury after prolonged ischemia in the canine model. Circulation **80:** 1388-1399.
31. KUSACHI, S., R. D. THOMPSON & R. A. OLSSON. 1983. Ligand selectivity in dog coronary adenosine receptor resembles that of adenylate cyclase stimulatory (R_a) receptors. J. Pharmacol. Exp. Ther. **227:** 316-321.
32. RAMAGOPAL, M. V., R. W. CHITWOOD, JR. & S. G. MUSTAFA. 1988. Evidence for an A_2 adenosine receptor in human coronary arteries. Eur. J. Pharmacol. **151:** 483-486.
33. NEWMAN. W. H., B. F. BECKER, M. HEIER, S. NEES & E. GERLACH. 1988. Endothelium-mediated coronary dilation by adenosine does not depend on endothelial adenylate cyclase activation: Studies in isolated guinea pig hearts. Pfluegers Arch. **413:** 1-7.
34. BURNSTOCK, G. 1988. Regulation of blood flow by neurohumoral substances released from perivascular nerves and endothelial cells. Acta Physiol. Scand. Suppl. **133:** 53-59.
35. SCHRADER, J., C. I. THOMPSON, G. HIENDLMAYER & E. GELACH. 1982. Role of purines in acetylcholine-induced coronary vasodilation. J. Mol. Cell. Cardiol. **14:** 427-430.
36. JAMES, T. N. 1965. The chronotropic action of ATP and related compounds studied by direct perfusion of the sinus node. J. Pharmacol. Exp. Ther. **149:** 233-247.
37. PELLEG, A., B. BELHASSEN, R. ILIA & S. LANIADO. 1985. Comparative electrophysiologic effects of ATP and adenosine in the canine heart: Influence of atropine, propranolol, vagotomy, dipyridamole and aminophylline. Am. J. Cardiol. **55:** 572-576.
38. HEADRICK, J. & R. WILLIS. 1988. Mediation by adenosine of bradycardia in rat heart during graded global ischemia. Pfluegers Arch. **412:** 618-623.
39. WEST, G. A. & L. BELARDINELLI. 1985. Correlation of sinus slowing and hyperpolarization caused by adenosine in sinus node. Pfluegers Arch. **403:** 75-81.
40. URTHALER, F. & T. N. JAMES. 1972. Effects of adenosine and ATP on AV conduction and on AV junctional rhythm. J. Lab. Clin. Med. **79:** 96-105.

41. MITSUOKA, T., T. HIRATA, M. HIRATA, Y. MATSUMOTO, K. YANO & K. HASHIBA. 1989. Comparison of effects of adenosine and ATP on sinus and AV junctional rhythms in the canine heart (abstr.). Jpn. Circ. J. **53:** 59.
42. HELLER, L. J. & R. A. OLSSON. 1985. Inhibition of rat ventricular automaticity by adenosine. Am. J. Physiol. **248:** H907-H913.
43. PELLEG, A., H. MITAMURA, T. MITSUOKA, E. L. MICHELSON & L. S. DREIFUS. 1986. Adenosine and ATP suppress ventricular escape rhythms in the canine heart. J. Am. Coll. Cardiol. **8:** 1145-1151.
44. HALEEN, S. J. & D. B. EVANS. 1985. Selective effects of adenosine receptor agonists upon coronary resistance and heart rate in isolated working rabbit hearts. Life Sci. **36:** 127-137.
45. KURAHASHI, K. & D. M. PATON. 1986. Negative chronotropic action of adenosine in rat atria: Evidence for action at A_1 receptors. Nucleosides & Nucleotides **5:** 493-501.
46. KURACHI, Y., T. NAKAJIMA & T. SUGIMOTO. 1986. On the mechanism of activation of muscarinic K^+ channels by adenosine in isolated atrial cells: Involvement of GTP-binding proteins. Pfluegers Arch. **407:** 264-274.
47. BELARDINELLI, L. & G. ISENBERG. 1983. Isolated atrial myocytes: Adenosine and acetylcholine increase potassium conductance. Am. J. Physiol. **224:** H734-H737.
48. BELARDINELLI, L., W. R. GILES & A. WEST. 1988. Ionic mechanisms of adenosine's actions in pacemaker cells from the rabbit heart. J. Physiol. **405:** 615-633.
49. PELLEG, A., T. MITSUOKA, T. MAZGALEV, E. L. MICHELSON & L. S. DREIFUS. 1988. Intermodulation of the negative chronotropic actions of the vagus and adenosine in the sinus node of the canine heart. Cardiovasc. Res. **22:** 55-61.
50. RONCA-TESTONI, S. & F. BORGHINI. 1982. Degradation of perfused adenine compounds up to uric acid in isolated rat heart. J. Mol. Cell. Cardiol. **14:** 177-180.
51. BELARDINELLI, L. *et al.* 1984. Effects of adenosine and adenine nucleotides on the atrioventricular node of isolated guinea pig hearts. Circulation **70:** 1083-1091.
52. PELLEG, A., T. MITSUOKA & E. L. MICHELSON. 1987. Adenosine mediates the negative chronotropic action of adenosine 5'-triphosphate in the canine sinus node. J. Pharmacol. Exp. Ther. **242:** 791-795.
53. COLLIS, M. G. & S. J. PETTINGER. 1982. Can ATP stimulate P_1-receptors in guinea pig atrium without conversion to adenosine? Eur. J. Pharmacol. **81:** 521-529.
54. KURACHI, Y., T. NAKAJIMA & R. TAKIKAWA. 1987. Induction of K^+ current by adenosine and adenine nucleotides in isolated atrial cells. Circulation **76**(Suppl. IV): 112.
55. PELLEG, A., T. MITSUOKA, T. MAZGALEV & E. L. MICHELSON. 1987. Vagal component in the chronotropic and dromotropic actions of adenosine and ATP. *In* Cardiac Electrophysiology and Pharmacology of Adenosine and ATP: Basic and Clinical Aspects. A. Pelleg, E. L. Michelson & L. S. Dreifus, Eds.: 375-384. Alan R. Liss. New York, NY.
56. SZENTMIKLOSI, A. J., M. NEMETH, J. SZEGI, J. G. PAPP & L. SZEKERES. 1980. Effect of adenosine on sinoatrial and ventricular automaticity of the guinea pig. Naunyn-Schmiedeberg's Arch. Pharmacol. **311:** 147-149.
57. ROSEN, M. R., P. DANILO & R. M. WEISS. 1983. Action of adenosine on normal and abnormal impulse initiation in canine ventricle. Am. J. Physiol. **244:** H715-H721.
58. BELARDINELLI, L., R. A. FENTON, A. WEST, J. LINDEN, J. S. ALTHAUS & R. M. BERNE. 1982. Extracellular action of adenosine and the antagonism by aminophylline on the atrioventricular conduction of isolated perfused guinea pig and rat hearts. Circ. Res. **51:** 569-579.
59. CLEMO, H. F., A. BOURASSA, J. LINDEN & L. BELARDINELLI. 1987. Antagonism of the effects of adenosine and hypoxia on atrioventricular conduction time by two novel alkylxanthines: Correlation with binding to adenosine A_1 receptors. J. Pharmacol. Exp. Ther. **242:** 478-484.
60. FAVALE, S., M. DIBIASE, U. RIZZO, L. BELARDINELLI & P. RIZZON. 1985. Effect of adenosine and adenosine 5'-triphosphate on atrioventricular conduction in patients. J. Am. Cardiol. **5:** 1212-1219.
61. WESLEY, R. C., B. B. LERMAN, J. P. DIMARCO, R. M. BERNE & L. BELARDINELLI. 1986. Mechanism of atropine-resistant atrioventricular block during inferior myocardial infarction: Possible role of adenosine. J. Am. Coll. Cardiol. **8:** 1232-1234.
62. SHAH, P. K., P. NALOS & T. PETER. 1987. Atropine resistant postinfarction complete AV block: Possible role of adenosine and improvement with aminophylline. Am. Heart J. **113:** 194-195.

63. DUNN, R. B. & B. M. SUMMERS. 1987. Effect of adenosine and adenine nucleotides on AV conduction in the ischemic canine heart (abstr.). Fed. Proc. **46:** 6530.
64. BELARDINELLI, L., G. A. WEST, R. CRAMPTON & R. M. BERNE. 1983. Chronotropic and dromotropic effects of adenosine. *In* Regulatory Function of Adenosine. R. M. Berne, T. W. Rall & R. Rubio, Eds.: 377-398. Martinus Nijhoff. The Hague.
65. PELLEG, A., C. HURT, A. MIYAGAWA, E. L. MICHELSON & L. S. DREIFUS. 1990. Differential sensitivity of cardiac pacemakers to exogenous adenosine *in vivo.* Am. J. Physiol.: in press.
66. DIFRANCESCO, D. 1985. The cardiac hyperpolarizing-activated current i_f: Origins and developments. Prog. Biophys. Mol. Biol. **6:** 163-183.
67. BELARDINELLI, L., W. R. GILES & A. WEST. 1988. Ionic mechanisms of adenosine's actions in pacemaker cells from the rabbit heart. J. Physiol. **405:** 615-633.
68. BROWN, H. F., D. DIFRANCESCO & S. J. NOBLE. 1979. How does adrenaline accelerate the heart? Nature **280:** 235-236.
69. HAGIWARA, N. & H. IRISAWA. 1989. Modulation by intracellular Ca^{2+} of the hyperpolarization-activated inward current in rabbit single sinoatrial node cells. J. Physiol. (London) **409:** 121-141.
70. YANAGIHARA, K. & H. IRISAWA. 1980. Inward current activated during hyperpolarization in the rabbit sinoatrial node cell. Pfluegers Arch. **385:** 11-19.
71. TSIEN, R. W. 1974. Effect of epinephrine on the pacemaker potassium current of cardiac Purkinje fibers. J. Gen. Physiol. **64:** 293-319.
72. CHIBA, S., K. KUBOTA & K. HASHIMOTO. 1973. Inhibition of the negative chronotropic action of adenosine by caffeine in the dog. Eur. J. Pharmacol. **21:** 281-285.
73. KURAHASHI, K. & D. M. PATON. 1986. Negative chronotropic action of adenosine in rat atria: Evidence for action at A_1 receptors. Nucleosides & Nucleotides **5:** 493-501.
74. MESZAROS, J., K. KELEMEN, V. KECSKEMETI & G. KARACS. 1986. Interaction between adenosine and antiarrhythmic agents in atrial myocardium of guinea pig. Arch. Int. Pharmacodyn. **280:** 84-96.
75. KURACHI, Y., T. NAKAJIMA & T. SUGIMOTO. 1987. Quinidine inhibition of the muscarinic receptor-activated K^+ channel current in atrial cells of guinea pig. Arch. Pharmacol. **335:** 216-218.
76. BERTOLINO, M. & D. DIFRANCESCO. 1983. L'adenosina antagonizza l'effetto delle catecolamine sul pace-maker cardiaco (abstr.). Atti del Congresso SIBS-SIF-SINU.
77. WESLEY, R. C., JR. & L. BERLARDINELLI. 1985. Role of adenosine on ventricular overdrive suppression in isolated guinea pig hearts and Purkinje fibers. Circ. Res. **57:** 517-531.
78. BROWN, H. F., D. DIFRANCESCO, J. KIMURA & D. NOBLE. 1981. Cesium: A useful tool of investigating sino-atrial pacemaking. J. Physiol. (London) **317:** 54.
79. DIFRANCESCO, D. 1981. A new interpretation of the pacemaker current in calf Purkinje fibers. J. Physiol. (London) **314:** 359-376.
80. NOMA, A., M. MORAD & H. IRISAWA. 1983. Does the "pacemaker current" generate the diastolic depolarization in the rabbit SA node cells? Pfluegers Arch. **397:** 190-194.
81. DIFRANCESCO, D., P. DUCOURET & R. ROBINSON. 1989. Muscarinic modulation of cardiac rate at low acetylcholine concentrations. Science **243:** 669-671.
82. DIFRANCESCO, D. & C. TROMBA. 1987. Acetylcholine inhibits activation of the cardiac hyperpolarizing-activated i_f. Pfluegers Arch. **410:** 139-142.
83. HEDQVIST, P. & B. B. FREDHOLM. 1979. Inhibitory effect of adenosine on adrenergic neuroeffector transmission in the rabbit heart. Acta Physiol. Scand. **105:** 120-122.
84. LOKHANDWALA, M. F. 1979. Inhibition of cardiac sympathetic neurotransmission by adenosine. Eur. J. Pharmacol. **60:** 353-357.
85. SEITELBERGER, R., W. SCHUTZ, O. SCHLAPPACK & G. RABERGER. 1984. Evidence against the adenosine-catecholamine antagonism under *in vivo* conditions. Arch. Pharmacol. **325:** 234-239.
86. RABERGER, G., G. FISCHER, G. KRUMPL, W. SCHNEIDER & H. STROIßNIG. 1987. Further evidence against adenosine-catecholamine antagonism *in vivo:* Investigations with treadmill exercise in dogs. *In* Topics and Perspectives in Adenosine Research. Proceedings of the Third International Symposium on Adenosine (Munich, June 1986). E. Gerlach & B. F. Becker, Eds.: 383-392. Springer-Verlag. New York, NY.
87. LERMAN, B. B., R. C. WESLEY, JR., J. P. DIMARCO, D. E. HAINES & L. BELARDINELLI. 1988. Antiadrenergic effects of adenosine on His-Purkinje automaticity. Evidence for accentuated antagonism. J. Clin. Invest. **82:** 2127-2135.

88. GORDON, J. L. 1986. Extracellular ATP: Effects, sources and rate. Biochem. J. **233:** 309-319.
89. NEEDHAM, L., N. J. CUSACK, J. D. PEARSON & J. L. GORDON. 1987. Characteristics of the P_2-purinoceptor that mediates prostacyclin production by pig aortic endothelial cells. Eur. J. Pharmacol. **134:** 199-209.
90. CATER, T. D., T. J. HALLAM, N. J. CUSACK & J. D. PEARSON. 1988. Regulation of P_{2Y}-purinoceptor-mediated prostacyclin release from human endothelial cells by cytoplasmic calcium concentration. Br. J. Pharmacol. **95:** 1181-1190.
91. KARMAZYN, M. & N. S. DHALLA. 1983. Physiological and pathophysiological aspects of cardiac prostaglandins. Can. J. Physiol. Pharmacol. **61:** 1207-1225.
92. TAKIKAWA, R., Y. KURACHI, S. MASHIMA, H. ITO & Y. ASSARO. 1989. ATP-induced sinus tachycardia mediated by prostaglandin synthesis via phospholipase C in rabbit heart. Circulation **80**(Suppl. II): 442.
93. PELLEG, A., H. MITAMURA, E. L. MICHELSON & L. S. DREIFUS. 1986. Evidence against prostaglandin mediation of the electrophysiologic effects of ATP versus adenosine in the canine heart. J. Cardiovasc. Pharmacol. **8:** 534-538.
94. CIABATTONI, G. & A. WENNMALM. 1985. Adenosine-induced coronary release of prostacyclin at normal and low pH in isolated heart of rabbit. Br. J. Pharmacol. **85:** 557-563.
95. KARWATOWSKA-PROKOPCZUK, E., G. CIABATTONI & A. WENNMALM. 1988. Effect of adenosine on the formation of prostacyclin in the rabbit isolated heart. Br. J. Pharmacol. **94:** 721-728.

DISCUSSION OF THE PAPER

M. H. MAGUIRE (*University of Kansas Medical Center, Kansas City, KS*): With regard to the cardioprotective role that you postulated for adenosine, how does this accord with the role of ischemia-induced adenosine as the precursor of hypoxanthine, which on reperfusion is the substrate for the generation of toxic oxygen radicals? These radicals are formed by action of xanthine oxidase on hypoxanthine and are believed to be responsible for reperfusion injury.

PELLEG: The overall cardioprotective effect of adenosine is based on several actions of the nucleoside that appeared to take place in a stepwise fashion depending on the amount of adenosine in the active compartments. Thus, the adenosine level threshold for each action is different.

Relatively small amounts of adenosine cause coronary vasodilation and increased blood flow and oxygen supply to the affected tissue. Higher levels of adenosine are required for electrophysiologic effects of the nucleoside, that is, its negative chronotropic action on the cardiac pacemakers and negative dromotropic action on AV nodal conduction. These effects reduce cardiac work and oxygen demand. Further increase in adenosine release from ischemic cells will result in extracellular levels sufficient for an indirect, antiadrenergic action that will also reduce cardiac work and, in turn, oxygen demand.

In addition, adenosine can cause a shift in cardiac metabolism from fatty acid to carbohydrate, which is beneficial under conditions of reduced oxygen supply.

Upon reperfusion, adenosine can reduce reperfusion damage by attenuating platelet aggregation and leukocyte activation, and by attenuating the release of superoxides. These latter effects of adenosine were shown in animal models as well as in humans using exogenous adenosine. Therefore, it is still to be determined whether the net effect of endogenous adenosine during reperfusion is beneficial as suggested by the above-mentioned studies or detrimental because of its metabolism to precursors of oxygen radicals.

Dual Control of Local Blood Flow by Purines[a]

G. BURNSTOCK

Department of Anatomy and Developmental Biology
University College London
London WC1E 6BT, England

NERVE, MUSCLE, ENDOTHELIAL CELL INTERRELATIONS IN VESSEL WALL

Before discussing the mechanisms by which purines control local blood flow, it is important to understand the geometry of cell relationships in the vessel wall and to define some of the terms that will be used in this paper. The wall of an artery consists of adventitial, medial, and intimal layers. Endothelial cells in the intima vary considerably in their fine structure and are often separated from the muscle coat by dense elastic tissue. The perivascular nerves are confined to the adventitial-medial border in most vessels, and whole-mount stretch preparations show that the perivascular plexus consists of an extensive network of bundles of branching varicose terminal fibers.[3] At the electron microscopic level, it can be seen that individual varicose fibers pass close to muscle fibers, but it is important to recognize that varicosities do not form synapses with muscle cells, which would, by definition, involve both pre- and postsynaptic specializations similar to those seen at the skeletal neuromuscular synapse or at ganglionic synapses. Rather, the varicosities in autonomic nerves continuously vary in their relationship to the muscle cells in the medial coat (FIG. 1), and the number of varicosities (and therefore the position of varicosities) varies during development and with the level of activity of the nerve. The vascular neuromuscular junction, therefore, involves "en passage" release of transmitters from specialized prejunctional sites in the varicosities, which then reach receptors located more or less homogeneously on a group of smooth muscle cells. The smooth muscle cells on which the receptors are located do not show any postjunctional specializations and are usually in electrical continuity with each other via gap junctions. These neuromuscular junctions are particularly suitable for the neuromodulatory actions of local agents, such as prostaglandin, histamine, or bradykinin, or of circulating neurohormones.[4] Neuromodulation may be either prejunctional, decreasing or increasing the amount of neurotransmitter released by the nerve, or postjunctional, modulating the time course or extent of action of the neurotransmitter.

[a] I refer the reader to my previous article in this volume (pp. 1-18), where I dealt in particular with the subclassification of purinoceptors. In this article, I am concerned with analyzing the various roles of purines in controlling local blood flow (see also references 1 and 2).

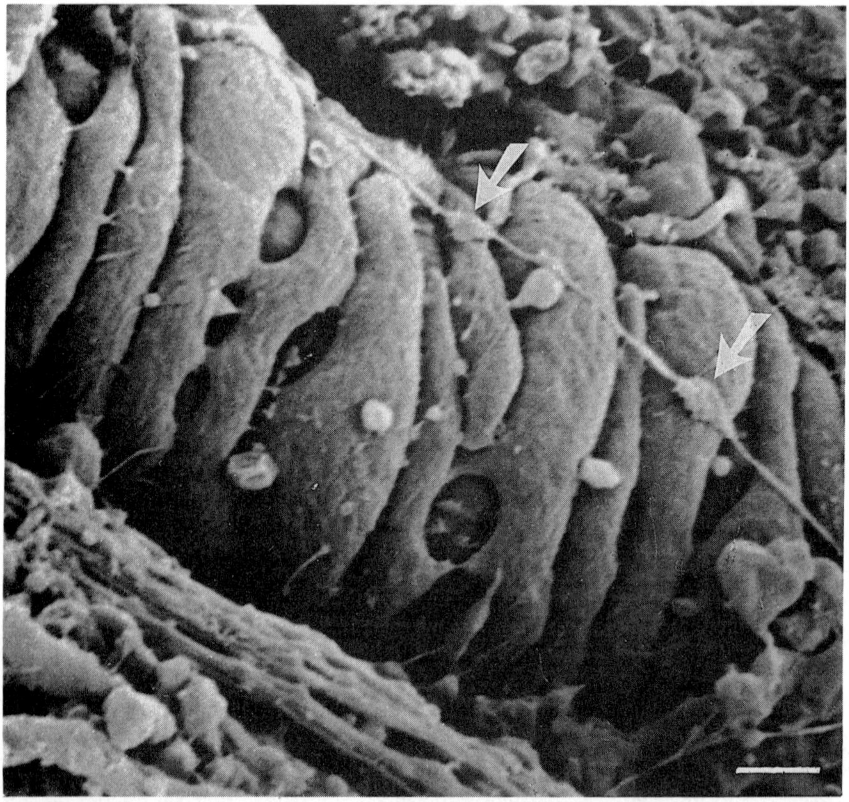

FIGURE 1. Scanning electron micrograph of the adventitial surface of the smooth muscle cells of the central arteriole of the rat retina digested by collagenase and trypsin before fixation. Arrows point to varicosities in a single nerve fiber. Bar: 3.2 μm. From Uehara and Suyama.[139]

PERIVASCULAR PURINERGIC NERVES

Nonsympathetic, Purinergic Transmission

In view of the potent actions of purines on the vasculature, it was perhaps surprising that few examples of vessels controlled by nerves containing ATP as the principal transmitter emerged in the 1970s. As we shall see later, however, ATP is now known to have a widespread role as a cotransmitter in sympathetic and sensory-motor nerves and has potent vasodilator effects via the endothelium (FIG. 2).

Rabbit Portal Vein

The best example of nonsympathetic, purinergic innervation of vessels is the innervation that supplies the rabbit portal vein. The evidence includes the following: 1) fluorescence histochemical localization of a fine perivascular plexus in the wall of the vein utilizing quinacrine, a fluorescent dye that is known to bind tightly to ATP (this plexus is unaffected by sympathectomy)[5–8]; 2) release of endogenous ATP during stimulation of perivascular nerves (the release of ATP is abolished by tetrodotoxin as is the vasodilator response to stimulation of perivascular nerves that persisted in the presence of atropine and guanethidine)[9–12]; 3) mimicry by ATP of the rapid vasodilatation produced by stimulation of the perivascular nerves in the presence of atropine and guanethidine[9,13–15]; 4) reduction of vasodilatation mediated by perivascular nerves and vasodilatation produced by ATP in the presence of reactive blue 2, a drug that appears to selectively antagonize the P_{2Y}-purinoceptor.[15,16]

Rat Intrapulmonary Arteries

Excitatory junction potentials recorded in the smooth muscle of these vessels were abolished by α,β-methylene ATP, a selective desensitizer of the P_{2X}-purinoceptor, but

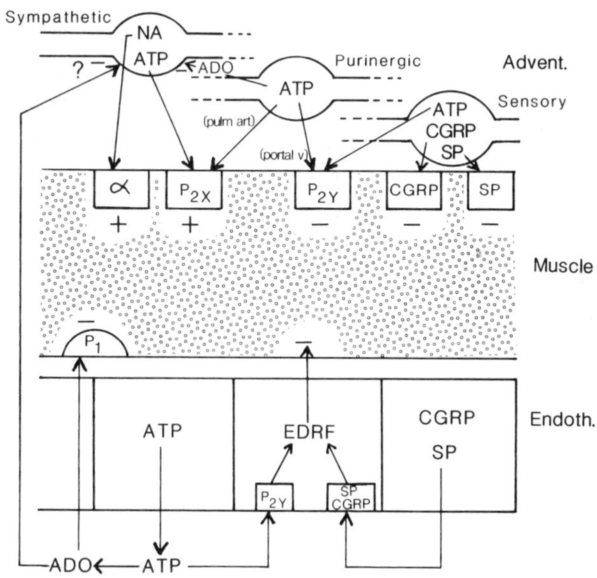

FIGURE 2. A schematic representation of the possible roles of purines in local control of blood flow. It is suggested that ATP is released from endothelial cells (Endoth.) during hypoxia or increased flow to act on endothelial P_{2Y}-purinoceptors leading to production of EDRF and subsequent vasodilatation (−). In contrast, ATP release from perivascular sympathetic nerves at the adventitial (Advent.)-muscle border produces vasoconstriction (+) via P_{2X}-purinoceptors located in the muscle cells. Adenosine (ADO), resulting from rapid breakdown of ATP by ectoenzymes, produces vasodilatation by direct action on the muscle via P_1-purinoceptors. Adenosine also acts on P_1-purinoceptors on the perivascular nerves to inhibit release of excitatory transmitters. Also included in the scheme are nonsympathetic purinergic nerves known to release ATP in two vessels, namely rabbit portal vein (producing vasodilatation) and rat intrapulmonary arteries (producing vasoconstriction), and sensory-motor nerves that may release ATP as a vasodilatory cotransmitter with substance P (SP) and calcitonin gene-related peptide (CGRP) during "axon reflexes." From Burnstock.[140]

were unaffected by guanethidine, which inhibits release of ATP as well as noradrenaline (NA) from sympathetic nerves.[17] This is good evidence for nonsympathetic, purinergic excitatory innervation of these vessels. In view of the quinacrine staining of a subpopulation of intramural neurons in the airways,[18] it seems likely that the perivascular purinergic nerves originate in these ganglion cells.

Other Vessels

Other examples may be found where there is some evidence for direct purinergic vasodilator innervation of smooth muscle (that is, vasodilatation is not via endothelial purinergic receptors). Vessels with such innervation include vessels in skeletal muscle,[19] and possibly some coronary and cerebral resistance vessels.[20]

Sympathetic Purinergic Transmission

Early evidence for ATP and NA as sympathetic cotransmitters comes largely from studies of the vas deferens.[21–25] There are now a number of reports of sympathetic, purinergic excitatory cotransmission to various blood vessels, although there is considerable variation in the proportions of ATP and NA utilized.[2,10] In most of the vessels studied so far, neuropeptide Y also coexists with NA in the sympathetic nerves; however, it does not always appear to act as a cotransmitter, but rather as a neuromodulator of NA release and/or activation.[26] The purinergic component of the vasopressor response to stimulation of the sympathetic outflow in the pithed rat is blocked by nifedipine, whereas the α-adrenoceptor-mediated response to the cotransmitter, NA, is relatively resistant.[27] Colocalization and synthesis of ATP with NA in cultured sympathetic neurons has also been demonstrated.[28,29]

Rabbit Saphenous Artery

This is a vessel where there is a substantial ATP component involved in sympathetic neurotransmission. In most blood vessels, the contributions of ATP and NA to the mechanical response to sympathetic nerve stimulation are not so clearly separated as they are in the vas deferens, but prazosin reduces the response of the rabbit saphenous artery by less than 30% even at high frequencies of sympathetic nerve stimulation (which favors the NA response), while the remaining response is blocked by α,β-methylene ATP[30,31] or ANAPP$_3$.[30]

Mesenteric Artery

The mesenteric artery appears to be another vessel where the ATP contribution to sympathetic cotransmission is high. This effect is seen in the dog,[32–35] rat,[36–38] guinea pig,[39] and rabbit[40–42] mesenteric artery.

Rat Tail Artery

Excitatory junction potentials in the rat tail artery are resistant to prazosin,[43] but are blocked by α,β-methylene ATP.[44–46] The ATP component, however, appears to be smaller relative to NA in the sympathetic nerves supplying this vessel, so that it is more difficult to demonstrate a mechanical prazosin-resistant component, except in tail artery taken from spontaneously hypertensive rats, where it has been suggested that the ATP component in the sympathetic nerves is more prominent.[47,48]

Rabbit Ear Artery

Evidence has been presented for cotransmission involving ATP and NA in the ear artery,[49–53] but, as in the rat tail artery, it is difficult to demonstrate a prazosin-resistant (purinergic) mechanical component of the response to perivascular nerve stimulation, except with short bursts of pulses lasting about a second.[54] Excitatory junction potentials recorded in the rabbit ear artery are clearly prazosin resistant.[55,56]

Rabbit Coronary Artery

In the rabbit coronary artery, though not in other vessels, NA has been shown to cause dilatation via β-adrenoceptors, while its sympathetic cotransmitter, ATP, produces direct vasodilatation of smooth muscle via P_{2Y}-purinoceptors.[57] This situation is consistent with the synergistic nature of sympathetic cotransmission (albeit excitatory) seen in other vessels.

Pulmonary Artery

Sympathetic purinergic cotransmission has been demonstrated in pulmonary arteries of rabbit,[58,59] and P_2-purinoceptors of both P_{2X} and P_{2Y} subclasses have been detected in pulmonary arteries of cats[60] and man.[61] Recently, we have used $[^3H]\alpha,\beta$-methylene ATP as a radioligand for labeling the P_{2X}-purinoceptor.[62] The localization of this receptor in the rabbit pulmonary artery is illustrated in FIGURE 3.

Intestinal Circulation

Evidence for purinergic cotransmission in rabbit ileocolonic artery[63] and arterial mesenteric vessels in cat intestine[64] has been presented.

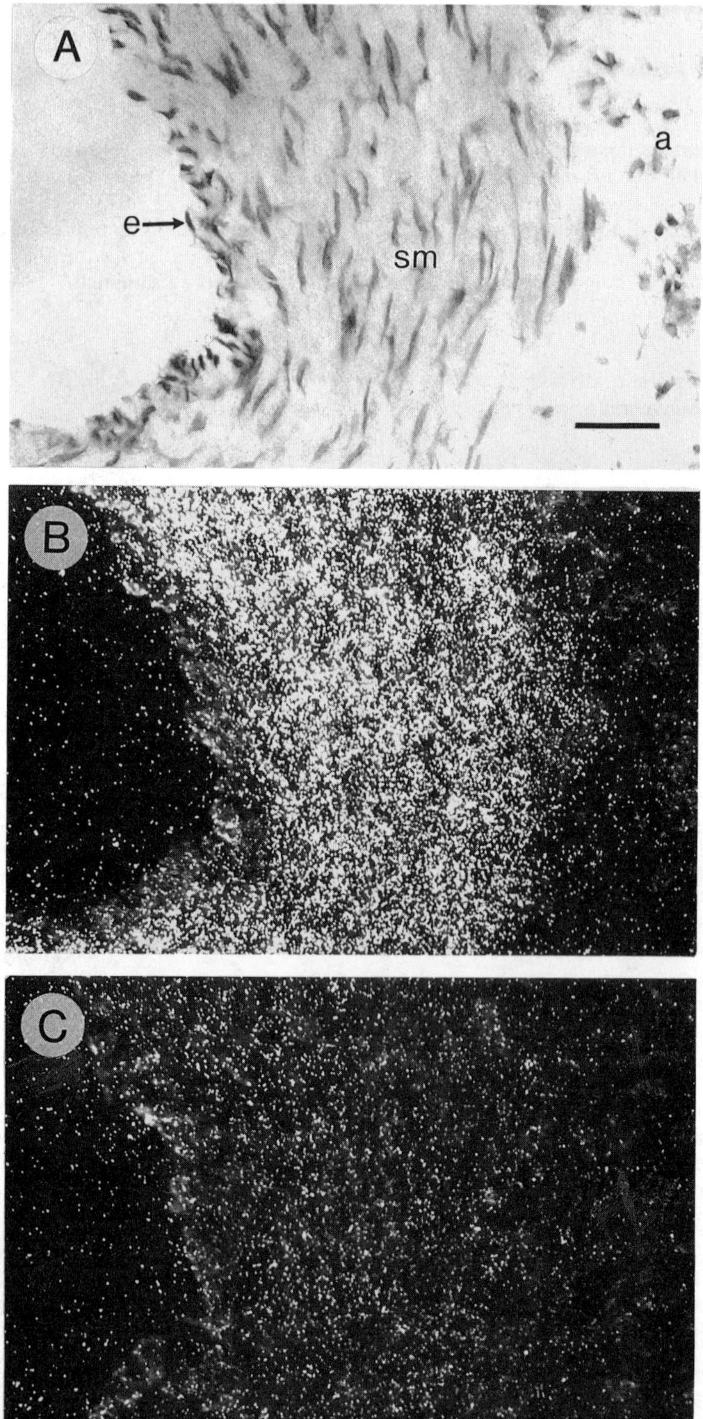
A
a
e
sm
B
C

Renal Vascular Bed

Renal periarterial nerve-stimulation-induced vasoconstriction at low frequencies is primarily due to release of a purinergic transmitter in the rat.[65]

Cerebral Arteries

Coexistence of NA and ATP has been demonstrated in dog basilar artery[66] and middle cerebral artery.[67]

Rabbit Aorta

Su[12,68] used tritium-labeled adenosine and NA to indicate that ATP is released together with NA from sympathetic nerves supplying the rabbit aorta.

Skeletal Muscle Circulation

Vasodilatation produced by hypothalamic stimulation in anesthetized rabbits has been claimed to be mediated by ATP released from sympathetic nerves.[69] A contribution of ATP to sympathetic vasopressor responses of the pithed rabbit has also been demonstrated.[13]

Sensory-Motor Purinergic Transmission

As far back as 1954, Pamela Holton[70] showed that ATP might be released during antidromic impulses in sensory nerves supplying the rabbit ear artery. More recently,

---

FIGURE 3. (**A**) Rabbit pulmonary artery viewed using bright-field optics (cryostat section stained with 0.5% toluidine blue). (**B**) A dark-field view of **A** showing the distribution of [^{3}H]α,β-meATP binding sites (the section was incubated with 10 nM [^{3}H]α,β-meATP at 30 °C for 30 min). (**C**) A dark-field view of a section adjacent to **A** showing the nonspecific binding sites of [^{3}H]α,β-meATP (the section was incubated with 10 nM [^{3}H]α,β-meATP in the presence of 10 μM β,γ-meATP). Abbreviations: a: adventitia; e: endothelium; sm: smooth muscle. Bar: 30 μm.

ATP has been claimed as a neurotransmitter in some small primary afferent neurons that impinge on neurons in sensory nuclei in spinal cord and brain.[71-74] It has been suggested that ATP may be coreleased with substance P during "axon reflex" vasodilatation of skin vessels,[75] and because there are P_{2Y}-purinoceptors directly on the smooth muscle of some vessels (rather than exclusively on the endothelium), sensory-motor purinergic transmission may occur in these vessels too, including rabbit mesenteric artery[76] and rabbit hepatic artery.[77]

PURINERGIC NEUROMODULATION

Prejunctional Modulation

Adenine nucleotides and nucleosides have been shown to inhibit NA release from sympathetic nerves that supply a number of blood vessels, including saphenous, tibial, portal, pulmonary, and mesenteric vessels.[68,78-85] The prejunctional receptor that mediates these actions is almost always the P_1-purinoceptor because the inhibitory actions of both ATP and adenosine are blocked by methylxanthines and because the stable analogues of ATP, such as β,γ-methylene ATP, are ineffective.[86] It has been suggested that occupation of P_1-purinoceptors leads to decreases in Ca^{2+} influx and subsequent reductions in NA release.[87] In most tissues, the prejunctional P_1-purinoceptor has been shown to be of the A_1 subtype,[79] but the A_2 subtype appears to be dominant in the rat portal vein.[88] Autoinhibition of release of ATP from sympathetic nerves by negative feedback of purines on prejunctional receptors has also been described.[89,90] Adenosine enhancement of sympathetic neurotransmission has been claimed in the guinea pig pulmonary artery.[91]

Although ATP and adenosine have been shown to act on prejunctional purinergic receptors, and to lead to modulation of the release of acetylcholine from cholinergic nerves in skeletal muscle, brain, and intestine,[92,93] specific studies on cholinergic transmission to blood vessels do not appear to have been carried out. Because the modulatory effects on acetylcholine release at motor nerve endings are blocked by methylxanthines, it seems likely that they are mediated by P_1-purinoceptors.[94,95] Silinsky[96] has suggested that two types of adenosine receptor may be present at cholinergic nerve endings—one reducing and the other enhancing acetylcholine release.

Postjunctional Modulation

Purine nucleotides and/or nucleosides can have modulatory effects on the postjunctional actions of both NA and acetylcholine. For example, ATP, AMP, and adenosine have been shown to potentiate the contractile actions of NA on the vas deferens.[97,98] Postjunctional synergism of NA and ATP has also been demonstrated in the mesenteric arterial bed of the rat.[99] Increases in acetylcholine receptor sensitivity in response to ATP have been demonstrated, particularly at the motor end plate and sympathetic ganglia.[12,100-104]

PURINERGIC VASODILATATION VIA ENDOTHELIUM

Furchgott and Zawadzki,[105] while exploring the action of acetylcholine on endothelial cells of the rabbit aorta, were the first to introduce the concept of endothelium-mediated vasodilatation. Soon afterward, it was reported that ATP and ADP (but rarely adenosine) receptors were present on vascular endothelial cells and produced potent vasodilatation via the same mechanism by which acetylcholine participates in the release of endothelium-derived relaxing factor (EDRF).[106–112] Recently ATP has been shown to directly release nitric oxide from cultured aortic endothelial cells.[113]

It has been shown in several different vascular preparations that the P_{2X}-purinoceptor on the smooth muscle mediates vasoconstriction produced by ATP released as a cotransmitter with NA, and that the P_{2Y}-purinoceptor on endothelial cells mediates vasodilatation produced by EDRF.[16,37,109,114,115] In the Langendorff heart preparation, increases in perfusion pressure are produced by α,β-methylene ATP, a selective P_{2X} agonist, whereas decreases in pressure are produced by 2-methylthio-ATP, which is selective for the P_{2Y}-purinoceptor. Prolonged exposure to α,β-methylene ATP selectively blocks the vasoconstriction, whereas reactive blue 2 selectively blocks the vasodilatation.[115,116]

The P_{2Y}-purinoceptor on vascular endothelial cells has been characterized. The P_{2Y}-purinoceptors mediate not only endothelium-dependent relaxation, but also prostacyclin synthesis[117,118] and cell mitogenesis.[119] Mitogenic actions are probably involved in the accelerated repair of lesions; prostacyclin controls the platelet aggregation following vascular damage.[120] The P_{2Y}-purinoceptors on endothelial cells are coupled to a phospholipase C by a GTP-binding protein, and ATP or ADP induces a rapid and transient accumulation of inositol 1,4,5-trisphosphate in cultured endothelial cells.[121–123] This is followed by an increase in cytoplasmic Ca^{2+},[124,125] which is likely to mediate further cellular events.

PHYSIOLOGICAL AND PATHOPHYSIOLOGICAL SIGNIFICANCE OF PURINERGIC VASOMECHANISMS

It is suggested that following ischemia, there is a release of ATP (and other substances) from endothelial cells, which are known to contain high levels of ATP following high-affinity uptake of adenosine.[126,127] This ATP then acts locally on P_{2Y}-purinoceptors on endothelial cells (together perhaps with ADP released from aggregating platelets) to lead to release of EDRF and vasodilatation, which protects the delicate tissues of, for example, heart and brain, which are particularly sensitive to hypoxia. Because ATP is rapidly broken down by ectoenzymes, especially in the blood, reactive hyperemia mediated by ATP, although powerful, will be short-lived; also, the prolonged vasodilatation that occurs with hypoxia is probably mediated by its breakdown product, adenosine, acting directly on the vascular smooth muscle (FIG. 2). This hypothesis would provide an explanation of some of the anomalies posed by the Berne hypothesis of reactive hyperemia, in which adenosine is claimed to be the sole purine involved, released per se from myocytes.[128–130] For example, it would explain why there is only partial antagonism of vasodilatation of the coronary circulation in response to hypoxia by methylxanthines or adenosine deaminase; it would also explain why the initial response to hypoxia in the lung circulation is a vasoconstriction, even though adenosine is largely a vasodilator (ATP constricts pulmonary vessels). The lack of reduction in hypoxic vasodilatation produced by the adenosine transport inhibitor, dipyridamole, could also be explained if ATP rather than adenosine per se was released from endothelial cells and was broken down by ectoenzymes to adenosine later, with its uptake prevented.

To put the role of purines in producing vasodilatation in perspective, it is important to recognize that ATP is not the only agent stored and released from endothelial cells during hypoxia. The acetylcholine-synthesizing enzyme choline acetyltransferase has been localized in endothelial cells in brain vessels[131]; also, Loesch and Burnstock[132] have demonstrated, with electron-microscopic immunocytochemical techniques, that serotonin and substance P also show positive staining in some endothelial cells in the mesenteric and femoral arteries of the rabbit. We have also collected acetylcholine, serotonin, and substance P, as well as ATP, in the venous effluent of the Langendorff heart preparation during hypoxia.[116,133,134]

The physiological conditions under which ATP is released from sympathetic nerves, as well as the conditions under which purines modulate the release of NA, are not known, but some speculation is tempting. For example, during vasodilatation produced via endothelial cells during ischemia, the adenosine resulting from breakdown of released endothelial ATP would diffuse through the vessel wall to act directly on the muscle (reinforcing and prolonging vasodilatation); also, since adenosine breaks down relatively slowly, it might reach sympathetic nerves at the adventitial-medial border, where it would reinforce this protective response further by inhibiting release of the vasoconstrictor transmitters.

The existence of different mechanisms of postjunctional actions of ATP and NA—these mechanisms being electromechanical and pharmacomechanical, respectively—suggests that they play different roles in the neurotransmission process. For example, one could speculate that long-lasting, low-frequency discharge in sympathetic nerves, which favors the NA component of the response, could be involved in the kind of homeostatic regulation involved in, say, gentle exercise, whereas short bursts of high-frequency pulses in sympathetic nerves, which favor the ATP component, might be involved in the rapid, short-lasting, powerful vasoconstriction involved in the defense reaction.

It has been claimed that in spontaneously hypertensive rats, purinergic mechanisms are affected, including reduction in prejunctional adenosine-mediated neuromodulation of NA release,[135,136] diminished endothelium-mediated vasodilatation by ADP,[137] and an increased role for ATP as a sympathetic cotransmitter in rat tail artery.[47] Recent studies also indicate that sympathetic-mediated forearm vasoconstriction persists after α-adrenoceptor blockade in hypertensive patients.[138]

SUMMARY

The potent and widespread vascular actions of purine nucleotides and nucleosides have long been recognized. A dual function for ATP in the regulation of vascular tone is considered. ATP acts as an excitatory cotransmitter with noradrenaline from sympathetic perivascular nerves, to cause vasoconstriction via P_{2X}-purinoceptors located on vascular smooth muscle. In contrast, ATP can act via P_{2Y}-purinoceptors located on vascular endothelial cells to release EDRF, which diffuses to the vascular smooth muscle and produces vasodilatation. The main source of intraluminal ATP is likely to be endothelial cells, and its release can be measured during conditions such as changes in flow and hypoxia, in amounts sufficient to activate endothelial P_{2Y}-purinoceptors. In some vessels, ATP acts directly on P_{2Y}-purinoceptors located in the vascular smooth muscle to produce vasodilatation; the possibility that the origins of this ATP are nonsympathetic purinergic or sensory-motor nerves is discussed. ATP can also be released during intravascular platelet aggregation and from intact and damaged vascular smooth muscle cells, and so may play a role in the complex physiological mechanisms controlling local vascular tone under normoxic conditions, during changes in blood flow and during vessel injury.

REFERENCES

1. BURNSTOCK, G. & C. KENNEDY. 1986. Circ. Res. **58:** 319-330.
2. BURNSTOCK, G. 1988. Trends Pharmacol. Sci. **9:** 116-117.
3. BURNSTOCK, G. 1986. J. Anat. **146:** 1-30.
4. DHITAL, K. K. & G. BURNSTOCK. 1989. *In* Diseases of the Arterial Wall. J.-P. Camilleri, C. L. Berry, J.-N. Fiessinger & J. Bariéty, Eds.: 97-126. Springer. London.
5. BURNSTOCK, G., T. COCKS, R. CROWE & L. KASAKOV. 1978. Br. J. Pharmacol. **63:** 125-138.
6. CROWE, R. & G. BURNSTOCK. 1981. Cell Tissue Res. **221:** 93-107.
7. CROWE, R. & G. BURNSTOCK. 1982. Cardiovasc. Res. **16:** 384-390.
8. DA PRADA, M., J. G. RICHARDS & H. P. LOREZ. 1978. *In* Platelets: A Multidisciplinary Approach. G. de Gaetano & S. Garattini, Eds.: 331-353. Raven Press. New York, NY.
9. BURNSTOCK, G., R. CROWE & H. K. WONG. 1979. Br. J. Pharmacol. **65:** 377-388.
10. BURNSTOCK, G., R. CROWE, C. KENNEDY & J. TÖRÖK. 1984. Br. J. Pharmacol. **82:** 359-368.
11. LEVITT, B. & D. P. WESTFALL. 1982. Blood Vessels **19:** 30-40.
12. SU, C. 1975. J. Pharmacol. Exp. Ther. **195:** 159-166.
13. HUGHES, J. & J. R. VANE. 1970. Br. J. Pharmacol. Chemother. **39:** 476-489.
14. KENNEDY, C. & G. BURNSTOCK. 1985. Eur. J. Pharmacol. **111:** 49-56.
15. REILLY, W. M., V. L. SAVILLE & G. BURNSTOCK. 1987. Eur. J. Pharmacol. **140:** 47-53.
16. BURNSTOCK, G., A. M. HOPWOOD, C. H. V. HOYLE, W. M. REILLY, V. L. SAVILLE, M. D. A. STANLEY & J. J. I. WARLAND. 1986. Br. J. Pharmacol. **89:** 857P.
17. INOUE, T. & M. S. KANNAN. 1988. Am. J. Physiol. **254:** H1142-H1148.
18. BELAI & HASSALL. Unpublished data.
19. SHIMADA, S. G. & J. T. STITT. 1984. Br. J. Pharmacol. **83:** 577-589.
20. BURNSTOCK, G. 1980. *In* Handbook of Physiology. Section 2: The Cardiovascular System. Vol. II: Vascular Smooth Muscle. D. F. Bohr, A. D. Somlyo, H. W. Sparks & S. R. Geiger, Eds.: 567-612. American Physiological Society/Waverly Press. Baltimore, MD.
21. SNEDDON, P. & D. P. WESTFALL. 1984. J. Physiol. (London) **347:** 561-580.
22. SNEDDON, P. & G. BURNSTOCK. 1984. Eur. J. Pharmacol. **100:** 85-90.
23. STJÄRNE, L. & P. ÅSTRAND. 1985. Neuroscience **14:** 929-946.
24. KIRKPATRICK, K. & G. BURNSTOCK. 1987. Eur. J. Pharmacol. **138:** 207-214.
25. KASAKOV, L., J. ELLIS, K. KIRKPATRICK, P. MILNER & G. BURNSTOCK. 1988. J. Auton. Nerv. Syst. **22:** 75-82.
26. LUNDBERG, J. M., J. PERNOW, C. DAHLÖF & K. TATEMOTO. 1985. Acta Physiol. Scand. **125:** 511-513.
27. BULLOCH, J. M. & J. C. MCGRATH. 1988. Br. J. Pharmacol. **95:** 695-700.
28. POTTER, D. D., E. J. FURSHPAN & S. C. LANDIS. 1983. Fed. Proc. **42:** 1626-1632.
29. TOLKOVSKY, A. M. & H. S. SUIDAN. 1987. Neuroscience **23:** 1133-1142.
30. CHEUNG, D. W. & M. FUJIOKA. 1986. Br. J. Pharmacol. **89:** 3-5.
31. BURNSTOCK, G. & J. J. I. WARLAND. 1987. Br. J. Pharmacol. **90:** 111-120.
32. MURAMATSU, I. 1986. Br. J. Pharmacol. **87:** 478-480.

33. OMOTE, S., S. KIGOSHI & I. MURAMATSU. 1989. Eur. J. Pharmacol. **160:** 239-245.
34. MURAMATSU, I., T. OHMURA & M. OSHITA. 1989. J. Physiol. (London) **411:** 227-243.
35. MACHALY, M., H. H. DALZIEL & P. SNEDDON. 1988. Eur. J. Pharmacol. **147:** 83-91.
36. YAMAMOTO, R., W. H. CLINE, JR. & K. TAKASAKI. 1988. J. Auton. Pharmacol. **8:** 303-309.
37. RALEVIC, V. & G. BURNSTOCK. 1988. Br. J. Pharmacol. **95:** 637-645.
38. ÅSTRAND, P. & L. STJÄRNE. 1989. Acta Physiol. Scand. **136:** 355-365.
39. ISHIKAWA, S. 1985. Br. J. Pharmacol. **86:** 777-787.
40. RAMME, D., J. T. REGENOLD, K. STARKE, R. BUSSE & P. ILLES. 1987. Naunyn-Schmiedeberg's Arch. Pharmacol. **336:** 267-273.
41. VON KÜGELGEN, I. & K. STARKE. 1985. J. Physiol. (London) **367:** 435-455.
42. KRISHNAMURTY, V. S. R. & P. J. KADOWITZ. 1983. Can. J. Physiol. Pharmacol. **61:** 1409-1417.
43. CHEUNG, D. W. 1982. J. Physiol. (London) **328:** 461-468.
44. SNEDDON, P. & G. BURNSTOCK. 1984. Eur. J. Pharmacol. **106:** 149-152.
45. STJÄRNE, L. & E. STJÄRNE. 1989. Acta Physiol. Scand. **135:** 227-239.
46. ÅSTRAND, P., J. A. BROCK & T. C. CUNNANE. 1988. J. Physiol. (London) **401:** 657-670.
47. VIDAL, M., P. E. HICKS & S. Z. LANGER. 1986. Naunyn-Schmiedeberg's Arch. Pharmacol. **332:** 384-390.
48. BAO, J. X., I. E. ERIKSSON & L. STJÄRNE. 1989. Acta Physiol. Scand. **136:** 287-288.
49. HEAD, R. J., R. E. STITZEL, I. S. DE LA LANDE & S. M. JOHNSON. 1977. Blood Vessels **14:** 229-239.
50. SUZUKI, H. 1985. J. Physiol. (London) **359:** 401-415.
51. MUIR, T. C. & K. A. WARDLE. 1988. J. Auton. Pharmacol. **8:** 203-218.
52. BENHAM, C. D., T. B. BOLTON, N. G. BYRNE & W. A. LARGE. 1987. J. Physiol. (London) **387:** 473-488.
53. SAVILLE, V. L. & G. BURNSTOCK. 1988. Eur. J. Pharmacol. **155:** 271-277.
54. KENNEDY, C., V. L. SAVILLE & G. BURNSTOCK. 1986. Eur. J. Pharmacol. **122:** 291-300.
55. HOLMAN, M. E. & A. SURPRENANT. 1980. Br. J. Pharmacol. **71:** 651-661.
56. SUZUKI, H. & K. KOU. 1983. Jpn. J. Physiol. **33:** 743-756.
57. CORR & BURNSTOCK. Unpublished data.
58. KATSURAGI, T. & C. SU. 1982. J. Pharmacol. Exp. Ther. **220:** 152-156.
59. KATSURAGI, T., R. MORI & T. FURUKAWA. 1987. Gen. Pharmacol. **18:** 485-490.
60. NEELY, C. F., P. J. KADOWITZ, H. LIPPTON, M. NEIMAN & A. L. HYMAN. 1989. J. Pharmacol. Exp. Ther. **250:** 170-176.
61. LIU, S. F., D. G. MCCORMACK, T. W. EVANS & P. J. BARNES. 1989. Br. J. Pharmacol. **98:** 1014-1020.
62. BO, X. & G. BURNSTOCK. 1989. J. Auton. Nerv. Syst. **28:** 85-88.
63. BULLOCH, J. M. 1989. Paper presented at a conference: Purine Nucleosides and Nucleotides in Cell Signalling: Target for New Drugs. September 17-20, 1989. Bethesda, MD.
64. TAYLOR, E. M. & M. E. PARSONS. 1989. Eur. J. Pharmacol. **164:** 23-33.
65. SCHWARTZ, D. D. & K. U. MALIK. 1989. J. Pharmacol. Exp. Ther. **250:** 764-771.
66. MURAMATSU, I., M. FUJIWARA, A. MIURA & Y. SAKAKIBARA. 1981. J. Pharmacol. Exp. Ther. **216:** 401-409.
67. MURAMATSU, I. & S. KOGISHI. 1987. Br. J. Pharmacol. **92:** 901-908.
68. SU, C. 1978. J. Pharmacol. Exp. Ther. **204:** 351-361.
69. GRANT, T. L., N. A. FLAVAHAN, J. GRIEG, J. C. MCGRATH, C. E. MCKEAN & J. L. REID. 1985. Clin. Sci. **68**(Suppl. 10): 25s-30s.
70. HOLTON, F. A. & P. HOLTON. 1954. J. Physiol. (London) **126:** 124-140.
71. FYFFE, R. E. W. & E. R. PERL. 1984. Proc. Natl. Acad. Sci. USA **81:** 6890-6893.
72. JAHR, C. E. & T. M. JESSEL. 1983. Nature **304:** 730-733.
73. SALT, T. E. & R. G. HILL. 1983. Neurosci. Lett. **35:** 53-57.
74. SALTER, M. W. & J. L. HENRY. 1985. Neuroscience **15:** 815-825.
75. BURNSTOCK, G. 1977. J. Invest. Dermatol. **69:** 47-57.
76. MATHIESON, J. J. I. & G. BURNSTOCK. 1985. Eur. J. Pharmacol. **118:** 221-229.
77. BRIZZOLARA, A. L. & G. BURNSTOCK. 1990. Br. J. Pharmacol. **99:** 835-839.
78. ENERO, M. A. & B. O. SAIDMAN. 1977. Naunyn-Schmiedeberg's Arch. Pharmacol. **297:** 39-46.

79. Fredholm, B. B., L. E. Gustafsson, P. Hedqvist & A. Sollevi. 1983. *In* Regulatory Function of Adenosine. R. M. Berne, T. W. Rall & R. Rubio, Eds.: 479-495. Martinus Nijhoff. The Hague.
80. Moylan, R. D. & T. C. Westfall. 1979. Blood Vessels **16:** 302-310.
81. Nedergaard, O. A., S. Husted & J. Schrold. 1980. *In* Vascular Neuroeffector Mechanisms. J. A. Bevan, Ed.: 139-146. Raven Press. New York, NY.
82. Paton, D. M. 1981. *In* Purinergic Receptors: Receptors and Recognition. Series B. G. Burnstock, Ed. Vol. 12: 199-219. Chapman & Hall. London.
83. Ribeiro, J. A. 1979. J. Theor. Biol. **80:** 259-270.
84. Snyder, S. H. 1985. Annu. Rev. Neurosci. **8:** 103-124.
85. Vizi, E. S. 1979. Prog. Neurobiol. **12:** 181-290.
86. De Mey, J., G. Burnstock & P. M. Vanhoutte. 1979. Eur. J. Pharmacol. **55:** 401-405.
87. Wakade, A. R. & T. D. Wakade. 1978. J. Physiol. (London) **282:** 35-49.
88. Kennedy, C. & G. Burnstock. 1984. Eur. J. Pharmacol. **100:** 363-368.
89. Sneddon, P., L. Meldrum & G. Burnstock. 1984. Eur. J. Pharmacol. **105:** 293-299.
90. Lautt, W. W. 1986. Can. J. Physiol. Pharmacol. **64:** 1291-1295.
91. Wiklund, N. P., B. Cederqvist & L. E. Gustafsson. 1989. Br. J. Pharmacol. **96:** 425-433.
92. Burnstock, G. 1980. Gen. Pharmacol. **11:** 15-18.
93. Burnstock, G. 1982. *In* Co-transmission. A. C. Cuello, Ed.: 151-163. Macmillan. London.
94. Moody, C. & G. Burnstock. 1982. Eur. J. Pharmacol. **77:** 1-9.
95. Ribeiro, J. A. & M. L. Dominguez. 1978. J. Physiol. (Paris) **74:** 491-496.
96. Silinsky, E. M. 1980. Br. J. Pharmacol. **71:** 191-194.
97. Hedqvist, P. & B. B. Fredholm. 1976. Naunyn-Schmiedeberg's Arch. Pharmacol. **293:** 217-223.
98. Holck, M. I. & B. H. Marks. 1978. J. Pharmacol. Exp. Ther. **205:** 104-117.
99. Ralevic, V. & G. Burnstock. 1990. Eur. J. Pharmacol. **175:** 291-299.
100. Akasu, T., K. Hirai & K. Koketsu. 1981. Br. J. Pharmacol. **74:** 505-507.
101. Buchthal, F. & G. Kahlson. 1944. Acta Physiol. Scand. **8:** 325-334.
102. Ewald, D. A. 1976. J. Membr. Biol. **29:** 47-65.
103. Ribeiro, J. A. 1977. J. Membr. Biol. **33:** 401-402.
104. Saji, Y., G. Escalona de Motta & J. del Castillo. 1975. Life Sci. **16:** 945-954.
105. Furchgott, R. F. & J. V. Zawadzki. 1980. Nature **288:** 373-376.
106. De Mey, J. G. & P. M. Vanhoutte. 1981. J. Physiol. (London) **316:** 347-355.
107. Furchgott, R. F. 1981. Trends Pharmacol. Sci. **2:** 173-176.
108. Gordon, J. L. & W. Martin. 1983. Br. J. Pharmacol. **79:** 531-541.
109. Houston, D. A., G. Burnstock & P. M. Vanhoutte. 1987. J. Pharmacol. Exp. Ther. **241:** 501-506.
110. Kennedy, C., D. Delbro & G. Burnstock. 1985. Eur. J. Pharmacol. **107:** 161-168.
111. Martin, W., N. J. Cusack, J. S. Carleton & J. L. Gordon. 1985. Eur. J. Pharmacol. **108:** 295-299.
112. Sauve, R., L. Parent, C. Simoneau & G. Roy. 1988. Pfluegers Arch. **412:** 469-481.
113. Kelm, M., M. Feelisch, R. Spahr, H. M. Piper, E. Noack & J. Schrader. 1988. Biochem. Biophys. Res. Commun. **154:** 236-244.
114. Burnstock, G. & J. J. I. Warland. 1987. Br. J. Pharmacol. **90:** 383-391.
115. Hopwood, A. M. & G. Burnstock. 1987. Eur. J. Pharmacol. **136:** 49-54.
116. Hopwood, A. M., J. Lincoln, K. A. Kirkpatrick & G. Burnstock. 1989. Eur. J. Pharmacol. **165:** 323-326.
117. Boeynaems, J. M. & N. Galand. 1983. Biochem. Biophys. Res. Commun. **112:** 290-296.
118. Pearson, J. D., L. L. Slakey & J. L. Gordon. 1983. Biochem. J. **214:** 273-276.
119. Van Coevorden, A., P. P. Roger & J.-M. Boeynaems. 1989. Thromb. Haemostas. **62:** 190.
120. Radomski, M. W., R. M. J. Palmer & S. Moncada. 1987. Br. J. Pharmacol. **92:** 639-646.
121. Pirotton, S., E. Raspe, D. Demolle, C. Erneux & J.-M Boeynaems. 1987. J. Biol. Chem. **262:** 17461-17466.

122. Forsberg, E. J., G. Feuerstein, E. Shohami & H. B. Pollard. 1987. Proc. Natl. Acad. Sci. USA **84:** 5630-5634.
123. Brock, T. A., P. A. Dennis, K. K. Griendling, T. S. Diehl & P. F. Davies. 1988. Am. J. Physiol. **255:** C667-C673.
124. Carter, T. D., T. J. Hallam, N. J. Cusack & J. D. Pearson. 1988. Br. J. Pharmacol. **95:** 1181-1190.
125. Carter, T. D., T. J. Hallam & J. D. Pearson. 1989. Biochem. J. **262:** 431-437.
126. Pearson, J. D. & J. L. Gordon. 1985. Annu. Rev. Physiol. **47:** 617-627.
127. Schubert, P. & G. W. Kreutzberg. 1976. *In* The Cerebral Vessel Wall. J. Cervós-Navarro, E. Betz, F. Matakas & R. Wüllenweber, Eds.: 207-213. Raven Press. New York, NY.
128. Berne, R. M. 1963. Am. J. Physiol. **204:** 317-322.
129. Berne, R. M., T. W. Rall & R. Rubio. 1983. Regulatory Function of Adenosine. Martinus Nijhoff. The Hague.
130. Sparks, H. V., Jr. & M. W. Gorman. 1987. *In* Topics and Perspectives in Adenosine Research. E. Gerlach & B. F. Becker, Eds.: 406-415. Springer. Berlin.
131. Parnavelas, J., W. Kelly & G. Burnstock. 1985. Nature **316:** 724-725.
132. Loesch, A. & G. Burnstock. 1988. Anat. Embryol. **178:** 137-142.
133. Burnstock, G., J. Lincoln, E. Fehér, A. M. Hopwood, K. Kirkpatrick, P. Milner & V. Ralevic. 1988. Experientia **44:** 705-707.
134. Milner, P., V. Ralevic, A. M. Hopwood, E. Fehér, J. Lincoln, K. A. Kirkpatrick & G. Burnstock. 1989. Experientia **45:** 121-125.
135. Kamikawa, Y., W. H. Cline & C. Su. 1980. Eur. J. Pharmacol. **66:** 347-353.
136. Jackson, E. K. 1987. Am. J. Physiol. **253:** H909-H918.
137. Lüscher, T. F. & P. M. Vanhoutte. 1986. Hypertension **8**(Suppl. II): II-55-II-60.
138. Taddei, S., A. Salvetti & R. Pedrinelli. 1989. Circulation **80:** 485-490.
139. Uehara, Y. & K. Suyama. 1978. J. Electron Microsc. **27:** 157-159.
140. Burnstock, G. 1989. Eur. Heart J. **10**(Suppl. F): 15-21.

DISCUSSION OF THE PAPER

J. S. Fedan (*National Institute for Occupational Safety and Health, Morgantown, WV*): Your comment about sympathetic varicosities has implications for the view that adrenoceptor density is not uniform on (at least) smooth muscle cells. Can you comment on the relationship?

Burnstock: If you are referring to the c-adrenoceptor hypothesis, I do not believe that this can be regarded as secure evidence for an unhomogenous distribution of receptors on smooth muscle. Autoradiographic studies of receptor localization suggest a homogenous distribution for most smooth muscles.

I. L. O. Buxton (*University of Nevada School of Medicine, Reno, NV*): Dr. Burnstock, you have described the release of ATP from blood vessels in response to hypoxia and sheer stress. Do you believe that ATP is responsible for moment-to-moment regulation of blood vessel tone, or do you see ATP's role as being important in pathophysiological states?

Burnstock: My guess is that endothelium-mediated vasodilation triggered by ATP is operating in normal homeostatic conditions where it is in dynamic balance with perivascular nerve activity as part of dual control of local blood flow. In pathophysiological conditions such as ischemia, hypoxia, or rapid change in blood flow, however, endothelium-mediated vasodilation may play a dominant role.

R. I. HUME (*University of Michigan, Ann Arbor, MI*): How would both P_{2X} and P_{2Y} receptors on vessels be selectively activated if nerves can move?

BURNSTOCK: I do not believe that movement of varicosities in extended perivascular nerve fibers would make much difference to the effects of ATP release. It is true, however, that we know too little about the regional distribution of perivascular sympathetic and nonsympathetic nerves releasing ATP in different vascular beds to predict which response would dominate in high- or low-tone conditions in a given vessel. A similar problem exists for the receptors of vessels containing both α- and β-adrenoceptors.

J. S. WILEY (*Austin Hospital, Heidelberg, Australia*): I am curious about the source of plasma ATP. In the past, we have focused on platelet-dense granules as a source of plasma ATP, but your comment about an endothelial source of plasma ATP does raise the question of whether this secreted compound is vesicular (granular) or cytoplasmic in origin. Do Wiebel-Palade bodies or other endothelial granules contain ATP as well as factor 8?

BURNSTOCK: We have been carrying out a series of studies of endothelial storage and release of not only ATP, but also of acetylcholine, substance P, and serotonin. Release is Ca^{2+} dependent, and my guess is that it is largely by exocytosis from vascular stores, comparable to release from nerve terminals. We still need hard evidence for this, however. As for nerves, we cannot exclude the possibility of nonvesicular release of ATP.

The Effects of ATP on Endothelium

JOHN L. GORDON

British Biotechnology, Ltd.
Oxford OX4 5LY, England

INTRODUCTION

When considering the significance of extracellular ATP in relation to a particular cell type or tissue, we should keep in mind certain fundamental questions:

- What are the opportunities for the cells to encounter extracellular ATP?
- From what source(s) and in what concentrations?
- How do the cells respond to ATP?
- What are the biological consequences of such responses?
- How and where is the extracellular ATP metabolized?
- Do the metabolic pathways themselves exert biological effects?

In some situations where extracellular ATP may be encountered, the answers to most of the questions listed above are a matter for conjecture, but in the case of the vascular endothelium, a considerable amount of relevant information is available (see Gordon[1] for a review).

ANATOMICAL LOCATION OF ENDOTHELIUM

The opportunities for vascular endothelial cells to interact with extracellular ATP are considerable, partly because of the anatomical location of the endothelium. Because it is the interface between the blood and the rest of the body, the vascular endothelium is the first tissue exposed to any ATP released into the plasma, and there are several potential sources of plasmatic ATP.

ATP FROM BLOOD CELLS

ATP is a virtually ubiquitous intracellular constituent, and therefore it is no surprise that erythrocytes, leukocytes, and platelets all contain significant amounts of ATP.

As is the case with other tissues in the body, ATP can be released from blood cells exposed to damaging stimuli, even if the damage is sublethal—for example, release of ATP from erythrocytes in response to hemodynamic and other stresses has been demonstrated. In addition, however, the blood platelets are a particularly rich source of plasmatic ATP because their dense bodies, in which serotonin is stored, also contain ATP in remarkably high concentrations—of the order of 1 M. The contents of these dense storage granules are released from the platelets when they aggregate and undergo a secretion reaction. In consequence, the concentration of ATP plus ADP in the bulk phase of the plasma can reach concentrations of around 20 μM or so,[1,2] and the ATP concentration immediately around a platelet aggregate can be even higher.

ATP FROM THE VESSEL WALL AND OTHER TISSUES

Research into the effects of trauma during and shortly after the Second World War showed that ATP was released into the plasma following physical damage to tissues, or ischemia.[3] Subsequent work quantified the ATP released into the blood perfusing ischaemic vascular beds[4] or working muscle.[5] The ATP released from individual small blood vessels subjected to physical trauma has also been measured,[6] and studies with vascular endothelial cells in culture showed that they could release ATP in biologically significant amounts when exposed to stimuli such as cationic proteins or proteolytic enzymes.[7–9]

RESPONSES OF ENDOTHELIAL CELLS TO ATP

Prostacyclin

When endothelial cells are grown to confluence on microcarrier beads, and these beads are packed in columns, continuously perfused, and briefly exposed to ATP, there is a rapid production of prostacyclin.[10] The production of prostacyclin is dose dependent, with the maximum secretion rate being reached only seconds after the initial exposure. This represents an increase of several hundred fold over basal production, and prostacyclin secretion returns to basal level within a few minutes, even in the continuing presence of the stimulus. Repeat exposures to extracellular ATP result in the repeated production of prostacyclin,[11] unless the intervals between the exposures are too short, in which case desensitization occurs.[12]

The receptor involved in this response is of the P_{2Y} subclass.[13] Studies with a range of analogues showed that substitution in the 2-position increases potency, that the stereochemistry of the sugar moiety is critical, and that metabolism of the ATP is unrelated to stimulatory potency—analogues poorly metabolized are of similar potency to those that are readily susceptible to metabolism.[11]

Endothelium-Derived Relaxing Factor

Shortly after the discovery of endothelium-derived relaxing factor (EDRF) induced by vasodilators such as acetylcholine,[14] it was established that ATP could also induce endothelium-dependent vasodilation,[15] even when the capacity of the endothelial cells to synthesize prostacyclin was blocked.[16] The receptor involved in this response is also of the P_{2Y} subclass and shows broad similarities to the receptor responsible for prostacyclin production, in terms of the rank order of potency of a range of ATP analogues, but there are some intriguing differences.[11,17] It is too early to conclude that subtypes of the P_{2Y} receptor may exist, but the possibility of further subdivision of purinoceptors should be kept in mind.

The experimental system involving columns of microcarrier beads covered in endothelial cells and subjected to continuous perfusion, described above for studies on prostacyclin production, was also used to establish that EDRF released from stimulated endothelial cells was nitric oxide,[18] derived from L-arginine.[19] This pathway has now been identified in other cell types.[20]

INTRACELLULAR REGULATION

It has become clear that the levels of intracellular calcium in endothelium are central to the control of prostacyclin production in response to various stimuli, including ATP.[21] This rise in intracellular calcium begins first by liberation from intracellular stores and second by influx from the surrounding medium.[22] The full details of the metabolic pathways responsible for EDRF production in response to stimulation by ATP remain to be elucidated.

Membrane Permeability

In common with many other cell types,[23,24] endothelial cells respond to high concentrations of ATP (around millimolar) with an increase in membrane permeability.[25] Such concentrations of extracellular ATP are probably seldom encountered *in vivo,* although theoretical calculations indicate that they can arise in the pericellular environment surrounding an aggregate of platelets that are undergoing maximum degranulation.

BIOLOGICAL CONSEQUENCES OF STIMULATION OF ENDOTHELIUM BY ATP

The actions of extracellular ATP on vascular endothelium have two main effects: one on vascular tone (and therefore blood flow), and one on platelet aggregation.

Vascular Tone

The appearance of ATP in the plasma usually results in vasodilation. This vasodilation could arise (in theory) either through the actions of prostacyclin or of EDRF, but studies in which ATP was injected into perfused vascular beds (for example, the coronary circulation) have shown that although prostacyclin and EDRF can both induce vasodilation and therefore increase blood flow, EDRF is the agent responsible under most conditions. The vasodilator effects of prostacyclin are only seen in response to larger amounts of ATP.[26] Although ATP can cause vasoconstriction in some blood vessels by an action via P_{2X} receptors on vascular smooth muscle cells,[27] the net effect is usually vasodilation—presumably because the actions are mediated via the P_{2Y} receptors on endothelial cells, which predominate under normal conditions when the endothelium is intact. ATP-induced vasodilation has been seen following bolus injections, continuous infusion, or topical application to the external surface of some vascular beds (for example, the cat pial arteries).[28]

Platelet Function

Both prostacyclin and nitric oxide can also act as powerful inhibitors of platelet aggregation, the former by stimulating adenylate cyclase and thus raising intracellular levels of cyclic AMP, and the latter by stimulating guanylate cyclase and thus raising cyclic GMP. Where both stimuli are present together, and therefore both pathways are activated, there is a synergistic inhibitory effect (see Moncada *et al.*[29] for a review).

EFFECTS OF ATP METABOLITES

In addition to the effects of prostacyclin and nitric oxide, which are released from endothelium stimulated by ATP, the metabolites of ATP also affect vascular tone and platelet function. When ATP in the circulation is metabolized (*vide infra*), ADP, AMP, and adenosine are formed. ADP acts like ATP on P_{2Y} receptors, and can thus reinforce ATP's vasodilating effects, but it also induces platelet aggregation via a distinct class of purinoceptor on blood platelets. Adenosine, the eventual metabolite, can induce vasodilation by direct action on vascular smooth muscle cells, and it can also inhibit platelet aggregation by stimulating adenylate cyclase—like prostacyclin, although less effectively.

METABOLISM OF ATP

ATP is not rapidly metabolized in isolated blood samples: the half-time is 20–30 min in plasma at 37 °C, and 10-15 min in whole blood. In the circulation, however, metabolism is very efficient: the half-time for adenine nucleotides in the pulmonary vasculature has been calculated to be less than 0.6 sec.[30] ATP is metabolized in the circulation mainly by ectoenzymes on the luminal surface of endothelium.[31] The speed of ATP metabolism derives not only from the characteristics of these endothelial ectonucleotidases, but also from the enormous surface area of endothelium in microcirculatory beds. In consequence, ATP metabolism takes place mainly in the microcirculation, not in large blood vessels. The metabolic pathway involves sequential dephosphorylation, with ATP being converted to ADP, then to AMP, then to adenosine, which is taken up by active transport and rephosphorylated intracellularly.[32]

Ectonucleotidases have also been described on other cells in various tissues of the body, and on circulating blood cells, but any ATP released into the plasma is metabolized primarily by endothelial ectonucleotidases, largely because of the position of endothelium as the biological interface with the blood (q.v.).

The endothelial ectonucleotidases that metabolize ATP, ADP, and AMP are three discrete enzymes with complex interactive control.[33] Their characteristics are in some respects different from those of the ectonucleotidases present on other cell types (see Slakey *et al.*[34] for full details of ATP metabolism by endothelial and vascular smooth muscle cells).

REFERENCES

1. GORDON, J. L. 1986. Biochem. J. **233:** 309-319.
2. INGERMAN, C. M., J. B. SMITH & M. J. SILVER. 1979. Thromb. Res. **16:** 335-344.
3. GREEN, H. N. & H. B. STONER. 1950. Biological Actions of the Adenine Nucleotides. H. K. Lewis. London.
4. CLEMENS, M. G. & T. FORRESTER. 1981. J. Physiol. **312:** 143-158.
5. FORRESTER, T. 1972. J. Physiol. **224:** 611-628.
6. BORN, G. V. R. & M. A. A. KRATZER. 1984. J. Physiol. **354:** 419-429.
7. PEARSON, J. D. & J. L. GORDON. 1979. Nature **281:** 384-386.
8. LOLLAR, P. & W. G. OWEN. 1981. Ann. N.Y. Acad. Sci. **370:** 51-56.
9. LEROY, E. C., A. AGER & J. L. GORDON. 1984. J. Clin. Invest. **74:** 1003-1010.
10. PEARSON, J. D., L. L. SLAKEY & J. L. GORDON. 1983. Biochem. J. **214:** 273-276.
11. NEEDHAM, L. A., N. J. CUSACK, J. D. PEARSON & J. L. GORDON. 1987. Eur. J. Pharmacol. **134:** 199-209.
12. TOOTHILL, V. J., L. A. NEEDHAM, J. L. GORDON & J. D. PEARSON. 1988. Eur. J. Pharmacol. **157:** 189-196.
13. BURNSTOCK, G. & C. KENNEDY. 1985. Gen. Pharmacol. **16:** 433-440.
14. FURCHGOTT, R. F. & J. V. ZAWADSKI. 1980. Nature **288:** 373-376.
15. DEMEY, J. G. & P. M. VANHOUTTE. 1981. J. Physiol. (London) **316:** 437-455.
16. GORDON, J. L. & W. MARTIN. 1983. Br. J. Pharmacol. **79:** 531-542.
17. MARTIN, W., N. J. CUSACK, J. S. CARLETON & J. L. GORDON. 1985. Eur. J. Pharmacol. **108:** 295-299.
18. PALMER, R. M. J., A. G. FERRIGE & S. MONCADA. 1987. Nature **327:** 524-526.
19. PALMER, R. M. J., D. S. ASHTON & S. MONCADA. 1988. Nature **333:** 664-666.
20. MONCADA, S., R. M. J. PALMER & A. HIGGS. 1989. Commentary Biochem. Pharmacol. **38:** 1709-1715.
21. HALLAM, T. J. & J. D. PEARSON. 1986. FEBS Lett. **207:** 95-99.
22. PEARSON, J. D. & T. D. CARTER. 1990. Ann. N.Y. Acad. Sci. This volume.
23. DICKER, P., L. A. HEPPEL & E. ROZENGURT. 1980. Proc. Natl. Acad. Sci. USA **77:** 2103-2107.
24. COCKCROFT, S. & B. D. GOMPERTS. 1980. Biochem J. **188:** 789-798.
25. PEARSON, J. D. & J. L. GORDON. Unpublished results.
26. FLEETWOOD, G. & J. L. GORDON. 1987. Br. J. Pharmacol. **90:** 219-227.

27. KENNEDY, C., D. DELBRO & G. BURNSTOCK. 1985. Eur. J. Pharmacol. **107:** 161-168.
28. FORRESTER, T., A. M. HARPER, E. T. MACKENZIE & E. M. THOMSON. 1979. J. Physiol. (London) **296:** 343-355.
29. MONCADA, S., R. M. J. PALMER & A. HIGGS. 1988. Hypertension **12:** 365-372.
30. RYAN, U. & J. W. SMITH. 1971. Trans. Assoc. Am. Physicians **84:** 297-306.
31. PEARSON, J. D. & J. L. GORDON. 1985. Annu. Rev. Physiol. **47:** 617-627.
32. PEARSON, J. D., J. S. CARLETON, A. L. HUTCHINGS & J. L. GORDON. Biochem. J. **170:** 265-271.
33. GORDON, E. L., J. D. PEARSON & L. L. SLAKEY. 1986. J. Biol. Chem. **261:** 15496-15504.
34. SLAKEY, L. L., E. L. GORDON & J. D. PEARSON. 1990. Ann. N.Y. Acad. Sci. This volume.

DISCUSSION OF THE PAPER

S. C. SILVERSTEIN (*Columbia University, New York, NY*): Can you distinguish the receptors on the endothelial cells that recognize ATP from those that recognize UTP? Does ATP desensitize endothelial cell responses to UTP? Peter Ward has said that in neutrophils UTP desensitizes the ATP effect on Ca^{2+} mobilization.

GORDON: We have not done that experiment. Peter Ward's group, however, by measuring calcium transients in human neutrophils, has shown that ATP and UTP cross-desensitize, and that the response to FMLP is unaffected.

P. A. WARD (*University of Michigan Medical School, Ann Arbor, MI*): In data to be presented by Dr. B. Walker, in human neutrophils ATP and UTP mutually desensitize to each other with respect to calcium transients, but the response to FMLP is entirely intact, suggesting that signal transduction pathways related to nucleotides are linked but independent of these pathways related to FMLP.

B. B. FREDHOLM (*Karolinska Institutet, Stockholm, Sweden*): We (Fredholm & Lerner. *Biochem. Pharmocol.* **34:** 937-940, 1985) found that 2-chloro-adenosine can cause PG release and PG-dependent bone resorption that is not mimicked by other adenosine analogues or blocked by adenosine receptor antagonists. Could 2-Cl-adenosine act on the same site as 2-Cl-ATP that you described? Or is it possible that 2-Cl-adenosine could be metabolized to 2-Cl-ATP and be active in that form?

GORDON: 2-Cl-Adenosine was ineffective at stimulating PGI_2 production in endothelial cells.

C. F. NEELY: What contributions does prostacyclin release make in the ATP-induced relaxation of precontracted pig aortic rings? I have demonstrated that under conditions of controlled pulmonary blood flow the ATP-induced vasodilation of precontracted vessels in the feline pulmonary vascular bed is not affected by cyclooxygenase inhibition. It would depend on the response of a given vascular bed to respond to nitric oxide; it most likely would vary according to the vascular bed.

GORDON: PGI_2 did not contribute to relaxation of the piglet aorta induced by ATP. ATP probably induces endothelium-dependent vasodilation in most vascular beds by stimulating the production of nitrous oxide.

J. S. DAVIDSON (*University of Cape Town Medical School, Cape Town, South Africa*): Although you are reluctant to propose two P_{2Y} subtypes, your data would appear to demand it. It is very difficult to envisage how a similar receptor could produce the different responses to UTP and to 2-methylthio-ATP you have observed.

GORDON: Subdivision of the P_{2Y} receptor class may eventually be necessary, but until antagonists that discriminate between subtypes are available, attempts at formal subclassification are probably premature at present.

Y. H. EHRLICH (*College of Staten Island, New York, NY*): Based on your data of ATP content and percentage release, what is the estimate of the ATP concentration at the cell surface during the cytoplasmic release of ATP or during the vesicular exocytosis of ATP?

GORDON: The cytoplasmic concentration of ATP is about 1 mM in most cells; therefore, if most of this is released, and diffuses into a pericellular unstirred layer of, say, 10 times greater volume than the intracellular compartment, the pericellular concentration of ATP could initially be up to 100 μM or so. The pericellular concentration around a platelet aggregate that has undergone maximum degranulation could initially be even higher, because the ATP concentration in the dense granules (from which ATP is selectively released by exocytosis) approaches 1 M.

I. FRIEDBERG (*Tel Aviv University, Tel Aviv, Israel*): I am referring to the first part of your lecture, about the release of ATP from endothelial cells by trypsin and other stimulators. Is this release specific for nucleotides, or is the release due to a nonspecific increase in the membrane permeability?

GORDON: The release of ATP is selective, inasmuch as no concomitant release of LDH from the cytoplasm was observed.

S. E. O'CONNOR (*Fisons plc, Loughborough, Leicestershire, England*): Continuing the theme of possible subtypes of P_{2Y} receptors, you showed that 2-Me-S-ATP had very low efficacy for PGI_2 release from endothelial cells, suggesting that it is a partial agonist. Have you tested the antagonist properties of 2-Me-S-ATP against full agonists in this system, that is, 2-Cl-ATP, ATP, and particularly UTP?

GORDON: Using ATP as the agonist, we tested 2-Me-S-ATP, but—to our disappointment—we found no antagonist effects.

The Effects of ATP and Related Nucleotides on Visceral Smooth Muscle

DAVID SATCHELL

Department of Zoology
University of Melbourne
Parkville, Victoria, Australia 3052

INTRODUCTION

Sixty years ago the effects of adenine nucleotides on smooth muscle were reported.[1] Since this time a large volume of data has accumulated, first on the actions of the naturally occurring adenine nucleotides on smooth muscle, and more recently on the effects of adenine nucleotide analogues, that is, adenine nucleotides having modifications in the purine and ribose moieties and in the phosphate chain.

In order to gain a comprehension of the actions of ATP and its analogues on visceral smooth muscle, the actions of ATP on *in vitro* preparations of a wide variety of tissues have been determined and the results examined. The guinea pig taenia coli preparation has been selected for detailed studies on the interpretation of the action of ATP, and it will be shown that the action of this nucleotide is complicated not only by its rapid breakdown during the contact time with the preparation, but by actions of the breakdown products themselves.

Studies have also been carried out on the effects of a short series of analogues of ATP on selected preparations from the gut, the reproductive system, and the airways, in order to further examine the concept[2] of different receptor subtypes for ATP.

METHODOLOGY

Lung strips were dissected from along the pointed margins of lung lobes of guinea pigs. These strips were 15 mm in length and 1.5 mm in diameter. All other preparations were dissected as described previously[3] and placed in organ baths containing a Krebs solution of the following composition (millimolar concentrations): NaCl, 133; $CaCl_2$, 2.5; $MgSO_4$, 1.2; $NaHCO_3$, 16.3; NaH_2PO_4, 1.5; dextrose, 7.8; KCl, 4.7. The solution was bubbled with a 95% oxygen-5% carbon dioxide mixture and maintained at 37 °C.

Muscle activity was registered on Grass polygraphs recorded isometrically with Grass FT03 transducers or Gould-Statham UC2 transducers.

Adenine nucleotide analogues were obtained from the Sigma Chemical Company, St Louis, MO.

At the conclusion of each organ bath experiment, tetrodotoxin (1 μg/ml) was added to the organ bath and the effects of agonists tested, to rule out the possibility of nerve-mediated responses.

RESULTS AND DISCUSSION

Actions of ATP on Various Preparations of Visceral Smooth Muscle

ATP and usually ADP have the potential to cause relaxation or contraction of most preparations containing visceral smooth muscle. In some cases both responses are seen. For example, in preparations of guinea pig ileum, ATP (5 μM) consistently caused biphasic responses in which relaxations were followed by contractions (FIG. 1 & TABLE 1). The shape of the response was influenced by the tone of the preparation. Contractions were pronounced in low-tone preparations, and relaxations were enhanced by raising the tone of preparations (FIG. 1).

When the individual circular and longitudinal muscle layers of guinea pig ileum were surgically isolated and mounted in organ baths, preparations containing the circular layer consistently relaxed in response to ATP, while preparations containing the longitudinal layer consistently contracted and failed to relax even at high tone (FIG. 1). It is not clear, then, how Finkleman-type preparations, that is, whole-segment preparations of guinea pig ileum, elicited their response. One possibility is that the relaxation of the circular muscle layer preceded contraction of the longitudinal layer and that the initial relaxation of the segment was due to relaxation of the circular muscle. Another possibility is that longitudinal muscle that is in segments behaves differently from longitudinal muscle that is dissected free.

These results emphasize the advisability of conducting experiments on individual muscle layers rather than whole segments. There is a large amount of information,

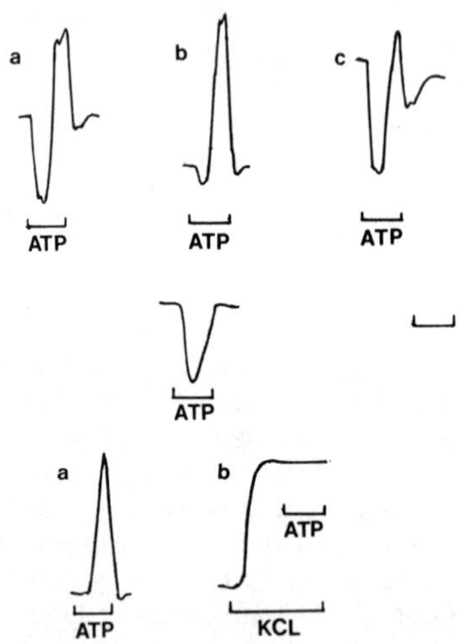

FIGURE 1. Effects of ATP (5 μM) on preparations of guinea pig ileum over 30-sec intervals. The traces in the upper panel show responses of whole-segment preparations: (**a**) a preparation with normal tone; (**b**) a preparation with low tone; (**c**) a preparation with high tone. The trace in the middle panel shows the response of the circular muscle layer dissected free of the longitudinal muscle. The traces in the lower panel show responses of preparations of longitudinal layer dissected free of the circular layer: (**a**) a preparation with normal tone; (**b**) a preparation in which the tone had been raised by adjusting the K^+ concentration of the Krebs solution to 17.5 mM. Time marker: 30 sec.

TABLE 1. Effect of ATP on Preparations of Visceral Smooth Muscle[a]

Tissue	Concentration (μM)	Response
Guinea pig taenia coli	0.2-10	Relaxation
Guinea pig and mouse colon	5-150	Relaxation
Mouse rectum	10-50	Relaxation
Mouse and rabbit ileum	1-100	Relaxation
Guinea pig ileum	5	Relaxation then contraction
Guinea pig internal anal sphincter	1-200	Relaxation
Rabbit anococcygeus	10-500	Relaxation
Stomach of various mammalian species	1-200	Relaxations most common—contractions and biphasic responses also, depending on muscle layer and species
Guinea pig ileum longitudinal muscle	1-30	Contraction
Guinea pig ileum circular muscle	1-30	Relaxation
Rabbit colon and rectum longitudinal muscle	100-300	Contraction—sometimes biphasic (that is, relaxation then contraction)
Rabbit colon circular muscle	100-300	Relaxation
Guinea pig bladder	0.2-10	Contraction
Guinea pig vas deferens	0.05-5	Contraction
Guinea pig seminal vesicle	30-100	Contraction
Guinea pig trachea	50-200	Relaxation

[a] Intestinal preparations were as segments except where stated

however, on the effects of ATP on various gut segments from a variety of mammalian species. The usual effect of ATP (1-150 μM) is a relaxation (TABLE 1). It is likely that the powerful relaxations that are seen in these preparations are the result of inhibition of the longitudinal muscle layer. It is well established that ATP causes relaxations in other longitudinal muscle layers associated with the intestine, for example, guinea pig taenia coli and muscles to the coccyx (see Maguire and Satchell[4]) (TABLE 1).

The effects of ATP in preparations of esophagus, stomach, and uterus were variable, depending on the species and the site and disposition of the muscle layer (TABLE 1) (see also Maguire and Satchell[4]). In other areas of the urinogenital system ATP caused rapid contractions, and its actions in vas deferens and bladder have been widely studied in relation to the concept of its role as a cotransmitter with norepinephrine in sympathetic nerve terminals.[5] The effects of ATP on isolated vas deferens and bladder preparations may be attenuated by tissue barriers limiting access to the smooth muscle cells. In guinea pig bladder, preparations responded with contractions over a range from 0.2 to 10 μM ATP (TABLE 1). In the guinea pig vas deferens preparation, responses to ATP were enhanced by dissecting the preparation free of its adventitial coat. In preparations that had been opened by a longitudinal cut along the wall opposite to the blood vessels, careful removal of the epithelium and the underlying lamina propria provided preparations that were particularly sensitive to ATP. They responded to threshold concentrations of 0.05 μM, although such preparations usually exhibited a decreased force of contraction due to dissection procedures.

Analysis of the Actions of ATP on Isolated Taenia Coli Preparations

Adenosine (25 μM) caused a slowly developing relaxation that took approximately 30 sec to reach maximum. In contrast to this, ATP (7 μM) caused a relaxation that was much more rapid in onset, taking 11 to 15 sec to reach maximum (FIG. 2). The response to ATP returned toward the baseline even while ATP was in the bath, but there was a delay, even after washout, before the response returned to baseline. When dipyridamole (1 μM) was present, the initial response to ATP remained unchanged, but the delayed relaxation was enhanced, suggesting that the delayed component of the relaxation could be due to adenosine. In smaller baths (that is, baths reduced from 10 to 3 ml, each bath containing four preparations), the delayed component of the response was further enhanced, suggesting a further accumulation of adenosine formed by the breakdown of ATP. When the preparations were placed in a 50-ml bath, applied ATP still caused a markedly delayed relaxation in the presence of dipyridamole (FIG. 2). Presumably this is because adenosine formed from ATP only partially exchanged with the bath medium.

To further elucidate the actions of ATP on guinea pig taenia coli preparations, a strip was mounted in air and perfused dropwise with Krebs solution. ATP (5.5 μM) was introduced to the bath and allowed to perfuse over the preparation for 6 min. This perfusate was collected and stored, and in the interim normal Krebs solution was perfused over the preparation for 15 min. At this time the stored perfusate was reintroduced to the preparation and reperfused over the surface. During the first perfusion ATP induced a rapid relaxation. This relaxation, however, was not maintained over the 6 min allowed for the perfusion even though fresh ATP was being supplied by the perfusion process. The preparation had returned almost 80% of the way back to the baseline by the time normal Krebs solution was restored after the 6 min (FIG. 3). When the collected perfusate was reintroduced to the preparation, the pattern of the relaxation was the same as in the previous perfusion. Clearly no significant breakdown of ATP had occurred during the first perfusion. The possibility remained that ATP may have been causing autoinhibition or tachyphylaxis. This seemed unlikely, however, because perfusion with the more stable α,β-methylene analogue of ATP (henceforth AOPCPOP)(3.5 μM) for 6 min caused a relaxation that was sustained over the 6 min (FIG. 3). The differences between the actions of ATP and AOPCPOP in these experiments were not readily explained. One possibility was that the perfused ATP was accumulating metabolites in the extracellular space that were not exchanging readily with the incoming ATP from the perfusate. Also, it seemed that the concentration of metabolites formed would need to be small relative to the total perfused ATP because stored recycled perfusate was as effective as the initial perfusate.

Minimal Structural Requirements for Actions at P_2 Purinoceptors

The presence of a 5′-polyphosphate was mandatory for actions at P_2 sites. There was not much difference in potency between two and three phosphates. Four phosphates were more active than three in causing contraction of guinea pig vas deferens strips (FIG. 4) and causing relaxation of taenia coli strips (see Satchell and Maguire[6]). Five phosphates in the form of a diadenosine pentaphosphate were most potent in contracting and causing desensitization of guinea pig vas deferens strips.[7]

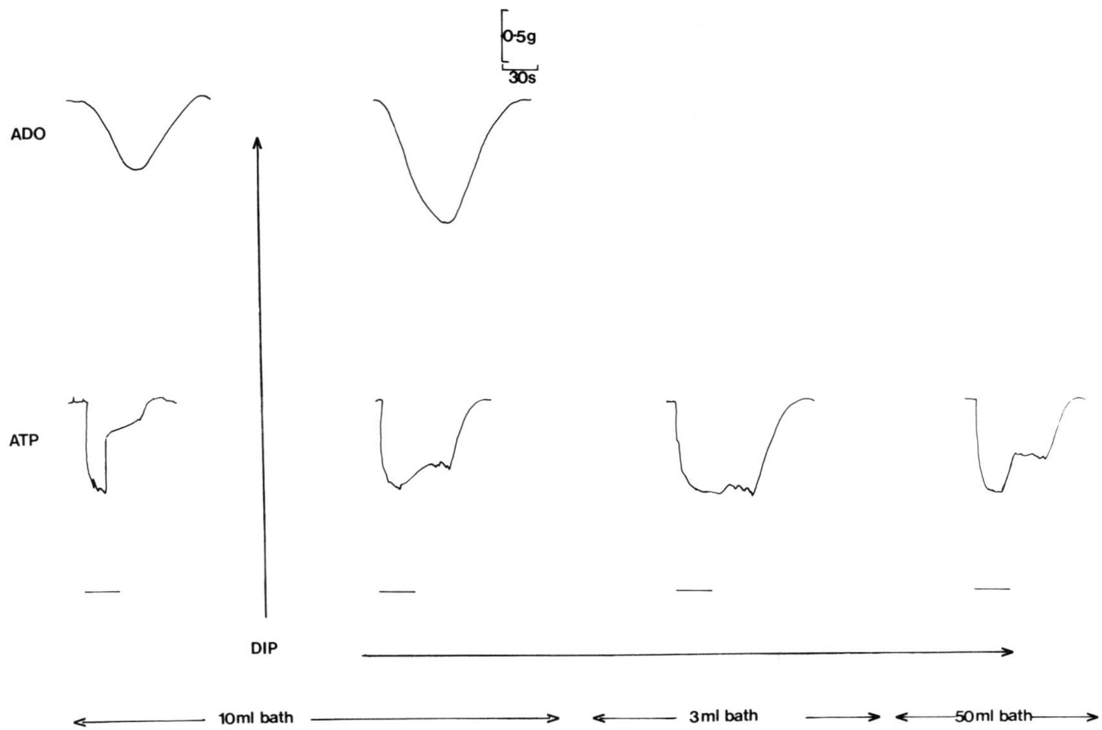

FIGURE 2. Relaxations of guinea pig taenia coli strips. Upper panel: adenosine (ADO, 25 μM), before and 20 min after treatment with dipyridamole (DIP, 1 μM). Lower panel: ATP (7 μM), before and after dipyridamole (DIP, 1 μM) in a 10-ml bath; ATP (7 μM), after dipyridamole (DIP, 1 μM) in a 3-ml bath and a 50-ml bath. Ordinate: 0.5 g; abscissa: 30 sec.

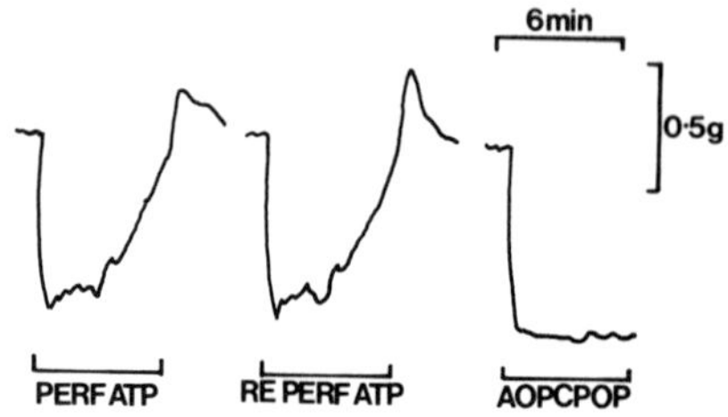

FIGURE 3. Left panel: response of guinea pig taenia coli strips perfused for 6 min at 1.5 ml/min with Krebs solution containing ATP (5.5 μM). This perfusate was collected and stored, then reapplied after the strips had been perfused with normal Krebs solution for 15 min. Middle panel: response of strips after reapplication of the perfusate. Third panel: response of strips perfused for 6 min at 1.5 ml/min with Krebs solution containing AOPCPOP (3.5 μM). Ordinate: 0.5 g; abscissa: 6 min.

Earlier studies had shown that changes in the ribose moiety caused a small decrease in potency or, in some cases, failed to cause any significant decrease in potency in taenia coli preparations. Modifications to the ribose moiety of adenosine, however, caused a marked reduction or an entire loss of potency.[8] Thus the ribose, which is a central and integral part of the ATP molecule, may be changed without a substantial loss in potency.

The 6-amino group is essential for actions at P_2 receptors because ITP and GTP exhibited reduced activity in taenia coli strips.[4] GTP, however, was found to be more potent than ITP,[4] suggesting perhaps that the 2-amino group could partially make up for the loss of the 6-amino group, at least at P_{2Y} receptors of the taenia coli. Nevertheless, Fedan *et al.*[9] demonstrated that ITP was significantly more potent than GTP on isolated preparations guinea pig vas deferens. The difference in potency is perhaps due to differences in requirements for binding at P_{2X} receptors of vas deferens.

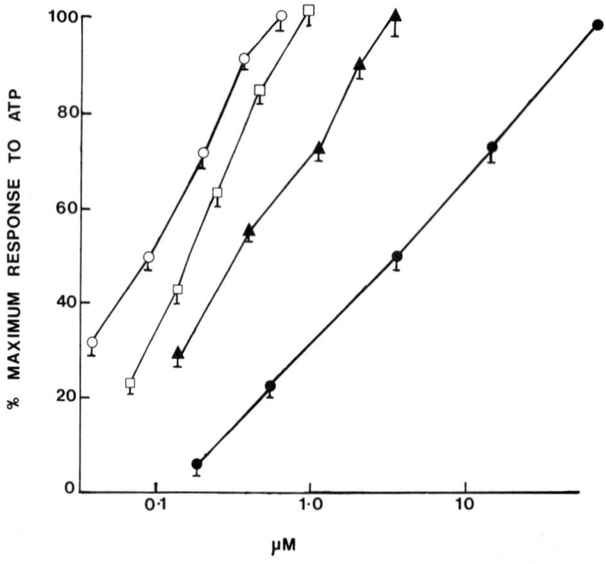

FIGURE 4. Contractions of guinea pig vas deferens preparations, induced by the following: ATP (●), 2-methylthio-ATP (▲), 2-chloro-ATP (□), and adenosine tetraphosphate tricyclohexyl ammonium salt (○). Each point is the mean ± SEM of at least six values from at least two preparations and is expressed as a percentage of the maximum response to ATP.

Effects of ATP Analogues on Isolated Preparations of Taenia Coli and Vas Deferens of Guinea Pig

The taenia coli and vas deferens preparations are reported to contain P_{2Y} and P_{2X} purinoceptor subtypes, respectively.[2] These preparations were selected for the study of the actions of four analogues of ATP that had proved to be of considerable value in differentiating agonist properties at P_1 and P_2 purinoceptors.[8]

In taenia coli preparations all four analogues were potent in causing relaxations. 2′-Deoxy-ATP, 8-bromo-ATP, and ara-ATP all produced concentration-response curves that were for the most part to the right of the concentration-response curves

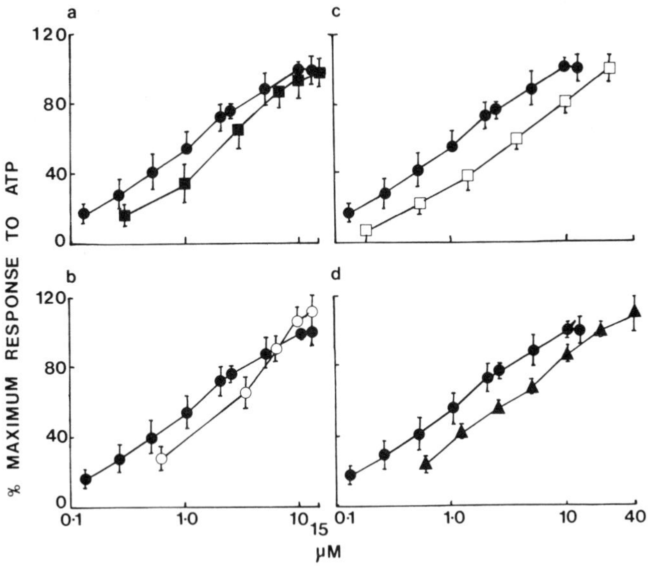

FIGURE 5. Relaxations of the guinea pig taenia coli preparations, induced by the following: (**a**) ATP (●) and 2′-deoxy-ATP (■); (**b**) ATP (●) and 8-bromo-ATP (○); (**c**) ATP (●) and 3′-deoxy-ATP (□); (**d**) ATP (●) and ara-ATP (▲). Each point is the mean ± SEM of at least six values obtained from at least two preparations and is expressed as a percentage of the maximum response to ATP.

for ATP. The curve for 3′-deoxy-ATP was also to the right of the curve for ATP, but the displacement was significant (FIG. 5). In vas deferens preparations ara-ATP and 3′-deoxy-ATP exhibited a considerably greater potency than ATP in causing contraction, and their curves were placed well to the left of the curves for ATP (FIG. 6). These results are only in partial agreement with earlier studies by Fedan *et al.*,[9] who observed some displacement to the left of the curves for ara-ATP and 3′-deoxy-ATP in their central regions while the curves of the lower and higher concentrations were the same, or to the right of the curves for ATP. 2′-Deoxy-ATP exhibited little potency, and its curve was placed far to the right. 8-Bromo-ATP exhibited a concentration-response curve that was similar to that of ATP, as in the taenia coli study (FIG. 5).

The above results demonstrated that there was a considerable variation in the potencies of the analogues, relative to ATP, between the two preparations studied. A further insight into the significance of these results was obtained from studies of the same four analogues in causing contractions of rat portal vein and rabbit femoral artery spiral strips (presumably by acting on P_{2X} receptors) and in causing relaxation of rabbit portal vein.[3]

The potencies of the four analogues relative to ATP in the taenia coli study were the same for relaxations of rabbit portal vein, except for ara-ATP, the curve for which was significantly displaced to the right of that for ATP in rabbit portal vein (TABLE 2).[3]

In rabbit femoral artery and rat portal vein the potencies of all four analogues relative to ATP in causing contractions[3] were similar to those obtained in the vas deferens (TABLE 2).

Although the four analogues studied varied considerably in the potencies relative to ATP, a consistent pattern of their actions was observed—on one hand in preparations that contracted in response to ATP, and on the other hand in preparations that relaxed in response to ATP. The results support the concept[2] of different P_2 receptor subtypes mediating contraction and relaxation. Further studies are necessary, including the use of desensitizing analogues and receptor blocking drugs to further evaluate this view.

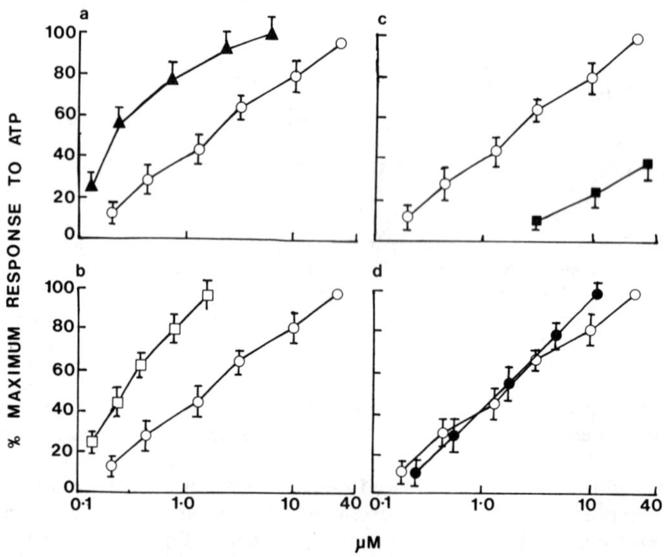

FIGURE 6. Contractions of guinea pig vas deferens preparations, induced by the following: (**a**) ATP (○) and ara-ATP (▲); (**b**) ATP (○) and 3′-deoxy-ATP (□); (**c**) ATP (○) and 2′-deoxy-ATP (■); (**d**) ATP (○) and 8-bromo-ATP (●). Each point is the mean ± SEM of at least six values from at least two preparations and is expressed as a percentage of the maximum response to ATP.

TABLE 2. Potency of Analogues Relative to ATP

	Ara-ATP	3′-Deoxy-ATP	2′-Deoxy-ATP	8-Bromo-ATP
Relaxations/ Guinea pig taenia coli	=	<	=	=
Relaxations/ Rabbit portal vein[3]	<	<	=	=
Contractions/ Guinea pig vas deferens	> >	> >	< <	=
Contractions/Rat portal vein[3]	> >	> >	<	=
Contractions/ Rabbit femoral artery[3]	> >	> >	< <	=

Further Studies on Preparations of Vas Deferens of Guinea Pig

Three further analogues were found to be more potent than ATP in causing contraction of guinea pig vas deferens preparations. 2-Methylthio-ATP, which in an earlier study had been considered to be equipotent with ATP,[2] was found to exhibit a significantly greater potency than ATP (FIG. 4). 2-Chloro-ATP also exhibited a high potency in contracting the vas deferens. Adenosine tetraphosphate was the most potent of the three analogues studied, and its curve was displaced to the left of that for ATP by more than one order of magnitude (FIG. 4). The ammonium salt of adenosine tetraphosphate was more potent than the sodium salt (results not shown). The ammonium salt, but not the sodium salt, caused marked desensitization in vas deferens preparations. This finding is consistent with the recent findings of Mackenzie *et al.* on other extended phosphate chain analogues.[7]

Actions of ATP and Related Analogues on Airway Smooth Muscle

ATP and adenosine were equally effective in causing relaxation of the guinea pig trachealis muscle. The responses were potentiated equally by dipyridamole, the adenosine transport inhibitor, suggesting that the actions of ATP were due at least in part to rapidly formed adenosine.[10] Analogues of ATP resistant to metabolic degradation were inactive in this preparation, suggesting that receptors for ATP were absent and that ATP elicits its responses entirely by rapid formation of adenosine.

Recent studies on peripheral airway smooth muscle have shown that strips of lung parenchyma contract in response to ATP (100-1000 μM). They also exhibit weak contractions in response to adenosine (200-1000 μM). Analogues of ATP were much more potent. Adenosine tetraphosphate (50-200 μM) elicited substantially greater contractions, as did 2-methylthio-ATP, which exhibited powerful contractions even at 0.1 μM (FIG. 7).

Further studies are necessary to characterize the receptor at this site. Moreover, experiments should be carried out on the smooth muscle of the fine airways. For example, one could look for the presence of receptors at different bronchioles.

ACKNOWLEDGMENTS

The initial experiments on lung parenchyma were carried out by Miss Peta Burns.

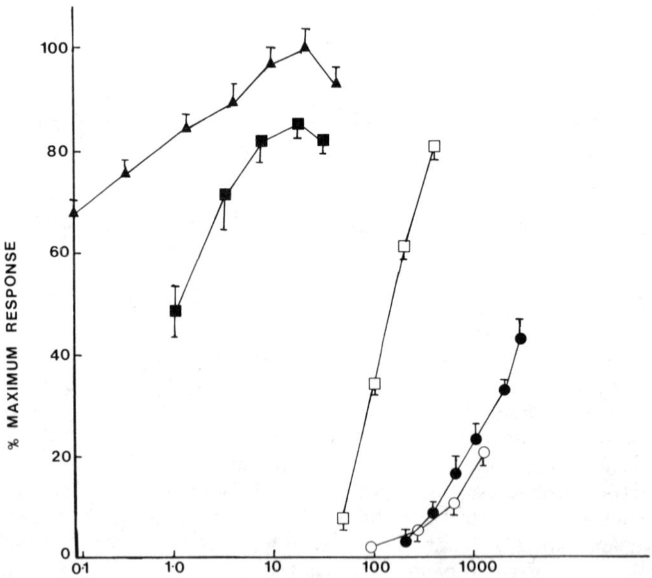

FIGURE 7. Contraction of strips of guinea pig lung parenchyma induced by the following: adenosine (○), ATP (●), adenosine tetraphosphate tricyclohexyl ammonium salt (□), carbachol (■), and 2-methylthio-ATP (▲). Each point is the mean ± SEM of at least five values from at least two preparations and is expressed as a percentage of the contraction induced by Krebs solution containing 50 mM KCl.

REFERENCES

1. DRURY, A. N. & A. SZENT-GYÖRGYI. 1929. The physiological activity of adenosine compounds with especial reference to their action upon the mammalian heart. J. Physiol. (London) **68:** 213-237.
2. BURNSTOCK, G. & C. KENNEDY. 1985. Is there a basis for distinguishing two types of P_2-purinoceptor? Gen. Pharmacol. **16:** 433-440.
3. SATCHELL, D. G. 1988. Differences in the structural requirements for agonist properties at P_1 and P_2 receptors in smooth muscle. *In* Adenosine and Adenine Nucleotides. D. M. Paton, Ed.: 85-92. Taylor & Francis. London.

4. MAGUIRE, M. H. & D. G. SATCHELL. 1981. Purinergic receptors in visceral smooth muscle. *In* Purinergic Receptors. G. Burnstock, Ed.: 49-91. Chapman Hall. London.
5. BURNSTOCK, G. 1986. The changing face of autonomic neurotransmission. Acta Physiol. Scand. **126:** 67-91.
6. SATCHELL, D. G. & M. H. MAGUIRE. 1975. Inhibitory effects of adenine nucleotide analogues on the isolated guinea-pig taenia coli. J. Pharmacol. Exp. Ther. **195:** 540-548.
7. MACKENZIE, I. & K. A. KIRKPATRICK & G. BURNSTOCK. 1988. Comparative study of the actions of AP5A and α,β-methylene ATP on nonadrenergic, noncholinergic, neurogenic exitation in the guinea-pig vas deferens. Br. J. Pharmacol. **94:** 699-706.
8. SATCHELL, D. G. & M. H. MAGUIRE. 1982. Evidence for separate receptors for ATP and adenosine in the guinea-pig taenia coli. Eur. J. Pharmacol. **81:** 669-672.
9. FEDAN, J. S., G. K. HOGABOOM & J. P. O'DONNELL. 1986. Further comparison of contractions of the smooth muscle of the guinea-pig isolated vas deferens induced by ATP and related analogues. Eur. J. Pharmacol. **129:** 279-291.
10. CHRISTIE, J. & D. G. SATCHELL. 1980. Purine receptors in the trachea: Is there a receptor for ATP? Br. J. Pharmacol. **70:** 512-514.

Effects of Extracellular ATP on Surfactant Secretion[a]

WARD R. RICE

Division of Neonatology
Department of Pediatrics
College of Medicine
University of Cincinnati
Cincinnati, Ohio 45267

INTRODUCTION

Pulmonary surfactant that contains both lipids and proteins is synthesized, stored as lamellar bodies, and secreted by alveolar type II cells.[1–4] Surfactant maintains the stability of the alveolus, insuring normal ventilation and gas exchange by the lungs. Inability to synthesize and secrete surfactant leads to hypoxemia and hypoventilation in certain clinical conditions like respiratory distress syndrome of the premature neonate and adult respiratory distress syndrome. Although the type II cell of the lung is known to be responsible for synthesis and secretion of surfactant, the mechanisms regulating these processes are less well understood.[1–4]

Secretion of surfactant from alveolar type II cells is being actively studied in many laboratories and is known to be augmented by agents affecting cAMP levels (β-adrenoceptor agonists,[5,6] cholera toxin,[7] and forskolin[8]), prostaglandins,[9] protein kinase C agonists (phorbol esters[10]), agents that mobilize cytosolic calcium (A23187[11] and ionomycin[12]), and agents that alter cytoskeletal associations (cytochalasins[13]). One of the most potent agonists for surfactant secretion identified to date is exogenous ATP, which results in a three- to sixfold stimulation of surfactant secretion *in vitro.*[14–16] In the present work, the purinoceptor regulating surfactant secretion is characterized and the possible second messenger systems involved in regulation of purinoceptor-mediated surfactant secretion are described.

MATERIALS AND METHODS

Materials

Methyl[^{3}H]choline chloride, dipalmitoyl[^{14}C]phosphatidylcholine, [^{3}H]phorbol-12,13-dibutyrate ([^{3}H]PDBu), and the radioimmunoassay kit for 6-keto-prostaglandin

[a]This work was supported by Grant HL38764 from the National Institutes of Health.

$F_{1\alpha}$ (6-keto-PG$F_{1\alpha}$) were purchased from New England Nuclear (Boston, MA). Chloroform, methanol, and scintillation fluid were from Fisher (Cincinnati, OH). Sphingosine, α,β-methylene adenosine 5′-triphosphate (AMPCPP), β,γ-methylene adenosine 5′-triphosphate (AMPPCP), and reactive blue 2 were obtained from Sigma (St. Louis, MO). Bis-(2-amino-5-methylphenoxy)ethane-*N,N,N′,N′*-tetraacetic acid tetraacetoxymethyl ester (MAPTAM) was from Calbiochem Behring (La Jolla, CA). Egg phosphatidylcholine was from Avanti Polar Lipids (Birmingham, AL). Adenosine 5′-*O*-(3-thiotriphosphate) (γS-ATP) and β,γ-imidoadenosine 5′-triphosphate (AMPPNP) were from Boehringer Mannheim Biochemicals (Indianapolis, IN). Elastase was obtained from Cooper Biomedicals (Malvern, PA), and 1-[2,15-carboxyoxazol-2-yl)-6-aminobenzo-furan-5-oxy]-2-(2′-amino-5′-methylphenoxy)-ethane-*N,N,N′,N′*-tetraacetic acid methyl ester (fura-2-AM) was from Molecular Probes (Eugene, OR).

Animals

Pathogen-free, male Sprague-Dawley rats (200-250 g) were obtained from Charles River (Wilmington, MA).

Isolation and Culture of Type II Cells

Type II cells were isolated from lungs by methodology previously described.[17] Briefly, type II cells were obtained following elastase digestion of rat lungs and panning cells on petri dishes coated with immunoglobulin G. This procedure is readily reproduced and routinely yields 90% viable type II cells.

Secretion of Phosphatidylcholine

Secretion of phosphatidylcholine by cultured type II cells was determined as previously described.[14] Cells were incubated overnight with 1 μCi [^{3}H]choline/500,000 cells/ml, and after this incubation were washed three times with media. The cells were allowed to equilibrate for 30 min, agents were added, and the [^{3}H]phosphatidylcholine released was determined by aspirating the medium, washing the cells with fresh medium, and counting the disintegrations per minute (dpm) in cells and medium in a β-scintillation counter. [^{3}H]Phosphatidylcholine release was expressed as (dpm in medium/dpm in cells + dpm in medium) $\times$ 100%. Lactate dehydrogenase activity was determined in each sample as a measure of cytotoxicity.

Mobilization of Intracellular Calcium

Intracellular calcium concentrations were determined by a method published previously.[18] Fresh type II cells or cells that had plated down overnight were loaded for

10 min with 5 μM fura-2-AM, centrifuged, and resuspended in fresh medium for an additional 10-min period at 37 °C. Loaded cells were then washed three times with ice-cold medium and stored on ice prior to use. Fluorescence was monitored utilizing an excitation wavelength of 339 nm and an emission wavelength of 500 nm.

6-Keto-PGF$_{1\alpha}$ Determination

Concentrations of 6-keto-PGF$_{1\alpha}$ in cells or media were determined using a commercially available radioimmunoassay kit and the instructions provided. Results are expressed as percentages of the control level.

[^{3}H]PDBu Binding to Intact Cells

Type II cells were plated down overnight in 24-well plates (approximately 1 × 10^6 cells/well) as previously described.[14] Cells were washed and incubated for 30 min at 37 °C with fresh medium. At the end of this time, medium was removed and replaced with fresh medium containing 20 nM [^{3}H]PDBu for 5 min at 37 °C as previously described.[19] Agents were then added for 5 min, after which cells were rapidly washed with fresh medium and counts extracted with 0.5 ml of 1 N sodium hydroxide. Nonspecific binding was determined in the presence of 10 μM PDBu. Binding was expressed relative to control binding, which was subtracted from all values shown.

Statistical Analyses

Analysis of variance for both parametric and nonparametric data was performed on a microcomputer using commercially available statistical packages. Each value was expressed as a mean ± SEM for three to six experiments, each in triplicate, and $p < .05$ was utilized to assess statistical significance.

RESULTS

ATP and ATP analogues were examined for their ability to stimulate surfactant phospholipid secretion. ATP, the most potent analogue examined (FIG. 1), stimulated [^{3}H]phosphatidylcholine secretion in a dose-dependent fashion. ADP also stimulated [^{3}H]phosphatidylcholine secretion in a dose-dependent manner, whereas AMP and adenosine did not have significant effects on surfactant phospholipid secretion. The rank order of agonist potency is consistent with a P$_2$-purinoceptor mediating surfactant phosphatidylcholine secretion. Other ATP analogues were also examined for an effect on [^{3}H]phosphatidylcholine secretion (FIG. 2A). Methylene analogues of ATP were less potent at stimulating surfactant phospholipid secretion than ATP, consistent with a P_{2Y}-purinoceptor mediating the secretory effect. In support of this hypothesis, reactive blue 2 also selectively inhibited ATP-stimulated phospholipid secretion (FIG. 3). Reactive blue 2 inhibited ATP-induced surfactant phospholipid secretion in a dose-dependent fashion, but was without effect on β-adrenoceptor-induced surfactant phospholipid secretion.

Several second messenger systems are activated in type II cells following agonist binding. Calcium is mobilized from an intracellular compartment in a dose-dependent fashion, and the dose response for calcium mobilization is similar to the dose response for ATP-induced surfactant phospholipid secretion, the EC_{50} (concentration producing

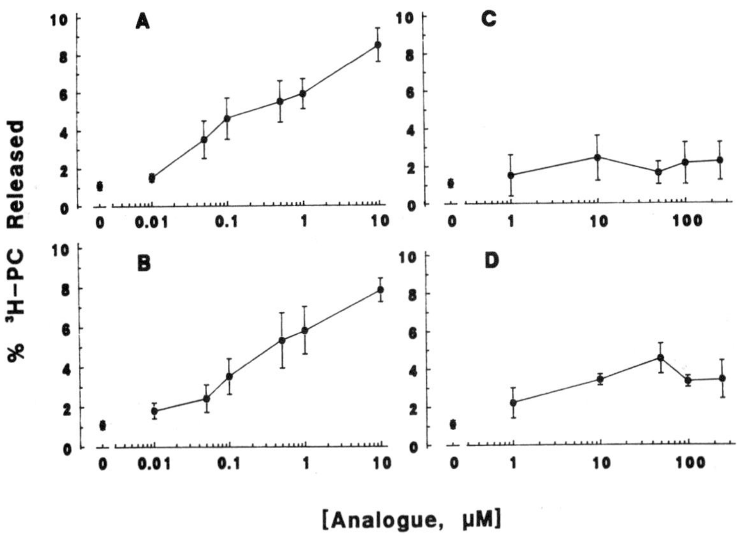

FIGURE 1. Effect of ATP, ADP, AMP, and adenosine on [^{3}H]phosphatidylcholine ([^{3}H]PC) release. Type II cells were exposed to ATP (**A**), ADP (**B**), AMP (**C**), or adenosine (**D**) for 3 hr at the indicated concentrations. Release of [^{3}H]PC was then determined and expressed as stated in the methods section. Data are shown for four experiments. Both ATP and ADP significantly stimulated [^{3}H]PC secretion.

one-half the maximal response) in each case being 2 μM (FIG. 4). However, using MAPTAM that was trapped in the intracellular compartment as a calcium chelator, we were able to block calcium mobilization in a dose-dependent fashion without affecting ATP-induced surfactant secretion (FIG. 5), arguing against a role for calcium mobilization in regulation of ATP-induced surfactant secretion.

Because P_2-purinoceptors are known to be coupled to phospholipase A_2 and prostacyclin production in other systems,[20,21] and because prostaglandins have been reported to augment surfactant secretion,[9] we next examined whether prostaglandins mediate ATP-induced surfactant phospholipid secretion. The dose response for prostacyclin production (measured as the stable prostacyclin metabolite, 6-keto-$PGF_{1\alpha}$) was similar to the dose response for ATP-induced surfactant secretion (FIG. 4). We attempted

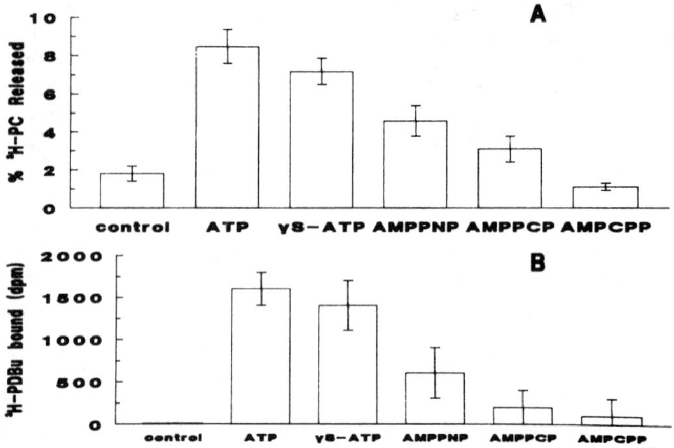

FIGURE 2. Effect of ATP and ATP analogues on [^{3}H]PC secretion and [^{3}H]PDBu binding by isolated type II cells. Type II cells were exposed to ATP or the indicated analogues (10 μM) for 3 hr (**A**) or 5 min (**B**), and [^{3}H]PC released (**A**) or [^{3}H]PDBu bound (**B**), determined as stated in the methods section. The analogues utilized were adenosine 5′-*O*-(3-thiotriphosphate) (γS-ATP); β,γ-imidoadenosine 5′-triphosphate (AMPPNP); α,β-methylene adenosine 5′-triphosphate (AMPCPP); and β,γ-methylene adenosine 5′-triphosphate (AMPPCP). Data are shown from four to six experiments.

to block ATP-induced surfactant phospholipid secretion with the prostaglandin synthesis inhibitor, indomethacin (FIG. 6). Although indomethacin effectively blocked production of 6-keto-PGF$_{1\alpha}$, it was without effect on surfactant phospholipid secretion, arguing against a role for prostacyclin in regulating ATP-induced surfactant secretion.

We next examined whether protein kinase C might mediate ATP-induced surfactant phospholipid secretion and examined protein kinase C activation by following PDBu binding in intact cells. This method was chosen because it does not involve fractionation of cells and therefore allows an examination of C kinase translocation

FIGURE 3. Inhibition of purinoceptor-induced [^{3}H]PC secretion by reactive blue 2. Type II cells were exposed to ATP (10 μM, **A**) or terbutaline (10 μM, **B**) for 3 hr in the presence or absence of the indicated concentrations of reactive blue 2, and the [^{3}H]PC released was determined as stated in the methods section. Data are shown from four to six experiments, each in triplicate. Reactive blue 2 significantly inhibited the purinoceptor-induced, but not the adrenoceptor-induced [^{3}H]PC secretion.

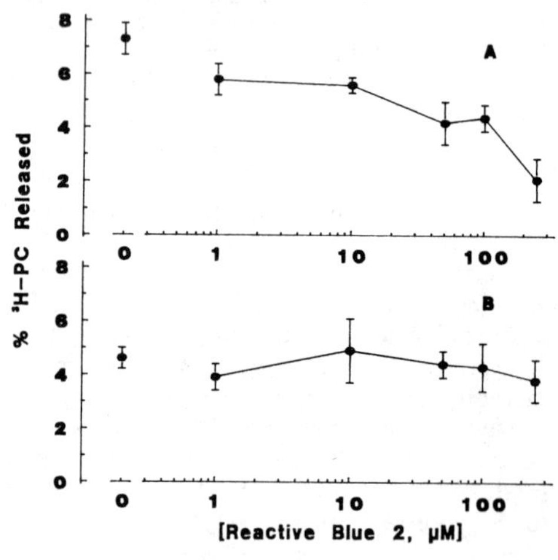

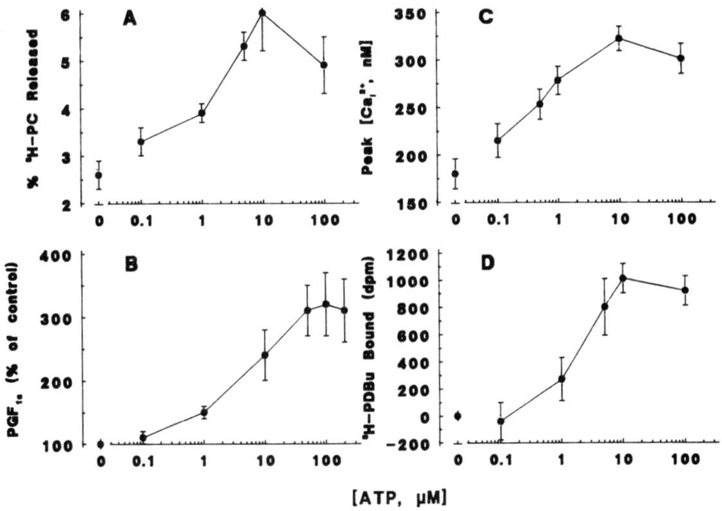

FIGURE 4. Effect of ATP on phosphatidylcholine secretion, calcium mobilization, [^{3}H]PDBu binding, and prostacyclin production in isolated type II cells. Type II cells were exposed to the indicated concentrations of ATP for 3 hr (**A**), 30 min (**B**), 10 sec (**C**), or 5 min (**D**), and then the [^{3}H]PC released (**A**), 6-keto-$PGF_{1\alpha}$ (**B**), peak $[Ca^{2+}]_i$ (**C**), or [^{3}H]PDBu bound (**D**) was determined. ATP had a statistically significant dose-dependent effect in each case. Data are shown from four or five experiments.

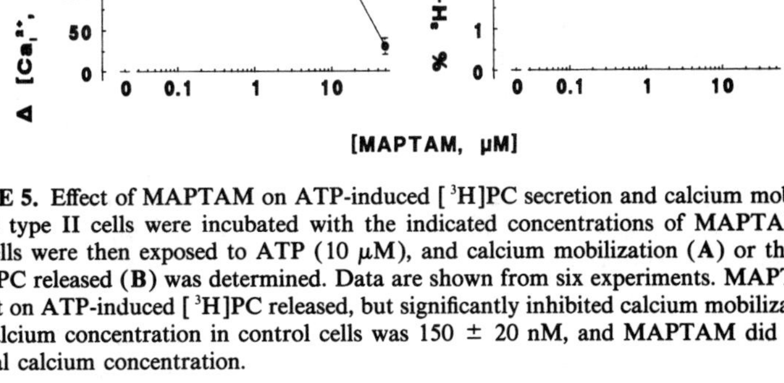

FIGURE 5. Effect of MAPTAM on ATP-induced [^{3}H]PC secretion and calcium mobilization. Isolated type II cells were incubated with the indicated concentrations of MAPTAM for 30 min. Cells were then exposed to ATP (10 μM), and calcium mobilization (**A**) or the amount of [^{3}H]PC released (**B**) was determined. Data are shown from six experiments. MAPTAM had no effect on ATP-induced [^{3}H]PC released, but significantly inhibited calcium mobilization. The basal calcium concentration in control cells was 150 ± 20 nM, and MAPTAM did not affect the basal calcium concentration.

in intact cells.[19] We found the dose response for PDBu binding to be analogous to the dose response for ATP-induced surfactant secretion (FIG. 4), the EC_{50} in each case being approximately 2 μM. We next attempted to block surfactant phospholipid secretion with the protein kinase C inhibitor, sphingosine. In this case, the dose response for inhibition of C kinase translocation was similar to the dose response for inhibition of surfactant phospholipid secretion (FIG. 7). In each case, the IC_{50} (concentration producing one-half the maximal inhibition) was 5-10 μM. We also examined the effect of various ATP analogues on PDBu binding by type II cells (FIG. 2B) and found the rank order of agonist potency for [^{3}H]phosphatidylcholine secretion was the same as the rank order for [^{3}H]PDBu binding. These data support a role for protein kinase C as a mediator of P_2-purinoceptor-induced surfactant phospholipid secretion.

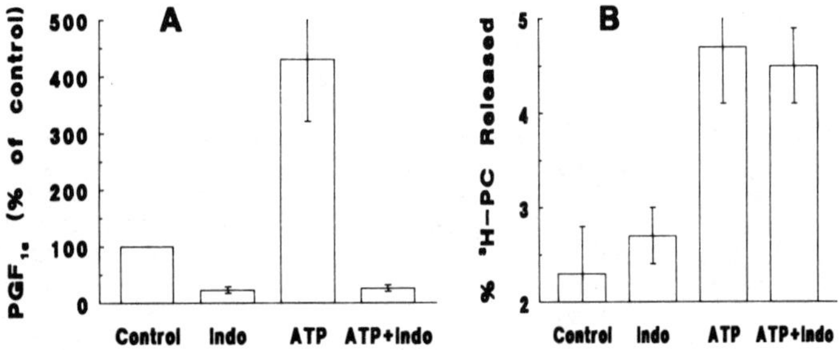

FIGURE 6. Effect of indomethacin and ATP on 6-keto-$PGF_{1\alpha}$ production and [^{3}H]PC release by type II cells. Type II cells were exposed to buffer, ATP (10 μM), indomethacin (20 μM), or both agents. Determinations were then made for 6-keto-$PGF_{1\alpha}$ levels after 30 min (**A**) or for [^{3}H]PC released after 3 hr (**B**). Indomethacin significantly inhibited 6-keto-$PGF_{1\alpha}$ production by type II cells, but had no effect on ATP-induced [^{3}H]PC release. Data are shown from four to six experiments.

DISCUSSION

ATP is the most potent endogenous stimulus for surfactant phospholipid secretion from alveolar type II cells identified to date.[14-16] The purinoceptor responsible for mediating this process appears to belong to the P_{2Y} subclass for the following reasons:

1. The rank order of agonist potency is ATP > ADP > AMP = adenosine.
2. Methylene analogues of ATP are less potent secretagogues than ATP itself.
3. Reactive blue 2 inhibits ATP-induced surfactant phospholipid secretion in a dose-dependent fashion.

Multiple second messenger systems are activated following binding of ATP to the alveolar type II cell (FIG. 8). Of these second messengers, protein kinase C appears

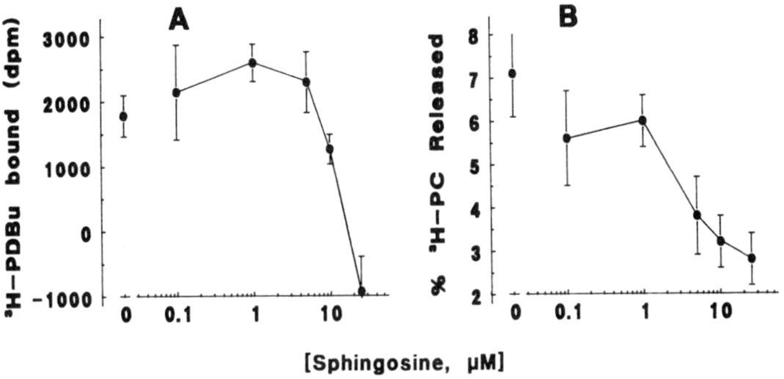

FIGURE 7. The effect of sphingosine on ATP-induced [^{3}H]PDBu binding and [^{3}H]PC release. Type II cells were exposed to the indicated concentrations of sphingosine, and [^{3}H]PDBu binding was determined after 5 min (**A**), or the amount of [^{3}H]PC released was determined after 3 hr (**B**) following exposure to ATP (10 μM). Sphingosine significantly inhibited both [^{3}H]PDBu binding and [^{3}H]PC released. The IC_{50} in each case was 5-10 μM.

to mediate the signal responsible for secretion of phospholipid. Although receptor occupancy also resulted in calcium mobilization and prostacyclin production, we were able to block both calcium mobilization and prostacyclin production without affecting surfactant phospholipid secretion. On the other hand, both surfactant phospholipid secretion and protein kinase C translocation were blocked by the protein kinase C inhibitor, sphingosine, with similar IC_{50} values (5-10 μM). In addition, the dose response for ATP-induced C kinase activation and [^{3}H]phosphatidylcholine secretion were similar (EC_{50} = 2 μM), and the rank order of agonist potency among various ATP analogues was the same as that for ATP-induced [^{3}H]phosphatidylcholine secretion and [^{3}H]PDBu binding by type II cells. Previous work with the phorbol ester,

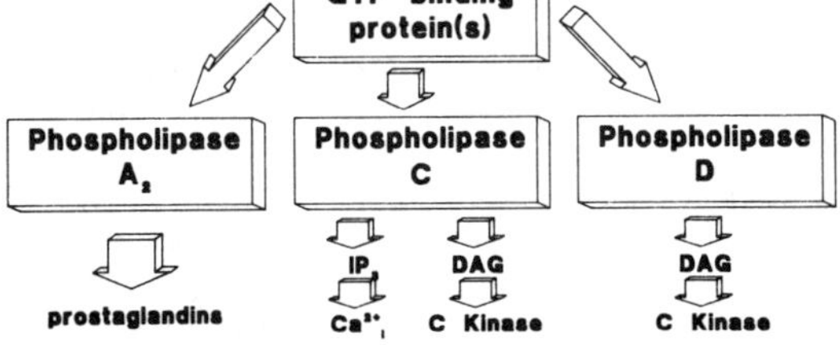

FIGURE 8. Schematic representation of P_{2Y}-purinoceptor coupling to second messenger generation. Abbreviations: IP_3: inositol 1,4,5-trisphosphate; DAG: diaylglycerol; C kinase: calcium- and phospholipid-dependent protein kinase. The scheme is adapted from reference 18 and references 23-30.

TPA, suggested a role for protein kinase C in regulation of surfactant secretion,[22] and the present data are consistent with that hypothesis.

Although the present work suggests that protein kinase C plays a prominent role in mediating ATP-induced surfactant secretion, other second messengers are also generated following binding of ATP to the P_{2Y}-purinoceptor. This may indicate coupling of the P_{2Y}-purinoceptor to effector phospholipases by multiple G proteins (FIG. 8). Prostacyclin production in response to exogenous ATP has previously been reported in adrenal medullary endothelial cells[20] and porcine endothelial cells.[21] Prostacyclin production may result from direct activation of phospholipase A_2[23,24] by the G-protein-coupled P_{2Y}-purinoceptor or occur indirectly following mobilization of intracellular calcium and subsequent phospholipase A_2 activation.[25,26] Other G proteins may couple purinoceptors to phospholipase C resulting inositol trisphosphate and diacylglycerol production with subsequent C kinase activation[27–29] whereas another G protein may activate phospholipase D and generate diacylglycerol from phosphatidylcholine (FIG. 8).[30]

The role of these other second messengers in type II cell function is unclear. Local production of prostaglandins in response to ATP may mediate airway reactivity or vascular permeability in the area of the alveolus. Mobilization of calcium along with protein kinase C activation may mediate gene transcription in the type II cell and thus control growth or differentiation in this progenitor cell for the alveolar lining during periods of lung growth in the fetus or following cell damage after exposure to environmental toxins. In support of this hypothesis, exogenous ATP was recently reported to function as a mitogen in 3T3, 3T6, and A431 cell lines.[31] The source of ATP that could provide such growth regulation is unclear, but it is possible extracellular ATP may soon be added to the growing list of growth factors whose signals are transduced to the nucleus via calcium and protein kinase C.

In summary, these data are consistent with protein kinase C playing a major role in regulation of purinoceptor-induced surfactant phospholipid secretion from the alveolar type II cell. Further work designed to identify substrates for protein kinase C in the type II cell should help unravel the mystery of how phospholipid secretion is controlled by the cell. Although ATP has been identified in the airways in concentrations sufficient to cause surfactant phospholipid secretion *in vitro,* the source of ATP remains unclear, and it remains to be determined whether ATP plays a similar role *in vivo.*[32]

SUMMARY

Extracellular ATP is the most potent endogenous stimulus for surfactant phospholipid secretion from alveolar type II cells identified to date. This effect of ATP appears mediated via a P_2-purinoceptor because the rank order of agonist potency is ATP > ADP > AMP = adenosine. Examination of other ATP analogues demonstrates a rank order of agonist potency of ATP = γS-ATP > AMPPNP > AMPPCP > AMPCPP for surfactant secretion, consistent with a P_{2Y}-purinoceptor mediating this effect. This hypothesis is further supported by experiments with reactive blue 2, which selectively inhibits ATP-stimulated surfactant phospholipid secretion and has been purported as a specific inhibitor at P_{2Y}-purinoceptors.

Several second messenger systems are activated in the type II cell following agonist binding: intracellular Ca^{2+} is mobilized, prostaglandin levels increase, and protein kinase C is activated. Of these three second messengers, protein kinase C appears to be the most important for surfactant secretion because inhibition of protein kinase C activation blocks ATP-induced surfactant secretion whereas inhibition of Ca^{2+} mobilization and prostaglandin production does not affect ATP-induced surfactant secretion.

ACKNOWLEDGMENTS

I appreciate the outstanding technical support of Susie Singleton, Pat Walker, Jeryl Appleby, and Curt Dorn.

REFERENCES

1. DOBBS, L. G. 1989. Pulmonary surfactant. Annu. Rev. Med. **40:** 431-446.
2. WRIGHT, J. R. & J. A. CLEMENTS. 1987. Metabolism and turnover of lung surfactant. Am. Rev. Respir. Dis. **135:** 426-444.
3. ROONEY, S. A. 1985. The surfactant system and lung phospholipid biochemistry. Annu. Rev. Respir. Dis. **131:** 439-460.
4. MASON, R. J. 1987. Surfactant synthesis, secretion and function in alveoli and small airways. Respiration **51**(Suppl. 1): 3-9.
5. DOBBS, L. G. & R. J. MASON. 1979. Pulmonary alveolar type II cells isolated from rat. J. Clin. Invest. **63:** 378-387.
6. METTLER, N. R., N. E. GRAY, S. SCHUFFMAN & V. S. LEQUIRE. 1981. β-Adrenergic-induced synthesis and secretion of phosphatidylcholine by isolated pulmonary alveolar type II cells. Lab. Invest. **45:** 575-586.
7. MESCHER, E. J., L. G. DOBBS & R. J. MASON. 1983. Cholera toxin stimulates secretion of phosphatidylcholine and increases cellular cAMP in isolated rat alveolar type II cells. Exp. Lung Res. **5:** 173-182.
8. RICE, W. R., W. M. HULL, C. A. DION, B. A. HOLLINGER & J. A. WHITSETT. 1985. Activation of cyclic AMP-dependent protein kinase during surfactant release from type II pneumocytes. Exp. Lung Res. **9:** 135-149.
9. GILFILLAN, A. M. & S. A. ROONEY. 1985. Arachidonic acid metabolites stimulate phosphatidylcholine secretion in primary cultures of type II pneumocytes. Biochim. Biophys. Acta **833:** 336-341.
10. DOBBS, L. G. & R. J. MASON. 1978. Stimulation of secretion of disaturated phosphatidylcholine from isolated alveolar type II cells by 12-*O*-tetradecanoyl-13-phorbol acetate. Am. Rev. Respir. Dis. **118:** 705-713.
11. MARINO, P. A. & S. A. ROONEY. 1980. Surfactant secretion in the newborn rabbit lung slice model. Biochim. Biophys. Acta **620:** 509-519.
12. PIAN, M. S., L. G. DOBBS & N. DÜZGÜNES. 1988. Positive correlation between cytosolic free calcium and surfactant secretion in cultured rat alveolar type II cells. Biochim. Biophys. Acta **960:** 43-53.
13. RICE, W. R., K. C. OSTERHOUDT & J. A. WHITSETT. 1984. Effect of cytochalasins on surfactant release from alveolar type II cells. Biochim. Biophys. Acta **805:** 12-18.
14. RICE, W. R. & F. M. SINGLETON. 1986. P_2-Purinoceptors regulate surfactant secretion from rat isolated alveolar type II cells. Br. J. Pharmacol. **89:** 485-491.
15. GILFILLAN, A. M. & S. A. ROONEY. 1987. Purinoceptor agonists stimulate phosphatidylcholine secretion in primary cultures of adult rat type II pneumocytes. Biochim. Biophys. Acta **917:** 18-23.

16. WARBURTON, D., S. BUCKLEY & L. COSICO. 1989. P_1 and P_2 purinergic receptor signal transduction in rat type II pneumocytes. J. Appl. Physiol. **66:** 901-905.
17. DOBBS, L. G., R. GONZALES & M. C. WILLIAMS. 1986. An improved method for isolating type II cells in high yield and purity. Am. Rev. Respir. Dis. **134:** 141-145.
18. RICE, W. R. & F. M. SINGLETON. 1987. P_{2Y}-Purinoceptor regulation of surfactant secretion from rat isolated alveolar type II cells is associated with mobilization of intracellular calcium. Br. J. Pharmacol. **91:** 833-838.
19. TRILIVAS, I. & J. H. BROWN. 1989. Increases in intracellular Ca^{2+} regulate the binding of [^{3}H]phorbol-12,13-dibutyrate to intact 1321N1 astrocytoma cells. J. Biol. Chem. **264:** 3102-3207.
20. FARSBERG, E. J., G. FEUERSTEIN, E. SHOHAMI & H. B. POLLARD. 1987. Adenosine triphosphate stimulates inositol phospholipid metabolism and prostacyclin formation in adrenal medullary endothelial cells by means of P_2-purinergic receptors. Proc. Natl. Acad. Sci. USA **84:** 5630-5634.
21. PEARSON, J. D., L. L. SLAKEY & J. L. GORDON. 1983. Stimulation of prostaglandin production through purinoceptors on cultured porcine endothelial cells. Biochem. J. **214:** 273-276.
22. SANO, K., D. R. VOELKER & R. J. MASON. 1985. Involvement of protein kinase C in pulmonary surfactant secretion from alveolar type II cells. J. Biol. Chem. **260:** 12725-12729.
23. BURCH, R. M. & J. AXELROD. 1987. Dissociation of bradykinin-induced prostaglandin formation from phosphatidylinositol turnover in Swiss 3T3 fibroblasts: Evidence for G protein regulation of phospholipase A_2. Proc. Natl. Acad. Sci. USA **84:** 6374-6378.
24. PEARCE, B., S. MURPHY, J. JEREMY, C. MORROW & P. DANDONA. 1989. ATP-evoked Ca^{2+} mobilization and prostanoid release from astrocytes: P_2-Purinergic receptors linked to phosphoinositide hydrolysis. J. Neurochem. **52:** 971-977.
25. DUBYAK, G. R. 1986. Extracellular ATP activates polyphosphoinositide breakdown and Ca^{2+} mobilization in Ehrlich ascites tumor cells. Arch. Biochem. Biophys. **245:** 84-95.
26. PIROTTON, S., E. RASPE, D. DEMOLLE, C. ERNEUX & J. M. BOEYNAEMS. 1987. Involvement of inositol 1,4,5-trisphosphate and calcium in the action of adenine nucleotides on aortic endothelial cells. J. Biol. Chem. **262:** 17461-17466.
27. HARDEN, T. K., P. T. HAWKINS, L. STEPHENS, J. L. BOYER & C. P. DOWNES. 1988. Phosphoinositide hydrolylsis by guanosine 5′-[γ-thio]triphosphate-activated phospholipase C of turkey erythrocyte membranes. Biochem. J. **252:** 583-543.
28. VANDERMERNE, P. A., I. K. WAKEFIELD, J. FINE, R. P. MILLAR & J. S. DAVIDSON. 1989. Extracellular adenosine triphosphate activates phospholipase C and mobilizes intracellular calcium in primary cultures of sheep anterior pituitary cells. FEBS Lett. **243:** 333-336.
29. OKAJIMA, F., K. SATO & Y. KONDO. 1989. P_2-Purinergic agonists activate phospholipase C in a guanine nucleotide and Ca^{2+}-dependent manner in FRTL-5 thyroid cell membranes. FEBS Lett. **253:** 132-136.
30. DUBAYAK, G. R., D. S. COWEN & L. M. MEULLER. 1988. Activation of inositol phospholipid breakdown in HL60 cells by P_2-purinergic receptors for extracellular ATP. J. Biol. Chem. **263:** 18108-18117.
31. HUANG, N., D. WANG & L. A. HEPPEL. 1989. Extracellular ATP is a mitogen for 3T3, 3T6 and A431 cells and acts synergistically with other growth factors. Proc. Natl. Acad. Sci. USA **86:** 7904-7908.
32. RICE, W. R., M. BURHANS & J. R. WISPÉ. 1989. Effect of oxygen exposure on ATP content of rat bronchoalveolar lavage. Pediatr. Res. **25:** 396-398.

DISCUSSION OF THE PAPER

S. C. SILVERSTEIN (*Columbia University, New York, NY*): Have you tried to potentiate surfactant secretion *in vivo* by placing ATP into the bronchial tree?

RICE: We have not, although we plan to perform such experiments in an animal model.

J. S. DAVIDSON (*University of Cape Town Medical School, Cape Town, South Africa*): I believe your experiments with the intracellular calcium chelator are inconclusive. The reason is that exocytosis is triggered by a rise in calcium in a narrow zone immediately beneath the plasma membrane. This zone may not contribute much to the Fura-2 signal observed. Thus, although the apparent rise in cytosol calcium may have been dampened, there may well have been elevated calcium levels at the exocytotic site beneath the plasma membrane, and these elevated calcium levels may have been sufficient to swamp the chelator.

RICE: Although that is possible, we have also performed experiments in the presence of extracellular EGTA. Extracellular EGTA would inhibit a calcium rise in the narrow band beneath the plasma membrane as well, and under these conditions, phospholipid secretion in response to ATP, TPA (12-*O*-tetradecanoylphorbol 13-acetate), and terbutaline still occurs in a normal fashion, consistent with the hypothesis that calcium mobilization is not necessary for ATP-induced surfactant phospholipid secretion.

F. DI VIRGILIO (*Institute of General Pathology, Padua, Italy*): Sphingosine is not only an inhibitor of PKC (protein kinase C) but is also a damaging agent to several cell types. Have you tried PKC down-regulation as an alternative to probe the hypothesis that ATP causes activation of PKC?

RICE: We are at present conducting such experiments utilizing TPA treatment to down-regulate type II cell C kinase. The effect of sphingosine in our system does not appear to be due to cytotoxicity because lactate dehydrogenase release is not different in the presence or absence of sphingosine, and because terbutaline-induced surfactant phospholipid secretion is unaffected by the same dose of sphingosine.

I. L. O. BUXTON (*University of Nevada School of Medicine, Reno, NV*): Dr. Rice, I believe that PAF (platelet-activating factor) release is associated with the appearance of surfactant in the fetal lung. Is it possible then that the origin of ATP in this case is the blood platelet?

RICE: Although the PAF may come from platelets, there may also be lung cells that produce PAF or a PAF-immunoreactive molecule. The source of bronchoalveolar lavage ATP is unknown at present, although platelets would be one possible source.

A. S. DAHMS (*San Diego State University, San Diego, CA*): Do the 4-alpha phorbol esters (which do not affect protein kinase C) have any effect on surfactant secretion?

RICE: We have examined 4-alpha phorbol esters for an effect, and they do not affect surfactant secretion at the concentrations examined ($< 10\ \mu M$). Bob Mason's group in Denver has utilized oleoylacetylglycerol, and we have used dioctanoylglycerol. Both compounds stimulate surfactant phosphotidylcholine secretion.

B. WALKER: Do you have any experience with 2-Me-S-ATP (2-methylthio-ATP) or UTP?

RICE: We have examined the effect of 2-Me-S-ATP on surfactant phospholipid secretion and calcium mobilization. 2-Me-S-ATP is a much weaker agonist than ATP, requiring concentrations of 100-250 μM to produce a response.

Elevation of $[Ca^{2+}]_i$ and the Activation of Ion Channels and Fluxes by Extracellular ATP and Phospholipase C-Linked Agonists in Rat Parotid Acinar Cells

STEPHEN P. SOLTOFF, MICHAEL K. McMILLIAN, JAMES D. LECHLEITER, LEWIS C. CANTLEY, AND BARBARA R. TALAMO

Department of Physiology and Neuroscience Program
Tufts University School of Medicine
Boston, Massachusetts 02111

INTRODUCTION

Extracellular ATP produces a myriad of effects on normal and transformed cells, including alterations in ion fluxes, phospholipid metabolism, growth and differentiation, and secretion. Many of these changes appear to be mediated by specific receptors on the plasma membrane. Burnstock[1] proposed that ATP might be released from nerves as a neurotransmitter, and that "purinergic" receptors could be subclassified into two groups: P_1, those that recognized adenosine, and P_2, those that recognized ATP and other nucleotides. The P_2 group was further subdivided by Burnstock and his collaborators into P_{2X} and P_{2Y} based on the pharmacological potencies of different adenine nucleotides.[2] Gordon[3] suggested the further expansion to P_{2T}, the purinoceptor on platelets, and P_{2Z}, the purinoceptor on a number of cells that respond to ATP^{4-}.

One of the main effects of extracellular ATP is the activation of ion channels (TABLE 1).[4–12] Some of the ATP-activated channels, including K^+ channels,[7,11] are Ca^{2+} sensitive, and these appear to be indirectly activated by the effects of ATP on $[Ca^{2+}]_i$. In several cells, extracellular ATP can stimulate phosphoinositol turnover, resulting in the production of inositol 1,4,5-trisphosphate ($InsP_3$), the elevation of $[Ca^{2+}]_i$, and the activation of Ca^{2+}-sensitive ion channels. In contrast, ATP activates cation-selective channels in a number of systems by a mechanism that does not appear to involve $InsP_3$ production. In general, these channels are not sensitive to $[Ca^{2+}]_i$, and may be directly activated by ATP without the involvement of a second messenger.

We have been studying the effects of classical neurotransmitter agonists and extracellular ATP on rat parotid cells,[13–17] which provide a useful model for studying receptor-mediated changes in intracellular events (that is, second messengers), $[Ca^{2+}]_i$, and ion transport systems. The constitutive activity of ion channels in these cells is

TABLE 1. A Partial List of Ion Channels Activated by Extracellular ATP

Cell	Current(s)[a]	Selectivity	Single Channel Conductance (pS)	Depolarizing	Desensitizing	Na_o-Sensitive Current	Ca_i-Sensitive Current	Reference
Smooth muscle								
Rat vas deferens	Inward	Cation	2.5-4.0		Yes	Yes		4
Rat vas deferens	Inward	Cation	20	Yes	Yes	Yes		5
Rabbit ear artery	Inward[b]	$3Ca^{2+}:1Na^{+}$	5	Yes	Yes	Yes	No	6
Bullfrog atrial	Inward	Cation	< 1		Yes	Yes	No	7
	Outward	K^{+}			No		Yes	
Skeletal								
Chick myoblast/ myotube	Inward[c]	Cation		Yes	Yes			8
Oocyte								
Xenopus	Inward[d]	Cl^{-}		Yes	Yes			9
	Outward	K^{+}						
Neurons								
Rat sensory	Inward	Cation			Yes	Yes		10
Epithelia								
MDCK cells	Outward	K^{+}	66			No	Yes	11
Rat parotid	Inward	Cation	31			Yes	Yes	15
	Outward	K^{+}	130			No	Yes	
Other								
Macrophage	Inward	Cation/Anion	≤ 20	Yes	No			12

[a] Direction when voltage clamped near physiological level.
[b] Observable in outside-out patches.
[c] Not observable in outside-out patches.
[d] Inward activated by P_2 receptor, outward activated by P_1 receptor.

very low. Neurotransmitters that activate muscarinic, substance P, and α-adrenergic receptors greatly increase the membrane permeability and stimulate fluid secretion by the parotid gland. Gallacher[18] reported that extracellular ATP produced changes in the membrane conductance of mouse parotid acinar cells similar to alterations produced by acetylcholine and α-adrenergic agonists. Like Gallacher, we have found that ATP promotes similar changes to those produced by other agonists, but we find that ATP appears to have unique effects as well. This paper summarizes some of our findings. Our results suggest that ATP directly activates a Ca^{2+}-permeable channel, and that the resultant elevation of $[Ca^{2+}]_i$ activates additional channels that are Ca^{2+} sensitive. ATP, which is stored and cosecreted with muscarinic or adrenergic neurotransmitters in some systems,[3,19] may play a physiological role in the stimulation of fluid secretion by the parotid gland.

RESULTS AND DISCUSSION

Effects of Extracellular ATP and Neurotransmitter Agonists on Ca^{2+} Mobilization

The effects of ATP and other agonists on $[Ca^{2+}]_i$ are shown in FIGURE 1. Carbachol and substance P rapidly elevated $[Ca^{2+}]_i$. Carbachol produced a sustained elevation that was reversed by the muscarinic antagonist atropine. The substance P-initiated elevation spontaneously returned to baseline levels because of desensitization of the substance P receptor.[14] ATP usually elevated $[Ca^{2+}]_i$ to greater levels than those produced by other agonists. The $K_{0.5}$ for ATP in elevating $[Ca_2^{+}]_i$ to maximal levels was > 100 μM. Our recent studies, however, indicate that there is a biphasic $[Ca^{2+}]_i$ response to ATP, consisting of a saturable response at relatively low (< 10 μM) ATP concentrations and a second saturable response at higher ATP concentrations (McMillian *et al.*, in preparation).

In the absence of extracellular Ca^{2+}, the peak $[Ca^{2+}]_i$ response to carbachol was identical to that measured in the presence of Ca^{2+}, but the elevation rapidly returned to baseline levels.[17] This indicated that the initial muscarinic response was due to the release of Ca^{2+} from intracellular stores and that an influx of Ca^{2+} across the plasma membrane was required to maintain the elevation. Similar findings have been reported previously in the parotid cell as well as many other cells. In contrast, we found that the response to ATP was greatly diminished (FIG. 1B) in the absence of extracellular Ca^{2+}, indicating that the initial response to ATP relied primarily on the influx of Ca^{2+}. This was verified by measuring Ca^{2+} influx using $^{45}Ca^{2+}$.

In the absence of agonist, $^{45}Ca^{2+}$ entered relatively slowly. The rate was slightly enhanced by carbachol, but ATP promoted a marked acceleration of extracellular Ca^{2+} influx (FIG. 2). The ATP-stimulated rate of Ca^{2+} entry was more than 20 times the basal rate. The effect of ATP was intermediate between that of the calcium ionophore ionomycin and that of carbachol.

Unlike the other agonists that elevate $[Ca^{2+}]_i$ in parotid cells, ATP did not act via phospholipase C. The relative effects of carbachol, substance P, and phenylephrine on $InsP_3$ production and on the elevation of $[Ca^{2}]_i$ were about the same.[15] Although ATP produced a large elevation of $[Ca^{2+}]_i$, it produced only a small increase in $InsP_3$, similar to ionomycin (FIG. 3). The small effect of ATP on $InsP_3$ was reduced by about 75% by removal of extracellular Ca^{2+}, but these conditions barely altered the effects of carbachol and the other agonists.[15] These results, in combination with those described above, indicated that the effects of ATP were not mediated by $InsP_3$, and that the receptors of the other agonists were linked to phospholipase C. Thus, extracellular ATP and phospholipase C-linked agonists elevated $[Ca^{2+}]_i$ by markedly different mechanisms.

In the absence of extracellular Mg^{2+}, ATP was about 10-fold more potent in its effects on $[Ca^{2+}]_i$ and other physiological responses.[13,17] As shown in FIGURE 4, in

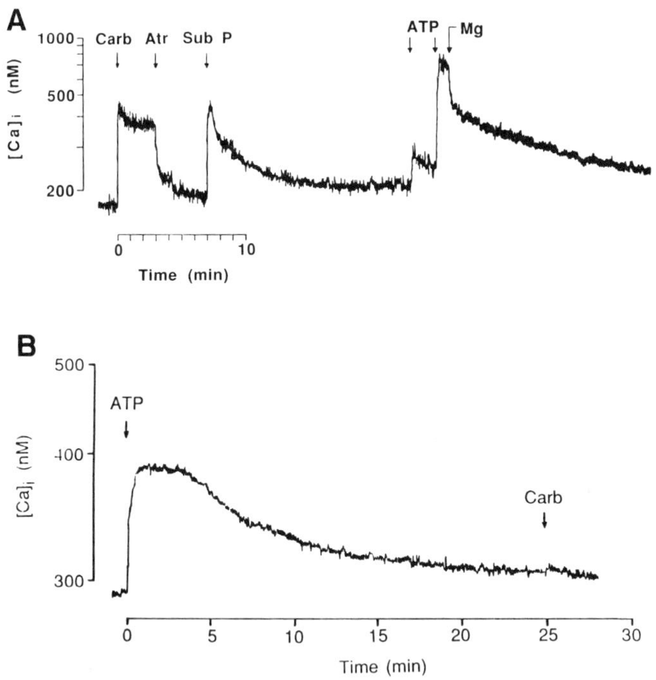

FIGURE 1. The effects of carbachol, substance P, and extracellular ATP on $[Ca^{2+}]_i$. The cells were loaded with Fura 2, and $[Ca^{2+}]_i$ was measured as reported.[14] (**A**) The effects of carbachol (20 μM), atropine (1 μM), substance P (10 nM), and two concentrations of ATP (30 μM, 1 mM) on cells suspended in solution A, a Ca^{2+}-containing medium (constituents, in mM: NaCl, 120; KCl, 5; $MgCl_2$, 2.2; $CaCl_2$, 1; HEPES, 20; betahydroxybutyrate, 5; glucose, 10; bovine serum albumin, 0.1%) (pH 7.4). (**B**) The effect of ATP (1 mM) in the absence of extracellular Ca^{2+}. From Soltoff *et al.*,[17] reprinted with permission.

the presence of Mg^{2+} more than 300 μM ATP was required to elevate $[Ca^{2+}]_i$ to a level comparable to or greater than that produced by carbachol, but much less ATP produced this effect in the absence of Mg^{2+}. These experiments demonstrated that the effects of ATP were not mediated by ectokinases or ecto-ATPases, which require $Mg\text{-}ATP^{2-}$ as a substrate. Instead, it appeared that ATP^{4-} is the form that acts on parotid cells. This suggested that the parotid cell purinoceptor is of the P_{2Z} class. In addition, α,β-methylene ATP and 2-methylthio-ATP, selective agonists of the P_{2X} and P_{2Y} receptors, respectively, were less potent than ATP in eliciting alterations in

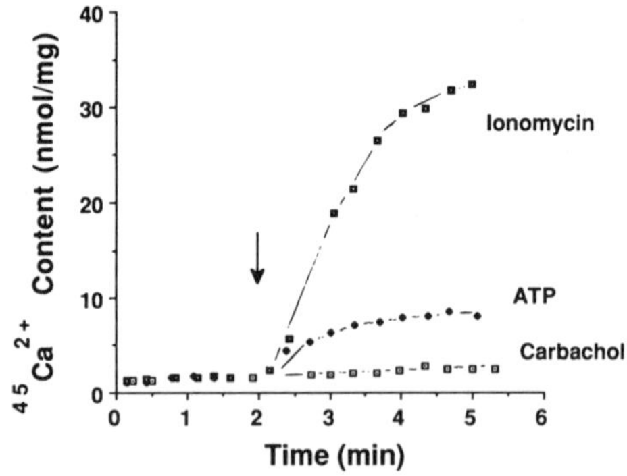

FIGURE 2. The effect of different agonists on $^{45}Ca^{2+}$ entry into parotid cells. Cells were suspended in solution B (constituents, in mM: NaCl, 116.4; KCl, 5.4; $MgSO_4$, 0.8; NaH_2PO_4, 1; NaHEPES, 25; $CaCl_2$, 1.8; butyrate, 1; glucose, 5.6) (pH 7.4). Isotope was added at time 0, and samples were collected as previously described[16] to measure the cellular $^{45}Ca^{2+}$ content. Carbachol (20 μM), ATP (1 mM, with 1 mM $MgCl_2$), or ionomycin (1 μM) was added at the time indicated by the arrow.

$[Ca^{2+}]_i$ or other physiological responses. Recently, we observed that 3′-*O*-(4-benzoyl)benzoyl-ATP (Bz-ATP) was more potent than ATP in promoting responses in parotid cells (Soltoff *et al.*, in preparation). Although this agonist was highly potent in activating a P_{2Z} receptor in transformed mouse fibroblasts,[20] it was also a potent agonist of the P_{2Y} receptor on turkey erythrocytes,[21] and thus may not discriminate P_2 receptor subtypes. The potency order for the parotid cell (Bz-ATP > ATP >

FIGURE 3. Comparison of various agonists on the accumulation of [^{3}H]$InsP_3$ and the elevation of $[Ca^{2+}]_i$ in rat parotid acinar cells. Cells were loaded with [^{3}H]inositol or Fura 2, and the $InsP_3$ levels (accumulated after a 15-sec exposure) or peak values of $[Ca^{2+}]_i$ in response to carbachol (10 μM), substance P (1 nM), norepinephrine (10 μM) or phenylephrine (100 μM), ATP (1 mM), and ionomycin (1 μM) were measured as described.[15,16] The results are presented in terms of the basal levels of $InsP_3$ and $[Ca^{2+}]_i$ measured in the absence of agonist. The $InsP_3$ results were presented in tabular form in McMillian *et al.*,[15] and the $[Ca^{2+}]_i$ results were reported in Soltoff *et al.*[16,17]

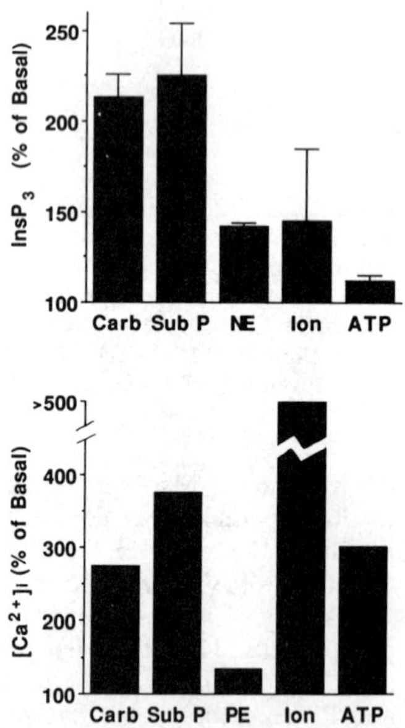

ATP-γS > 2-methylthio-ATP > AMP-PNP, ADP, α,β-methylene ATP) is consistent with the parotid receptor being of the P_{2Z} class. Other cells with receptors in the P_{2Z} class are lymphocytes, macrophages, and mast cells.[3,12,22–24] For all of these cells, ATP is more potent in the absence of extracellular Mg^{2+}. One characteristic that is not shared by parotid cells and several of the other cells in this category, however, is an ATP-promoted increase in permeability to normally impermeant compounds.[24] For example, ATP caused the loss of intracellular nucleotides from transformed fibroblasts[19] and the uptake of ethidium bromide (M_r = 314) into rat mast cells.[23] ATP, however, did not promote the uptake of ethidium bromide into parotid cells (data not shown). In addition, the effects of ATP on $[Ca^{2+}]_i$ (measured using Quin 2 or Fura 2) were reversible (FIG. 4), demonstrating that Quin 2 (M_r = 388)

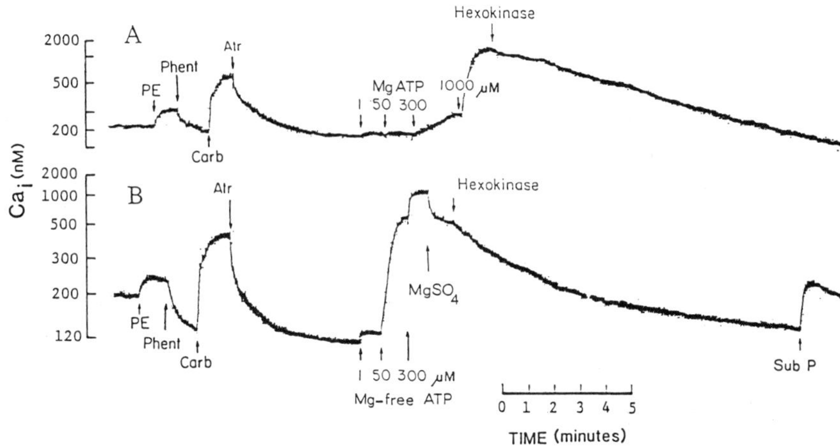

FIGURE 4. The effect of carbachol (10 μM) and ATP on $[Ca^{2+}]_i$ in parotid cells suspended in the presence (**A**) and absence (**B**) of extracellular Mg^{2+}. Note that ATP was much more potent in the absence of Mg^{2+}. Measurements of $[Ca^{2+}]_i$ were made using Quin 2. Cells in **A** were suspended in solution A, and ATP was added in equimolar amounts with $MgSO_4$. Cells in **B** were suspended in Mg^{2+}-free solution A, and ATP was added without Mg^{2+}. The addition of hexokinase (2.5 units/ml), which consumed the added ATP, reversed the ATP-induced elevation of $[Ca^{2+}]_i$. In **B**, $MgSO_4$ (2.2 mM) was added as a cofactor for hexokinase. From McMillian *et al.*,[13] reprinted with permission.

and Fura 2 (M_r = 831) did not leak out of the cytosolic space. (In the presence of extracellular Ca^{2+}, the efflux of Quin 2 or Fura 2 would have resulted in an irreversible increase in fluorescence.)

DIDS Blocks the Effects of ATP and the Binding of [α-^{32}P]ATP

The effects of ATP on parotid cells were blocked by 4,4′-diisothiocyano-2,2′-stilbene disulfonate (DIDS). Pretreatment of the cells with DIDS concentrations of 100-150 μM completely blocked the ATP-stimulated elevation of $[Ca^{2+}]_i$, but did not block the $[Ca^{2+}]_i$ rise produced by the phospholipase C-linked agonists.[15] Thus, DIDS discriminated between the purinoceptors and the phospholipase C-linked receptors. Initially, we exposed the cells to DIDS at room temperature for ≥ 30 min, and found that the exposure resulted in the irreversible blockade of the responses to ATP. More recently, we have observed that an exposure to DIDS at 37 °C for as little as 5 min was sufficient to produce irreversible inhibition (Soltoff, unpublished results). Because DIDS is fluorescent, $[Ca^{2+}]_i$ measurements were performed using cells that were pretreated with DIDS, and then washed. We have also found that dihydro-DIDS (H_2-DIDS), the reduced form of DIDS, is nearly as effective as DIDS in blocking ATP-stimulated alterations in $[Ca^{2+}]_i$. Unlike DIDS, H_2-DIDS does not greatly in-

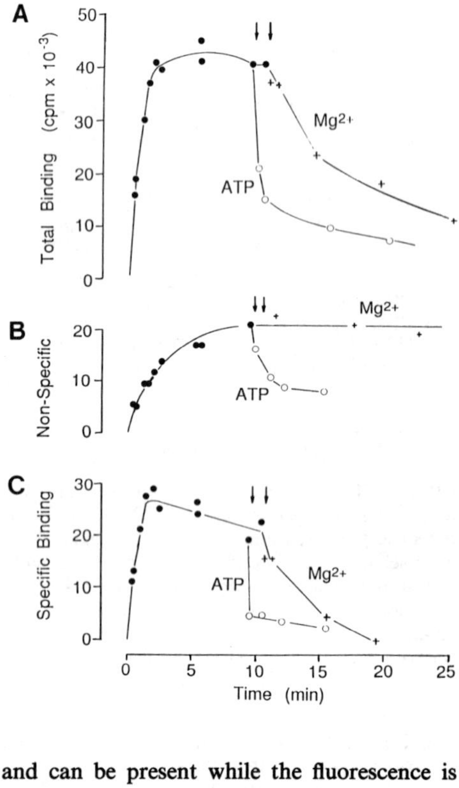

FIGURE 5. The binding of [α-^{32}P]ATP to intact parotid cells is reduced by DIDS. Radiolabeled ATP (0.1 μM) was added in the absence (**A**) or presence (**B**) of 100 μM DIDS to parotid cells suspended at 37 °C in phosphate-buffered saline (constituents, in mM: NaCl, 136.9; NaH_2PO_4, 15.65; KH_2PO_4, 1.47; KCl, 2.68) containing 1 mM EDTA. After about 10 min, either 5 mM $MgSO_4$ or 1 mM unlabeled ATP was added. The DIDS-sensitive binding (**A** minus **B**) is shown in **C**. From McMillian *et al.*,[15] reprinted with permission.

terfere with the Fura 2 fluorescence, and can be present while the fluorescence is being monitored.

[α-^{32}P]ATP bound rapidly to intact parotid cells.[15] It was displaced by excess unlabeled ATP or by Mg^{2+}, and DIDS substantially reduced this binding (FIG. 5). These results suggested that the effectiveness of DIDS in blocking the physiological effects of ATP was due to its blockage of ATP binding, rather than at a site beyond this step. In preliminary results, the number of DIDS-sensitive ATP-binding sites was estimated as 1.2 pmol/mg total cell protein. If we use a value of 2.58 × 10^6 cells/mg protein (derived from reference 25), the estimate represents 0.47 pmol/10^6 cells, or about 2.9 × 10^5 sites/cell. Presumably, this fraction of ATP-binding sites identifies the P_{2Z} receptor, and represents an upper limit for the number of purinoceptors found on these cells. Additional studies, however, are necessary for a more complete characterization, which would include a biochemical description of this receptor. Keppens and De Wulf[26] found about 10 times as many P_2-binding sites on hepatocytes using [^{35}S]ATP-α-S as a ligand. [^{35}S]ADP-β-S was used as a ligand to characterize the turkey erthyrocyte P_{2Y} receptor,[27] which is linked to phospholipase C and stimulates the production of inositol phosphates.[28] Membranes prepared from these cells had about 3 pmol binding sites/mg protein. This value was similar to that obtained for the parotid cell, but was normalized to plasma membrane protein rather than total cellular protein.

DIDS blocked all of the ATP-stimulated ion fluxes that we examined, including Ca^{2+} influx and Na^+ influx.[15,17] In contrast, Chahwala and Cantley[29] reported that DIDS blocked the ATP-stimulated influx of Ca^{2+} and Cl^- into Friend erythroleukemia cells, but did not affect the ATP-stimulated Na^+ uptake. Kimmich and Randles[30] found that 4-acetamino-4′-isothiocyanostilbene-2,2′-disulfonate (SITS), a stilbene monoisothiocyanate compound, blocked ATP-stimulated Na^+ entry but not Ca^{2+} entry into intestinal epithelial cells. It is curious that these isothiocyanate compounds do not appear to block the ATP effects in a consistent fashion in different cells. To our knowledge, our findings are the first to report that DIDS may block the effects mediated by purinoceptors because of its effects on ATP binding, and that these effects may be partly due to the similarities of the sulfonate and phosphate residues. DIDS has been used extensively as an inhibitor of the anion exchanger of red cells and other cells, where it binds to a lysine residue on the transport protein.[31] It also appears to bind to a lysine residue on the ATP-binding region of the Na^+,K^+-ATPase.[32] It remains to be determined whether DIDS is an effective inhibitor of the P_{2Z} class of receptors on other cells, and whether it blocks other classes of purinoceptors.

Agonist-Induced Activation of Ca^{2+}-Sensitive Ion Transport Systems

Among the initial ionic events involved in the stimulation of fluid secretion are the activation of Ca^{2+}-sensitive basolateral K^+ channels and the activation of luminal Cl^- channels.[33,34] These events promote the loss of large amounts of K^+ and Cl^- from the acinar cell, and consequently the cells shrink. Subsequently, Na^+ enters the cell through multiple pathways. Both extracellular ATP and phospholipase C-linked agonists initiate this sequence of events by the elevation of $[Ca^{2+}]_i$.[16,17] The extent of these changes are shown on FIGURE 6, which compares the effects of carbachol and ATP.

To characterize the Ca^{2+} dependence of the K^+ efflux, we measured the rate of K^+ efflux initiated by carbachol and that initiated by ATP under several different conditions (FIG. 7). Loading the cells with BAPTA, a Ca^{2+} buffer that slows the rise in $[Ca^{2+}]_i$ in both ATP- and carbachol-stimulated cells, slowed the agonist-stimulated K^+ efflux.[17] This indicated that an elevation of $[Ca^{2+}]_i$ was required. In the absence of extracellular Ca^{2+}, the stimulation by carbachol was inhibited only slightly, whereas the response to ATP was severely reduced. These results are consistent with findings that the initial rise in $[Ca^{2+}]_i$ promoted by carbachol was due to the mobilization of an intracellular pool of Ca^{2+}, and that the rise promoted by ATP was due primarily to the mobilization of extracellular Ca^{2+}.

Ion channels were characterized electrophysiologically using patch-clamp techniques.[15] Single-channel recordings of ionic currents were made in the cell-attached

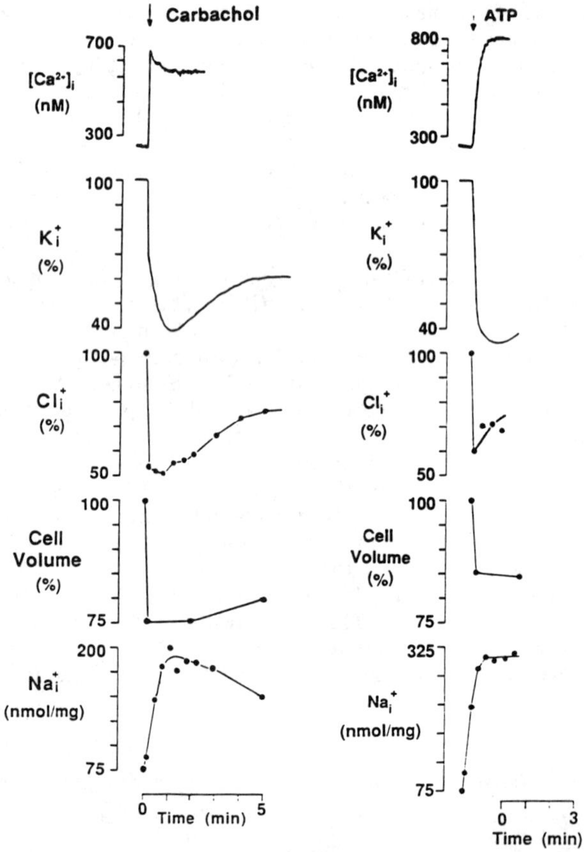

FIGURE 6. Comparison of the physiological effects of carbachol (30 μM) and ATP (1 mM) on rat parotid acinar cells. Measurements of $[Ca^{2+}]_i$ were made using Fura 2. Alterations in the intracellular Cl^- and Na^+ contents were measured using $^{36}Cl^-$ and $^{22}Na^+$, respectively. Alterations in the intracellular K^+ content were obtained using a K^+-sensitive extracellular electrode. Cell volume was measured using [^{3}H]H_2O and [^{14}C]sucrose. For additional details on these techniques, see Soltoff *et al.*[16,17]

mode in response to the addition of carbachol or ATP to the bath solution. Both agonists activated two classes of single channels: a small conductance channel of 31 pS and a larger conductance channel of 130 pS. The larger channel was blocked by 10 mM tetraethylammonium (TEA), suggesting that it was the BK (or maxi-K^+) K^+-sensitive channel. Ionomycin also activated the two channel types, indicating that both channels were activated by Ca^{2+}. When the cell was penetrated with the patch electrode to make whole-cell recordings, the application of either agonist produced a small inward current followed by a larger outward current. The outward current was not observed when cells were dialyzed with a zero K^+/high Na^+ solution in the patch electrode, indicating that the larger current was normally carried by K^+.

To study the possible involvement of GTP-dependent proteins in the activation of membrane currents, patch electrodes were filled with GDP-β-S, and the above experiments were repeated.[15] After a positive response was obtained upon application of carbachol or ATP under cell-attached conditions, the cell was penetrated with the electrode to make whole-cell recordings. After waiting 1 min to allow for GDP-β-S to dialyze into the cell interior, the agonist was applied again (FIG. 8). Under these conditions, ATP elicited its normal response, but the cells did not respond to carbachol. Therefore, it is likely that the muscarinic agonist acts through a G protein (presumably G_p, which links the receptor to phospholipase C). ATP acted by a different mechanism, consistent with the phosphoinositide results (FIG. 3) and with the hypothesis that ATP may directly activate a Ca^+-permeable channel.

The effects of different K^+ channel blockers on agonist-stimulated K^+ efflux were examined (FIG. 7) to investigate whether activation of the BK channel was sufficient to account for the net loss of K^+ from the cell (FIG. 6).[17] TEA reduced the carbachol-stimulated efflux by about 60%, but reduced the ATP-stimulated efflux by only about 30%. Because TEA effectively blocked the BK K^+ channel in the patch-clamp experiments,[15] these results indicated 1) that multiple types of K^+-permeable channels must contribute to the agonist-initiated loss of K^+, and 2) that the relative contribution of these channels varies with the agonist. Charybdotoxin (CTX), another inhibitor of various types of K^+ channels,[35] was more effective than TEA in blocking the effects of both ATP and carbachol. Thus, in the parotid gland, CTX appears to inhibit additional K^+-permeable channels that are not blocked by TEA.

ATP Increases $[Na^+]_i$ through Multiple Na^+ Entry Pathways

Extracellular ATP and carbachol maximally increased the intracellular Na^+ content to about 4 times and 2.5 times, respectively, the basal content (FIG. 6). The relative increase in the Na^+ concentration was even greater than the relative increase

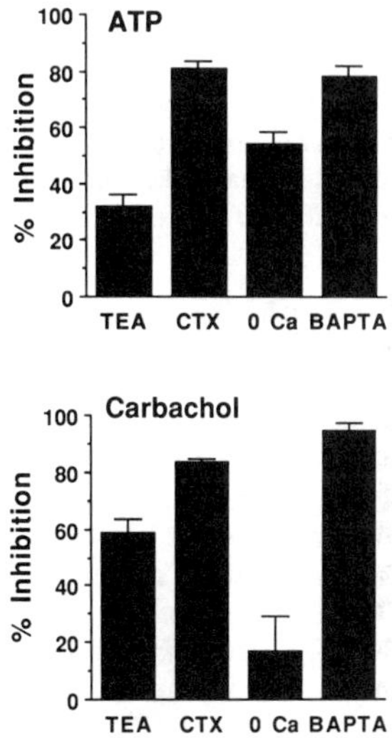

FIGURE 7. Comparison of the effects of carbachol (30 μM) and extracellular ATP (1 mM) on the initial rate of K^+ efflux from parotid cells. The rate of K^+ efflux was measured using a K^+-sensitive extracellular electrode. The cells were suspended in solution B and were exposed to TEA (tetraethylammonium, 15 mM) or CTX (charybdotoxin, 100 nM) about 2 min prior to the addition of agonist. For "O Ca^{2+}" experiments, cells were suspended in the absence of extracellular Ca^{2+} and with 500 μM EGTA. BAPTA-treated cells were exposed to 25 μM BAPTA-AM for at least 20 min prior to the addition of agonist. The data were normalized to measurements obtained under control conditions. These results were presented in tabular form in Soltoff *et al.*[17]

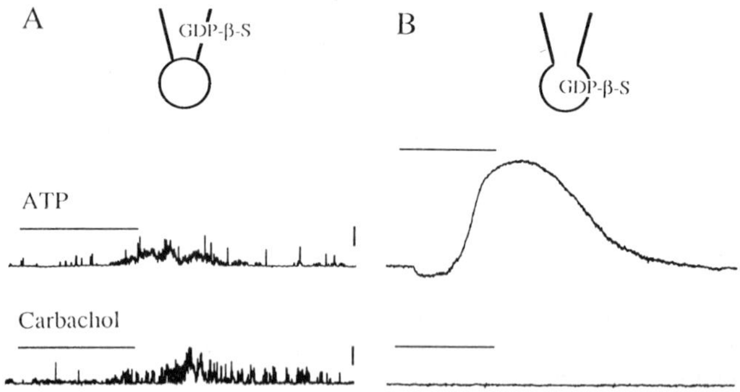

FIGURE 8. GDP-β-S blocks the membrane current initiated by carbachol but not that initiated by ATP. (**A**) Single-channel patch recordings were made in the cell-attached configuration with 10 mM GDP-β-S in the patch pipette (0 mV patch potential) in response to a 2-sec exposure (horizontal bar) to agonist. Vertical calibration bar: 15 pA. (**B**) Whole-cell recordings (at −50 mV) from the same cell as that shown in **A.** After a 1-min dialysis with the contents of the patch pipette, the cells were again exposed to the agonist for 2 sec (horizontal bar). Vertical bar: 50 pA. From McMillian *et al.*,[15] used with permission.

in the Na^+ content, because the parotid cells rapidly decreased in cell volume in response to the agonists (FIG. 6). The bulk of the carbachol-stimulated entry was via the Na^+-K^+-$2Cl^-$ cotransporter, and about 20% of the entry was via the Na^+-H^+ exchanger.[16] In addition to promoting the entry of Na^+ through these two pathways, ATP stimulated Na^+ uptake via an additional pathway. This may be an ATP-gated Na^+-permeable ion channel, similar to that described in other cells (TABLE 1). Such a channel may be permeable to both Na^+ and Ca^{2+}, accounting for the Ca^{2+}uptake as well as the additional Na^+ uptake promoted by ATP. In support of this, measurements of the parotid cell membrane potential using a voltage-sensitive dye showed that ATP promoted a biphasic change: an initial hyperpolarization followed by a depolarization (FIG. 9). In contrast, carbachol promoted a hyperpolarization only. The hyperpolarization appeared to be due to the activation of Ca^{2+}-sensitive K^+ channels by both agonists, whereas the ATP-promoted depolarization appeared to be due to the activation of an ATP-gated Na^+- and Ca^{2+}-permeable channel (Soltoff *et al.,* in preparation). Unlike the other ion channels characterized in this cell (TABLE 1),[15] this pathway was not activated by an elevation in $[Ca^{2+}]_i$.

FIGURE 9. Comparison between the effects of extracellular ATP (300 μM) and the effects of carbachol (30 μM) on the membrane potential of rat parotid acinar cells. Alterations in the membrane potential (bis-oxonol fluorescence measured at 540/580 nm emission/excitation wavelength pair) were measured using cells suspended in Mg^{2+}-free solution B containing 100 nM bis-oxonol. An increase (or decrease) in the fluorescence represents a depolarization (or hyperpolarization) of the membrane potential.

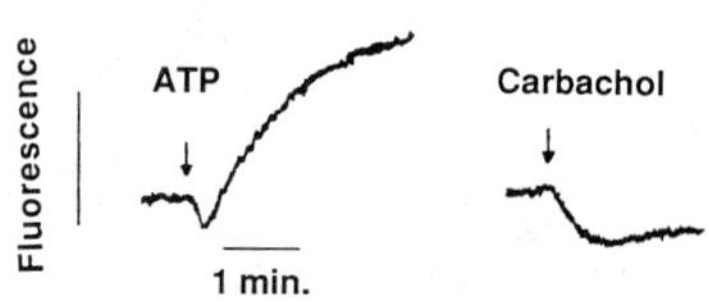

Activation of the Na^+ Pump

The Na^+ pump activity was rapidly activated by ATP and other agonists. This was easily detected by measuring the ouabain-sensitive oxygen consumption of the cells, because there is a tight coupling between mitochondrial oxidative phosphorylation and the Na^+ pump (Na^+,K^+-ATPase) activity.[36] In response to ATP (FIG. 10) and other agonists, the rate of oxygen consumption (QO_2) was greatly increased. The stimulatory effects of the agonists were blocked by ouabain, which inhibits the Na^+,-K^+-ATPase. The ouabain-sensitive rate of the ATP-stimulated or carbachol-stimulated cells was more than seven times the ouabain-sensitive basal QO_2.[16,17]

In the parotid cell, the stimulation of the Na^+ pump was mediated by the increase in intracellular Na^+ as well as the decrease in intracellular K^+ (FIG. 6). The effect

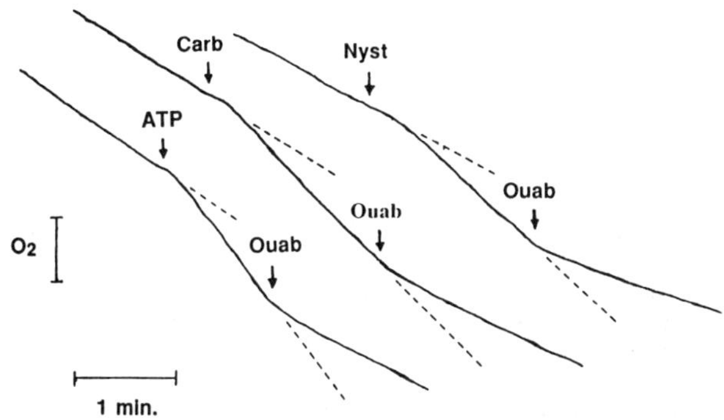

FIGURE 10. Stimulation of the oxygen consumption of parotid cells by ATP, carbachol, and nystatin. Shown is the disappearance of oxygen from a closed chamber containing parotid cells suspended in solution B. ATP (1 mM, added with 1 mM $MgCl_2$), carbachol (20 μM), nystatin (0.4 mM), and ouabain (3 mM) were added where indicated. When ouabain was added first (not shown), it blocked the effects of all three stimuli, demonstrating that their effects were due to stimulation of the Na^+ pump.

of the Na^+/K^+ ratio on the Na^+,K^+-ATPase hydrolytic activity in parotid cell membranes is shown in FIG. 11A. At a given concentration of Na^+, the Na^+,K^+-ATPase activity was greater when the K^+ concentration was relatively low. The $K_{0.5}$ for Na^+ was ~10 mM when the K^+ concentration was fixed at 10 mM and was ~35 mM when the K^+ concentration was over 100 mM.

To characterize the response of the Na^+ pump in intact cells at various concentrations of intracellular Na^+ and low intracellular K^+, the ouabain-sensitive QO_2 of parotid cells exposed to different concentrations of extracellular Na^+ was measured in response to nystatin. Nystatin is a monovalent cationophore that collapses both the Na^+ and K^+ gradients across the plasma membrane. The response of the Na^+ pump to different concentrations of intracellular Na^+ at low concentrations of intracellular K^+ was examined by measuring the nystatin-stimulated ouabain-sensitive QO_2 of cells suspended in various concentrations of Na^+ (FIG. 11B). The nystatin-stimulated QO_2 of parotid cells was very sensitive to the Na^+ concentration under these conditions. Similar results were observed in renal cells.[37] The sensitivity of the ATP-stimulated ouabain-sensitive QO_2 to Na^+ was nearly identical to that of the nystatin-treated cells, suggesting that ATP promoted a similar alteration of the intracellular Na^+/K^+ ratio. In contrast, the carbachol-stimulated QO_2 was markedly less sensitive to the extracellular Na^+ concentration. Presumably, the relatively poor stimulation by carbachol at low Na^+ concentrations was due, at least in part, to the relatively lower increases in intracellular Na^+ produced by carbachol, which was restricted to the combined activities of the Na^+-K^+-$2Cl^-$ cotransporter and the Na^+-H^+ exchanger.

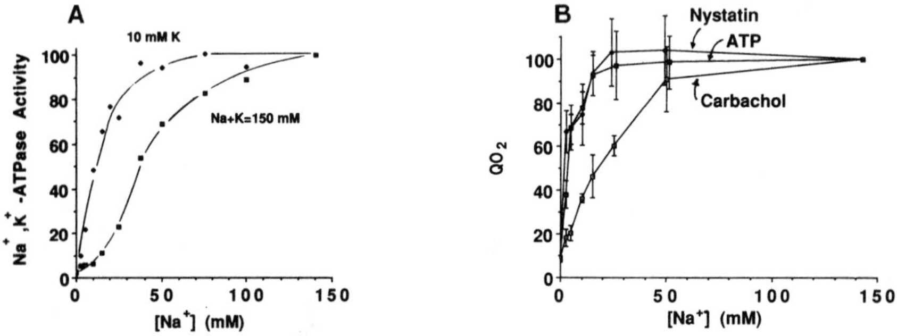

FIGURE 11. The dependence of the Na^+,K^+-ATPase (**A**) and the Na^+ pump (**B**) on the Na^+/K^+ ratio. (**A**) The Na^+, K^+-ATPase hydrolytic activity in lysed parotid acinar cells was measured in a coupled assay system that monitored the disappearance of NADH at 340 nm. The membranes were incubated in a solution that contained varying amounts of Na^+ with either a fixed or varying amount of K^+. In one case, the K^+ concentration was maintained at 10 mM, and the Na^+ concentration was varied between 0 and 140 mM by replacement of NaCl with tetramethylammonium chloride. In the other case, NaCl was varied by replacement with KCl, such that the combined sum of Na^+ and K^+ was 140 mM. In addition, the solutions contained the following: 2.5 mM $MgCl_2$, 20 mM HEPES, 1.4 mM phospho(enol)pyruvate, 20 U/ml lactate dehydrogenase, 5 mg/ml NADH, 3.5 U/ml pyruvate kinase, and 2 mM Tris-ATP-$MgCl_2$. For each condition, membranes were assayed in the presence and absence of 4 mM ouabain. The results are presented as the ouabain-sensitive ATPase, and are normalized to the activity measured at the highest Na^+ concentration. (**B**) The rate of oxygen consumption (QO_2) was measured using parotid cells suspended in varying concentrations of Na^+, in which *N*-methyl-D-glucamine was used to replace Na^+. Otherwise, the solution resembled solution B. The increase in the stimulated ouabain-sensitive QO_2 values in response to nystatin (0.4 mM), ATP (1 mM, added with 1 mM $MgCl_2$), and carbachol (30 μM) were normalized to the maximum increases (20.4 $\pm$ 3.4 (3), 18.1 $\pm$ 2.6 (3), and 20.0 $\pm$ 1.8 (4) nmol O_2/mg protein/min, respectively) produced by these stimuli at 142.4 mM Na^+.

SUMMARY

Extracellular ATP initiates a variety of changes in the parotid acinar cell. The initial effect appears to be the entry of Ca^{2+} (and perhaps Na^+), and a series of ion transport events result from the subsequent elevation of $[Ca^{2+}]_i$ (FIG. 6). Agonists of phospholipase C-linked receptors elevate $[Ca^{2+}]_i$ by a different pathway, involving the generation of inositol polyphosphate compounds, but share in the subsequent initiation of the ion transport events. Although the maintenance of the physiological changes may depend on specific inositol polyphosphate intermediates,[38] the critical initiating factor is the elevation of $[Ca^{2+}]_i$. Fluid secretion by the parotid gland is triggered by the action of neurotransmitters, which alter the membrane permeability of the acinar cell. The similarities between the two receptor-mediated activation pathways suggests that ATP may act as a neurotransmitter and play a role in the control of fluid secretion. Basing our analysis on the purinoceptor characteristics outlined by Gordon,[3] we suggest that the parotid receptor belongs to the P_{2Z} class, which is highly sensitive to ATP^{4-}. Basing his analysis on the earlier report by Gallacher[18] of the effects of ATP on mouse parotid cells, Gordon placed the parotid purinoceptor in a different P_2 subclass (P_{2Y}). However, our findings of an increased potency of ATP in the absence of Mg^{2+} (FIG. 4), as well as the potency order of different nucleotides, indicate that the P_{2Z} class is a more appropriate category.

REFERENCES

1. BURNSTOCK, G. 1978. A basis for distinguishing two types of purinergic receptor. *In* Cell Membrane Receptors for Drugs and Hormones: A Multidisciplinary Approach. R. W. Straub & L. Bolis, Eds.: 107-118. Raven Press. New York, NY.
2. BURNSTOCK, G. & C. KENNEDY. 1985. Is there a basis for distinguishing two types of P_2-purinoceptor? Gen. Pharmacol. **16:** 433-440.
3. GORDON, J. L. 1986. Extracellular ATP: Effects, sources and fate. Biochem. J. **233:** 309-319.
4. FRIEL, D. D. 1988. An ATP-sensitive conductance in single smooth muscle cells from the rat vas deferens. J. Physiol. **401:** 361-380.
5. NAKAZAWA, K. & N. MATSUKI. 1987. Adenosine triphosphate-activated inward current in isolated smooth muscle cells from rat vas deferens. Pfluegers Arch. **409:** 644-646.
6. BENHAM, C. D. & R. W. TSIEN. 1987. A novel receptor-operated Ca^{2+}-permeable channel activated by ATP in smooth muscle. Nature **328:** 275-278.
7. FRIEL, D. D. & B. P. BEAN. 1988. Two ATP-activated conductances in bullfrog atrial cells. J. Gen. Physiol. **91:** 1-27.
8. HUME, R. I. & M. G. HONIG. 1986. Excitatory action of ATP on embryonic chick muscle. J. Neurosci. **6:** 681-690.
9. LOTAN, I., N. DASCAL, S. COHEN & Y. LASS. 1986. ATP-evoked membrane responses in *Xenopus* oocytes. Pfluegers Arch. **406:** 158-162.
10. KRISHTAL, O. A., S. M. MARCHENKO & V. I. PIDOPLICHKO. 1983. Receptors for ATP in the membrane of mammalian sensory neurones. Neurosci. Lett. **35:** 41-45.
11. FRIEDRICH, F., H. WEISS, M. PAULMICHL & F. LANG. 1989. Activation of potassium channels in renal epitheliod cells (MDCK) by extracellular ATP. Am. J. Physiol. **256:** C1016-C1021.
12. BUISMAN, H. P., T. H. STEINBERG, J. FISCHBARG, S. C. SILVERSTEIN, S. A. VOGELZANG, C. INCE, D. L. YPEY & P. C. J. LEIJH. 1988. Extracellular ATP induces a large nonselective conductance in macrophage plasma membranes. Proc. Natl. Acad. Sci. USA **85:** 7988-7992.
13. MCMILLIAN, M. K., S. P. SOLTOFF, L. C. CANTLEY & B. R. TALAMO. 1987. Extracellular ATP elevates intracellular free calcium in rat parotid acinar cells. Biochem. Biophys. Res. Commun. **149:** 523-530.
14. MCMILLIAN, M. K., S. P. SOLTOFF & B. R. TALAMO. 1987. Rapid desensitization of substance P- but not carbachol-induced increases in inositol trisphosphate and intracellular Ca^{2+} in rat parotid acinar cells. Biochem. Biophys. Res. Commun. **148:** 1017-1024.

15. McMillian, M. K., S. P. Soltoff, J. D. Lechleiter, L. C. Cantley & B. R. Talamo. 1988. Extracellular ATP increases free cytosolic calcium in rat parotid acinar cells: Differences from phospholipase C-linked receptor agonists. Biochem. J. **255:** 291-300.
16. Soltoff, S. P., M. K. McMillian, L. C. Cantley, E. J. Cragoe, Jr. & B. R. Talamo. 1989. The effects of muscarinic, alpha-adrenergic, and substance P agonists and ionomycin on ion transport mechanisms in the rat parotid acinar cell. J. Gen. Physiol. **93:** 285-319.
17. Soltoff, S. P., M. K. McMillian, L. C. Cantley, E. J. Cragoe, Jr. & B. R. Talamo. 1990. The effects of extracellular ATP on ion fluxes and $[Ca^{2+}]_i$ in rat parotid acinar cells: Comparison to the muscarinic agonist carbachol. J. Gen. Physiol. **95:** in press.
18. Gallacher, D. V. 1982. Are there purinergic receptors on parotid acinar cells? Nature **296:** 83-86.
19. Stone, T. W. 1981. Physiological roles for adenosine and adenosine 5′-triphosphate in the nervous system. Neuroscience **6:** 523-555.
20. Gonzales, F. A., A. H. Ahmed, K. D. Lustig, L. Erb & G. A. Weisman. 1989. Permeabilization of transformed mouse fibroblasts by 3′-*0*-(4-benzoyl)benzoyl adenosine 5′-triphosphate and the densensitization of the process. J. Cell. Physiol. **139:** 109-115.
21. Boyer, J. L. & T. K. Harden. 1989. Mol. Pharmacol.: in press.
22. Cockcroft, S. & B. D. Gomperts. 1980. The ATP^{4-} receptor of rat mast cells. Biochem. J. **188:** 789-798.
23. Tatham, P. E. R., N. J. Cusack & B. D. Gomperts. 1988. Characterization of the ATP^{4-} receptor that mediates permeabilisation of rat mast cells. Eur. J. Pharmacol **147:** 13-21.
24. Steinberg, T. H., A. S. Newman, J. A. Swanson & S. C. Silverstein. 1987. ATP^{4-} permeabilizes the plasma membrane of mouse macrophages to fluorescent dyes. J. Biol. Chem. **262:** 8884-8888.
25. Kanagasuntheram, P. & P. J. Randle. 1976. Calcium metabolism and amylase release in rat parotid acinar cells. Biochem. J. **160:** 547-564.
26. Keppens, J. & H. De Wulf. 1986. Characterization of the liver P_2-purinoceptor involved in the activation of glycogen phosphorylase. Biochem. J. **240:** 367-371.
27. Cooper, C. L., A. J. Morris & T. K. Harden. 1989. Guanine nucleotide-sensitive interaction of a radiolabeled agonist with a phospholipase C-linked P_{2Y}-purinergic receptor. J. Biol. Chem. **264:** 6202-6206.
28. Boyer, J. L., C. P. Downes & T. K. Harden. 1989. Kinetics of activation of phospholipase C by P_{2Y}-purinergic receptor agonists and guanine nucleotides. J. Biol. Chem. **264:** 884-890.
29. Chahwala, S. B. & L. C. Cantley. 1984. Extracellular ATP induces ion fluxes and inhibits growth of Friend erythroleukemia cells. J. Biol. Chem. **259:** 13717-13722.
30. Kimmich, G. A. & J. Randles. 1982. An ATP- and Ca^{2+}-regulated Na^+ channel in isolated intestinal epithelial cells. Am. J. Physiol. **243:** C116-C123.
31. Ramjeesingh, M., A. Gaarn & A. Rothstein. 1981. The amino acid conjugate formed by the interaction of the anion transport inhibitor 4,4′-diisothiocyano-2,2′-stilbenedisulfonic acid (DIDS) with band 3 protein from human red blood cell membranes. Biochem. Biophys. Acta **641:** 173-182.
32. Pedemonte, C. H. & J. H. Kaplan. 1988. Inhibition and derivitization of the renal Na,K-ATPase by dihydro-4,4′-diisothiocyanatostilbene-2,2′-disulfonate. Biochemistry **27:** 7966-7973.
33. Petersen, O. H. & D. V. Gallacher. 1988. Electrophysiology of pancreatic and salivary cells. Annu. Rev. Physiol. **50:** 65-80.
34. Marty, A., Y. P. Tan & A. Trautmann. 1984. Three types of calcium-dependent channel in rat lacrimal glands. J. Physiol. **357:** 293-325.
35. Moczydlowski, E., K. Lucchesi & A. Ravindran. 1988. An emerging pharmacology of peptide toxins targeted against potassium channels. J. Membr. Biol. **105:** 95-111.
36. Mandel, L. J. & R. S. Balaban. 1981. Stoichiometry and coupling of active transport to oxidative metabolism in epithelial cells. Am. J. Physiol. **240:** F357-F371.
37. Soltoff, S. P. & L. J. Mandel. 1984. Active ion transport in the renal proximal tubule. II. Ionic dependence of the Na pump. J. Gen. Physiol. **84:** 623-642.
38. Morris, A. P., D. V. Gallacher, R. F. Irvine & O. H. Petersen. 1987. Synergism of inositol trisphosphate and tetrakisphosphate in activating Ca^{2+}-dependent K^+ channels. Nature **330:** 653-655.

DISCUSSION OF THE PAPER

P. A. WARD (*University of Michigan Medical School, Ann Arbor, MI*): It is possible that a phospholipase C pathway using PC as a substrate is involved? Is your response sensitive to pertussis toxin?

SOLTOFF: We were unable to find any evidence for phosphatidylcholine breakdown in these cells. Pertussis toxin did not appear to affect the responses to ATP or other agonists.

T. D. WHITE (*Dalhousie University, Halifax, Nova Scotia, Canada*): Do you anticipate that ATP^{4-} will have effects on parotid cells *in vivo,* where there is presumably extracellular Mg^{2+} that could prevent the ATP^{4-} effects?

SOLTOFF: Yes, I think that there will be an ATP^{4-} effect *in vivo.* I imagine the nerve terminal to be closely opposed to the basolateral membrane of the acinar cell. Large concentrations of ATP are costored with neurotransmitters. The release of ATP might initially "overwhelm" the extracellular Mg^{2+}, allowing for a substantial amount of ATP^{4-}. Eventually, because of formation of Mg-ATP^{2-}, a substrate for ecto-ATPases and kinases, and because of diffusion, the ATP^{4-} would be diminished. Of course, this scenario presents a stop-flow view of these events rather than a dynamic one.

R. I. HUME (*University of Michigan, Ann Arbor, MI*): I want to point out the striking similarity of the work you presented to the results we have found on chick skeletal muscle. [See pp. 484-485 and 486-488 in this volume—Ed.] In particular, in chick muscle, DIDS blocks the ATP responses, and ATP activates K^+ channels without the mediation of a G protein. However, our evidence indicates that these are *not* Ca^{2+}-dependent K^+ channels, and we think it very likely that there is a membrane-bound second messenger.

SOLTOFF: We are interested in hearing whether other groups have found DIDS to be inhibitory and whether it is specific to any particular P_2 receptor subtype. Because DIDS binds to lysine groups, and has been observed to bind to the ATP-binding site of various enzymes, we would predict that it might be effective on purinoceptors in a number of cells.

F. DI VIRGILIO (*University of Padua, Italy*): The experiment showing a larger increase in intracellular Ca^{2+} in the absence of Mg^{2+} does not prove formally that the active species in ATP^{4-}, since Mg^{2+} may compete with Ca^{2+} for translocation across the channel.

SOLTOFF: Elevating the extracellular Ca^{2+} concentration also decreases the potency of extracellular ATP. Under these conditions, the entry of Ca^{2+} would not be blocked; rather, the formation of Ca-ATP^{2-} diminishes the concentration of ATP^{4-}.

DI VIRGILIO: Did you test whether other nucleotides, such as UTP, CTP, and GTP, can also trigger the small Ca^{2+} rise caused by ATP?

SOLTOFF: The agonist potency series for the small response was ATP > ATP-γ-S > 2-methylthio-ATP (a P_{2Y}-selective agonist) > > ADP, ITP, and α,β-methylene ATP (a P_{2X}-selective agonist). GTP had no effect. We did not examine UTP or CTP on this response.

G. A. WEISMAN (*University of Missouri, Columbia, MO*): With respect to your findings of two subclasses of P_2 receptor in parotid cells, are both subclasses activated by ATP^{4-}? We have found evidence for two subclasses (P_{2Y} and P_{2Z}) in mouse fibroblasts and also indications that P_{2Y} receptors, like P_{2Z} receptors, may be activated by ATP^{4-}. Also, in the presence of Ca^{2+} and Mg^{2+}, small amount of ATP^{4-} would still be present. Have you directly titrated the small Ca^{2+} response with the ATP^{4-} concentration?

SOLTOFF: In the absence of extracellular Mg^{2+}, we find a biphasic concentration-response to extracellular ATP. The larger response (ATP concentrations above 30 μM) is activated by ATP^{4-}, and appears to involve a P_{2Z}-type receptor. Although we have not directly titrated the small response (observed between 0.1 and 10 μM ATP), it is not affected by large concentrations (up to 10 mM) of Mg^{2+} or Ca^{2+}, and may not be due to the activation of a P_{2Z}-type receptor. However, 2-methylthio-ATP (a potent P_{2Y} agonist) and α,β-methylene ATP (a potent P_{2X} agonist) were much less effective than ATP, so the small response does not appear to be either a P_{2Y} or P_{2X} receptor, either.

A. CHRISTIE (*Case Western Reserve University, Cleveland, OH*): What type of current did you depict with the cell-attached patch-clamp experiments—the Ca^{2+}-activated K^+ current or the ATP-activation nonselective current?

SOLTOFF: In the cell-attached mode, when the cell membrane outside of the patch electrode was exposed to ATP, we found two channels (31 pS and 130 pS). Carbachol also activates these channels under these conditions. These Ca^{2+}-sensitive channels appear to be a nonselective cation channel and the BK (maxi-K^+) K^+ channel.

CHRISTIE: Do you have any evidence for single-channel events induced by ATP, that is, single-channel data on the ATP-activated nonselective current?

SOLTOFF: Although our ion flux measurements (for which we used ^{22}Na and ^{45}Ca) indicate that ATP also activates a Ca^{2+}-insensitive, nonselective cation channel, we have not yet characterized this channel using electrophysiology.

S. C. SILVERSTEIN (*Columbia University, New York, NY)*: Could it be that all cells secrete ATP and that ATP is an autocoid?

SOLTOFF: Clearly there are some cells (most notably, platelets) that secrete ATP during the exocytosis of storage granules. Although ATP may function as an autocoid in some cellular systems, I think that these would be limited in number.

SILVERSTEIN: Are ATP receptors only on the basolateral surface or might they also be on the apical membrane, as suggested in type II pneumocytes by Dr. Rice?

SOLTOFF: Because ATP appears to have neurotransmitter-like effects on the parotid acinar cell, we predict that ATP receptors are on the basolateral membrane along with the other receptors. However, we have not yet performed any studies to demonstrate this.

W. R. RICE (*University of Cincinnati Medical Center, Cincinnati, OH*): In response to Dr. Silverstein's comment, we have measured ATP levels in type II cell media and have found that the levels are nondetectable. Lamellar bodies have not been examined for ATP content, but appear to contain SP-A, which down-regulates surfactant phospholipid secretion *in vivo.*

B. R. TALAMO (*Tufts University School of Medicine, Boston, MA*): In response to Dr. Silverstein's suggestion that ATP may be released from epithelial (and endothelial) cells and have an autocrine effect, I do not have an answer for parotid cells, but would like to point out that parotid secretory granules contain high concentrations of Ca^{2+}, which would be likely to complex ATP and make it less effective in activating ATP^{4-} receptors.

Purine Nucleosides and Nucleotides as Central Nervous System Modulators

Adenosine as the Prototypic Paracrine Neuroactive Substance

MICHAEL WILLIAMS

Neuroscience Research
Abbott Laboratories
Abbott Park, Illinois 60064

INTRODUCTION

Studies on the role of adenosine and adenosine triphosphate in mammalian tissue function can be divided historically into two general types: biochemical (that is, predominately pharmacological) and physiological. Biochemical studies have focused on extracellular recognition sites, both receptors and nucleoside transporters, for adenosine, second messenger systems and effects on neurotransmitter release and turnover. Physiological studies have been concerned with studying the effect of adenosine and its related nucleotides in the context of "nonadrenergic, noncholinergic" (NANC) neurotransmission in peripheral tissues[1]; local regulation of coronary and cerebral blood flow[2,3]; and behavior in animal models.[4] Because of the extensive metabolism of ATP by ectoenzyme systems in the synaptic cleft,[5] ascribing functionality to ATP as opposed to ADP, AMP, or adenosine has been difficult, especially in the complex and confined milieu of the mammalian peripheral and central nervous systems. This has been further compounded not only by a lack of potent and/or selective ATP antagonists to aid in delineating nucleotide effects from those of the nucleoside, but also by a limited availability of nucleotide analogues. Of all tissues studied, the mammalian brain has the highest levels of adenosine plus the richest density of adenosine receptors.[4]

RECEPTOR SUBCLASSES

Adenosine

Adenosine elicits its cellular effects via interaction with two major receptor classes, A_1 and A_2. The latter can be further divided into A_{2a} and A_{2b}. The A_1 receptor is involved in regulation of transmitter release and is selectively activated by adenosine analogues substituted in the N^6-position. These include N^6-cyclohexyladenosine (CHA), N^6-cyclopentyladenosine (CPA) (FIG. 1), 2-chloro-N^6-cyclopentyladenosine[6] (CCPA), and (*S*)-N^6-endonorbornyladenosine ((*S*)-ENBA).[7] In contrast, A_{2a} receptors can be selectively activated by adenosine analogues substituted in the 2- and 5'-positions including 5'-*N*-ethylcarboxamidoadenosine (NECA) (FIG. 1), a potent (K_i = 15), nonselective agonist while 2-phenylaminoadenosine (CV 1808),[8] although weak (K_i ~ 100 nM), is 5-fold A_2 selective.[9] 2-*p*-((Carboxyethyl)phenethylamino)-5-carboxamidoadenosine (CGS 21680) (FIG. 1)[10] is the most potent (K_i ~ 14 nM) and selective (140-fold) A_2 agonist yet reported.

A_{2b} receptors represent a low-affinity form of the A_{2a} receptor,[9] with no potent, selective ligands yet described. Other adenosine receptor subtypes include an A_3 receptor, which is involved in the regulation of neurotransmitter release.[11] A_{1a} and A_{1b} receptors[12] have also been proposed.

Adenosine receptors can be characterized by their antagonism by the xanthines. While theophylline and caffeine were instrumental in providing evidence for the existence of adenosine receptors,[13] many 8-substituted xanthines are available which are highly potent (K_i values in the nanomolar range) and usually A_1 selective. These include xanthine amine congener (XAC),[14] cyclopentylxanthine (CPX), and cyclopentyltheophylline (CPT) (FIG. 2).[15] PD 115,199, a sulfonamide congener of 1,3-dipropyl-8-phenylxanthine, is a potent (K_i = 16 nM), nonselective A_2 receptor antagonist,[15] as is the triazoloquinazoline, CGS 15943 (FIG. 2).[16]

Biochemical studies have shown that both the A_1 and A_2 receptor subtypes are heterogeneously distributed in the CNS.[4] Autoradiographic binding studies show that A_1 receptors labeled by [^{3}H]CHA have the highest binding in the molecular layer of the cerebellum and in the hippocampus. Intermediate binding is observed in the thalamus, caudate-putamen, septum, and cerebral cortex. Minimal binding is observed in the hypothalamus and brain stem.[17] Similar studies on A_2 receptor distribution using [^{3}H]NECA in the presence of CPA, to selectively label A_2 receptors,[18] and the A_2-selective ligand [^{3}H]CGS 21680[19] have shown that A_2 receptors are selectively localized to the striatum and olfactory tubercle. Based on comparisons with biochemical effects elicited through A_2 receptor activation, it would appear that the receptors identified with NECA/CPA and CGS 21680 are of the A_{2a} subtype. Marked species differences in both A_1 and A_2 receptors have been documented.[20,21]

Adenine Nucleotide

ATP and ADP can alter mammalian cell function via interaction with their own discrete set of receptors. Initially designated by Burnstock in 1978[1] as P_2 receptors (with the adenosine receptor(s) being classified as P_1), at least four adenine nucleotide receptor subclasses have been described: P_{2T}, P_{2X}, P_{2Y}, and P_{2Z}.[22,23] Whereas the X, Y, and Z receptors can be activated by various ATP analogues (TABLE 1), the T receptor is most sensitive to 2-methylthio-ADP. Hence the designation of nucleotide as opposed to ATP receptor subclasses.

Many analogues of the various adenine nucleotides have been made and tested in a variety of different tissues.[24] The structure-activity relationship for the various receptors has been limited because of a lack of comparison of a series of these ligands across different tissues and a lack of confirmation of reported activities in separate laboratories. Studies to date may therefore be confounded by what Black[25] has termed circular reasoning, especially in regard to the possibility that the rank activities for these entities may reflect differences in their efficacies in different tissues.

A comprehensive study of a series of ATP analogues in rat nodose ganglion[26] led to the conclusion that the nature of the polyphosphate side chain as well as C-6

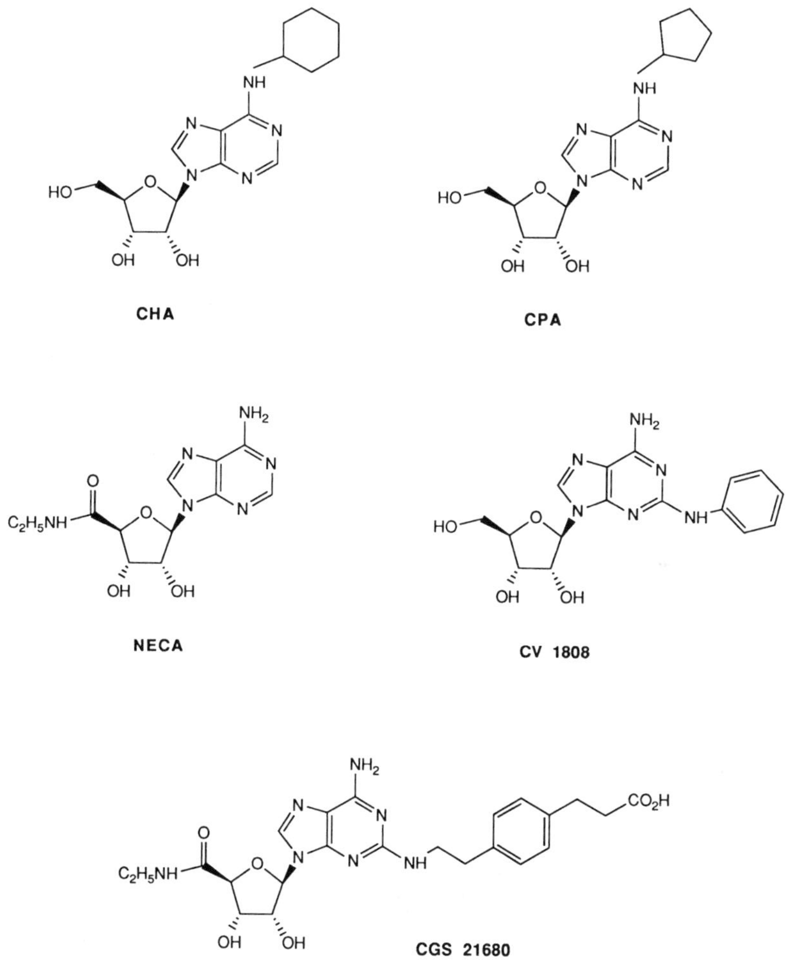

FIGURE 1. Adenosine analogues.

substituents determined activity at ATP receptors in this tissue. Interestingly, whereas adenosine 5′- (β,γ-methylene)-triphosphonate (APPCP) was found to be an agonist, the difluoro analogue, APPCF2P, was found to be a partial agonist, and the dichloro analogue, adenosine 5′-(β,γ-dichloromethylene)-triphosphonate (APPCC12P), a rather potent (K_i = 21 μM) competitive antagonist.

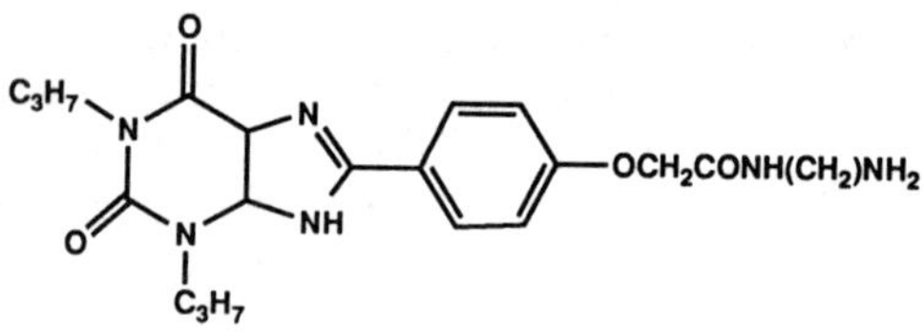

XAC

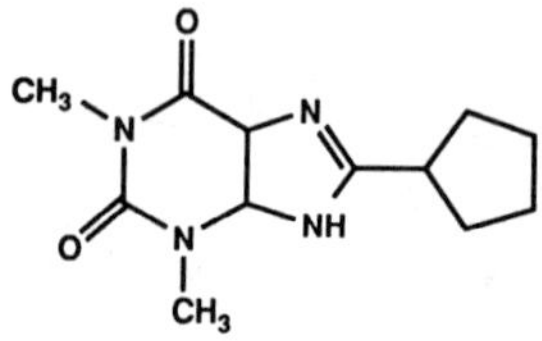

CPT

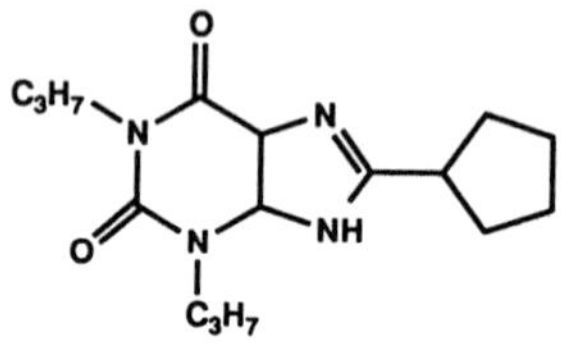

CPX

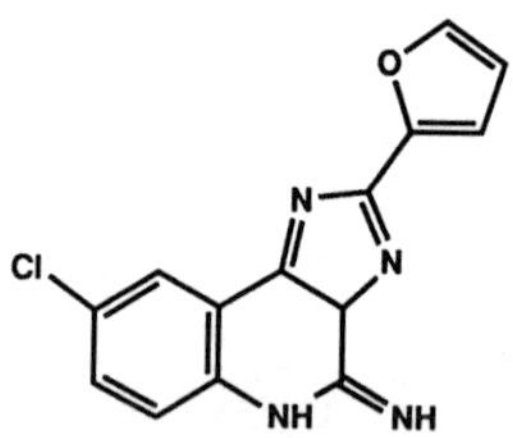

CGS 15943

FIGURE 2. Xanthine analogues.

As the various adenine nucleotide analogues synthesized by Cusack[24] and Blackburn[26] become more readily available, it is to be hoped that the data base used to delineate adenosine nucleotide receptors will increase to the point that some generalizations about receptor subtype functionality in various tissues will be possible.

PURINE RELEASE IN THE CNS

The mechanisms responsible for modulating the release of adenosine and ATP in nervous tissue have been extensively studied[27] since the initial observation of Holton in 1959.[28] In essence, although both depolarization and electrical stimulation have been reported to effect the release of adenosine in both calcium-dependent and independent modes, it is still unresolved as to whether purines are released from neurons, glia, or both. Fredholm and colleagues[29] have suggested that adenosine release is a carrier-mediated phenomenon. Based on the physiological effects of endogenous adenosine, it is probable that the purine nucleoside functions as a paracrine homeostatic neuromodulatory agent as defined by Schmitt.[30] The ramifications of this role in regard to the concept of "dynamical diseases" related to chaos theory[31] have been discussed elsewhere.[32,33]

Evidence for ATP being stored with acetylcholine and norepinephrine and for ATP having a functional role as a cotransmitter has also been extensively documented.[23,34] Release can be evoked by depolarization and is calcium dependent. Although ATP release from caudate synaptosomes[35] and hippocampal slices[36] has been reported, evidence for the "unequivocal release of endogenous ATP has yet to be demonstrated."[34]

BIOCHEMICAL EFFECTS OF ADENOSINE AND ATP

The role of ATP in mammalian CNS function has generally been assumed to be due to adenosine formed by ectonuclotidase action. As a consequence, few studies have focused on the role of ATP in the CNS beyond the issue of release.

Based on recent evidence for a rapidly expanding role in a variety of tissues,[23] an increased focus on the role of ATP, as distinct from that of adenosine, appears well justified. In addition to being a primary source of adenosine, and functioning as an agonist via interactions with its own discrete receptors, ATP has the potential to function as a substrate for ectokinases involved in long-term potentiation.[37] Purine nucleotides can modulate binding to A_1 receptors in rat brain membranes in a concentration-dependent, biphasic manner.[38] An initial inhibition of binding due to ATP-dependent adenosine formation was followed at higher concentrations ($> 100\ \mu M$) by a stimulation of binding. This stimulation may involve a P_{2Y} receptor.

Adenosine can modulate cyclic AMP formation, phosphatidyl inositol turnover, and calcium mobilization, as well as directly regulate potassium channel activity.[39,40] The consequences of these activities are alterations in nerve tissue responsiveness as

TABLE 1. Purine Receptor Types

Receptor Type	Selective Agonist(s)	Antagonist(s)
A_1	CCPA[a] > CPA = CHA	XAC, CPX, CPT
A_{2a}	CGS 21680	CGS 15943,[b] PD 115,199[c]
P_{2T}	2-Methylthio-ADP > ADP	ATP, AMP
P_{2X}	α,β-Methylene ATP > ATP	
	L-Adenosine-5′-(β,γ-methylene) triphosphate	$ANAPP_3$[d]
P_{2Y}	2-Methylthio-ATP > > ATP > Adenosine-5′-(2-fluorodiphosphate)	Reactive blue 2
P_{2Z}	ATP^{4-} > ATP	

[a] CCPA: 2-Chloro-N^6-cyclopentyladenosine.
[b] CGS 15943: 9-Chloro-2-(2-furyl)[1,2,4]triazolo[1,5-*c*]quinazolin-5-amine.
[c] PD 115,119: 1,3-Dipropyl-8-phenylxanthine sulfonamide congener.
[d] $ANAPP_3$: 3′-*O*-{3[*N*-(4-Azido-2-nitrophenyl)amino]propionyl} ATP.

well as a generalized inhibition of neurotransmitter release via A_1 receptor activation. A decrease in calcium availability for stimulus secretion coupling or an effect on potassium conductance may account for the ability of adenosine to modulate transmitter release. Among the transmitters whose release is affected by adenosine are norepinephrine, dopamine, serotonin, acetylcholine, GABA, and glutamate. Given the putative paracrine homeostatic role of the purine nucleoside,[32] the ability of adenosine to modulate the release of a number of transmitters with apparently conflicting actions (that is, glutamate as an excitatory transmitter and GABA as an inhibitory effector), as well as function within the context of autoreceptor regulation of transmitter release, has been difficult to place in perspective.[36] Many of these studies were conducted either *in vitro* or using exogenously administered purine nucleoside, and it is probable that endogenous adenosine availability subserves a far more discrete function. A discrete effect of adenosine on glutamate and aspartate, but not GABA release, has been found in rat hippocampal slices.[42]

ATP can also alter phosphatidyl inositol turnover, potassium conductance, and the opening of voltage-dependent calcium channels.[23] P_{2Y} receptors in the endothelium mediate vasorelaxation via the release of endothelium-derived relaxing factor (EDRF).[43] In the liver, ATP has glycogenolytic activity.[44]

BEHAVIORAL ACTIONS OF ADENOSINE AND ATP

The initial studies of Feldberg and Sherwood in the 1950s[45] demonstrated the sedative/hypnotic actions of ATP when administered into the brain, a finding replicated by many workers using both peripheral and central routes of administration. ATP has been shown to have excitatory effects, however, when applied to cortex and spinal cord neurons.[46,47] In the initial studies of Phillis and co-workers,[46] a biphasic response was noted, an initial excitatory response ascribed to an effect of the polyphosphate side chain followed by a depression of cell firing presumed to be due to adenosine formation. This action was also accompanied by a decrease in locomotor activity. Because of the profound effects of adenosine in decreasing blood pressure, a decade-old controversy has addressed whether the *in vivo* actions of adenosine on CNS function are indirectly mediated via peripheral effects on cardiovascular and respiratory function. Despite many elegant studies predicated on comparisons of systemic versus central administration of adenosine and various stable analogues and the use of xanthine antagonists that are unable to cross the blood-brain barrier, this issue has yet to be convincingly resolved. It is further compounded by the central actions of adenosine in regulating cardiac function and respiration.[48–50] An analogue of CGS 21680, the A_2-selective agonist, has been found that is less acidic and hence more likely to cross the blood-brain barrier. This analogue has pronounced effects on cardiac output that can only be explained in terms of a direct CNS action.[10] Adenosine can alter scheduled-controlled responding[51] and can elicit a discrimitive cue.[52] Much of the behavioral work on adenosine, however, has derived from studies on the psychostimulant actions of the methylxanthine, caffeine.[4] Caffeine is a locomotor stimulant and can be self-administered. Like adenosine, many of the actions of the alkylxanthines can be interpreted in terms of effects on neurotransmitter release, in this instance the xanthine acting to antagonize the inhibitory tone imposed by the purine nucleoside and increase transmitter release.

ADENOSINE IN CNS FUNCTION

Many of the studies implicating adenosine-related systems in a wide range of CNS pathophysiologies have been dependent on observations of the effects of known drugs on adenosine systems (uptake, receptor interactions) or the effects of exogenous adenosine agonists and antagonists on the actions of known drugs. Because many CNS-active entities have poorly understood mechanisms of action, both acutely and in a chronic context, a confusing picture has emerged, especially in light of the many effects of adenosine that may occur indirectly via modulation of transmitter release. This has represented a major problem in regard to the CNS as a target for drugs acting via adenosine systems except in instances where major traumatic insults (stroke, convulsive episodes) have caused dramatic changes in the extracellular availability of adenosine.

Epilepsy

At high doses, caffeine has convulsant and proconvulsant activity.[53] In contrast, adenosine can prevent audiogenic,[54] chemically, and electrically induced[55–57] convulsant activity. As might be anticipated in view of the hypoxia associated with convulsive episodes, adenosine levels are markedly increased,[58] and it has been proposed that the purine acts as an endogenous anticonvulsant.[59] Adenosine has also been implicated in the mechanism of action of the barbiturate anticonvulsants,[60] carbamazepine,[61] as well as the benzodiazepine (BZ) anticonvulsants.[62]

Anxiety

Caffeine is a well-known anxiogenic. Following the discovery of the central BZ receptor, the metabolic product of adenosine, inosine, was identified as a ligand, albeit weak, for the BZ receptor.[63] Xanthines can antagonize the electrophysiological and behavioral actions of the BZs as well as their sedative actions.[64,65] Chronic treatment with the methylxanthines can upregulate A_1 receptors[66] and uncouple the functional state of the central BZ receptor complex.[67] Conversely, diazepam can potentiate adenosine-mediated depression of cell firing,[64] and has been reported to block adenosine uptake,[68] whereas the anxiogenic β-carbolines can block the depressant actions of adenosine.[65,69] The finding that the BZ receptor antagonist, anexate (Ro 15-1788), can block caffeine-induced seizures[70] reinforces the relationship between adenosine and the BZs, as does the observation that the BZ inverse agonist, CGS 8216, is also an adenosine antagonist[71] and in fact led to CGS 15943. The finding that the peripheral BZ receptor ligand, Ro 5-4864, can also interact with purine-modulated systems does cast doubt on the robustness of the BZ/adenosine hypothesis, but there is a convincing body of evidence that adenosine and the centrally active BZs share some degree of commonality in their actions. Both the BZs and adenosine interact with the oldest known anxiolytic of all, alcohol,[32] and adenosine-stimulated adenylate cyclase activity in the NG 108-15 neuroblastoma cell line has been used as a model of acute alcohol intoxication, tolerance, and dependence.[72] These data are summarized in TABLE 2. Continued work in elaborating this relationship, either in regard to adenosine systems mediating sedative actions of the BZs or in the context of a discrete anticonvulsant effect of the BZs distinct from their anxiolytic actions, may clarify the existing body of evidence. However, based on the effects of Ro 5-4864 and the involvement of the peripheral BZ receptor in the modulation of cardiac function,[73] it may be premature to limit such studies to the CNS.

Sleep

Following from the work of Feldberg and Sherwood,[45] the hypnotic/sedative actions of adenosine and ATP have been extensively studied.[74] Adenosine agonists suppressed REM (rapid eye movement) sleep, decreased waking, increased deep slow-wave sleep, and increased total sleep time. Sleep deprivation for 48 hr caused an upregulation in brain A_1 receptors.[74] The nucleoside transport inhibitor, mioflazine, which increases the concentration of adenosine in the extracellular milieu, is currently in clinical trials as a hypnotic agent.[75]

TABLE 2. Comparison of Actions of BZs and Adenosine

Pharmacological Action	BZs	Adenosine
Anxiolytic	+++	+?
Anticonvulsant	+++	++
Stress	++	++
Sedative	++	+++
Alcohol interactions	+++	++
Muscle relaxation	+++	+++
Inhibitory role in CNS	+++[a]	++++

[a] BZs act via potentiation of the actions of GABA.

Pain

Studies implicating adenosine as a mediator of the pain process have been controversial. In some instances, adenosine has been implicated, via the effects of caffeine, as a nociceptive factor.[76] In others,[77–79] adenosine appears to be an antinociceptive agent. The central effects of morphine appear to be dependent on adenosine release,[80] whereas in the spinal cord[81] the antinociceptive actions of the opiate involve adenosine release from primary afferents in the dorsal horn. An A_2 receptor mechanism apparently mediates the spinal actions of the purine.[82] Increases in A_1 receptor density have been reported in morphine-dependent mice.[83]

Depression

In animal models of depression, adenosine receptors are upregulated,[84] whereas in the behavioral despair test, potentiation of adenosine actions via inhibition of uptake can increase the immobilization period[85]—an action blocked by administration of either methylxanthines or tricyclic antidepressants. The latter agents, at high concentrations, have been reported to facilitate adenosine release[86] as well as potentiate the effects of exogenous adenosine.[87] Chronic electroconvulsive treatment or ECT, a routine measure used to treat chronically depressed patients, can upregulate A_1 receptors.[88] Chronic antidepressant treatment,[89] however, has no effect on A_1 receptor binding. A novel nonxanthine, aminotriazolol[4,3-*a*]quinoxaline adenosine antagonist has recently been in clinical trials as an antidepressant,[90] but for unspecified reasons appears unlikely to be developed as a therapeutic entity.

Basal Nucleus Function

The relationship between adenosine and central dopamine systems is complex and may involve effects of the purine on dopamine release as well as distinct postsynaptic effects. As with many of the central actions of adenosine, the initial studies linking purines and dopamine occurred with studies on methylxanthine actions. Thus these prototypic adenosine antagonists were found to enhance dopamine release[91] and either induce or potentiate rotational behavior in the classical dopaminergic striatally lesioned rat.[92,93] Subsequent studies[94,95] demonstrated that adenosine analogues could decrease dopamine synthesis and release as well as decrease serum prolactin dopamine levels,[96] the latter a direct measure of dopamine receptor activation. Chemical lesions of central dopaminergic pathways have been reported to both decrease[97] and increase[98] caffeine-stimulated motor activity, a paradigm also blocked by the dopamine antagonists, haloperidol and pimozide.[99] Apomorphine-induced rotation in unilaterally lesioned rats can be modulated by both NECA and theophylline.[100]

Psychomotor stimulants including L-dopa, pemoline, amphetamine, apomorphine, SKF 38393, and the methylxanthines can induce a self-mutilation behavior[4] that is reduced by intrastriatal injection of adenosine analogues via an A_2 receptor mechanism.[98] A malfunction in purine metabolism is the basis of the self-mutilation behavior characterized as Lesch-Nyhan syndrome,[101] an X-linked recessive disorder. Interestingly, this disease, like methamphetamine-induced neurotoxicity, is characterized by a marked decrease in striatal dopamine.[102]

Adenosine agonists have putative antipsychotic agents in animal models,[103] a reflection no doubt of their ability to functionally antagonize dopamine function by inhibiting release. The difference in dose between these central effects and the ability of the agonists to decrease blood pressure was somewhat low. In addition, the tests used to evaluate their antipsychotic potential were those used for classical dopamine receptor antagonists/autoreceptor agonists suggesting that the same limitations occurring with these latter agents would also be seen for adenosine agonists.

Stroke

A reduction in blood flow to the brain can lead to cell death due to the release of excitatory amino acids and a consequent irreversible and fatal influx of calcium into neurons. Agents to treat the consequences of stroke have become a major focus in neuroscience research.[104] Such agents include free radical scavengers (such as superoxide dismutase and the lazaroids),[105] *N*-methyl-D-aspartate receptor antagonists (such as MK 801, CGS 19755, and CGP 37849), and calcium entry blockers.[106] An immediate consequence of the ischemia resulting from the reduction in cerebral blood flow is an increase in central adenosine levels.[104] Accordingly, a number of groups[107–109] have evaluated adenosine agonists for their effectiveness in both preventing or reversing hippocampal cell death and increasing survival in a variety of animal models with encouraging results. Presumably the purine is acting to inhibit the release of the excitotoxin, glutamate, as has been shown in model systems.[42] The use of an adenosine agonist in stroke therapy, in addition to addressing the sequence of events preceding cell death at a point proximal to the initial tissue insult, also has as an advantage the ability to rapidly lower blood pressure. If this latter effect could be brought about without a reflex effect on cardiac output,[9] an adenosine agonist or potentiator such as AICA-riboside[110] could be the most beneficial agent in preventing or restricting stroke-related cell damage. As an additional benefit, the role of adenosine in preventing neutrophil-associated reperfusion injury[111] could aid significantly in reducing the damage potential from overactivation of the brain's own defense systems.

Cognition

The central stimulant actions of caffeine have long been thought of as a model for cognition enhancement, although there is currently little evidence to suggest that impaired mental functioning in old age is due to an increase in purinergic tone. Xanthines increase cerebral blood flow,[112,113] thereby increasing oxygen and glucose supply to the brain.[113] Caffeine causes a global increase in functional activity,[114] whereas the 1-substituted xanthine, HWA 285, has cognition-enhancing properties in human trials.[115] A limitation to the evaluation of adenosine antagonists in humans with a major incidence of cognitive deficits, that is, the elderly, has been the complication with the cardiotonic actions of adenosine antagonists—these actions may be expected to compromise cardiovascular function. On a lighter note, it is of interest that the only evidence to suggest that an excess of adenosine can lead to marked performance deficits is anecdotal, and comes from the guinea pig, which various experimental paradigms have indicated has the highest functional activity in terms of central adenosine systems. For those who have worked with this animal, its behavior leaves much to be desired, and might be considered to be an inappropriate model for studying cognitive impairment.

Substance Abuse

The potential involvement of adenosine in the central actions of ethanol, opiates, psychostimulants, and antipsychotics suggests that central purine systems may be intimately involved in the mechanisms underlying substance abuse. The involvement of the purine in regulating the limbic dopaminergic reward systems can be considered consistent with this hypothesis. Although initial reports[116] showing that the adenosine agonist, phenylisopropyladenosine (PIA), could stereoselectively antagonize phencyclidine (PCP) discrimination were shown to be due to an effect on PCP pharmacokinetics,[117] the major crisis in containing substance abuse should aid in focusing effort on establishing whether purinergic systems are involved in the substrates of such behavior.

ADENOSINE AND ATP AS CNS MODULATORS

Although it is broadly based in terms of the number of CNS behaviors, the role of adenosine (and by inference that of ATP) is unequivocal. In terms of a globalized homeostatic role, it is hardly surprising that the purine appears to be involved in a variety of CNS behaviors. This has been taken to be a negative in developing CNS-active drugs based on adenosine agonism or antagonism,[118] but given the still limited understanding of the interactions between the various neurotransmitter/neuromodulatory systems that contribute to the highly subtle nuances of behavior, it would not be surprising that central purinergic systems act as the master network for cellular integration, not only neuronal but also glial and systemic. As a comparative example, the centrally active BZ agonists are effective in the treatment of anxiety and epilepsy, whereas the BZ antagonists, anexate and CGS 8216, are central stimulants. The ubiquitous role of adenosine in interfacing the cellular elements of the diverse organ systems of the body, all of which are necessary for survival under adverse conditions, is a factor that has been somewhat underemphasized in studies related to the role in the CNS of both adenine nucleosides and nucleotides.

PURINE LIGANDS AS CNS DRUGS

As noted, a considerable number of the therapeutic agents used to treat CNS conditions are limited both in terms of efficacy and safety margins. A major effort in industrial CNS-related research is to derive novel entities for the treatment of depression, epilepsy, schizophrenia, anxiety, pain, alcohol and substance abuse, sleep disorders, and cognitive and neurodegenerative disorders. In several instances, effective treatment is either inadequate or lacking. Accordingly, every neurotransmitter/neuromodulator identified in the brain has been considered as a potential new drug target. Together with the neuropeptides, adenosine has a wide array of potential actions on CNS function none of which have been realistically amenable to evaluation due to both a lack of potent and selective receptor agonists and a profound effect of these on cardiovascular function. A novel entity, however effective in treating a CNS condition without the limiting side effects seen with currently available agents, would be severely compromised by a blood pressure lowering effect in the range of 20-50 mm Hg.[10]

A major strategy in drug discovery, however, is to identify receptor subtypes and design selective ligands for these.[119] Experience has shown that for the β-adrenoceptor, the histamine receptor, and a variety of others, it is possible to differentiate the beneficial actions resulting from modulation of a transmitter system from side effects or liabilities by targeting receptor subtypes. In the adenosine arena, work has only just begun in the isolation of the A_1 and A_2 receptors.[120,121] Based on the identification (using molecular biological techniques) of a number of receptor superfamilies, however, it may be anticipated that the number of adenosine receptors may increase by an order of magnitude by the end of the century.

Furthermore, it is only in the past year or so that effective and selective adenosine A_2 receptor agonists have become available. Accordingly, many of the studies reviewed in the present monograph which that have focused on the A_1 receptor because of the compounds available will require reevaluation in terms of the potential role of the A_2 receptor. An important goal in this area will be the identification of nonpurine agonists as well as the further elucidation of the role of "site- and event-specific" adenosine potentiators, including nucleoside transport inhibitors[75] and novel entities such as AICA riboside.[110]

For the ATP receptor, it is only in the past five years that the persistence and foresight of Burnstock and a myriad of co-workers has led to an increasing pharmacological perspective on the role of ATP in cell function. A major limitation to defining this role in greater detail has been the scarcity of the various isosteric and isopolar phosphonate analogues of the adenine nucleotides. An effort to redress this situation with more detailed studies in various target tissues will do much to bring ATP in line with what is known about adenosine. Significant progress has been made in the last decade,[23,122] and it is hoped that purinergic drugs will be the result of research in the next.

REFERENCES

1. BURNSTOCK, G. 1978. *In* Cell Membrane Receptors for Drugs and Hormones: A Multidisciplinary Approach. L. Bolis & R. W. Straub, Eds.: 107-118. Raven Press. New York, NY.
2. BERNE, R. M. 1963. J. Physiol. (London) **204:** 317-322.
3. PHILLIS, J. W. 1989. Cerebrovasc. Brain Metab. Rev. **1:** 26-54.
4. JARVIS, M. F. & M. WILLIAMS. 1990. *In* Adenosine and Adenosine Receptors. M. Williams, Ed.: 423-476. Humana. Clifton, NJ.
5. PEARSON, J. & L. L. SLAKEY. 1990. *In* Purine Nucleosides and Nucleotides in Cell Signalling: Targets for New Drugs. K. A. Jacobson, J. W. Daly & V. C. Manganiello, Eds.: 13-19. Springer-Verlag. New York, NY.
6. LOHSE, M. J., K.-N. KLOTZ, U. SCHWABE, G. CRISTALLI, S. VITTORI & M. GRIFANTINI. 1988. Naunyn-Schmiedeberg's Arch. Pharmacol. **337:** 687-689.
7. TRIVEDI, B. K., A. J. BRIDGES, W. C. PATT, S. R. PRIEBE & R. F. BRUNS. 1988. J. Med. Chem. **32:** 8-11.
8. KAWAZOE, K., N. MATSUMOTO, M. TANABE, S. FUJIWARA, M. YANAGIMOTO, M. HIRATA & K. KAKIUCHI. 1980. Arzneim. Forsch. **30:** 1083-1091.
9. BRUNS, R. F., G. H. LU. & T. A. PUGSLEY. 1986. Mol. Pharmacol. **29:** 331-346.
10. HUTCHISON, A. J., R. L. WEBB, H. H. OEI, G. R. GHAI, M. B. ZIMMERMAN & M. WILLIAMS. 1989. J. Pharmacol. Exp. Ther. **251:** 47-55.
11. RIBERIO, J. A. & A. M. SEBASTIAO. 1986. Prog. Neurobiol. **26:** 179-209.
12. GUSTAFFSON, L. E., C. U. WIKLUND, N. P. WIKLUND & L. STELIUS. 1989. *In* Adenosine Receptors in the Nervous System. J. A. Riberio, Ed.: 194. Taylor & Francis. London.
13. SATTIN, A. & T. W. RALL. 1970. Mol. Pharmacol. **6:** 13-23.
14. JACOBSON, K. A., K. L. KIRK, W. L. PAGETT & J. W. DALY. 1985. J. Med. Chem. **28:** 1334-1340.

15. BRUNS, R. F., R. E. DAVIS, F. W. NINTEMAN, B. P. H. PORSCHEL, J. N. WILEY & T. G. HEFFNER. 1988. *In* Adenosine and Adenine Nucleotides: Physiology and Pharmacology. D. M. Paton, Ed.: 39-49. Taylor & Francis. London.
16. WILLIAMS, M., J. E. FRANCIS, G. R. GHAI, A. BRAUNWALDER, S. PSYCHOYOS, G. A. STONE & W. D. CASH. 1987. J. Pharmacol. Exp. Ther. **241:** 415-420.
17. JARVIS, M. F. 1988. *In* Receptor Localization: Ligand Autoradiography. F. Leslie & C. A. Altar, Eds.: 95-113. Alan R. Liss. New York, NY.
18. JARVIS, M. F., R. H. JACKSON & M. WILLIAMS. 1989. Brain Res. **484:** 111-118.
19. JARVIS, M. F. & M. WILLIAMS. 1989. Eur. J. Pharmacol. **268:** 172-176.
20. FERKANY, J. W., H. VALENTINE, G. A. STONE & M. WILLIAMS. 1986. Drug Dev. Res. **9:** 85-93.
21. STONE, G. A., M. F. JARVIS, M. A. SILLS, B. WEEKS, E. A. SNOWHILL & M. WILLIAMS. 1988. Drug Dev. Res. **15:** 31-46.
22. GORDON, J. L. 1986. Biochem. J. **233:** 309-319.
23. BURNSTOCK, G. 1989. *In* Adenosine Receptors in the Nervous System. J. A. Riberio, Ed.: 1-14. Taylor & Francis. London.
24. CUSACK, N. & S. M. HOURANI. 1990. *In* Purine Nucleosides and Nucleotides in Cell Signalling: Targets for New Drugs. L. A. Jacobson, J. W. Daly & V. C. Manganiello, Eds.: 254-259. Springer-Verlag. New York, NY.
25. BLACK, J. W. 1987. *In* Perspectives on Receptor Classification. J. W. Black, D. H. Jenkinson & V. P. Gerskowitch, Eds.: 11-15. Alan R. Liss. New York, NY.
26. KRISHTAL, O. A., S. M. MARCHENKO, A. G. OBUKHOV & T. M. VOLKOVA. 1988. Br. J. Pharmacol. **95:** 1057-1062.
27. STONE, T. W., A. C. NEWBY & H. G. E. LLOYD. 1990. *In* Adenosine and Adenosine Receptors. M. Williams, Ed.: 173-224. Humana. Clifton, NJ.
28. HOLTON, P. 1959. J. Physiol. (London) **145:** 494-504.
29. JONZON, B. & B. B. FREDHOLM. 1985. J. Neurochem. **44:** 217-224.
30. SCHMITT, F. O. 1982. *In* Molecular Genetic Neuroscience. F. O. Schmitt, S. E. Bird & F. E. Bloom, Eds.: 1-9. Raven Press. New York, NY.
31. POOL, R. 1989. Science **243:** 604-607.
32. WILLIAMS, M. 1989. Neurochem. Int. **14:** 249-264.
33. WILLIAMS, M. 1990. *In* Adenosine and Adenosine Receptors. M. Williams, Ed.: 501-508. Humana. Clifton, NJ.
34. WHITE, T. D. 1988. *In* Adenosine and Adenine Nucleotides: Physiology and Pharmacology. D. M. Paton, Ed.: 205-215. Taylor & Francis. London.
35. RICHARDSON, P. J. & S. J. BROWN. 1987. J. Neurochem. **48:** 622-628.
36. WIERASZKO, A., G. GOLDSMITH & T. N. SEYFRIED. 1989. Brain Res. **485:** 244-250.
37. EHRLICH, Y. H., R. M. SNYDER, E. KORNECKI, M. G. GARFIELD & R. H. LENOX. 1988. J. Neurochem. **50:** 295-301.
38. WILLIAMS, M. & A. BRAUNWALDER. 1986. J. Neurochem. **47:** 88-97.
39. COOPER, D. M. F. & K. CALDWELL. 1990. *In* Adenosine and Adenosine Receptors. M. Williams, Ed.: 105-142. Humana. Clifton, NJ.
40. SILINSKY, E. M. 1990. Semin. Neurosci. **1:** in press.
41. WILLIAMS, M. 1987. Annu. Rev. Pharmacol. Toxicol. **27:** 315-345.
42. BURKE, S. P. & D. J. V. NADLER. 1988. J. Neurochem. **51:** 1541-1551.
43. BURNSTOCK, G. 1988. *In* Vasodilitation: Vascular Smooth Muscle, Peptides, Autonomic Nerves and Endothelium. P. M. Vanhoutte, Ed.: 1-14. Raven Press. New York, NY.
44. HAUSSINGER, D., T. STEHLE, W. GEROK, T.-A. TRAN-THI & K. DECKER. 1987. Eur. J. Biochem. **169:** 645-650.
45. FELDBERG, W. & S. L. SHERWOOD. 1954. J. Physiol. (London) **123:** 148-167.
46. PHILLIS, J. W., J. P. EDSTROM, G. K. KOSTOPOULOS & J. P. KIRKPATRICK. 1979. Can. J. Physiol. Pharmacol. **57:** 1289-1312.
47. JAHR, C. F. & T. M. JESSEL. 1983. Nature (London) **304:** 730-732.
48. BARRACO, R. F., A. K. AGARAWAL, J. W. PHILLIS, M. A. MORAN & P. H. WU. 1984. Neurosci. Lett. **48:** 139-143.
49. SOLLEVI, A. 1986. Prog. Neurobiol. **27:** 319-349.
50. ELDRIDGE, F. L. & D. E. MILLHORN. 1987. *In* Topics and Perspectives in Adenosine Research. E. Gerlach & B. F. Becker, Eds.: 586-596. Springer-Verlag. Berlin.

51. COFFIN, V. C. & R. SPEALMAN. 1987. J. Pharmacol. Exp. Ther. **241:** 76-83.
52. HOLLOWAY, F. A., H. E. MODROW & R. C. MICHAELIS. 1985. Pharmacol. Biochem. Behav. **22:** 815-824.
53. SNYDER, S. H. & P. SKLAR. 1984. J. Psychiatr. Res. **18:** 91-106.
54. MAITRE, M., L. CIESIELSKI, A. LEHMANN, E. KEMPF & P. MANDEL. 1974. Biochem. Pharmacol. **23:** 2807-2816.
55. DUNWIDDIE, T. V. & T. WORTH. 1982. J. Pharmacol. Exp. Ther. **220:** 70-76.
56. BARRACO, R. A., T. H. SWANSON, J. W. PHILLIS & R. F. BERMAN. 1984. Neurosci. Lett. **46:** 317-322.
57. MURRAY, T. F., D. SYLVESTER, C. S. SCHULZ & P. SZOT. 1985. Neuropharmacology **24:** 761-766.
58. WINN, R. H., R. RUBIO & R. M. BERNE. 1981. J. Cereb. Blood Flow Metab. **1:** 239-247.
59. DRAGUNOW, M. 1988. Prog. Neurobiol. **31:** 85-107.
60. LOHSE, M. J., K.-N. KLOTZ, K. JACOBS & U. SCHWABE. 1985. J. Neurochem. **45:** 1761-1770.
61. MARANGOS, P. J., R. M. POST, J. PATEL, A. ZANDER, A. PARMA & S. WEISS. 1983. Eur. J. Pharmacol. **93:** 175-182.
62. PHILLIS, J. W. & M. O'REGAN. 1988. Trends Pharmacol. Sci. **9:** 153-154.
63. MARANGOS, P. J., S. M. PAUL, A. M. PARMA, F. K. GOODWIN, P. SYAPIN & P. SKOLNICK. 1979. Life Sci. **24:** 851-858.
64. PHILLIS, J. W. & P. H. WU. 1981. Prog. Neurobiol. **16:** 187-193.
65. PHILLIS, J. W. & M. O'REGAN. 1988. Prog. Neuro-Psychopharmacol. Biol. Psychiatry **12:** 389-404.
66. BOULENGER, J. P., J. PATEL, R. M. POST, A. M. PARMA & P. J. MARANGOS. 1983. Life Sci. **32:** 1135-1142
67. ROCA, D. J., G. D. SCHILLER & D. H. FARB. 1988. Mol. Pharmacol. **30:** 481-485.
68. PHILLIS, J. W. 1985. *In* Purines: Pharmacology and Physiological Roles. T. W. Stone, Ed.: 45-56. VCH. Deerfield Beach, FL.
69. ALBERTSON, T. E., R. M. JOY & L. G. STARK. 1982. Pharmacol. Biochem. Behav. **19:** 339-343.
70. ALBERTSON, T. E., J. F. BOWYER & M. G. PAULE. 1982. Life Sci. **31:** 1597-1601.
71. CZERNIK, A., B. PETRACK, H. J. KALINSKY, S. PSYCHOYOS, W. D. CASH, C. TSAI, R. K. RINEHART, F. R. GRANAT, R. A. LOVELL, D. E. BRUNDISH & R. WADE. 1982. Life Sci. **30:** 363-370.
72. GORDON, A. S., K. COLLIER & I. DIAMOND. 1986. Proc. Natl. Acad. USA **83:** 2105-2108.
73. WILLIAMS, M. & R. W. OLSEN. 1988. *In* Receptor Pharmacology and Function. M. Williams, R. A. Glennon & P. B. M. W. M. Timmermans, Eds.: 385-413. Marcel Dekker. New York, NY.
74. RADULOVACKI, M. 1985. *In* Adenosine: Receptors and Modulation of Cell Function. V. Stefanovich, K. Rudolphi & P. Schubert, Eds.: 211-219. IRL Press. Oxford.
75. WAUQUIER, A. *et al.* 1987. Psychopharmacology **91:** 434-439.
76. GOURLEY, D. R. H. & S. K. BECKNER. 1973. Proc. Soc. Exp. Biol. Med. **144:** 774-780.
77. HO, I. K., H. H. LOH & E. L. WAY. 1973. J. Pharmacol. Exp. Ther. **185:** 336-346.
78. YARBROUGH, G. G. & J. C. MCGUFFIN-CLINESCHMIDT. 1981. Eur. J. Pharmacol. **76:** 137-144.
79. JURNA, I. 1984. Naunyn-Schmiedeberg's Arch. Pharmacol. **327:** 23-28.
80. STONE, T. W., B. B. FREDHOLM & J. W. PHILLIS. 1989. Trends Pharmacol. Sci. **10:** 316.
81. SAWYNOK, J., M. I. SWEENEY & T. D. WHITE. 1989. Trends Pharmacol. Sci. **10:** 186-189.
82. DELANDER, G. E. & C. J. HOPKINS 1987. Eur. J. Pharmacol. **139:** 215-223.
83. AHILIJANIAN, M. & A. E. TAKEMORI. 1986. J. Pharmacol. Exp. Ther. **236:** 615-620.
84. ANDERSON, S. M., J. R. LEU & G. J. KANT. 1988. Pharmacol. Biochem. Behav. **30:** 169-175.
85. KULKARNI, S. K. & A. K. MEHTA. 1985. Psychopharmacology **85:** 460-463.
86. STONE, T. W. 1983. *In* Regulatory Function of Adenosine. R. M. Berne, T. W. Rall & R. Rubio, Eds.: 467-477. Martinus Nijhoff. Boston, MA.
87. SATTIN, A., T. W. STONE & D. A. TAYLOR. 1979. Life Sci. **23:** 2621-2626.
88. NEWMAN, M. E., J. ZOHAR, M. KALIAN & R. F. BELMAKER. 1984. Brain Res. **291:** 188-192.
89. WILLIAMS, M., E. A. RISLEY & J. A. ROBINSON. 1983. Neurosci. Lett. **35:** 47-51.

90. SARGES, R., H. R. HOWARD, R. G. BROWNE & B. K. KOE. 1990. *In* Purine Nucleosides and Nucleotides in Cell Signalling: Targets for New Drugs. K. A. Jacobson, J. W. Daly & V. C. Manganiello, Eds.: 417-418. Springer-Verlag. New York, NY.
91. BERKOWITZ, B., J. H. TARVER & S. SPECTOR. 1970. Eur. J. Pharmacol. **10:** 64-71.
92. FUXE, K. & U. UNGERSTEDT. 1974. Med. Biol. **52:** 48-54.
93. FREDHOLM, B. B., M. HERRERA-MARSCHWITZ, B. JONZON, K, LINDSTROM & U. UNGERSTEDT. 1983. Pharmacol. Biochem. Behav. **19:** 535-541.
94. MICHAELIS, M. K., E. K. MICHAELIS & S. L. MYERS. 1979. Life Sci. **24:** 2083-2092.
95. MYERS, S. L. & T. A. PUGSLEY. 1986. Brain Res. **375:** 193-197.
96. STEWART, S. F. & T. A. PUGSLEY. 1985. Naunyn-Schiedeberg's Arch. Pharmacol. **331:** 140-145.
97. ERINOFF, L. & S. R. SNODGRASS. 1986. Pharmacol. Biochem. Behav. **24:** 1039-1045.
98. CRISWELL, H., R. A. MULLER & G. R. BREESE. 1988. J. Pharmacol. Exp. Ther. **244:** 493-500.
99. WALDECK, B. 1975. Acta Pharmacol. Toxicol. **36**(Suppl. 4): 1-23.
100. GREEN, R. D., H. K. PROUDFIT & S.-M. H. YEUNG. 1982. Science **218:** 58-61.
101. NYHAN, W. L., W. J. OLIVER & M. LESCH. 1985. J. Pediatr. **67:** 257-263.
102. LLOYD, K. G., O. HORNYKIEWICZ, L. DAVIDSON, K. SHANNAK, I. FARLEY, M. GOLDSTEIN, M. SHIBUYA, W. N. KELLY & I. H. FOX. 1981. N. Engl. J. Med. **305:** 1106-1111.
103. HEFFNER, T. G. & R. F. BRUNS. 1989. Psychopharmacology **98:** 31-38.
104. MELDRUM, B. S. 1990. *In* Current and Future Trends in Anticonvulsant, Antianxiety and Stroke Therapy. B. S. Meldrum & M. Williams, Eds.: 275-290. Alan R. Liss. New York, NY.
105. HALL, E. D., J. M. BRAUGHLER & J. M. MCCALL. 1990. *In* Current and Future Trends in Anticonvulsant, Antianxiety and Stroke Therapy. B. S. Meldrum & M. Williams, Eds. Alan R. Liss. New York, NY.
106. FOSTER, A. C., R. GILL, L. L. IVERSEN, J. A. KEMP, E. H. F. WONG & G. N. WOODRUFF. 1990. *In* Current and Future Trends in Anticonvulsant, Antianxiety and Stroke Therapy. B. S. Meldrum & M. Williams, Eds.: 351-362. Alan R. Liss. New York, NY.
107. EVANS, M. C., J. H. SWAN & B. S. MELDRUM. 1987. Neurosci. Lett. **83:** 287-292.
108. GOLDBERG, M. P., H. MONYER, J. H. WEISS & D. W. CHOI. 1988. Neurosci. Lett. **89:** 323-327.
109. VON LUBITZ, D., J. M. DAMBROSIA, O. KEMPSKI & D. J. REDMOND. 1988. Stroke **19:** 1133-1139.
110. ENGLER, R. 1987. Fed. Proc. **46:** 2407-2412.
111. CRONSTEIN, B. N., S. B. KRAMER, E. D. ROSENSTEIN, G. WEISMANN & R. HIRSCHHORN. 1983. J. Exp. Med. **158:** 1160-1177.
112. GROME, J. J. & V. STEFANOVICH. 1986. Nauyn Schmiedeberg's Arch. Pharmacol. **333:** 172-179.
113. NEHLIG, A., G. LUCIGNANI, L. KADEKARO, L. J. PORRINO & L. SOKOLOFF. 1984. Eur. J. Pharmacol. **101:** 91-100.
114. LOUIE, G. L., P. G. PROKOCIMER, E. A. NICHOLLS & M. MAZE. 1986. Brain Res. **383:** 377-381.
115. HINDMARCH, I. & Z. SUBHAN. 1985. Drug Dev. Res. **5:** 379-386.
116. BROWNE, R. G. & W. M. WELCH. 1982. Science **217:** 1157-1158.
117. BROWNE, R. G., W. M. WELCH, M. R. KOZLOWSKI & G. DUTHU. 1983. *In* Phencyclidiene and Related Arylcyclohexylamines: Present and Future Applications. J. M. Kamenka, E. F. Domino & G. Erneste, Eds.: 639-636. Xerox Books. Ann Arbor, MI.
118. WILLIAMS, M. 1990. *In* Purine Nucleosides and Nucleotides in Cell Signalling: Targets for New Drugs. K. A. Jacobson, J. W. Daly & V. C. Manganiello, Eds.: 174-183. Springer-Verlag. New York, NY.
119. WILLIAMS, M. & A. M. NADZAN. 1991. *In* Medicinal Chemistry. P. Krogsgaard-Larsen, P. Bundgaard & F. Jorgansen, Eds.: in press. Ellis Harwood. London.
120. RAMKUMAR, V., G. PIERSON & G. L. STILES. 1988. Prog. Drug Res. **32:** 195-247.
121. BARRINGTON, W. W., K. A. JACOBSON, A. J. HUTCHISON, M. WILLIAMS & G. L. STILES. 1989. Proc. Natl. Acad. Sci. USA **86:** 6572-6576.
122. BURNSTOCK, G. 1978. *In* Physiological and Regulatory Functions of Adenosine and Adenine Nucleotides. H. P. Baer & G. I. Drummond, Eds.: 3-32. Raven Press. New York, NY.

Functional Consequences of Interactions between Human Neutrophils and ATP, ATPγS, and Adenosine[a]

PETER A. WARD, BLAIR A. M. WALKER, AND BRIAN E. HAGENLOCKER

Department of Pathology
University of Michigan Medical School
Ann Arbor, Michigan 48109

INTRODUCTION

There is an increasing body of evidence suggesting that nucleotides such as ATP can serve as important extracellular factors to bring about significant functional changes in phagocytic cells. For instance, ATP and its analogues (for example, adenosine-5′-*O*-(3-thiotriphosphate), or ATPγS) can interact with neutrophils to cause increases in intracellular calcium.[1–4] This appears to be associated with "priming" of these cells for enhanced superoxide (O_2^-) responses to agonists such as chemotactic peptides.[1,2,5,6] Because platelets contain and, when activated, release substantial amounts of ATP and ADP, mammalian systems have the potential for delivering significant amounts of adenine nucleotides to sites of inflammation. It may well be that *in vivo* secretion products of platelets will bring about an amplification of oxygen radical formation by phagocytic cells, and so intensify tissue injury at the inflammatory site. This concept is supported both by *in vitro* and *in vivo* observations.[6–8] In contrast to ATP with respect to phagocytic cells, adenosine has an opposing effect: suppression of O_2^- responses to chemotactic peptides.[5,9] In this study, we have assessed the effects of ATP (and/or ATPγS) and adenosine on chemotactic peptide receptor numbers and affinities, the cell membrane content of CR3 (complement receptor III, or, as defined by monoclonal antibodies, Mo1 antigen (CD11b)), and the intracellular changes in calcium induced by exposure of neutrophils to chemotactic peptide.

[a]This work was supported in part by Grants HL-31963, GM-29507, and HL-34635 from the National Institutes of Health.

MATERIALS AND METHODS

Reagents

Except as noted, reagents were purchased from Sigma Chemical Company (St. Louis, MO). ATPγS was obtained from Boehringer Mannheim Biochemicals (Indianapolis, IN). *N'*-formyl-Met-Leu-[^{3}H]Phe ([^{3}H]fMLP) (56.9 Ci/mmol) was purchased from DuPont-New England Nuclear Research Products (Boston, MA).

Preparation of Neutrophils

For most studies, neutrophils were prepared as previously described.[10] Briefly, blood was anticoagulated with $\frac{1}{10}$ volume of acid citrate dextrose and subjected to centrifugation (400 × g, 15 min, room temperature). The upper layer of platelet-rich plasma was then removed. The remaining pellet containing red blood cells and leukocytes was diluted 1:1 with phosphate-buffered saline (140 mM NaCl, 1 mM KH_2PO_4, 5 mM Na_2HPO_4, pH 7.3), and 25-ml aliquots were layered over 15 ml of Ficoll-Hypaque (specific gravity of 1.077). Centrifugation was carried out (400 × g, 30 min), and the upper layers were aspirated, leaving the red cell pellet containing neutrophils. Red cells in the cell pellets were then lysed with equal volumes of 150 mM NH_4Cl containing 1 mM EDTA. The remaining neutrophils were washed with 50 ml of phosphate-buffered saline and resuspended in Tris-buffered Hanks' Balanced Saline Solution (HBSS) and used immediately.

The second method of neutrophil preparation was similar to that described previously,[1] with the exception that lysis of the red cell pellet was omitted. Briefly, heparinized whole blood was diluted 1:1 with 6% (w/v) T-500 dextran (Pharmacia, Piscataway, NJ) and sedimented a 1 × g for 40 min. Twenty-five aliquots of the upper leukocyte-rich layer were layered over 15 ml of Ficoll-Hypaque (specific gravity of 1.077) and centrifuged (400 × g, 30 min). The upper layers were removed by aspiration and the neutrophil pellet was washed once with phosphate-buffered saline and used immediately.

N'-*Formyl-Met-Leu-[^{3}H]-Phe Binding*

Neutrophils suspended in binding buffer (140 mM NaCl, 1 mM KH_2PO_4, 5mM Na_2HPO_4, 0.5 mM $MgSO_4$, 1.8 mM $CaCl_2$, pH 7.3) were prewarmed for 3 min and then exposed to either 10 μM ATPγS, 10 μM adenosine, or buffer for 5 min. Incubations were terminated by placing the cells on ice followed by addition of phenylmethylsulfonyl fluoride (1 mM). [^{3}H]fMLP binding was carried out at 4 °C by incubating 100 μl of cells (8 × 10^5 cells) with 25 μl of [^{3}H]fMLP (3-300 nM) and 25 μl of either buffer or cold fMLP (360 μM). After 1-1.5 hr, cells were harvested by vacuum filtration over glass fiber filters (Whatman GF/C, Whatman International, Maidstone, United Kingdom) and washed with binding buffer (4 × 4 ml). Filters were dried overnight, scintillation fluid (Safety Count, Research Products International, Mount Pleasant, IL) was added, and radioactivity was assessed in a Beckman LS 5801 scintillation counter (Fullerton, CA). Nonspecific binding (uptake of [^{3}H]fMLP in the presence of unlabeled 60 μM fMLP) was usually less than 13% of the total binding. Scatchard plots were obtained with the aid of the program EBDA/ligand (Elsevier-Biosoft, Cambridge, United Kingdom). Analysis of [^{3}H]fMLP binding using concentrations of 2.5-200 nM showed no significant improvement of fit for the Scatchard plot using a two receptor site model over a single site (F test, $p > .05$).

O_2^- Response of Neutrophils

Generation of O_2^- was determined by the superoxide dismutase-inhibitable reduction of ferricytochrome *c* as described previously.[2] Enhancement studies were performed by the addition of prewarmed cells to reaction mixtures containing either ATP, ATPγS, or adenosine. After 2 min, fMLP was added. For experiments involving cytoplasts, a preliminary assay of O_2^- production was performed to determine the number of cells necessary to generate O_2^- in amounts that would be the same for cytoplasts and intact neutrophils. This adjustment of cell number allowed for comparison of the relative enhancement of these two groups. The O_2^- responses were comparable between intact neutrophils kept on ice and intact neutrophils exposed to Ficoll and cytochalasin B under conditions similar to those involved in cytoplast preparation (data not shown).

Neutroplast Preparation

Neutrophils (10^8) were suspended in 10 ml of 12.5% (w/v) Ficoll solution containing 17.4 μM cytochalasin B at 37 °C. This was layered onto a discontinuous gradient (10 ml of 17% Ficoll and 10 ml of 25% Ficoll, each containing 17.4 μM cytochalasin B), and was subjected to ultracentrifugation at 34 °C at 79,500 $\times$ g for 30 min. The band of neutroplasts at the interface of the 12.5% and 17% Ficoll layers was aspirated, diluted to 50 ml with phosphate-buffered saline, and centrifuged at 725 $\times$ g for 6 min. The pellet was resuspended in HBSS and used. Enzyme content of neutroplasts produced in this manner showed less than 3.5% of β-glucuronidase or vitamin B_{12} binding protein by standard assay in conformity with other reports.[11] Transmission electron microscopy also confirmed granule depletion (data not shown).

Statistical Analysis

The data was expressed as the mean ($\bar{x}$) and standard error of the mean (SEM). A paired t test was used to compare the response between two treatments. Because of the variability of specific fluorescence for control samples in different experiments, CR3 (Mo1) fluorescence data was compared using a t test of the mean log values of specific fluorescence. Statistical significance was defined at $p < .05$.

Immunofluorescence Analysis

The generation and characterization of murine monoclonal antibodies anti-Mo1 (anti-CD11b, IgG2a; clone 44) and anti-I3 (IgG2a; clone 9-4) have been described.[10] Neutrophils were subjected to indirect immunofluorescence staining for the expression of the Mo1 determinant relative to background staining by an isotype-identical negative control antibody (anti-I3) and subjected to flow cytometric analysis. For each experiment, specific fluorescence intensity represents the computed mean channel number (0-230 channels, linear scale) of cells stained with the anti-Mo1 antibody minus the mean channel number of cells stained with the anti-I3 reagent. The photomultiplier tube setting for all experiments was 1000.

Changes in the Concentration of Intracellular Calcium

Measurements of $[Ca^{2+}]_i$ using the fluorescent probe fura-2 were described previously.[2] Briefly, neutrophils were suspended in Hepes buffered saline (2 mM $CaCl_2$, 1 mM $MgCl_2$, 10 mM glucose), loaded for 15 min at 37 °C with 20 μM fura-2-AM (the acetoxymethyl ester of fura-2, Calbiochem, La Jolla, CA), and then loaded for an additional 15 min after a 10-fold dilution. The cells were centrifuged and resuspended in buffer and stored on ice until use. Changes in $[Ca^{2+}]_i$-dependent fluorescence were measured in a Perkin-Elmer LS-5B luminescence spectrometer using a thermally equilibrated cuvette holder at 37 °C. Changes in $[Ca^{2+}]_i$ were determined as previously described.[2]

RESULTS

Effects of ATP and ATPγS on Superoxide Response of Neutrophils Prepared by Two Different Methods

Because the usual procedure of isolating neutrophils from human blood includes the lysis of red cells by addition of NH_4Cl, the ability of ATPγS to enhance O_2^- responses in fMLP-stimulated cells was evaluated in neutrophils prepared by addition or omission of the fluid used to lyse red cells (150 mM NH_4Cl). Neutrophils (1×10^6/ml) were then stimulated with 100 nM fMLP in the presence of increasing concentrations of ATPγS, and the O_2^- responses were measured (FIG. 1). As is evident from the data, the addition of ATPγS caused the production of O_2^- to be enhanced in a dose-dependent manner, but there was a remarkably amplified effect on O_2^- production in cells that had not been exposed to the lysing solution. In this case, there was a >100% increase in the production of O_2^- as compared to a 27% increase in cells that had been in contact with the lysing solution. Reasons for the differences are not understood. In all subsequent experiments to be reported, values were obtained using neutrophils that had been isolated in the conventional manner (using the lysing procedure), because contamination of neutrophils with red cells was incompatible with some of the analytical procedures.

As shown in FIGURE 2, the ability of ATP and ATPγS to prime neutrophils for subsequently enhanced O_2^- responses was proportional to the concentration of nucleotide employed. Enhancement in the O_2^- responses of neutrophils subsequently stimulated with 100 nM fMLP was found in the low μM dose range of nucleotide. Although the ATP effect tended to reach a plateau concentration of 10 μM, the effect of ATPγS was maintained and was proportional to the concentration of nucleotide, even at 100 μM (FIG. 1).

The well-known ability of adenosine to inhibit O_2^- responses of human neutrophils is shown in FIGURE 3 as a dose-response relationship. Neutrophils were exposed to varying concentrations of adenosine at 37 °C for 2 min, and these exposures were followed by the addition of 100 nM fMLP. Production of O_2^- over a 30-min period was then determined. As is apparent from the data, the inhibitory effects of adenosine were dose related, with maximal inhibition found at a concentration of 1 μM.

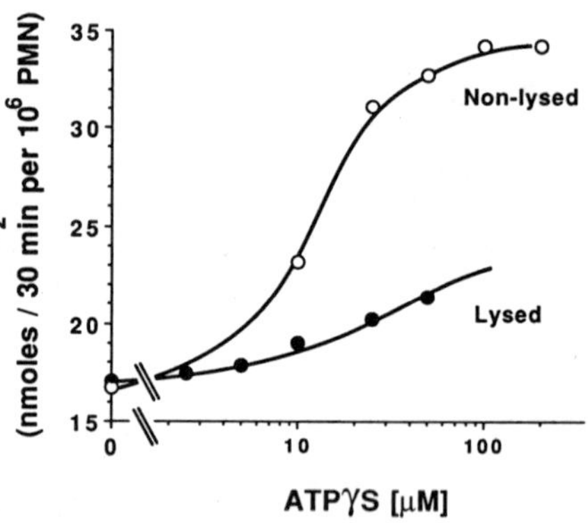

FIGURE 1. Enhancement by ATPγS of O_2^- responses in fMLP-stimulated neutrophils as a function of whether cells had been previously exposed to 0.15 M NH_4Cl, the lysing solution for RBCs.

Effects of ATPγS on Calcium Changes Induced by fMLP

In order to determine if the priming effects of ATPγS might be related to a greater increase in intracellular calcium after cell stimulation with fMLP, human neutrophils (5×10^6) that had been preloaded with fura-2 were first exposed to 25 μM ATPγS for 5 min at 37 °C, and then 1 μM fMLP was added. The intracellular calcium levels were continuously monitored in a spectrofluorometer. For convenience, computed concentrations of intracellular calcium at selected time points were used for graphic expression. As is apparent from the data in FIGURE 4, the intracellular changes in calcium in response to cell contact with fMLP were indistinguishable from the calcium changes induced in fMLP-stimulated cells that had been previously exposed to ATPγS. When the two sets of data (from ATPγS- or buffer-pretreated cells) were compared, there was no statistically significant differences at any time point. Thus, cells that are primed with ATPγS do not demonstrate a different pattern of increase in intracellular calcium when the cells are subsequently stimulated with fMLP.

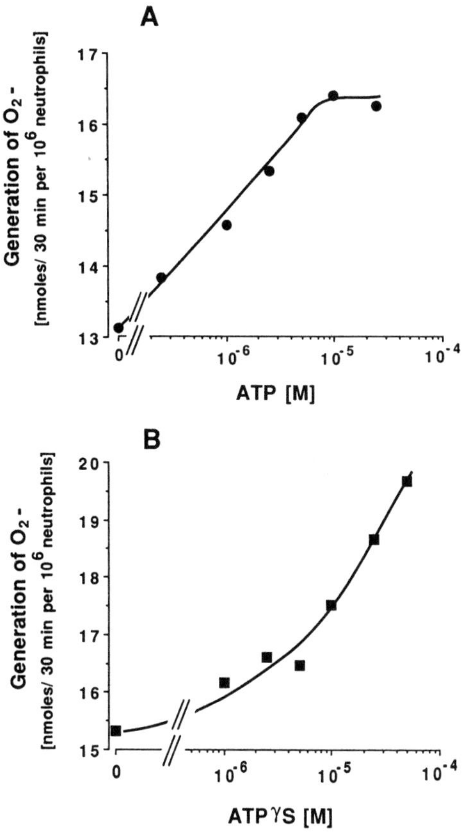

FIGURE 2. Dose-response relationships between O_2^- production in fMLP-stimulated neutrophils and the concentration of (**A**) ATP or (**B**) ATPγS.

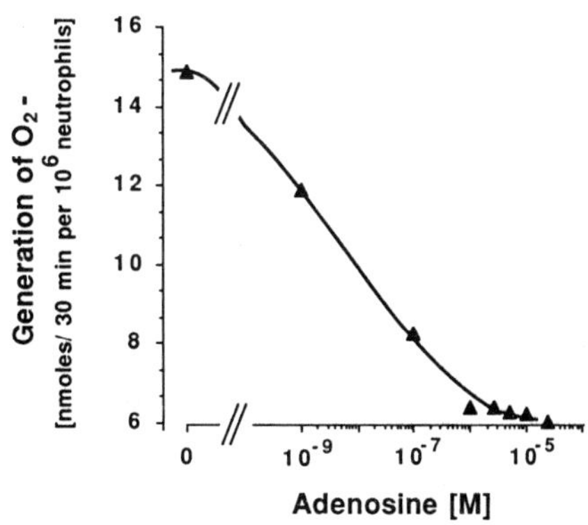

FIGURE 3. Dose-response relationship of inhibition by adenosine of O_2^- in fMLP-stimulated neutrophils.

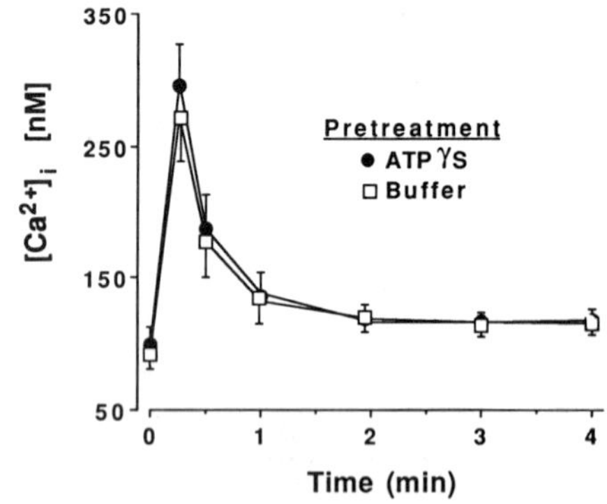

FIGURE 4. Increases in intracellular calcium in fMLP-stimulated neutrophils that had been pretreated with ATPγS (25 μM) or buffered salt solution.

In order to determine if, under the experimental conditions employed, calcium changes might be at a plateau and, therefore, any ATPγS effects on calcium levels might be masked, the experiment described by the data in FIGURE 5 was carried out. Cells were exposed to either 25 μM ATPγS or buffered salt solution for 5 min at 37 °C, and then fMLP over a wide range of concentrations was added. Changes in intracellular calcium were measured. In cells incubated 5 min at 37 °C, the "basal" concentrations of intracellular calcium were 61.5 ± 3.3 and 62.3 ± 2.6 nM in primed (ATPγS-treated) and unprimed cells, respectively. Thus, the levels of intracellular calcium in these two preparations of cells prior to stimulation with fMLP were indistinguishable. When cells were exposed to a range of fMLP concentrations (10^{-11}-10^{-6} M), the increases in intracellular calcium were the same in the two sets of cells whether they had been previously (5 min earlier) exposed to 25 μM ATPγS or to buffered salt solution. Thus, at no concentration of fMLP does prior cell contact with ATPγS alter the maximal increase in intracellular calcium induced by fMLP. In whatever manner ATPγS brings about enhancement of O_2^- in fMLP-stimulated neutrophils, this does not appear to be related to exaggerated increases in intracellular levels of calcium induced by cell contact with fMLP.

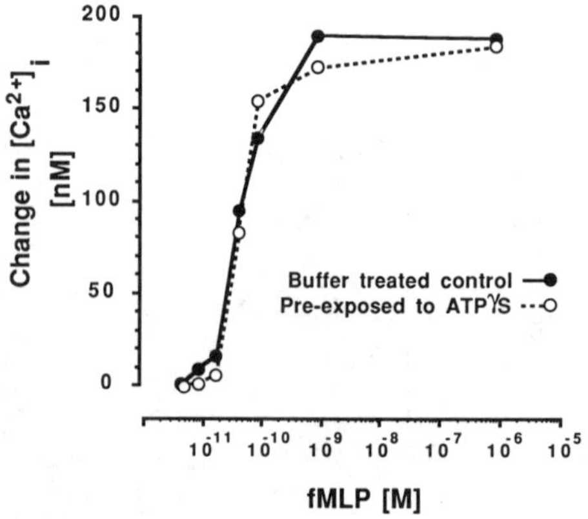

FIGURE 5. Peak changes in intracellular calcium in neutrophils exposed to a range of concentrations of fMLP, as a function of prior exposure of cells (5 min earlier) to buffered salt solution or 25 μM ATPγS.

Effects of ATPγS and Adenosine on fMLP Receptors and CR3 Content

Consideration was given to the possibility that ATPγS and adenosine may alter neutrophil responses to fMLP by causing changes in fMLP receptor content on the surfaces of intact neutrophils. Accordingly, after the neutrophils were pretreated with 10 μM adenosine or ATPγS, the cells were allowed to equilibrate with [^{3}H]fMLP, and the specific binding of the radioactive ligand was determined. From these measurements, the number of receptors/cell and the binding affinities (K_d) for [^{3}H]fMLP were calculated. The results of these studies are described in TABLE 1, where it is evident that cell contact with either adenosine or ATPγS resulted in a small (10-17%) drop in the mean number of receptors/cell without a change in the affinity of the binding of [^{3}H]fMLP to the neutrophils. Because it is known that agonists (including fMLP) for neutrophils will cause changes in the content of the cell membrane resulting from fusion of secretory granules, cells were exposed to ATPγS and both the content of CR3 and the fMLP receptor number were measured in cells from the same donor pool. The results are described in TABLE 2. CR3 cell content (Mo1 antigen, CD11b), as assessed by the use of fluorescent antibody and flow cytometry, was greatly increased. Consistent with the data in TABLE 1, the number of fMLP receptors fell by 27% when cells were exposed to ATPγS (TABLE 2). These data indicate that ATPγS has paradoxical effects on CR3 and fMLP receptors of human neutrophils. Because it is known that secretory granules contain both fMLP and CR3 receptors and that granule fusion to the cell membrane can result in adduction of these receptors to the surface of the neutrophil,[12,13] the data in TABLE 2 suggest that adduction of these two different receptors to the cell membrane may be separately regulated. The findings described in TABLES 1 and 2 do not point to a consistent pattern of change in fMLP receptor content as an explanation for why ATPγS and adenosine alter O_2^- responses of neutrophils stimulated with fMLP.

TABLE 1. Effects of ATPγS and Adenosine on [^{3}H]fMLP Binding to Neutrophils[a]

Experiment	Material Added to Neutrophils	[^{3}H]fMLP Receptor Number/Cell (mean ± SEM)	K_d (nM, mean ± SEM)
1	Buffer	13,183 ± 1790	4.64 ± 0.65
	Adenosine (10 μM)	11,974 ± 1559[b]	4.14 ± 0.05[c]
2	Buffer	16,530 ± 1438	4.46 ± 0.49
	ATPγS (10 μM)	13,717 ± 1193[d]	4.13 ± 0.49[c]

[a] Derived from reference 10.
[b] As compared to the data line above, $p < .007$, $N = 7$.
[c] As compared to the data line above, p is not significant.
[d] As compared to the data line above, $p < .02$, $N = 9$.

Effects of ATPγS and Adenosine on O_2^- Production: Independence of a Requirement for Cytoplasmic Granules

Because, as discussed above, fusion of cytoplasmic granules to the cell membrane of the neutrophil can change the composition of that membrane, the question was explored as to whether the availability of cytoplasmic granules was required for the enhancing effects of ATPγS and the inhibitory effects of adenosine on O_2^- responses of fMLP-stimulated cells. Accordingly, neutroplasts, which lack cytoplasmic granules but still respond to fMLP or to phorbol ester with O_2^- responses, were prepared and used in amounts that produced O_2^- responses similar to those of 500,000/ml neutrophils stimulated with 100 nM fMLP. It should be emphasized that these studies employed neutrophils and neutroplasts that had been prepared from the same pool of cells, allowing simultaneous comparison of O_2^- responses of neutroplasts and neutrophils. As is evident from the data in TABLE 3, the O_2^- responses of fMLP-stimulated neutrophils and neutroplasts in the absence of ATPγS or adenosine were similar. The enhancing effects of 10 μM ATPγS and the inhibitory effects of 10 μM adenosine were quantitatively similar in the two cell preparations. These data prove beyond reasonable doubt that however ATPγS and adenosine alter O_2^- responses in fMLP-stimulated cells, the presence of cytoplasmic granules is not required.

DISCUSSION

There is an increasing body of evidence suggesting that adenine nucleotides and adenosine may be important modulators of cellular responses. ATP (and its analogues) and to a lesser extent ADP will cause transient increases in levels of intracellular calcium immediately after cell exposure. In some systems it is known that this phenomenon is related to synthesis of inositol phosphates. The functional significance of these changes has been defined in neutrophils where it has been demonstrated that for cells first treated with ATP, and then exposed to the chemotactic peptide fMLP, the production of O_2^- and the extracellular release of lysosomal enzymes are increased. The biological relevance of these observations may be linked to the finding that for human neutrophils stimulated with fMLP or with immune complexes, in the presence of intact platelets, platelet lysates, or platelet secretion products, the generation of

TABLE 2. Effects of ATPγS on CR3 Content and fMLP Receptors of Neutrophils[a]

Material Added	CR3 Content[b] (mean ± SEM)	[^{3}H]fMLP Receptor Number/Cell (mean ± SEM)
Buffer	3.20 ± 0.60	11,068 ± 2745
ATPγS (10 μM)	9.66 ± 3.54[c]	8,111 ± 1828[d]

[a] Summarized from reference 10 ($N = 8$).
[b] Specific fluorescence intensity measured by flow cytometry, expressed on linear scale.
[c] As compared to the data line above, $p < .03$.
[d] As compared to the data line above, $p < .05$.

TABLE 3. O_2^- Responses of Neutrophils and Neutroplasts Stimulated with fMLP[a]

Preparation Employed	Addition	O_2^- Response (nmol/30 min, mean ± SEM)	Significance[b]
Neutrophils	None	7.1 ± 1.7	
	ATPγS (10 μM)	9.1 ± 2.2	$p < .04$
	Adenosine (10 μM)	3.7 ± 1.1	$p < .03$
Neutroplasts	None	7.4 ± 1.0	
	ATPγS (10 μM)	10.1 ± 1.4	$p < .002$
	Adenosine (10 μM)	2.7 ± 0.6	$p < .001$

[a] Summarized from reference 10.
[b] All tests for significance were by comparison to the cell preparation to which there were no additions.

O_2^- is significantly amplified.[6] Also, when complement activation is brought about *in vivo* by infusion of cobra venom factor, the resulting injury of pulmonary microvascular endothelial cells is greatly attenuated by prior platelet depletion.[7] Because the vascular damage in this model is known to be directly attributable to toxic oxygen products of activated neutrophils, it may well be that the role of platelets in this inflammatory model is in part linked to the release of adenine nucleotides, which enhance oxygen radical production by neutrophils.

Although ATPγS causes increased CR3 expression on the plasma membrane of neutrophils, it has the opposite effect on fMLP receptor number (with no change in receptor affinity). It seems likely that the effects of ATP, ATPγS, and adenosine on fMLP-induced O_2^- responses in neutrophils cannot be attributed to changes in fMLP receptor content. This is further underscored by the fact that fMLP-induced O_2^- responses in granule-deficient neutroplasts are fully responsive to ATPγS and adenosine, even though granule fusion events are precluded.

It is possible that ATP, ATPγS, and adenosine may be altering a late step in the signal transduction pathway, such as the activity, translocation, or proteolysis of protein kinase C, or perhaps a direct effect on NADPH oxidase. Alternatively, one or more of these adenine compounds could be causing selective enhancement (by ATP or ATPγS) or inhibition (by adenosine) of diacylglycerol production from a phosphatidylglyceride other than phosphatidylinositol, such as phosphatidylcholine. In this way, there would be no change in the amount of inositol trisphosphate formed, but there could be an asymmetrical generation of diacylglycerol, higher in the case of ATP (or ATPγS) followed by fMLP, and lower in the case of adenosine followed by fMLP. The evidence assembled to date does not allow a distinction to be made between the effects of the adenine compounds on early or late steps in signal transduction.

REFERENCES

1. KUHNS, D. B., D. G. WRIGHT, J. NATH, S. S. KAPLAN & R. E. BASFORD. 1988. ATP induces transient elevations of $[Ca^{2+}]_i$ in human neutrophils and primes these cells for enhanced O_2^- generation. Lab. Invest. **48:** 448.
2. WARD, P. A., T. W. CUNNINGHAM & K. J. JOHNSON. 1989. Signal transduction events

in stimulated rat neutrophils: Effects of adenine nucleotides. Clin. Immunol. Immunopathol. **50:** 30.
3. COWEN, D. S., H. M. LAZARUS, S. B. SURIN, S. E. STOLL & G. DUBYAK. 1989. Extracellular adenosine triphosphate activates calcium mobilization in human phagocytic leukocytes and neutrophils/monocyte progenitor cells. J. Clin. Invest. **83:** 1651.
4. COCKCROFT, S. & J. STUTCHFIELD. 1989. ATP stimulates secretion in human neutrophils and HL60 cells via a pertussis toxin-sensitive quanine nucleotide-binding protein coupled to phospholipase C. FEBS Lett. **245:** 25.
5. WARD, P. A., T. W. CUNNINGHAM, K. K. MCCULLOCH & K. J. JOHNSON. 1988. Regulatory effects of adenosine and adenine nucleotides on oxygen radical responses of neutrophils. Lab. Invest. **58:** 438.
6. WARD, P. A., T. W. CUNNINGHAM, K. K. MCCULLOCH, S. H. PHAN, J. POWELL & K. J. JOHNSON. 1988. Platelet enhancement of O_2^- responses in stimulated human neutrophils: Identification of platelet factor as adenine nucleotides. Lab. Invest. **58:** 37.
7. WARD, P. A., D. MACCONI, M. C. SULAVIK, G. O. TILL, J. S. WARREN, K. J. JOHNSON & J. POWELL. 1988. Rat neutrophil-platelet interactions in oxygen radical-mediated lung injury. *In* Oxy-Radicals in Molecular Biology and Pathology. P. A. Cerutti, I. Fridovich & J. M. McCord, Eds.: 83-98. Alan R. Liss. New York, NY.
8. JOHNSON, R. J., C. E. ALPERS, P. PRITZL, M. SCHULZE, P. BAKER, C. PRUCHNO & W. G. COUSER. 1988. Platelets mediate neutrophil-dependent immune complex nephritis in the rat. J. Clin. Invest. **82:** 1225.
9. CRONSTEIN, B. N., E. D. ROSENSTEIN, S. B. KRAMER, G. WEISSMAN & R. HIRSCHHORN. 1985. Adenosine: A physiological modulation of superoxide anion generation by human neutrophils. Adenosine acts via an A2 receptor on human neutrophils. J. Immunol. **135:** 1366.
10. WALKER, B. A. M., T. W. CUNNINGHAM, D. R. FREYER, R. F. TODD III, K. J. JOHNSON & P. A. WARD. 1989. Regulation of the superoxide responses of human neutrophils by adenine compounds: Independence of requirement for cytoplasmic granule. Lab. Invest. **61:** 515.
11. ROOS, D., A. A. VOETMAN & L. J. MEERHOF. 1983. Functional activity of enucleated human polymorphonuclear leukocytes. J. Cell Biol. **97:** 368.
12. FLETCHER, M. P., B. E. SELIGMANN & J. I. GALLIN. 1982. Correlation of human neutrophil secretion, chemotactic receptor mobilization and enhanced functional capacity. J. Immunol. **128:** 941.
13. PETREQUIN, P. R., R. F. TODD III, L. J. DEVALL, L. A. BOXER & J. T. CURNUTTE III. 1987. Association between gelatinase release and increased plasma membrane expression of the Mo1 glycoprotein. Blood **69:** 605.

DISCUSSION OF THE PAPER

L. L. SLAKEY (*University of Massachusetts, Amherst, MA*): I am struck by the similarity between the relationship of nucleotides and adenosine to platelet responsiveness, and the results that you report. That is, in the platelet, ADP is proaggregatory, and adenosine is antiaggregatory, and you find that ATP and ADP enhance the neutrophil O_2^- response and that adenosine inhibits it. My comment springs from the hypothesis that the pathway of nucleotide hydrolysis at cell surfaces can serve as a timer. Early after ATP/ADP release, both platelet and neutrophil responses are enhanced. As adenosine is produced, the response is inhibited. Both the platelet and the neutrophil would be in the neighborhood of endothelial cells, and the endothelial ectonucleotidases, unlike those on some other cell types, do tend to create a distinct time gap between the appearance of a bolus of ATP or ADP, and the generation of an adenosine-rich milieu.

WARD: The brief appearance and subsequent disappearance of intravascular aggregates of platelets may well be due to the release by the platelets of ATP and ADP. ATP and ADP are metabolized very rapidly to adenosine, which then results in the disaggregation and disappearance of the platelets.

Y. H. EHRLICH (*College of Staten Island, New York, NY*): My questions regard the mechanism of action. What is the effect of ATP analogues other than ATP[S]? Could the measured binding of ATP[^{35}S] be actually or partly this phosphorylation of proteins rather than binding?

WARD: Human neutrophils respond with 1) calcium transients and 2) enhanced O_2^- response to other nucleotide triphosphates with the following rank orders of potency: 1) ATP > > 2-Me-S-ATP > AMP-PCP>; 2) UTP $\geq$ ATP > ITP > GTP > CTP. Our studies with ATP[^{35}S] suggest that the binding is reversible (in an excess of cold ATP[S]) and shows no covalent thiophosphorylation (as determined by SDS-PAGE). Further, examination of label eluted from neutrophils shows that all of the label is in the form of ATP[^{35}S] without evidence of hydrolysis.

R. F. COLMAN (*University of Delaware, Newark, DE*): Is the protein kinase C system involved? Have you looked at specific inhibitors of the enzyme for effects on the potentiating effect of ATP[S]?

WARD: The endpoint is O_2^- production, which seemed to be blocked by inhibitors of protein kinase C. Thus, we have not used these inhibitors. There is still no evidence either for or against the ability of ATP[S] to modify protein kinase C.

J. S. WILEY (*Austin Hospital, Heidelberg, Australia*): The increase in CR3 expression on neutrophils due to exposure to ATP[S] is impressive. Do other agonists give increased CR3 expression of the same magnitude?

WARD: Like other neutrophil agonists (such as formyl chemostatic peptide, C5a, phorbol ester, and calcium ionophore A23187), ATP causes increased expression of CR3 by fusing secondary granules to the cell membrane. ATP carries out this action in a dose-dependent manner.

Effects of Extracellular ATP on Mononuclear Phagocytes

THOMAS H. STEINBERG,[a,b] HENK P. BUISMAN,[c] STEVEN GREENBERG,[d] FRANCESCO DI VIRGILIO,[e] AND SAMUEL C. SILVERSTEIN [d]

[b] *Washington University School of Medicine*
St. Louis, Missouri 63110

[c] *University of Leiden*
2333 Al Leiden, the Netherlands

[d] *Columbia University College of Physicians and Surgeons*
New York, New York 10032

[e] *University of Padua*
I-35131 Padua, Italy

INTRODUCTION

Macrophages are involved in a wide variety of activities: they ingest invading microorganisms, senescent erythrocytes, and other cellular debris[1]; secrete over one hundred products that mediate and modulate inflammatory and immune responses[2]; and initiate specific recognition of many antigens by lymphocytes.[3] Given the diversity of tasks that macrophages perform, it may not be too surprising that murine macrophages respond to extracellular nucleotides in a variety of ways, possibly making more different uses of these extracellular signaling molecules than any other cell. Macrophages therefore provide an excellent system in which the effects of extracellular nucleotides can be studied. At present we can measure four distinct activities elicited by extracellular nucleotides:

1. ATP^{4-}-induced pore.
2. Nucleotide-activated receptor mediating calcium mobilization.
3. Ectonucleotidase.
4. Ectokinase.

This paper summarizes our current understanding of these different ATP-mediated activities. The data suggest that these activities are mediated by distinct plasma

[a] Address for correspondence: Washington University School of Medicine, Department of Medicine, Division of Infectious Diseases, Box 8051, 660 South Euclid Avenue, St. Louis, Missouri 63110.

membrane nucleotide receptors or enzymes. The selection of variant J774 cells that specifically lack one of these activities, ATP-mediated permeabilization, has provided additional support for the hypothesis that each of these activities reside in distinct molecules. The cells used in most of these experiments were taken from the J774 mouse macrophage-like cell line.[4] Qualitatively similar results were obtained with thioglycollate-elicited mouse peritoneal macrophages.

EFFECTS OF ATP ON J774 CELLS

In 1967 Cohn and Parks demonstrated that exposure of macrophages to ATP caused an increase in the number of phase-lucent vacuoles within the cells.[5] This report led Sung *et al.* to explore the effects of ATP and other nucleotides on membrane potential and ion fluxes in J774 cells.[6] In these experiments, millimolar quantities of extracellular MgATP caused membrane depolarization, influx of extracellular sodium, efflux of intracellular potassium or rubidium, an increase in $[Ca^{2+}]_i$, and inhibition of Fc receptor-mediated phagocytosis. The ATP-induced permeability of the plasma membrane to monovalent cations was profound: in cells exposed to 10 mM MgATP for 10 min, the intracellular Na^+ and K^+ concentrations approached those of the extracellular medium. Neither adenosine, AMP, ADP, nor nucleoside triphosphates other than ATP altered membrane potential or permeability of the plasma membrane to monovalent cations.

Most intracellular activities that utilize ATP require the presence of divalent cations, or, put another way, $MgATP^{2-}$ is the active species of ATP in most of these reactions. When we assessed the role of divalent cations in ATP-induced $^{86}Rb^+$ efflux, however, we found that Mg^{2+} and Ca^{2+} need not be present in the extracellular medium for ATP to induce $^{86}Rb^+$ efflux, and that divalent cations were in fact inhibitory, as depicted in FIGURE 1. By calculating the concentrations of the various ATP species present in the medium, we found that $^{86}Rb^+$ efflux correlates with the concentration of ATP^{4-}, and that $^{86}Rb^+$ efflux is elicited by less than 100 μM ATP^{4-}. Cockcroft and Gomperts had shown that ATP^{4-} is the species of ATP responsible for permeabilization of rat mast cells.[7] The dependence of monovalent cation fluxes on ATP^{4-} explains why millimolar quantities of MgATP were required in the studies mentioned above, as only a small fraction of the extracellular ATP was present as the relevant species.

We further characterized ATP-mediated plasma membrane depolarization using electrophysiologic techniques.[8] Membrane potential in single cells was assessed in the whole cell patch clamp configuration. ATP was introduced into the medium surrounding the patch-clamped cell from a second pipette placed near the cell. FIGURE 2 shows the membrane depolarization that occurs when ATP is applied to the cell for varying lengths of time. Membrane depolarization occurred within 40 msec of ATP application, and approached 0 mV as ATP was applied for increasing lengths of time. The membrane remained depolarized as long as the concentration of ATP in the medium remained elevated. As the ATP diffused away from the patched cell the membrane potential returned to -75 mV. The whole cell conductance in the presence of 5 mM ATP was 10 nS, and had a reversal potential of 0 mV, showing that the ATP-induced increase in ionic permeability was not selective for specific ions.

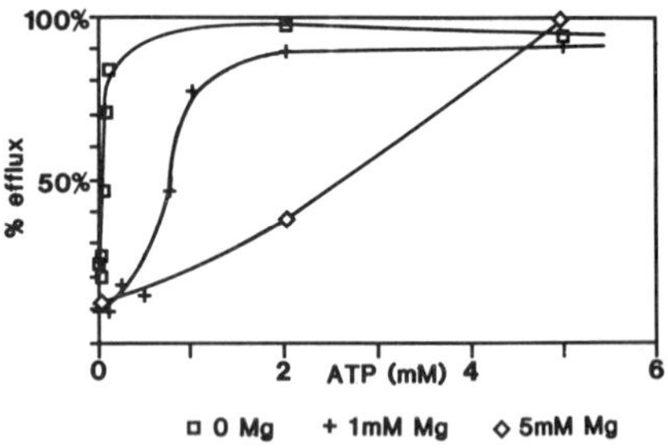

FIGURE 1. Effect of extracellular Mg^{2+} on ATP-induced $^{86}Rb^{+}$ efflux in J774 cells. Cells were incubated in medium containing 5 μM $^{86}RbCl$ for 90 min, washed, and then incubated in phosphate-buffered saline containing varying concentrations of $MgSO_4$. The $^{86}Rb^{+}$ efflux was measured and expressed as a percentage of total intracellular $^{86}Rb^{+}$. Reproduced from reference 20, by permission.

ATP-MEDIATED PERMEABILIZATION OF MACROPHAGES

It had previously been shown that rat mast cells[7,9] and transformed mouse fibroblasts[10,11] become permeable to nucleotides and sugar phosphates in the presence of ATP. Because of these findings and our own observations that ATP^{4-} induced both monovalent and divalent cation fluxes in J774 cells, we tested the hypothesis that ATP^{4-} caused a more general membrane permeabilization in mouse macrophages. We found that a variety of membrane-impermeant fluorescent dyes were able to gain entrance to the cytoplasmic matrix of these cells in the presence of extracellular ATP. We have primarily used the dye Lucifer Yellow to study ATP-induced permeabilization (FIG. 3), although the largest molecule we have been able to introduce into J774 cells by this means is fura-2.[12] Thus monovalent cation fluxes and membrane depolarization induced by ATP in mouse macrophages are not caused by activation of specific ion channels, but rather are due to the more general permeabilization of the plasma membrane to a variety of molecules by the formation of a ligand-operated pore.

The general features of ATP-mediated pore formation are similar in all cells in which this phenomenon has been described, including transformed mouse fibroblasts,[10,11] rat mast cells,[7,9] Chinese hamster ovary cells,[13] and mouse neuroblastoma cells[14] as well as mouse macrophages. Only ATP and a few nonhydrolyzable ATP analogues induce pore formation, whereas other nucleotide triphosphates are ineffective. Because the active ATP species is ATP^{4-}, pore formation is enhanced by lowering the extracellular divalent cation and hydrogen ion concentrations. Permeabilization

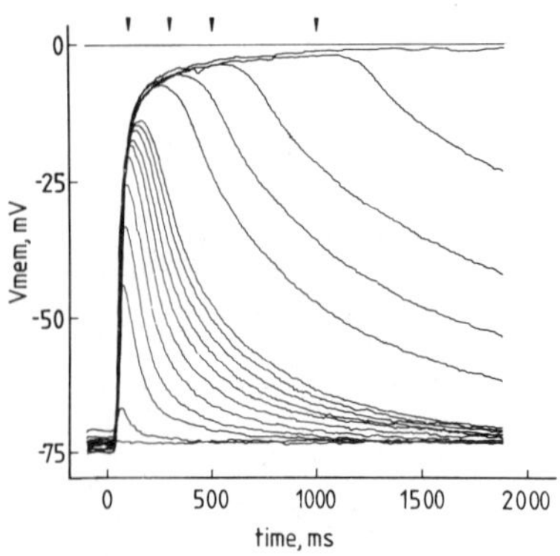

FIGURE 2. ATP-induced depolarization of J774 cells. ATP was pulsed for varying periods of time from an ejection pipette onto a J774 cell maintained in the whole cell patch configuration. Reproduced from reference 8, by permission.

is rapidly reversed when ATP is removed from the medium or when sufficient divalent cations are added to bind ATP^{4-}. In cells permeabilized by ATP, a variety of small molecules can be introduced into the cytoplasmic matrix, and small molecules within the cytoplasmic matrix can escape into the extracellular medium.

The function of the ATP-induced pore in macrophages remains a mystery. It would seem highly counterproductive for cells to permit the major perturbation of intracellular homeostasis that occurs during ATP permeabilization of the plasma membrane. One need only consider that ATP permeabilization allows the release of the intracellular pool of nucleoside triphosphates, which the cell generated at great expense. A clue to a possible role of this phenomenon may be the size of the pore,

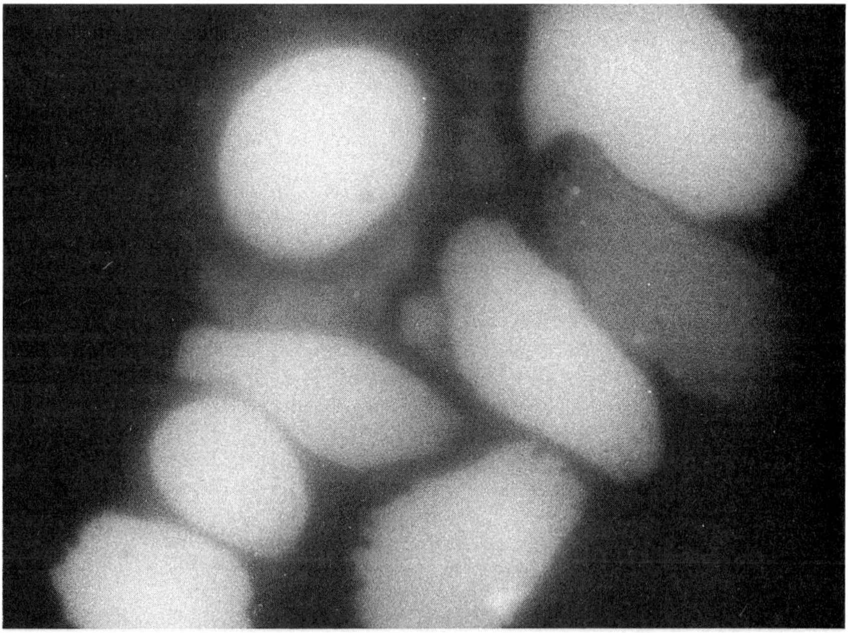

FIGURE 3. Introduction of Lucifer Yellow into the cytoplasm of J774 cells by ATP permeabilization. J774 cells adherent to glass coverslips were incubated in medium containing 0.5 mg/ml Lucifer Yellow and 5 mM ATP for 5 min. These cells, after being washed, were then viewed by fluorescence microscopy. Reproduced from reference 12, by permission.

which functionally appears to be about the size of a gap junction. Gap junctions electrically and chemically couple adjacent cells, allowing them to function in concert. Macrophages perform many roles that involve cell-cell contact, such as antigen presentation and cytotoxicity. In response to organisms and foreign materials that are not readily degradable, they may fuse to form giant cells, and it may be in the context of such activities that the ATP pore makes physiological sense. Because these cells can form tight seals and strictly delimited compartments both with substrates to which they are adherent and to other cells, the opening of a pore such as that formed in the presence of ATP does not necessarily expose the cytosol to a vast extracellular space. Instead, the ATP pore may form only in a strictly delimited compartment formed by the apposition of two cells. If both cells possessed such a pore, exchange of cytosolic constituents might ensue.

NUCLEOTIDE-INDUCED CALCIUM TRANSIENTS

Although permeabilization of the plasma membrane as described above causes a rise in $[Ca^{2+}]_i$ when the extracellular medium contains calcium, upon further investigation we found that extracellular ATP causes calcium transients by an additional mechanism that is independent of the ATP-operated pore described above.[15] This rise in $[Ca^{2+}]_i$ is caused both by a release of calcium from intracellular stores and by an influx of calcium from the extracellular medium. Calcium rises induced by this second mechanism can be distinguished from calcium influx through ATP-induced pores by several criteria (TABLE 1). First, the quantity of ATP^{4-} required to initiate calcium mobilization is significantly lower than the quantity of ATP^{4-} required to mediate ATP permeabilization. Increases in $[Ca^{2+}]_i$ via the calcium-mobilizing receptor occur

TABLE 1. Characteristics of ATP-Induced Permeabilization and Nucleotide-Activated Calcium Mobilization in J774 cells.

	ATP-Induced Pore	Ca-Mobilizing Receptor
[ATP] required	10^{-4} M	10^{-6} M
Agonists	ATP > ATPγS > AMP-PNP	UTP > ATP > ATPγS > ADP > other nucleotides
ATP species	ATP^{4-}	ATP^{4-}
Activity in ATPR cells	Absent	Present

at concentrations of ATP that do not cause membrane depolarization or monovalent cation fluxes. Second, while the ATP-induced pore responds only to ATP and a few ATP analogues, calcium mobilization can be elicited by many other nucleoside di- and triphosphates. UTP is the most potent agonist in this system. As with the receptor that induces ATP-mediated permeabilization, mobilization of $[Ca^{2+}]_i$ is most effective in the absence of Mg^{2+}, and appears to be caused by ATP^{4-}. Third, certain ATP-resistant cells described below are not permeabilized by ATP, but do retain the ability to respond to extracellular ATP by calcium mobilization.

Although the putative receptor that mediates ATP permeabilization is present on a limited number of cell types, the receptor that mobilizes calcium in response to ATP appears on a wider range of cells, and appears to correspond to the P_{2Y} or P_{2U} receptors described in other cells. We have but a small body of data that addresses the physiologic significance of this receptor on macrophages; however, several groups of investigators have studied the effects of adenine nucleotides on monocytes and polymorphonuclear leukocytes, and have shed some light on the potential uses of this receptor by cells of the immune system. Altieri and Edgington have shown that exposure of human monocytes to ADP induces the expression of high-affinity binding sites for coagulation factor X, probably through upregulation of complement receptor 3 (CR3), the receptor that binds the C3bi component of complement, by these cells.[16]

Similarly, Freyer and colleagues have shown that ATP and nonhydrolyzable ATP analogues induce increased expression of this complement receptor on human neutrophils, and that this is accompanied by increased neutrophil aggregation.[17] These findings suggest that purinergic receptors on phagocytic cells may regulate a variety of important surface adhesive properties on these cells, such as adherence of monocytes to vascular endothelium, regulation of intravascular coagulant activity, and binding of phagocytes to pathogenic organisms. In addition to these effects of adenine nucleotides, Ward and co-workers have demonstrated that adenine nucleotides enhance the generation of superoxide anion by human neutrophils in response to bacterial chemoattractants or immune complexes.[18] These workers also demonstrated that ATP and ADP released by platelets had a similar effect.[19]

Thus there are several effects of adenine nucleotides on phagocytic cells. Although the receptors that transduce these effects have not been identified, it is reasonable to speculate that the calcium mobilizing nucleotide receptor described in J774 cells is similar or identical to those that modulate neutrophil function. The potential roles of these receptors in mediating and modulating immune and inflammatory responses remain to be clarified.

ATP-RESISTANT CELL LINES

ATP-induced membrane permeabilization and the membrane depolarization and ion fluxes that occur as a consequence of this change in permeability can be reversed if J774 cells are exposed to the nucleotide for less than 30 min. Exposure of J774 cells for longer periods of time, however, causes cell death, presumably due to the massive alterations in the intracellular ionic composition. We exposed J774 cells to sufficient ATP to kill >95% of the cells (10 mM ATP for 45 min at 37 °C), allowed the remaining cells to grow to confluence, and repeated these steps until the cells were no longer killed by ATP.[20] Clones of cells derived from the resulting population, called ATP-resistant (ATPR) cells, do not become permeable to the membrane-impermeant dye Lucifer Yellow when exposed to concentrations of ATP as high as 20 mM.[12] ATP does not cause membrane depolarization[8] or $^{86}Rb^+$ efflux[20] in these cells. These findings show that ATPR cells do not respond to ATP by pore formation, which further supports the hypothesis that ATP mediates pore formation through ligation of a specific membrane protein and not via a nonspecific effect of ATP on the plasma membrane.

ATPR cells are not deficient in other ATP-induced responses. ATP causes an increase in $[Ca^{2+}]_i$ in ATPR cells that is indistinguishable from the nucleotide-mediated calcium mobilization in wild-type J774 cells.[15] FIGURE 4 compares the calcium transients induced by 500 μM ATP in J774 cells and ATPR cells. In both the wild-type J774 cells and the ATPR cells, ATP induces a rapid and brief increase in $[Ca^{2+}]_i$ by causing a release of calcium from intracellular stores. In ATPR cells this calcium spike is followed by a second, less pronounced rise in $[Ca^{2+}]_i$, producing a biphasic calcium response. This pattern is similar to the pattern seen in wild-type cells exposed to concentrations of ATP that are sufficient to mobilize calcium but not high enough to permeabilize the cells. In wild-type cells, 500 μM ATP causes a second calcium spike that saturates the reporter dye, indicating that plasma membrane permeabilization has occurred. In addition to possessing functional calcium-mobilizing receptors, ATPR cells also express ecto-ATPase activity.[20]

ATPR cells therefore appear to be deficient in only one of the activities mediated by extracellular ATP in J774 cells. The characteristics of these cells lend support to the hypothesis that the different actions of extracellular ATP on J774 cells are mediated by distinct plasma membrane receptors. Therefore, these cells may serve as useful tools in further characterizing these receptors.

THE ECTONUCLEOTIDASE OF MACROPHAGES

Ectoenzymes in general, and ecto-ATPases in particular, have been found on many cells; the ecto-ATPase of guinea pig polymorphonuclear leukocytes was studied in detail fifteen years ago by DePierre and Karnovsky; they defined the criteria for ectoenzymes.[21,22] Ecto-ATPase activity has subsequently been described on many other cells, including human polymorphonuclear leukocytes[23] and mouse macrophages.[6] As with most other ectoenzymes, biochemical characterization has not been accompanied by an understanding of the physiological function of these enzymes.

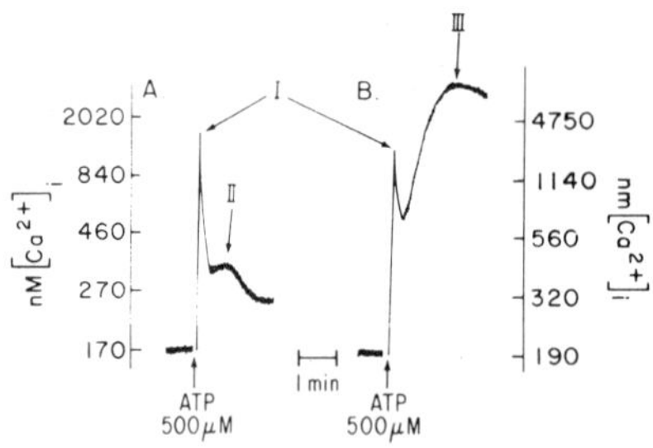

FIGURE 4. Calcium transients induced by 500 μM ATP in ATPR G6 variants (trace A) and wild-type J774 cells (trace B). Reproduced from reference 15, by permission.

J774 cells possess ecto-ATPase activity (apparent K_m of 350 μM). It was therefore natural to speculate that the hydrolysis of extracellular ATP by this enzyme was linked to the other membrane events that were induced by this nucleotide. We tested this hypothesis by comparing the requirements for ecto-ATPase activity with those required for the ATP-mediated efflux of $^{86}Rb^+$. These studies showed that the ecto-ATPase activity was unrelated to the ATP-mediated generation of monovalent cation fluxes.[20]

First, the pharmacologic profiles of these two activities were different. Ecto-ATPase activity of J774 cells was inhibited by the addition of other nucleotide triphosphates, including ITP, GTP, and CTP. This finding implied that the ectoenzyme was capable of hydrolyzing other nucleoside triphosphates in addition to ATP, and indeed we showed directly that GTP was as good a substrate for the ecto-ATPase as was ATP. This enzyme is therefore actually an ectonucleotidase. As mentioned above, of all nucleotide triphosphates tested, only ATP induced $^{86}Rb^+$ efflux.

Second, ecto-ATPase activity and ATP-induced cation fluxes also differed in their requirements for divalent cations. As with most ATP-utilizing enzymes, ecto-ATPase activity had an absolute requirement for Mg^{2+} or Ca^{2+}, in contrast to the inhibitory effect of divalent cations on ATP-induced ion fluxes and membrane depolarization.

Third, the ATPR clones mentioned above expressed ATPase activity in spite of the inability of ATP to elicit membrane depolarization or monovalent cation fluxes in these cells. These studies provided persuasive evidence for at least two different membrane molecules on mouse macrophages that responded to ATP; they suggested that the function of the ectonucleotidase may be to limit the access of ATP to membrane ATP receptors, analogous to the role of the acetylcholinesterase in cholinergic synapses.

ECTO-PROTEIN KINASE ACTIVITY OF MACROPHAGES

Macrophages also possess a fourth ATP-utilizing activity, an ectoprotein kinase. Remold-O'Donnell described this activity in guinea pig macrophages, and showed that both exogenous proteins and plasma membrane proteins were phosphorylated upon the addition of exogenous ATP.[24] Although we have no information on the nature of these phosphorylated membrane proteins, it is likely that phosphorylation of membrane proteins affects macrophage function. In a similar fashion, phosphorylation of exogenous proteins provides another mechanism by which macrophages can interact with extracellular matrix proteins in their environment.

CONCLUSIONS

These investigations of the effects of extracellular adenine nucleotides on macrophages have revealed four different cell surface activities that utilize extracellular nucleotides; they suggest that macrophages possess different membrane proteins that mediate each of these responses. We have little evidence at present for the functions of these diverse activities: the ectonucleotidase of these cells may limit the access of nucleoside triphosphates to the other receptors; ATP-induced calcium mobilization may regulate surface adhesive properties. Because cells with which macrophages interact, especially platelets[19,25] and endothelial cells,[26] not only secrete but respond to extracellular ATP, dissecting the role of ATP in regulating the function of macrophages and other cells of the immune system may offer many insights into inflammatory and immune processes, and provide new opportunities for therapeutic interventions.

REFERENCES

1. Silverstein, S. C., S. Greenberg, F. Di Virgilio & T. H. Steinberg. 1989. Phagocytosis. *In* Fundamental Immunology. 2nd edit. W. E. Paul, Ed.: 703-720. Raven Press. New York, NY.
2. Nathan, C. F. 1987. Secretory products of macrophages. J. Clin. Invest. **79:** 319-326.
3. Unanue, E. R. & P. M. Allen. 1986. Biochemistry and biology of antigen presentation by macrophages. Cell. Immunol. **99:** 3-6.
4. Ralph, P., J. Prichard & M. Cohn. 1975. Reticulum cell sarcoma: An effector cell in antibody-dependent cell-mediated immunity. J. Immunol. **114:** 898-905.

5. COHN, Z. A. & E. PARKS. 1967. The regulation of pinocytosis in mouse macrophages. III. The induction of vesicle formation by nucleosides and nucleotides. J. Exp. Med. **125:** 457-466.
6. SUNG, S. S., J. D. YOUNG, A. M. ORIGLIO, J. M. HEIPLE, H. R. KABACK & S. C. SILVERSTEIN. 1985. Extracellular ATP perturbs transmembrane ion fluxes, elevates cytosolic [Ca^{2+}], and inhibits phagocytosis in mouse macrophages. J. Biol. Chem. **260:** 13442-13449.
7. COCKCROFT, S. & B. D. GOMPERTS. 1980. The ATP^{4-} receptor of rat mast cells. Biochem. J. **188:** 789-798.
8. BUISMAN, H. P., T. H. STEINBERG, J. FISCHBARG, S. C. SILVERSTEIN, S. A. VOGELZANG, C. INCE, D. L. YPEY & P. C. J. LEIJH. 1988. Extracellular ATP induces a large, nonselective conductance in macrophage plasma membranes. Proc. Natl. Acad. Sci. USA **85:** 7988-7992.
9. COCKCROFT, S. & B. D. GOMPERTS. 1979. ATP induces nucleotide permeability in rat mast cells. Nature **279:** 541-542.
10. HEPPEL, L. A., G. A. WEISMAN & I. FRIEDBERG. 1985. Permeabilization of transformed cells in culture by external ATP. J. Membr. Biol. **86:** 189-196.
11. ROZENGURT, E., L. A. HEPPEL & I. FRIEDBERG. 1977. Effect of exogenous ATP on the permeability properties of transformed cultures of mouse cell lines. J. Biol. Chem. **252:** 4584-4590.
12. STEINBERG, T. H., A. S. NEWMAN, J. A. SWANSON & S. C. SILVERSTEIN. 1987. ATP^{4-} permeabilizes the plasma membrane of mouse macrophages to fluorescent dyes. J. Biol. Chem. **262:** 8884-8888.
13. KITAGAWA, T. & Y. AKAMATSU. 1986. Control of membrane permeability by external ATP in mammalian cells: Isolation of an ATP-resistant variant from Chinese hamster ovary cells. Biochim. Biophys. Acta **860:** 185-193.
14. DI VIRGILIO, F., C. FASOLATO & T. H. STEINBERG. 1988. Inhibitors of membrane transport system for organic anions block fura-2 excretion from PC12 and N2A cells. Biochem. J. **256:** 959-963.
15. GREENBERG, S., F. DI VIRGILIO, T. H. STEINBERG & S. C. SILVERSTEIN. 1988. Extracellular nucleotides mediate Ca^{2+} fluxes in J774 macrophages by two distinct mechanisms. J. Biol. Chem. **263:** 10337-10343.
16. ALTIERI, D. C. & T. S. EDGINGTON. 1988. The saturable high-affinity association of factor X to ADP-stimulated monocytes defines a novel function of the Mac-1 receptor. J. Biol. Chem. **263:** 7007-7015.
17. FREYER, D. R., L. A. BOXER, R. A. AXTELL & R. F. TODD III. 1988. Stimulation of human neutrophil adhesive properties by adenine nucleotides. J. Immunol. **141:** 580-586.
18. WARD, P. A., T. W. CUNNINGHAM, K. K. MCCULLOCH & K. J. JOHNSON. 1988. Regulatory effects of adenosine and adenine nucleotides on oxygen radical responses of neutrophils. Lab. Invest. **58:** 438-447.
19. WARD, P. A., T. W. CUNNINGHAM, K. K. MCCULLOCH, S. H. PHAN, J. POWELL & K. J. JOHNSON. 1988. Platelet enhancement of O_2^- responses in stimulated human neutrophils. Lab. Invest. **58:** 37-47.
20. STEINBERG, T. H. & S. C. SILVERSTEIN. 1987. Extracellular ATP^{4-} promotes cation fluxes in the J774 mouse macrophage cell line. J. Biol. Chem. **262:** 3118-3122.
21. DEPIERRE, J. & M. L. KARNOVSKY. 1974. Ectoenzymes of the guinea-pig polymorphonuclear leukocyte. I. Evidence for an ecto-adenosine monophosphatase, adenosine triphosphatase, and *p*-nitropenyl phosphatase. J. Biol. Chem. **249:** 7111-7120.
22. DEPIERRE, J. & M. L. KARNOVSKY. 1974. Ecto-enzymes of the guinea-pig polymorphonuclear leukocyte. II. Properties and suitability as markers for the plasma membrane. J. Biol. Chem. **249:** 7121-7129.
23. HARLAN, J., L. R. DECHATELET, D. B. IVERSON & C. E. MCCALL. 1977. Magnesium-dependent adenosine triphosphatase as a marker enzyme for the plasma membrane of human polymorphonuclear leukocytes. Infect. Immun. **15:** 436-443.
24. REMOLD-O'DONNELL, E. 1978. Protein kinase activity associated with the surface of guinea pig macrophages. J. Exp. Med. **148:** 1099-1104.

25. MEYERS, K. M., H. HOLMSEN & C. L. SEACHORD. 1982. Comparative study of platelet-dense granule constituents. Am. J. Physiol. **243:** R454-R461.
26. PEARSON, J. D. & J. L. GORDON. 1985. Nucleotide metabolism by endothelium. Annu. Rev. Physiol. **47:** 617-627.

DISCUSSION OF THE PAPER

S. P. SOLTOFF (*Tufts University, Boston, MA*): Regarding the selection pressure used to generate the mutant cell line, how can you discriminate between the idea that the selection pressure (long-term exposure to ATP) may down-regulate an ATP-gated channel/pore rather than down-regulate a receptor that may promote these changes?

STEINBERG: As you suggest,the selection pressure only selects for resistance to permeabilization. We do not know whether the resistant cells lack the receptor that mediates permeabilization or whether the defect resides elsewhere. The salient point is that calcium release and influx via the nucleotide-mediated receptor is intact, and this provides a way for us to distinguish between the nucleotide receptor and the ATP-induced pore.

I. FRIEDBERG (*Tel Aviv University, Tel Aviv, Israel*): In our studies, in collaboration with Dr. Leon Heppel and his group, we found that transformed mouse fibroblasts were permeabilized by ionophores, indicating that dissipation of the membrane potential influences ATP-induced cell permeabilization. Did you try the effect of ionophores on the membrane permeability in J774 mouse macrophages?

STEINBERG: We have not specifically used ionophores for this purpose. Permeabilization in J774 cells, however, can be carried out in both Na^+- and K^+-containing medium, suggesting that membrane potential is not a critical factor.

G. WEISMAN (*University of Missouri, Columbia, MO*): I am interested in the similarity of the ATP-activated pore in macrophages to an ATP-activated pore in transformed mouse fibroblasts (3T6 cells). Leon Heppel's group and my group have shown that cross-linking benzoyl benzoic-ATP (Bz-ATP) to 3T6 cells caused subsequent formation of pores in these cells. Have you used this probe in macrophages?

STEINBERG: We have not used Bz-ATP. The only ATP analogues that induce permeabilization are ATPγS and AMP-PNP.

E. HEILBRONN (*University of Stockholm, Stockholm, Sweden*): Could you shut off the second Ca^{2+} increase peak, that is, the influx of Ca^{2+}, with any known Ca^{2+} channel blocker?

STEINBERG: We have not performed these studies. Would Dr. Di Virgilio like to comment?

F. DI VIRGILIO (*Institute of General Pathology, Padua, Italy*): Verapamil, a Ca^{2+} channel blocker, has been shown to decrease Ca^{2+} influx through the ATP-operated channel in human lymphocytes. We have made the same observation; however, other Ca^{2+} channel blockers, such as nifedipine, did not have the same effect. This suggests that the effects of verapamil are not specifically due to its known action on the voltage-gated channels.

Use of ATP following Shock and Ischemia[a]

IRSHAD H. CHAUDRY

Departments of Surgery and Physiology
Michigan State University
East Lansing, Michigan 48824

INTRODUCTION

Although shock has been recognized for over a century, a clear definition of this complex pathophysiological condition has evolved slowly.[1] Most of the earliest definitions of shock, although accurate to a certain extent, suffered from imprecision.[1] A current, precise definition describes shock as "inadequate blood flow to vital organs or the inability of the body cell mass to metabolize nutrients normally."[2] Thus, shock results from a sustained reduction in perfusion of capillaries and therefore of tissues and organs. Trauma, myocardial diseases, major operations, and severe infections all may lead to shock or circulatory collapse. Additionally, repair or transplantation of an organ, vessel injury, or atherosclerosis may lead to ischemia of one or more organs. Under such conditions, what happens to an organ with regard to function of its component cells and what happens to the cell of an organ with sufficiently diminished blood flow becomes a critical series of events in terms of survival of the organ and the individual.

Because of prolonged and sustained reduction in tissue perfusion during shock, various alterations in tissue metabolism, structure, and function occur at the systemic, cellular, and subcellular levels. The numerous alterations in cellular and subcellular function that occur following shock and ischemia have been dealt with in detail previously[3] and therefore will not be described here. In addition, a scheme of progressive cell injury with insults such as shock and ischemia has been previously described.[3] To summarize, it is sufficient to state that the major changes during shock occur in the microcirculation, affecting cell membrane transport and function, energy metabolism, and mitochondrial function. Such progressive and interrelated events eventually lead to cell swelling. When such swelling has developed, blood flow cannot rapidly return to normal despite fluid resuscitation. Part of the insult under such conditions may be caused by continued ischemia related to endothelial and parenchymal cell swelling along with sludging of blood. Pharmacologic interventions which improve blood flow, microcirculation, and cell function after such conditions may be necessary when volume resuscitation and conventional means of circulatory support have failed.[3] In this regard, many pharmacologic agents for the treatment of shock and ischemia are being evaluated.[3] As ATP in the cell exists as Mg-ATP, however, this article will focus on the use of ATP complexed with $MgCl_2$ as an adjunct for the treatment of low-flow conditions.

[a]This work was supported by Grant GM RO1 39519 from the National Institutes of Health.

TREATMENT OF SHOCK AND ISCHEMIA

Blood or fluid administration to provide or maintain an adequate vascular volume during the early stages of shock is often sufficient to correct the problem. If fluids alone are not successful, then the use of inotropic agents to increase cardiac output and improve blood flow may be required.[3] Along with these, various adjunctive agents and approaches such as administering buffering agents, vasoactive agents, steroids, and other means of circulatory support have been used.[3] If the foregoing nonspecific treatment of circulation becomes ineffective, then correction of specific alterations in cell function may be necessary.

According to Stoner and Threlfall,[4] shock is "the state following injury which is characterized by a fall in energy production." It is now clear that metabolic problems and alterations in cell function occur following shock and low-flow conditions. In view of this, an exciting area of clinical and animal research for the treatment of shock, now and in the future, is the correction of various metabolic problems and alterations in cell function that occur following shock and low-flow conditions. If a depressed circulation, such as that occurring with hypovolemic shock, alters cell functions by decreasing membrane capability and energy (ATP) production within the cell, then this can be considered a kind of energy crisis for the cell and various organ systems. Thus, attempts are being made to provide for these needs by giving substrates or compounds that improve energy production or supply within the cell following adverse circulatory conditions. These include the use of hypertonic glucose, increased glycogen stores, and other approaches such as administration of fructose 1,6-diphosphate and inosine, to name a few.[5]

Before proceeding further, it is pertinent to discuss why substrates and energy factors are important following shock. In this regard, a common feature of all forms of shock is thought to be an inadequate circulation with diminished blood flow to tissues.[3] As a result of this decreased blood flow to tissues, hypoxic conditions prevail and the production of ATP is significantly depressed at a time when more energy is needed to protect the cell against the insult. In addition, with reduced flow and perfusion, the availability and the delivery of substrate to target tissues decreases, and energy production is further compromised.

The essential role of ATP in membrane function, carbohydrate metabolism, tissue respiration, muscle contraction, and supplying energy for various intracellular reactions is well recognized.[3] It is clear that in order for ATP levels to be maintained in the tissues, oxidative phosphorylation must continue. This, however, is not the case during low-flow conditions because oxygen delivery as well as substrate delivery to tissues is decreased.[3] Moreover, because glycolysis cannot match ATP production in relation to its utilization during shock, the cellular adenine nucleotide levels decrease under such conditions.[3]

Although provision of a number of the above-mentioned substrates during shock may augment cellular energy levels, such substrates may not necessarily improve the microcirculation. Other agents that have been and are being evaluated include angiotensin-converting enzyme inhibitors, aprotinin, calcium channel blocking agents, coenzyme Q_{10}, naloxone, prostacycline, reduced glutathione, ribose, stroma-free hemoglobulin, free radical scavengers, and thromboxane inhibitors.[3] Although sporadic beneficial effects of the above agents have been reported, detailed studies of most of these agents on cellular function and microcirculation after shock have not been carried out.

A number of studies have shown that tissue ATP levels decrease following shock and ischemia and, coincident with this decrease, that there is altered cellular function.[1,3]

The provision of substrates during shock may be helpful; however, the basic problem during low-flow conditions is that an adequate amount of substrates cannot reach the target tissues because of reduced blood flow. Moreover, because the metabolism of substrates (for example, glucose) is decreased during shock,[5] it appears unlikely that provision of substrates per se could correct the cellular and metabolic alterations until blood flow increases. In this regard, it could be postulated that the provision of energy (that is, ATP or creatine phosphate) that does not have to pass through the glycolytic pathway, the Kreb cycle, and the electron transport chain to produce ATP may be the most advantageous and direct method for the treatment of shock and ischemia when fluid and vasoactive agents are not effective.

Because the resynthesis of ATP is a major rate-limiting factor following adverse circulatory conditions,[5] it seemed logical that the most direct approach for raising tissue ATP levels under these conditions would be to infuse ATP rather than administer agents that would synthesize it. Such an approach was used by Talaat and associates,[6] who showed that ATP infusion before or during shock improved the survival of animals. Subsequent studies by Sharma and Eiseman[7] showed that ATP was protective if given prior to hemorrhage but that it was not protective if it was administered following severe hemorrhagic shock. Our studies extended the work of these investigators[6,7] by introducing one essential difference: we gave $MgCl_2$ along with ATP.[3] ATP, which is a biological complexing agent,[8] may, when given alone, chelate divalent cations from the vascular system and produce a different hemodynamic effect. Such undesirable effects may be eliminated by giving an equimolar amount of $MgCl_2$ along with ATP.[8] Moreover, studies have shown that magnesium *in vitro* and *in vivo* inhibits the deamination and dephosphorylation of ATP by tissues.[8] Thus, if ATP is given along with $MgCl_2$, a higher concentration of ATP could be available to tissues than when ATP is given alone.

Although many pharmacologic agents are being evaluated,[3] the focus here will be on the use of ATP-$MgCl_2$ as an adjunct in the treatment of shock and ischemia. Our hypothesis has been that ATP-$MgCl_2$ can help parenchymal cells, both directly and through an improvement in microcirculation. The net result would thus be an improvement in organ blood flow and an amelioration of the depressed cell membrane function and energy metabolism following adverse circulatory conditions.

EFFECT OF ATP-$MgCl_2$ ON SURVIVAL AND ORGAN FUNCTION FOLLOWING SHOCK AND ISCHEMIA

Studies from a number of laboratories, including our own, have shown that infusion of ATP-$MgCl_2$ following hemorrhagic shock,[1,3,9–11] severe burns,[12] sepsis-peritonitis,[3] postischemic hepatic failure,[3] bowel ischemia,[13] and endotoxic shock[14,15] significantly improved the survival of animals. The beneficial effects on survival following shock were not observed if ATP or $MgCl_2$ alone or adenosine-$MgCl_2$ were infused.[3] From these results, it could be concluded that ATP together with $MgCl_2$ is required for the treatment to be effective.

ATP-$MgCl_2$ has been shown to accelerate the recovery of hepatic and renal function following acute hepatic[3] and renal failure,[3,16] respectively. ATP-$MgCl_2$ has also been shown to improve skin flap survival.[17,18] Moreover, studies have demonstrated that kidneys subjected to episodes of warm ischemia could be salvaged by the addition of ATP-$MgCl_2$ to the perfusate.[19] ATP-$MgCl_2$ has been effective in hastening renal recovery from various toxic injuries.[20,21] Studies have also suggested that ATP-$MgCl_2$ may be useful in preventing ischemic damage.[22] ATP-$MgCl_2$ appears to be a promising adjunct to cardioplegia during extended periods of aortic cross-clamping necessitated by complex cardiac procedures.[23] Additional experiments indicated that infusion of ATP-$MgCl_2$ during reperfusion following hypothermic myocardial ischemia may help ameliorate the reperfusion injury.[24]

Studies were also conducted to compare the effects of ATP-$MgCl_2$ and adenosine-$MgCl_2$ on renal function following ischemia.[25] The results indicated that although adenosine-$MgCl_2$ treatment provided a transient improvement in renal function, it failed to provide a sustained improvement in renal function or attain control values compared with ATP-$MgCl_2$.[25] Such findings led to the conclusion that the salutary effects of ATP-$MgCl_2$ following warm ischemia in the kidney also are not mediated by adenosine.

EFFECTS OF ATP-$MgCl_2$ ON ALTERED CELLULAR FUNCTIONS FOLLOWING SHOCK AND ISCHEMIA

Although the above-mentioned studies indicate that ATP-$MgCl_2$ administration improves the survival of animals following shock and ischemia, the issue that should be discussed is whether infused ATP-$MgCl_2$ corrects the abnormalities in cell function after shock or ischemia. The studies of Kraven *et al.*[26] have shown that infused ATP-$MgCl_2$ decreased tissue lactate production. These studies have also suggested that the decrease was due to a direct intracellular effect of administered ATP. Moreover, the work of Machiedo *et al.*[27] has indicated that treatment of animals following shock with ATP-$MgCl_2$ returns the altered membrane permeability toward normal.[27] Additional studies indicated that exogenously administered ATP-$MgCl_2$ can reverse the inhibition of ornithine metabolism and the changes in tissue lactate levels during shock.[27] Because both of these are intracellular ATP-dependent reactions, Machiedo *et al.*[27] were led to conclude that ATP-$MgCl_2$ administration after hemorrhagic shock either replenishes intracellular ATP levels or returns the altered cell membrane permeability toward normal or both. In this regard, results from our laboratory indicated that administration of ATP-$MgCl_2$ following hemorrhagic shock significantly increased tissue ATP levels.[3] Additional work by Garvin *et al.*[28] indicated that cortical ATP levels were restored after 24 hr in both perfused and cold storage kidneys when an intraaortic infusion of ATP-$MgCl_2$ was given after ischemia. Moreover, it has been shown that ATP-$MgCl_2$ reverses ischemically induced hepatic cellular edema,[3,29] improves the electrolyte balance,[30] and restores the depressed hepatocellular function after ischemia.[3]

Studies from our laboratory have shown that following hemorrhagic shock, the mitochondrial respiratory control ratio was significantly depressed and treatment of animals after hemorrhagic shock with ATP-$MgCl_2$ resulted in restoration of the respiratory control ratios.[31] Infusion of ATP-$MgCl_2$ after hepatic ischemia and reflow also resulted in a progressive and significantly higher adenine nucleotide translocase activity in liver mitochondria, but neither dopamine nor papaverine (vasodilatory controls) was effective.[32] In addition, mitochondrial free fatty acids and calcium levels increased after ischemia and reflow, and treatment with ATP-$MgCl_2$ resulted in significantly lower free fatty acid levels and normalized mitochondrial calcium levels.[32]

Thus, mitochondrial function can be significantly improved by treatment of animals with ATP-$MgCl_2$ after shock or ischemia. This was also associated with significantly higher hepatic ATP levels and energy charge.[33]

At the cellular level, infused ATP-$MgCl_2$ has been shown to reverse the tissue insulin resistance in skeletal muscle following hemorrhagic shock,[3] and to improve the sodium-potassium balance in the liver following ischemia.[29,30] These functional changes were accompanied by changes in hepatic ultrastructure, which deteriorated progressively after ischemia and reflow in saline-treated animals.[33] With ATP-$MgCl_2$ treatment, however, the hepatic ultrastructure was normal at 1 and 20 hr after reflow and treatment.[33] Studies have indicated that ATP-$MgCl_2$ improves hepatic blood flow following hepatic ischemia[33] and endotoxemia.[34] The effects of ATP-$MgCl_2$ administration on hepatic surface microvasculature after ischemia were investigated using intravital fluorescence microscopy. Such studies demonstrated that ATP-$MgCl_2$ improves hepatic blood flow by ameliorating the progressive loss of perfused capillaries seen during the reperfusion.[35] Thus, ATP-$MgCl_2$ improves microcirculatory blood flow.

Nuclear magnetic resonance studies of isolated kidneys indicated that postischemic administration of ATP-$MgCl_2$ corrects the intracellular acidosis, restores the tissue ATP stores, and prevents the deterioration in perfusate flow that occurs following ischemia and reflow.[36] Thus, it can be concluded that ATP-$MgCl_2$ prevents the reperfusion injury.

Recent studies by Korthuis *et al.*[37] have indicated that vascular injury in dogs during ischemia and reperfusion can be prevented by pretreatment of the dogs with ATP-$MgCl_2$. Studies have also indicated that administration of ATP-$MgCl_2$ not only restores immunoresponsiveness after burn injury,[3] but also restores depressed reticuloendothelial function after hepatic ischemia[3] and sepsis-peritonitis.[3] More recent experiments from our laboratory have indicated that the decreased interleukin-2 levels following hemorrhage and resuscitation can be restored if ATP-$MgCl_2$ is given to animals following hemorrhagic shock.[38]

MECHANISM OF THE BENEFICIAL EFFECTS OF ATP-$MgCl_2$

Although a number of studies have clearly shown that ATP-$MgCl_2$ increases intracellular ATP levels and improves cellular function following shock and ischemia,[3] it is pertinent to discuss whether infused ATP-$MgCl_2$ crosses the cell plasma membranes or whether the crossing of these membranes is required for ATP-$MgCl_2$ to produce its various beneficial effects. Although a large number of studies have shown that ATP can cross the cell plasma membrane,[39–41] this agent may not produce beneficial effects by providing a large amount of energy, but perhaps by catalyzing improvement of energy production, by improving the microcirculation, or by exerting actions through extracellular effects mediated by cell surface ATP receptors[3] or phosphorylation of membrane proteins.[3] Although the precise mechanisms of action of ATP-$MgCl_2$ are not entirely clear, the available information clearly indicates that infused ATP-$MgCl_2$ improves cellular function directly, through long-term improvement in the microcirculation, not simply through vasodilation, as produced by dopamine or papaverine.[32,33]

IS ATP WITHOUT $MgCl_2$ EFFECTIVE?

Studies have shown that administration of ATP or $MgCl_2$ alone following hemorrhagic shock or hepatic ischemia failed to improve the survival of animals and the recovery of renal function following renal ischemia.[3,8,42] Based on this information, it would appear that ATP together with $MgCl_2$ treatment following shock and ischemia is required for this agent to be effective. Because ATP and ADP in the cell exist as magnesium complexes, and because most ATP reactions require not only ATP as a substrate but also magnesium as a cofactor, it is possible that parallel changes in the levels of these two substances may contribute to the depression of cellular functions observed following shock and low-flow conditions. Additional studies from our laboratory have indicated that tissue and mitochondrial magnesium levels decreased significantly following ischemia and reflow and that ATP-$MgCl_2$, but not ATP or $MgCl_2$ alone, after ischemia significantly increased tissue and mitochondrial magnesium levels.[8] Tissue ATP and ADP levels also follow the same trend. Thus, it could be concluded that 1) a significant loss of tissue ATP and ADP occurs, together with decreased tissue and mitochondrial magnesium levels, after ischemia and reflow and that 2) ATP together with $MgCl_2$ treatment is required for increasing cellular and mitochondrial magnesium levels and for improving cellular functions under such conditions.[42]

EFFECT OF ATP-$MgCl_2$ ON MYOCARDIAL AND TOTAL BODY OXYGEN CONSUMPTION

Studies from our laboratory as well as from others have indicated that administration of ATP-$MgCl_2$ increases cardiac output and produces peripheral vasodilatation.[43,44] It could be argued, however, that if such an agent is administered under conditions of compromised cardiac function, the increased myocardial performance together with peripheral vasodilatation might create a deleterious situation in which oxygen demand exceeds oxygen supply. To investigate this possibility, additional studies were conducted which indicated that infusion of ATP-$MgCl_2$ decreased not only total body oxygen consumption,[45] but myocardial oxygen consumption during normovolemic as well as hypovolemic conditions.[46] These decreases occurred in spite of increased coronary flow and cardiac output.[46] Because the decreased oxygen consumption was greater than what would be predicted from the observed afterload reduction, the results indicated that ATP-$MgCl_2$ exerts effects on myocardial oxygen demand over and above the effects exerted through its peripheral vasodilatory action. The combination of increased cardiac output, increased coronary flow, and decreased myocardial oxygen consumption clearly supports a role for the therapeutic use of ATP-$MgCl_2$ during low-flow conditions and during clinical instances of coronary insufficiency.

PRECAUTIONS IN ADMINISTERING ATP-$MgCl_2$

The precautions, hazards, and controversies in using ATP-$MgCl_2$ have been reviewed in detail in recent publications[3,47] and will not be repeated here. Nonetheless, some of the important precautions that should be taken in administering this agent will be mentioned. As with any potent vasodilator, this agent should not be given as a bolus because doing so could lead to severe tachycardia and even cardiac arrest. Careful monitoring of blood pressure should be undertaken during infusion to avoid significant hypotension. If this agent is given following shock, efforts should be made to restore the blood pressure to normal with volume resuscitation prior to the initiation of ATP-$MgCl_2$ administration. If ATP-$MgCl_2$ is given at a rate that decreases blood pressure precipitously, then deleterious rather than beneficial effects would be expected to occur because a secondary hypotension would be created under these conditions. Support for this comes from the studies in which beneficial effects of ATP-$MgCl_2$ were not observed.[47] If the blood pressure drops rapidly, the infusion of ATP-$MgCl_2$ should be stopped for a few minutes to let the blood pressure recover. Once the blood pressure recovers, the administration can be restarted at a lower rate of infusion. The doses of ATP-$MgCl_2$ required for effective therapy of different adverse circulatory conditions have been described in detail previously.[47,48]

CLINICAL TRIALS OF ATP-$MgCl_2$

The safety and hemodynamic responses of ATP-$MgCl_2$ administration in man have been determined. The results indicated that intravenous infusion of this agent in normal human volunteers increased cardiac output by as much as 131% without affecting mean blood pressure.[49] The increased cardiac output was dependent on the ATP-$MgCl_2$ infusion rate but not on the total dose of ATP-$MgCl_2$ infused. Stroke volume index also increased with ATP-$MgCl_2$ infusion, but there were no changes in blood glucose or chemistries during or after the infusion of ATP-$MgCl_2$. Based on these studies, it was concluded that ATP-$MgCl_2$ administration is safe in healthy humans and increases cardiac output without producing hypotension. After the administration of ATP-$MgCl_2$ in normal volunteers was found to be safe, the effects of intracoronary ATP-$MgCl_2$ administration on coronary sinus blood flow and myocardial oxygen consumption in patients with coronary artery disease were investigated.[50] The results indicated that during the infusion of ATP-$MgCl_2$, at rates that did not alter the heart rate, cardiac output, or wedge pressure, there was a 65% increase in coronary sinus blood flow with a concomitant 27% reduction in myocardial oxygen consumption. This would indicate that ATP-$MgCl_2$ is a demand-independent coronary vasodilator.[50] The reduction in myocardial oxygen consumption in the absence of changes in the measured determinants of myocardial oxygen demand suggest a possible oxygen-sparing effect of ATP-$MgCl_2$. Thus, ATP-$MgCl_2$ shows favorable characteristics for potential applications in patients with coronary artery disease.

Hirasawa and co-workers[51,52] have been using ATP-$MgCl_2$ in Japan in patients with acute renal failure as well as in patients with multiple organ failure,[53] and the reported results appear very encouraging. In this country, ATP-$MgCl_2$ has also been given to some patients with acute renal failure, acute hepatic failure, trauma, and low cardiac output, and the results obtained so far are very promising (unpublished observations). A clinical trial of ATP-$MgCl_2$ treatment in patients with various adverse circulatory conditions is underway at several institutions in our country.

SUMMARY

The available information indicates that shock and ischemia are associated with such phenomena as diminished microcirculatory blood flow and diminished metabolic (including ATP levels) and cellular capabilities, and that these phenomena are associated with altered cellular functions. Infusion of ATP-$MgCl_2$ as an adjunct following shock or ischemia significantly improves microcirculatory blood flow, tissue and mitochondrial magnesium levels, tissue ATP levels, cellular functions, and overall survival of animals. Administration of ATP or $MgCl_2$ alone after such conditions was ineffective in improving cellular functions or the survival of animals. Thus, it could be concluded that ATP together with $MgCl_2$ is required for an effective treatment. ATP-$MgCl_2$ can be administered safely in normal volunteers as well as in patients following various adverse circulatory conditions. Administration of this agent in humans produces positive inotropic, negative chronotropic, and peripheral vasodilatory actions, which clearly suggests the potential use of this agent in patients with low-flow conditions or organ ischemia. Clinical trials of ATP-$MgCl_2$ treatment in patients with various adverse circulatory conditions are underway at several institutions in this country as well as in Japan.

REFERENCES

1. CHAUDRY, I. H. & A. E. BAUE. 1982. Overview of hemorrhagic shock. *In* Pathophysiology of Shock, Anoxia and Ischemia. R. A. Cowley & B. F. Trump, Eds.: 203-219. William & Wilkins. Baltimore, MD.
2. MACLEAN, L. D. 1977. Causes and management of circulatory collapse. *In* Davis-Christopher Textbook of Surgery. D. C. Sebastian, Ed.: 65-94. Saunders. Philadelphia, Pa.
3. CHAUDRY, I. H. 1983. Cellular mechanisms in shock and ischemia and their correction. Am. J. Physiol. **245:** R117-R134.
4. STONER, H. B. & C. J. THRELFALL. 1954. Effect of nucleotide and ischemic shock on the level of energy-rich phosphates in the tissues. Biochem. J. **58:** 115-121.
5. CHAUDRY, I. H. & A. E. BAUE. 1980. The use of substrates and energy in the treatment of shock. *In* Advances in Shock Research. A. M. Lefer, T. M. Saba & L. M. Mela, Eds. Vol. 3: 27-46. Alan R. Liss. New York, NY.
6. TALAAT, S. M., W. H. MISSION & J. A. SCHILLING. 1964. Effects of adenosine triphosphate administration in reversible hemorrhagic shock. Surgery **55:** 813-819.
7. SHARMA, G. P. & B. EISEMAN. 1966. Protective effect of ATP in experimental hemorrhagic shock. Surgery **59:** 66-75.
8. CHAUDRY, I. H., M. G. CLEMENS & A. E. BAUE. 1986. The role of ATP-magnesium in ischemia and shock. Magnesium **5:** 211-220.
9. DISTAZIO, J., W. B. MALEY, H. THOMPSON, S. SEMBRAT & J. STREMPLE. 1980. Effect of ATP-$MgCl_2$-glucose administration during hemorrhagic shock on cardiac metabolism function and survival. *In* Advances in Shock Research. A. M. Lefer, T. M. Saba & L. M. Mela, Eds. Vol. 3: 153-166. Alan R. Liss. New York, NY.

10. HIRASAWA, H., S. ODA, H. HAYASHI, Y. OHTAKE, M. ODAKA & H. SATO. 1983. Improved survival and reticuloendothelial function with intravenous ATP-$MgCl_2$ following hemorrhagic shock. Circ. Shock **11:** 141-148.
11. KRAVEN, T., B. F. RUSH, A. GHOSH & M. ADAMS-GRIFFIN. 1979. Improved survival and metabolic changes in a rat shock model produced by ATP-$MgCl_2$. Curr. Surg. **36:** 435-437.
12. ZAKI, M. S., J. L. BURKE & R. L. TRELSTAD. 1978. Protective effects of adenosine triphosphate administration in burns. Arch. Surg. **113:** 605-610.
13. CIKRIT, D., K. GROSS, S. KATZ, D. PLAGER, D. ROSS, M. WOLFE, T. R. WEBER & J. L. GROSFELD. 1983. Comparative effects of cytoprotective agents in bowel ischemia. Surg. Forum **34:** 208-210.
14. FULTON, R. L. 1974. Prevention of endotoxin death with nicotinamide and adenosine triphosphate. Surg. Forum **25:** 17-19.
15. FILKINS, J. P. & B. J. BUCHANAN. 1977. Protection against endotoxin shock and impaired glucose homeostasis with ATP. Circ. Shock **4:** 253-258.
16. ANDREWS, P. M. & A. K. COFFEY. 1983. Protection of kidneys from acute renal failure resulting from normothermic ischemia. Lab. Invest. **49:** 87-98.
17. BOSS, W. K., S. FUSI, I. H. CHAUDRY, M. G. CLEMENS, G. CHATER, D. LAUB, C. B. CUONO & S. ARIYAN. 1984. Improved skin flap survival with adenosine triphosphate-magnesium chloride. Surg. Forum **35:** 576-578.
18. CIKRIT, D., D. BILLMORE, M. R. LOPEZ, R. W. WEST, J. L. GROSFELD & D. SMITH. 1984. Beneficial effect of exogenous adenosine triphosphate on skin flap viability. Surg. Forum **35:** 378-480.
19. LYTTON, B., V. R. VAISBORT, W. B. GLAZIER, I. H. CHAUDRY & A. E. BAUE. 1981. Improved renal function using ATP-$MgCl_2$ in preservation of canine kidneys subjected to warm ischemia. Transplantation **31:** 187-189.
20. SUMPIO, B. E., I. H. CHAUDRY & A. E. BAUE. 1985. Reduction of drug-induced nephrotoxicity by ATP-$MgCl_2$. I. Effects on the *cis*-diammine-dichloroplatinum-treated isolated perfused kidneys. J. Surg. Res. **38:** 429-437.
21. SUMPIO, B. E., I. H. CHAUDRY & A. E. BAUE. 1985. Reduction of drug-induced nephrotoxicity by ATP-$MgCl_2$. II. Effects on gentamicin-treated isolated perfused kidneys. J. Surg. Res. **38:** 438-445.
22. MCDONAGH, P. F., H. LAKS, I. H. CHAUDRY & A. E. BAUE. 1984. Improved myocardial recovery from ischemia: Treatment with low-dose adenosine triphosphate-magnesium chloride. Arch. Surg. **119:** 1378-1384.
23. KOPF, G. S., I. H. CHAUDRY, S. CONDOS & A. E. BAUE. 1986. Improved myocardial performance following prolonged ischemia with ATP-$MgCl_2$ cardioplegia. Surg. Forum **37:** 234-236.
24. KOPF, G. S., I. H. CHAUDRY, S. CONDOS & A. E. BAUE. 1987. Reperfusion with ATP-$MgCl_2$ following prolonged ischemia improves myocardial performance. J. Surg. Res. **43:** 114-117.
25. SUMPIO, B. E., M. J. HULL, A. E. BAUE & I. H. CHAUDRY. 1987. Comparison of effects of ATP-$MgCl_2$ and adenosine-$MgCl_2$ on renal function following ischemia. Am. J. Physiol. **252:** R388-R393.
26. KRAVEN, T., B. RUSH, G. J. SLOTMAN & M. ADAMS-GRIFFIN. 1979. Permeability of the shock cell to ATP-$MgCl_2$. Surg. Forum **30:** 7-9.
27. MACHIEDO, G. W., S. GHUMAN, B. F. RUSH, T. KRAVEN & G. DIKDAN. 1981. The effect of ATP-$MgCl_2$ infusion on hepatic cell permeability and metabolism following hemorrhagic shock. Surgery **90:** 328-335.
28. GARVIN, P. J., M. JELLINEK, R. MORGAN & J. E. CODD. 1981. Renal cortical levels of adenosine triphosphate: Restoration after prolonged ischemia by *in situ* perfusion of ATP-$MgCl_2$. Arch. Surg. **116:** 221-224.
29. HIRASAWA, H., M. OHKAWA, H. KOBAYASHI, M. ODAKA & H. SATO. 1980. Reversal of ischemia-induced hepatic cellular edema by administration of ATP-$MgCl_2$. Surg. Forum **31:** 10-12.
30. OHKAWA, M., I. H. CHAUDRY, M. G. CLEMENS & A. E. BAUE. 1982. Improvement in water and electrolyte homeostasis by ATP-$MgCl_2$ after ischemia. Circ. Shock **9:** 162-163.

31. CHAUDRY, I. H., M. OHKAWA, M. G. CLEMENS & A. E. BAUE. 1983. Alterations in electron transport and cellular metabolism with shock and trauma. *In* Molecular and Cellular Aspects of Shock and Trauma. A. M. Lefer & W. Schumer, Eds.: 67-88. Alan R. Liss. New York, NY.
32. CHAUDRY, I. H., M. OHKAWA & M. G. CLEMENS. 1984. Improved mitochondrial function following ischemia and reflow by ATP-$MgCl_2$. Am. J. Physiol. **246:** R799-R804.
33. OHKAWA, M., M. G. CLEMENS & I. H. CHAUDRY. 1983. Studies on the mechanism of beneficial effects of ATP-$MgCl_2$ following hepatic ischemia. Am. J. Physiol. **244:** R695-R702.
34. ASHER, E. F., R. N. GALLISON & D. E. FRY. 1983. Improvement of hepatic nutrient blood flow with ATP-$MgCl_2$ in endotoxin rats. Circ. Shock **10:** 248.
35. CLEMENS, M. G., P. F. MCDONAGH, I. H. CHAUDRY & A. E. BAUE. 1985. Hepatic microcirculatory failure after ischemia and reperfusion: Improvement with ATP-$MgCl_2$ treatment. Am. J. Physiol. **248:** H804-H811.
36. SUMPIO, B. E., I. H. CHAUDRY & A. E. BAUE. 1985. Adenosine triphosphate-magnesium chloride ameliorates reperfusion injury following ischemia as determined by phosphorus nuclear magnetic resonance. Arch. Surg. **120:** 233-240.
37. KORTHUIS, R. J., M. B. GRISHAM, B. J. ZIMMERMAN, D. N. GRANGER & A. E. TAYLOR. 1988. Vascular injury in dogs during ischemia-reperfusion: Improvement with ATP-$MgCl_2$ pretreatment. Am. J. Physiol. **254:** H702-H708.
38. STEPHAN, R. N., C. A. JANEWAY, R. E. DEAN & I. H. CHAUDRY. 1987. Decreased interleukin-2 production following simple hemorrhage and its restoration with ATP-$MgCl_2$ treatment. Magnesium **6:** 163.
39. CHAUDRY, I. H. 1982. Does ATP cross the cell plasma membrane? Yale J. Biol. Med. **55:** 1-10.
40. CHAUDRY, I. H., M. G. CLEMENS & A. E. BAUE. 1985. Uptake of ATP by tissues. *In* Purines: Pharmacology and Physiological Roles. T. W. Stone, Ed.: 115-124. Macmillan. Hampshire.
41. PANT, H. C., S. TERAKAWA, T. YOSHIOKI, I. TASAKI & H. GAINER. 1979. Evidence for the utilization of extracellular [γ-^{32}P]ATP for the phosphorylation of intracellular proteins in the squid giant axon. Biochim. Biophys. Acta **582:** 102-114.
42. CHAUDRY, I. H. 1989. ATP-$MgCl_2$ and liver blood flow following shock and ischemia. *In* Progress in Clinical and Biological Research. Perspectives in Shock Research, Metabolism, Immunology, Mediators, and Models. J. C. Passmore, S. M. Reichard, D. G. Reynolds & D. E. Traber, Eds. Vol. 299: 19-31. Alan R. Liss. New York, NY.
43. KLAY, J. W., I. H. CHAUDRY, A. S. GEHA & A. E. BAUE. 1980. Improved myocardial performance with adenosine triphosphate-$MgCl_2$ infusion. Surg. Forum **31:** 260-262.
44. MCGOVERN, P. J., B. F. RUSH, B. W. MACHIDIO & T. KRAVEN. 1981. ATP-$MgCl_2$ improves cardiac output after shock. Surg. Forum **32:** 1-3.
45. KRINSLEY, J., I. H. CHAUDRY, J. LOKE, P. SNYDER, J. VIRGULTO, A. GEHA & A. BAUE. 1985. Effect of adenosine triphosphate-magnesium chloride infusion on oxygen consumption and hemodynamics in anesthetized dogs. Am. Rev. Respir. Dis. **131:** A310.
46. CLEMENS, M. G., I. H. CHAUDRY & A. E. BAUE. 1985. Increased coronary flow and myocardial efficiency with systemic infusion of ATP-$MgCl_2$. Surg. Forum **36:** 244-246.
47. HAYASHI, I. H., I. H. CHAUDRY, M. G. CLEMENS & A. E. BAUE. 1986. Hepatic ischemia models for determining the effects of ATP-$MgCl_2$. J. Surg. Res. **40:** 167-175.
48. CHAUDRY, I. H. & A. E. BAUE. 1982. Energy problems during shock and their correction with adenosine triphosphate-magnesium chloride. *In* Role of chemical mediators in the pathophysiology of acute illness and injury. R. McConn, Ed.: 361-376. Raven Press. New York, NY.
49. CHAUDRY, I. H., J. R. KEEFER, P. BARASH, M. G. CLEMENS, G. KOPF & A. E. BAUE. 1984. ATP-$MgCl_2$ infusion in man: Increased cardiac output without adverse systemic hemodynamic effects. Surg. Forum **35:** 13-15.
50. WOHLGELERNTER, D., C. JAFFE, M. CLEMEN, L. YOUNG, M. G. CLEMENS & I. H. CHAUDRY. 1985. Effects of ATP-$MgCl_2$ on coronary blood flow and myocardial oxygen consumption. Circulation **72**(III): 315.
51. HIRASAWA, H., M. ODAKA, K. SOEDA, H. KOBAYASHI, Y. OHTAKE, S. ODA, S. KO-

BAYASHI & H. SATO. 1983. Experimental and clinical study of ATP-$MgCl_2$ administration for postischemic acute renal failure. Clin. Exp. Dialysis Apheresis **7:** 37-47.
52. HIRASAWA, H., K. SOEDA, M. OHKAWA, S. KOBAYASHI, N. MUROTANI, M. ODAKA & H. SATO. 1985. A randomized clinical trial of ATP-$MgCl_2$ for postischemic acute renal failure. Circ. Shock **13:** 66.
53. HIRASAWA, H., T. SUGAI, Y. OHTAKE, S. ODA, H. SHIGA, T. AOE & M. OHKAWA. 1989. Effect of impaired hepatic mitochondrial function (HMF) on systemic metabolism in multiple organ failure (MOF) patients and its treatment with ATP-$MgCl_2$. Circ. Shock **27:** 361.

DISCUSSION OF THE PAPER

C. NEELY: 1) What is the longest period of time that you have measured survival in animals treated with ATP-$MgCl_2$? 2) Have you measured serum Mg^{2+} levels in animals or humans during ATP-$MgCl_2$ infusion?

CHAUDRY: 1) Two weeks. 2) No.

S. C. SILVERSTEIN (*Columbia University, New York, NY*): Could high levels of ADP appear in the blood after the infusion of ATP-$MgCl_2$?

T. FORRESTER (*St. Louis University Medical Center, St. Louis, MO*): D. C. B. Mills showed in 1966 that the ATP in blood was broken down directly by a dipyrophosphorylase to AMP. The development of high levels of ADP are thus avoided.

P. A. WARD (*University of Michigan Medical School, Ann Arbor, MI*): Dr. Chaudry, is it possible that these data could be explained by conversion of ATP-$MgCl_2$ to adenosine?

CHAUDRY: I do not believe so. We infused equimolar adenosine-$MgCl_2$ in separate studies and did not find any beneficial effects on survival or for a sustained improvement of renal function following renal ischemia.

E. KORNECKI (*SUNY Health Science Center at Brooklyn, New York, NY*): The conversion of ATP to ADP may result in concentrations of ADP capable of interacting with platelets to produce a refractory state resulting in loss of platelet function to ADP. Have you tested whether the infusion of Mg-ATP in human volunteers resulted in platelets that became refractory to ADP *in vivo?*

CHAUDRY: We have not tested whether the platelets are refractory to ADP. Nonetheless, we did check whether ATP-$MgCl_2$ administration caused any platelet aggregation. It did not.

Y. H. EHRLICH (*College of Staten Island, New York, NY*): What is the experimental evidence for your definitive conclusion that the beneficial effect of ATP-$MgCl_2$ infusion is the provision of an intracellular energy source? What is the mechanism involved?

CHAUDRY: We are not saying that ATP-$MgCl_2$ acts solely as an intracellular energy source. It does, however, cross the plasma cell membrane as ATP, and it does increase intracellular ATP. Nonetheless, ATP-$MgCl_2$ also phosphorylates cell surface receptors, improves the microcirculation, and it also stimulates the resynthesis of ATP. Thus, the mechanism of action of ATP-$MgCl_2$ following shock and ischemia is due to a combination of the following: 1) provision of ATP, 2) improvement in microcirculation, 3) reduction of cell swelling, and 4) priming of the resynthesis of ATP.

J. BARANKIEWICZ (*Gensia Pharmaceuticals, San Diego, CA*): Did you observe uric acid overproduction after ATP infusion?

CHAUDRY: Uric acid levels did not increase significantly after ATP-$MgCl_2$ was given continuously for approximately 2 hr. If, however, ATP-$MgCl_2$ was given continuously for 7 days in patients with cancer, the uric acid levels increased. The dose of ATP-$MgCl_2$ that we used for the treatment of shock and ischemia, however, did not produce any significant elevation in uric acid.

Mechanisms of Anticancer Activities of Adenine Nucleotides in Tumor-Bearing Hosts

ELIEZER RAPAPORT

Department of Microbiology
Boston University School of Medicine
Boston, Massachusetts 02118

INTRODUCTION

The role of ATP as an extracellular mediator of a variety of physiological functions, including those originating intraluminally as well as those originating extraluminally, was established during the 1970s by Burnstock and his colleagues.[1-4] The involvement of ATP in intracellular energetics as well as in the allosteric regulation of intracellular protein functions is widely known. Acid-soluble adenine and guanine nucleotides have been implicated in the regulation of specific sets of intracellular metabolic reactions in bacterial and animal cells.[5-7] Because of their metabolic lability, which allows their intracellular pools (steady state concentrations) to fluctuate rapidly in response to extracellular conditions that affect growth or cellular metabolism, these compounds are suitable for their role as intracellular signals.[5-8] Polypeptides, however, because of their higher metabolic stability, evolved into intercellular signals or mediators affecting responder cells primarily by interacting with specific membrane receptors.[5,9] The emergence of ATP as an extracellular signal was commensurable with the concept requiring extracellular mediation by a class of metabolically labile compounds, a concept based on the accumulating data suggesting pinpoint accuracy in the physiological execution of vascular and extravascular functions. More importantly, two types of purinergic receptors (purinoceptors) were identified and distinguished on a variety of cells and tissues.[4,10]

More recently we[11,12] and others[13-16] have demonstrated that the growth of human and animal tumor cells exposed to extracellular ATP, under several *in vitro* conditions, was markedly inhibited. The cytostatic and cytotoxic effects of extracellular ATP against tumor cells in serum-containing media were much more pronounced than the effects observed after similar treatments of untransformed cells.[11,13,16] Furthermore, after the ectoenzymatic activities of tumor cells in catalyzing the degradation of extracellular ATP were taken in account, together with the soluble catabolic enzymatic activities present in the serum-supplemented growth media and the lack of growth-inhibitory activities of the degradation products, adenosine and AMP, it was concluded that very low levels of extracellular ATP were actually sufficient to yield substantial tumor cell growth inhibition *in vitro*.[12] Studies with live animals were then initiated to determine the distribution of ATP administered intravenously to tumor-bearing rats and rabbits. Because we have previously demonstrated that low levels of adenine nucleotides can permeate the plasma membrane of certain human and animal tumor cells, an attempt was made to utilize ^{99m}Tc-radiolabeled-chelated forms of adenine nucleotides for tumor imaging.[17] Low levels of ^{99m}Tc-labeled ATP and ^{99m}Tc-labeled diadenosine 5′,5‴-P^1,P^4-tetraphosphate (Ap_4A) injected intravenously into rats or rabbits were suitable for the *in vivo* detection of tumors by radionuclide imaging.[17] The incorporation of the radionuclide (^{99m}Tc) as a chelate into the tumor was dependent on the entry of low levels of nucleotides into the tumor cells without prior degradation of these nucleotides, because the nonchelated ^{99m}Tc or nonspecific ^{99m}Tc chelates were not effective as tumor-seeking agents. These studies also provided data regarding the tissue distribution of ^{99m}Tc-ATP, which presumably indicates the biodistribution of ATP after intravenous injections. The results demonstrated an extremely slow clearance of ^{99m}Tc-ATP from the blood (cellular) in comparison to another ^{99m}Tc-chelated adenine nucleotide, ^{99m}Tc-Ap_4A, or in comparison to a nonspecifically chelated ^{99m}Tc, ^{99m}Tc-labeled diethylenetriaminepentaacetic acid (DTPA).[17] A suggestion that ATP could semipermeabilize red blood cells (RBCs) was made earlier by another group, which showed that under certain conditions extracellular ATP could induce influx and mostly efflux of adenine nucleotides associated with RBCs.[18]

After these initial studies, the effects of adenine nucleotides *in vivo* in murine tumor models were elucidated. Adenine nucleotides (AMP, ADP, or ATP), but not adenosine, were demonstrated to yield elevated blood (total cellular) and plasma compartment (extracellular) pools of ATP after their intraperitoneal administration into mice.[19–21] Elevated RBC and blood plasma ATP pools were correlated with the inhibition of tumor growth and the inhibition of host weight loss in tumor-bearing murine hosts.[19–21] The two anticancer activities of adenine nucleotides do not exhibit a cause-effect relationship, although both are mediated by the expansion of RBC and blood plasma ATP pools.[19–21] The expansions of RBC and plasma compartment pools of ATP were unexpected because of the rapid degradation of adenine nucleotides in the vascular bed. The catabolic activities, which are catalyzed by a variety of ectoenzymatic and soluble enzymatic activities present intraluminally, led to the assumption that the pharmacological activities of externally introduced adenine nucleotides were due to the major breakdown product of these nucleotides, adenosine. The mechanism of expansion of RBC ATP pools was elucidated and was shown to involve prior expansion of total hepatic ATP pools after intraperitoneal adminstration of AMP, ADP, or ATP.[21] The adenosine precursor for enhanced RBC ATP synthesis originates in the turnover of expanded liver ATP pools and is taken up by circulating RBCs in the hepatic sinusoids.[21] The two- to fourfold expansion of RBC ATP pools lasts for several hours with specific continuous slow release of micromolar amounts of ATP from RBCs into the blood plasma compartment. Although ATP degrades rapidly in the blood plasma compartment, it is sufficient to elevate the steady state extracellular ATP levels.[20,21] The anticancer activities of adenine nucleotides may be the result of the direct effects of extracellular ATP on the tumor cells or may be mediated by the effects of extracellular ATP on host functions, because many of the functions that could conceivably affect tumor growth are known to be influenced by elevated blood plasma ATP pools.[22–24] Effects of extracellular ATP on the functions of a variety of immune and nonimmune cells that contribute to host defense mechanisms have also been established,[25–30] and could favorably influence the response of these cells after the administration of adenine nucleotides and elevation of blood plasma ATP pools in tumor-bearing hosts. The data reviewed in this paper indicate that *in vivo* administration of ATP and other adenine nucleotides yields a sustained "secondary wave" of extracellular ATP resulting from the immediate rapid degradation of the administered adenine nucleotide followed by expansion of RBC ATP pools and release of micromolar levels of ATP from these RBCs into the extracellular compartment. These findings explain the different systemic and cardiovascular effects observed after the administration of ATP versus the administration of adenosine in humans.[24] The different physiological and pharmacological activities of the administered adenosine or ATP were not commensurable with the widely accepted notion that the activity of ATP is due to its primary catabolic product, adenosine, which is rapidly produced from ATP by powerful soluble enzymatic and ectoenzymatic activities present in the vascular bed. The results reviewed here also explain why the beneficial effects of bolus injections or infusions of ATP as compared to adenosine are more pronounced in humans than what has been observed in animal studies.[24] The reasons relate to the enzymatic and ectoenzymatic catabolic activities that catalyze the degradation of extracellular ATP: the activities present in humans are much weaker than the activities present in animals. Thus, for animals and humans to achieve the same levels of ATP in blood plasma, animals must receive relatively large doses of ATP, whereas humans may receive relatively small doses of ATP. Humans may, as a consequence, experience fewer and less severe side effects.[18–20,24]

Recent reports from my laboratory[19–21] have established the following results in murine models:

1. Under normal pathophysiological conditions, the vast majority of extracellular, blood plasma compartment pools of ATP originate in RBCs, as opposed to vascular endothelium or blood platelets.
2. Administration of AMP, ADP, or ATP—but not adenosine—into murine hosts results in the unexpected expansion of total liver and RBC ATP pools and in the elevation of blood plasma ATP levels. The elevated extracellular levels of ATP are the result of slow release of ATP from RBCs containing expanded ATP pools.
3. Expanded liver, RBC, and plasma compartment ATP pools yield inhibition of tumor growth in murine models and a marked inhibition of host weight loss in tumor-bearing hosts.[19–21]

Studies demonstrating the improvement in some hepatic functions after the intraperitoneal administration of adenine nucleotides to tumor-bearing hosts are discussed in this report along with the implications of the expansions of hepatic ATP pools with regard to the inhibition of host weight loss in tumor-bearing hosts.

METHODS

CT26 tumors were grown in the footpads of $CB6F_1$ mice according to previously published procedures.[19–21] On day 1, 8-week-old males were inoculated with CT26 cells. On day 9, when 100% of the animals carried palpable tumors, treatment schedules were initiated. Treatment consisted of daily intraperitoneal injections of 2 ml of saline or 25 mM of AMP or ATP in 2 ml of saline (AMP or ATP solutions were adjusted to pH 6.4) for 10 consecutive days. Blood was withdrawn from the inferior vena cava into 1-ml syringes (500 μl per mouse) while the animals were anesthetized with ether. Blood was collected on day 19, that is, on the first day after the last injection. Animals were weighed and tumor sizes were determined every 3 days. On day 19, animals were weighed before blood collection and tumors were excised and weighed after blood collection. Blood was allowed to clot and retract, and serum was collected by centrifugation. Serum phosphodiesterase activities were determined using a modification of the published thin-layer chromatographic assay.[31] Incubation mixtures contained 10 μl of serum in a total volume of 25 μl containing 100 mM HEPES (pH 8.0), 5 mM $MgCl_2$, and 0.05 mM of [^{3}H]Ap_4A (of 1 Ci/mmol specific radioactivity). Incubations were performed for 30 min. The rate of degradation of [^{3}H]Ap_4A was linear during the 30-min incubation at 37 °C. Thin-layer chromatography on poly(ethylene)imine-cellulose was carried out according to the published procedures.[31]

RESULTS AND DISCUSSION

Treatment of tumor-bearing murine hosts with adenine nucleotides results in significant anticancer effects. The activity of the administered adenine nucleotides is mediated by the expansions of liver, RBC, and blood plasma compartment (extracellular) ATP pools.[19–21] This anticancer activity consists of direct effects of the extracellular ATP on the tumor cells, as has been observed in *in vitro* systems,[11–16] and host-mediated responses, which are induced by the expansion of total liver, RBC, and plasma compartment ATP pools. Our previous results demonstrate that intraperitoneal injections of adenine nucleotides yield expansions of total hepatic ATP pools. Circulating RBCs take up the adenosine precursor needed for the expansion of RBC ATP pools in the hepatic sinusoids.[21,32] The increased supply of adenosine as a salvage precursor for increased RBC ATP synthesis is the result of the turnover of expanded liver ATP pools.[21] RBCs containing expanded ATP pools slowly release micromolar amounts of ATP into the extracellular plasma compartment.[20] The levels of ATP in the extracellular plasma compartment resulting from this release can sustain elevated steady state plasma ATP concentrations (pools) in spite of the strong catabolic ectoenzymatic and soluble enzymatic activities that catalyze the degradation of ATP in the vascular bed.[18,20,22] As noted earlier, in addition to the inhibition of tumor growth, host weight loss was markedly inhibited in tumor-bearing hosts. Although tumor size is related to the rate of weight loss, the inhibition of tumor growth and host weight loss in tumor-bearing hosts did not exhibit a cause-effect relationship.[20,21] The component of the inhibition of weight loss due to decreased tumor size in treated animals could be separated from the direct effects of adenine nucleotides on host functions. These direct effects, which constitute another component of the inhibition of weight loss, presumably account for the inhibition of weight loss in tumor-bearing murine hosts.[20,21] Fluid retention (see the data for AMP- or ATP-treated mice[21]) or alterations in food consumption by the treated animals (data not shown) were eliminated as possible causes of the inhibition of weight loss in tumor-bearing hosts.

Weight loss is a frequent and poorly understood aspect of cancer.[33] The prognostic effect of weight loss on the response rate to chemotherapy in patients with a variety of tumor types has been established. Chemotherapy response rates were lower in patients with weight loss and reversal of poor nutritional state was shown to affect the outcome in cases where the anticancer therapy was inherently effective.[34] Intravenous infusions of nutritional substrate for support of tumor-bearing hosts (total parenteral nutritional support) were shown to be ineffective. Tumor growth is generally increased to a greater extent than is host tissue growth.[35] Although cancer cachexia in man as well as in experimental animals produces not only host weight loss but also anorexia, hormonal aberrations, and depletion and redistribution of host components, all of which lead to a progressive decline in vital host functions,[36] much of the attention has been focused on the alterations in energy metabolism in cancer patients.[37] Gluconeogenesis from lactate and amino acids was proposed to account for lactate recycling and the enormous expenditures in energy that occur in the liver and kidney cortex during cancer cachexia.[38,39] Whereas gluconeogenesis is costly in terms of the number of ATP molecules required for glucose synthesis from lactate in the liver and kidney cortex, anaerobic glycolysis of the newly synthesized glucose at another site, namely the tumor, occurs at a rapid rate and produces fewer molecules of ATP. This type of host-tumor interplay was suggested as the main reason for the progressive weight loss associated with cancer cachexia.[38,39] Aberrations in glucose metabolism showing marked increases in Cori cycle activity (the cyclic metabolic pathway in which glucose is converted to lactic acid which is in turn utilized for glucose resynthesis in the liver via gluconeogenesis) have been demonstrated using radioactively labeled

TABLE 1. Serum Phosphodiesterase Activities in Non-Tumor-Bearing (NTB) and Tumor-Bearing (TB) Mice after Treatment with Adenine Nucleotides[a]

Animal Treatment[b]	Phosphodiesterase Activity (pmols/min·μl serum)	Total Serum Protein (g/dl)	Final Body Weight[c] (g)	Final Tumor Weight (g)
NTB/Saline	1.13 ± 0.19	4.8 ± 0.3	27.2 ± 2.3	
NTB/ATP	1.21 ± 0.20	4.7 ± 0.3	25.9 ± 2.4	
TB/No treatment	0.63 ± 0.09 ($p < .001$[d])	4.7 ± 0.4	21.1 ± 2.1 ($p < .001$[d])	2.1 ± 0.4
TB/Saline	0.65 ± 0.11 ($p < .001$[d])	4.9 ± 0.3	20.9 ± 2.6 ($p < .001$[d])	2.1 ± 0.5
TB/AMP	1.12 ± 0.22 ($p < .001$[e])	4.9 ± 0.4	25.9 ± 2.8 ($p < .01$[e])	1.2 ± 0.3 ($p < .005$[e])
TB/ATP	0.83 ± 0.15 ($p < .05$[e])	4.8 ± 0.3	25.6 ± 2.8 ($p < .01$[e])	1.0 ± 0.4 ($p < .005$[e])

[a] Experimental protocols are described in the text.
[b] Each group included 10 animals. No deaths occurred before the end of the experiment.
[c] Total body weight including the tumor.
[d] Statistical significance as compared to the NTB/Saline group (Student's *t* test).
[e] Statistical significance as compared to the TB/Saline group (Student's *t* test).

precursors in human cancer patients suffering from progressive weight loss.[40] Furthermore, the significant depletion of host visceral energy stores by a progressively growing tumor was recently demonstrated in experimental animals.[41] The utilization of ^{31}P nuclear magnetic resonance spectroscopy *in vivo* demonstrated a marked reduction in total rat liver ATP pools and increases in total liver inorganic phosphate levels at the point where the growing tumor began to adversely affect the weight of tumor-bearing animals.[41]

Data outlined in this paper show that the intraperitoneal administration of AMP or ATP to mice ($CB6F_1$) bearing syngeneic palpable tumors (CT26) resulted in marked inhibitions of tumor growth and host weight loss in tumor-bearing mice (TABLE 1). In order to assess the improvement in hepatic functions after adenine nucleotide treatments of tumor-bearing mice, the activity of a soluble serum protein that is synthesized in the liver and acts in catalyzing the degradation of extracellular nucleotides in the blood was determined. This protein, serum phosphodiesterase, utilizes Ap_4A, in addition to a wide range of adenine nucleotides, as a substrate yielding AMP and ATP.[31,42,43] Isoenzymes of this protein were reported to be elevated in cases of clinically diagnosed liver cancer,[44,45] but not in serum of patients suffering from cancers other than liver cancers.[45] Significant decreases in serum phosphodiesterase were observed in tumor-bearing hosts, and treatment with adenine nucleotides restored the activity of the serum protein almost to the levels found in non-tumor-bearing animals (TABLE 1). Synthesis of this enzyme is most likely induced in the liver by adenine nucleotides, which are the substrates for the catabolic activity of the enzyme. Thus the increased activity of the enzyme in the sera of tumor-bearing mice after treatment with adenine nucleotides is indicative of the improvement in the energy state and metabolic functions of the liver.[41] Hepatic gluconeogenesis was proposed as a target for inhibition in previous attempts to devise a treatment for cancer cachexia and progressive weight loss.[38,39,46] In this regard, it is important to note that elevated ATP pools were shown to effectively inhibit gluconeogenesis in isolated rat liver hepatocytes.[47]

In conclusion, data presented in this report and elsewhere[21] suggest that the marked inhibition of host weight loss in tumor-bearing animals after the intraperitoneal administration of adenine nucleotides is in part the result of the expansion of total hepatic ATP pools. These elevated ATP pools improve liver functions by abolishing the deterioration in the energy state of the liver, a deterioration known to occur after the onset of cancer cachexia.[38,39,41] The expansion of hepatic ATP pools, which was also shown to inhibit gluconeogenesis in isolated hepatocytes,[47] is likely to provide a significant inhibition of the Cori cycle in tumor-bearing hosts. The Cori cycle is viewed by many as the reason for the depletion of visceral energy stores in cancer patients and a likely cause of weight loss in these patients.[37–41]

REFERENCES

1. BURNSTOCK, G. 1972. Purinergic nerves. Pharmacol. Rev. **24:** 509-581.
2. BURNSTOCK, G. 1976. Purinergic receptors. J. Theor. Biol. **62:** 491-503.
3. BROWN, C., G. BURNSTOCK & T. COCKS. 1979. Effect of adenosine 5′-triphosphate (ATP) and beta-gamma-methylene ATP on the rat urinary bladder. Br. J. Pharmacol. **65:** 97-102.
4. BURNSTOCK, G. & C. KENNEDY. 1986. A dual function for adenosine 5′-triphosphate in the regulation of vascular tone. Circ. Res. **58:** 319-330.
5. TOMKINS, G. M. 1975. The metabolic code. Science **189:** 760-763.
6. RAPAPORT, E. & P. C. ZAMECNIK. 1976. Presence of diadenosine 5′,5‴-P^1,P^4-tetraphosphate (Ap_4A) in mammalian cells in levels varying widely with proliferative activity of the tissue: A possible positive "pleiotypic activator." Proc. Natl. Acad. Sci. USA **73:** 3984-3988.
7. FARR, S. B., D. N. ARNOSTI, M. J. CHAMBERLIN & B. N. AMES. 1989. An *apaH* mutation causes AppppA to accumulate and affects motility and catabolite repression in *Escherichia coli.* Proc. Natl. Acad. Sci. USA **86:** 5010-5014.
8. ALPER, M. D. & B. N. AMES. 1978. Transport of antibiotics and metabolite analogs by systems under cyclic AMP control: Positive selection of *Salmonella typhimurium cya* and *crp* mutants. J. Bacteriol. **133:** 149-157.
9. BURNSTOCK, G. 1982. Neuropeptides as trophic factors. *In* Systemic Role of Regulatory Peptides. S. R. Bloom, J. M. Polak & E. Lindenlaub, Eds.: 423-441. F. K. Schattauer Verlag.
10. BURNSTOCK, G. & C. M. BROWN. 1981. An introduction to purinergic receptors: Receptors and recognition. Series B. G. Burnstock, Ed. Vol. 12: 1-45. Chapman & Hall. London.

11. RAPAPORT, E. 1983. Treatment of human tumor cells with ADP or ATP yields arrest of growth in the S phase of the cell cycle. J. Cell. Physiol. **114:** 279-283.
12. RAPAPORT, E., R. F. FISHMAN & C. GERCEL. 1983. Growth inhibition of human tumor cells in soft-agar cultures by treatment with low levels of adenosine 5′-triphosphate. Cancer Res. **43:** 4402-4405.
13. CHAHWALA, S. B. & L. CANTLEY. 1984. Extracellular ATP induces ion fluxes and inhibits growth of Friend erythroleukemia cells. J. Biol. Chem. **259:** 13717-13722.
14. WEISMAN, G. A., K. D. LUSTIG, E. LANE, N. HUANG, I. BELZER & I. FRIEDBERG. 1988. Growth inhibition of transformed mouse fibroblasts by adenine nucleotides occurs via generation of extracellular adenosine. J. Biol. Chem. **263:** 12367-12372.
15. KITAGAWA, T., F. AMANO & Y. AKAMATSU. 1988. External ATP-induced passive permeability change and cell lysis of cultured transformed cells: Action in serum-containing growth media. Biochim. Biophys. Acta **941:** 257-263.
16. BELZER, I. & I. FRIEDBERG. 1989. ATP-resistant variants of transformed mouse fibroblasts. J. Cell. Physiol. **140:** 524-529.
17. ELMALEH, D. R., P. C. ZAMECNIK, F. P. CASTRONOVO, H. W. STRAUSS & E. RAPAPORT. 1984. ^{99m}Tc-labeled nucleotides as tumor-seeking radiodiagnostic agents. Proc. Natl. Acad. Sci. USA **81:** 918-921.
18. TRAMS, E. G., H. KAUFMAN & G. BURNSTOCK. 1980. A proposal for the role of ectoenzymes and adenylates in traumatic shock. J. Theor. Biol. **87:** 609-621.
19. RAPAPORT, E. 1988. Experimental cancer therapy in mice by adenine nucleotides. Eur. J. Cancer Clin. Oncol. **24:** 1491-1497.
20. RAPAPORT, E. & J. FONTAINE. 1989. Anticancer activities of adenine nucleotides in mice are mediated through expansion of erythrocyte ATP pools. Proc. Natl. Acad. Sci. USA **86:** 1662-1666.
21. RAPAPORT, E. & J. FONTAINE. 1989. Generation of extracellular ATP in blood and its mediated inhibition of host weight loss in tumor-bearing mice. Biochem. Pharmacol. **38:** 4261-4266.
22. GORDON, J. L. 1986. Extracellular ATP: Effects, sources and fate. Biochem. J. **223:** 309-319.
23. CHAUDRY, I. H., R. N. STEPHAN, R. E. DEAN, M. G. CLEMENS & A. E. BAUE. 1988. Use of magnesium-ATP following liver ischemia. Magnesium **7:** 68-77.
24. FLACKE, W. E. 1988. Adenosine and adenosine triphosphate for acute blood pressure control. Semin. Anesth. **7:** 216-225.
25. BUISMAN, H. P., T. H. STEINBERG, J. FISCHBARG, S. C. SILVERSTEIN, S. A. VOGELZANG, C. INCE, D. L. YPEY & P. C. LEIGH. 1988. Extracellular ATP induces a large nonselective conductance in macrophage plasma membrane. Proc. Natl. Acad. Sci. USA **85:** 7988-7992.
26. LIN, J., R. K. KRISHNARAJ & R. G. KEMP. 1985. Exogenous ATP enhances calcium influx in intact thymocytes. J. Immunol. **135:** 3403-3410.
27. EL-MOATASSIN, C., J. DORNAND & J. C. MANI. 1987. Extracellular ATP increases cytosolic free calcium in thymocytes and initiates the blastogenesis of the phorbol 12-myristate 13-acetate-treated medullary population. Biochim. Biophys. Acta **927:** 437-444.
28. KUROKI, M., K. TAKESHIGE & S. MINAKAMI. 1989. ATP-induced calcium mobilization in human neutrophils. Biochim. Biophys. Acta **1012:** 103-106.
29. WILEY, J. S. & G. R. DUBYAK. 1989. Extracellular adenosine triphosphate increases cation permeability of chronic lymphocytic leukemic lymphocytes. Blood **73:** 1316-1323.
30. SOSLAU, G. & J. PARKER. 1989. Modulation of platelet function by extracellular adenosine triphosphate. Blood **74:** 984-993.
31. LÜTHJE, J. & A. OGILVIE. 1985. Catabolism of Ap_3A and Ap_4A in human plasma: Purification and characterization of a glycoprotein complex with 5′-nucleotide phosphodiesterase activity. Eur. J. Biochem. **149:** 119-127.
32. LERNER, M. H. & B. A. LOWY. 1974. The formation of adenosine in rabbit liver and its possible role as a direct precursor of erythrocyte adenine nucleotides. J. Biol. Chem. **249:** 959-966.
33. DEWYS, W. D. & Eastern Cooperative Oncology Group. 1980. Prognostic effect of weight loss prior to chemotherapy in cancer patients. Am. J. Med. **69:** 491-497.
34. VAN EYS, J. 1982. Effect of nutritional status on response to therapy. Cancer Res. **42**(Suppl.): 747S-753S.

35. POPP, M. B., S. C. WAGNER & O. J. BRITO. 1983. Host and tumor responses to increasing levels of intravenous nutritional support. Surgery **94:** 300-308.
36. COSTA, G. 1977. Cachexia, the metabolic component of neoplastic diseases. Cancer Res. **37:** 2327-2335.
37. YOUNG, V. R. 1977. Energy metabolism and requirements in the cancer patient. Cancer Res. **37:** 2336-2347.
38. GOLD, J. 1974. Cancer cachexia and gluconeogenesis. Ann. N.Y. Acad. Sci. **230:** 103-110.
39. STEIN, T. P. 1978. Cachexia, gluconeogenesis and progressive weight loss in cancer patients. J. Theor. Biol. **73:** 51-59.
40. HOLROYDE, C. P., T. G. GABUZDA, R. C. PUTNAM, P. PAUL & G. A. REICHARD. 1975. Altered glucose metabolism in metastatic carcinoma. Cancer Res. **35:** 3710-3714.
41. SCHNEEBERGER, A. L., R. T. THOMPSON, A. A. DRIEDGER, R. J. FINLEY & R. I. INCULET. 1989. Effects of cancer on the *in vivo* energy state of rat liver and skeletal muscle. Cancer Res. **49:** 1160-1164.
42. LÜTHJE, J. & A. OGILVIE. 1987. Catabolism of Ap_4A and Ap_3A in human serum. Eur. J. Biochem. **169:** 385-388.
43. LÜTHJE, J., A. HOLLER, A. OGILVIE, W. SIEGFRIED, A. WALDHERR & W. DOMSCHKE. 1987. Catalytic activity concentrations of diadenosine tetraphosphate hydrolase in normal and pathological sera. J. Clin. Chem. Clin. Biochem. **25:** 493-498.
44. TSOU, K. C., S. LEDIS & M. G. MCCOY. 1973. 5′-Nucleotide phosphodiesterase isoenzyme pattern in the serum of human hepatoma. Cancer Res. **33:** 2215-2217.
45. LIN, K., S. YE & Q. LIU. 1988. The significance of new abnormal isoenzymes of 5′-nucleotide phosphodiesterase in the diagnosis of human liver cancer. Int. J. Cancer **41:** 30-32.
46. GOLD, J. 1976. Hydrazine sulfate in the treatment of cancer. Cancer Treat. Rep. **60:** 964-965.
47. LUND, P., N. W. CORNELL & H. A. KREBS. 1975. Effect of adenosine on the adenine nucleotide content and metabolism of hepatocytes. Biochem. J. **152:** 593-599.

DISCUSSION OF THE PAPER

I. FRIEDBERG (*Tel Aviv University, Tel Aviv, Israel*): Cockcroft and Gomperts (*Biochem. J.* **188:** 789, 1980) have shown that ATP induces the release of histamine from mast cells. On the other hand, side effects due to this phenomenon were not detected in your study or in Chaudry's study. Do you have an explanation for the lack of side effects?

RAPAPORT: The most probable explanation is that conditions are different in a whole animal as compared to isolated mast cells treated with low levels of ATP^{4-}. In a host, there are probably several factors that affect mast cells—factors that are themselves influenced by the elevated levels of extracellular plasma compartment ATP.

I. H. CHAUDRY (*Michigan State University, East Lansing, MI*): With regard to the potential side effects of ATP-$MgCl_2$, I would like to mention that with the infusion of ATP-$MgCl_2$ there are no significant side effects, at least if this agent is infused up to 0.3 mg/kg/min. There are, however, subjective feelings of warmth and dryness of the throat that disappear as soon as the infusion is stopped. If one administers ATP-$MgCl_2$ above 0.4 mg/kg/min, there is hypotension and an associated feeling of nausea.

G. WEISMAN (*University of Missouri, Columbia, MO*): Have you considered that the antitumor effects of ATP *in vitro* may be due to generation of adenosine from extracellular ATP and uptake of adenosine by the cells? In Heppel's group, after hearing of your results, we also found that ATP (but not adenosine) inhibited the growth of transformed mouse fibroblasts. We also showed that adenosine translocator inhibitors, such as NBTI and dipyridamole, blocked the inhibitory effect of ATP, and that suboptimal doses of adenosine added to the cells over a long period (16 hr) were as effective as ATP in inhibiting cell growth. Our evidence supported the role of adenosine in the induction of pyrimidine starvation, as shown by Ishii and Green (*J. Cell Sci.* **13:** 429, 1973).

RAPAPORT: I am aware of your studies with regard to the *in situ*-produced adenosine (*J. Biol. Chem.* **263:** 12367, 1988) exhibiting cytotoxicity against fibroblasts. My *in vitro* studies, however, demonstrate that tumor cells of epithelioid morphology are resistant to cytotoxicity or cytostatic effects of adenosine or AMP at a variety of levels and schedules of administration. These cells, however are very sensitive to low micromolar levels of ADP or ATP. Adenosine toxicity in fibroblasts induces, as you mentioned, pyrimidine starvation. The cytostatic and cytotoxic effects of ADP and ATP against tumor cells of epithelioid morphology were not the result of pyrimidine starvation.

Structural and Chemical Properties of ATP and Its Metal Complexes in Solution

MILDRED COHN

Department of Biochemistry and Biophysics
University of Pennsylvania School of Medicine
Philadelphia, Pennsylvania 19104

INTRODUCTION

ATP is ubiquitous in cells and tissues and is a most versatile molecule both functionally and conformationally. Many of its functions may be inferred from FIGURE 1. As indicated, it is the central compound for storing chemical energy—energy that is transduced in many essential biological processes. Nucleoside triphosphates such as ATP serve in the first step of all macromolecular biosyntheses by enzymatic transfer of a nucleotidyl group concomitant with the formation of inorganic pyrophosphate. ATP also serves as a source of phosphorylating metabolites and proteins by enzymatic transfer of the γ-phosphoryl group with the concomitant formation of adenosine diphosphate. In binding to some proteins, ATP acts as an effector of activity.

The question arises which inherent features of its structure are correlated with its various functions. Some aspects of this question have been considered by Westheimer in his paper entitled, "Why Nature Chose Phosphates."[1] Of primary importance is the negative charge of ATP at physiological pH. It is always ionized because the first pH of a phosphate group is about pH 2. One consequence is the usual retention of ATP within the cell with significant exceptions of particular relevance to this volume. The negative charge on ATP is crucial for two other properties of the nucleotide, its kinetic stability to hydrolysis and its binding to metal ions and to positively charged moieties of proteins. The negative charge makes phosphoanhydrides much less vulnerable to nucleophilic attack by hydroxide ion or water than, for example, carboxylic anhydrides. The retention of thermodynamic instability coupled to kinetic stability in an aqueous medium is indispensable for the role of ATP in bioenergetics. Binding to proteins as substrate or effector depends not only on charge but also on the formation of many types of intermolecular hydrogen bonds and the ability of its ring system to stack with aromatic groups of the protein.

Some structural parameters of ATP and other nucleotides remain invariant whether in crystals or solution or whether liganded to metals or proteins. Thus the ribose is always D-ribose, and the adenine is always linked by a β-N-glycosidic linkage. In solution many conformations may coexist, but in a crystal or in a protein complex, there is only one conformation.

For example, ATP may exist in two conformations about the glycosidic C(1′)-N bond, *syn* if N-3 of adenine lies above the plane of the sugar group, and *anti* if N-3

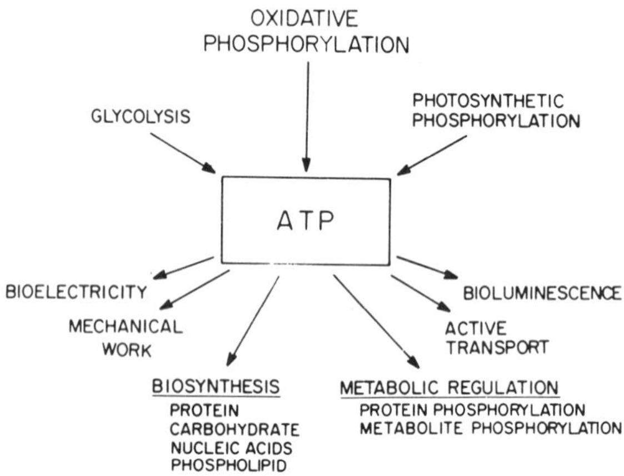

FIGURE 1. Formation and utilization of ATP in cells.

points away from the sugar group. NMR studies have shown that ATP in solution, in an equilibrium mixture, lies predominantly but not exclusively in the *anti* conformation. Which conformation ATP adopts, *syn* or *anti,* on binding to a protein is unpredictable because intermolecular interactions such as hydrogen bonding and stacking effects are often critical. Even in solution, the equilibrium may be shifted to the *syn* form by structural modification. In 8-Br-ATP, the *syn* conformation is assumed, but 2-Cl-ATP remains in the *anti* form.[2] In tubercidine triphosphate (N-7 is replaced by CH), the purine can rotate around the glycosidic bond, and no preference exists for *syn* or *anti.*[3]

STRUCTURE OF ATP

The direct methods for determining structure, X-ray diffraction for crystals and NMR for solutions, both have their limitations. Only Na_2-ATP·$3H_2O$ has been successfully crystallized and analyzed by X-ray diffraction.[4,5] Evaporation of solutions of divalent metal ion salts and nucleoside polyphosphates invariably leads to mixtures of metal phosphates and nucleoside monophosphates. Divalent metal ions have been shown to catalyze hydrolysis of ATP.[6] Several adenosine diphosphate species have been successfully crystallized and have had their structures determined: the free acid,[7] the isomorphous crystals of Rb-ADP·$3H_2O$,[8,9] Rb-ADP·H_2O,[10] K-ADP·$2H_2O$,[11,12] and Tris-(hydroxymethyl)methylammonium-ADP·$2H_2O$.[13]

In the Na_2-ATP·H_2O crystal, ATP exists as an asymmetric dimer with two sodium ions coordinating the two ATP molecules through the phosphate oxygens and N-7. There is no evidence for a dimeric species of Na_2-ATP in solution. In the crystal, the triphosphate chain is in the folded conformation, but it forms a left-handed helix in one ATP of the dimer and a right-handed helix in the other. The conformation of the ribose rings also differs with C-3′ *endo* in one ATP and C-2′ *endo* in the other.

Sodium ATP in solution has been investigated with ^{31}P and ^{23}Na by titration with NaCl at constant pH 6.7.[14] With NMR, there is no difficulty in obtaining data for

ATP in any form; however, there is difficulty in interpreting the data in structural terms. The pure spectra of two species, Na-ATP^{3-} and NaH-ATP^{2-}, could be observed, and to reconcile the spectra of mixtures, it was necessary to postulate equilibria involving two other species, ATP^{4-} and H-ATP^{3-}. The constants for the equilibria between the pairs of species were calculated, as were some dynamic parameters of the system.

BINARY Mg-ATP COMPLEX

It should be pointed out that Mg^{2+}-ATP or another divalent ion complex is always the relevant ATP species at the catalytic site of enzymes and often at other protein-binding sites. For many cells and tissues, the metal-coordinated form is the predominant species; in extracellular fluid with 5 mM Ca^{2+} and 3 mM Mg^{2+}, ATP would be almost exclusively coordinated to Ca^{2+} and/or Mg^{2+}. Phosphorus-31 NMR is particularly useful in delineating the conformation of the polyphosphate chain. In solution, there is only a small effect of Mg^{2+} coordination on the adenine ring as inferred from the small chemical shifts of H-1 and H-8 in ^{1}H NMR.[15] It is unnecessary to invoke binding of Mg^{2+} to N on the adenine ring because the small protein chemical shifts can be explained by changes in base stacking. Further presumptive evidence indicating that Mg^{2+} is not bound to the adenine ring arises from comparisons of ^{31}P NMR spectra of the magnesium complexes of ATP, ribose triphosphate, and inorganic tripolyphosphate.[16] All three complexes gave essentially the same chemical shifts and spin-lattice relaxation times for the three ^{31}P nuclei. The ^{1}H NMR parameters of the ribose moiety of ATP were not significantly affected by Mg^{2+} complexation, although one possible exception was observed —a small, unexplained chemical shift of H-3′.[15] The coordination complex with the three phosphate groups of ATP supplying ligands for Mg^{2+} is the predominant species. The β-P shows the largest shift, as first reported in 1962,[17] because its P-O bond angles change the most,[18] as shown in FIGURE 2. In

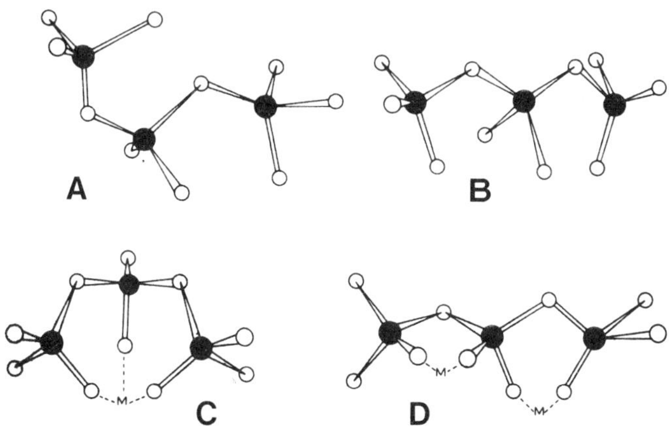

FIGURE 2. Conformations of (**A**) ATP and (**B**) Mg-ATP deduced from ^{31}P NMR. Probable conformations in solution of (**C**) Mg-ATP and (**D**) Mg_2-ATP.[16]

Mn^{2+}-ATP, the paramagnetic effect of Mn^{2+} on the relaxation rates of α-P, β-P, and γ-P indicates that all three phosphates are liganded to Mg^{2+}, but Cu^{2+} only affects β-P and γ-P, forming a bidentate complex.[17]

Analogues of ATP with the nonbridging oxygen substituted by sulfur on the α-, β-, or γ-phosphate may be synthesized.[19] When the α- or β- phosphate is so modified, a new chiral center is created on the corresponding phosphorous and two diastereomers designated R_p and S_p exist for the phosphorothioates ATPαS and ATPβS (FIG. 3). The diastereomers may be distinguished by ^{31}P NMR spectroscopy.[20] The Mg^{2+} complexes of the R_p and S_p diastereomers are designated Λ and Δ, respectively. It should be pointed out that the coordination of Mg^{2+} to ATP also creates chiral centers at α-P and β-P when their phosphates are liganded to Mg^{2+}. The use of Mn(II) electron paramagnetic resonance (EPR) spectroscopy has allowed the diastereomers of Mn(II)-ADP to be distinguished stereospecifically with ^{17}O at α-P.[21] A number of stable diastereomers of metal-ATP complexes with Cr(III) and Co(III), that is, substitution-inert chiral complexes with bidentate (β-γ) coordination or tridentate (α-β-γ) coordination, were synthesized by Cleland and co-workers.[22] All diastereoisomers of ATP and their metal complexes serve as stereospecificity probes of the protein sites to which they bind.

TERNARY AND HIGHER COMPLEXES OF ATP

A number of model ternary complexes have been crystallized and have had their crystal structures determined by X-ray diffraction, including the dimeric species of bis[(ATP)(2,2′-bipyridine)Zn(II)]·$4H_2O$[23] and the isomorphous Mg^{2+}, Ca^{2+}, Mn^{2+}, and CO^{2+} complexes with ATP and 2,2′-dipyridylamine.[24] The relevance of these structures to metal-ATP-protein complexes is somewhat dubious. The conformation ATP or Mg^{2+}-ATP assumes upon binding to a protein is always monomeric, the details of the conformation being largely determined by the binding domain on the protein.

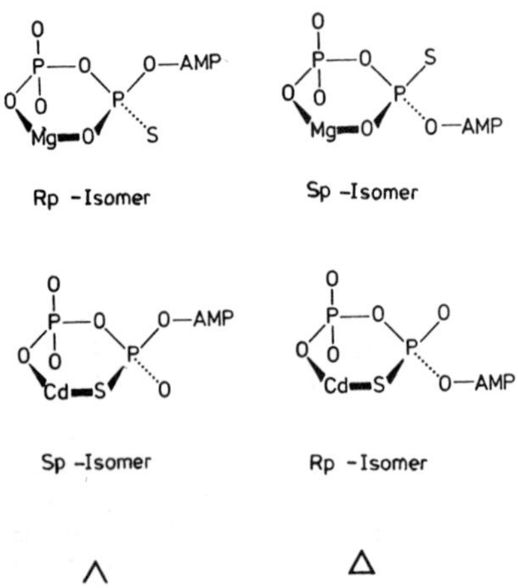

FIGURE 3. The Mg^{2+} complexes of the R_p and S_p isomers of ATPβS (Λ and Δ, respectively), and the Cd^{2+} complexes of the R_p and S_p isomers of ATPβS (Δ and Λ, respectively).

TABLE 1. Dissociation Constants from Yeast 3-Phosphoglycerate Kinase[25]

Compound	Catalytic Site (μM)	Secondary Site (mM)
ATP	70	0.5
Mg-ATP	65	5.0

A number of different methods have been used to investigate the higher complexes of ATP or metal-ATP with enzymes and other proteins. Both direct methods, which include X-ray diffraction in the crystalline state and NMR and EPR in solution, and indirect methods, which utilize ATP analogues modified in the adenine or ribose or phosphate moieties to determine specificity and stereochemistry of binding, have contributed to the elucidation of the structures of bound ATP. The most striking conclusion from an overwhelming number of studies is that ATP is capable of assuming different conformations when bound to different proteins even in a group of enzymes that catalyze the same type of reaction.

SPECIFICITY

The simplest aspect of structural specificity in an ATP-protein complex is that of the metal ion. In almost all enzymatic reactions of ATP, the substrate is a divalent ion complex of ATP, but uncoordinated ATP may bind at other sites. An example is the enzyme 3-phosphoglycerate kinase, which requires Mg-ATP at the catalytic site but binds ATP at a secondary site.[25] The binding constants of ATP and Mg-ATP[25] are given in TABLE 1; note that the secondary site prefers ATP to Mg-ATP by a factor of 10.

In the reaction catalyzed by adenylate kinase, that is, Mg-ATP + AMP $\rightleftarrows$ Mg-ADP + ADP, where ATP is the phosphoryl donor in the forward direction and ADP is the phosphoryl donor in the reverse direction, only the ATP donor is coordinated to Mg^{2+}. Thus, in the reverse reaction, bound ADP at the acceptor site (ATP site) is in the form of Mg-ADP, but the bound donor ADP (AMP site) is not coordinated to Mg^{2+}, as demonstrated by ^{31}P NMR.[26]

In most enzymatic reactions of ATP, Mn^{2+}, Co^{2+}, and Ca^{2+} can substitute for Mg^{2+}, although the catalytic rate is somewhat affected. In those complexes where the metal ion binds to the protein, however, as in pyruvate kinase, Ca^{2+} cannot substitute for Mg^{2+}, probably because of size. One dynamic property that differs 1000-fold between Mg^{2+} and Ca^{2+} is the rate of complex formation. As shown in FIGURE 4,[27] the fast rate of exchange on the NMR time scale between free ATP and Ca-ATP leads the ^{31}P NMR spectrum (at a 2.5:1 ATP:Ca^{2+} ratio) to show a narrow β-P resonance at a chemical shift intermediate between ATP and Ca-ATP. For the analogous exchange between free ATP and Mg-ATP (at a 2:1 ATP:Mg^{2+} ratio), the β-P resonance is broadened because the exchange rate is not fast enough on the NMR time scale.

Examining the base specificity for enzyme activity and/or binding reveals that there are some enzymes that will accept only ATP or another single nucleoside

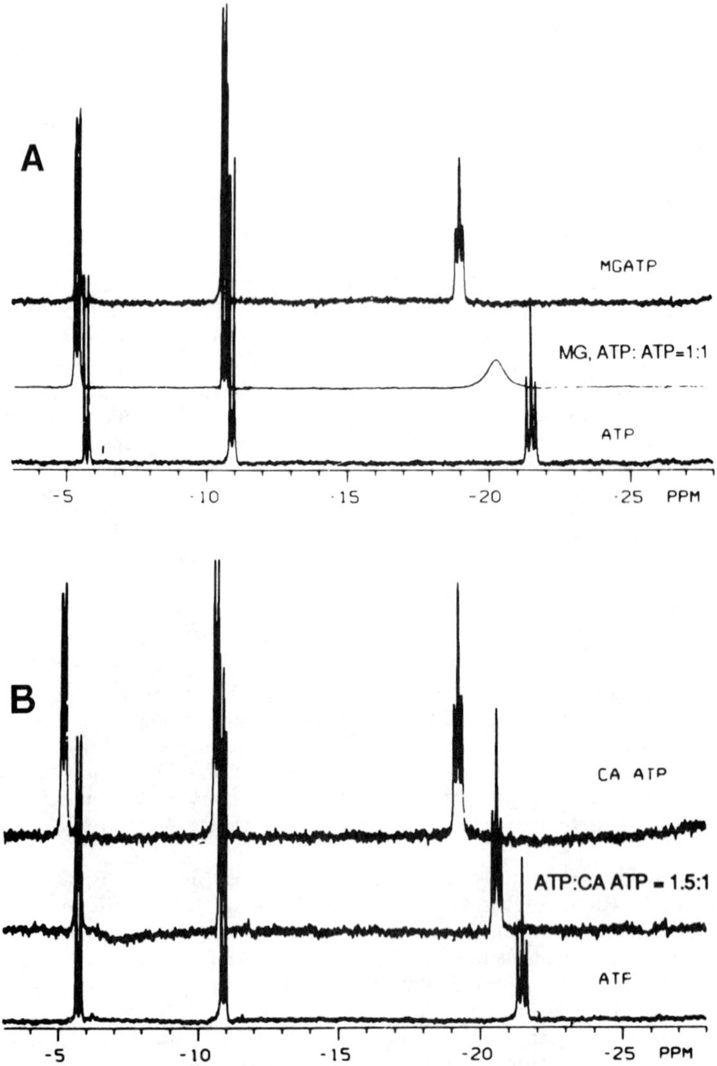

FIGURE 4. Spectra obtained using ^{31}P NMR. (**A**) Mg-ATP, Mg-ATP/ATP in a 1/1 mixture, and ATP. (**B**) Ca-ATP, Ca-ATP/ATP in a 1/1.5 mixture, and ATP.[27]

triphosphate, some that will accept only ATP and GTP, and some that will accept adenosine, cytidine, guanosine, or uridine triphosphates.

The aminoacyl-t-RNA synthetases have been investigated extensively for substrate specificity. Freist[2] has reported on the activities of 20 enzymes from *E. coli* and 13 enzymes from yeast with eight analogues of ATP. Four of these analogues had modifications at the adenine (*N*^{6}-methyl, 2-chloro, 8-bromo, and 7-deaza (tubercidine)), and four had modifications at the ribose (2′-deoxy, 3′-deoxy, 2′-methoxy, and 3′-methoxy). Only the 8-Br-ATP analogue (*syn* conformation) was never a substrate, but it was an inhibitor for a number of enzymes. There was no analogue that was a substrate for the whole group of enzymes. The most active analogue, the 2-chloro derivative, was a substrate for 23 of the 33 enzymes. The other analogues varied with each member of the group, and acted as substrates for some, were inhibitors for others, or had no effect. The only conclusion that can be drawn is that the structural requirements for ATP binding to a group of functionally similar proteins differ greatly in detail.

Experiments with analogues of ATP can at best give qualitative information about the bound structure of Mg-ATP. A most ingenious modification of ATP was introduced by Leonard and co-workers[28] that in favorable cases can yield dimensionally quantitative information. An excellent inhibitor of porcine muscle adenylate kinase is diadenosine pentaphosphate, Ap_5A ($K_i = 1.2 \times 10^{-7}$ M).[29] The corresponding tetraphosphate, Ap_4A, which differs in its phosphate chain length by 2.7 Å, is not a good inhibitor ($K_i = 1.2 \times 10^{-5}$ M).[28] The modification introduced yielded a fused ring 2.4 Å in length between the two rings of adenine at one end of the molecule (P^1-(lin benzo-5′-adenosyl)-P^4-(5′-adenosyl)tetraphosphate) and the corresponding pentaphosphate. The tetraphosphate thus lengthened was a more effective inhibitor ($K_i = 5 \times 10^{-7}$ M) than the pentaphosphate ($K_i = 5 \times 10^{-6}$ M). Furthermore, the modified form is fluorescent, and it could be shown that the compounds are intramolecularly stacked in aqueous solution. When bound to the enzyme, however, the phosphate chain assumes an "open" or "extended" structure. The finding that Mg-Ap_5A bound to adenylate kinase in an extended form had been shown directly by ^{31}P NMR: Ap_5A in solution is a symmetrical molecule with two NMR peaks, but the ^{31}P spectrum of bound Mg-Ap_5A has five peaks.[30]

The specificity of the phosphate chain has been investigated with a number of analogues. The bridging oxygens have been substituted with a CH_2 group or an NH group. The methylene modification differs more in structure from ATP than the imido modification. AMPPNP binds to all ATP sites, but it is generally stable to hydrolases and often less active or inactive as substrate for other ATP enzymes.[31] A nonbridging oxygen may be substituted by a sulfur at α-, β-, or γ-P to form the respective phosphorothioates, which are generally substrates for ATP-utilizing enzymes, except for ATPγS, which is not susceptible to enzymatic hydrolysis. The stability of the phosphorothioate ATPγS makes it useful, particularly in those systems where ATP hydrolysis is coupled to activation. In these systems, it becomes possible to determine whether binding is sufficient or whether hydrolysis is necessary.[32]

METAL COORDINATION SCHEME OF METAL-ATP-PROTEIN COMPLEXES

Four methods are available in solution to determine the ligands of ATP to the metal ion: 1) Apparent reversal of stereospecificity of phosphorothioate complexes of Mg^{2+} versus Cd^{2+}. 2) The measurement of enzyme activity of stable Cr(III) and Co(III) bidentate (β-γ) or tridentate (α-β-γ) complexes of ATP. 3) Paramagnetic effect of Mn(II) or Co(II) on the NMR relaxation rates of phosphorus nuclei of ATP in metal-ATP-protein complexes. 4) The hyperfine splitting due to ATP labeled with

^{17}O in the nonbridging oxygens on the EPR spectrum of Mn(II)-ATP-protein. Of these, only the third, the NMR method, can yield distances.

The phosphorothioate approach depends on findings from ^{31}P the NMR spectroscopy that Mg^{2+} coordinates preferentially to oxygen and that Cd^{2+} coordinates preferentially to sulfur.[33] Thus the Δ isomers of Mg complexes are formed from the S_p isomers of ATPαS and ATPβS, but the Δ isomers of Cd complexes are formed from the R_p isomers (FIG. 3). If an enzyme is stereoselective for the Δ isomer of the metal complex, then a reversal of stereoselectivity of S_p to R_p may be expected when Mg^{2+} and Cd^{2+} complexes respectively are used as substrates. Such a reversal was very marked for ATPβS in the hexokinase reaction,[33] but no reversal was seen with ATPαS, and it was concluded that the bidentate β-γ coordination structure held for this enzyme. Those enzymes that show reversal with both ATPαS and ATPβS are inferred to use tridentate metal ATP coordination in their substrates. Not all results are clear-cut, and limitations may arise if the α-P is not liganded in the rate-determining step in the reaction or if constraints of the protein interaction force Cd-O rather then Co-S ligands to form. The stereoselectivity reversal is thoroughly reviewed by Eckstein.[32]

The enzymatic activity or lack thereof of various Cr(III) and Co(III) bidentate and tridentate complexes has been used extensively[34,35] to determine both the coordination scheme as well as the stereochemistry of enzyme-bound metal-ATP. The paramagnetic Cr(III) complexes have also been utilized as distance probes in NMR experiments. The metal coordination scheme inferred from this approach is not always unequivocal. At best, one is usually observing a single turnover because the complex is substitution inert, and, when the true substrate is the tridentate α-β-γ complex, the complex may not show activity because of kinetic and structural constraints in this species.

The NMR method of determining distances from the metal ion to each of the three P atoms has been extensively used by Mildvan and his co-workers[36,37] and by Nageswara Rao and his co-workers.[38] The effect of a paramagnetic center on the nuclear spin relaxation time of nuclei in the system is a function of the inverse sixth power of the distance between the paramagnetic center (such as Mn(II) to a nucleus such as α-P of ATP) and also of τ_c, the correlation time, and τ_e, the exchange time between bound and free species. Measurements of the frequency dependence and temperature dependence of the nuclear spin relaxation times serve to distinguish the magnitudes of the various contributions. The creatine kinase metal-ATP-enzyme complex was investigated using first Mn(II) and then Co(II) as the paramagnetic species.[39] The relaxation rates of the Mn(II) complex were found to be limited by the exchange rate of the free and bound forms of Mn-ATP. The Co(II) complexes were not exchange limited and yielded a range of distances for α-, β-, and γ-P, indicating that all three are within the first coordination sphere of the metal ion. Therefore the tridentate complex is the substrate. The corresponding metal-ADP bidentate α-β complex is the substrate in the reverse direction. The metal-ATP structure for adenylate kinase,[40] on the other hand, proved to be the bidentate β-γ complex of metal-ATP. The corresponding metal-ADP, monodentate β complex is the substrate in the reverse direction. The limitations and pitfalls of this method have been discussed.[39,41] The most frequently encountered problem is that the relaxation rate is primarily determined by the exchange rate rather than the distance. It should be noted that the NMR method is the only one that yields distances between atoms.

The EPR method, with which a hyperfine effect may be observed in the spectrum of a Mn(II) complex when the ligand is labeled with ^{17}O, allows unequivocal assignment of the phosphate ligands of protein-bound Mn-ATP, as demonstrated by Reed and co-workers.[42,43] Thus, in agreement with the NMR experiments, the coordination of metal-ATP bound to creatine kinase was found to be tridentate. In addition, the number of water molecules liganded to Mn(II) can be ascertained by observing the EPR spectrum in $H_2{}^{17}O$. All six ligands of creatine kinase were accounted for. The chief limitation of this technique is the necessity for the EPR spectrum to have sufficiently narrow lines to make the ^{17}O hyperfine effect observable.

The four methods do not always give the same result. The disagreements are usually due to inherent ambiguities and limitations of some of the methods, as discussed in the references cited. The totality of the evidence, however, indicates convincingly that for some proteins the tridentate α-β-γ metal coordination complex is the active species, and that for others it is the bidentate β-γ complex.

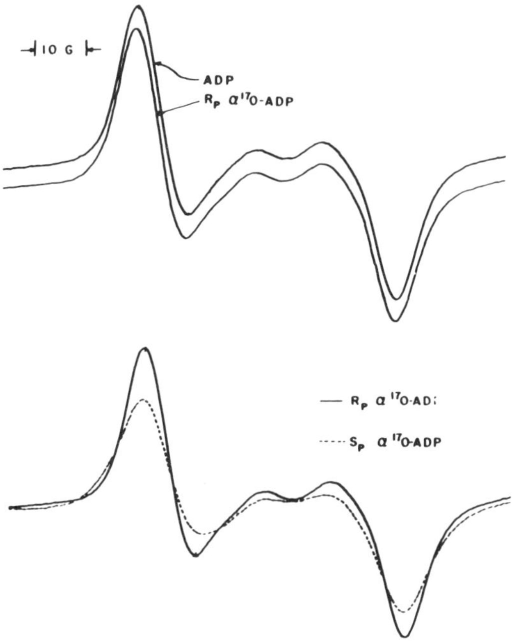

FIGURE 5. Partial EPR spectra of the Mn-ADP-3-phosphoglycerate kinase complex. Upper curve: comparison of unlabeled ADP and the R_p isomer of Mn-ADP (α-^{17}O); lower curve: comparison of the R_p and S_p isomers of Mn-ADP (α-$_{17}O$).[44]

STEREOSELECTIVITY

As described above, the bidentate β-γ Mg^{2+} complex has a chiral center on the β-P and the tridentate α-β-γ complex has chiral centers at both the α-P and the β-P. The criterion of the efficacy of a substrate, the V_{max}/K_m value, has been obtained for S_p and R_p diastereoisomers of the Mg^{2+} complexes of ATPαS and ATPβS for a number of enzymes.[32]

The stereoselectivity of Mg-ATP itself for the R_p and S_p diastereoisomers, on α-P and β-P, respectively, has only been detected by labeling each nonbridging oxygen on α-P or β-P stereospecifically, that is, $R_p(^{17}O)$ and $S_p(^{17}O)$, and observing the effect of each diastereoisomer on the EPR spectrum of bound Mn-ATP, α-^{17}O, or β-^{17}O. The EPR method of determining stereoselectivity of Mn(II)-ADP is exemplified by a study of the interaction of Mn-ADP stereospecifically labeled ^{17}O with the 3-phosphoglycerate kinase enzyme.[44] As shown in FIGURE 5, for one portion of the

Mn-ADP (α-^{17}O)-enzyme spectrum, the R_p isomer is identical with that of unlabeled ADP, but the S_p isomer exhibits broadening due to the interaction of Mn with ^{17}O in the α position.

BINDING DOMAINS OF ATP

In considerations of the nature of the binding site for ATP on proteins, it must be emphasized that the binding site is not a static structure. The ATP binding site structure changes as other ligands bind to the protein. For the kinases, the progressive changes in the nucleotide binding site as each ligand is added are reflected in the effects of enzyme-bound Mn-ADP or Mn-ATP on the enhancement of the proton relaxation rate of water in NMR, and in the progressive changes in EPR spectra of the same complexes.[45] Similar progressive changes have been observed in these systems in the ^{31}P NMR chemical shifts of enzyme-bound nucleotides, again indicating a change in the conformation in the ATP binding domain of the protein as the second substrate or cofactors are added. The geometric arrangement of the rigid Co(III)-ATP bound to PP-ribose-P synthetase changes when ribose-5-P is bound to the enzyme, as shown in FIGURE 6.[46]

The electronic structures and/or geometric structures of both the nucleoside and phosphate moieties of protein-bound nucleoside triphosphates differ considerably—as discussed for the aminoacyl tRNA synthetases and the kinases, respectively. For adenylate kinase, the binding of the two nucleotides at the active site differ both for the phosphate chain[30] and the adenine moieties,[47] as shown by the NMR (^{1}H and ^{31}P) of the bound bisubstrate analogue Ap_5A.

Unexpected amino acid sequence homologues are sometimes encountered for apparently unrelated proteins. For example, 48% homology[48] and 71% hydropathy[49]

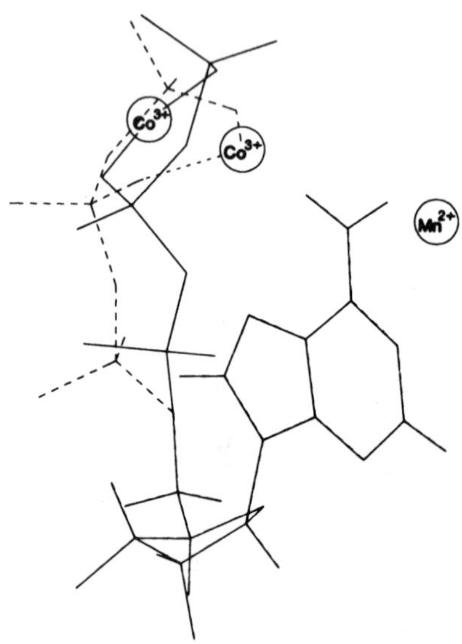

FIGURE 6. Structure deduced from ^{31}P NMR of Co(III)ATP bound to PP-ribose-P synthetase. The same complex with ribose-5-P bound to the enzyme.[46]

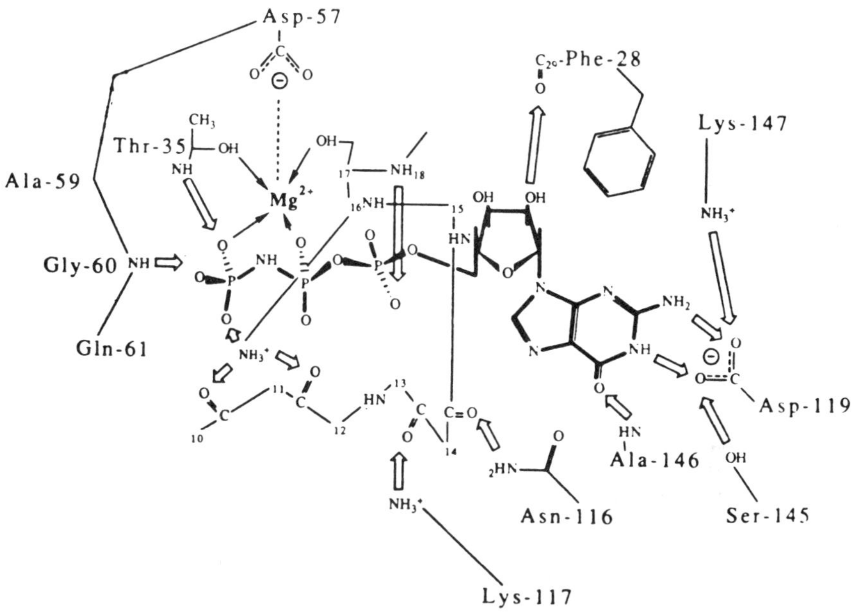

FIGURE 7. Structure of Mg-GTP and its binding domain on p21 deduced from X-ray diffraction.[50]

have recently been found between *E. coli* tyrosyl tRNA synthetase and human estrogen receptor. A similarity in the GTP binding domains of the elongation factor EF-Tu and the GTP binding site of the *Ha ras* oncogene product, p21,[50] is not surprising, but the close similarity of the ATP binding domain of adenylate kinase to that of the GTP binding domain of p21[51,52] was unanticipated. In addition to evidence from structural studies of adenylate kinase by NMR[37,51] and X-ray diffraction,[53] point mutations in the nucleotide binding domain of *E. coli* adenylate kinase greatly affects the kinetics and structure of the enzyme,[54] as has been found for p21. The crystal structure of the GTP binding domain of p21 *Ha ras* in its conformation has been determined in greater detail (FIG. 7) than any other nucleoside triphosphate-binding protein. Every structural interpretation is bolstered, for example, by activity changes in solution with analogues, and by mutations (see Baker[48] and the references therein). The investigators used a fully active truncated form of p21 (1-166 amino acids) complexed with the slowly hydrolyzable analogue GppNp. The structure has the same topology and shares a high degree of amino acid sequence homology with the GDP binding domain of EF-Tu, the only other guanine nucleotide-binding protein with a known three-dimensional structure. The crystal structure of the GDP-bound form of p21 has been determined,[55] and significant differences between the diphosphate and triphosphate complexes have been noted. Their similarity to the differences between the EF-Tu complexes of GDP and GTP led the authors to suggest that perhaps the same structural elements are involved for all guanine nucleotide-binding proteins in their transition between GDP- and GTP-bound forms.

As shown in FIGURE 7, the three main-chain nitrogens of residues 13 to 16 (in the loop) and those of 17 and 18 (in the following helix) point toward and are within

3.5 Å of the phosphate oxygens. These nitrogens can form hydrogen bonds. For example, the amide nitrogen of residue 18 forms a hydrogen bond to the *pro-R* oxygen of the α-phosphate. Supporting evidence arises from experiments with GDPαS and GTPαS showing that replacement of the S_p oxygen by sulfur leads to greater activity than replacement of the R_p oxygen. The amino group of Lys 18 lies close enough to the γ-phosphate to allow an ion-pair interaction.

The guanine base is situated between the aromatic side chain of Phe 28 on one side and the aliphatic part of the side chain of Lys 117 on the other. The binding of the guanine base is greatly facilitated by the strong hydrogen bond between O-6 of the guanine ring and the amide nitrogen of Ala 146. When the O-6 of guanine is removed (h6-GDP), the affinity for p21 is decreased 25-fold. The modification has an even greater effect for EF-Tu. Asp 119, which lies close to the N-1 and the N-2 of guanine, is part of the N-K-X-D sequence, which is highly conserved in all guanine necleotide-binding proteins. The ribose moiety of GppNp is in a 2′-*endo* conformation. The 2′ and 3′-hydroxyl groups are exposed to the solvent, in agreement with the finding that at least one of the hydroxyl groups can be substituted in GDP and GTP without changing the binding affinity.

The phosphate-binding loop in both the GDP and GTP structures is wrapped around the β-phosphate group. In addition to the four main-chain peptide nitrogens in the loop, there are other main-chain amide nitrogens that interact with phosphates: NH-17, NH-18, and NH-60. As expected from the work in solution, Mg^{2+} binds in a bidentate β-γ complex as it does in the EF-Tu complex. In addition to the two phosphate ligands of Mg^{2+}, at least two of the remaining four, Ser 17 and Thr 35, are close enough to the metal ion to be in the first coordination sphere, and Asp 57 may be close enough as well. Thr 35 also interacts with an oxygen of γ-phosphate. Mutations of Thr 35 → Ser or, even worse, Thr 35 → Ala affect the interaction with the GTPase-activating protein. Further detailed relationships between structure and function can be found in Baker,[48] but the ones discussed above illustrate the importance of the protein on the conformation of the bound metal nucleotide.

In conclusion, as more detailed quantitative data is collected from X-ray diffraction and NMR, together with experiments on chemically modified nucleotides, which could reveal protein sequence homologies and identify mutationally modified proteins, some generalizations will be possible concerning the effect of the structure of the binding domain on the conformation of the bound metal nucleotide. Perhaps it will be found that one criterion of topographical similarity of the binding domains may well be the metal coordination scheme, that is, the existence of β-γ bidentate coordination or α-β-γ tridentate coordination of nucleoside triphosphates.

REFERENCES

1. WESTHEIMER, F. H. 1987. Science **235:** 1173-1178.
2. FREIST, W. 1988. Angew. Chem. Int. Ed. Engl. **27:** 773-788.
3. EVANS, F. E. & R. H. SARMA. 1975. Cancer Res. **35:** 1458-1463.
4. KENNARD, O., N. W. ISAACS, W. D. S. MOTHERWELL, J. C. KOPPOLA, D. L. WAMPLER, A. C. LARSON & D. G. WATSON. 1971. Proc. R. Soc. London Ser. A **235:** 401-436.
5. LARSON, A. C. 1978. Acta Crystallogr. Sect. B **34:** 3601-3604.
6. TETAS, M. & J. M. LOWENSTEIN. 1963. Biochemistry **2:** 350-357.
7. VISWAMITRA, M. A. & M. V. HOSUR. 1977. Abstracts of the Fourth European Crystallographic Meeting (Oxford): 265. Cotswold Press. Oxford.

8. MULLER, F. A. & A. B. DULUKE. 1971. Abstracts of the American Crystallographic Association (winter meeting): 26.
9. SUNDRALINGAM, M. 1973. Fifth Jerusalem Symposium in Quantum Chemistry and Biochemistry. E. D. Bergmann & B. Pullman, Eds.: 417. Academic Press. New York, NY.
10. VISWAMITRA, M. A., M. V. HOSUR, Z. SHAKKED & O. KENNARD. 1976. Nature (London) **262:** 234-236.
11. ADAMIAK, D. A. & W. SAENGER. 1980. Acta Crystallogr. Sect. B **36:** 2585-2589.
12. SWAMINATHAN, P. & M. SUNDRALINGAM. 1980. Acta Crystallogr. Sect. B **36:** 2590-2597.
13. SHAKKED, Z., M. A. VISWAMITRA & O. KENNARD. 1980. Biochemistry **19:** 2567-2571.
14. PILATUS, V., A. MAYER, W. OFFERMAN & D. LEIBFRITZ. 1987. Biochim. Biophys. Acta **926:** 106-113.
15. BOCK, J. L. 1980. J. Inorg. Biochem. **12:** 119-130.
16. BISHOP, E. O., S. J. KIMBER, D. ORCHARD & B. E. SMITH. 1981. Biochem. Biophys. Acta **635:** 63-72.
17. COHN, M. & T. R. HUGHES. 1962. J. Biol. Chem. **237:** 176-181.
18. GORENSTEIN, D. G. 1981. Annu. Rev. Biophys. Bioeng. **10:** 355-386.
19. ECKSTEIN, F. & R. S. GOODY. 1976. Biochemistry **15:** 1685-1691.
20. JAFFE, E. & M. COHN. 1978. Biochemistry **17:** 652-657.
21. LEYH, T. S., R. D. SAMMONS, P. A. FREY & G. H. REED. 1982. J. Biol. Chem. **257:** 15047-15053.
22. DUNAWAY-MARIANO, D. & W. W. CLELAND. 1980. Biochemistry **19:** 1496-1505.
23. ORIOLI, P., R. CINI, D. DONATI & S. MANGANI. 1981. J. Am. Chem. Soc. **103:** 4446-4452.
24. CINI, R., M. SABAT, M. SUNDRALINGAM, M. C. BURLA, A. NUNGI, G. POLIDORI & P. F. ZANAZZI. 1983. J. Biomol. Struct. Dyn. **1:** 633-637.
25. RAY, B. D. & B. D. NAGESWARA RAO. 1988. Biochemistry **27:** 5574-5579.
26. ROA, B. D. N., M. COHN & L. NODA. 1978. J. Biol. Chem. **253:** 1149-1158.
27. VASAVADA, K. V., B. D. RAY & B. D. NAGESWARA RAO. 1984. J. Inorg. Biochem. **21:** 323-335.
28. VAN DER LIJN, P., J. R. BARRIO & N. J. LEONARD. 1979. Biochemistry **18:** 5557-5561.
29. LIENHARD, G. E. & I. I. SECEMSKI. 1973. J. Biol. Chem. **248:** 1121-1123.
30. NAGESWARA RAO, B. D. & M. COHN. 1977. Proc. Natl. Acad. Sci. USA **74:** 5355-5357.
31. YOUNT, R. G. 1975. Adv. Enzymol. Relat. Areas Mol. Biol. **43:** 1-56.
32. ECKSTEIN, F. 1985. Annu. Rev. Biochem. **54:** 367-402.
33. JAFFE, E. K. & M. COHN. 1979. J. Biol. Chem. **256:** 9450-9456.
34. CLELAND, W. W. & A. S. MILDVAN. 1979. Adv. Inorg. Biochem. **1:** 163-191.
35. DUNAWAY-MARIANO, D. & W. W. CLELAND. 1980. Biochemistry **19:** 1506-1515.
36. MILDVAN, A. S. & R. K. GUPTA. 1978. Methods Enzymol. **49:** 332-358.
37. MILDVAN, A. S. 1989. FASEB J. **3:** 1705-1714.
38. NAGESWARA RAO, B. D. 1987. *In* Biophosphates and Their Analogues, Synthesis, Structure, Metabolism and Activity. K. S. Bruizik & W. J. Stec, Eds.: 547-559. Elsevier. Amsterdam.
39. JARORI, G. K., B. D. RAY & B. D. NAGESWARA RAO. 1985. Biochemistry **24:** 3487-3494.
40. RAY, B. D., P. ROSCH & B. D. NAGESWARA RAO. 1988. Biochemistry **27:** 8669-8675.
41. MCLAUGHLIN, A., J. S. LEIGH & M. COHN.. 1976. J. Biol. Chem. **251:** 2777-2787.
42. LEYH, T. S., P. J. GOODHART, A. C. NGUYEN, G. L. KENYON & G. H. REED.. 1985. Biochemistry **24:** 308-316.
43. LODATO, D. T. & G. H. REED. 1987. Biochemistry **26:** 2243-2250.
44. MOORE, J. M. & G.H. REED. 1985. Biochemistry **24:** 3487-3494.
45. COHN, M. & B. D. NAGESWARA RAO. 1978. *In* Frontiers in Physicochemical Biology. B. Pullman, Ed.: 193-211. Academic Press. New York, NY.
46. GRANOT, J., K. J. GIBSON, R. L. SWITZER & A. S. MILDVAN. 1980. J. Biol. Chem. **255:** 10931-10937.
47. ROSCH, P., W. KLAUS, M. AUER & R. S. GOODY. 1989. Biochemistry **28:** 4318-4325.
48. BAKER, M. E. 1989. FASEB J. **3:** 2086-2088.
49. TRITSCH, G. L. 1989. FASEB J. **3:** 2554.
50. PAI, E. F., W. KABSCH, V. KRENGEL, K. C. HOLMES, J. JOHN & A. WITTINGHOFER. 1989. Nature **341:** 209-214.

51. FRY, D. C., S. A. KUBY & A. S. MILDVAN. 1986. Proc. Natl. Acad. Sci. USA **83:** 907-911.
52. DREUSICKE, D. & G. E. SCHULZ. 1986. FEBS Lett. **208:** 301-304.
53. MULLER, C. W. & G. E. SCHULZ. 1988. J. Mol. Biol. **202:** 909-912.
54. REINSTEIN, J., M. BRUNE & A. WITTINGHOFER. 1988. Biochemistry **27:** 4712-4720.
55. DEVOS, A. M., L. TONG, M. V. MILBURN, P. M. MATIUS, J. JANCARIK, S. NOGUCHI, S. NISHIMURA, K. MIURA, E. OHTSUKA & S. H. KIM. 1988. Science **239:** 888-893.

DISCUSSION OF THE PAPER

E. RAPAPORT (*Boston University School of Medicine, Boston, MA*): Have you ever observed a folded conformation of ATP that could exist at around pH 5 where the purine ring acquires a partial positive charge and may interact electrostatically with the partially negatively charged phosphate moieties? We have seen such interactions in our studies of the preferred conformations (in solutions) of Ap_4A and Ap_5A. These compounds possess a stacked conformation at pH 7 and 37 °C, but when the pH is dropped to 4-5, they switch to a folded, unstacked conformation because of the type of interactions mentioned above.

COHN: In solution, ATP molecules are stacked intermolecularly, and Ap_4A or Ap_5A are stacked intramolecularly. As far as I know, the effect of pH on the stacking of ATP has not been examined by NMR. The binding of Ap_5A to adenylate kinase unstacks the intramolecular stacking as evidenced by the appearance of five different ^{31}P peaks in enzyme-bound Ap_5A as opposed to the two peaks seen for Ap_5A in solution.

T. FORRESTER (*St. Louis University Medical Center, St. Louis, MO*): Two questions: Can you comment upon the nature of the disodium salt of ATP? You stated that the high negative charge on ATP kept the molecule inside the cell. How does that agree with the inside of the cell being negative? Is the protein binding of ATP intracellular enough to overcome the tendency for ATP to be repulsed out of the cell?

COHN: The high charge on ATP makes it difficult to go through the lipid membrane. Extracellular ATP does not enter cells. Nor, for that matter, do other charged species. Specific mechanisms are required to release ATP from cells.

A Comparison of P_1- and P_2-Purinoceptors

D. M. PATON [a] AND T. TAERUM [b]

[a]*Department of Oral Biology*
and
[b]*University Computing Systems*
University of Alberta
Edmonton, Alberta, Canada T6G 2N8

INTRODUCTION

In 1978, Burnstock[1] proposed a system for the classification of the receptors at which adenosine and adenine nucleotides act to produce their diverse and numerous effects. This proposal was based on a review of the extensive literature concerning the actions of purine nucleosides and nucleotides on a wide variety of tissues. This classification system termed the receptors at which the purine nucleosides and nucleotides act purinoceptors and proposed that there were two major subtypes, the P_1- and P_2-purinoceptors. This classification system was based on four criteria:

1. The relative potencies as agonists of adenosine, AMP, ADP, and ATP
2. The selective actions of antagonists, particularly methylxanthines such as theophylline.
3. The activation of adenylate cyclase by adenosine but not by ATP.
4. The induction of prostaglandin synthesis by ATP but not by adenosine.

According to this classification, P_1-purinoceptors are more responsive to adenosine and AMP than to ADP and ATP. Methylxanthines such as caffeine and theophylline are selective antagonists at these P_1-purinoceptors, and occupation of these receptors alters the activity of adenylate cyclase. P_2-purinoceptors are more responsive to ADP and ATP than to adenosine and AMP. Methylxanthines are not antagonists at such receptors, and occupation of these receptors may result in increased prostaglandin synthesis.

In this review, the present understanding of P_1- and P_2-purinoceptors will be discussed in terms of the original criteria proposed by Burnstock for their classification.

CLASSIFICATION OF PURINOCEPTORS UTILIZING THE RELATIVE POTENCIES OF ADENOSINE AND ADENINE NUCLEOTIDES AS AGONISTS

Classification systems for the receptors for adenosine were independently proposed by Van Calker *et al.*[2] and Londos *et al.*[3] These classifications were based on the relative potencies of adenosine analogues as agonists, and on whether such agonists increased or decreased the activity of adenylate cyclase. It was proposed that occupation of A_1 (R_i) receptors results in inhibition of adenylate cyclase whereas occupation of A_2 (R_a) receptors results in activation of the enzyme.

The actions of adenosine at both A_1 and A_2 receptors are antagonized by methylxanthines such as caffeine and theophylline. Because of this finding, A_1 and A_2 receptors can be regarded as subtypes of the P_1-purinoceptor.

A problem in the use of adenosine in such studies is that it is subject to transport into tissues and to hydrolysis by adenosine deaminase. As a consequence, the potency of adenosine as an agonist in peripheral tissues is significantly potentiated by inhibitors of adenosine transport such as dipyridamole and, to a lesser extent, by inhibitors of adenosine deaminase such as deoxycoformycin.[4] It is essential therefore that the uptake of adenosine is inhibited if the true potency of the nucleoside is to be established.

In peripheral tissues, AMP, ADP, and ATP are subject to extracellular breakdown by ectonucleotidases to adenosine.[5] As a consequence, it is often difficult to decide whether the effects produced by the adenine nucleotides are produced by the nucleotides per se or are produced following the formation of adenosine.[6–8] The use of theophylline as an adenosine antagonist or of dipyridamole as an inhibitor of adenosine transport in such studies has not always been helpful. In rat vas deferens, for example, a series of 2′-substituted adenine nucleotide analogues were not hydrolyzed to adenosine as determined by HPLC analysis but their presynaptic inhibitory effects were antagonized by theophylline and potentiated by dipyridamole.[6] In guinea pig atria, through the use of exogenous adenosine deaminase, 5′-nucleotidase, and inhibitors of 5′-nucleotidase, evidence was obtained that ATP could apparently act directly at P_1-purinoceptors.[7,8]

These considerations have led to the realization that the potencies of adenosine and adenine nucleotides as agonists are subject to a number of important modifying factors. Because the classification of P_1- and P_2-purinoceptors could be complicated by these modifying factors, increasing reliance has been placed on the actions and potencies of adenosine and adenine nucleotide analogues less susceptible to enzymatic breakdown and/or tissue uptake.[9,10]

CLASSIFICATION OF P_1-PURINOCEPTOR SUBTYPES UTILIZING ADENOSINE AGONISTS

Adenosine analogues have been used extensively to determine which P_1-purinoceptors belong to the A_1 subtype and which belong to the A_2 subtype.[9,10] Particular reliance has been placed on the relative potencies of N^6-(*R*)-(phenylisopropyl)adenosine (*R*-PIA), N^6-(*S*)-(phenylisopropyl)adenosine (*S*-PIA), N^6-cyclohexyladenosine (CHA), 2-chloroadenosine (2-CADO), and 5′-*N*-ethylcarboxamidoadenosine (NECA). It has been concluded, for example, that at A_1 receptors *R*-PIA, CPA > NECA, *S*-PIA > adenosine, whereas at A_2 receptors NECA > 2-CADO > *S*-PIA, CHA > adenosine.[10]

The relative potencies of adenosine analogues reported in the literature for 15 different tissues were recently reviewed by one of us,[11] with particular attention being paid to the potencies of *R*-PIA, *S*-PIA, CHA, N^6-cyclopentyladenosine (CPA), NECA, and 2-(phenylamino)adenosine (CV-1808). It was concluded that the potencies of *R*-PIA, CHA, and CPA are generally so similar that there is little to be gained by using more than one of these three N^6-substituted analogues. It was proposed that the following represent reasonable working criteria for the classification of adenosine or P_1-purinoceptors:

1. Rank order of potency of CPA (or CHA), NECA, and CV-1808. The order at A_1 receptors is CPA (or CHA) > NECA > > CV-1808, and the order at A_2 receptors is NECA > CV-1808 > CPA (or CHA).
2. Ratio of potencies of *R*-PIA and *S*-PIA. A ratio of greater than 20 suggests that the receptor is A_1 in type, and a ratio of less than 10 suggests that it is A_2 in type. The potency of *R*-PIA should be greater than that of *S*-PIA at both receptors.
3. Ratio of potencies of NECA and CV-1808. A ratio of greater than 40 suggests that the receptor is A_1 in type, and a ratio of less than 20 suggests that it is A_2 in type. The potency of NECA should be greater than that of CV-1808 at both receptors.

We have recently examined the utility of a multidimensional scaling procedure[12] for the classification of adenosine receptors. The aim of this analysis, employing ALSCAL, was to objectively determine the relative importance of different analogues in distinguishing between the adenosine receptors in different tissues. Distances between samples were expressed as the differences in potency (nanomolar) between samples. Data were then scaled using the INSCAL model assuming interval data and individual differences. Samples were given weights by ALSCAL on two dimensions.

In this analysis the following adenosine agonists were selected for study: CHA, CPA, *R*-PIA, *S*-PIA, 2-CADO, CV-1808, and NECA—analogues that have been extensively used in studies of adenosine receptor subtypes. These analogues also appeared to have potential utility for this purpose as they have been considered to be helpful in the classification of adenosine subtypes. Published results were included in the study if the potencies for at least four of these analogues were reported, as this is the minimum number of values required by the ALSCAL program. Results of published studies of 89 different tissues, parameters, or processes were recorded and examined.

For 15 tissues or preparations, potency values for all seven adenosine analogues were available. These tissues were examined in detail using the multidimensional scaling procedure, which provided values for two dimensions. It was found that in the absence of values for either *S*-PIA or CV-1808 the scaling procedure did not allow the assignment of tissues to two groups or subtypes. When both *S*-PIA and CV-1808 were included in such scaling, however, it was found that tissues could be readily assigned to one of two groups.

In group 1, dimension 1 was >0.9 while dimension 2 was <0.1; in group 2, dimension 1 was generally <0.1 while dimension 2 was >0.9. All 10 tissues or preparations in group 1 had been assigned to the A_1 subtype by the authors of the papers reporting the original data used in this analysis, while 5 tissues or preparations, with one exception, in group 2 had been originally assigned to the A_2 subtype. The single exception was the presynaptic inhibitory receptor in rat vas deferens, which had been originally assigned to the A_1 subtype.[13] Subsequent studies have indicated that the actions of some analogues in this tissue (for example, NECA) are more complex than at first thought, and may reflect actions at both A_1 and A_2 receptors.

As a result of these preliminary studies, the following two combinations of analogues were selected for further study utilizing this multidimensional scaling procedure: 1) *R*-PIA (or CHA), *S*-PIA, CV-1808, and NECA; 2) *R*-PIA (or CHA), *S*-PIA, 2-CADO, and NECA. These combinations were then used in a study of the larger group of 89 tissues or preparations.

It was found that only one combination of analogues (*R*-PIA, *S*-PIA, CV-1808, and NECA) allowed the clear separation into two groups of 23 tissues or preparations for which values for the four analogues had been reported. These corresponded to the original designation of such tissues as being A_1 or A_2 in type, with the exception of rat vas deferens, as noted previously. Additional details of the method and the results obtained will be published separately. In these tissues at A_1 receptors, *R*-PIA > (or =) NECA > *S*-PIA > CV-1808 (or *R*-PIA > *S*-PIA > NECA > CV-1808); at A_2 receptors, NECA > CV-1808 = *R*-PIA > *S*-PIA.

It seems reasonable to conclude that this multidimensional scaling procedure has potential in the classification of P_1-purinoceptor subtypes and indeed of all receptor subtypes. As a result of this statistical approach, it is proposed that a study of the relative potencies of *R*-PIA, *S*-PIA, CV-1808, and NECA is likely to be particularly helpful in the classification of P_1-purinoceptors as being A_1 or A_2 in subtype. The addition of CHA and/or CPA to such studies is unlikely to add very much as their potencies are so similar to that of *R*-PIA.

Recent detailed structure-activity studies have resulted in the synthesis of adenosine analogues with increased selectivity in radioligand binding studies. N^6-[(1*R*,2*S*,4*S*)-2-*Endo*-norbornyl]adenosine (*S*-ENBA) has been reported to have > 1800-fold selectivity for A_1 receptors,[14] whereas N^6-[(*RS*)-2-(3,5-dimethoxyphenyl)-2-(2-methylphenyl)ethyl]adenosine (PD 125,944) and 2-(2-phenylethylamino)-5′-*N*-carboxamidoadenosine (CGS 21577) have been reported to have about 30-fold[15] and 60-fold[16] selectivity, respectively, for A_2 receptors. It is obvious that these analogues should be included in future studies designed to classify the adenosine receptor subtype present in different tissues and that their potencies should be compared to the potencies of *R*-PIA, *S*-PIA, CV-1808, and NECA. It seems very likely that the addition of such analogues, provided that their reported A_1 or A_2 selectivity can be confirmed in other tissues, will greatly facilitate the classification of P_1-purinoceptor subtypes.

Another approach that has been used in the study of P_1-purinoceptor subtypes is an examination of the effect of hypothermia on responses of peripheral tissues to adenosine and 2-CADO.[17] It was found that in tissues with adenosine receptors of the A_1 subtype (rat vas deferens, guinea pig ileum, rat atria), lowering the temperature of the external medium from 37 °C to 27 °C increased the sensitivity of tissues to the adenosine analogues. By contrast, in guinea pig trachea this degree of hypothermia did not alter tissue sensitivity to the analogues. It was suggested that there may be a fundamental temperature-dependent difference between the two adenosine receptor subtypes.

The L-enantiomers of adenosine, NECA, 2-CADO, and 2-azidoadenosine, had no inhibitory effects in guinea pig atria and ileum (tissues with A_1 receptors), and had only weak potency in guinea pig trachea (a tissue with A_2 receptors), indicating that the P_1-purinoceptors in these tissues are stereospecific for the D-enantiomers of adenosine and its analogues.[18] By contrast, the P_2-purinoceptors have been shown to lack this degree of stereoselectivity.[19]

CLASSIFICATION OF P_2-PURINOCEPTORS UTILIZING ADENINE NUCLEOTIDE ANALOGUES

The development and utilization of a number of adenine nucleotide analogues, particularly by Burnstock, Cusack, and their colleagues, has provided considerable evidence for a number of P_2-purinoceptor subtypes.[10,19]

Burnstock and Kennedy[20] proposed that P_2-purinoceptors could be divided into two subtypes based on the relative orders of potencies of ATP, α,β-methylene-ATP (AMP-CPP), β,γ-methylene-ATP (AMP-PCP), and 2-methylthio-ATP. At P_{2X} receptors, AMP-CPP, AMP-PCP > ATP, 2-methylthio-ATP. At P_{2Y} receptors, 2-methylthio-ATP > > ATP > AMP-CPP, AMP-PCP.

Cusack and his colleagues have synthesized and examined the actions of a number of adenine nucleotides that have provided further evidence for the P_{2X} and P_{2Y} subtypes.[19] At the P_{2Y} receptor, 2-methylthio-ATP is 700-fold more potent than 2-methylthio-L-ATP, whereas the P_{2X} receptor shows no stereoselectivity toward the D- and L-enantiomers of ATP, ADP, 2-chloro-ATP, or 2-methylthio-ATP. These workers also demonstrated that the P_1-purinoceptor shows absolute stereoselectivity for the natural enantiomer. Based on these observations and the knowledge that AMP-PCP is a potent agonist at P_{2X} receptors, they synthesized β,γ-methylene-L-ATP (L-AMP-PCP). This compound was found to be the most potent known agonist at P_{2X} receptors while having no agonist activity at P_{2Y} receptors. Adenosine 5′-(2-fluorodiphosphate) (ADP-β-F) may be a selective agonist at P_{2Y} receptors.

The receptors for ATP on platelets and mast cells do not fit the P_{2X} or P_{2Y} classification and have therefore been termed P_{2T}- and P_{2Z}-purinoceptors, respectively.[21]

CLASSIFICATION OF P_1-PURINOCEPTOR SUBTYPES UTILIZING ADENOSINE ANTAGONISTS

The utility of the methylxanthines, caffeine, and theophylline as adenosine antagonists is limited because these agents are not selective for A_1 or A_2 adenosine receptor subtypes, they are not very potent as antagonists, and they also inhibit phosphodiesterase.[22] Consequently, considerable effort has been devoted to the search for adenosine antagonists that are A_1 or A_2 selective, are more potent, and do not inhibit the enzyme, phosphodiesterase. A number of promising compounds have proved to have very limited solubility, however, and this has limited their usefulness. Recent studies in this area have been reviewed by a number of authors.[9,22]

8-Phenyltheophylline is about 100-fold more potent than theophylline and does not inhibit phosphodiesterase. It is also nonselective, however, and has a greatly reduced solubility.[22] To increase the solubility of this agent, charged side chains have been added to the basic structure. The xanthine congeners XCC and XAC have resulted from this approach.[9] XAC, the xanthine amine congener of 8-phenyl-1,3-dipropylxanthine, has proved to be a valuable antagonist radioligand for the study of A_1 receptors with a degree of A_1 selectivity in some tissues and species.[9]

8-Cyclopentyltheophylline (CPT) has about 100-fold A_1 selectivity in the rat, has high A_1 affinity, has moderately good water solubility, and is only a weak phosphodiesterase inhibitor.[22] 1,3-Dipropyl substitution led to the antagonist 8-cyclopentyl-1,3-dipropylxanthine (DPCPX or PD 116,948), which proved to have a greater than 20-fold increase in A_1 affinity and a 6-fold increase in A_1 selectivity compared to the parent compound, CPT.[22]

CGS 15943A, a triazoloquinazoline, has been reported to be a potent adenosine antagonist with a limited degree of A_2 selectivity.[23] A nonxanthine, CGS 15943A does not inhibit phosphodiesterase.

Thus for the classification of P_1-purinoceptor subtypes, there are now a number of xanthine antagonists (CPT, DPCPX) with a significant degree of A_1 selectivity, and these should certainly be utilized in classification studies.

TRANSDUCTION MECHANISMS FOR P_1-PURINOCEPTORS

In the original classification scheme for purinoceptors, the P_1 subtype referred to those purinoceptors whose occupation by adenosine resulted in activity changes for adenylate cyclase.[1] The subclassification of P_1-purinoceptors (or adenosine receptors) into A_1 (R_i) or A_2 (R_a) subtypes was also based on the ability of adenosine and adenosine analogues to respectively inhibit or activate adenylate cyclase.[2,3] As Burnstock has pointed out,[1] however, few of the actions of adenosine have been shown conclusively to result from changes in cyclic AMP.

In relation to the presynaptic inhibitory actions of adenosine, Fredholm and Dunwiddie[24] have summarized the evidence that there may be only one type of A_1 adenosine receptor that can couple with different types of G proteins, each of which then interacts with a different effector system. These appear to include not only the enzyme, adenylate cyclase, but also Ca^{2+} and K^+ channels. In atria, the A_1 adenosine receptor may be directly coupled to a K^+ channel via a pertussis toxin-sensitive N protein.[25]

It was also originally proposed that ATP, but not adenosine, could induce prostaglandin synthesis.[1] In endothelial cells, for example, Boeynaems and colleagues found that there is a burst of prostacyclin synthesis in response to ADP and ATP, mediated by P_{2Y}-purinoceptors.[26] These workers have suggested that this may result from a rapid and transient increase of cytosolic Ca^{2+} due to Ca^{2+} release from the endoplasmic reticulum by inositol 1,4,5-trisphosphate, though other mechanisms are possible.[26] In the isolated rabbit heart, adenosine also appeared to increase the formation of prostacyclin.[27]

In view of such considerations, Burnstock[10] has concluded that purinoceptors should not be classified according to their effect on adenylate cyclase.

CONCLUSIONS AND FUTURE DIRECTIONS

A decade has passed since Burnstock's proposal that receptors for adenosine and adenine nucleotides be classified as P_1- and P_2-purinoceptors,[1] and it is now clear that this proposal has represented an important advance and has served as an impetus for a considerable amount of research.

There is now considerable evidence for at least two subtypes of the P_1-purinoceptor, the A_1 and A_2 adenosine receptors. It is suggested that classification of such receptors should involve a study of the relative potencies of the agonists *R*-PIA, *S*-PIA, CV-1808, and NECA, together with the relative potencies of the newer agonists *S*-EBNA, PD 125,944, and CGS 21577, as these agonists have been reported to have significant selectivity for A_1 or A_2 receptors. Antagonists should also be employed in such studies, and CPT and DPCPX would be important to include in view of their reported selectivity at A_1 receptors.

For the classification of the P_{2X} and P_{2Y} subtypes of the P_2-purinoceptor it is clearly important to use L-AMO-PCP, a selective agonist, as well as ATP, AMP-PCP, AMP-CPP, and 2-methylthio-ATP. There is clearly a continuing need, however, for more potent and selective antagonists at P_2-purinoceptors.

REFERENCES

1. BURNSTOCK, G. 1978. *In* Cell Membrane Receptors for Drugs and Hormones: A Multidisciplinary Approach. L. Bolis & R. W. Straub, Eds.: 107-118. Raven Press. New York, NY.
2. VAN CALKER, D., M. MULLER & B. HAMPRECHT. 1978. Nature **276:** 839-841.
3. LONDOS, C., D. M. F. COOPER & J. WOLFF. 1980. Proc. Natl. Acad. Sci. USA **77:** 2551-2554.
4. MULLER, M. J. & D. M. PATON. 1979. Naunyn-Schmiedeberg's Arch. Pharmacol. **306:** 23-28.
5. CUSACK, N. J., S. M. O. HOURANI & L. A. WELFORD. 1988. *In* Adenosine and Adenine Nucleotides: Physiology and Pharmacology. D. M. Paton, Ed.: 93-100. Taylor & Francis. London.
6. WEBSTER, D. R., G. D. BOSTON, N. H. G. HOLFORD & D. M. PATON. 1986. Naunyn-Schmiedeberg's Arch. Pharmacol. **333:** 163-167.
7. BURNSTOCK, G. & P. MEGHJI. 1981. Br. J. Pharmacol. **73:** 879-885.
8. COLLIS, M. G. & S. J. PETTINGER. 1983. Eur. J. Pharmacol. **81:** 521-529.
9. DALY, J. W. & K. A. JACOBSON. 1989. *In* Adenosine Receptors in the Nervous System. J. A. Ribeiro, Ed.: 41-52. Taylor & Francis. London.
10. BURNSTOCK, G. 1989. *In* Adenosine Receptors in the Nervous System. J. A. Ribeiro, Ed.: 1-14. Taylor & Francis. London.
11. PATON, D. M. 1989. *In* Adenosine Receptors in the Nervous System. J. A. Ribeiro, Ed.: 79-86. Taylor & Francis. London.
12. SCHIFFMAN, S., M. L. REYNOLDS & F. W. YOUNG. 1981. Introduction to Multidimensional Scaling Theory: Theory, Methods and Applications. Academic Press. Toronto.
13. PATON, D. M., R. A. OLSSON & R. D. THOMPSON. 1986. Naunyn-Schmiedeberg's Arch. Pharmacol. **333:** 313-322.
14. TRIVEDI, B. K., A. J. BRIDGES, W. C. PATT, S. R. PRIEBE & R. F. BRUNS. 1989. J. Med. Chem. **32:** 8-10.
15. BRIDGES, A. J., R. F. BRUNS, D. G. ORTWINE, S. R. PRIEBE, D. L. SZOTEK & B. K. TRIVEDI. 1988. J. Med. Chem. **31:** 1282-1285.
16. WILLIAMS, M., R. L. WEBB, H. H. OEI, M. F. JARVIS & A. J. HUTCHISON. 1989. *In* Adenosine Receptors in the Nervous System. J. A. Ribeiro, Ed.: 61-68. Taylor & Francis. London.
17. BROADLEY, K. J., S. BROOME & D. M. PATON. 1985. Br. J. Pharmacol. **84:** 407-415.
18. BROWN, C., G. BURNSTOCK, N. J. CUSACK, P. MEGHJI & C. J. MOODY. 1982. Br. J. Pharmacol. **75:** 101-107.
19. CUSACK, N. J., L. A. WELFORD & S. M. O. HOURANI. 1988. *In* Adenosine and Adenine Nucleotides: Physiology and Pharmacology. D. M. Paton, Ed.: 73-84. Taylor & Francis. London.
20. BURNSTOCK, G. & C. KENNEDY. 1985. Gen. Pharmacol. **16:** 433-440.
21. GORDON, J. L. 1986. Biochem. J. **233:** 309-319.
22. BRUNS, R. F., R. E. DAVIS, F. W. NINTEMAN, B. P. H. POSCHEL, J. N. WILEY & T. G. HEFFNER. 1988. *In* Adenosine and Adenine Nucleotides: Physiology and Pharmacology. D. M. Paton, Ed.: 39-49. Taylor & Francis. London.
23. WILLIAMS, M., J. FRANCIS, G. GHAI, A. BRAUNWALDER, S. PSYCHOYOS, G. A. STONE & W. D. CASH. 1987. J. Pharmacol. Exp. Ther. **241:** 415-420.
24. FREDHOLM, B. B. & T. V. DUNWIDDIE. 1988. TIPS **9:** 130-134.
25. SCHMITZ, W., T. HELLER, M. KOCHER, J. NEUMANN, H. SCHOLZ, J. SCHOLZ, V. STEMMILDT & K. STORTZEL. 1988 *In* Adenosine and Adenine Nucleotides: Physiology and Pharmacology. D. M. Paton, Ed.: 167-176. Taylor & Francis. London.
26. BOEYNAEMS, J. M., E. RASPE, S. PIROTTON, D. DEMOLLE, A. VAN COEVORDEN & C. ERNEUX. 1988. *In* Adenosine and Adenine Nucleotides: Physiology and Pharmacology. D. M. Paton, Ed.: 177-184. Taylor & Francis. London.
27. KARWATOWSKA-PROKOPCZUK, E., G. CIABATTONI & A. WENNWALM. 1988. Br. J. Pharmacol. **94:** 721-728.

Subtypes of P_2-Purinoceptors

Studies Using Analogues of ATP

NOEL J. CUSACK

Whitby Research, Inc.
Irvine, California 92715

SUSANNA M. O. HOURANI

University of Surrey
Guildford, Surrey GU2 5XH, England

INTRODUCTION

The design, synthesis, and pharmacology of analogues of ATP have proved extremely useful in the classification of P_2-purinoceptors,[1] and provide clear evidence for at least four subtypes,[2] which have been termed P_{2Z}, P_{2X}, P_{2Y}, and P_{2T}.[3,4] The P_{2Z}-purinoceptor is found on mast cells, macrophages, lymphocytes, and epithelial cells; the excitatory P_{2X}-purinoceptor, on visceral and vascular smooth muscle and sensory neurons; the inhibitory P_{2Y}-purinoceptor, on visceral and vascular smooth muscle, endothelial cells, hepatocytes, parotid acini, type II alveoli, and pancreatic β cells; and the P_{2T}-purinoceptor, uniquely, on blood platelets.[4] Systematic alterations to the ATP molecule define closely the structure-activity relationships for agonist potencies at each subtype of the P_2-purinoceptor.[2] In particular, 2-chloro-ATP, 2-methylthio-ATP, N^6-phenyl-ATP, and 8-bromo-ATP are used to determine the effects of substituents at various positions of the adenine base.[5–7] Replacement of the D-ribose sugar by its unnatural enantiomer L-ribose (to generate L-ATP) enables the stereoselectivity (enantioselectivity) of P_2-purinoceptor subtypes to be compared.[8–10] Replacement of an ionized oxygen on the inner or middle phosphate of the 5′-triphosphate chain by ionized sulfur so as to produce the Rp and Sp diastereoisomers of ATP-α-S and ATP-β-S, respectively, enables the stereoselectivity (diastereoselectivity) of this important region of ATP to be determined.[11–15] The requirement for bridging oxygens linking the phosphates is examined by their replacement by methylene at the innermost position as in homo-ATP, between the innermost and middle phosphates as in α,β-methylene-ATP, and between the middle and outer phosphates as in β,γ-methylene-ATP.[7,16–20] The necessity for electronegativity at the β,γ-position is assessed by replacement of that bridging oxygen by an imido group as in β,γ-imido-ATP, and by dihalomethylene as in β,γ-difluoromethylene-ATP and β,γ-dichloromethylene-ATP.[21,22]

P_{2Z}-PURINOCEPTORS

Extracellular ATP renders certain cells permeable by inducing the formation of pores or lesions in the plasma membrane, apparently by an interaction with a surface receptor, which has been termed a P_{2Z}-purinoceptor.[4] Cells susceptible to such permeabilization include mast cells, macrophages, lymphocytes, and some transformed epithelial cells in culture. Studies in the presence of chelating agents have shown that ATP^{4-}, present as a minor component in solutions containing magnesium to which ATP is usually complexed, is the actual agonist.[23] On mast cells, where the actions of ATP analogues have been most extensively studied (FIG. 1), permeabilization by ATP can be monitored by the uptake of normally impermeant dyes such as ethidium bromide, whose fluorescence emission becomes enhanced upon intercalcating with nuclear DNA.[24] Of the naturally occurring nucleoside triphosphates, only ATP is active, whereas GTP, CTP, and UTP, as well as ADP and AMP, are without activity.[23] The P_{2Z}-purinoceptor is exquisitely sensitive to alterations in the structure of ATP, as measured by uptake of ethidium by rat mast cells, and only very few ATP analogues can induce permeabilization at a concentration (10 μM) at which ATP itself is

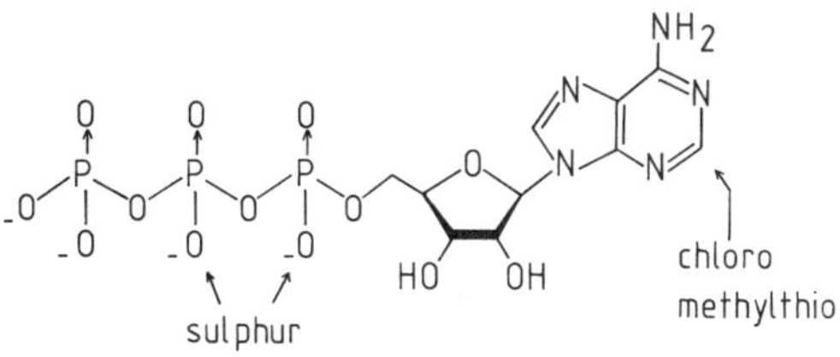

FIGURE 1. Structure-activity relationships for P_{2Z}-purinoceptor agonists on rat mast cells.

maximally active. Analogues with substituents on the C-2 position of the adenine base, such as 2-chloro-ATP, 2-methylthio-ATP, and 2-ethylthio-ATP, are at least as potent as ATP, but substitution at the C-8 position as in 8-bromo-ATP, or at the N^6 position as in N^6-phenyl-ATP, generates inactive analogues.[24] The P_{2Z}-purinoceptor exhibits absolute stereoselectivity for the naturally occurring enantiomer of ATP, as the unnatural enantiomer L-ATP is without activity, as are the C-2-substituted enantiomers of L-ATP,[24] 2-chloro-L-ATP and 2-methylthio-L-ATP. The 5′-triphosphate chain is an absolute requirement, but replacement of one of the ionized oxygens by ionized sulfur, as in ATP-α-S and ATP-β-S, generates agonists more potent than ATP, although the P_{2Z}-purinoceptor does not distinguish between the Rp or Sp diastereoisomers of those phosphorothioate analogues. Replacement of any of the bridging phosphate oxygens by methylene, as in homo-ATP, α,β-methylene-ATP, or β,γ-methylene-ATP, results in loss of agonist potency, and the unnatural L-enantiomers, α,β-methylene-L-ATP and β,γ-methylene-L-ATP (L-AMP-PCP), as well as their electronegative dichloro- and difluoromethylene versions, are also inactive.[24] Competitive antagonists of the P_{2Z}-purinoceptor are unknown, but 2-methylthio-L-ATP (100 μM) added simultaneously with ATP (10 μM) inhibits permeabilization of rat mast cells by about 50%,[24] and may provide a lead to the design of selective antagonists.

P_{2X}-PURINOCEPTORS

Extracellular ATP, acting at excitatory P_{2X}-purinoceptors, induces contraction of visceral and vascular smooth muscle such as the urinary bladder, vas deferens, anococcygeus, and portal vein. Studies of structure-activity relationships at the P_{2X}-purinoceptor have been carried out most extensively on the guinea pig urinary bladder (FIG. 2).[2] All of the naturally occurring nucleoside triphosphates, GTP, CTP, and UTP, as well as ADP, are approximately as active as ATP, but GDP, CDP, and AMP fail to induce contraction of the bladder.[25] 2-Chloro-ATP, 2-methylthio-ATP, and 8-bromo-ATP have potencies similar to the potency for ATP, but N^6-phenyl-ATP is inactive.[7] L-ATP is as potent as ATP (replacement of D-ribose by L-ribose having no effect), and so—unlike the P_{2Z}-, P_{2T}-, and P_{2Y}-purinoceptors—the P_{2X}-purinoceptor exhibits no stereoselectivity whatever toward the enantiomers of ATP or the enantiomers of 2-chloro-ATP or of 2-methylthio-ATP.[10] No stereoselectivity is exhibited either toward the diastereoisomers of the phosphorothioate analogues of ATP, but ATP-β-S is much more potent than ATP, whereas ATP-α-S is considerably less potent than ATP at inducing contraction.[12] Among the methylene analogues,

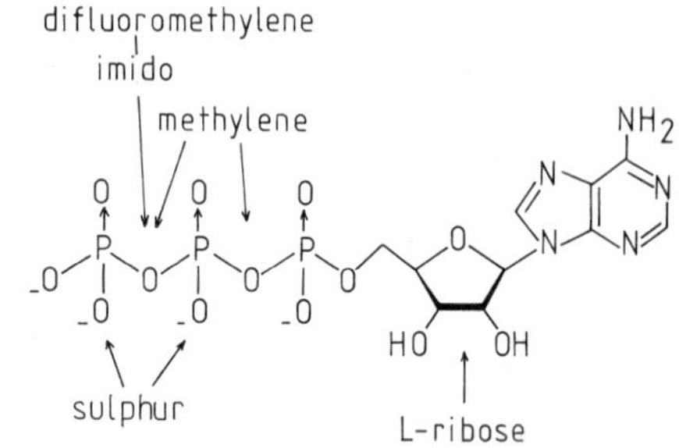

FIGURE 2. Structure-activity relationships for P_{2X}-purinoceptor agonists having potency greater than ATP at contracting the guinea pig urinary bladder.

homo-ATP is as potent as ATP, but α,β-methylene-ATP and β,γ-methylene-ATP are much more potent.[7,17,18] Substitution of the adenine base, as in 2-chloro-β,γ-methylene-ATP and 2-methylthio-β,γ-methylene-ATP, does not lead to further increases in potency.[21] Similarly, the electronegative versions, β,γ-difluoromethylene-ATP and β,γ-dichloromethylene-ATP, and their 2-methylthio analogues, 2-methylthio-β,γ-difluoromethylene-ATP and 2-methylthio-β,γ-dichloromethylene-ATP, as well as β,γ-imido-ATP,[21] are no more potent than β,γ-methylene-ATP, although all of them are much more potent than ATP itself at contracting the urinary bladder. A series of specific agonists for the P_{2X}-purinoceptor have been designed based on the unnatural enantiomer L-AMP-PCP (FIG. 3).[18] L-AMP-PCP is the most potent known agonist acting on P_{2X}-purinoceptors in a variety of tissues that respond to ATP,[17,18,26,27] yet is completely without agonist or antagonist potency at the P_{2Y}, P_{2Z}, and P_{2T} subtypes.[19,24,27] Use of L-AMP-PCP has important advantages over α,β-methylene-ATP and β,γ-methylene-ATP, which are commonly employed in studies of P_{2X}-purinoceptors. L-AMP-PCP is completely resistant to dephosphorylation,[17] and in any event the resultant nucleoside would be the unnatural enantiomer L-adenosine, which is inactive at adenosine receptors.[8,9] Unlike α,β-methylene-ATP (which is of similar potency and efficacy as ATP in some P_{2Y}-mediated effects such as relaxation of the

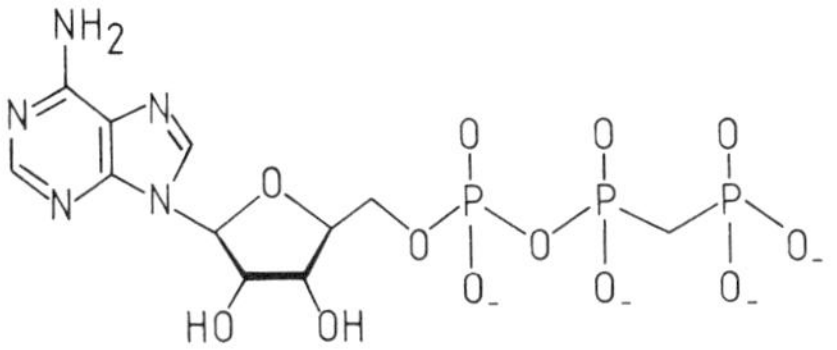

FIGURE 3. L-AMP-PCP, and its derivatives, specific P_{2X}-purinoceptor agonists.

taenia coli[16]) and β,γ-methylene-ATP (which, although less potent than ATP in taenia coli P_{2Y}-purinoceptors, nonetheless has the same efficacy[16]), L-AMP-PCP is utterly without agonist (or antagonist) activity, except at the P_{2X} subtype.[18,19] This absolute selectivity of L-AMP-PCP for the P_{2X}-purinoceptor is shared by the 2-substituted analogues, 2-chloro-L-AMP-PCP and 2-methylthio-L-AMP-PCP, and the dihalomethylene versions, β,γ-difluoromethylene-L-ATP (L-AMP-PCF_2P) and β,γ-dichloromethylene-L-ATP (L-AMP-$PCCl_2P$).[21] Each of the dichloro analogues, β,γ-dichloromethylene-ATP, 2-methylthio-β,γ-dichloromethylene-ATP, and L-AMP-$PCCl_2P$, are less efficacious than β,γ-methylene-ATP itself, with intrinsic activities of approximately 0.65, and these partial agonists could give a lead to the design of competitive antagonists at the P_{2X} subtype.[21] Indeed, in a detailed study of the effects of ATP analogues on mammalian isolated sensory neurons, β,γ-dichloromethylene-ATP is reported to antagonize ATP-induced inward currents.[28]

P_{2Y}-PURINOCEPTORS

Extracellular ATP, acting at inhibitory P_{2Y}-purinoceptors, induces relaxation of visceral smooth muscle such as the taenia coli, and vascular smooth muscle such as the aorta by an endothelial cell-dependent process. Most studies of structure-activity relationships at the P_{2Y}-purinoceptor have been carried out on the guinea pig taenia coli (FIG. 4), and because the same species supplied the urinary bladder, this facilitates comparison[2] between the most controversial divisions, the P_{2X} and P_{2Y} subtypes.[3] All of the naturally occurring 5′-triphosphates as well as ADP are approximately as potent as ATP, but CDP, GDP, UDP, and AMP are less than one-tenth as potent as ATP at inducing relaxation of the taenia coli. The effects of substituents on the adenine ring vary as to the point of substitution, N^6-phenyl-ATP being inactive, 8-bromo-ATP having the same potency as ATP, but C-2-substituted analogues such as 2-azido-ATP, 2-chloro-ATP, 2-methylthio-ATP having considerably enhanced potencies, being 22-fold, 35-fold, and 195-fold more potent, respectively, than ATP.[8,10,16] This increase

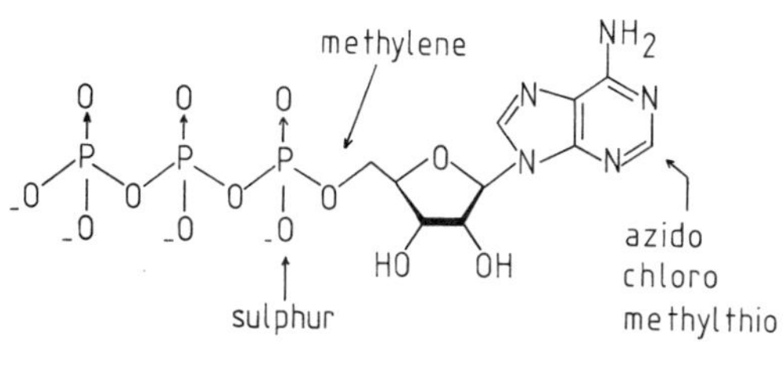

FIGURE 4. Structure-activity relationships for P_{2Y}-purinoceptor agonists having potency greater than ATP at relaxing the guinea pig taenia coli.

in potency following 2-substitution is similar to the behavior of P_{2T}-purinoceptor agonists, but dissimilar to P_{2Z} and P_{2X} agonists where no such increases are seen. Only poor stereoselectivity is exhibited toward the ribose enantionmers, so that ATP is 3- to 6-fold more potent than L-ATP.[8] However, C-2 substitution enhances this stereoselectivity considerably because of the lack of effect of such substitution on the L-enantiomers compared to the increased potency of the natural enantiomers. The photoaffinity analogue 2-azido-ATP is 124-fold more potent than 2-azido-L-ATP,[8] 2-chloro-ATP is 35-fold more potent than 2-chloro-L-ATP, and 2-methylthio-ATP is 724-fold more potent than 2-methylthio-L-ATP.[10] Unlike the P_{2T} and P_{2Z} subtypes, absolute stereoselectivity is not displayed toward any of these pairs because the L-enantiomers are still as fully efficacious as ATP. Some stereoselectivity is also displayed toward the phosphorothioate analogue ATP-α-S, where the Rp diastereoisomer is approximately 7-fold more potent than the Sp diastereoisomer, and the Rp and Sp diastereoisomers are 50-fold and 9-fold more potent, respectively, than ATP at inducing relaxation of the taenia coli.[12] ATP-β-S is only a little more potent than ATP, and no stereoselectivity is displayed toward its Rp and Sp diastereoisomers.[12] The potency of the methylenephosphonate analogues depends on where the methylene linkage has been placed, so that homo-ATP is 70-fold more potent than ATP, α,β-methylene-ATP is approximately equipotent with ATP, and β,γ-methylene-ATP has about one-

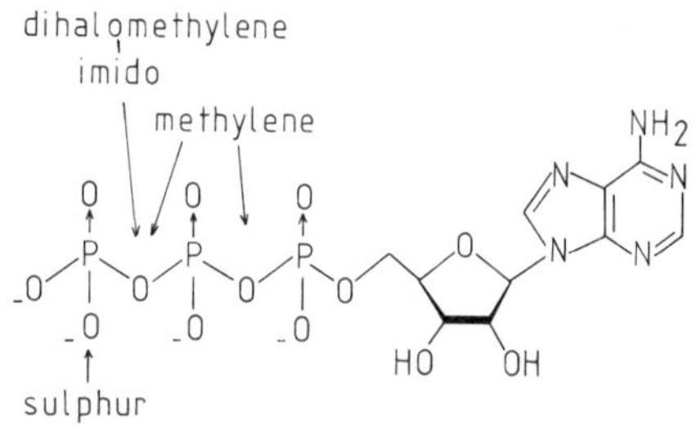

FIGURE 5. Structure-activity relationships for resistance to dephosphorylation by tissue ectonucleotidases of ATP analogues.

tenth the potency of ATP, with all of these analogues having the same or even a greater efficacy than ATP at inducing relaxation of the taenia coli.[16,20] The potency of β,γ-methylene-ATP is slightly improved if the electropositive methylene group is replaced by dihalomethylene, and reflects the order of electronegativity, so that β,γ-dichloromethylene-ATP and β,γ-difluoromethylene-ATP are 1.4-fold and 2.6-fold more potent, respectively.[21] The effect on C-2-substituted analogues is much greater, and 2-methylthio-β,γ-dichloromethylene-ATP and 2-methylthio-β,γ-difluoromethylene-ATP are 6-fold and 67-fold as potent, respectively, as 2-methylthio-β,γ-methylene-ATP, and indeed the difluoromethylene analogue is about twice as potent as ATP itself.[21] This behavior is in contrast to that at P_{2X}-purinoceptors, where the dihalomethylene analogues are no more potent than methylene and, in the case of the dichloromethylene analogues, are less potent and less efficacious.[21] The imido analogue, β,γ-imido-ATP, has less than one-tenth the potency of ATP, but is considerably more potent than ATP at P_{2X}-purinoceptors.[7] These differences in potencies between the P_{2Y} and P_{2X} responses are real as, unlike ATP and 2-methylthio-ATP, these β,γ-methylene analogues are not dephosphorylated by tissue ectonucleotidases (FIG. 5).[7,20,21] No competitive antagonists at the P_{2Y}-purinoceptor are known, and an earlier claim[29] that reactive blue 2 is a selective antagonist for the P_{2Y} subtype has been discounted.[30] Adenosine 5′-(2-fluorodiphosphate) (ADP-β-F) (FIG. 6) has been found

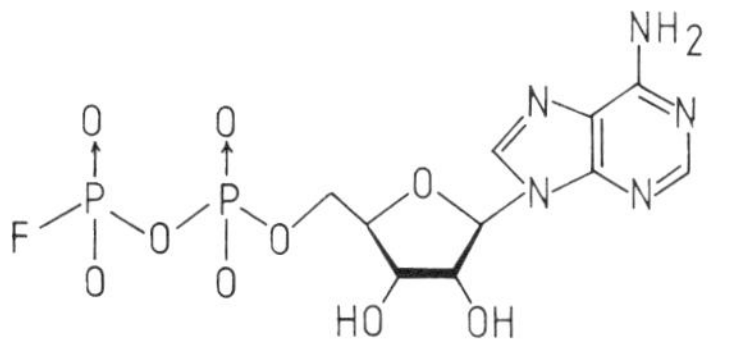

FIGURE 6. ADP-β-F, a specific P_{2Y}-purinoceptor agonist.

to be a specific agonist at the P_{2Y} subtype, having no agonist or antagonist activity at P_{2T}, P_{2Z}, or P_{2X} receptors.[31] ADP-β-F, although less potent than ATP, induces P_{2Y}-mediated relaxation of the taenia coli and aorta, but not P_{2X}-mediated contraction of the urinary bladder or vas deferens.[31] ADP-β-F is not an ideal P_{2Y}-purinoceptor agonist because the terminal fluorophosphate, despite conferring some resistance to ectonucleotidase activity, is not immune to dephosphorylation with the eventual formation of adenosine, although so far that has been shown not to be a problem.[31]

P_{2T}-PURINOCEPTORS

Extracellular ADP, acting at P_{2T}-purinoceptors, activates human blood platelets (with an associated release into the cytosol of calcium from intracellular sequestrations, as well as an influx of extracellular calcium) and a potent inhibition of stimulated adenylate cyclase.[4] ATP and AMP are not agonists at the P_{2T} subtype, but instead are competitive antagonists of the action of ADP,[11,32] an arrangement that seems to be unique to the platelet P_{2T}-purinoceptor.[4] The existence of competitive antagonists means that, unlike the situation with P_{2Z}, P_{2X}, and P_{2Y} subtypes, it is possible to perform classical pharmacology on the P_{2T}-purinoceptor with a variety of adenine nucleotide analogues that are agonists, partial agonists, and antagonists.[9,11,15,33–36]

Of the naturally occurring diphosphates, only ADP (FIG. 7) is active as an agonist, CDP, UDP, and GDP being inactive or giving slight antagonist activity. Small substituents on the N^6 position reduce, and larger ones abolish, agonist potency, and substituents on the C-8 position, as in 8-bromo-ADP, also abolish agonist potency.[5] A variety of substituents, however, may be placed at the C-2 position so as to enhance potencies.[5] 2-Azido-ADP, a photoaffinity analogue of ADP, is 5-fold,[9,37] and 2-chloro-ADP and 2-methylthio-ADP are 10-fold more potent than ADP at inducing human platelet aggregation.[9,35] The C-2-substituted analogues were even more potent than ADP at inhibiting stimulated adenylate cyclase,[35] and these differences in potencies

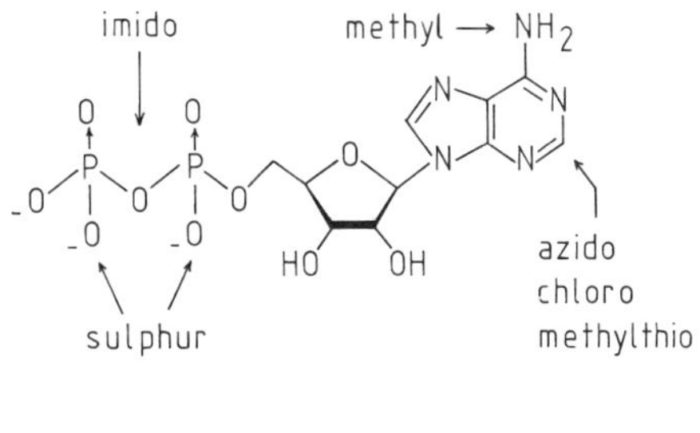

FIGURE 7. Structure-activity relationships for P_{2T}-purinoceptor agonists having the property of inducing platelet aggregation.

could be accounted for by these 2-substituted analogues having actions in addition to one at a single P_{2T}-purinoceptor. All of the C-2 analogues, however, were competitively inhibited by ATP, with apparent dissociation constants the same as that obtained for inhibition by ATP of these actions of ADP itself.[35] Their differences in agonist potencies are therefore due to differences in efficacies at one P_{2T}-purinoceptor rather than additional actions elsewhere or at another class of P_{2T}-purinoceptor. Very large substituents can be placed at the C-2 position with little or no loss of potency, suggesting that this region of the ADP molecule is orientated away from the receptor. The P_{2T}-purinoceptor shows absolute stereoselectivity toward the ribose enantiomers, and L-ADP and its C-2-substituted analogues 2-azido-L-ADP and 2-chloro-L-ADP are completely inactive as agonists or as antagonists.[9] The phosphorothioate analogues of ADP, ADP-α-S and ADP-β-S, are both partial agonists with the same efficacy as aggregating agents, having intrinsic activities relative to ADP of 0.75, and exhibiting a 5-fold stereoselectivity toward the Sp diastereoisomer of ADP-α-S.[33,34] ADP-β-S also inhibits stimulated adenylate cyclase, still as a partial agonist, but ADP-α-S does not, and instead antagonizes competitively ADP-induced inhibition of adenylate cyclase, and again 5-fold stereoselectivity for the Sp diastereoisomer is exhibited.[33,34] None of the methylenephosphonate analogues of ADP are agonists, homo-ADP having virtually no activity, whereas α,β-methylene-ADP is a very weak antagonist both of ADP-induced platelet aggregation and of inhibition by ADP of stimulated adenylate cyclase.[22] The dihalomethylene versions are more potent, both as antagonists of aggregation and as antagonists of the effect of ADP on adenylate cyclase, but α,β-difluoromethylene-ADP is not more potent than $\alpha\beta$-dichloromethylene-ADP. The imido analogue, α,β-imido-ADP, is an agonist but has only one-hundredth the potency of ADP.[22]

A variety of analogues of AMP and of ATP are antagonists at the P_{2T}-purinoceptor, and structure-activity relationships show that in addition to a 5′-mono- or polyphosphate chain, the ribose sugar and adenine base are required.[11] ATP, 2-chloro-ATP, 2-chloroadenosine 5′-monophosphorothioate, β,γ-methylene-ATP, the diastereoisomers of ATP-α-S, P_1P_5-diadenosine pentaphosphate, and γ-fluoro-ATP have all been shown by Schild analysis to be competitive inhibitors of the actions of ADP.[11] A comparison of their apparent dissociation constants for antagonism of ATP-induced aggregation with those for antagonism of ADP-induced inhibition of adenylate cyclase confirms that one P_{2T}-purinoceptor mediates both actions of ADP, 2-chloro-ADP, 2-azido-ADP, and 2-methylthio-ADP.[11,35] 2-Alkylthio-analogues of AMP and of ATP are specific but noncompetitive antagonists of the actions of ADP at P_{2T}-purinoceptors.[15,36] Radioligand binding studies of intact human platelets using β-[^{32}P]-2-methylthio-ADP detects one population of binding sites with an affinity constant of 5 to 20 nM, and 400 to 1200 sites per platelet.[38]

SUMMARY

Studies using analogues of ATP have allowed a comparison of structure-activity relationships that has provided ample evidence for the existence of at least four subtypes of P_2-purinoceptors responsive to adenine nucleotides (TABLE 1). Competitive antagonists have defined clearly the P_{2T} subtype, specific agonists are now available for the P_{2X} and P_{2Y} subtypes, and leads have been given for development of reversible specific antagonists for the P_{2X} and P_{2Z} subtypes. The availability especially of inhibitors of

TABLE 1. Structure-Activity Relationships at Subtypes of P_2-Purinoceptors

	P_{2Z} (Mast Cells)	P_{2X} (Urinary Bladder)	P_{2Y} (Taenia Coli)	P_{2T} (Platelets)
Agonists (endogenous)	ATP	ATP/ADP	ATP/ADP/AMP	ADP
Antagonists (endogenous)	—	—	—	ATP/AMP
C-2 analogues	=ATP	=ATP	>ATP	>ADP
Stereoselectivity (D & L enantiomers)	L inactive	L = D	L < D	L inactive
Phosphorothioates				
ATP-α-S	R = S	R = S	R > S	S > R
ATP-β-S	R = S	R = S	R = S	—
Methylene phosphonates				
α,β-CH_2	Inactive	>ATP	=ATP	Antagonist
β,γ-CH_2	Inactive	>ATP	<ATP	Antagonist
β,γ-CF_2	Inactive	$CF_2 = CH_2$	$CF_2 > CH_2$	Antagonist $CF_2 > CH_2$
Specific agonist	—	L-AMP-PCP	ADP-β-F	—
Competitive antagonist	2-Me-S-L-ATP ?	AMP-$PCCl_2P$?	—	2-Cl-ATP

ectonucleotidases and of specific antagonists for each P_2-purinoceptor subtype would enable the roles of endogenous extracellular nucleotides to be ascertained, as has already been shown for the interaction of platelets with the vasculature.[39,40]

REFERENCES

1. Cusack, N. J. 1985. Synthesis of adenosine and adenine nucleotide analogs. *In* Methods in Adenosine Research: Methods in Pharmacology. D. M. Paton, Ed. Vol. 6: 3-20. Plenum. New York, NY.
2. Cusack, N. J. & S. M. O. HOURANI. 1990. Structure-activity relationships for adenine nucleotide receptors on mast cells, human platelets, and smooth muscle. *In* Purines in Cellular Signalling: Targets for New Drugs. K. A. Jacobson, J. W. Daly & V. Manganiello, Eds.: 254-259. Springer-Verlag. New York, NY.
3. Burnstock, G. & C. Kennedy. 1985. Is there a basis for distinguishing two types of P_2-purinoceptor? Gen. Pharmacol. **17:** 433-440.
4. Gordon, J. L. 1986. Extracellular ATP: Effects, sources and fate. Biochemistry **233:** 309-319.
5. Gough, G., M. H. Maguire & F. Penglis. 1972. Analogues of adenosine 5′-diphosphate: New platelet aggregators. Mol. Pharmacol. **8:** 170-177.
6. Satchell, D. G. & M. H. Maguire. 1982. Evidence for separate receptors for ATP and adenosine in the guinea-pig taenia coli. Eur. J. Pharmacol. **81:** 669-672.
7. Welford, L. A., N. J. Cusack & S. M. O. Hourani. 1987. The structure-activity relationships of ectonucleotidases and of excitatory P_2-purinoceptors: Evidence that the dephosphorylation of ATP analogues reduces pharmacological potency. Br. J. Pharmacol. **141:** 123-130.
8. Cusack, N. J. & M. Planker. 1979. Relaxation of isolated taenia coli of guinea-pig by enantiomers of 2-azido analogues of adenosine and adenine nucleotides. Br. J. Pharmacol. **67:** 153-158.
9. Cusack, N. J., M. E. Hickman & G. V. R. Born. 1979. Effects of D- and L-enantiomers of adenosine, AMP and ADP and their 2-chloro and 2-azido analogues on human platelets. Proc. R. Soc. London Ser. B **206:** 139-144.
10. Burnstock, G., N. J. Cusack, J. M. Hills, I. Mackenzie & P. Meghji. 1983. Studies on the stereoselectivity of the P_2-purinoceptor. Br. J. Pharmacol. **79:** 907-913.
11. Cusack, N. J. & S. M. O. Hourani. 1982. Adenosine 5′-diphosphate antagonists and human platelets: No evidence that aggregation and inhibition of stimulated adenylate cyclase are mediated by different receptors. Br. J. Pharmacol. **76:** 221-227.
12. Burnstock, G., N. J. Cusack & L. A. Meldrum. 1984. Effects of phosphorothioate analogues of ATP, ADP and AMP on guinea-pig taenia coli and urinary bladder. Br. J. Pharmacol. **82:** 369-374.
13. Needham, L., N. J. Cusack, J. D. Pearson & J. L. Gordon. 1987. Characteristics of the P_2-purinoceptor that mediates prostacyclin production by pig aortic endothelial cells. Eur. J. Pharmacol. **134:** 199-209.
14. Burnstock, G., N. J. Cusack & L. A. Meldrum. 1985. Studies on the stereoselectivity of the P_2-purinoceptor on the guinea-pig vas deferens. Br. J. Pharmacol. **84:** 431-434.
15. Hourani, S. M. O., L. A. Welford & N. J. Cusack. 1986. 2-MeS-AMP-PCP and human platelets: Implications for the role of adenylate cyclase in ADP-induced aggregation? Br. J. Pharmacol. **87:** 84P.
16. Satchell, D. E. & M. H. Maguire. 1975. Inhibitory effects of adenine nucleotide analogs on the isolated guinea-pig taenia coli. J. Pharmacol. Exp. Ther. **195:** 540-548.
17. Cusack, N. J. & S. M. O. Hourani. 1984. Some pharmacological and biochemical interactions of the enantiomers of adenylyl 5′-(β,γ-methylene)diphosponate with the guinea-pig urinary bladder. Br. J. Pharmacol. **82:** 155-159.
18. Hourani, S. M. O. 1984. Desensitization of the guinea-pig urinary bladder by the enantiomers of adenylyl 5′-(β,γ-methylene)diphosphonate and by substance P. Br. J. Pharmacol. **82:** 161-164.

19. HOURANI, S. M. O., N. J. CUSACK & L. A. WELFORD. 1985. L-AMP-PCP, an ATP receptor agonist in guinea-pig bladder, is inactive on taenia coli. Eur. J. Pharmacol. **108:** 197-200.

20. WELFORD, L. A., N. J. CUSACK & S. M. O. HOURANI. 1986. ATP analogues and the guinea-pig taenia coli: A comparison of the structure-activity relationships of ectonucleotidases with those of the P_2-purinoceptor. Eur. J. Pharmacol. **129:** 217-224.

21. CUSACK, N. J., S. M. O. HOURANI, G. D. LOIZOU & L. A. WELFORD. 1987. Effects of isopolar phosphonate analogues of ATP on guinea-pig taenia coli and urinary bladder. Br. J. Pharmacol. **90:** 791-795.

22. CUSACK, N. J. & C. J. PETTEY. 1988. Effects of isopolar isosteric phosphonate analogues of adenosine 5′-diphosphate (ADP) on human platelets. *In* Adenosine and Adenine Nucleotides: Physiology and Pharmacology. D. M. Paton, Ed.: 287. Taylor & Francis. London.

23. COCKCROFT, S. & B. D. GOMPERTS. 1980. The ATP^{4-} receptor of rat mast cells. Biochem. J. **188:** 789-798.

24. TATHAM, P. E. R., N. J. CUSACK & B. D. GOMPERTS. 1988. Characterization of the ATP^{4-} receptor that mediates permeabilization of rat mast cells. Eur. J. Pharmacol. **147:** 13-21.

25. LUKACSKO, P. & R. D. KRELL. 1982. Responses of the guinea-pig urinary bladder to purine and pyrimidine nucleotides. Eur. J. Pharmacol. **80:** 401-406.

26. WELFORD, L. A., N. J. CUSACK & S. M. O. HOURANI. 1986. ATP analogues and the guinea-pig urinary bladder: Pharmacological potency is related to resistance to breakdown by ectoenzymes. Br. J. Pharmacol. **89:** 514P.

27. HOURANI, S. M. O., G. D. LOIZOU & N. J. CUSACK. 1986. Pharmacological effects of L-AMP-PCP on ATP receptors in smooth muscle. Eur. J. Pharmacol. **131:** 99-103.

28. KRISHTAL, O. A., S. M. MARCHENKO, A. G. OBUKHOV & T. M. VOLKOVA. 1988. Receptors for ATP in rat sensory neurones: The structure-function relationship for ligands. Br. J. Pharmacol. **95:** 1057-1062.

29. BURNSTOCK, G. & J. J. I. WARLAND. 1987. P_2-purinoceptors of two subtypes in the rabbit mesentery artery: Reactive blue 2 selectively inhibits responses mediated via P_{2Y}- but not the P_{2X}-purinoceptors. Br. J. Pharmacol. **90:** 383-391.

30. WHITE, T. D. 1988. Role of adenine compounds in autonomic neurotransmission. Pharmacol. Ther. **38:** 129-168.

31. HOURANI, S. M. O., L. A. WELFORD, G. D. LOIZOU & N. J. CUSACK. 1988. Adenosine 5′-(2-fluorodiphosphate) is a selective agonist at P_2-purinoceptors mediating relaxation of smooth muscle. Eur. J. Pharmacol. **147:** 131-136.

32. MACFARLANE, D. E. & D. C. B. MILLS. 1975. The effects of ATP on platelets: Evidence against the central role of released ADP in primary aggregation. Blood **46:** 309-320.

33. CUSACK, N. J. & S. M. O. HOURANI. 1981. Partial agonist behavior of adenosine 5′-O-(2-thiodiphosphate). Br. J. Pharmacol. **73:** 405-408.

34. CUSACK, N. J. & S. M. O. HOURANI. 1981. Effects of Rp and Sp diastereoisomers of adenosine 5′-O-(1-thiodiphosphate) on human platelets. Br. J. Pharmacol. **73:** 409-412.

35. CUSACK, N. J. & S. M. O. HOURANI. 1982. Competitive inhibition by adenosine 5′-triphosphate of the actions on human platelets of 2-chloroadenosine 5′-diphosphate, 2-azidoadenosine 5′-diphosphate, and 2-meththioadenosine 5′-diphosphate. Br. J. Pharmacol. **77:** 329-333.

36. CUSACK, N. J. & S. M. O. HOURANI. 1982. Specific but noncompetitive inhibition by 2-alkylthio analogues of adenosine 5′-monophosphate and adenosine 5′-triphosphate of human platelet aggregation by adenosine 5′-diphosphate. Br. J. Pharmacol. **75:** 397-400.

37. CUSACK, N. J. & G. V. R. BORN. 1977. Effects of photolysable 2-azido analogues of adenosine, AMP and ADP on human platelets. Proc. R. Soc. London Ser. B **197:** 515-520.

38. MACFARLANE, D. E., P. C. SRIVASTAVA & D. C. B. MILLS. 1983. 2-Methylthioadenosine [β-^{32}P]diphosphate: An agonist and radioligand for the receptor that inhibits the accumulation of cyclic AMP in intact platelets. J. Clin. Invest. **71:** 420-428.

39. BORN, G. V. R. & M. A. A. KRATZER. 1984. Source and concentration of extracellular adenosine triphosphate during haemostasis in rats, rabbits and man. J. Physiol. (London) **354:** 419-429.

40. MCCLURE, M. O., A. KAKKER, N. J. CUSACK & G. V. R. BORN. 1988. Evidence for dependence of arterial haemostasis on ADP. Proc. R. Soc. London Ser. B **234:** 255-262.

P_2-Purinoceptor Antagonists[a]

JEFFREY S. FEDAN

Physiology Section
Division of Respiratory Disease Studies
National Institute for Occupational Safety and Health
Morgantown, West Virginia 26505

SHEILA J. LAMPORT

Department of Pharmacology and Toxicology
West Virginia University
Morgantown, West Virginia 26506

INTRODUCTION

An understanding of the antagonist compounds that block the actions of ATP[1-4] are required for the elucidation of the physiologic roles and mechanisms of responses of cells to extracellular ATP, and for the classification of P_2-purinoceptor subtypes. We will summarize characteristics of agents that block P_2-purinoceptor-mediated responses. Included in this topic are agents thought to exert their inhibitory effects at the purinoceptors, as well as others that interfere with transduction via mechanisms that are as yet but poorly understood.

It has been with difficulty that P_2-purinoceptor antagonists have become available. Equilibrium competitive antagonists that act specifically and reversibly at the receptor, and that are devoid of agonist or nonspecific activity, have not been described for either P_{2X}- or P_{2Y}-purinoceptors. Conventional approaches to antagonist design, which rely upon chemical substitutions to decrease agonist efficacy and retain affinity, have not been realized in this area.

Agents that have been studied for their P_2-purinoceptor antagonist activities are listed in TABLE 1. For the purpose of discussion, they are divided into two operational groups. The first group consists of agents that structurally resemble ATP, and are thus able to induce responses prior to the antagonistic effect. The second group is diverse and consists of agents that bear no obvious structural or mechanistic relationship to each other or to ATP. We will dwell on investigations involving the use of P_2-purinoceptor antagonists in the autonomic nervous system and in smooth muscle systems—after all, it was in an attempt to define transmitter roles for ATP that led to the usage of many of these compounds. The application of these compounds in transmitter identification is discussed elsewhere in this volume.

[a] This paper was supported, in part, by Grant PTR 5 T32 GM07039-13 from the National Institutes of Health.

AGENTS WITH INITIAL AGONIST ACTIVITY

Arylazido Aminopropionyl ATP ($ANAPP_3$)

Arylazido aminopropionyl ATP, or $ANAPP_3$ (3′-*O*-{3[*N*-(4-azido-2-nitrophenyl)amino]propionyl}adenosine 5′-triphosphate), is a photoaffinity label that was first described with respect to its covalent insertion, after photolysis, at or near the ATP-binding sites in myosin and mitochondrial ATPases.[5,6] Because such a process occurring at P_2-purinoceptors in intact tissues could result in an antagonism of ATP-induced responses, experiments with $ANAPP_3$ were performed using vas deferens isolated from guinea pig.[7] Being an analogue of ATP, $ANAPP_3$ elicits upon first addition an ATP-like, transient contraction. Photolysis of $ANAPP_3$ in the presence of the tissue resulted, after washout of the agent, in an irreversible antagonism of ATP-induced contractile responses (FIG. 1). The antagonism was inhibited if ATP was present during photolysis, and it was specific in that responses to norepinephrine

TABLE 1. P_2-Purinoceptor Antagonists

Agents with initial agonist activity
Arylazido aminopropionyl ATP ($ANAPP_3$)
ATP or α,β-methylene ATP (APCPP)
ATP-2′,3′-dialdehyde (periodate-oxidized ATP; P-ATP)
Agents without agonist activity
Isatogens
Apamin
Reactive blue 2
p-Chloromercuribenzene sulfonic acid (PCMBS)
4,4′-Diisothiocyano-2,2′-disulfonic acid stilbene (DIDS)
5′-*p*-Fluorosulfonylbenzoyl adenosine (FSBA)

and K^+ were not affected. The antagonism by $ANAPP_3$ is known to occur at a cell surface site because contractions to agarose-ATP (ATP linked to solid agarose beads) were also inhibited.[8] In P_{2X}-purinoceptor systems, $ANAPP_3$ is noteworthy in its specificity. For example, in guinea pig urinary bladder[9] $ANAPP_3$ treatment inhibited contractile responses to ATP but not to acetylcholine and K^+ (FIG. 2).

The effects of $ANAPP_3$ have not been studied extensively in preparations in which ATP elicits inhibitory responses. Relaxation responses of taenia coli to ATP were antagonized 6-fold by $ANAPP_3$; responses to ADP, AMP, and adenosine were blocked to lesser degrees, that is, 2.4-, 3.8-, and 2.4-fold, respectively.[10] The effect on adenosine responses could reflect an interaction of $ANAPP_3$ with adenosine receptors, or an interaction of adenosine with P_{2Y}-purinoceptors. The latter case is more likely (see reference 10 for a discussion). In the annococcygeus, extensive conversion of ATP to adenosine appears to explain the ineffectiveness of $ANAPP_3$ as an ATP antagonist in this preparation[11–13]; in ileum, it has been suggested that conversion of $ANAPP_3$ to an adenosine congener renders it an adenosine receptor agonist devoid of P_2-purinoceptor antagonist activity.[14] It has been tempting to refer to $ANAPP_3$ as a P_{2X}-purinoceptor antagonist.[2–4] Although a benefit would be derived from unequivocal

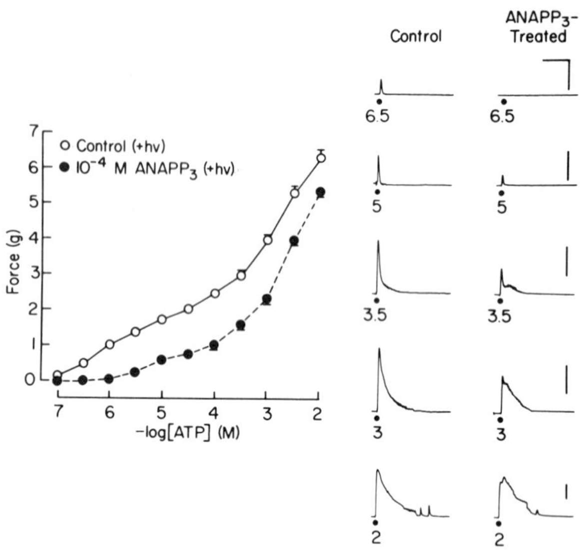

FIGURE 1. Effect of ANAPP$_3$ following photolysis ($+h\nu$; 15 min) in the presence of the guinea pig isolated vas deferens on contractile responses to ATP. The protocols for this experiment have been described.[1,15,20,22] The numbers below the individual responses are the negative log molar ATP concentrations. Horizontal bar: 30 sec; vertical bar: 2 g. Modified from Fedan and Lamport[37a] with permission.

P_{2X}-purinoceptor selectivity, the effects of ANAPP$_3$ on P_{2Y}-purinoceptor responses have not been studied sufficiently to establish whether such selectivity exists.

The antagonistic effect of ANAPP$_3$ following photolysis results from true, as opposed to pseudo-, photoaffinity labeling.[15–17] An analogue of ANAPP$_3$, ANABP$_3$, containing a larger, aminobutyryl "spacer" between ribose and the photoprobe, yielded a greater antagonistic effect than ANAPP$_3$.[18]

An advantage of nonequilibrium competitive antagonists, such as ANAPP$_3$, over competitive antagonists is that the covalent bond formed affords an opportunity to biochemically characterize the insertion site. We examined the incorporation of ^3H after intact vasa deferentia were treated with [^3H]ANAPP$_3$ (while monitoring func-

FIGURE 2. Effect of ANAPP$_3$ treatment on acetylcholine, ATP, and K^+ concentration-response curves of urinary bladder isolated from guinea pig. Open circles: irradiated control preparations; filled circles: ANAPP$_3$-treated preparations. Reproduced from Westfall *et al.*[9] with permission.

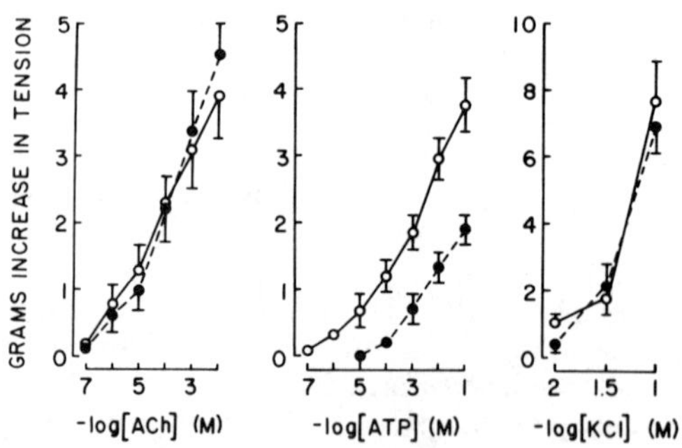

tional responses), homogenized, and processed for SDS-PAGE (FIG. 3).[19] Incorporation of [^{3}H]ANAPP$_3$ was observed into two regions, of molecular weights 43-57 and 54-66 kD, and was inhibited by ATP present during photolysis and by prior treatment with nonradiolabeled ANAPP$_3$. The nucleoside transport inhibitor dipyridamole had no effect. In fractionation studies, the incorporation of ^{3}H was associated with a crude membrane fraction and not cell cytosol. These findings provided additional evidence for a cell surface localization of the P_2-purinoceptor.[8]

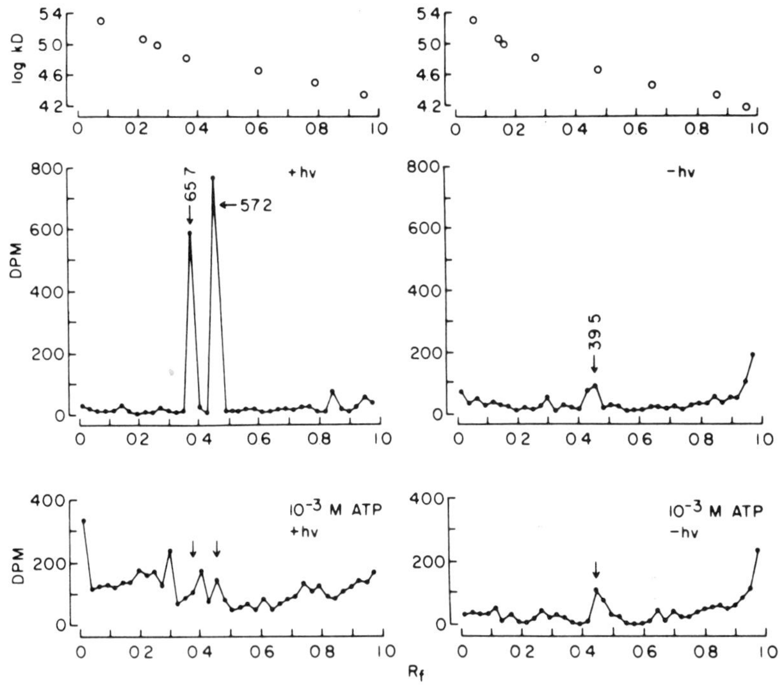

FIGURE 3. Distribution of ^{3}H in samples of homogenates after photolysis of intact vasa deferentia in the presence of [^{3}H]ANAPP$_3$ (10^{-5} M). Shown is the incorporation of ^{3}H after SDS-PAGE. The topmost panels indicate the locations of the protein standards. Left half of figure: paired vasa deferentia were incubated with [^{3}H]ANAPP$_3$ and irradiated (+hν) in the absence (middle panel) or presence (bottom panel) of 10^{-3} M ATP ("protection experiment"). Right half of figure: paired vasa deferentia were incubated in the absence (middle panel) or presence (bottom panel) of ATP but were not irradiated. Arrows point to interpolated molecular weights. Reproduced from Fedan *et al.*[19] with permission.

ANAPP$_3$ has allowed probing of the complex interaction of ATP with the vas deferens.[20-22] In this smooth muscle, the ATP concentration-response curve is biphasic (FIG. 1). Responses to ATP in intermediate to high concentrations (3×10^{-5} to 10^{-2} M) are also biphasic. After ANAPP$_3$ treatment, the curve becomes monophasic. The antagonism by ANAPP$_3$ shown in the curves is an expression of inhibition of the first phase of contraction. Because the second phase is relatively resistant to ANAPP$_3$, the peak force developed is not appreciably affected.

α,β-Methylene ATP (APCPP)

If ATP is allowed to remain in contact with smooth muscle, the contractile response to added ATP is diminished.[2,7,23] This effect, referred to as desensitization or autoinhibition, is reversible upon removal of the nucleotide. The extent of desensitization depends upon the nucleotide concentration,[7,24,25] and is produced by a large number of purine nucleotides.[25] Desensitization may occur upon repeated challenge with nucleotide, and the decline in responsiveness is related inversely to the interval between challenges. The biochemical mechanism of desensitization is poorly understood. The phenomenon has been observed in whole-cell voltage clamp experiments on ATP-activated conductance changes in frog atrium and rat vas deferens.[26,27] Recovery of channel conductance after desensitization with ATP follows a time course with a half-life of about 2 min.[26]

The slowly degradable 5′-anhydride-substituted ATP analogue, APCPP, was introduced[28,29] as a desensitizing agent to circumvent the problem of ATP metabolism by ectoenzymes to adenosine and the activation of adenosine receptors. Although the conductance changes it evokes in vas deferens differ from those evoked by ATP,[26] APCPP, nevertheless, produces a more profound and longer lasting desensitization than ATP.[28] As such, APCPP has found wide usage as an approach for the subdivision of P_2-purinoceptors into the P_{2X} and P_{2Y} classes. Desensitization occurs in association with the P_{2X}-purinoceptor-mediated excitatory effects of adenine nucleotides, such as contraction of smooth muscle.[4,30,31]

Desensitization by APCPP is generally specific for P_{2X}-purinoceptor-mediated responses; that is, responses to agents acting via other receptors are usually unaffected.[4,31,32] In immature rat basilar artery, however, desensitization with APCPP inhibited depolarizations to norepinephrine as well as ATP.[33] APCPP inhibited responses of taenia coli to epinephrine (acting via α_1-adrenoceptors).[34] These findings indicate that APCPP may cause nonspecific inhibitory effects, perhaps by affecting transduction mechanisms shared by purinoceptors and adrenoceptors.[33,34]

ANAPP$_3$ versus APCPP

The mechanism of inhibition of ATP-induced responses following treatment with ANAPP$_3$ could involve the covalent insertion of ANAPP$_3$ into the P_2-purinoceptor per se, or occlusion of the receptor resulting from insertion distant to the recognition site. Nonphotolyzed ANAPP$_3$ shares with ATP and other adenine nucleotides the ability to desensitize the P_2-purinoceptor specifically.[7] It is thus alternatively possible that the covalent insertion of ANAPP$_3$ at the receptor induces a prolonged agonist desensitization signal. The consequences of either mechanism would be functionally identical. This could explain the reported similarities in the antagonistic effects of ANAPP$_3$ and APCPP.[4,22,24] That is, ANAPP$_3$ and APCPP might desensitize cells by similar mechanisms, but the desensitization by ANAPP$_3$ is rendered irreversible by photoaffinity labeling of the receptor. The experiments done with [^{3}H]ANAPP$_3$ (FIG. 3) do not allow a finer distinction to be made between these possibilities.

ATP-2′,3′-Dialdehyde (Periodate-Oxidized ATP; P-ATP)

The high degree of pharmacological specificity exhibited by $ANAPP_3$, coupled with the biochemical advantages of labeling receptors covalently, has focused our attention on site-directed affinity labeling as an approach to the development of P_2-purinoceptor antagonists. Our motivation to follow this approach is to eventually find a replacement for $ANAPP_3$, the photolysis required for covalent insertion being inconvenient in organ bath studies. The desire to avoid photolysis has also led us to avoid other available ATP photoaffinity labels. One criterion that has guided the selection of agents for study is that the P_2-purinoceptor, in contrast to adenosine receptors, is not tolerant of N^6-substitutions.[8,20,22] Because N^6-substituted analogues have weak activities as agonists, it is unlikely that they would bind to the receptor tightly enough to favor covalent interactions. In contrast, ribose substitutions, as in the case of $ANAPP_3$ and $ANABP_3$ and other 2′- and 3′-deoxy ATP analogues, are well tolerated.[18,20–22] These considerations led us to investigate P-ATP as an ATP antagonist.[35–37a]

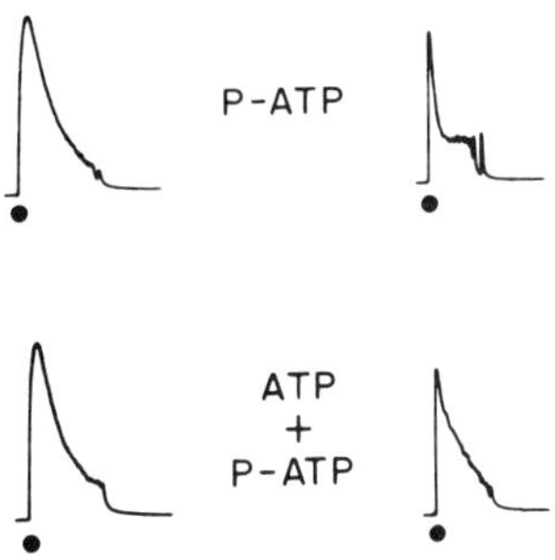

FIGURE 4. Effect of P-ATP (10^{-2} M; 5-min incubation) on contractions of paired guinea pig vasa deferentia to ATP (10^{-2} M; dots) following incubation in the absence (top row) or presence (bottom row) of ATP (10^{-2} M; "protection experiment"). The left column of responses are the controls. Modified from Fedan and Lamport[37a] with permission.

The ATP affinity label P-ATP, which was introduced by Lowe and Beechey,[38] binds covalently and irreversibly to inhibit mitochondrial ATPase in a time- and concentration-dependent manner. In vas deferens isolated from guinea pig, incubation of the preparations with 10^{-2} M P-ATP for 5 min, followed by washout, resulted in a selective inhibition of the second phase of contractile responses to ATP; this is opposite to the effect of $ANAPP_3$ treatment (FIG. 4). This treatment protocol had no effect on norepinephrine, acetylcholine, histamine, and K^+ concentration-response curves; that is, the effect was specific for ATP-induced responses. Longer incubation periods (10 and 15 min) led to potentiation of norepinephrine and acetylcholine responses, and to inhibition to a "saturation point" of the first phase of the ATP responses. The inhibitory effect of P-ATP on the second phase of ATP-induced responses was lessened after treatment was conducted in the presence of ATP in a "protection experiment" (FIG. 4). These results indicated that the inhibitory effect of P-ATP resulted from affinity labeling.

$ANAPP_3$ and P-ATP act at separate and independent sites to produce their differing antagonistic effects (FIG. 5). $ANAPP_3$ alone resulted in selective inhibition of the first phase of response. Subsequent P-ATP treatment then inhibited the second phase. When used in reverse order, the second phase was inhibited first by P-ATP, followed

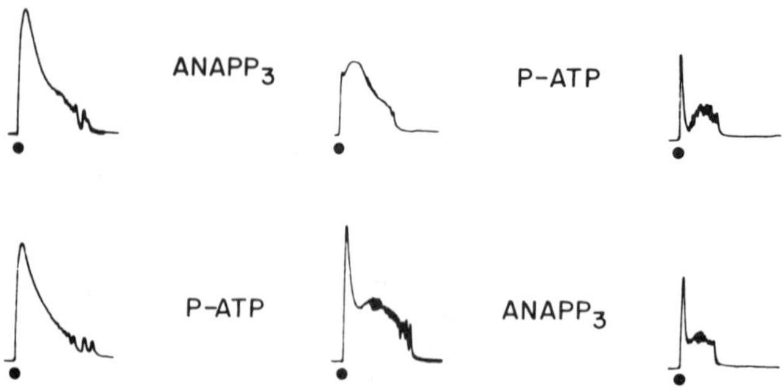

FIGURE 5. Effect of $ANAPP_3$ and P-ATP treatments alone and in combination on contractions of vas deferens (isolated from guinea pig) to ATP (10^{-2} M; dots). Top row: preparations treated first with $ANAPP_3$, then with P-ATP; bottom row: order of treatments reversed. Modified from Fedan and Lamport[37a] with permission.

by inhibition of the second phase by $ANAPP_3$. The residual responses were equivalent, regardless of the order of treatment. Thus, the phase-selective effects of these two agents indicate that at least two distinct and dissociable mechanisms mediate contractions of the vas deferens to ATP.

We have hypothesized[20–22] that the first phase of contraction involves a conventional, reversible interaction between ATP and $ANAPP_3$-sensitive P_2-purinoceptors, and have suggested that the presence of two P_{2X}-purinoceptor subtypes may explain the biphasic ATP concentration-response curve. We have suggested further that the second contractile phase is initiated by the enzymatic cleavage of the β,γ-anhydride oxygen bond by a cell surface hydrolytic or phosphoryl transfer (for example, kinase) enzyme. These hypotheses were based on the following observations: 1) ATP analogues without 5′-anhydride oxygen substitutions initiate responses containing a second phase of contraction; 2) nucleotides with anhydride-oxygen substitutions (such as APCPP, β,γ-methylene ATP (APPCP), and β,γ-imido ATP (APPNP)), which render them poor substrates for ATPases, initiate responses containing abbreviated second phases; and 3) adenosine 5′-*O*-(3-thiotriphosphate) (ATPγS) gives rise to responses with prolonged second phases (FIGS. 6 & 7). Compared to its effect on ATP-induced responses, P-ATP had little effect on the already abbreviated second phases of APPNP

FIGURE 6. Effect of P-ATP treatment on contractions of vas deferens (isolated from guinea pig) to ATP, β,γ-imido-ATP (APPNP) and APCPP (all 3×10^{-3} M; dots). Differences in the profiles of response to the three nucleotides are shown in this figure. Modified from Fedan and Lamport[37a] with permission.

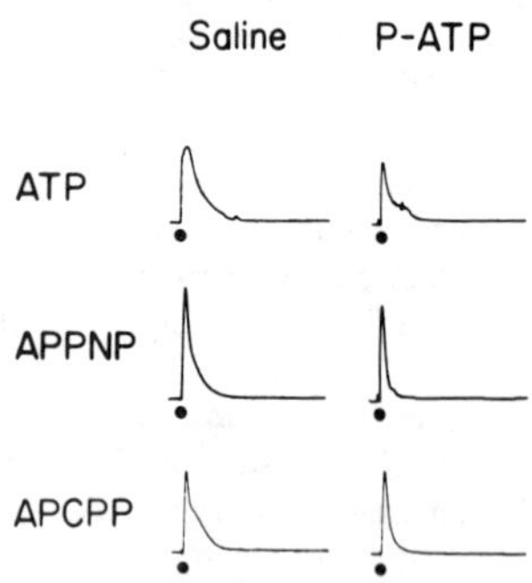

and APCPP responses, whereas the ATPγS second phase was markedly inhibited by P-ATP. We interpret these results to indicate that P-ATP inhibits an ectophosphoryltransfer transduction mechanism. In support of this idea are two lines of evidence,[39] obtained with the intact guinea pig vas deferens. First, ^{35}S from [^{35}S]ATPγS and ^{32}P from [γ-^{32}P]ATP are incorporated into cellular proteins along a time course that mimics that of contraction (FIG. 7). Incorporation of ^{35}S is inhibited in the presence of competing ATP, or following prior treatment with P-ATP. The relationship of phosphoryltransfer-mediated transduction to the P_{2X}- and P_{2Y}-classification system is difficult to reconcile because chemical stability of the agonist during receptor interactions is assumed. The ubiquity of phosphoryltransfer transduction, that is, its possible relation to P_{2Y}-purinoceptor systems, should be examined. This mechanism, having been reported to occur in a neural cell line, is not likely to be unique to the vas deferens.[40]

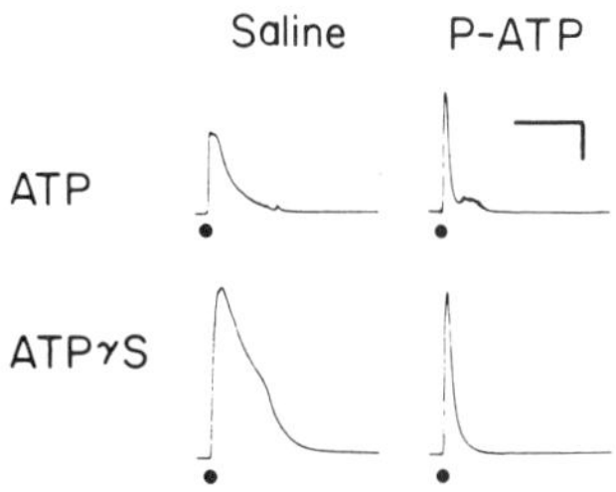

FIGURE 7. Effect of P-ATP treatment on contractions of vas deferens (isolated from guinea pig) to ATP and ATPγS (both 3×10^{-3} M; dots). Differences in the profiles of response to ATP and ATPγS are shown in this figure. Modified from Fedan and Lamport[37a] with permission.

AGENTS WITHOUT AGONIST ACTIVITY

The prototype compound for structure-activity studies that could lead to the development of competitive antagonists at P_{2X}- and P_{2Y}-purinoceptors has not been defined. ATP, or possibly an unrelated, undescribed compound, may be used. One limitation to P_2-purinoceptor competitive antagonist design is that ATP analogues lacking agonist activity are, generally, not antagonistic as well. They simply have no activity. The correlation between lack of agonist activity in an ATP analogue and its antagonist activity, however, has not been vigorously explored.

Isatogens: 2,2′-Pyridylisatogen (PIT)

These agents, exemplified by PIT, were among the first to be evaluated for their selective purinoceptor blocking activity.[41–43] Relaxant responses of guinea pig taenia coli to ATP are antagonized with reasonable specificity by PIT. The muscle, however, is relaxed by PIT directly. Such a baseline effect makes it difficult to interpret the antagonism of the relaxant effect of ATP. A series of isatogen compounds were tested in an attempt to eliminate relaxant activity and concomitantly to increase antagonist activity.[44] Despite enhanced relative ATP inhibitory activity, the relaxant effect of the isatogens was still evident. The specificity of the isatogen derivatives for a purinoceptor effect was not reported.

Apamin

Apamin, a neurotoxin from bee venom, inhibits ATP-induced relaxation of gastrointestinal smooth muscle preparations (see reference 45 for review) by inhibiting Ca^+-dependent increases in K^+ conductance. Because transduction by several receptor systems, in addition to P_{2Y}-purinoceptors, is linked to activation of Ca^+-dependent K^+ channels, apamin is relatively nonspecific in its antagonist action.[34] Apamin, nevertheless, is useful in clarifying transduction. For example, the possibility that $ANAPP_3$ could exert its antagonistic effects by interfering with Ca^+-dependent K^+ permeability changes was examined in the guinea pig vas deferens.[45] Rather than inhibition, apamin caused a nonspecific potentiation of contractions to ATP, norepinephrine, histamine, and acetylcholine, but not K^+, which is indicative of an involvement of the channel in responses to the receptor agonists. After $ANAPP_3$ treatment, apamin did not potentiate the residual responses to ATP, possibly because receptor transduction had been inhibited. Ectophosphoryltransfer transduction may not involve apamin-sensitive K^+ channels.

Reactive Blue 2

Introduced for this purpose by Kerr and Krantis,[46] the anthraquinone sulfonic acid dye, reactive blue 2, has been used as an ATP antagonist in several preparations.[4,47–49] In guinea pig cecum, reactive blue 2 inhibited hyperpolarization induced by APCPP, but not by norepinephrine,[49] and inhibited spike formation in response to carbachol without affecting depolarization and developed force, although the rate of force development was slowed.[49] In some studies, reactive blue 2 has been found to possess selectivity for P_{2Y}-purinoceptors. For example, endothelial cell-mediated relaxation responses of blood vessels to ATP were inhibited by reactive blue 2, not by APCPP or $ANAPP_3$, whereas the direct contractile responses of the muscles were antagonized by $ANAPP_3$ or APCPP desensitization.[4,31,50] Reactive blue 2 did not inhibit contractile responses of guinea pig ileum to ATP or APCPP.[51] Nonspecific inhibitory effects of reactive blue 2, however, have been reported. For example, endothelial-dependent relaxations of coronary artery to acetylcholine[31] and adenosine-induced dilation of coronary vessels[52] were inhibited appreciably by reactive blue 2. Commercially available reactive blue 2 is only approximately 60% pure, and it varies chemically with the supplier. Its nonspecific effects could be due to impurities.

Because reactive blue 2 is reasonably specific and effective in antagonizing the inhibitory effects of ATP, the view is being espoused that reactive blue 2 is a P_{2Y}-purinoceptor antagonist.[2–4] The effects of reactive blue 2 on excitatory, P_{2X}-purinoceptor-mediated responses have not been studied to a great degree. We have examined the effects of reactive blue 2 in the guinea pig vas deferens (FIG. 8). After a 30-min incubation with 2×10^{-5} M reactive blue 2 (a low concentration[31] in comparison with higher ones that have been employed[49,51,53]), an $ANAPP_3$-like effect on the ATP concentration-response curve was observed (FIG. 1). The greatest shift to the right occurred in low ATP concentrations. The biphasic curve became monophasic, and the maximum response was not affected. Reactive blue 2 also affected reactivity to norepinephrine and acetylcholine. The norepinephrine curve acquired a novel inflection, and the maximum response was reduced. The maximum response to acetylcholine was reduced substantially. The response profiles were altered in the presence of reactive

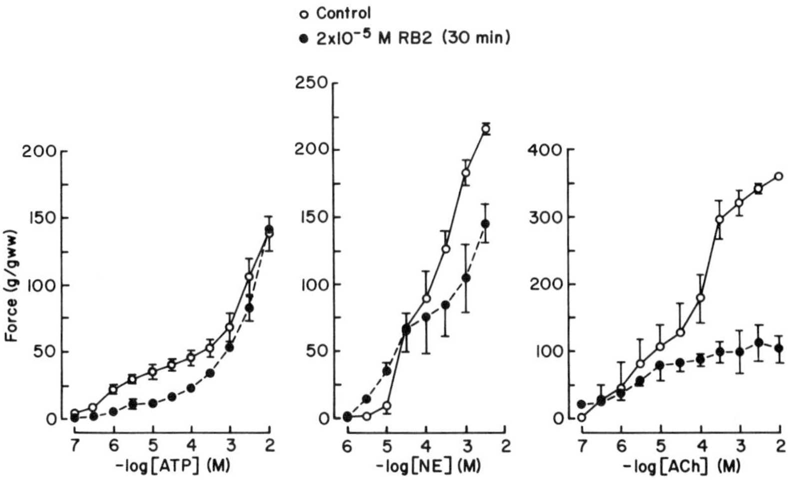

FIGURE 8. Effect of reactive blue 2 (RB2; 2×10^{-5} M) on the ATP ($N = 4$), norepinephrine (NE; $N = 2$), and acetylcholine (ACh; $N = 2$) concentration-response curves of vasa deferentia isolated from guinea pig. The reactive blue 2 remained present after a 30-min incubation prior to the addition of agonist.

blue 2 (FIG. 9). The second phase of contraction to ATP was delayed, and appreciable spontaneous contractions followed the peak, initial responses to acetylcholine and norepinephrine.

It is uncertain whether these effects of reactive blue 2 reflect the actions of impurities or the nonselectivity reactive blue 2 has on several receptors. The similarity in the antagonistic effects of $ANAPP_3$ and reactive blue 2 on the ATP concentration-response curve (FIGS. 1 & 9), however, is suggestive of a P_{2X}-purinoceptor interaction. Our findings would indicate that reactive blue 2 is not a selective antagonist in the vas

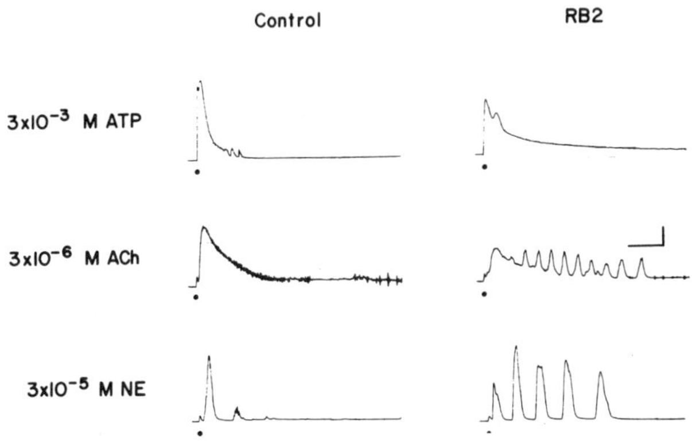

FIGURE 9. Effect of reactive blue 2 (RB2; 2×10^{-5} M) on the responses of vasa deferentia (isolated from guinea pig) to ATP, acetylcholine (ACh), and norepinephrine (NE). Horizontal bar: 30 sec; vertical bar: 2 g.

deferens, and that it is not P_{2Y} selective, either, providing its effect on ATP responses reflects a true receptor antagonism.

p-*Chloromercurbenzene Sulfonic Acid (PCMBS)*

Wiklund & Gustafsson[51,53] examined whether PCMBS, a sulfhydryl reagent, could act as a purinoceptor antagonist in smooth muscles of the guinea pig. Contractions of ileum to ADP, ATP, and APCPP were inhibited irreversibly and noncompetitively, and were concentration-dependent with respect to PCMBS. PCMBS inhibited norepinephrine- and ATP-induced contractions of the vas deferens,[51] but the investigators' results were not shown. In contrast, PCMBS did not inhibit relaxation responses of taenia coli to ADP or ATP. PCMBS also relaxed the taenia coli directly.

We extended these observations in vas deferens isolated from the guinea pig.[51] Our observations for the vas deferens, in contrast to those for the taenia coli,[51] revealed no effect of 3×10^{-5} M PCMBS on resting force. After a 30-min exposure to PCMBS, responses to ATP and norepinephrine were virtually abolished, and responses to acetylcholine were markedly reduced (FIG. 10). Contractions in response to 120 mM K^+, which was added at the end of these experiments, were also greatly attenuated (not shown). Our findings are illustrative of noncompetitive antagonistic effects of PCMBS in the vas deferens.

4'-Diisothiocyano-2,2'-Disulfonic Acid Stilbene (DIDS)

DIDS is a homobifunctional crosslinking reagent that inhibits anion exchange. McMillian *et al.*[54] examined the effects of DIDS on ATP-induced elevations of in-

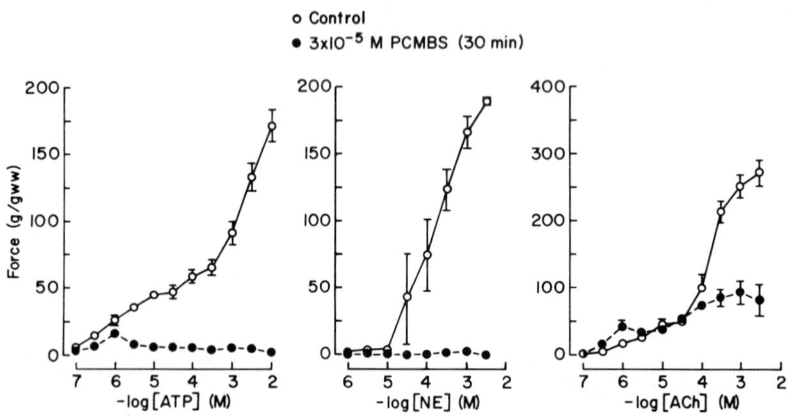

FIGURE 10. Effect of PCMBS (3×10^{-5} M) on the ATP ($N = 4$), norepinephrine (NE; $N = 2$), and acetylcholine (ACh; $N = 2$) concentration-response curves of vasa deferentia isolated from guinea pig. The PCMBS remained present after a 30-min incubation prior to the addition of agonist.

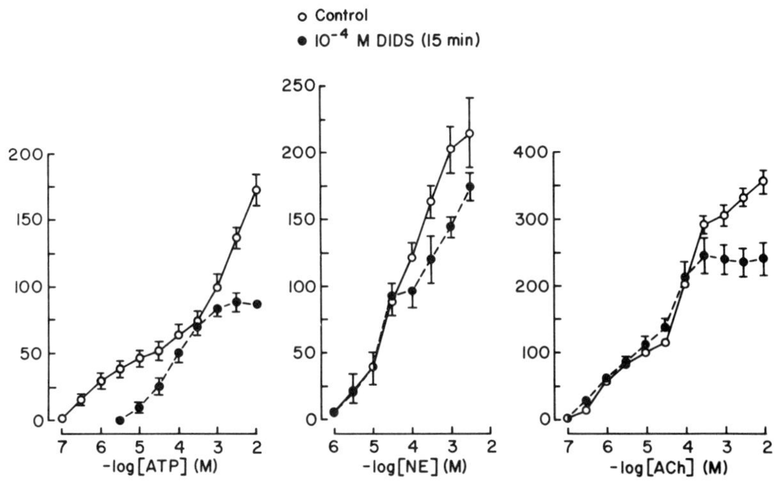

FIGURE 11. Effect of DIDS (100 μM) on the ATP (N = 4), norepinephrine (NE; N = 4), and acetylcholine (ACh; N = 4) concentration-response curves of vasa deferentia isolated from guinea pig. The DIDS (dissolved in DMSO; 0.1% v/v final concentration; DMSO present in controls) remained present after a 15-min incubation prior to the addition of agonist.

tracellular Ca^{2+} in rat parotid cells. These investigators provided qualitative evidence that DIDS inhibited the ATP-induced rise in intracellular Ca^{2+}, while having no effect on the response of the cells to substance P. Likewise, DIDS inhibited ATP-induced conductance increases in whole-cell patch-clamp experiments, but not those elicited with carbachol. The binding of [α-^{32}P]ATP to parotid cells was inhibited by DIDS. Unpublished work cited by McMillian *et al.*[54] indicated that these effects were not mimicked by the replacement of Cl^- with the impermeant anion, isethionate. It was suggested that the blockade of these effects of ATP provided evidence for the involvement of a purinoceptor that is blocked by DIDS.

In view of the significance of these findings, we examined the pharmacologic profile of DIDS in the vas deferens isolated from the guinea pig. DIDS (100 μM[54]) produced some novel effects (FIGS. 11 & 12). DIDS antagonized responses to ATP in concentrations of the nucleotide less or greater than 3×10^{-4} M, but had no effect at this specific ATP concentration. The norepinephrine and acetylcholine concentration-response curves in the presence of DIDS were superimposable upon the control curves, up to a point, after which the maximum responses became reduced. DIDS altered the profiles of response to these agonists. Although it contained oscillations, the second phase of the response to ATP was maintained at a reasonably steady level during exposure to the nucleotide. This is the first instance where we have observed that the contraction in response to ATP did not to return to baseline. Contractions to norepinephrine became rhythmic after the peak force was generated. The contractions to acetylcholine did not fade to basal force, but acquired a long-lasting tonic component. The reasons for these changes are unknown. If they do not result from nonspecific interactions owing to the chemical reactivity of DIDS, these changes may result secondarily from interference in anion permeability. Although isethionate substitution for Cl^- increases membrane resistance in smooth muscles,[55,56] the electophysiological role of Cl^- in smooth muscle of guinea pig vas deferens, to our knowledge, has not been studied. The altered response profiles could be expressions of increased membrane resistance and/or diminished rectification. Experiments with Cl^- substitution could allow a distinction to be made between these alternatives. Experiments to evaluate the specificity of DIDS for P_{2X}- and P_{2Y}-purinoceptor-mediated responses are needed.

5′-Fluorosulfonylbenzoyl Adenosine (FSBA)

This agent, introduced by R. F. Colman,[57-60] has been used to affinity label via the sulfonylfluoride group a multitude of nucleotide-binding proteins. The agent can be viewed as being an ATP analogue in which the 5′-polyphosphate chain is substituted with the bulky 5′-fluorosulfonylbenzoyl moiety. We examined FSBA for its ability to antagonize ATP-induced contractions in vas deferens isolated from the guinea pig. After incubation of the vas deferens for 6 min ($N = 2$) or 10 min ($N = 2$) with 10^{-4}

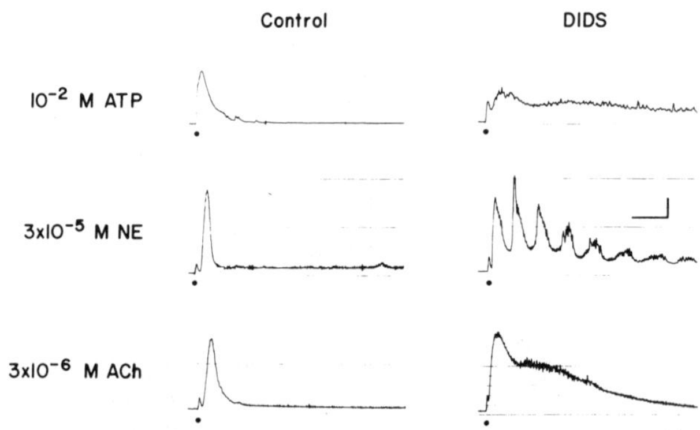

FIGURE 12. Effect of 100 μM DIDS on responses of vasa deferentia (isolated from guinea pig) to ATP, acetylcholine (ACh), and norepinephrine (NE). Horizontal bar: 30 sec; vertical bar: 2 g.

M FSBA, there was no effect on the contractions in response to 3×10^{-3} M ATP. The lack of an antagonistic effect suggests that FSBA does not interact with the receptor to form covalent bonds. The inactivity of FSBA is in keeping with the weak activity in vas deferens of ATP analogues with 5′-substitutions or deletions, other than anhydride oxygen or thiophosphate substitutions. Regardless of bulk, many such 5′-substituted analogues have less activity than AMP.[20,22] These experiments suggest, at least for P_{2X}-purinoceptors, that affinity or photoaffinity labels with the probe at the 5′-position will be devoid of antagonist activity.

CONCLUDING REMARKS

There are at least two P_2-purinoceptors that are recognized as distinct entities. Evidence for their existence has been obtained based, in part, on the effects of antagonists that are inherently unconventional in their chemistry (being distant from the endogenous agonist, ATP) or in their application.

Many aspects of the mechanisms of cellular responses to ATP remain obscure. Furthermore, we remain ignorant of the mechanisms of desensitization, as well as of the exact species of ATP in divalent cation-containing solutions that activates both types of receptors. We do not know why ATP cannot be rendered an antagonist with simple chemical substitutions, and we do not know why the responses of some cells to ATP appear to involve the breakdown of the agonist. In the face of all this uncertainty, the development of ideal P_2-purinoceptor antagonists will be an ambitious undertaking in receptor pharmacology. Without such agents it will be very difficult to define the important *in vivo* roles of extracellular ATP in health, and to understand and apply nucleotide agonist or antagonist therapy in disease. The differences between receptor and physiological antagonists should always be recognized.

REFERENCES

1. FEDAN, J. S., G. K. HOGABOOM, J. P. O'DONNELL & D. P. WESTFALL. 1985. *In* Methods in Pharmacology, Vol. 6. Methods Used in Adenosine Research. D. M. Paton, Ed.: 279-292. Plenum. New York, NY.
2. BURNSTOCK, G. & C. KENNEDY. 1985. Gen. Pharmacol. **16:** 433-440.
3. GORDON, J. L. 1986. Biochem. J. **233:** 309-319.
4. WHITE, T. D. 1988. Pharmacol. Ther. **38:** 129-168.
5. JENG, S. J. & R. J. GUILLORY. 1975. J. Supramol. Struct. **3:** 448-468.
6. RUSSELL, J., S. J. JENG & R. J. GUILLORY. 1976. Biochem. Biophys. Res. Commun. **70:** 1225-1234.
7. HOGABOOM, G. K., J. P. O'DONNELL & J. S. FEDAN. 1980. Science **208:** 1273-1276.
8. HEAD, R. J., J. P. O'DONNELL, G. K. HOGABOOM & J. S. FEDAN. 1983. Biochem. Pharmacol. **32:** 563-565.
9. WESTFALL, D. P., J. S. FEDAN, J. COLBY, G. K. HOGABOOM & J. P. O'DONNELL. 1983. Eur. J. Pharmacol. **87:** 415-422.
10. WESTFALL, D. P., G. K. HOGABOOM, J. COLBY, J. P. O'DONNELL & J. S. FEDAN. 1982. Proc. Natl. Acad. Sci. USA **79:** 7041-7045.
11. SNEDDON, P., D. P. WESTFALL & J. S. FEDAN. 1982. Eur. J. Pharmacol. **80:** 93-98.
12. FREW, R. F. & P. M. LUNDY. 1982. Life Sci. **30:** 259-267.
13. FREW, R. F. & P. M. LUNDY. 1982. Eur. J. Pharmacol. **81:** 333-336.
14. FREW, R. F. & P. M. LUNDY. 1986. Eur. J. Pharmacol. **123:** 395-400.
15. FEDAN, J. S., G. K. HOGABOOM & J. P. O'DONNELL. 1982. Life Sci. **31:** 921-928.
16. FEDAN, J. S., G. K. HOGABOOM & J. P. O'DONNELL. 1984. Biochem. Pharmacol. **33:** 1167-1180.
17. FEDAN, J. S., G. K. HOGABOOM, D. P. WESTFALL & J. P. O'DONNELL. 1983. Fed. Proc. **42:** 2846-2850.
18. O'DONNELL, J. P., G. K. HOGABOOM & J. S. FEDAN. 1983. Eur. J. Pharmacol. **86:** 435-440.
19. FEDAN, J. S., G. K. HOGABOOM, J. P. O'DONNELL, S. J. JENG & R. J. GUILLORY. 1985. Eur. J. Pharmacol. **108:** 49-61.
20. FEDAN, J. S., G. K. HOGABOOM, D. P. WESTFALL & J. P. O'DONNELL. 1982. Eur. J. Pharmacol. **81:** 193-204.
21. FEDAN, J. S., G. K. HOGABOOM, D. P. WESTFALL & J. P. O'DONNELL. 1982. Eur. J. Pharmacol. **85:** 277-290.
22. FEDAN, J. S., G. K. HOGABOOM & J. P. O'DONNELL. 1986. Eur. J. Pharmacol. **129:** 279-291.
23. FEDAN, J. S., G. K. HOGABOOM, J. P. O'DONNELL, J. COLBY & D. P. WESTFALL. 1981. Eur. J. Pharmacol. **69:** 41-53.

24. SNEDDON, P. & G. BURNSTOCK. 1984. Eur. J. Pharmacol. **100:** 85-90.
25. FEDAN, J. S. 1987. Br. J. Pharmacol. **91:** 633-639.
26. FRIEL, D. D. 1988. J. Physiol. (London) **401:** 361-380.
27. FRIEL, D. D. & B. P. BEAN. 1988. J. Gen. Physiol. **91:** 1-27.
28. KASAKOV, L. & G. BURNSTOCK. 1983. Eur. J. Pharmacol. **86:** 291-294.
29. MELDRUM, M. & G. BURNSTOCK. 1983. Eur. J. Pharmacol. **92:** 161-163.
30. BURNSTOCK, G. & C. KENNEDY. 1986. Circ. Res. **58:** 319-330.
31. HOUSTON, D. A., G. BURNSTOCK & P. M. VANHOUTTE. 1987. J. Pharmacol. Exp. Ther. **241:** 501-506.
32. FLAVAHAN, N. A. & P. M. VANHOUTTE. 1986. J. Pharmacol. Exp. Ther. **239:** 784-789.
33. BYRNE, N. G. & W. A. LARGE. 1986. Br. J. Pharmacol. **88:** 6-8.
34. DEN HERTOG, A. & J. VAN DEN AKKER. 1986. Eur. J. Pharmacol. **122:** 117-121.
35. FEDAN, J. S. 1986. Pharmacologist **28:** 238.
36. FEDAN, J. S. 1987. Fed. Proc. **46:** 376.
37. LAMPORT, S. J. & J. S. FEDAN. 1989. FASEB J. **3:** A1198.
37a. FEDAN, J. S. & S. J. LAMPORT. 1990. J. Pharmacol. Exp. Ther.: in press.
38. LOWE, P. N. & R. B. BEECHEY. 1982. Biochemistry **21:** 4073-4082.
39. LAMPORT, S. J. & J. S. FEDAN. 1990. Ann. N.Y. Acad. Sci. This volume.
40. EHRLICH, Y. H., T. B. DAVIES, E. BOCK, E. KORNECKI & R. H. LENOX. 1986. Nature **320:** 67-70.
41. SPEDDING, M., A. J. SWEETMAN & D. F. WEETMAN. 1975. Br. J. Pharmacol. **53:** 575-583.
42. SPEDDING, M. & D. F. WEETMAN. 1978. J. Pharm. Pharmacol. **30:** 335-336.
43. FOSTER, H., M. HOOPER, M. SPEDDING, A. J. SWEETMAN & D. F. WEETMAN. 1978. Br. J. Pharmacol. **63:** 309-314.
44. FOSTER, H., M. HOOPER, S. H. INMAN, G. S. LOVETT, J. NICHOLSON, C. J. SWAIN, A. J. SWEETMAN & D. F. WEETMAN. 1983. Br. J. Pharmacol. **79:** 273-278.
45. FEDAN, J. S., G. K. HOGABOOM & J. P. O'DONNELL. 1984. Eur. J. Pharmacol. **104:** 327-334.
46. KERR, D. I. B. & A. KRANTIS. 1979. Proc. Aust. Physiol. Pharmacol. Soc. **10:** 156P.
47. CHOO, L. K. 1981. J. Pharm. Pharmacol. **33:** 248-250.
48. MANZINI, S., C. A. MAGGI & A. MELI. 1985. Eur. J. Pharmacol. **113:** 399-408.
49. MANZINI, S., C. H. V. HOYLE & G. BURNSTOCK. 1986. Eur. J. Pharmacol. **127:** 197-204.
50. BURNSTOCK, G. & J. J. I. WARLAND. 1987. Br. J. Pharmacol. **90:** 111-120.
51. WIKLUND, N. P. & L. E. GUSTAFSSON. 1988. Eur. J. Pharmacol. **148:** 361-370.
52. HOPWOOD, A. M. & G. BURNSTOCK. 1987. Eur. J. Pharmacol. **136:** 49-54.
53. WIKLUND, N. P. & L. E. GUSTAFSSON. 1988. Acta Physiol. Scand. **132:** 15-21.
54. MCMILLIAN, M. K., S. P. SOLTOFF, J. D. LECHLEITER, L. C. CANTLEY & B. R. TALAMO. 1988. Biochem J. **255:** 291-300.
55. CASTEELS, R. 1970. *In* Smooth Muscle. E. Bülbring, A. F. Brading, A. W. Jones & T. Tomita, Eds.: 70-99. Williams & Wilkins. Baltimore, MD.
56. CASTEELS, R. 1981. *In* Smooth Muscle: An Assessment of Current Knowledge. E. Bülbring, A. F. Brading, A. W. Jones & T. Tomita, Eds.: 105-126. University of Texas Press. Austin, TX.
57. PAL, P. K., W. J. WECHTER & R. F. COLMAN. 1975. J. Biol. Chem. **250:** 8140-8147.
58. COLMAN, R. F. 1983. Annu. Rev. Biochem. **52:** 67-91.
59. COLMAN, R. F. 1987. *In* Proteins. J. J. L'Italien, Ed.: 569-580. Plenum. New York, NY.
60. COLMAN, R. F. 1990. Ann. N.Y. Acad. Sci. This volume.

DISCUSSION OF THE PAPER

S. E. O'CONNOR (*Fisons plc, Loughborough, Leicestershire, England*): I was interested in your ability to separate out the first and second phases of the response of the vas deferens to ATP using $ANAPP_3$ and P-ATP, respectively. If you desensitize with Me-ATP, what happens to these two phases? Based on your suggestion that the first phase is receptor mediated, I would predict that it would selectively abolish the first phase.

FEDAN: That is a very interesting suggestion. We have been considering that experiment for some time, but have not yet done it. It all depends on what you think is the mechanism of $ANAPP_3$ at the receptor, that is, whether $ANAPP_3$ causes a conventional antagonism or simply an irreversible agonist signal. The experiment you suggest would go some way to clarifying this situation.

T. D. WHITE (*Dalhousie University, Halifax, Nova Scotia, Canada*): Does P-ATP antagonize the second phase of contraction of the guinea pig vas deferens observed during sympathetic nerve stimulation?

FEDAN: We have not yet done this experiment. Only at intermediate to high concentrations of ATP ($> 3 \times 10^{-5}$ M) does the second phase of contraction to ATP become apparent. The possible effect of P-ATP on neurogenic responses might allow one to determine a wide range of concentrations at which neurally released ATP is acting.

ADP Receptors in Platelets[a]

ROBERT W. COLMAN

Thrombosis Research Center
Temple University School of Medicine
Philadelphia, Pennsylvania 19140

INTRODUCTION

Blood platelets were the first cells—or, more precisely, anucleate megakaryocyte fragments—demonstrated to respond to extracellular adenine nucleotides. Gaarder *et al.*[1] identified ADP as the factor that was released from red cells and that stimulated platelets to adhere to glass. Platelets circulate in blood without adhering to other platelets or to the endothelium lining the vessels. Following exposure to ADP and other physiological agonists such as collagen or thrombin, platelets undergo adhesion, shape change (from discs to spiculated spheres), aggregation, and secretion. ADP released from injured endothelial or red cells is not only an agonist, but functions as an autocrine regulator because it is also released from platelet-dense granules during secretion. ADP is also involved in platelet activation by thromboxane A_2, platelet activating factor, collagen, and thrombin. Exposure of platelets to ADP results in ionized calcium transport into the cell as well as intracellular mobilization of calcium. Calmodulin-dependent phosphorylation of myosin light chain then occurs, followed by secretion of alpha and dense granule contents. ADP initiates two separate pathways that unite to produce platelet activation (FIG. 1). At low concentrations, ADP (0.1-0.5 μM) produces shape change, followed by reversible aggregation.[2] Secondary aggregation and release of storage granule contents occur when ADP reaches 2-5 μM in the extracellular environment. ADP also inhibits adenylate cyclase when it is stimulated by such agonists as prostaglandin I_2 and adenosine.[3] The conversion of ATP to AMP results in an increase of cyclic AMP, which inhibits platelet activation, including aggregation and secretion. Thus, by inhibiting stimulated adenylate cyclase, ADP reduces cyclic AMP, further potentiating the activation of platelets.

FUNCTIONAL AND BINDING STUDIES OF THE ADP RECEPTOR

The platelet stands alone among cells with surface receptors for adenine nucleotides by preferring ADP to AMP, ATP, and adenosine. Functional studies have been performed by measuring biologic response, such as aggregation over a variety of ADP concentrations. These studies have established that the ADP receptor for platelet

[a]The preparation of this review was supported by Grant HL36579 from the National Institutes of Health.

shape change and aggregation displays a high degree of specificity. On the purine ring, only the 2-position can be modified, and the substitution of a chloro or a methylthiol group results in more active agonists.[4] Substitutions at the 6- or 8-positions of the purine ring or in the ribose ring result in a decrease in activity.[5] The pyrophosphate in the 5′-position is required, but ATP serves as a competitive antagonist.[6] Adenosine does not directly inhibit the effect of ADP, but rather inhibits aggregation by increasing intracellular cyclic AMP. When platelet membranes have been used, formal binding experiments have been frustrated by the presence of ADP-binding proteins, including actin and myosin, on the cytoplasmic face of the membranes.[7] Binding to intact platelets is hampered by the presence of ecto-ADPases that degrade ADP to AMP or kinases that convert ADP to ATP. Jefferson *et al.*[8] have recently measured the binding of [2-^{3}H]ADP to paraformaldehyde-fixed platelets, but this approach has the disadvantage that chemical changes in the binding sites induced by the fixative could affect the accessibility of the ligand. In addition, correlation of function with the number of sites is difficult.

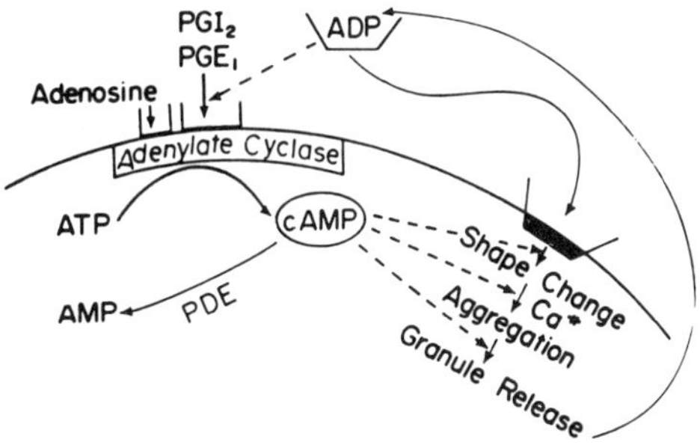

FIGURE 1. Dual role of ADP in platelet activation. See text for explanation.

IDENTIFICATION OF ADP RECEPTOR (AGGREGIN) BY AFFINITY LABELING

Another approach is that of affinity labeling, which allows the introduction of a covalent bond between the ligand and the receptor, and which thus facilitates definitive identification of particular proteins in the complex mixture that is characteristic of the platelet membrane. Bennett *et al.*[9] introduced such a reagent to label a putative ADP receptor, 5′-*p*-fluorosulfonylbenzoyl adenosine (FSBA) (FIG. 2). This compound is similar in structure to ADP or ATP in the adenosine moiety, and, in its extended configuration, has a carbonyl group at the 5′-position of the ribose (similar to the first phosphoryl group of ADP or ATP) and a sulfonyl group (similar to the terminal phosphate of ADP or ATP). The sulfonyl fluoride group has been demonstrated to react covalently with several nucleophilic amino acids. Reactions between the sulfonyl fluoride group and these amino acids are accompanied by displacement of the fluoride ion and concomitant covalent labeling of multiple enzymes.[10] The first event after the binding of ADP, platelet shape change, as detected by either changes in absorbance or transmission electron microscopy, is inhibited by FSBA. Inhibition due to FSBA parallels the covalent incorporation of SBA into platelets in both a time- and concentration-dependent manner.[11,12] Phosphorylation of the myosin light chain, the bio-

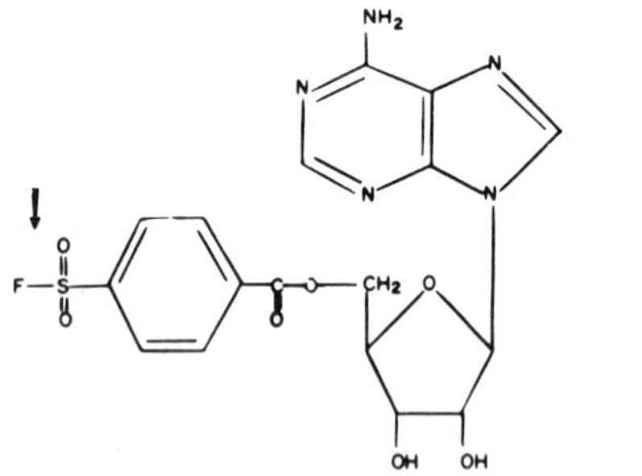

5′-p-fluorosulfonylbenzoyl adenosine (5′FSBA)

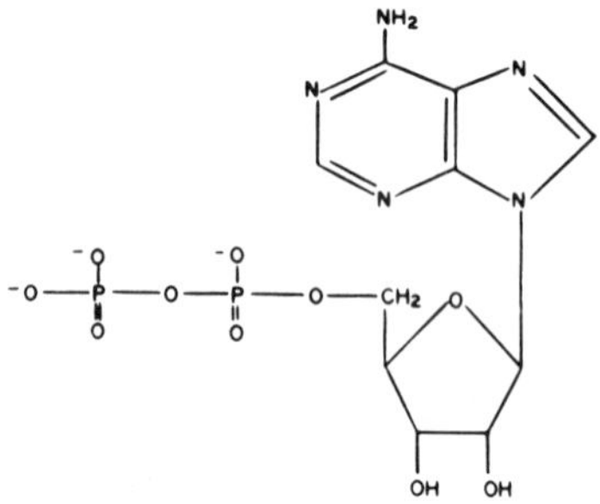

Adenosine-5′-diphosphate (ADP)

FIGURE 2. Comparison of the structure of FSBA and ADP. Reproduced from reference 37 with permission.

chemical correlate of ADP-induced shape change, is also inhibited by FSBA.[12] Following shape change, platelets undergo aggregation when exposed to ADP. For aggregation to occur, ADP must first expose fibrinogen-binding sites.[13] Fibrinogen then binds to a calcium complex of platelet membrane glycoproteins IIb and IIIa[14] that serves as a receptor on the platelet surface. Exposure of washed platelets[15] to FSBA prevented ADP-induced aggregation and inhibited the exposure of fibrinogen receptors by ADP in proportion to the covalent incorporation of the affinity analogue.

Intact platelets incubated with [^{3}H]FSBA display on SDS electrophoresis a single component having a relative molecular mass of 100,000 Da, aggregin, which is covalently modified by the affinity analogue.[7,9,15] The specificity is indicated by the observation that the incorporation is prevented by ADP or ATP, but not by adenosine. FSBA, like ADP, cannot penetrate the intact cell without prior hydrolysis to adenosine.[9] Because modification of a single surface membrane component, aggregin, by FSBA, renders platelets refractory to activation by ADP, aggregin becomes a strong candidate for assignment as an ADP receptor responsible for activation.

FSBA also serves as a partial agonist for platelet activation. ADP-induced shape change has an EC_{50} of 0.4 μM, but we found that FSBA stimulated platelet shape change with an EC_{50} of 220 μM,[20] a 550-fold difference (FIG. 3). The specificity is confirmed by the absence of any measurable effect of the guanosine analogue of FSBA (500 μM), on platelets. Because the amount of SBA covalently incorporated into aggregin at the time of maximum platelet shape change (2 min) was only 3% of that incorporated when ADP-induced shape change was totally inhibited, the prior noncovalent binding of FSBA to platelets was responsible for the activation of platelets.

Similar observations have been made for the phosphorylation of the myosin light chain, a biochemical correlate of shape change.[16] High concentrations of FSBA induce rapid myosin phosphorylation,[12] whereas the incubation of platelets with 40 μM FSBA for 1 hr results in the inhibition of this response.

DISTINCTION BETWEEN AGGREGIN AND GPIIIa

GPIIIa, the β component of the heterodimer fibrinogen receptor, has a molecular mass (100 kDa) similar to that for aggregin on nonreduced gels. Moreover, thrombasthenic platelets lacking GPIIIa resemble FSBA-treated platelets in failing to aggregate or bind fibrinogen after ADP stimulation, and both proteins are degraded by chymotrypsin to 70-kDa fragments.[17] A recent study,[18] however, has assembled overwhelming evidence that aggregin and GPIIIa are distinct proteins. Thrombasthenic platelets display normal ADP-induced shape change, and although thrombasthenic platelets from the two patients examined contained less than 5% of the GPIIIa present in normal controls, the amounts of SBA incorporated into the platelets of the two patients were similar to that incorporated into the controls. Moreover, aggregin was not precipitated by murine monoclonal antibodies directed against the GPIIb-GPIIIa complex or against GPIIIa or by a monospecific rabbit polyclonal antibody to GPIIIa. Finally, on reduction, the electrophoretic migration of GPIIIa decreases because of the cleavage of intrachain disulfide bonds, whereas that of aggregin remains unchanged.

DISTINCTION BETWEEN AGGREGIN AND THE RECEPTOR COUPLED TO ADENYLATE CYCLASE

Is the ADP receptor mediating shape change distinct or the same as that modulating adenylate cyclase (FIG. 1)? We[11] found that increasing concentrations of ADP pro-

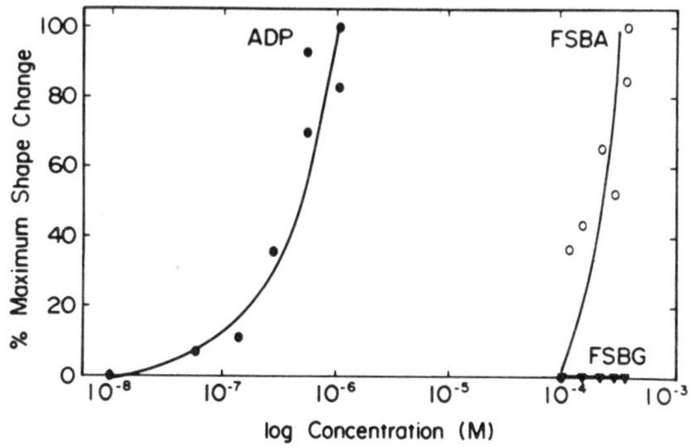

FIGURE 3. Effect of various concentrations of ADP, FSBA, and FSBG on induced platelet shape change. Each compound was tested at a wide range of concentrations for its ability to produce shape change. For comparison, each is normalized to the percentage of maximum shape change and plotted as a log of concentration. The data are from a single experiment and are representative of three such experiments. Reproduced from reference 12 with permission.

gressively decreased the prostaglandin I_2-stimulated platelet cyclic AMP formation. Neither the basal nor the stimulated levels of cyclic AMP were affected by FSBA. Further, 2-methylthio-ADP (MeSADP) was 75-fold more potent as an inhibitor of stimulated adenylate cyclase than as a stimulator of platelet aggregation. Radiolabeled MeSADP bound to intact platelets at concentrations that do not affect aggregation but cause marked inhibition of adenylate cyclase. FSBA at a concentration 25-fold greater than that needed to block aggregation did not change the number of binding sites for MeSADP. Finally, the nonpenetrating thiol reagent, *p*-chloromercuribenzene-sulfonate, failed to inhibit shape change induced by ADP but reversed the effect of ADP in inhibiting cyclic AMP accumulation. This finding further indicates the distinct nature of the two receptors.

RELATIONSHIP OF AGGREGIN TO RECEPTORS FOR OTHER PLATELET AGONISTS

Prior to the use of FSBA, involvement of ADP was evaluated by converting endogenous or exogenous ADP to ATP with creatine phosphokinase (K_m = 50 μM) or by hydrolyzing ADP to AMP with potato apyrase (K_m = 240 μM). ADP potentiates other agonists at concentrations as low as 50 nM, and thus the results with these enzymes may not be definitive. Using FSBA to block the receptor has made it possible to reassess the requirement for ADP for activation by other platelet agonists.

EPINEPHRINE

Epinephrine was able to induce platelet aggregation and bind fibrinogen, but only when incubated at 1-5 μM, a concentration 100-fold higher than that which exists *in vivo* during stress.[19] Thus, epinephrine must act in synergy with other agonists such as ADP to have physiological significance in platelet activation. Both platelet aggregation and inhibition of adenylate cyclase are mediated by an α_2-adrenergic receptor, the gene for which has been cloned.[20] To study the "cross-talk" between this receptor and aggregin, FSBA was used to render the platelet insensitive even to submicromolar concentrations of ADP.[21] FSBA progressively inhibited epinephrine-induced platelet aggregation and fibrinogen binding, indicating the absolute dependence of epinephrine-induced platelet aggregation on the ADP receptor (aggregin). FSBA did not change the number or affinity of α_2-adrenergic receptors[21] as measured by [^{3}H]yohimbine binding.[22]

Epinephrine is able to potentiate ADP-induced aggregation by increasing by 10-fold the affinity of binding of ADP to platelets without altering the total number of ADP binding sites as measured by incorporation of [^{3}H]FSBA. Thus, epinephrine, by decreasing the concentration of ADP in this way, could protect against FSBA incorporation (FIG. 4). The rate of incorporation of FSBA may in fact be inhibited by much lower concentrations of ADP in the presence of epinephrine.[21] This synergistic effect appears to depend on α_2-adrenergic receptors because it is blocked by the competitive antagonist, yohimbine.

PROSTAGLANDIN H_2

Prostaglandin H_2 (PGH_2) is one of the predominant metabolites of arachidonic acid and a potent inducer of platelet activation. PGH_2 is synthesized from arachidonic acid by cyclooxygenase. The role of released ADP in the action of PGH_2 has been unclear. ADP was thought to be unnecessary for endoperoxide-stimulated platelet aggregation because PGH_2 can stimulate aggregation without detectable ADP release,[22] because arachidonic acid can induce aggregation of degranulated platelets,[23] and because endoperoxides induce aggregation of platelets with "storage pool disease." The

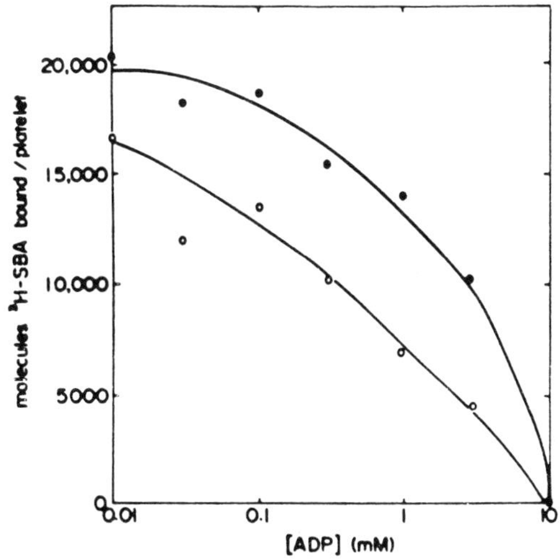

FIGURE 4. Effect of epinephrine on the ability of ADP to protect against covalent incorporation of [^{3}H]FSBA into intact platelets. Platelets (1 ml, 5 × 10^8 cells/ml) in Tyrode's buffer without added calcium were incubated with [^{3}H]FSBA (40 μM, 37 °C) in the presence (○——○) and absence (●——●) of epinephrine (10 μM) at various concentrations of ADP. The incubations were carried out in the presence of adenosine deaminase (2 units/ml). At the end of the incubation period (5 min), the suspensions were treated with dithiothreitol (0.2 M) to destroy the FSBA and stop the reaction. The suspensions were then washed and dialyzed. Reproduced from reference 21 with permission.

necessary ADP concentrations, however, may just be very low. In transfer experiments involving thromboxane, ADP appeared to be important in the observed responses.[24]

The use of FSBA has established that prostaglandin endoperoxide-induced aggregation of platelets requires ADP.[25] The platelet aggregation and fibrinogen binding induced by stable PGH_2 analogues (U44619 and azo-PGH_2) are inhibited by FSBA. In contrast, the shape change induced by high concentrations of prostaglandin endoperoxides is not inhibited by FSBA. Thus, ADP is required for the induction of aggregation by the prostanoid derivatives, but not for the shape change induced by high concentrations of prostanoids, the shape change being mediated by a distinct receptor.

COLLAGEN

Collagen is known to stimulate platelet aggregation and fibrinogen binding, but the role of ADP has been controversial. Investigators have suggested that ADP is essential,[26] that it is partially responsible,[22] and that it plays no role in collagen-induced aggregation.[27] We have studied[28] collagen-induced platelet activation using FSBA to block ADP actions by modifying aggregin covalently. The induction of platelet aggregation and exposure of fibrinogen receptors are completely inhibited. FSBA also inhibits collagen-induced shape change as well as shape change stimulated by low concentrations of azo-PGH_2. At higher concentrations of azo-PGH_2, however, the inhibition by FSBA can be overcome. Thus, activation of platelets by collagen is mediated directly by ADP binding to aggregin to induce aggregation and exposure of fibrinogen binding sites and indirectly by ADP through potentiating PGH_2-induced shape change.

THROMBIN

Thrombin appears to stimulate platelets by multiple mechanisms. At low concentrations of thrombin (≤ 1 nM), secretion of ADP and fibrinogen may contribute to platelet stimulation. Thrombin, unlike collagen, epinephrine, and prostaglandin endoperoxides, however, stimulates platelets by a mechanism independent of ADP. Thus, as predicted,[29] aggregation and exposure of fibrinogen binding sites are little affected by FSBA when stimulated by moderate concentrations of thrombin (>2 nM, 0.2 NIH units/ml). Harmon and Jamieson[30] summarized the evidence for two types of receptors, and for each of these receptors being coupled to different platelet responses. The first type, R_1, is ADP independent, involves a small number of high-affinity sites, and requires only low concentrations of thrombin (<1 nM). This receptor mediates the inhibition of stimulated adenylate cyclase, the secretion of acid hydrolases, and the activation of phospholipase A_2, and is not stimulated by γ-thrombin. The second type, R_2, is dependent on occupancy of an intermediate number of binding sites with moderate affinity and requires a moderate concentration of thrombin (>2 nM). This receptor is linked to phospholipase C, probably through a G protein, and thus produces diacylglycerol, which stimulates protein kinase C and is closely linked to secretion, and inositol trisphosphate, which plays a major role in mobilizing intracellular calcium. These responses, which are also mediated by γ-thrombin, result in platelet aggregation and exposure of fibrinogen binding sites.

We have shown that aggregin is cleaved after exposure of intact platelets to thrombin. Aggregin is not cleaved, however, if thrombin is incubated with platelet membranes containing aggregin (FIG. 5) or with intact platelets depleted of ATP.[31] These findings suggested that thrombin may act indirectly by activating an intracellular proteolytic enzyme dependent on metabolic energy. Because the cleavage was inhibited by thiol protease inhibitors, leupeptin and antipain (20 μM), it was suggested that a thiol protease cleaved aggregin.[31] Moreover, at 20 μM leupeptin, there is no inhibition of the ability of thrombin to elevate intracellular Ca^{2+}, to produce shape change, to cleave tripeptide amidolytic substrates (TABLE 1), or to clot fibrinogen. Purified platelet calpain, unlike thrombin, is able to cleave aggregin in the platelet membrane as well as in intact platelets. If fibrinogen is added, calpain will aggregate washed

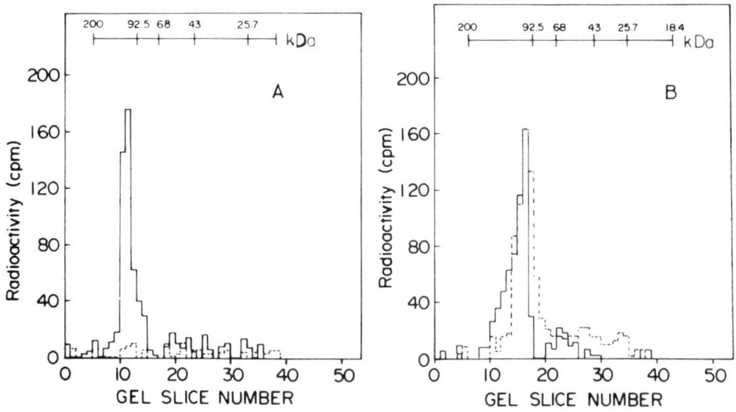

FIGURE 5. Effect of thrombin on the proteolytic cleavage of [^{3}H]FSBA-modified aggregin in (**A**) intact platelets and (**B**) platelet membrane. The solid lines and dashed lines indicate the distribution of radioactivity in gels corresponding to control and thrombin-treated samples, respectively. Platelets were labeled with [^{3}H]FSBA. The radiolabeled platelets (1×10^{10}/20 ml) were incubated in the absence or presence of thrombin (0.2 units/ml) at 37 °C for 30 min. Following incubation, thrombin was inactivated by the addition of 3 nM PPACK. Membrane preparations of [^{3}H]FSBA-labeled aggregin from control and thrombin-treated platelets were solubilized, dialyzed, and separated by electrophoresis. Membranes containing [^{3}H]FSBA-labeled aggregin were first prepared from intact [^{3}H]FSBA-labeled platelets (3.2×10^{10}). The radiolabeled membranes were suspended in 300 μl of 50 mM sodium phosphate buffer, pH 7.2, and divided into two equal halves. They were incubated with thrombin (0.2 units/ml) for 30 min at 37 °C. Thrombin activity was inhibited by the addition of hirudin (20 units/ml). In the control experiment, thrombin was omitted from the incubation mixture. The control and thrombin-treated membranes were solubilized, dialyzed, and separated by electrophoresis. Reproduced from reference 30 with permission.

TABLE 1. Effect of Leupeptin on Thrombin-Induced Platelet Aggregation and Shape Change, on $[Ca^{2+}]_i$, and on Thrombin Enzymatic Activity

Leupeptin	$[Ca^{2+}]_i$ (nM)	Extent of Aggregation (mV)	Extent of Shape Change (mV)	Amidolytic Activity (%)
0	869 ± 33	40 ± 2	14 ± 1	100
5	876 ± 36	34 ± 2	14 ± 1	—
10	889 ± 22	0	14 ± 1	98 ± 4
20	883 ± 28	0	15 ± 1	94 ± 5
30	475 ± 18	0	14 ± 1	—
40	208 ± 10	0	14 ± 2	84 ± 5

NOTE: Platelet shape change was monitored in a Lumi aggregometer in the presence of 2 mM EDTA. Intracellular Ca^{2+} concentrations, $[Ca^{2+}]_i$, and platelet aggregation were measured. In each case, platelets were preincubated with a given concentration of leupeptin for 30 sec at 37°C followed by the addition of thrombin (0.2 units/ml). Platelet concentrations were as follows: 5×10^8/ml for $[Ca^{2+}]_i$ and platelet aggregation and 1×10^8/ml for platelet shape change. Thrombin amidolytic activity was measured with 1 mM S2238 (*H*-D-phenylalanine pipecolyl-L-arginine-*p*-nitroanilide) as substrate. Each result is expressed as a mean ± SD of three experiments. The basal $[Ca^{2+}]_i$ in platelets in the absence of thrombin was 80 ± 8 nM.

platelets preincubated with FSBA, further supporting the thesis that thrombin-induced platelet aggregation and exposure of fibrinogen binding sites are mediated through the activation of calpain[31] and are independent of ADP.

Competition binding studies show that N^{α}-tosyl-L-lysine chloromethyl ketone (PPACK) modifies thrombin so that it competes for the R_2 site on platelets.[29] We have recently shown[32] that PPACK-thrombin inhibits thrombin-induced platelet aggregation and cleavage of aggregin, indicating that this thrombin response is receptor mediated. Iloprost (a stable PGI_2 analogue) and forskolin, both of which raise cAMP levels and inhibit thrombin binding,[33] inhibit thrombin-associated aggregation and cleavage of aggregin,[32] indicating that both aggregation and cleavage of aggregin depend on the binding of thrombin to a platelet receptor (R_2).

High-molecular-weight kininogen (HK), the most potent plasma inhibitor of platelet calpain,[34] was found to inhibit thrombin-induced aggregation and cleavage of aggregin,[35] further strengthening the involvement of calpain in the aggregation by platelets. About 10 times as much thrombin is needed to stimulate platelets in plasma as washed platelets. This difference may be due to the presence of HK, since plasma deficient in HK behaves similarly to washed platelets. This inhibition displays specificity because HK inhibits both α- and γ-thrombin but fails to inhibit aggregation by ADP; collagen; a calcium ionophore, A23187 (without added Ca^{2+}); the protein kinase C stimulator, phorbol myristrate acetate; a combination of A23187 and phorbol myristrate acetate; or the stable prostaglandin endoperoxide, U46619.

CONCLUSION

A working model first developed to explain chymotrypsin-induced platelet aggregation[36,37] has recently been expanded to explain thrombin-induced platelet aggregation (FIG. 6).[38] Aggregin, a platelet membrane protein with a molecular mass of 100,000 daltons, exhibits a nucleotide binding site capable of interacting with ADP and ATP. Binding at the ADP site results in exposure of latent fibrinogen binding sites necessary for platelet aggregation. FSBA prevents the binding of ADP to aggregin and thus inhibits exposure of fibrinogen binding sites, shape change, and aggregation of the cells. Aggregin is distinct from the ADP binding site that mediates the inhibition of adenylate cyclase as well as glycoprotein IIIa (GPIIIa). Binding of ADP to the receptor is pictured as introducing a conformational change in the receptor molecule, thereby facilitating the spatial approximation of glycoproteins IIb/IIIa to form the fibrinogen receptor. Thrombin binds to glycoprotein Ib (GPIb) on the platelet membrane and initiates a series of events that activate calpain. Metabolic energy is required to supply ATP for the synthesis of phosphatidyl inositol bisphosphate (PIP_2) from phosphatidyl inositol. Thrombin activates phospholipase C, which cleaves PIP_2 to diacylglycerol. The other product, inositol trisphosphate, raises intracellular Ca^{2+} and thus activates platelet calpain. Calpain then cleaves aggregin, resulting in the exposure of the GPIIb/GPIIIa complex to allow binding of fibrinogen and platelet aggregation. Thus, aggregin plays a major role in platelet activation not only by serving as the receptor for ADP-induced shape change and aggregation but also by reversing the latency of the fibrinogen receptor.

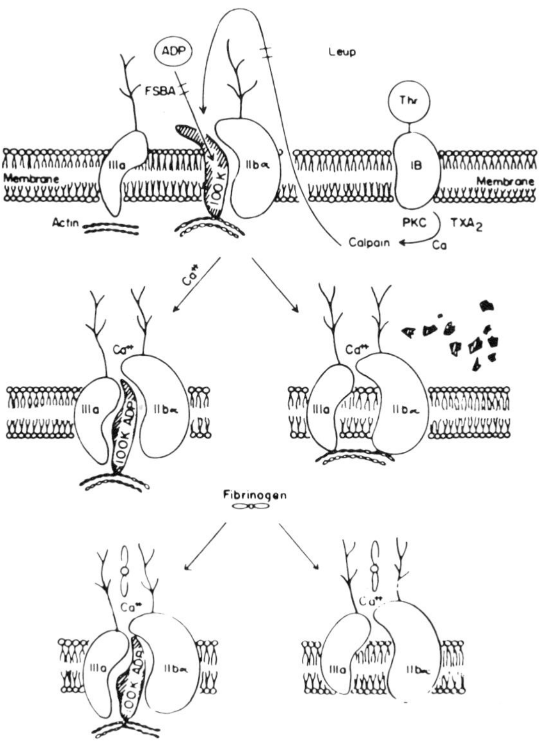

FIGURE 6. Schematic model for mechanisms of platelet aggregation by ADP and thrombin. See text for explanation. Reproduced from reference 35 with permission.

ACKNOWLEDGMENTS

The investigations reviewed in the article required substantial contributions, for which I am grateful, from my colleagues, Roberta F. Colman, of the University of Delaware, and Rajinder N. Puri and Stefan Niewiarowski, of the Temple University School of Medicine.

REFERENCES

1. GAARDER, A., J. JONSEN, S. LALAND, A. HELLEM & P. A. OWREN. 1961. Adenosine diphosphate in red cells as a factor in the adhesiveness of human blood platelets. Nature **193:** 531-532.
2. BORN, G. V. R. 1962. Aggregation of blood platelets by adenosine diphosphate and its reversal. Nature **194:** 927-929.
3. HASLAM, R. J. 1973. Interactions of the pharmacological receptors of blood platelets with adenylate cyclase. Semin. Hematol. **6:** 333-350.
4. MAGUIRE, M. H. & F. MICHAL. 1968. Powerful new aggregator of blood platelets—2-chloroadenosine-5′-diphosphate. Nature **217:** 571-572.
5. GRANT, J. A. & M. C. SCRUTTON. 1980. Positive interaction between agonists in the aggregation response of human blood platelets: Interaction between ADP, adrenaline and vasopressin. Br. J. Haematol. **44:** 109-125.

6. MacFarlane, D. E. & D. C. B. Mills. 1975. The effects of ATP on platelets: Evidence against the central role of released ADP in primary aggregation. Blood **46:** 309-320.
7. Bennett, J. S., G. Vilaire, R. F. Colman & R. W. Colman. 1981. Localization of human platelet membrane-associated actomyosin using the affinity label 5′-*p*-fluorosulfonylbenzoyl adenosine. J. Biol. Chem. **256:** 1185-1190.
8. Jefferson, J. R., J. T. Harmon & G. A. Jamieson. 1988. Identification of high-affinity (K_d = 0.35 μmol/L) and low-affinity (K_d = 7.9 μmol/L) platelet binding sites for ADP and competition by ADP analogues. Blood **71:** 110-116.
9. Bennett, J. S., R. F. Colman & R. W. Colman. 1978. Identification of adenine nucleotide binding proteins in human platelet membranes by affinity labeling with 5′-*p*-fluorosulfonylbenzoyl adenosine. J. Biol. Chem. **253:** 7346-7354.
10. Colman, R. F. 1987. Advances in affinity labeling of purine nucleotide binding sites. *In* Protein: Structure and Function. J. L'Italien, Ed.: 569-580. Plenum. New York, NY.
11. Mills, D. C. B., W. R. Figures, L. M. Scearce, R. F. Colman & R. W. Colman. 1985. Two mechanisms for inhibition of ADP-induced platelet shape change by 5′-*p*-fluorosulfonylbenzoyl adenosine: Conversion of adenosine and covalent modification at an ADP binding site distinct from that which inhibits adenylate cyclase. J. Biol. Chem. **260:** 8078-8083.
12. Figures, W. R., L. M. Scearce, P. DeFeo, G. Stewart, F. Zhou, J. Chen, J. Daniel, R. F. Colman & R. W. Colman. 1987. Direct evidence for the interaction of the nucleotide affinity analog 5′-*p*-fluorosulfonylbenzoyl adenosine with a platelet ADP receptor. Blood **70:** 796-803.
13. Bennett, J. S. & G. Vilaire. 1979. Exposure of platelet fibrinogen receptors by ADP and epinephrine. J. Clin. Invest. **64:** 1393-1401.
14. Nachman, R. L. & L. L. K. Leung. 1982. Complex formation of platelet membrane glycoproteins IIb and IIIa with fibrinogen. J. Clin. Invest. **69:** 263-269.
15. Figures, W. R., S. Niewiarowski, T. A. Morinelli, R. F. Colman & R. W. Colman. 1981. Affinity labeling of a human platelet membrane protein with 5′-*p*-fluorosulfonylbenzoyl adenosine. J. Biol. Chem. **256:** 7789-7795.
16. Daniel, J. L., I. R. Molish, M. Rigmaiden & G. Stewart. 1984 Evidence for a role of myosin phosphorylation in the initiation of the platelet shape change. J. Biol. Chem. **259:** 9826-9831.
17. Kornecki, E., S. Niewiarowski, T. A. Morinelli & M. Kloczewiak. 1981. Exposure of fibrinogen receptors by chymotrypsin and adenosine diphosphate on normal and Glanzmann's thrombasthenic platelets. J. Biol. Chem. **256:** 5696-5701.
18. Colman, R. W., W. R. Figures, Q.-X. Wu, T. A. Morinelli, G. P. Tuszynski, R. F. Colman & S. Niewiarowski. 1988. Distinction between glycoprotein IIIa and the 100-kDa membrane protein (aggregin) mediating ADP-induced platelet activation. Arch. Biochem. Biophys. **262:** 298-306.
19. Cryer, P. E. 1980. Physiology and pathophysiology of the human sympatho-adrenal neuroendocrine system. N. Engl. J. Med. **303:** 436-444.
20. Kobilka, B. K., H. Matsui, T. S. Kobilka, T. L. Yang-Feng, U. Francke, M. G. Caron, R. J. Lefkowitz & J. W. Regan. 1987. Cloning, sequencing and expression of the gene coding for the human platelet α_2-adrenergic receptor. Science **238:** 650-656.
21. Figures, W. R., L. M. Scearce, Y. Wachtfogel, J. Chen, R. F. Colman & R. W. Colman. 1986. Platelet ADP receptor and α_2-adrenergic receptor interaction: Evidence for an ADP requirement for epinephrine-induced platelet activation and an influence of epinephrine on ADP binding. J. Biol. Chem. **261:** 5981-5986.
22. Kinlough-Rathbone, R. L., M. A. Packham & J. F. Mustard. 1977. Synergism between platelet aggregating agents: The role of the arachidonate pathway. Thromb. Res. **11:** 567-580.
23. Kinlough-Rathbone, R., H. Reimers & J. Mustard. 1976. Sodium arachidonate can induce platelet shape change and aggregation which are independent of the release reaction. Science **192:** 1011-1012.
24. Rao, A. K., J. Willis & H. Holmsen. 1984. A major role of ADP in thromboxane transfer experiments: Studies in patients with platelet secretion defects. J. Lab. Clin. Med. **104:** 116-126.

25. MORINELLI, T. A., S. NIEWIAROWSKI, E. KORNECKI, W. R. FIGURES, Y. T. WACHTFOGEL & R. W. COLMAN. 1983. Platelet aggregation and exposure of fibrinogen receptors by prostaglandin endoperoxide analogues. Blood **61:** 41-49.
26. GERATZ, J. D., R. R. TIDWELL, K. M. BRINKHOUS, S. F. MOHAMMAD, O. DANN & H. LOEWE. 1978. Specific inhibition of platelet agglutination and aggregation by aromatic amido compounds. Thromb. Haemostas. **39:** 411-425.
27. NUNN, B. 1979. Platelet collagen induced aggregation: Evidence against the essential role of platelet adenosine diphosphate. Thromb. Haemostas. **42:** 1193-1206.
28. COLMAN, R. W., W. R. FIGURES, L. M. SCEARCE, A. M. STRIMPLER, F. ZHOU & A. K. RAO. 1986. Inhibition of collagen-induced platelet activation by 5′-*p*-fluorosulfonylbenzoyl adenosine: Evidence for an ADP requirement and a synergistic influence of prostaglandin endoperoxides. Blood **68:** 565-570.
29. FIGURES, W. R., R. F. COLMAN, S. NIEWIAROWSKI, T. A. MORINELLI, Y. WACHTFOGEL & R. W. COLMAN. 1981. New evidence for an ADP-independent mechanism of thrombin-induced platelet activation. Thromb. Haemostas. **46:** 94.
30. HARMON, J. T. & G. A. JAMIESON. 1986. Activation of platelets by α-thrombin is a receptor-mediated event: D-Phenylalanine-L-prolyl-L-arginine chloromethylketone thrombin, but not *N*-α-tosyl-L-lysine chloromethylketone thrombin, binds to the high-affinity thrombin receptor. J. Biol. Chem. **261:** 15928-15933.
31. PURI, R. N., F. ZHOU, H. BRADFORD, C.-J. HU, R. F. COLMAN & R. W. COLMAN. 1989. Thrombin-induced platelet aggregation involves an indirect proteolytic cleavage of aggregin by calpain. Arch. Biochem. Biophys. **271:** 346-358.
32. PURI, R. N., F. ZHOU, R. F. COLMAN & R. W. COLMAN. 1989. Cleavage of a 100-kDa membrane protein (aggregin) during thrombin-induced platelet aggregation is mediated by the high-affinity thrombin receptors. Biochem. Biophys. Res. Commun. **162:** 1017-1024.
33. LEREA, K. M., J. A. GLOMSET & E. G. KREBS. 1987. Agents that elevate cAMP levels in platelets decrease thrombin binding. J. Biol. Chem. **262:** 282-288.
34. SCHMAIER, A. H., H. BRADFORD, L. D. SILVER, A. FARBER, C. F. SCOTT, D. SCHUTSKY & R. W. COLMAN. 1986. High molecular weight kininogen is an inhibitor of platelet calpain. J. Clin. Invest. **77:** 1565-1573.
35. PURI, R. N., E. J. GUSTAFSON, F. ZHOU, H. BRADFORD, R. F. COLMAN & R. W. COLMAN. 1988. Inhibition of thrombin-induced platelet aggregation by high molecular weight kininogen. Trans. Assoc. Am. Physicians **100:** 232-240.
36. COLMAN, R. W. & W. R. FIGURES. 1984. Characteristics of an ADP receptor mediating platelet activation. Mol. Cell. Biochem. **59:** 101-111.
37. COLMAN, R. W. 1986. Platelet activation: Role of an ADP receptor. Semin. Hematol. **23:** 319-328.
38. COLMAN, R. W., R. N. PURI, F. ZHOU & R. F. COLMAN. 1988. Aggregin: The platelet receptor mediating activation by ADP. *In* Platelet Membrane Receptors: Molecular Biology, Immunology, Biochemistry, and Pathology. G. A. Jamison, Ed.: 263-277. Alan R. Liss. New York, NY.

DISCUSSION OF THE PAPER

J. S. WILEY (*Austin Hospital, Heidelberg, Australia*): Do proteolytic enzymes such as trypsin that stimulate platelet aggregation also cause digestion of the 95-kDa "aggregin" protein band? Conversely, do proteolytic enzymes that do not cause platelet aggregation alter the 95-kDa "aggregin" band?

R. W. COLMAN: We have shown that plasmin acts identically to thrombin and causes aggregation and degradation of aggregin by raising the intracellular level of Ca^{2+} and by activating calpain. We have also studied chymotrypsin, which can aggregate platelets but cannot cause secretion or, presumably, raise the level of Ca^{2+}. This enzyme cleave aggregin to a 70-kDa peptide, whereas calpain, thrombin, and plasmin cleave aggregin to small peptides. Thus, chymotrypsin works directly on aggregin rather than by a mechanism dependent on calpain.

E. KORNECKI (*SUNY Health Science Center at Brooklyn, New York, NY*): Early studies have shown that the platelet epinephrine and ADP receptors are distinct receptors on the platelet surface. In these studies, Dr. David Mills showed that the primary wave of aggregation induced by epinephrine was not inhibited by ATP. Could you explain from your studies whether in this system (that is, in the ability of FSBA to completely block epinephrine-induced platelet aggregation), ATP and FSBA are acting through different mechanisms and through separate receptors?

R. W. COLMAN: I agree that ADP and epinephrine are acting through different receptors—the former through aggregin, the latter through the cloned α_2-adrenergic receptor. This conclusion is borne out by our studies because, as I mentioned, FSBA modification of aggregin does not change the binding of the competitive adrenergic antagonist yohimbine. The inhibition of epinephrine-induced platelet aggregation by FSBA is due to a postreceptor effect and indicates that ADP is necessary to bind to aggregin to reverse the latency of the fibrinogen receptor and allow platelet aggregation. In the presence of ADP (without FSBA), epinephrine produces aggregation by enhancing the affinity of ADP for its receptor, aggregin. The use of ATP as a competitive antagonist is difficult. In our hands, we need 500-1000 μM ATP to block ADP aggregation by 1-2 μM ADP. This concentration of ATP is probably needed because of the lower affinity and rapid hydrolysis of ATP.

J. S. DAVIDSON (*University of Cape Town Medical School, Cape Town, South Africa*): If your scenario regarding calpain is correct, the effect of thrombin should be mimicked by calcium ionophore. Have you done this experiment?

R. W. COLMAN: Yes. We find that the calcium ionophore A23187 in the presence of 5 mM Ca^{2+} can cause both platelet aggregation and cleavage of aggregin, thus mimicking thrombin. Interestingly, A23187 alone does not result in cleavage of aggregin. This finding is consistent with the fact that A23187 plus Ca^{2+} raises the intracellular Ca^{2+} level to above 1 μM, as does thrombin, allowing activation of calpain. A23187 alone, however, only gives values of Ca^{2+} of about 400 nM, a concentration that does not activate calpain.

Adenosine Receptors

Roles and Pharmacology

ROBERT F. BRUNS

Biochemical Pharmacology Research
Lilly Research Laboratories
Eli Lilly and Company
Indianapolis, Indiana 46285

INTRODUCTION

Adenosine is the final product of the stepwise dephosphorylation of ATP. Like ATP, adenosine has numerous pharmacodynamic actions mediated by cell surface receptors. Because of its close structural and metabolic relationship to ATP, adenosine has a role that is closely intertwined with the roles of intracellular and extracellular ATP.

This article will attempt to answer three questions that are central to adenosine research. These questions are also relevant to the extracellular actions of ATP, and in fact analogous questions can be framed for ATP. Because the field of extracellular ATP research is still relatively young, clear answers to most of these questions are not yet available for ATP, and in fact these questions may form an agenda for ATP research in the 1990s.

WHAT IS THE PHYSIOLOGICAL ROLE OF ADENOSINE?

An understanding of the role of adenosine must encompass both the pharmacodynamic actions of adenosine and the stimuli that lead to its production. The major source of adenosine appears to be intracellular AMP, whose concentration in turn is a function of intracellular ATP concentration and energy charge. [Energy charge is (ATP + ½ ADP) / (ATP + ADP + AMP).] If energy charge drops, because of an increase in the hydrolysis of ATP or a decrease in the rate of synthesis of ATP, then the AMP concentration rises (FIG. 1). A portion of the AMP is then converted to adenosine by an intracellular 5′-nucleotidase.[1] The localization, regulation, and general properties of this enzyme are probably the least understood area of adenosine research. Nevertheless, it is clear that physiological stimuli that lower energy charge (for example, hypoxia, ischemia, exercise) greatly increase adenosine production.[2–4]

In most cases, the limiting factor for maintenance of cellular energy charge is the availability of oxygen. Thus, adenosine levels act as a sort of oxygen sensor, with high levels of adenosine denoting inadequate tissue oxygenation.

Adenosine probably acts as a local hormone rather than a circulating hormone or neurotransmitter. Unlike classical neurotransmitters, adenosine can be produced by virtually any cell. Adenosine per se does not appear to be stored in exocytotic vesicles, but rather is produced on demand, much like the prostaglandins and leukotrienes. The extracellular half-life of a few seconds for adenosine implies that it should be capable of diffusing several millimeters from its source, indicating a wider sphere of action than a single synapse, but a much more localized role than would be seen with a circulating hormone.

Adenosine can also arise from release of ATP, which is rapidly broken down by ectonucleotidases. ATP is released from nerve endings (where it is stored in vesicles along with biogenic amines or other classical neurotransmitters), from blood platelets

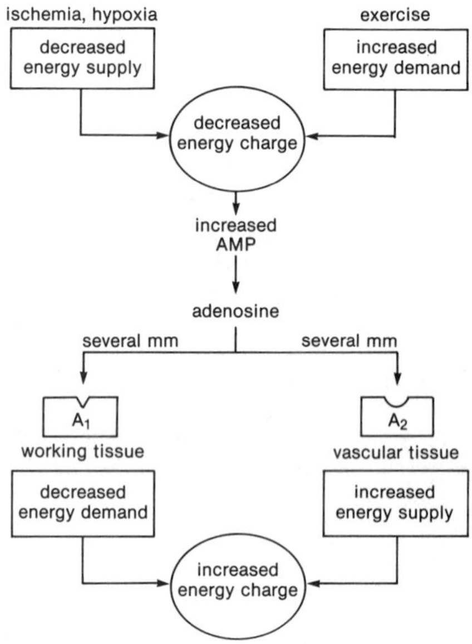

FIGURE 1. Role of adenosine in energy supply/demand balance. It should be noted that although most high-affinity A_2 responses occur in vascular elements (such as arterial smooth muscle and platelets), some responses (such as inhibition of locomotor activity) involve nonvascular tissues.

(where it is stored in secretory granules along with ADP), and from cells that are undergoing lysis. The adenosine that is generated from neuronal ATP release is involved in a negative feedback loop, as many nerve endings possess adenosine receptors that inhibit neurotransmitter release. These sources of adenosine probably are important under specialized circumstances (that is, at particular synapses or at sites of injury), whereas intracellularly generated adenosine probably predominates at a more macroscopic level. For instance, in brain slices ecto-5′-nucleotidase inhibitors block the elevation in cyclic AMP caused by AMP, but not that caused by depolarization, implying that the major source of the adenosine that is released by depolarizing agents is intracellular.[5]

The responses to adenosine are mediated by two major classes of extracellular receptors, called A_1 and A_2.[6] The two receptor subtypes were originally defined in terms of effects on adenylate cyclase and cyclic AMP levels, with A_1 receptors lowering cyclic AMP and A_2 receptors raising cyclic AMP.[6,7] In recent years, however, it has become apparent that at least the A_1 receptor can have effects that are independent of cyclic AMP, including opening potassium channels,[8] closing calcium channels,[9] and inhibiting[10] or stimulating[11] phosphatidylinositol turnover. Because of this, the receptors are now defined in terms of structure-activity relationships (see the following section for examples of selective agents).

A_1 and A_2 receptors appear to have distinct roles in protecting against the consequences of inadequate tissue oxygenation.[12] A_1 responses almost invariably bring about a decrease in oxygen demand, that is, a decrease in the amount of metabolic work carried out by a tissue (TABLE 1). In contrast, A_2 responses usually involve an increase in oxygen supply. In the heart, A_1 receptors in the cardiac muscle decrease heart rate, decrease force of contraction, and inhibit the conduction of action potentials.[13] All of these effects reduce the amount of work carried out by the heart. In adipocytes, A_1 receptor activation inhibits lipolysis,[14] reducing the amount of free fatty acids available for oxidation and concomitantly lowering oxygen consumption and thermogenesis. A_1 agonists injected into the brain decrease locomotor activity,[15] decreasing global oxygen consumption. Localized actions of adenosine to lower brain oxygen consumption include inhibition of neuronal firing[16] and inhibition of neurotransmitter release.[17,18] The inhibition of excitatory amino acid release by adenosine appears to be a powerful neuroprotective mechanism.[19] A_1 agonists lower body temperature, probably by both central and peripheral mechanisms.[20] One exception to the above generalization is the inhibition of breathing caused by intraventricular administration of A_1 agonists.[21] This would obviously cause a global decrease in the oxygen supply. It may be, however, that the teleological purpose of this effect is to protect the neurons of the respiratory center by limiting their oxygen consumption. In any case, the A_1-mediated inhibition of breathing is offset by a powerful A_2-mediated stimulation of breathing (see next paragraph).

Most A_2-mediated responses bring about an increase in the local oxygen supply (TABLE 1). A_2 agonists relax vascular smooth muscle, resulting in vasodilation and increased blood flow. A_2-receptor-mediated inhibition of platelet aggregation[22] increases the blood supply to tissues that are experiencing thrombosis. Adenosine also inhibits neutrophil activation, increasing blood flow by preventing adherence of neutrophils to capillary walls, as well as preventing reperfusion injury due to generation of superoxide radicals.[23] Adenosine increases breathing by activating A_2 receptors in the carotid artery.[24] Results with adenosine antagonists indicate that endogenous adenosine may have an important role in regulating breathing via the carotid chemoreceptors.[24] Interestingly, all of the above responses involve vascular tissue (construed in a broad sense), suggesting that the vasculature is probably the main target for high-affinity A_2 receptors.

Two other A_2 responses, relaxation of gut smooth muscle[25] and inhibition of locomotor activity,[26,27] should cause decreases in oxygen consumption. These responses are exceptions to the general rule that A_2 receptors mediate increases in oxygen supply; nevertheless, they are consistent with adenosine's overall role in preventing tissue damage from inadequate oxygenation. The inhibition of locomotor activity by A_2-selective adenosine agonists is of interest because of the very restricted distribution of high-affinity A_2 receptors in the brain: these receptors are found in the striatum, nucleus accumbens, and olfactory tubercle,[28] areas that have a high level of dopaminergic innervation and are part of the extrapyramidal motor system.

The role of A_1 and A_2 receptors in the kidney is of particular interest. Because the amount of work done by the kidney is proportional to renal blood flow (all other factors being equal), nonselective increases in renal blood flow will not improve oxygen supply/demand balance, and in fact may exacerbate a renal oxygen deficit.[29] The role of adenosine in the kidney reflects this dilemma. A_1 receptor activation selectively constricts the renal afferent arterioles, whereas A_2 receptor activation dilates the efferent arterioles. The net effect of these two actions is to reduce the pressure drop across the glomerular filter, reducing the amount of fluid that is filtered by the glomerulus and thereby reducing the amount of energy expended on reabsorption of solutes. The final result is an almost unchanged blood flow combined with a reduction in the metabolic demands of the kidney. In addition to the above hemodynamic effects, adenosine also lowers renin release via A_1 receptors, thereby lowering the amount of

TABLE 1. Role of Adenosine A_1 and A_2 Receptors in Oxygen Supply/Demand Balance

Response	Oxygen Demand	Oxygen Supply
	A_1 Receptor-Mediated Responses[a]	
Decrease heart rate	Decrease	
Decrease heart force	Decrease	
Decrease lipolysis	Decrease	
Constrict renal afferent arterioles	Decrease	(Decrease)
Decrease renin release	Decrease	
Decrease locomotor activity	Decrease	
Decrease neuronal firing	Decrease	
Decrease transmitter release	Decrease	
Decrease body temperature	Decrease	
Decrease breathing (CNS)	Decrease	(Decrease)
	A_2 Receptor-Mediated Responses	
Vasodilation		Increase
Decrease platelet aggregation		Increase
Decrease neutrophil activation		Increase
Increase breathing (carotid)		Increase
Dilate renal efferent arterioles	Decrease	Increase
Increase renin release	Increase	
Relax gut smooth muscle	Decrease	
Decrease locomotor activity	Decrease	

[a] The two examples in which A_1 responses lower oxygen supply are shown in parentheses because these responses are counteracted by A_2 effects.

work expended on retention of sodium. A_2 receptor activation, however, causes an increase in renin release,[30] which is difficult to reconcile with adenosine's role in maintaining adequate tissue oxygenation.

One can speculate that adenosine's role as a local hormone arose in evolution in the following way. Initially, an intracellular 5′-nucleotidase may have evolved as a mechanism to maintain energy charge during periods of rapid ATP hydrolysis. By breaking down AMP, such an enzyme would prevent AMP (and ADP) from accumulating, in this way maintaining ΔG for ATPase reactions and preventing such reactions from running in the reverse direction. Because adenosine levels would already be elevated during periods of oxygen deficit, it would be relatively simple for a hormonal role for adenosine to be added later in evolution.

It is not yet possible to outline a unified role for extracellular ATP. However, several hints about the role of ATP can be gleaned from existing information. The first source of extracellular ATP was probably cell lysis. It is easy to imagine that ATP receptors may have originally evolved in one-celled organisms as a means of chemotaxis toward potential food sources. In agreement with the idea that primordial ATP receptors may have evolved for the purpose of chemoreception, crustaceans can detect adenine nucleotides in the surrounding seawater via receptors located on their antennae.[31] In mammals, ATP derived from cell lysis may mediate some of the responses to injury. Other, more sophisticated, uses for ATP may have evolved opportunistically. For instance, ATP^{4-} may have been stored in vesicles as a counterion to positively charged biogenic amines, later evolving into a cotransmitter in its own right. The rapidity of hydrolysis of extracellular ATP implies that its role may be to mediate rapid, phasic responses. For instance, stimulation of the sympathetic nerves in the rat vas deferens results in a rapid, transient contraction mediated by ATP, followed by a sustained contraction that is mediated by norepinephrine.[32] Neuropeptide Y may mediate an even slower contractile response at some sympathetic junctions. Based on the considerations cited above, one is tempted to speculate that the overall role of ATP may be to mediate a rapid, transient response to injury.

WHAT PHARMACOLOGICAL TOOLS ARE AVAILABLE FOR STUDIES OF ADENOSINE RECEPTORS?

Although the overall role of adenosine is fairly well understood, many of the details remain to be worked out. Pharmacological tools such as selective agonists, antagonists, and enzyme inhibitors will be essential if we are to improve our understanding of adenosine's role *in vivo.* In addition, these tools can be used to distinguish the roles and actions of ATP from those of its breakdown product, adenosine.

Until their genes are cloned and sequenced, adenosine receptor subtypes will need to be defined by structure-activity relationships. Originally, A_1 and A_2 receptors were distinguished on the basis of the rank order of potency of the agonists N^6-(*R*)-(phenylisopropyl)adenosine (*R*-PIA) and 5′-*N*-ethyl-carboxamidoadenosine (NECA) (FIG. 2) and the magnitude of the difference in affinity between the *R* and *S* diastereomers of PIA.[7,17] Both of these criteria, however, have been found to sometimes result in ambiguities (for a discussion, see reference 33). Fortunately, more selective agonists have been reported in recent years (TABLE 2). CPA (FIG. 2) is the most A_1-selective (780-fold) agonist commercially available at present,[34] whereas N^6-[(1*R*,2*S*,4*S*)-2-*endo*-norbornyl]adenosine (*S*-ENBA) and its 5′-chloro derivative are the most selective overall (16,000-fold in the case of the latter).[35] On the A_2 side, PD 125,944, CGS 21577, and CGS 21680 are the first highly A_2-selective (up to 114-fold) agonists to be reported.[36,37] CPA, (*S*)-ENBA, PD 125,944, and CGS 21577 all have similar physicochemical properties (charge, partition coefficient, solubility), which should facilitate comparisons between these agents *in vivo.* Unlike the other selective agonists, all of which are uncharged, CGS 21680 carries a negative charge at physiological pH, a property that should improve solubility, but which may hinder comparisons with the other agonists.

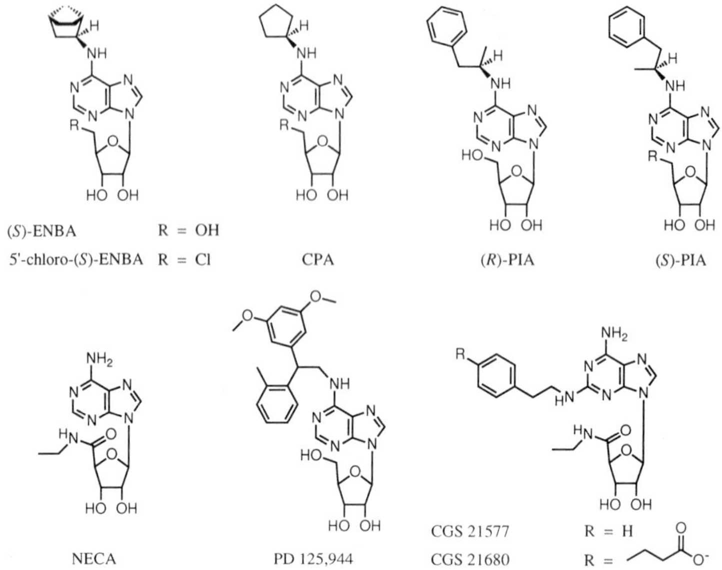

FIGURE 2. Structures of adenosine agonists.

Research on adenosine antagonists has proceeded with the twin goals of discovering subtype-selective antagonists and identifying potent antagonists that retain the excellent *in vivo* activity of theophylline, the first adenosine antagonist to be reported.[38] The latter goal has been impeded by the poor solubility of many derivatives of 8-phenyltheophylline, the first adenosine antagonist with nanomolar potency. Recently, it has become apparent that an important prerequisite for *in vivo* activity of adenosine antagonists is a sufficiently high ratio ($\geq 1{,}000$) between solubility and adenosine receptor affinity.[39] One strategy for achieving such a ratio has been to add a charged

TABLE 2. A_1 and A_2 Affinities of Adenosine Agonists[a]

Compound	K_i Values (nM)		
	A_1	A_2	A_2/A_1
5′-Chloro-*S*-ENBA	0.24	3,900	16,000
S-ENBA	0.30	1,390	4,700
CPA	0.59	460	780
R-PIA	1.17	124	106
S-PIA	49	1,820	37
NECA	6.3	10.3	1.64
PD 125,944	142	4.4	0.031
CGS 21577	480	9.7	0.020
CGS 21680	1,710	15	0.0088

[a] Structures are given in FIGURE 2. Affinities were determined in A_1 and A_2 binding in rat brain membranes as described.[42] Values for the two CGS compounds are from reference 37.

TABLE 3. Affinities and Solubilities of Adenosine Antagonists[a]

Compound	K_i or Solubility Values (nM)			Ratios		
	A_1	A_2	Sol	A_2/A_1	Sol/A_1	Sol/A_2
CPX	0.46	340	17,100	740	37,000	50
CPT	10.9	1,440	461,000	133	42,000	320
XAC	0.86	27	90,000	31	105,000	3,300
PD 113,297	5.6	70	945,000	12.5	169,000	13,500
CGS 15943	3.9	1.51	1,740	0.39	450	1,150
HTQZ	4,000	124	1,960	0.031	0.65	15.8

[a] Structures are given in FIGURE 3. Affinities were determined in A_1 and A_2 binding in rat brain membranes as described.[42] Solubilities were determined using a radioreceptor assay.[39]

side-chain to the 8-phenylxanthine core, for instance in PD 113,297 and xanthine amine congener (XAC) (TABLE 3, see FIG. 3 for structures).[40,41] PD 113,297 and XAC are potent adenosine antagonists *in vivo* when administered by the intravenous or intraperitoneal routes, but appear to have poor activity when given orally. Both compounds are peripherally selective adenosine blockers because they do not block behavioral effects of adenosine agonists and therefore probably do not cross the blood-brain barrier (personal communications from T. G. Heffner (PD 113,297) and K. A. Jacobson (XAC)).

A different strategy has provided antagonists that are active *in vivo* and highly selective for the A_1 receptor subtype. Replacement of the 8-phenyl moiety with cyclopentyl, as illustrated by 8-cyclopentyltheophylline (CPT) and its dipropyl homologue, CPX, has the dual effect of increasing solubility and greatly improving A_1

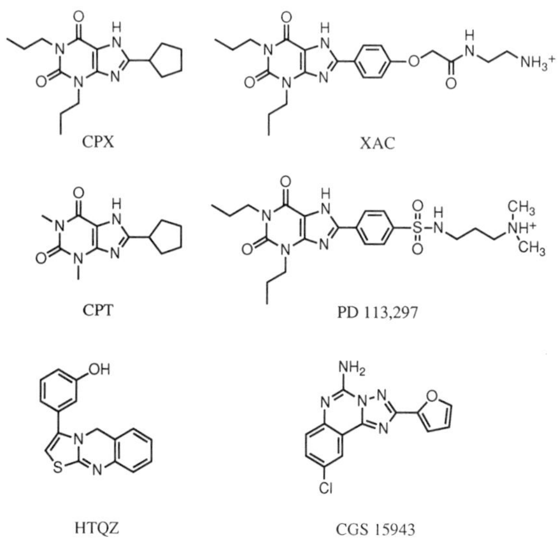

FIGURE 3. Structures of adenosine antagonists.

affinity.[42–45] Both compounds show A_1-selective adenosine antagonism *in vivo.*[26] CPT is orally active (CPX has not yet been tested for oral activity). Both compounds block behavioral effects of A_1 agonists, indicating that they cross the blood-brain barrier.

No highly A_2-selective antagonists are yet available. HTQZ (FIG. 3), the most A_2-selective antagonist to be reported, has a poor solubility/affinity ratio and therefore is unlikely to prove useful as a pharmacological tool.[39] CGS 15943 is only slightly selective for the A_2 receptor,[46] but has a better solubility/affinity ratio[39] and is orally active.[47]

Potent and selective adenosine antagonists are proving to be very useful tools for probing the roles of endogenous adenosine. For instance, CPT lacks the gross locomotor stimulant activity seen with caffeine and theophylline, implying that the locomotor stimulation shown by the latter compounds is not due solely to A_1 blockade.[26] CPT, however, does potentiate self-stimulation behavior, a more subtle indicator of stimulant or antidepressant activity. CPT, CPX, and 8-phenyltheophylline have interesting activity in the kidney, including natriuretic activity and protection against acute renal failure.[26,48]

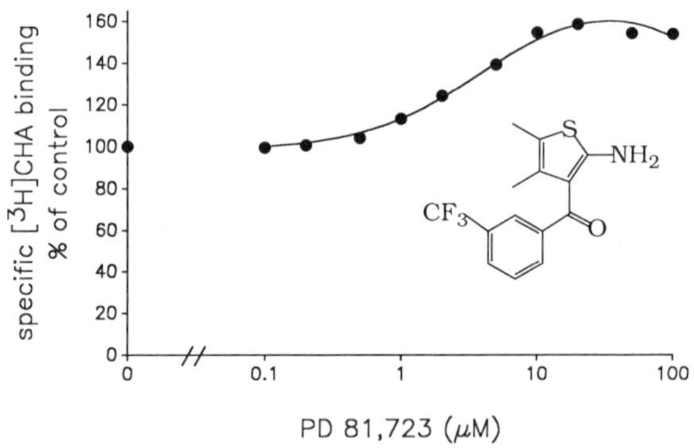

FIGURE 4. Enhancement of the binding of [^{3}H]N^6-cyclohexyladenosine to A_1 receptors in rat brain membranes by PD 81,723.

Recently, several 2-amino-3-benzoylthiophenes have been reported to show apparent allosteric enhancing activity at the A_1 receptor (FIG. 4).[49] The enhancement, which is seen in a functional assay as well as in binding, appears to be specific for the A_1 receptor because binding of agonists to A_2 receptors and to several other receptors was not enhanced. The compounds available to date, however, are limited in their potency, solubility, and stability. PD 81,723 appears to have the best overall profile, and may be worth examination as a potential pharmacological tool. Improved adenosine enhancers may be useful for treatment of ischemia.

Numerous other tools for adenosine research are available. Adenosine deaminase can be used to eliminate endogenous adenosine. Conversely, adenosine deaminase inhibitors such as 2′-deoxycoformycin (pentostatin, Parke-Davis) and *erythro*-2-hydroxy-3-nonyladenine (EHNA, Burroughs-Wellcome) can be used to potentiate endogenous adenosine by preventing its breakdown. 5-Iodotubercidin inhibits adenosine kinase, the other major adenosine-metabolizing enzyme. Potent inhibitors of ecto-5′-nucleotidase include α,β-methylene-ADP (APCP) and thymidine diphosphate (dTDP). Although dTDP is not as potent as APCP, it should be helpful in cases where the ATP receptor activity of the latter precludes its use. (The α,β-methylene derivative of dTDP or rTDP would be even more useful if commercially available.) No potent inhibitors of the intracellular 5′-nucleotidase that produces adenosine have been reported, although 5′-isobutylthioadenosine is reported to show weak inhibition of the enzyme.[1] Adenosine enters and leaves cells by a facilitated diffusion transporter than can be blocked by 6-(4-nitrobenzylthio)inosine, dipyridamole, and dilazep.

Efforts to find pharmacological tools for the study of extracellular ATP are mainly in the pioneering stages. Several selective ATP receptor agonists have been reported, including β,γ-methylene-L-ATP (specific for the P_{2X} receptor) and β-fluoro-ADP (specific for the P_{2Y} receptor).[50] As was the case for adenosine, the discovery of selective antagonists has lagged behind the discovery of selective agonists. Arylazido aminopropionyl adenosine triphosphate ($ANAPP_3$) is a relatively specific photoaffinity antagonist of ATP responses.[51] Its blocking activity, however, depends on photolysis, thus precluding its use *in vivo*. In the absence of photolysis, $ANAPP_3$ is an agonist. Other putative ATP antagonists such as reactive blue 2 and suramin are weak and nonspecific, although some of these compounds may represent chemical leads for development of improved blockers. Efforts to identify specific inhibitors of ecto-ATP-ases are continuing. Inhibitors of vesicular ATP storage would be very interesting pharmacological tools. The next ten years should bring many exciting developments in the pharmacology of extracellular ATP.

CAN ADENINE NUCLEOTIDES DIRECTLY ACTIVATE THE ADENOSINE RECEPTOR?

An area of continuing controversy for adenosine research is the question of whether adenine nucleotides can activate the adenosine receptor directly or whether they must first be converted to adenosine by ectonucleotidases. In many tissues, adenine nucleotides elicit responses that are blocked by theophylline and are qualitatively the same as responses to adenosine in such characteristics as time course and second messenger involvement. These responses are clearly mediated by adenosine receptors, yet in the last fifteen years repeated claims and counterclaims have been put forward with respect to whether adenine nucleotides act directly, or only indirectly via metabolism to adenosine. My own position for the last decade or so has favored the latter interpretation, and this section summarizes the case against direct action.

In VA13 human fibroblasts, adenosine causes a large increase in cyclic AMP accumulation. The response to AMP has an almost identical time course to that of adenosine, whereas ADP and ATP cause smaller, more slowly accumulating responses (FIG. 5).[52] Only about 15% of AMP is converted to adenosine during the course of a 2-min incubation, an amount that is insufficient to account for the response to 10 μM AMP. On the basis of similar observations, other investigators had previously concluded that AMP acted directly.[53] This argument, however, assumes that the concentration of adenosine in the bulk medium is equivalent to its concentration at the receptor. On closer investigation (see below), the validity of this assumption appears to be questionable.

The first argument against direct action is based on structure-activity relationships. The 5′-domain of the adenosine receptor has limited bulk tolerance and does not accept charged groups (for instance, adenosine 5′-uronic acid is completely inactive).[54]

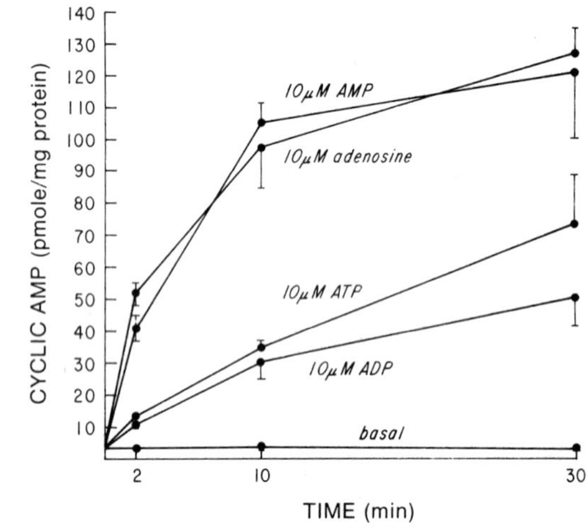

FIGURE 5. Time course of cyclic AMP increases in response to adenosine and adenine nucleotides in VA13 human fibroblasts. Taken with permission from reference 52.

It seems unlikely that a bulky, negatively charged phosphate group would be tolerated at this position. In agreement with this assessment, the hydrolysis-resistant nucleotides α,β-methylene-ADP (APCP) and 6′-deoxyhomoadenosine 6′-phosphonic acid (ACP) were inactive in VA13 cells. (ACP was previously shown to be inactive in the guinea pig taenia coli.[55])

Other experiments in VA13 cells also reveal a requirement for conversion of AMP to adenosine prior to activation of the adenosine receptor. Several inhibitors of 5′-nucleotidase blocked the response to AMP, but not the response to adenosine. There was an excellent correlation between inhibition of AMP breakdown and inhibition of the response to AMP (FIG. 6). In addition, adenosine deaminase blocked the response

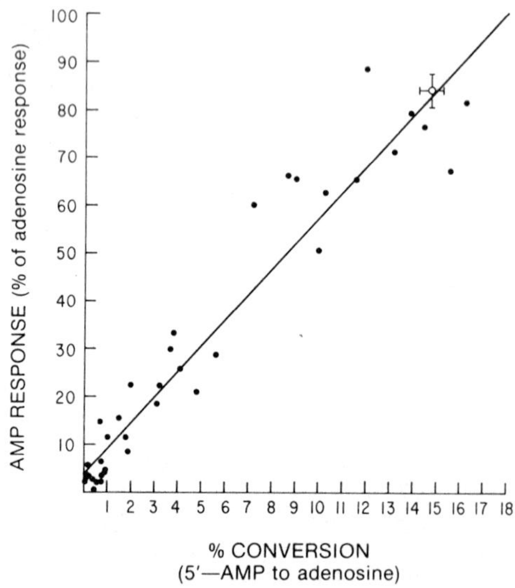

FIGURE 6. Correlation between inhibition of 5′-nucleotidase and blockade of the response to AMP in VA13 cells. The increase in cyclic AMP in response to 10 μM AMP is expressed as a percentage of the response to 10 μM adenosine. Each point represents a single concentration of a 5′-nucleotidase inhibitor; 12 compounds were tested. The open circle represents AMP alone (with standard errors). Reproduced from reference 52 with permission.

to AMP, as well as the response to adenosine. Because AMP is not a substrate for adenosine deaminase, these results imply that AMP must have been converted to adenosine prior to activating the receptor.

Interestingly, in the adenosine deaminase experiment described above, 3-10 times more enzyme was required to block the AMP response than the adenosine response. This suggests that the time interval between conversion of AMP to adenosine and activation of the adenosine receptor may have been very short. This in turn suggests an explanation for the lack of accumulation in the bulk medium: if adenosine is generated in close proximity to the receptor, the concentration of adenosine at the receptor could be high even though little adenosine might be seen in the bulk medium. This localized conversion hypothesis is illustrated in FIGURE 7. A simple mathematical model (see bottom of FIG. 7) indicates that, in a volume that is restricted by a bottleneck, the localized adenosine concentration will approach the concentration of AMP in the bulk medium if the rate of AMP hydrolysis exceeds the rate of diffusion through the bottleneck. Local conversion could occur in microvaginations, in endocytotic vesicles, or in the area underneath the cell.

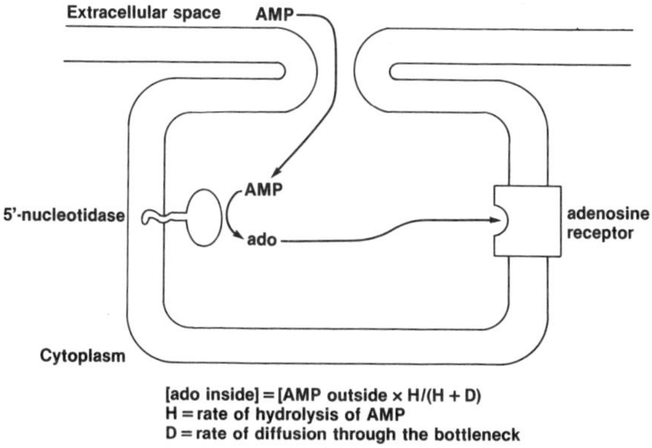

FIGURE 7. Local conversion of AMP to adenosine.

Experiments with cell lines expressing different levels of 5′-nucleotidase confirm the importance of local conversion (TABLE 4). In the neuro-2a cell line, as little as 0.5% conversion of AMP to adenosine is sufficient for a detectable response to AMP, whereas other cell lines with even lower 5′-nucleotidase activity do not respond to AMP.

The above data indicate that local conversion can explain the activation of adenosine receptors by AMP and other adenine nucleotides, at least in VA13 cells. It should be noted that intact tissue preparations such as brain slices and smooth muscle strips have an even greater potential for localized conversion because of the intricate, narrow passages through extracellular space that must be negotiated by a molecule in diffusing from the outside medium to the receptor site. Direct metabolic evidence for localized conversion of extracellular adenine nucleotides was recently reported in arterial smooth muscle cells.[56]

The specificity of adenosine receptors for adenosine (and not for adenine nucleotides) prompts a similar question with regard to ATP receptors: When two nucleotides both activate the same ATP receptor, is the activity of one nucleotide due to conversion to the other? For instance, ADP and ATP are equally active at some ATP receptors.

Is the activity of ATP due to breakdown to ADP? Unlike the situation that obtains at adenosine receptors, it appears that both nucleotides act directly at P_{2Y} receptors because hydrolysis-resistant analogues such as ATP-γ-S are active.[57] In C6 glioma cells, however, ADP and ATP both inhibit cyclic AMP accumulation, but the activity of ATP is due to its conversion to ADP.[58]

CONCLUSIONS

The role of adenosine in energy supply-demand balance is becoming increasingly well understood, in part because of the availability of selective pharmacological tools to study adenosine production and responses. Similar progress in the understanding of the role of extracellular ATP should follow as the physiology and pharmacology of ATP are explored over the next decade.

TABLE 4. Conversion of AMP to Adenosine and Increase in Cyclic AMP in Response to AMP in Various Cell Lines

Cell Line	Percentage of AMP Converted to Adenosine	Increase in cAMP in Response to AMP[a]
1321N1	26	90[b]
WI-38	24	73[b]
VA13	15	58[b]
B35	0.68	72[b]
B65	0.68	30
Neuro-2a	0.48	31[b]
L6	0.179	2.5
B11	0.076	1.8
BM1	0.038	−18

NOTE: These data are from reference 52.
[a] The response to 10 μM AMP is expressed as a percentage of the response to 10 μM adenosine.
[b] $p < .05$ (10 μM AMP versus basal, one-sided t test).

ACKNOWLEDGMENT

I thank David Mahoney for carrying out the experiment illustrated in FIGURE 4.

REFERENCES

1. SKLADANOWSKI, A. C., G. B. SALA & A. C. NEWBY. 1989. Inhibition of IMP-specific cytosolic 5′-nucleotidase and adenosine formation in rat polymorphonuclear leucocytes by 5′-deoxy-5′-isobutylthio derivatives of adenosine and inosine. Biochem. J. **262:** 203-208.

2. NEWBY, A. 1984. Adenosine and the concept of "retaliatory metabolites." Trends Biochem. Sci. **9:** 42-44.
3. SPARKS, H. V. & H. BARDENHEUER. 1986. Regulation of adenosine production by the heart. Circ. Res. **58:** 193-201.
4. BERNE, R. M. 1963. Cardiac nucleotides in hypoxia: Possible role in regulation of coronary blood flow. Am. J. Physiol. **204:** 317-322.
5. PONS, F., R. F. BRUNS & J. W. DALY. 1980. Depolarization-evoked accumulation of cyclic AMP in brain slices: The requisite intermediate adenosine is not derived from hydrolysis of released ATP. J. Neurochem. **34:** 1319-1323.
6. VAN CALKER, D., M. MÜLLER & B. HAMPRECHT. 1979. Adenosine regulates via two different types of receptors, the accumulation of cyclic AMP in cultured brain cells. J. Neurochem. **33:** 999-1005.
7. LONDOS, C., D. M. F. COOPER & J. WOLFF. 1980. Subclasses of external adenosine receptors. Proc. Natl. Acad. Sci. USA **77:** 2551-2554.
8. TRUSSELL, L. O. & M. B. JACKSON. 1987. Dependence of an adenosine-activated potassium current on a GTP-binding protein in mammalian central neurons. J. Neurosci. **7:** 3306-3316.
9. SCHUBERT, P. & G. W. KREUTZBERG. 1987. Pre- and postsynaptic effects of adenosine on neuronal calcium fluxes. *In* Topics and Perspectives in Adenosine Research. E. Gerlach & B. F. Becker, Eds.: 521-532. Springer-Verlag. Berlin.
10. DELAHUNTY, T. M., M. J. CRONIN & J. LINDEN. 1988. Regulation of GH_3-cell function via adenosine A_1 receptors: Inhibition of prolactin release, cyclic AMP production and inositol phosphate generation. Biochem. J. **255:** 69-77.
11. AREND, L. J., J. S. HANDLER, J. S. RHIM, F. GUSOVSKY & W. S. SPIELMAN. 1989. Adenosine-sensitive phosphoinositide turnover in a newly established renal cell line. Am. J. Physiol. **256:** F1067-F1074.
12. BRUNS, R. F. 1987. Adenosine and xanthines. *In* Role of Adenosine in Cerebral Metabolism and Blood Flow. V. Stefanovich & I. Okyayuz-Baklouti, Eds.: 57-80. VNU Boekengroep. Utrecht.
13. BELARDINELLI, L., J. LINDEN & R. M. BERNE. 1989. The cardiac effects of adenosine. Prog. Cardiovasc. Dis. **33:** 73-97.
14. FAIN, J. N., R. H. POINTER & W. F. WARD. 1972. Effect of adenosine nucleosides on adenylate cyclase, phosphodiesterase, cyclic adenosine monophosphate accumulation and lipolysis in fat cells. J. Biol. Chem. **247:** 6866-6872.
15. BARRACO, R. A., V. L. COFFIN, H. J. ALTMAN & J. W. PHILLIS. 1983. Central effects of adenosine analogs on locomotor activity in mice and antagonism by caffeine. Brain Res. **272:** 392-395.
16. PHILLIS, J. W., G. K. KOSTOPOULOS & J. J. LIMACHER. 1975. A potent depressant action of adenine derivatives on cerebral cortical neurones. Eur. J. Pharmacol. **30:** 125-129.
17. DALY, J. W., R. F. BRUNS & S. H. SNYDER. 1981. Adenosine receptors in the central nervous system: Relationship to the central actions of methylxanthines. Life Sci. **28:** 2083-2097.
18. PATON, D. M. 1981. Structure-activity relations for presynaptic inhibition of noradrenergic and cholinergic transmission by adenosine: Evidence for action on A_1 receptors. J. Autonom. Pharmacol. **1:** 287-290.
19. JANUSZEWICZ VON LUBITZ, D. K. E., J. M. DAMBROSIA & D. J. REDMOND. 1989. Protective effect of cyclohexyl adenosine in treatment of cerebral ischemia in gerbils. Neuroscience **30:** 451-462.
20. VAPAATALO, H., D. ONKEN, P. J. NEUVONEN & E. WESTERMANN. 1975. Stereospecificity in some central and circulatory effects of phenylisopropyladenosine (PIA). Arzneim. Forsch. **25:** 407-410.
21. HEDNER, T., J. HEDNER, P. WESSBERG & J. JONASON. 1982. Regulation of breathing in the rat: Indications for a role of central adenosine mechanisms. Neurosci. Lett. **33:** 147-151.
22. BORN, G. V. R., R. J. HASLAM, M. GOLDMAN & R. D. LOWE. 1965. Comparative effectiveness of adenosine analogues as inhibitors of blood-platelet aggregation and as vasodilators in man. Nature **205:** 678-680.

23. CRONSTEIN, B. N., R. I. LEVIN, J. BELANOFF, G. WEISSMANN & R. HIRSCHHORN. 1987. A new function for adenosine: Protection of vascular endothelial cells from neutrophil-mediated injury. *In* Topics and Perspectives in Adenosine Research. E. Gerlach & B. F. Becker, Eds.: 299-308. Springer-Verlag. Berlin.
24. MONTEIRO, E. C. & J. A. RIBEIRO. 1989. Inhibition by 1,3-dipropyl-8-(*p*-sulfophenyl)xanthine of the respiratory stimulation induced by common carotid artery occlusion in rats. Life Sci. **45:** 939-945.
25. BURNSTOCK, G., J. M. HILLS & C. H. V. HOYLE. 1984. Evidence that the P_1 purinoceptor in the guinea-pig taenia coli is an A_2 subtype. Br. J. Pharmacol. **81:** 533-541.
26. BRUNS, R. F., R. E. DAVIS, F. W. NINTEMAN, B. P. H. POSCHEL, J. N. WILEY & T. G. HEFFNER. 1988. Adenosine antagonists as pharmacologic tools. *In* Physiology and Pharmacology of Adenosine and Adenine Nucleotides. D. M. Paton, Ed.: 39-49. Taylor & Francis. London.
27. SPEALMAN, R. D. & V. L. COFFIN. 1986. Behavioral effects of adenosine analogs in squirrel monkeys: Relation to adenosine A_2 receptors. Psychopharmacology **90:** 419-421.
28. LEE, K. S. & M. REDDINGTON. 1986. Autoradiographic evidence for multiple CNS binding sites for adenosine derivatives. Neuroscience **19:** 535-549.
29. OSSWALD, H., H. H. HERMES & G. NABAKOWSKI. 1982. Role of adenosine in signal transmission of tubuloglomerular feedback. Kidney Int. **22**(Suppl. 12): S136-S142.
30. CHURCHILL, P. C. & M. C. CHURCHILL. 1985. A_1 and A_2 adenosine receptor activation inhibits and stimulates renin secretion of rat renal cortical slices. J. Pharmacol. Exp. Ther. **232:** 589-594.
31. CARR, W. E. S., R. A. GLEESON, B. W. ACHE & M. L. MILSTEAD. 1986. Olfactory receptors of the spiny lobster: ATP-sensitive cells with similarities to P_2-type purinoceptors of vertebrates. J. Comp. Physiol. A **158:** 331-338.
32. SNEDDON, P. & D. P. WESTFALL. 1984. Pharmacological evidence that adenosine triphosphate and noradrenaline are co-transmitters in the guinea-pig vas deferens. J. Physiol. **347:** 561-580.
33. BRUNS, R. F., G. H. LU & T. A. PUGSLEY. 1987. Adenosine receptor subtypes: Binding studies. *In* Topics and Perspectives in Adenosine Research. E. Gerlach & B. F. Becker, Eds.: 59-73. Springer-Verlag. Berlin.
34. MOOS, W. H., D. S. SZOTEK & R. F. BRUNS. 1985. N^6-Cycloalkyladenosines: Potent, A_1-selective adenosine antagonists. J. Med. Chem. **28:** 1383-1384.
35. TRIVEDI, B. K., A. J. BRIDGES, W. C. PATT, S. R. PRIEBE & R. F. BRUNS. 1989. N^6-Bicycloalkyladenosines with unusually high potency and selectivity for the A_1 adenosine receptor. J. Med. Chem. **32:** 8-11.
36. BRIDGES, A. J., R. F. BRUNS, D. F. ORTWINE, S. R. PRIEBE, D. S. SZOTEK & B. K. TRIVEDI. 1988. N^6-[2-(3,5-Dimethoxyphenyl)-2-(2-methylphenyl)ethyl]adenosine and its uronamide derivatives: Novel adenosine agonists with both high affinity and high selectivity for the adenosine A_2 receptor. J. Med. Chem. **31:** 1282-1285.
37. HUTCHISON, A. J., R. L. WEBB, H. H. OEI, G. R. GHAI, M. B. ZIMMERMAN & M. WILLIAMS. 1989. CGS 21680C, an A_2-selective adenosine receptor agonist with preferential hypotensive activity. J. Pharmacol. Exp. Ther. **251:** 47-55.
38. SATTIN, A. & T. W. RALL. 1970. The effect of adenosine and adenine nucleotides on the cyclic adenosine 3',5'-phosphate content of guinea pig cerebral cortex slices. Mol. Pharmacol. **6:** 13-23.
39. BRUNS, R. F. & J. H. FERGUS. 1989. Solubilities of adenosine antagonists determined by radioreceptor assay. J. Pharm. Pharmacol. **41:** 590-594.
40. HAMILTON, H. W., D. F. ORTWINE, D. F. WORTH, E. W. BADGER, J. A. BRISTOL, R. F. BRUNS, R. P. STEFFEN & S. J. HALEEN. 1985. Synthesis of xanthines as adenosine antagonists, a practical quantitative structure-activity relationship application. J. Med. Chem. **28:** 1071-1079.
41. JACOBSON, K. A., K. L. KIRK, W. L. PADGETT & J. W. DALY. 1985. Functionalized congeners of 1,3-dialkylxanthines: Preparation of analogues with high affinity for adenosine receptors. J. Med. Chem. **28:** 1334-1340.
42. BRUNS, R. F., G. H. LU & T. A. PUGSLEY. 1986. Characterization of the A_2 adenosine receptor labeled by [^{3}H]NECA in rat striatal membranes. Mol. Pharmacol. **29:** 331-346.
43. BRUNS, R. F., J. H. FERGUS, E. W. BADGER, J. A. BRISTOL, L. A. SANTAY, J. D.

HARTMAN, S. J. HAYS & C. C. HUANG. 1987. Binding of the A_1-selective adenosine antagonist 8-cyclopentyl-1,3-dipropylxanthine to rat brain membranes. Naunyn-Schmiedeberg's Arch. Pharmacol. **335:** 59-63.
44. MARTINSON, E. A., R. A. JOHNSON & J. N. WELLS. 1987. Potent adenosine receptor antagonists that are selective for the A_1 receptor subtype. Mol. Pharmacol. **31:** 247-252.
45. UKENA, D., K. A. JACOBSON, W. L. PADGETT, C. AYALA, M. T. SHAMIM, K. L. KIRK, R. A. OLSSON & J. W. DALY. 1986. Species differences in structure-activity relationships of adenosine agonists and xanthine antagonists at brain A_1 adenosine receptors. FEBS Lett. **209:** 122-128.
46. WILLIAMS, M., J. FRANCIS, G. GHAI, A. BRAUNWALDER, S. PSYCHOYOS, G. A. STONE & W. D. CASH. 1987. Biochemical characterization of the triazoloquinazoline, CGS 15943, a novel, non-xanthine adenosine antagonist. J. Pharmacol. Exp. Ther. **241:** 415-420.
47. GHAI, G., J. E. FRANCIS, M. WILLIAMS, R. A. DOTSON, M. F. HOPKINS, D. T. COTE, F. R. GOODMAN & M. B. ZIMMERMAN. 1987. Pharmacological characterization of CGS 15943A: A novel nonxanthine adenosine antagonist. J. Pharmacol. Exp. Ther. **242:** 784-790.
48. KELLETT, R., C. J. BOWMER, M. G. COLLIS & M. S. YATES. 1989. Amelioration of glycerol-induced acute renal failure in the rat with 8-cyclopentyl-1,3-dipropylxanthine. Br. J. Pharmacol. **98:** 1066-1074.
49. BRUNS, R. F. & J. H. FERGUS. 1989. Allosteric enhancers of adenosine A_1 receptor binding and function. *In* Adenosine Receptors in the Nervous System. J. A. Ribeiro, Ed.: 53-60. Taylor & Francis. London.
50. CUSACK, N. J., L. A. WELFORD & S. M. O. HOURANI. 1988. Studies on the P_2-purinoceptor using adenine nucleotide analogs. *In* Physiology and Pharmacology of Adenosine and Adenine Nucleotides. D. M. Paton, Ed.: 73-84. Taylor & Francis. London.
51. HOGABOOM, G. K., J. P. O'DONNELL & J. S. FEDAN. 1980. Purinergic receptors: Photoaffinity analog of adenosine triphosphate is a specific adenosine triphosphate antagonist. Science **208:** 1273-1276.
52. BRUNS, R. F. 1980. Adenosine receptor activation by adenine nucleotides requires conversion of the nucleotides to adenosine. Naunyn-Schmiedeberg's Arch. Pharmacol. **315:** 5-13.
53. CLARK, R. B., R. GROSS, Y. SU & J. P. PERKINS. 1974. Regulation of adenosine 3',5'-monophosphate content in human astrocytoma cells by adenosine and the adenine nucleotides. J. Biol. Chem. **248:** 5296-5303.
54. BRUNS, R. F. 1980. Adenosine receptor activation in human fibroblasts: Nucleoside agonists and antagonists. Can. J. Physiol. Pharmacol. **58:** 673-691.
55. SATCHELL, D. G. & M. H. MAGUIRE. 1975. Inhibitory effects of adenine nucleotide analogs on the isolated guinea-pig taenia coli. J. Pharmacol. Exp. Ther. **195:** 540-548.
56. GORDON, E. L., J. D. PEARSON, E. S. DICKINSON, D. MOREAU & L. L. SLAKEY. 1989. The hydrolysis of extracellular adenine nucleotides by arterial smooth muscle cells: Regulation of adenosine production at the cell surface. J. Biol. Chem. **264:** 18986-18992.
57. BOYER, J. L., C. P. DOWNES & T. K. HARDEN. 1989. Kinetics of activation of phospholipase C by P_{2Y} purinergic receptor agonists and guanine nucleotides. J. Biol. Chem. **264:** 884-890.
58. PIANET, I., M. MERLE & J. LABOUESSE. 1989. ADP and, indirectly, ATP are potent inhibitors of cAMP production in intact isoproterenol-stimulated C6 glioma cells. Biochem. Biophys. Res. Commun. **163:** 1150-1157.

DISCUSSION OF THE PAPER

T. D. WHITE (*Dalhousie University, Halifax, Nova Scotia, Canada*): Inhibition of adenosine receptors by drugs such as caffeine and theophylline stimulates the CNS even under apparently normal circumstances. I think this indicates that endogenous adenosine exerts tonic inhibitory effects in the CNS under normal circumstances, not just under pathological conditions of excessive neural activity in the CNS.

BRUNS: I agree that adenosine is involved in normal physiological regulation as well as in pathological circumstances. For instance, specific adenosine antagonists have significant effects in normal animals. Fortunately, adenosine is not the only modulator of oxygen supply/demand balance. Otherwise, a cup of coffee might be fatal!

E. RAPAPORT (*Boston University School of Medicine, Boston, MA*): I would like to comment with regard to the removal of adenosine from the vascular bed. Because the K_m for adenosine kinase is in the vicinity of 10^{-6} M, and because the K_m for adenosine deaminase is in the vicinity of 10^{-5} M, amounts of 1-2 μM of adenosine will be more than 95% phosphorylated after their transport into red blood cells. Only levels of intraluminal adenosine of more than 10 μM will be mostly deaminated. Therefore, red blood cells effectively remove intraluminal adenosine as well as release low levels of ATP into the extracellular plasma compartment.

BRUNS: Most metabolic studies indicate that phosphorylation is the primary pathway for recycling low concentrations of adenosine, and that deamination is more of a backup pathway for dealing with high adenosine concentrations.

M. WILLIAMS (*Abbott Laboratories, Abbott Park, IL*): In regard to the effects of adenosine on renin release lying outside your hypothesis of an energy-consuming role, Dr. Mark Miller, of Tulane University, has data showing that increased renin release following adenosine release is an indirect response to decreases in blood pressure.

J. BARANKEWICZ (*Gensia Pharmaceuticals, San Diego, CA*): I would be very cautious about saying that adenosine is produced intracellularly, because in many cells and tissues intracellular ATP degradation *in vitro* proceeds mainly via AMP deamination. Even in induced conditions, when ATP degradation is accelerated, no or very little adenosine is released. It is quite possible that the major source of extracellular adenosine is extracellular ATP.

BRUNS: Although more intracellular AMP may be broken down to IMP than adenosine, the amount of adenosine produced intracellularly during exercise or hypoxia is still sufficient to cause responses such as coronary vasodilation. Nucleoside transport inhibitors cause large accumulations of intracellular adenosine during hypoxia.

Activation of Inositol Phospholipid-Specific Phospholipase C by P_2-Purinergic Receptors in Human Phagocytic Leukocytes

Role of Pertussis Toxin-Sensitive G Proteins[a]

GEORGE R. DUBYAK[b] AND DANIEL S. COWEN

School of Medicine
Case Western Reserve University
Cleveland, Ohio 44106

INTRODUCTION

In many cell types, the functional effects elicited by exposure to extracellular ATP on cell function can be correlated with a rapid ATP-induced mobilization of intracellular Ca^{2+} stores. Consistent with this Ca^{2+}-mobilizing action, extracellular ATP (along with other nucleotides) has been shown to trigger rapid activation of inositol phospholipid hydrolysis in several cell types.[1-8] Thus, certain subtypes of P_2-purinergic receptors belong to the superfamily of cell surface receptors that are functionally coupled to inositol phospholipid-specific phospholipase C (PI-PLC) effector enzymes. Similar to the other receptor types that belong to this superfamily, P_2-purinergic receptors appear to indirectly activate PI-PLC effector enzymes via the mediation of GTP-binding regulatory proteins (G proteins). Two bodies of evidence support this latter hypothesis. First, the ability of P_{2Y}-purinergic agonists to activate PI-PLC in membranes isolated from turkey erythrocytes is strictly dependent on the presence of GTP or GTP-γ-S.[6] Second, in some (but not all) cell types, the ability of P_2-purinergic agonists to elicit inositol polyphosphate accumulation is inhibited or attenuated upon pretreatment of the cells with pertussis toxin.[3-5,9]

Fairly extensive agonist selectivity studies have been performed in a number of cell systems in order to define the ATP receptor subtype(s) involved in the activation of this particular transmembrane signaling pathway. Such studies suggest that P_2-

[a] This work was supported in part by Grant GM-36387 from the National Institutes of Health. G. R. D. is an Established Investigator of the American Heart Association. D. S. C. is a recipient of a Medical Scientist Training Program Award.

[b] To whom all correspondence and reprint requests should be addressed: Department of Physiology and Biophysics, School of Medicine, Case Western Reserve University, Cleveland, Ohio 44106.

purinergic activation of PI-PLC/Ca^{2+} mobilization in various cell types can be subcategorized into either of two broad nucleotide selectivity groups. In rat hepatocytes[1,2] bovine endothelial cells,[3] and turkey erythrocytes,[4,7] ADP is equipotent to ATP. In these cells, non-adenine nucleotides are generally inactive or much less potent than ATP/ADP. Cooper *et al.*,[7] after studying radioligand ([^{35}S]ADP-β-S) binding in turkey erythrocyte membranes, presented results consistent with a P_{2Y}-purinergic receptor subtype. Conversely, in cultured rat aortic smooth muscle cells,[6] HL-60 cells,[9] human neutrophils/monocytes,[10] and Ehrlich tumor cells,[8] ADP is several orders of magnitude less potent than ATP. Furthermore, these latter cell types can be activated by relatively low concentrations of certain non-adenine nucleotide triphosphates. In particular, UTP has been shown to be either equipotent to, or more potent than, ATP in such cells. These various functional studies suggest that there are multiple ATP receptor subtypes, and that these subtypes are functionally coupled to the PI-PLC transmembrane signaling system. The delineation of receptor subtype(s) is complicated, however, by the considerable likelihood that multiple types of G proteins and multiple PI-PLC isozymes are involved in the activation of inositol phospholipid hydrolysis by P_2-purinergic agonists. As is true for most biological actions of extracellular ATP, definitive characterization of the particular P_2-purinergic receptor subtypes(s) coupled to PI-PLC awaits the development of selective antagonists suitable for both functional and ligand-binding studies.

Our approach to the study of P_2-purinergic receptors coupled to PI-PLC has been to utilize cell types that fulfill the following criteria: 1) The cell type should be easily obtained in highly homogenous populations and in large numbers (10^9 cells). 2) The cell type should be well characterized as to the expression of other (nonpurinergic) PI-PLC-coupled receptors that can be readily assayed by conventional ligand-binding methods. 3) The cell type should be equally well characterized as to the expression of particular G protein subtypes and PI-PLC isozymes. With regard to these points, we have found the HL-60 human promyelocytic leukemia cell line to be a particularly useful model system. This easily cultured cell line has been extensively used as a model of human neutrophil/monocyte development and function. More germane to our purposes is the fact that the HL-60 cell line has been successfully employed in the study of PI-PLC activation by formylated chemotactic peptide receptors[11–13] as well as in studies of G protein expression/function.[14,15]

EXPRESSION OF Ca^{2+} MOBILIZING ATP RECEPTORS DURING DIFFERENTIATION OF HUMAN PHAGOCYTIC LEUKOCYTES

TABLE 1 summarizes our observations regarding the ability of extracellular ATP to trigger changes in cytosolic [Ca^{2+}] in a broad range of normal and leukemic human leukocytes, leukocyte progenitor cells, and established human white blood cell lines. As demonstrated by our studies and by other investigators,[16,17] micromolar concentrations of extracellular ATP trigger significant increases in the cytosolic [Ca^{2+}] of neutrophils and monocytes. In both cell types, Ca^{2+} transients induced by ATP are equal in magnitude to those elicited by supramaximal concentrations of fMet-Leu-Phe (FMLP) (30 μM), platelet activating factor (90 nM) (not shown), and leukotriene B_4 (300 nM) (not shown). In contrast, normal human lymphocytes are totally unresponsive to either extracellular ATP or FMLP. Similar studies have been performed with total mononuclear cell fractions isolated from the bone marrow of normal donors

TABLE 1. Transients of Ca^{2+} in Cells at Various Stages of Differentiation[a]

Cell Type	N	Basal $[Ca^{2+}]$ (nM)	Peak Change in Cytosolic $[Ca^{2+}]$ with Different Agonists (nM)		
			100 μM ATP	10 μM ATP	30 μM FMLP
Neutrophils	3	176 ± 17	1729 ± 226	1662 ± 74	1797 ± 258
Monocytes	3	197 ± 23	1388 ± 397	1136 ± 344	1447 ± 254
Lymphocytes	3	150 ± 12	156 ± 9		167 ± 3
Normal blood (mononuclear)	7	205 ± 16	282 ± 17	270 ± 20	280 ± 14
KG-1 (myeloblasts)	7	142 ± 19	265 ± 40	168 ± 18	142 ± 19
AML (FAB M1)	3	136 ± 17	263 ± 97	179 ± 30	138 ± 19
HL-60 (promyelocytes)	3	200 ± 20	>2000	>2000	303 ± 32
Promyelocytic leukemia (FAB M3)	1	67	1269	672	96
Chronic myelogenous leukemia (chronic phase)	4	180 ± 20	1173 ± 375	1062 ± 403	365 ± 93
Normal mononuclear bone marrow	6	170 ± 28	455 ± 75	357 ± 39	212 ± 22
U937 (promonocytes)	4	135 ± 5	872 ± 67	724 ± 98	159 ± 10
Myelomonocytic leukemia (FAB M4)	1	137	416	233	137
Monocytic leukemia (FAB M5)	2	123 ± 38	856 ± 580	817 ± 639	656 ± 339
B cell chronic lymphocytic leukemia	7	157 ± 9	166 ± 10	165 ± 10	164 ± 10
Acute lymphocytic leukemia (FAB L2)	1	156	176	156	156

[a] Each data point represents the mean ± standard error of the number of experiments indicated.

and from the marrow of nonleukemic patients undergoing autologous bone marrow transplantation. Such marrow mononuclear cell preparations consist of lymphocytes, monocytes, erythroid precursors, lymphocyte precursors, monocyte precursors, and those neutrophil precursors that are less mature than band neutrophils. Flow cytometric measurements of the cytosolic [Ca^{2+}] in single cells revealed that the percentage of ATP-responsive cells in these marrow-derived populations was greater than the percentage of FMLP-responsive cells. These results suggested the presence in bone marrow of leukocyte precursor cells that could be stimulated by ATP but not by FMLP.

In order to determine which types of hemopoietic progenitor cells expressed this Ca^{2+}-mobilizing response to ATP, a number of established human leukocyte cell lines were examined. HL-60 promyelocytes and U937 promonocytes exhibited large ATP-induced Ca^{2+} transients in contrast to only minor responses to FMLP (FIGS. 1C & 1D). Flow cytometric studies have demonstrated that >90% of undifferentiated HL-60 and U937 cells express these Ca^{2+}-mobilizing ATP receptors. After differentiation along the neutrophil/granulocyte pathway, HL-60 cells responded to both ATP and FMLP with equivalent large increases in [Ca^{2+}] (FIG. 1F). In contrast to these effects of ATP in HL-60 and U937 cells, cells from the less mature KG-1 myeloblastic line (FIG. 1B) exhibited only a small, 2-fold mean increase in cytosolic [Ca^{2+}] upon stimulation with 100 μM ATP; these cells were completely unresponsive to 30 μM FMLP. Because KG-1 cells did contain large intracellular Ca^{2+} stores (which could be released by the Ca^{2+} ionophore ionomycin), the modest Ca^{2+} mobilization triggered by ATP suggested 1) that most cells responded to ATP by mobilizing only a small fraction of their ionomycin-releasable Ca^{2+} stores, and/or 2) that only a minor subpopulation of the cells were responsive to ATP. In fact, flow cytometric measurements revealed that only 20 ± 1% ($N = 3$) of KG-1 cells responded to 100 μM ATP, and that the increases in cytosolic [Ca^{2+}] induced in these cells was small. The less mature KG-1A cell line (a variant of the KG-1 cell line)[17] was almost completely unresponsive to 100 μm ATP (as was observed with KG-1 cells, these KG-1A cells were also unresponsive to 30 μM FMLP). This suggests perhaps only a subpopulation of KG-1 cells, slightly more mature than the majority of cells in culture, had acquired responsiveness to ATP. Significantly, mobilization in response to stimulation with ATP was restricted to human leukemic cell lines exhibiting the phenotype of neutrophil/monocyte precursors. Cells from the K562 erythroblastic leukemia line (not shown), the Molt-4 T cell lymphoblastic leukemia line (FIG. 1E), and the HuT-78 T cell lymphoma line (not shown) were completely unresponsive to ATP (and FMLP).

NUCLEOTIDE SELECTIVITY OF THE PHOSPHOLIPASE C-COUPLED P_2-PURINERGIC RECEPTOR EXPRESSED BY HUMAN PHAGOCYTIC LEUKOCYTES

The stereoselectivity of the ATP effects on Ca^{2+} mobilization in HL-60 cells was also examined; these results are summarized in TABLE 2. ATP was the most potent of all tested adenine nucleotides/nucleosides. The EC_{50} for ADP action was 10 μM, whereas AMP and adenosine produced no significant Ca^{2+} mobilization at concentrations up to 300 μM. Modification of the triphosphate moiety greatly affected potency. The bridging oxygen between the β and γ phosphates appears critical because the nonhydrolyzable analogue, β,γ-methylene-ATP, was much less potent—a con-

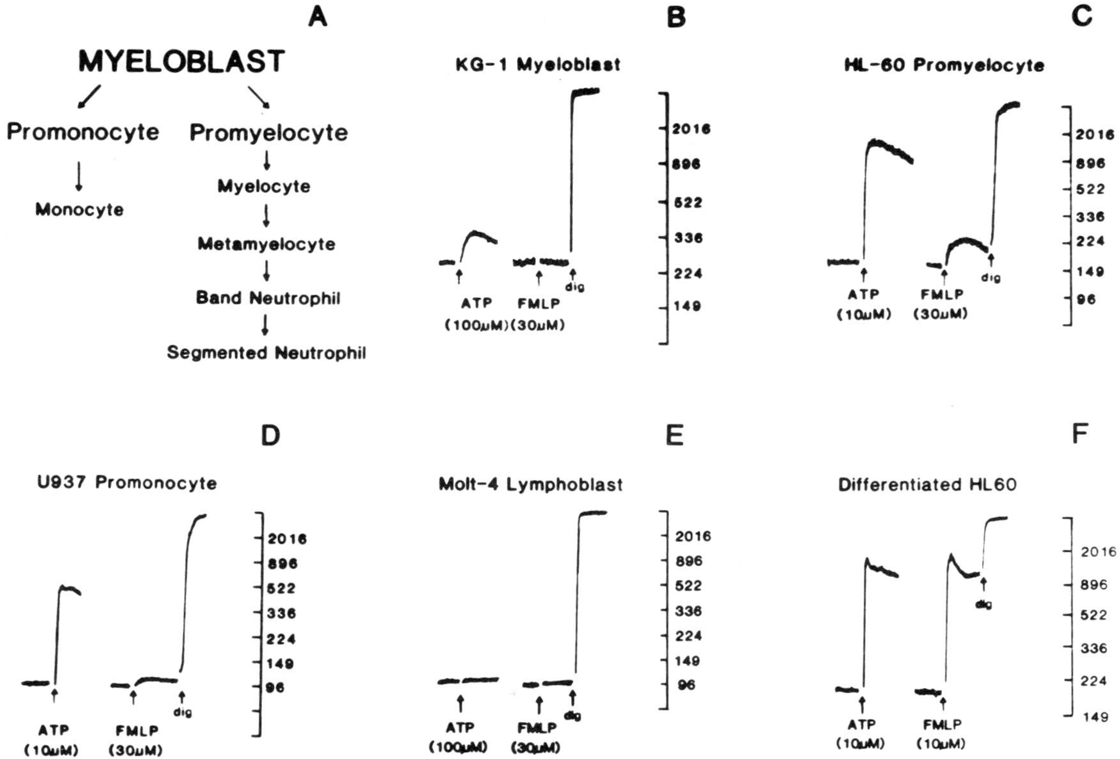

FIGURE 1. Comparative effects of ATP and chemotactic peptide on the cytosolic [Ca^{2+}] in human phagocyte progenitor cell lines at various stages of differentiation. Cultured cells from each of the indicated human leukemia cell lines were washed and loaded with fura 2 as described in reference 8. The Ca^{2}-dependent fura 2 fluorescence was recorded as described during incubation of cell aliquots at 3 °C. Cells were stimulated with either 10 μM ATP or 30 μM FMLP. After 1 min, 20 μg/ml of digitonin (dig) was added to lyse the cells. Free Ca^{2+} was calculated as described in reference 8.

TABLE 2. Relative Effects of Various Nucleotides on Ca^{2+} Mobilization in HL-60 Cells[a]

Nucleotide	EC_{50} for Ca^{2+} Mobilization (M)
Adenine Nucleotides	
ATP	1.5×10^{-7}
ADP	1×10^{-5}
AMP	$>3 \times 10^{-4}$
Adenosine	$>3 \times 10^{-4}$
ATP Analogues	
ATP-γ-S	1×10^{-6}
AMP-PCP[b]	$>1 \times 10^{-3}$
Purine Nucleotide Triphosphates	
ITP	2×10^{-6}
XTP	2×10^{-5}
GTP	3×10^{-5}
Pyrimidine Nucleotide Triphosphates	
UTP	3×10^{-7}
CTP	5×10^{-5}
TTP	8×10^{-5}

[a] The cytosolic [Ca^{2+}] in fura 2-loaded samples of undifferentiated HL-60 cells was measured at 37 °C. Separate aliquots of cells were exposed to various concentrations of the indicated nucleotides and the resultant Ca^{2+} transients were recorded as illustrated in FIGURE 1. The peak changes in cytosolic [Ca^{2+}] were calculated from these transients and plotted as a function of extracellular nucleotide concentration as illustrated in FIGURE 2; the EC_{50} values were estimated from these dose-response curves. For each nucleotide class, the listed EC_{50} values were obtained from the data collected from a single preparation of cells; that is, the adenine nucleotide data were obtained with one cell preparation, and the purine nucleotide data were obtained with a separate cell preparation. For each nucleotide, however, similar EC_{50} values (within a factor of 2) were obtained in least three separate experiments performed over a 12-month interval.

[b] AMP-PCP: β,γ-methylene-ATP.

centration as high as 1 mM produced less than a 2-fold change in [Ca^{2+}]. Conversely, another poorly hydrolyzable analogue, ATP-γ-S (characterized by thio-substitution of a nonbridging oxygen on the terminal phosphate), was only slightly less potent than ATP. As might be expected, potency was significantly affected by substitution of the adenine moiety. Of the other nonadenine purine nucleotide triphosphates tested, ITP ($EC_{50} = 2$ μM) was the most potent, whereas GTP and XTP were at least 100 times less potent that ATP. The pyrimidine triphosphates, CTP and TTP, were also much less potent, having EC_{50} values in the 30-100 μM region. Significantly, UTP was nearly equipotent with ATP ($EC_{50} = 300$ nM).

This basic rank order of potency was also observed when inositol trisphosphate (IP_3) accumulation, rather than Ca^{2+} mobilization, was the measured parameter. Stimulation of undifferentiated HL-60 cells with ATP evoked maximal 5-fold increases in IP_3 levels within 15 sec (FIG. 2). Significantly, treatment of these cells with UTP or ATP-γ-S triggered a maximal increase in IP_3 identical to that achieved with ATP. Although the potencies of ATP and UTP were equal ($EC_{50} = 2$ μM), ATP-γ-S was less potent ($EC_{50} = 20$ μM). All three nucleotides also elicited >10-fold maximal increases in IP_2 within 15 sec (data not shown). In contrast to the effects observed with extracellular adenine nucleotides and UTP, treatment of intact HL-60 cells with equivalent concentrations of guanine nucleotide triphosphates (both GTP and the poorly hydrolyzable analogue, GTP-γ-S) induced much smaller increases in IP_3 accumulation. Even at concentrations as high as 100 μM, these guanine nucleotides were only half as efficacious as ATP, ATP-γ-S, or UTP. Moreover, at concentrations below 10 μM, neither guanine nucleotide elicited significant activation of inositol phospholipid breakdown.

EFFECTS OF PERTUSSIS TOXIN TREATMENT ON P_2-PURINERGIC- VERSUS FMLP-STIMULATED INOSITOL POLYPHOSPHATE FORMATION IN INTACT HL-60 CELLS

Given the ability of pertussis toxin to substantially inhibit FMLP receptor-activated, GTP-dependent inositol phospholipid breakdown in neutrophils and differen-

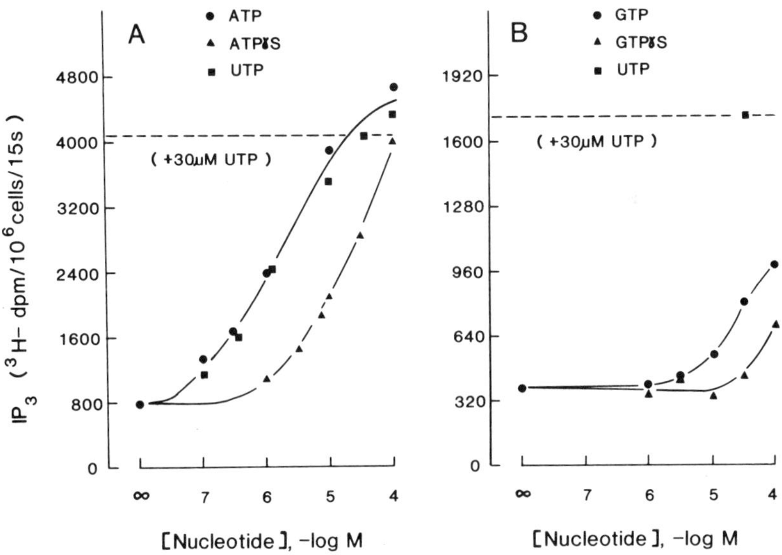

FIGURE 2. Dose-response relationships comparing the effects of various nucleotide triphosphates on IP_3 accumulation in intact, undifferentiated HL-60 cells. Undifferentiated HL-60 cells labeled with [^{3}H]myoinositol (1-2 μCi/ml) were incubated at 37 °C in 0.2-ml aliquots containing 1×10^6 cells. Individual aliquots were treated with the indicated concentrations of extracellular nucleotide for 15 sec. Samples were prepared for analysis of [^{3}H]inositol trisphosphate (IP_3) accumulation as described in reference 9. Data (expressed as ^{3}H dpm/106 cells/15 sec) represent the average of duplicate determinations from single experiments. Data in panels **A** and **B** were obtained using different cell preparations. Dotted lines denote the maximal rates of IP_3 accumulation (as triggered by 30 μM UTP) for each cell preparation. Each illustrated dose-response relationship is representative of data obtained in two or three similar experiments. (**A**) Effects of ATP (●), ATP-γ-S (▲), or UTP (■). (**B**) Effects of GTP (●), GTP-γ-S (▲), or UTP (■).

tiated HL-60 cells,[11,12] considerable attention has been focused on the role of pertussis toxin-sensitive G proteins in activating inositol phospholipid breakdown in these and other cell types. Polakis *et al.* have reported that the FMLP receptor copurifies with a 40-kDa GTP-binding protein that is the major pertussis toxin substrate in neutrophils and HL-60 cells.[13] Recent immunochemical and molecular cloning studies from several laboratories have identified this protein as the G_{i2}-subtype from the G_i superfamily of regulatory proteins.[14] Despite the large amount of evidence indicating that FMLP receptors are coupled to G_{i2}, important questions concerning the role of G_{i2} (or other pertussis toxin-sensitive G proteins) in the regulation of GTP-dependent inositol phospholipid breakdown remain unresolved. The mechanism(s) whereby pertussis toxin interferes with agonist-induced activation of the phospholipase is only partially understood. Although pertussis toxin treatment is known to uncouple G_{i2}/FMLP receptor interaction significantly,[13] it remains to be determined whether the interaction of the ADP-ribosylated G protein with the phospholipase effector (or some intermediary protein) is also inhibited or attenuated. Moreover, it has not been ascertained whether toxin-insensitive G proteins may also activate inositol phospholipid hydrolysis in neutrophils and other cells that exhibit pertussis toxin-sensitive receptor-phospholipase C coupling.

The activation of both inositol polyphosphate accumulation and Ca^{2+} mobilization by P_2-purinergic agonists is significantly, but not completely, inhibited in undifferentiated and differentiated HL-60 cells pretreated with pertussis toxin. This contrasts with the near-total inhibition of FMLP-stimulated phospholipase C activity and Ca^{2+} mobilization measured in the same pertussis toxin-treated cells. As shown in FIGURE 3, maximally activating concentrations of UTP or FMLP were equally efficacious in stimulating rapid accumulation of IP_3 (4-fold increases within 15 sec) and IP_2 (6-fold increases in 15 sec) in intact HL-60 granulocytes. The EC_{50} for UTP stimulation was about 3 μM, whereas that for FMLP was 30 nM. Treatment of HL-60 granulocytes with pertussis toxin significantly reduced the efficacies of both FMLP and UTP in stimulating inositol polyphosphate accumulation. The maximal extent of the pertussis toxin-induced inhibition was invariably smaller, however, when UTP (or ATP) rather than FMLP was the agonist. This differential sensitivity of P_2-purinergic receptors versus FMLP receptors to inhibition by pertussis toxin was very reproducible. It is important to note that varying the concentration of pertussis toxin (100-500 ng/ml) used to treat the cells had no effect on the magnitude of inhibition of the response to FMLP. Cells treated with this range of toxin concentrations were characterized by a 91 ± 1% (N = 7) inhibition of maximal FMLP-induced inositol polyphosphate accumulation. In these same cells there was an average 78 ± 2% (N = 8) inhibition of the maximal UTP-induced response. Treatment of HL-60 granulocytes with pertussis toxin, under the conditions used in these studies (100-1000 ng/ml; 3-4 hr), induces ADP ribosylation of 65-85% of the major 40-41-kDa toxin substrate present in isolated HL-60 cell membranes. This substrate presumably includes both the 41-kDa alpha subunit of G_{i3} and the 40-kDa subunit of G_{i2}, given the results of several immunochemical characterizations of G protein expression in HL-60 cells.[14] As noted in previous studies, it is difficult to resolve autoradiographically the relative distribution of incorporated ^{32}P between these subunits. Thus, we cannot ascertain the actual degrees of ADP ribosylation of G_{i2} versus G_{i3} in the pertussis toxin-treated cells and membranes. As previously noted, however, FMLP-induced activation and P_2-purinergic-induced activation of phospholipase C in intact toxin-treated cells were inhibited by 91% and 78%, respectively (FIG. 3). Likewise, the membranes isolated from these toxin-treated cells exhibited a 75% reduction in GTP-dependent inositol polyphosphate accumulation in the absence of receptor agonists and an 82% reduction in the extent of FMLP-induced potentiation of the GTP-dependent phospholipase C activity (see

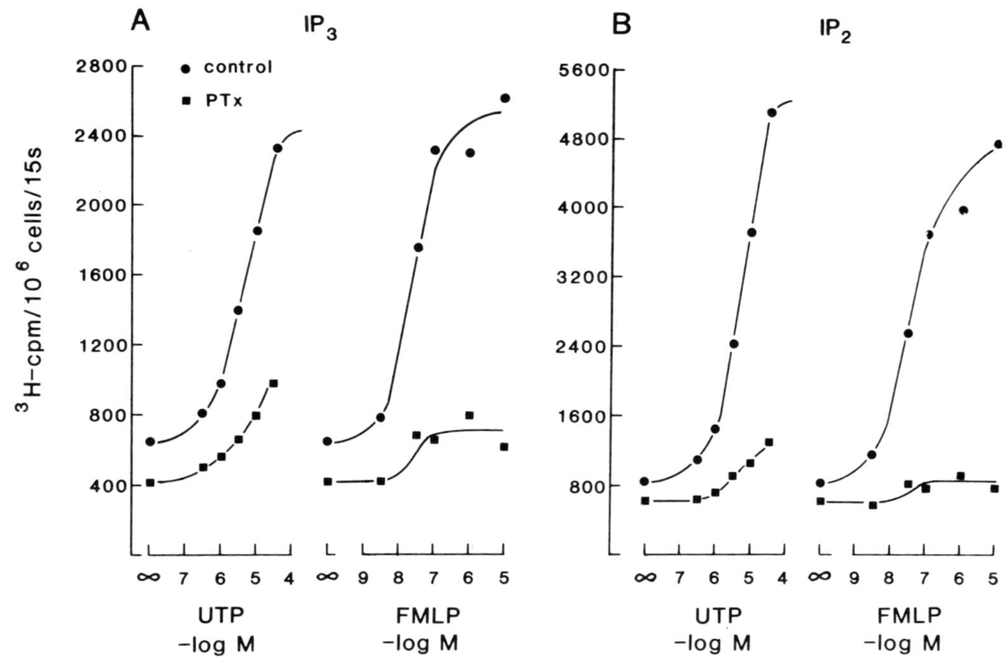

FIGURE 3. Inhibition of UTP- and FMLP-activated inositol polyphosphate accumulation in differentiated HL-60 granulocytes treated with pertussis toxin. HL-60 cells were cultured for 3 days in the presence of 2 μCi/ml [^{3}H]myoinositol and 0.5 mM dibutyryl cyclic AMP. The culture of labeled, differentiated cells was divided in two fractions, one of which was treated with pertussis toxin (500 ng/ml, 3 hr). The accumulation of [^{3}H]IP_3 (**A**) and [^{3}H]IP_2 (**B**) was quantified as described in reference 9. Data points represent the average of duplicate determinations from a single experiment. These data are representative of four separate experiments.

below). The magnitudes of these inhibitory effects strongly suggest that a substantial fraction of the particular G protein subpopulations that functionally interact with both the ATP receptors and the FMLP receptors has been modified by the toxin treatment. Indeed, there is a reasonable correlation between the degree of toxin-induced ADP ribosylation (65-85%) and the percentage inhibition (70-90%) of the various functional parameters in both intact HL-60 cells and isolated membranes. Okajima and Ui[18] noted a similar correlation between the degree of ADP ribosylation of the 40-41-kDa substrate and the relative inhibition of FMLP-induced functional responses in guinea pig neutrophils treated with increasing concentrations of pertussis toxin.

COMPARATIVE EFFECTS OF PERTUSSIS TOXIN ON P_2-PURINERGIC- VERSUS CHEMOTACTIC PEPTIDE-INDUCED ACTIVATION OF PHOSPHOLIPASE C IN HL-60 CELL MEMBRANES

The ability of P_2-purinergic receptor agonists to activate an inositol phospholipid-specific phospholipase C is retained in membranes isolated from these cells. FIGURE 4 compares the kinetics of inositol polyphosphate accumulation in membranes isolated from either control cells or pertussis toxin-treated cells. In control membranes, micromolar concentrations of GTP elicited 4-8-fold increases in the rate of inositol polyphosphate accumulation measured over a 4-min time period (TABLE 3). The inclusion of either 30 μM UTP or 10 μM FMLP produced an additional 2-fold increase in the rate. TABLE 3 shows that in six separate membrane preparations, UTP produced a mean 2.8 ($\pm$0.5)-fold potentiation of GTP-dependent phospholipase activity whereas FMLP elicited a mean 2.2 ($\pm$0.3)-fold increase. FIGURE 5 illustrates the dose-response curves characterizing UTP- and FMLP-induced potentiation of GTP-dependent inositol polyphosphate accumulation in membranes from control cells. We have previously verified that this effect of UTP (or other P_2-purinergic agonists) is not due to elevation of membrane polyphosphoinositide levels or to inhibition of GTP catabolism by membrane-associated phosphatases/nucleotidases. The threshold for UTP action was approximately 100 nM, the EC_{50} was about 1.8 μM, and near-maximal activation was obtained with 30-100 μM. These values are similar to those characterizing the effects of UTP on intact cells (FIGS. 2 & 3). Conversely, the FMLP dose-response relationship in membranes was different from that observed in intact cells. The threshold was 30 nM, and the EC_{50} was about 300 nM. FMLP in excess of 3 μM was required to achieve maximal activation.

Parallel studies performed on membranes isolated from cells pretreated with pertussis toxin revealed several differences. First, as shown in FIGURES 4 & 5 and TABLE 3, the basal rate of inositol polyphosphate formation as stimulated by 1 μM GTP (in the absence of receptor agonists) was significantly inhibited (mean inhibition = 76 $\pm$ 4%, N = 6) in the toxin-treated membranes. It should be emphasized that pertussis toxin treatment did not alter the levels of the various membrane [^{3}H]inositol phospholipids (data not shown). Nor did it significantly alter activation of the membrane-associated phospholipase C by 1 mM free Ca^{2+} (4 $\pm$ 7% increase over control membranes, N = 4). The finding that GTP-dependent phospholipase activity was inhibited in membranes prepared from pertussis toxin-pretreated cells was consistent with the observation that basal IP_2 and IP_3 levels were modestly reduced in intact cells pretreated with pertussis toxin (FIG. 3). As will be further discussed below, this

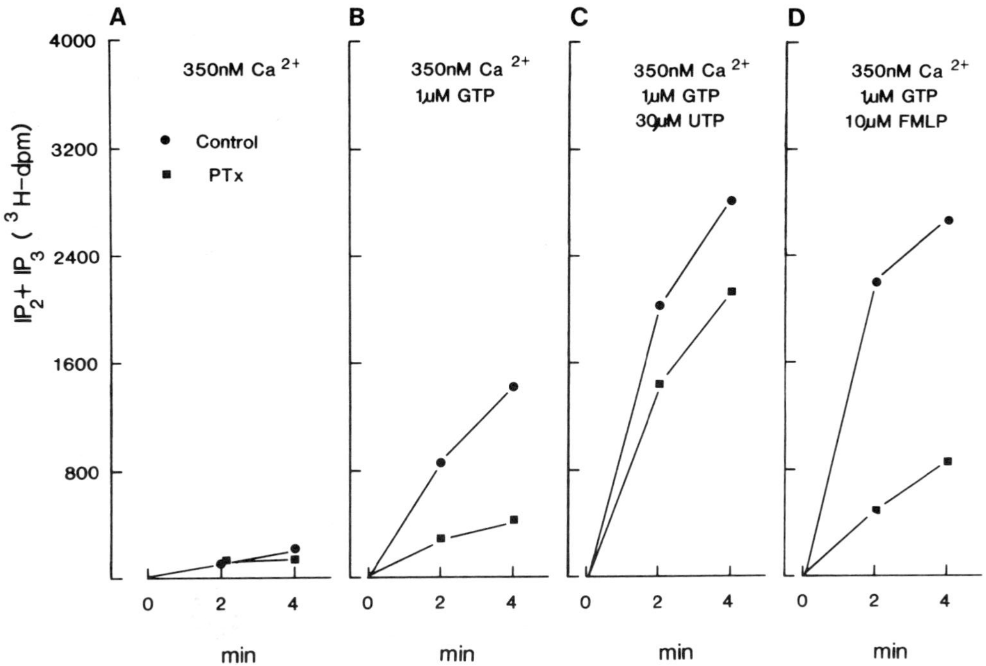

FIGURE 4. Differential inhibitory effects of pertussis toxin treatment on FMLP- and UTP-induced potentiation of GTP-dependent inositol polyphosphate accumulation in membranes isolated from HL-60 granulocytes—effects on the kinetics of accumulation. Membranes isolated from [^{3}H]inositol-labeled, differentiated HL-60 cells were incubated for the indicated times at 37 °C in basic assay medium containing 350 nM free Ca^{2+}. (**A**) Incubation media also contained the following: 1 μM GTP (**B**); 1 μM GTP plus 30 μM UTP (**C**); or 1 μM GTP plus 10 μM FMLP (**D**). Circles (●) represent data from membranes of control cells; squares (■) represent data from membranes of cells treated for 3 hr with 500 ng/ml pertussis toxin. Accumulation of [^{3}H]inositol polyphosphates was quantified as described in reference 9. Data points represent the average of duplicate determinations from a single experiment. These data are representative of six separate experiments.

TABLE 3. Effects of Pertussis Toxin on FMLP- and P_2-Purinergic-Induced Activation of GTP-Dependent Phospholipase C Activity in Membranes from HL-60 Granulocytes[a]

Toxin Treatment (ng/ml, hr)	Agonist	$IP_2 + IP_3$ Release (^{3}H dpm/sample) Control	PTX	Fractional Inhibition Total	GTP
250, 3	None	76	92		
	GTP	2008	456	0.81	
	GTP/UTP	4604	2772	0.43	0.11
	GTP/FMLP	3732	1240	0.68	0.55
	Ca^{2+} (1 mM)	9832	9400	0.05	
400, 3	None	104	132		
	GTP	892	296	0.79	
	GTP/UTP	2036	1460	0.31	−0.02
	GTP/FMLP	2228	512	0.82	0.84
	Ca^{2+} (1 mM)				
500, 3	None	128	112		
	GTP	440	216	0.67	
	GTP/UTP	1320	844	0.39	0.29
	GTP/FMLP	864	272	0.78	0.87
	Ca^{2+} (1 mM)	4332	4995	−0.15	
1000, 4	None	276	308		
	GTP	1116	560	0.71	
	GTP/UTP	2512	1704	0.38	0.18
	GTP/FMLP	1956	632	0.81	0.91
	Ca^{2+} (1 mM)	7224	8500	−0.13	
1000, 4	None	64	336		
	GTP	1112	440	0.90	
	GTP/UTP	2572	1728	0.56	0.12
	GTP/FMLP	1388	480	0.91	0.85
	Ca^{2+} (1 mM)	7508	7236	0.07	
100, 18	None	100	84		
	GTP	396	176	0.69	
	GTP/UTP	1844	1320	0.15	0.21
	GTP/FMLP	1328	240	0.87	0.93
	Ca^{2+} (1 mM)	6940	5488	0.21	
	GTP-γ-S (30 μM)	5288	2500	0.53	

[a] Aliquots from a single preparation of [^{3}H]inositol-labeled, differentiated HL-60 cells were incubated in the absence of pertussis toxin (PTX) or in the presence of the indicated concentrations of PTX for 3-18 hr (results obtained with the former incubations are listed under the Control heading; those obtained with the latter incubations, under the PTX heading). Membranes isolated from these cells were then incubated for 1-4 min at 37 °C in the basic assay medium containing either 350 nM free Ca^{2+} or 1 mM free Ca^{2+}. Parallel samples also contained the following (in addition to 350 nM free Ca^{2+}): no agonists (None); 1 μM GTP (GTP); 1 μM GTP plus 30 μM UTP (GTP/UTP); or 1 μM GTP plus 10 μM FMLP (GTP/FMLP). In the first experiment (PTX at 250 ng/ml for 3 hr), the [GTP] was 3 μM. Accumulation of [^{3}H]inositol polyphosphates was quantified. Each data point represents the average of duplicate determinations. Fractional inhibition of total activity was defined as follows: 1 - ($\{$Agonist - None$\}_{PTX}$ / $\{$Agonist - None$\}_{control}$). Fractional inhibition of GTP-dependent activity was defined as follows: 1 - ($\{$Agonist - GTP$\}_{PTX}$ / $\{$Agonist - GTP$\}_{control}$).

reduction in basal GTP-dependent phospholipase C activity in membranes from toxin-treated cells appeared to reflect a depressed rate of GTP-dependent activation of a toxin-sensitive G protein(s) that is functionally coupled to the phospholipase C.

The absolute rate of inositol polyphosphate release induced by the administration of both UTP and GTP in toxin-treated membranes was reduced by about 40% relative to that observed in control membranes (FIGS. 4 & 5, TABLE 3). We were surprised, however, to observe that the ability of UTP (or ATP) to potentiate GTP-dependent

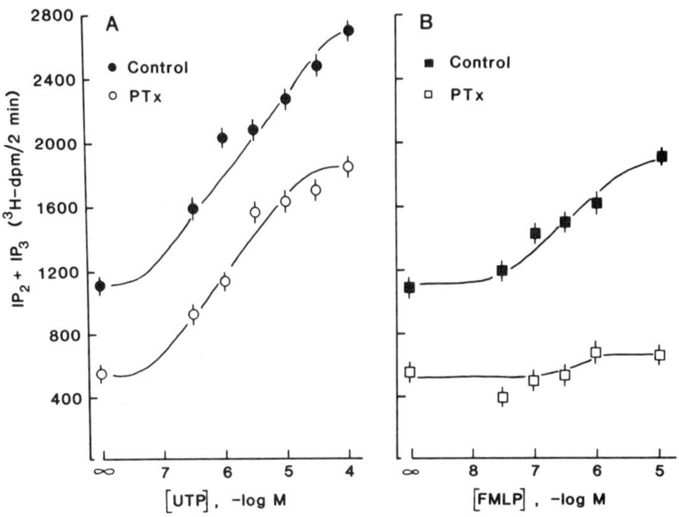

FIGURE 5. Differential inhibitory effects of pertussis toxin treatment on FMLP- and UTP-induced potentiation of GTP-dependent inositol polyphosphate accumulation in membranes isolated from HL-60 granulocytes—dose-response relationships. Membranes isolated from [^{3}H]inositol-labeled, differentiated HL-60 cells were incubated for 2 min at 37 °C in basic assay medium containing 350 nM free Ca^{2+} and 1 μM GTP. Incubation media also contained the indicated concentrations of UTP (●, panel **A**) or FMLP (■, panel **B**). Filled symbols (●,■) represent data from membranes of control cells; open symbols (○,□) represent data from membranes of cells treated for 4 hr with 1 μg/ml pertussis toxin (sufficient to induce 83% ADP ribosylation of the major 40-41-kDa substrate). Accumulation of [^{3}H]inositol polyphosphates was quantified. Data points represent the mean ± range of duplicate determinations from a single experiment. These data are representative of two separate experiments.

phospholipase C was very similar in isolated membranes prepared from either control or pertussis toxin-treated cells. Thus, the dose-response curve characterizing UTP-induced activity in pertussis toxin-treated membranes, though negatively shifted along the *y*-axis, paralleled that observed in control membranes. If it is assumed that P_2-purinergic receptors activate phospholipase C exclusively through the mediation of a toxin-sensitive G protein(s) (presumably G_{i2}), our results would suggest that ADP ribosylation does not preclude interaction of this G protein with occupied P_2-purinergic receptors. However, because ADP ribosylation appears to substantially reduce the absolute rate of GDP/GTP exchange by the relevant G protein in the absence of agonists, it is possible that the absolute rate of receptor-stimulated GTP/GDP exchange may be similarly reduced relative to that catalyzed by the unmodified G protein. Any attenuation of this guanine nucleotide exchange rate will actually limit activation of the phospholipase effector. Such an attenuation of G protein/effector coupling, rather than an absolute inhibition of receptor/G protein coupling, may explain the differential effects of toxin treatment on UTP- versus FMLP-induced phospholipase activation in both membranes and intact cells. Our results suggest that P_2-purinergic receptors, but not FMLP receptors, contain structural features that facilitate some degree of interaction with the ADP-ribosylated alpha subunits of G_i-type proteins. As noted in the Introduction, pertussis toxin treatment of aortic endothelial cells,[3] FRTL thyroid cells,[4] or rat aortic smooth muscle cells[5] induces an attenuation, rather than a complete blockade, of P_2-purinergic receptor-stimulated phospholipase C activity.

Because only 65-85% of the G_{i2+3} pool in these cells was ribosylated by ADP it may be argued that blockade of P_2-purinergic receptor-induced potentiation of inositol polyphosphate release in membranes requires modification of the residual pool of toxin substrate. In studying the ability of pertussis toxin to uncouple angiotensin II receptors from inhibition of hepatocyte adenylate cyclase, Pobiner *et al.*[19] found that at least 90% of the membrane G_i pool had to be ribosylated by ADP if a 50% attenuation of the angiotensin action on cyclase activity in hepatocyte membranes was to be achieved. Such results indicate an important role for receptor/G protein stoichiometry, particularly if $[G_i] > [\text{receptor}]$. It is important to note that nearly 100% ADP ribosylation was required to block the ability of the hormone to inhibit cyclic AMP accumulation in intact hepatocytes. Thus, there was no disparity between the effects of the toxin on agonist-induced signaling in intact cells or isolated membranes. This contrasts with our observation of a >75% inhibition of UTP-induced signaling in cells containing up to 35% of the normal content of unmodified G protein.

EFFECTS OF PERTUSSIS TOXIN ON THE BASAL GTP-DEPENDENT PHOSPHOLIPASE C ACTIVITY IN ISOLATED HL-60 CELL MEMBRANES

By preparing and assaying HL-60 cell membranes in the absence of ATP, it was possible to measure the ability of relevant G proteins to activate inositol phospholipid hydrolyisis in the absence of P_2-purinergic receptor-induced stimulation. Surprisingly, this basal GTP-dependent phospholipase C activity was significantly inhibited in membranes prepared from pertussis toxin-treated HL-60 granulocytes (FIGS. 4 & 5, TABLE 3). Therefore, pertussis toxin treatment appears to attenuate some step in G protein/phospholipase C coupling, in addition to the well-characterized inhibition of receptor/G protein coupling.[11,12] Our results suggest that ADP ribosylation of the relevant toxin-sensitive G protein(s) may slow the basal, agonist-independent cycle of G protein activation by guanine nucleotide triphosphates. This effect of pertussis toxin on GTP-dependent inositol polyphosphate formation in isolated membranes could be correlated with the 20% reduction in basal IP_2 and IP_3 levels measured in intact, pertussis-treated, differentiated cells (FIG. 3, TABLE 3). Such inhibition has not been reported in previous studies of GTP-dependent phospholipase C activity in membranes derived from differentiated HL-60 cells or neutrophils.[11,12] It is consistent, however, with results from a recent study by McLeish *et al.*[15] showing that membranes isolated from pertussis toxin-treated HL-60 granulocytes exhibit a 65% reduction in the rate of high-affinity GTPase measured in the absence of agonists. In these same membranes, the rate of high-affinity GTP-γ-S binding was reduced by about 70%. Similarly, Wilde *et al.*[20] have reported that pertussis toxin causes a 50% inhibition of basal GTPase activity in membranes from rabbit neutrophils. Okajima and Ui[18] and Okajima *et al.*[21] also observed a 20-30% inhibition of the GTPase assayed in membranes from toxin-treated guinea pig neutrophils. Such results suggest that ADP ribosylation of a predominant GTP-binding protein in HL-60 or neutrophil membranes significantly slows the rate of GTP/GDP exchange. Attenuation of this exchange rate would be expected to reduce the steady state level of activated G protein produced in response to incubation with GTP per se, and also to slow the rate, but not the extent, of G protein activation by the nonhydrolyzable GTP-γ-S. Consistent with this, we have observed that pertussis toxin inhibited only the rate, but not the extent of GTP-γ-S-induced phospholipase C activation in HL-60 cell membranes (FIGS. 6 & 7). Thus, ADP ribosylation appears to affect only the rate of G protein activation but not the ability of the activated G protein to stimulate phospholipase C.

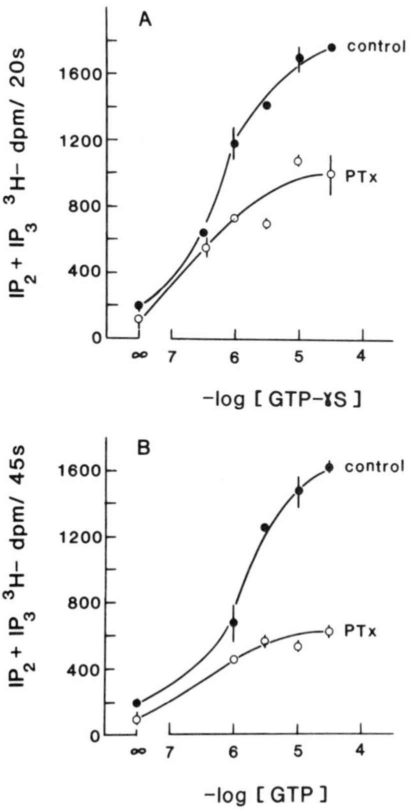

FIGURE 6. Inhibition of the initial rate of GTP- and GTP-γ-S-dependent inositol polyphosphate accumulation in membranes isolated from pertussis toxin-treated HL-60 granulocytes—dose-response relationships. Membranes isolated from [^{3}H]inositol-labeled, differentiated HL-60 cells were incubated at 37 °C in basic assay medium containing 350 nM free Ca^{2+}. (**A**) Membranes were incubated for 20 sec with the specified concentrations of GTP-γ-S. (**B**) Membranes were incubated for 45 sec with the specified concentrations of GTP. A GTP-regenerating system was included only for samples in which GTP was included (**B**). Filled symbols (●) represent data from the membranes of control cells; open symbols (○) represent data from the membranes of pertussis toxin-treated cells. Accumulation of [^{3}H]inositol polyphosphates was quantified as described in reference 9. Data points represent the mean ± range of duplicate determinations from a single experiment. These data are representative of two separate experiments.

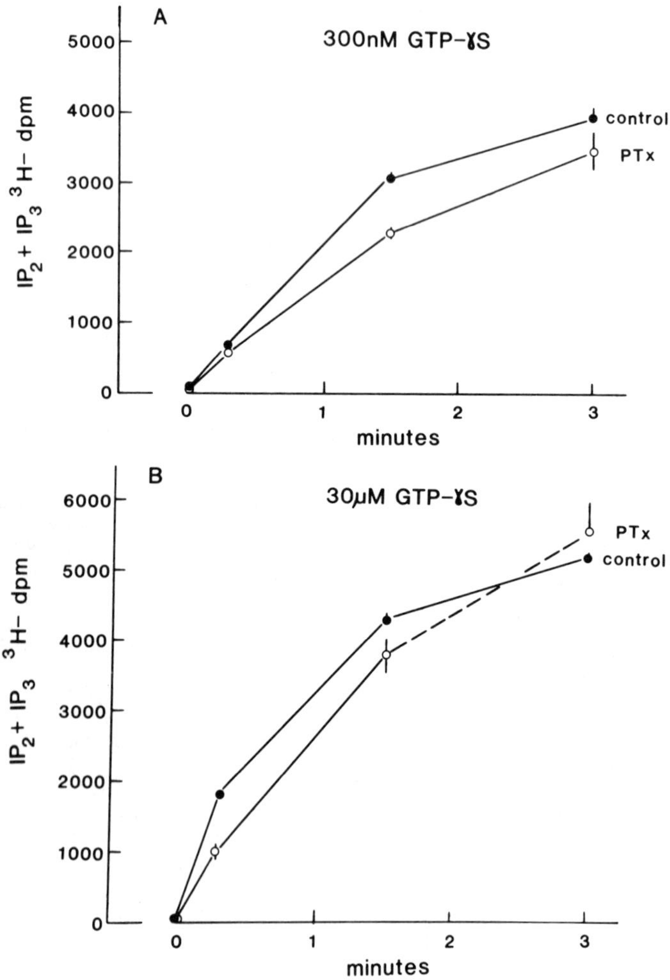

FIGURE 7. Inhibition of the GTP-γ-S-dependent inositol polyphosphate accumulation in membranes isolated from pertussis toxin-treated HL-60 granulocytes—effects on the kinetics of accumulation. Membranes isolated from [^{3}H]inositol-labeled, differentiated HL-60 cells were incubated for the indicated times at 37 °C in basic assay medium containing 350 nM free Ca^{2+}. No GTP-regenerating system was included. (**A**) Membranes were incubated with 300 nM GTP-γ-S. (**B**) Membranes were incubated with 30 μM GTP-γ-S. Filled symbols (●) represent data from membranes of control cells; open symbols (○) represent data from membranes of pertussis toxin-treated cells. Accumulation of [^{3}H]inositol polyphosphates was quantified as described in reference 9. Data points represent the mean ± range of duplicate determinations from a single experiment. These data are representative of two separate experiments.

FUNCTIONAL SIGNIFICANCE OF THE DIFFERENTIAL EFFECTS OF PERTUSSIS TOXIN ON G PROTEIN ACTIVATION BY P_2-PURINERGIC VERSUS FMLP RECEPTOR IN PHAGOCYTIC LEUKOCYTES

Our studies also do not rule out a role for pertussis toxin-insensitive G proteins in mediating the P_2-purinergic receptor-induced activation of phospholipase C in HL-60 cells or neutrophils. The apparent role of pertussis toxin-insensitive G proteins in mediating the activation of phospholipases by P_2-purinergic receptors has now been characterized in several cell types. In hepatocytes, P_2-purinergic stimulation of phosphatidylinositol-specific phospholipase C,[2] phospholipase D,[22] and phosphatidylcholine-specific phospholipase C[23] is not affected by pertussis toxin treatment. The P_2-purinergic receptors expressed by turkey erythrocytes also activate inositol phospholipid breakdown by a pertussis toxin-insensitive G protein.[24] It is possible that similar toxin-insensitve G proteins may the mediate the toxin-resistant portion of P_2-purinergic signaling in HL-60 granulocytes. In our own studies, we have consistently observed that pertussis toxin is significantly more efficacious in inhibiting the ATP-induced activation of inositol polyphosphate accumulation in differentiated HL-60 cells as opposed to undifferentiated cells. Such results suggest that differentiation of these cells induces a phenotypic shift in the G protein species available for coupling receptors to the phospholipase C effector enzymes. Indeed, previous studies[25] have demonstrated that the differentiation of HL-60 cells into neutrophils is accompanied by 1) 2-3-fold increases in the levels of mRNA transcripts for the alpha subunits of both G_{i2} and G_{i3}, and 2) at least 2-fold increases in the translation products of these alpha subunit transcripts.

McLeish *et al.*[15] have recently suggested that differences in the activation of G_{i2} (or additional G proteins) by FMLP receptors and leukotriene B_4 receptors may be responsible for differential activation of various neutrophil responses by these agonists. Similarly, FMLP, but not ATP, is an effective stimulus for neutrophil superoxide production and primary granule secretion.[16,17] That FMLP receptors and P_2-purinergic receptors also appear to activate a common G protein by different mechanisms may partially explain the differential capacities of FMLP and P_2-purinergic agonists to elicit functional changes in neutrophils.

REFERENCES

1. CHAREST, R., P. F. BLACKMORE & J. H. EXTOR. 1985. J. Biol. Chem. **260:** 15789-15794.
2. OKAJIMA, F., Y. TOKUMITSU, Y. KONDO & M. UI. 1987. J. Biol. Chem. **262:** 13483-13490.
3. PIROTTON, S., C. ERNEUX & J. M. BOEYNAEMS. 1987. Biochem. Biophys. Res. Commun. **147:** 1113-1120.
4. BOYER, J. L., G. L. WALDO, T. EVANS, J. K. NORTHUP, C. P. DOWNES & T. K. HARDEN. 1989. J. Biol. Chem. **264:** 13917-13922.
5. OKAJIMA, F., K. SHO & Y. KONDO. 1988. Endocrinology **123:** 1035-1043.
6. TAWADA, Y., K.-I. FURUKAWA & M. SHIGEKAWA. 1987. J. Biochem. **102:** 1499-1509.
7. COOPER, C. L., A. J. MORRIS & T. K. HARDEN. 1989. J. Biol. Chem. **264:** 6202-6206.
8. DUBYAK, G. R. & M. B. DEYOUNG. 1985. J. Biol. Chem. **260:** 10653-10661.
9. DUBYAK, G. R., D. S. COWEN & L. M. MUELLER. 1988. J. Biol. Chem. **263:** 18108-18117.
10. COWEN, D. S., H. M. LAZARUS, S. B. SHURIN & G. R. DUBYAK. 1989. J. Clin. Invest. **83:** 1651-1660.

11. SMITH, C. D., R. J. UHING & R. SNYDERMAN. 1987. J. Biol. Chem. **262:** 6121-6127.
12. KIKUCHI, A., O. KOZAWA, K. KAIBUCHI, M. UI. & Y. TAKAI. 1986. J. Biol. Chem. **261:** 11558-11562.
13. POLAKIS, P. G., R. J. UHING & R. SNYDERMAN. 1988. J. Biol. Chem. **263:** 4969-4976.
14. GOLDSMITH, P., P. GIERSCHIK, G. MILLIGAN, C. G. UNSON, R. VINITSKY, H. L. MALECH & A. M. SPIEGEL. 1987. J. Biol. Chem. **262:** 14683-14688.
15. MCLEISH, K. R., P. GIERSCHIK, T. SCHEPERS, D. SIDIROPOULOS & K. H. JACOBS. 1989. Biochem. J. **260:** 427-434.
16. WARD, P. A., T. W. CUNNINGHAM, K. K. MCCULLOCH & K. W. JOHNSON. 1988. Lab. Invest. **58:** 438-447.
17. KUHNS, D. B., D. G. WRIGHT, J. NATH, S. S. KAPLAN & R. E. BASFORD. 1988. Lab. Invest. **58:** 448-453.
18. OKAJIMA, F. & M. UI. 1984. J. Biol. Chem. **259:** 13863-13871.
19. POBINER, B. F., E. L. HEWLETT & J. C. GARRISON. 1985. J. Biol. Chem. **260:** 16200-16209.
20. WILDE, M. W., K. E. CARLSON, D. R. MANNING & S. H. ZIGMOND. 1989. J. Biol. Chem. **264:** 190-196.
21. OKAJIMA, F., T. KATADA & M. UI. 1985. J. Biol. Chem. **260:** 6761-6768.
22. BOCCKINO, S. B., P. F. BLACKMORE, P. B. WILSON & J. H. EXTON. 1987. J. Biol. Chem. **262:** 15309-15315.
23. IRVING, H. R. & J. H. EXTON. 1987. J. Biol. Chem. **262:** 3440-3443.
24. BOYER, J. L., G. L. WALDO, T. EVANS, J. K. NORTHRUP, C. P. DOWNES & T. K. HARDEN. 1989. J. Biol. Chem. **264:** 13917-13922.
25. MURPHY, P. M., B. EIDE, P. GOLDSMITH, M. BRANN, P. GIERSCHIK, A. SPIEGEL & H. L. MALECH. 1987. FEBS Lett. **221:** 81-86.

DISCUSSION OF THE PAPER

F. DI VIRGILIO (*Institute of General Pathology, Padua, Italy*): 1) It has been shown that ATP does not release Ca^{2+} from stores in mouse lymphocytes; however, it triggers Ca^{2+} influx across the plasma membrane. Do you have evidence that ATP, maybe at high concentrations, causes Ca^{2+} influx in human lymphocytes?

2) Does ATP cause plasma membrane depolarization in human neutrophils?

DUBYAK: 1) Certain types of human lymphocytes do indeed show enhanced Ca^{2+} influx in response to millimolar concentrations of extracellular ATP. Jim Wiley and I have recently reported (*Blood* **73:** 1316-1323, 1989) that ATP induces an increased permeability to both calcium and monovalent cations in chronic lymphocytic leukemic (CLL) lymphocytes (which are predominantly B cells in most cases). In contrast, we noted little ATP-induced ion influx in normal peripheral blood lymphocytes (which are predominantly T cells). It should be stressed that the observed response in CCL lymphocytes varied considerably between different patients. This suggests that only a particular subtype of B lymphocyte may express this responsiveness to ATP.

2) We have not tested whether ATP induces depolarization of human neutrophils. My speculation would be that ATP would induce depolarization in a manner analogous to that triggered by other Ca^{2+}-mobilizing agonists in these cells.

I. FRIEDBERG (*Tel Aviv University, Tel Aviv, Israel*): Do you have information about additional transmembrane signaling systems that are stimulated by extracellular ATP? If so, there "cross-talk" among these systems?

DUBYAK: Dr. Exton and his colleagues have demonstrated that ATP is a particularly efficacious activator of phosphatidylcholine-specific phospholipase C and phospholipase D in hepatocytes. The large amounts of phosphatidic acid and diacylglycerol generated from phosphatidylcholine undoubtedly can have marked effects on a variety of protein kinases, including protein kinase C. Activation of PKC can significantly attenuate the activation of inositol phospholipid-specific phospholipase C(s) by ATP and most other Ca^{2+}-mobilizing agonists. Thus, there would substantial "cross-talk" at this level. Ming-Sheng Xie, a graduate student in my lab, has similarly demonstrated that ATP (and UTP) can activate phospholipase D in differentiated HL-60 cells but, interestingly, not in undifferentiated HL-60 cells. Moreover, even though ATP and formlylated peptides (such as FMLP) are equally efficacious in activating the inositol phospholipid-specific phospholipase C in the differentiated cells, ATP is considerably less efficacious than FMLP in activating phospholipase D. This is intriguing given that that ATP cannot per se activate superoxide generation but can potentiate FMLP-induced superoxide release in these cells and in "real" neutrophils (*cf.* Ward *et al.*, this volume).

J. S. DAVIDSON (*University of Cape Town Medical School, Cape Town, South Africa*): Perhaps you might consider an alternative explanation for the difference you have observed between intact cells and the membrane preparations: Is it possible that part of the toxin-sensitive phospholipase C activity might be due to cytosolic phospholipase, which would have been lost in the membrane preparation?

DUBYAK: This is a very good question, and one that Dan Cowen and I have often contemplated. We know that the maximal FMLP + GTP-induced inositol polyphosphate release from membranes is less than the FMLP-induced inositol polyphosphate accumulation observed in intact cells or the ATP + GTP-induced inositol phosphate release from the same membranes. These membranes have lots of functional FMLP receptors that change their affinity state in the presence of GTP; that is, FMLP receptor/G protein interaction appears intact. The relatively low rates of PLC activation observed in membranes suggest that, in intact cells, FMLP may indeed be predominantly activating a PI-PLC isoform that is lost during membrane preparation. Conversely, ATP receptors may be equally capable of activating either the putative soluble PI-PLC or the membrane-associated PI-PLC.

Effects of Extracellular ATP on Phosphatidylcholine Phospholipase Signaling Systems

JOHN H. EXTON [a]

Howard Hughes Medical Institute
and
Vanderbilt University School of Medicine
Nashville, Tennessee 37232

INTRODUCTION

It is now well recognized that extracellular ATP can stimulate the breakdown of inositol phospholipids in certain mammalian cells and that the resulting increase in inositol 1,4,5-trisphosphate (IP_3) is responsible in part for the elevation of cytosolic Ca^{2+}.[1-7] The purinergic receptor that mediates these effects has been identified as the P_{2Y} type.[8,9] Studies with isolated plasma membranes have provided evidence that ATP and other adenine nucleotides that interact with the P_{2Y}-purinergic receptor stimulate the hydrolysis of endogenous phosphatidylinositol 4,5-bisphosphate (PIP_2) to IP_3, and that the response is dependent upon GTP and its analogues.[8,10] The characteristics of the guanine nucleotide effects on the activation of PIP_2 phospholipase C and on the binding of [^{35}S]ADP-β-S indicate the involvement of a G protein.[8-10]

Recent evidence has shown that agents that interact with P_{2Y}-purinergic receptors also stimulate the breakdown of phosphatidylcholine (PC) in cells.[11-14] As in the case of PIP_2 hydrolysis, the response appears to be mediated by a G protein, but both direct and indirect mechanisms are probably involved. In this report, the effects of ATP on PC breakdown will be described with reference to the enzymes and mechanisms involved and the possible physiological significance of agonist-induced PC breakdown.

EVIDENCE FOR ATP-INDUCED PC BREAKDOWN IN RAT HEPATOCYTES

The first clue that ATP and other Ca^{2+}-mobilizing agonists were causing the breakdown of another phospholipid besides PIP_2 in cells came from measurements of the amount of diacylglycerol (DAG) that accumulated when rat hepatocytes were

[a] Address for correspondence: Howard Hughes Medical Institute, 831 Light Hall, Vanderbilt University, Nashville, Tennessee 37232.

stimulated with ATP.[15] The amount of DAG that accumulated was found to be very much greater than the amount of IP_3 that accumulated.[1-16] Chemical measurements of the PIP_2 disappearing in hepatocytes stimulated with Ca^{2+}-mobilizing agonists, and of the DAG and phosphatidic acid (PA) accumulating,[12,15,17] confirmed the discrepancy.

When the molecular species of DAG formed in hepatocytes in response to ATP and other Ca^{2+}-mobilizing agonists were analyzed by high-performance liquid chromatography, the fatty acid compositions of the major DAG species were very different from those one would expect if one assumed inositol phospholipid to be the sole source (TABLE 1).[14] Comparison of the profile of DAG species with the profiles of the phospholipids of rat liver[18] strongly indicated that PC was a major source. This was confirmed when the molecular species of the phospholipids of isolated rat hepatocytes were determined.[14]

More direct evidence that ATP elicited the breakdown of PC in hepatocytes was obtained when the cells were incubated with 1-*O*-[^{3}H]alkyl-2-lyso-*sn*-glycero-3-phosphocholine[19] or were prepared from rats injected with this labeled compound 20 hr previously. This compound becomes incorporated into the PC fraction relatively selectively.[14] When cells labeled *in vitro* or *in vivo* were incubated with ATP, there was about a 2-fold stimulation of 1-*O*-[^{3}H]alkyl-2-acylglycerol production (TABLE 2)—a result consistent with the stimulation of a phospholipase C active on PC. In cells labeled *in vitro,* however, there was negligible stimulation of 1-*O*-[^{3}H]alkyl-PA formation in response to Ca^{2+}-mobilizing agonists. In contrast, in cells obtained from animals injected with the label 20 hr previously, there was about a 2-fold increase in [^{3}H]alkyl-PA (TABLE 2). Similar data were obtained when the cells were labeled with [^{14}C]lyso-PC.[14] These data indicate that both rapidly and slowly labeled PC pools are substrates for an agonist-activated phospholipase C, whereas only the slowly labeled pool is a substrate for a stimulated phospholipase D. The activation of a PC phospholipase D explains the rapid formation of PA in hepatocytes stimulated by Ca^{2+}-mobilizing agonists.[12]

TABLE 1. Molecular Species of DAG Accumulating in Hepatocytes Stimulated with ATP[a]

Molecular Species of DAG	Fold Increase Induced by ATP
$C_{16:0}/C_{20:4}$	4.5 ± 0.4
$C_{16:0}/C_{18:2}$	2.8 ± 0.3
$C_{18:0}/C_{20:4}$	4.5 ± 0.7
$C_{16:0}/C_{18:1}$, $C_{18:0}/C_{18:1}$	3.3 ± 0.3

[a] After hepatocytes were incubated for 10 min with saline (control) or 1 mM ATP, the molecular species of DAG were analyzed.[14] Each value is a mean ± S.E. from three experiments. Data are from reference 14.

ATP STIMULATION OF PC HYDROLYSIS IN LIVER PLASMA MEMBRANES

Further evidence that ATP and other P_{2Y}-purinergic agonists stimulate PC breakdown by phospholipase C and phospholipase D comes from experiments with isolated

TABLE 2. ATP Stimulation of the Formation of 1-*O*-[^{3}H]Alkyl-2-acyl-glycerol and 1-*O*-[^{3}H]Alkyl-PA in Hepatocytes Labeled with 1-*O*-[^{3}H]Alkyl-2-lyso-glycero-3-phosphocholine *in Vitro* and *in Vivo*[a]

	Fold Changes over Control			
	Hepatocytes Labeled *in Vitro*		Hepatocytes Labeled *in Vivo*	
	[^{3}H]AAG[b]	[^{3}H]APA[c]	[^{3}H]AAG[b]	[^{3}H]APA[c]
ATP	1.7 ± 0.1	1.2 ± 0.1		
Vasopressin	1.8 ± 0.1	1.1 ± 0.04	2.1 ± 0.1	1.7 ± 0.2

[a] Hepatocytes were either incubated for 30 min with 1-*O*-[^{3}H]alkyl-2-lyso-glycero-3-phosphocholine or prepared from rats injected intraperitoneally with this compound 20 hr previously. The hepatocytes were incubated for 7-10 min with saline (control), 1 mM ATP, or 100 nM vasopressin. Data are from reference 14.

[b] [^{3}H]AAG: 1-*O*-[^{3}H]alkyl-2-acyl-glycerol.

[c] [^{3}H]APA: 1-*O*-[^{3}H]alkyl-PA.

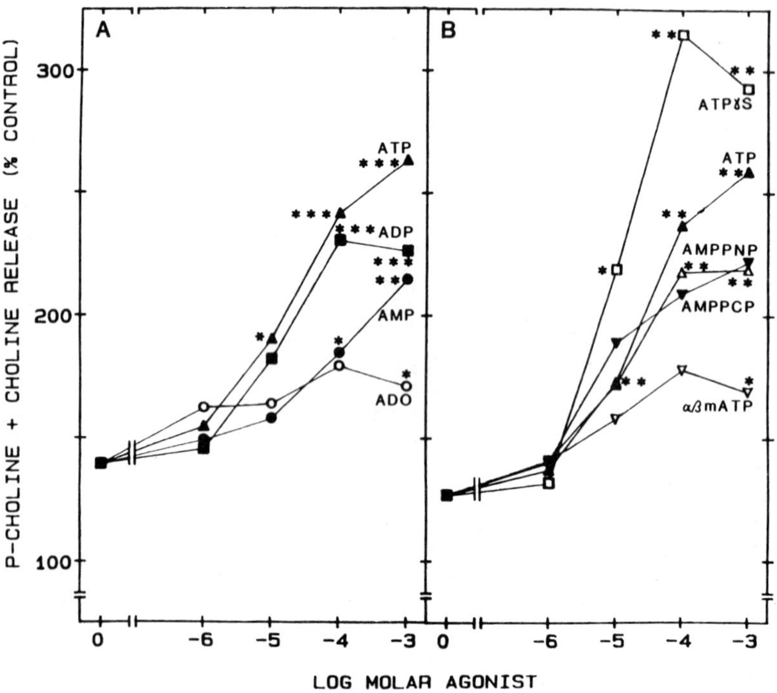

FIGURE 1. Effects of purinergic agonists on the release of phosphocholine and choline from rat liver plasma membranes. Plasma membranes were prepared from rats injected with [*methyl*-^{3}H]choline 18 hr previously, and incubated in the presence of 0.3 μM GTP-γ-S and the indicated concentrations of agonists for 5 min at 30 °C. [^{3}H]Phosphocholine plus choline were assayed as described in reference 11. (**A**) ○: adenosine (ADO); ●: AMP; ■: ADP; ▲: ATP. (**B**) ▲: ATP; □: ATP-γ-S; △: AMP-PNP; ▼: AMP-PCP; ▽: αβ-methylene-ATP.

rat liver plasma membranes.[11,12] In membranes prepared from rats previously injected with [^{3}H]choline, the addition of micromolar or lower concentrations of GTP analogues caused the breakdown of [^{3}H]PC to labeled phosphocholine and choline,[11] consistent with the activation of PC phospholipases C and D mediated by a G protein. When ATP, ADP, and certain ATP analogues were added in the presence of a low concentration (0.3 μM) of GTP-γ-S, there was a further stimulation of PC hydrolysis (FIG. 1). This stimulation was not observed with adenosine or AMP, except at very high concentrations (0.1 and 1 mM). The stimulation was blocked by P_2-purinergic antagonists such as α,β-methylene-ATP and 2,2′-pyridylisatogen (FIG. 1),[11] indicating the involvement of these receptors.

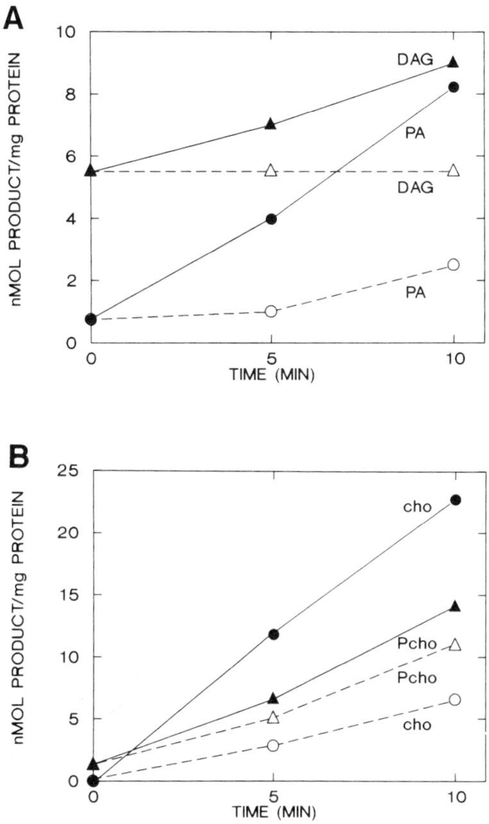

FIGURE 2. Effect of GTP-γ-S on the production of (**A**) diacylglycerol (DAG) and phosphatidic acid (PA) and on the production of (**B**) choline (cho) and phosphocholine (Pcho) by rat liver plasma membranes. Plasma membranes were incubated with (●——●, ▲——▲) or without (○----○, △----△) GTP-γ-S (20 μM) and assayed for DAG, PA, choline, and phosphocholine as described in reference 12.

In further experiments, the effects of GTP-γ-S on the release of choline and phosphocholine from rat liver plasma membranes were examined.[12] As shown in FIGURE 2, the nucleotide increased the release of choline about 4-fold, but caused only a small increase ($\sim$40%) in phosphocholine production. These data suggested that stimulation of the G protein(s) resulted mainly in the activation of PC phospholipase D. Because ATP was absent from the incubation medium, the enhanced formation of phosphocholine indicates that there was some stimulation of PC phospholipase C.

In agreement with the conclusion that GTP-γ-S activated both phospholipase D and phospholipase C in these experiments, there was increased formation of both PA and DAG (FIG. 2). The idea that PC was the major source of these lipids was supported by the finding that PC was the only liver plasma membrane phospholipid that decreased significantly in response to GTP-γ-S.[12] The proposal that P_2-purinergic receptors could be coupled to a G protein that controls PC phospholipase D was supported by the observation that addition of ATP and ADP enhanced the stimulatory effect of 0.3 μM GTP-γ-S on phosphatidate formation in isolated liver plasma membranes.[12]

The involvement of phospholipase D in the actions of hormones and other agonists on cells can also be explored utilizing the fact that the enzyme catalyzes a transphosphatidylation reaction in which primary alcohols substitute for water as the phosphatidyl acceptor.[20] As shown in TABLE 3, when hepatocytes were incubated with Ca^{2+}-mobilizing agonists in the presence of ethanol, the formation of phosphatidylethanol was enhanced, consistent with the activation of phospholipase D by these agonists. Stimulation of the formation of phosphatidylethanol was also shown when isolated liver plasma membranes were incubated with GTP-γ-S plus ethanol.[20]

DISCUSSION

There is now widespread evidence that Ca^{2+}-mobilizing agonists stimulate PC breakdown in a variety of tissues and cell lines (TABLE 4).[21] As a member of this group of agonists, extracellular ATP has been shown to elicit PC breakdown in hepatocytes[11,12,14] and endothelial cells[13] acting through the P_{2Y}-purinergic receptor. The mechanisms by which P_{2Y}-purinergic receptor activation causes PC hydrolysis may be manifold. As shown by our findings[11,12,14] and those of another laboratory,[13] there is evidence that the receptors activate a G protein that regulates plasma membrane PC phospholipases D and C. Whether the same G protein controls both of these phospholipases and also the PIP_2 phospholipase C in a given cell type remains unknown.

Findings with other agonists and other cell types indicate that PC hydrolysis can be controlled by other mechanisms besides direct interaction of the G protein with the phospholipases.[21] For example, Ca^{2+} depletion in hepatocytes leads to a partial inhibition of the stimulation of PC hydrolysis to PA and DAG by vasopressin,[12,14] and addition of the Ca^{2+} ionophore A23187 alone can stimulate PC breakdown and

TABLE 3. Effect of Ca^{2+}-Mobilizing Agonists on Phosphatidylethanol Accumulation in Rat Hepatocytes[a]

Additions	Phosphatidylethanol (pmol/mg)
None	5.8 ± 2.8
Vasopressin (100 nM)	50.4 ± 3.1
Angiotensin II (100 nM)	33.1 ± 4.0
Epinephrine (10 μM)	24.2 ± 2.3

TABLE 4. Cells or tissues showing PC breakdown in response to hormones, growth factors, phorbol esters, and other stimuli.

Cell or Tissue	Stimuli
Hepatocyte	Vasopressin, angiotensin II, adrenergic agonists (α_1), P_2-purinergic agonists, phorbol ester
Hepatoma (HepG2)	Serum factor(s), phorbol ester
Kidney cell (MDCK-D1)	Adrenergic agonists (α_1), phorbol ester
Promyelocyte (HL-60)	f-Met-Leu-Phe, phorbol ester, DAG
Neuroblastoma (N4TG1, NG108-15)	Serum factor(s), phorbol ester
Pheochromocytoma (PC-12)	Serum factor(s), phorbol ester
Pre-adipocyte (3T3-L1)	PDGF, phorbol ester
Fibroblast (Swiss 3T3, 3T3-A31, 11C9)	Bombesin, phorbol ester, thrombin, EGF
Aortic smooth muscle cell (A10)	Vasopressin
Rat embryo cell (REF52)	Vasopressin
T lymphocyte (Jurkat)	Interleukin-1
Neutrophil	f-Met-Leu-Phe, phorbol ester
Endothelial cell	Thrombin, bradykinin, P_2-purinergic agonists
Heart	Muscarinic agonists
Brain	Muscarinic agonists
Pancreas	Muscarinic agonists, bombesin, cholecystokinin
Platelet	Thrombin
Myoblast	Phorbol ester
HeLa cell	Phorbol ester
Uterine decidua cell	Phorbol ester
Pituitary cell (GH3)	DAG
Mast cell	Antigen
Ovarian granulosa cell	Gonadotropin-releasing hormone
Retina	Light
Electroplax	Acetylcholine, electrical stimulation
Spermatozoon (Sea urchin)	Egg jelly (fucose-sulfate glycoconjugate)
Hemopoietic stem cell (FDCP-Mix1)	Interleukin-3
Macrophage	Platelet-activating factor, interferon-γ, phorbol ester
Glioma (C_6)	Phorbol ester
Astrocytoma (1321N1)	Muscarinic agonists, phorbol ester

DAG and PA accumulation.[12,14] These findings suggest that part of the PC hydrolysis is attributable to the increase in cytosolic Ca^{2+} induced by ATP and other Ca^{2+}-mobilizing agonists in cells. Presumably, Ca^{2+} could act on the PC phospholipases either directly or through the mediation of a Ca^{2+}-binding protein or a Ca^{2+}-stimulated protein kinase.

Another mechanism that appears to operate in many cells is activation of the PC phospholipases directly or indirectly by protein kinase C. Thus the addition of synthetic forms of DAG or active phorbol esters promotes PC hydrolysis in many tissues (TABLE 4). One hypothesis explaining the role of protein kinase C is that the activation of PIP_2 hydrolysis by Ca^{2+}-mobilizing agonists generates a small ("trigger") amount of DAG that activates protein kinase C initially. Activation of the kinase, in turn, leads to the activation of PC phospholipases D and C.[21] Because activation of these phospholipases generates DAG either directly through phospholipase C action or through a coupled phospholipase D/PA phosphohydrolase reaction, the scheme would result in a positive feedback on protein kinase C leading to continuous activation of the two phospholipases, and hence in sustained production of PA and DAG.

Although not relevant to the P_{2Y}-purinergic system, there are several reports showing that PC hydrolysis can be induced by growth factors.[21] The mechanisms involved are not known, but could involve the phosphorylation of PC phospholipases or regulatory proteins, by growth factor receptor tyrosine kinases.

The physiological significance of agonist-induced PC breakdown remains unclear, principally because the phenomenon is of relatively recent discovery. Observations in several cell types, however, indicate that PC is the major source of the DAG that accumulates in stimulated cells, especially after the first few minutes of agonist action.[14,15,17,22–26] Because the molecular species of DAG formed from PC are capable of activating protein kinase C,[27] PC hydrolysis could play a major role in the regulation of this enzyme.

An interesting point relates to the fact that PC is the major phospholipid of all mammalian cells and is present at a cellular concentration several hundred-fold higher than PIP_2.[17,18] This means that it could provide a larger amount of DAG for a longer period of time than could PIP_2. Although both phospholipids can be regenerated by resynthesis, the cellular rates are unknown. Another point of interest is that PIP_2 appears to be localized to the plasma membrane,[17] whereas PC is present in all cell membranes.[28] Thus part of the hydrolysis of PC induced by agonists could be occurring in membranes other than the plasma membrane, leading to interesting possible mechanisms of cellular control.

There are some indications that PC hydrolysis is involved in long-term cellular regulation. For example, in fibroblasts stimulated with epidermal growth factor, PC breakdown with attendant elevation of DAG can persist for up to 4 hr.[26] In other situations, the elevation of DAG or PA or the breakdown of PC has been observed to proceed for 1 hr or more.[23,25,29–33] In all these situations, the experiments could have been continued longer. In summary, these observations suggest that PC hydrolysis may be important for cellular changes involving long-term activation of protein kinase C or sustained elevation of DAG or PA.

In addition to activating protein kinase C, PC-derived DAG may act as a source of arachidonic acid for eicosanoid production. In contrast to the known cellular actions of DAG, the physiological functions of PA remain obscure. Although several roles have been ascribed to this lipid,[21] none of these seems entirely convincing. In many cells, PA acts as an important source of DAG through the action of PA phosphohydrolase,[21] but it is unclear why DAG should be produced from PC by a two-step reaction involving phospholipase D and PA phosphohydrolase rather than directly by phospholipase C. Perhaps DAG produced via PA arises at different cellular sites and has different functions compared with DAG produced directly. In any event, PA accumulation is very rapid and substantial in most cells stimulated by Ca^{2+}-mobilizing agonists, and it seems reasonable to propose that PA has some significant signaling function(s). It is also possible that the other products of PC hydrolysis, choline and phosphocholine, act as cellular signals in some cells. To date there have been few studies of the cellular changes in these compounds, and no physiological roles have been suggested, apart from choline acting as a substrate for acetylcholine synthesis.

In summary, PC hydrolysis induced by certain hormones, neurotransmitters, and growth factors has now been shown to be a widespread phenomenon in mammalian tissues. Specific functions for PC hydrolysis in cellular regulation remain elusive. It seems likely, however, that PC hydrolysis plays an important role in certain responses to these agonists, especially those occurring on a longer time scale.

REFERENCES

1. CHAREST, R., P. F. BLACKMORE & J. H. EXTON. 1985. Characterization of responses of isolated rat hepatocytes to ATP and ADP. J. Biol. Chem. **260:** 15789-15794.
2. DUBYAK, G. R. 1986. Extracellular ATP activates polyphosphoinositide breakdown and Ca^{2+} mobilization in Ehrlich ascites tumor cells. Arch. Biochem. Biophys. **245:** 84-95.
3. PHANEUF, S., B. PHILIPPE & J. CASANOVA. 1987. ATP stimulates inositol phosphate accumulation and calcium mobilization in a primary culture of rat aortic myocytes. Biochem. Biophys. Res. Commun. **143:** 454-460.
4. PIROTTON, S., E. RASPE, D. DEMOLLE, C. ERNEUX & J.-M. BOEYNAEMS. 1987. Involvement of inositol 1,4,5-trisphosphate and calcium in the action of adenine nucleotides on aortic endothelial cells. J. Biol. Chem. **262:** 17461-17466.
5. OKAJIMA, F., Y. TOKUMITSU, Y. KONDO & M. UI. 1987. P_2-Purinergic receptors are coupled to two signal transduction systems leading to inhibition of cAMP generation and to production of inositol trisphosphate in rat hepatocytes. J. Biol. Chem. **262:** 13483-13490.
6. DUBYAK, G. R., D. S. COWEN & L. M. MEULLER. 1988. Activation of inositol phospholipid breakdown in HL-60 cells by P_2-purinergic receptors for extracellular ATP: Evidence for mediation by both pertussis toxin-sensitive and pertussis toxin-insensitive mechanisms. J. Biol. Chem. **263:** 18108-18117.
7. SASAKAWA, N., T. NAKAKI, S. YAMAMOTO & R. KATO. 1989. Stimulation by ATP of inositol trisphosphate accumulation and calcium mobilization in cultured adrenal chromaffin cells. J. Neurochem. **52:** 441-447.
8. BOYER, J. L., C. P. DOWNES & T. K. HARDEN. 1989. Kinetics of activation of phospholipase C by P_{2Y}-purinergic receptor agonists and guanine nucleotides. J. Biol. Chem. **264:** 884-890.
9. COOPER, C. L., A. J. MORRIS & T. K. HARDEN. 1989. Guanine nucleotide-sensitive interaction of a radiolabeled agonist with a phospholipase C-linked P_{2Y}-purinergic receptor. J. Biol. Chem. **264:** 6206-6206.
10. OKAJIMA, F., K. SATO & Y. KONDO. 1989. P_2-Purinergic agonists activate phospholipase C in a guanine nucleotide- and Ca^{2+}-dependent manner in FRTL-5 thyroid cell membranes. FEBS Lett. **253:** 132-136.
11. IRVING, H. & J. H. EXTON. 1987. Phosphatidylcholine breakdown in rat liver plasma membranes: Roles of guanine nucleotides and P_2-purinergic agonists. J. Biol. Chem. **262:** 3440-3443.
12. BOCCKINO, S. B., P. F. BLACKMORE, P. B. WILSON & J. H. EXTON. 1987. Phosphatidate accumulation in hormone-treated hepatocytes via a phospholipase D mechanism. J. Biol. Chem. **262:** 15309-15315.
13. MARTIN, T. W. & K. MICHAELIS. 1989. P_2-Purinergic agonists stimulate phophodiesteratic cleavage of phosphatidylcholine in endothelial cells: Evidence for activation of phospholipase D. J. Biol. Chem. **264:** 8847-8856.
14. AUGERT, G., S. B. BOCCKINO, P. F. BLACKMORE & J. H. EXTON. 1989. Hormonal stimulation of diacylglycerol formation in hepatocytes: Evidence for phosphatidylcholine breakdown. J. Biol. Chem. **264:** 21689-21698.
15. BOCCKINO, S. B., P. F. BLACKMORE & J. H. EXTON. 1985. Stimulation of 1,2-diacylglycerol accumulation in hepatocytes by vasopressin, epinephrine and angiotensin II. J. Biol. Chem. **260:** 14201-14207.
16. CHAREST, R., V. PRPIC, J. H. EXTON & P. F. BLACKMORE. 1985. Stimulation of myoinositol-1,4,5-triphosphate formation in hepatocytes by vasopressin and epinephrine and its relationship to changes in free cytosolic Ca^{2+}. Biochem. J. **227:** 79-90.
17. AUGERT, G., P. F. BLACKMORE & J. H. EXTON. 1989. Changes in the concentration and fatty acid composition of phosphoinositides induced by hormones in hepatocytes. J. Biol. Chem. **264:** 2574-2580.
18. PATTON, G. M., J. M. FASULO & S. J. ROBINS. 1982. Separation of phospholipids and individual molecular species of phospholipids by high-performance liquid chromatography. J. Lipid Res. **23:** 190-196.

19. Daniel, L. W., M. Waite & R. L. Wykle. 1986. A novel mechanism for diglyceride formation: 12-*O*-Tetradecanoylphorbol-13-acetate stimulates the cyclic breakdown and resynthesis of phosphatidylcholine. J. Biol. Chem. **261:** 9128-9132.
20. Bocckino, S. B., P. B. Wilson & J. H. Exton. 1987. Ca^{2+}-mobilizing hormones elicit phosphatidylethanol accumulation via phospholipase D activation. FEBS Lett. **225:** 201-204.
21. Exton, J. H. 1989. Signaling through phosphatidylcholine breakdown (minireview). J. Biol. Chem. **265:** 1-4.
22. Wright, T. M., L. A. Rangan, H. S. Shin & D. M. Raben. 1988. Kinetic analysis of 1,2-diacylglycerol mass levels in cultured fibroblasts: Comparison of stimulation of α-thrombin and epidermal growth factor. J. Biol. Chem. **263:** 9374-9380.
23. Pessin, M. S. & D. M. Raben. 1989. Molecular species analysis of 1,2-diglycerides stimulated by α-thrombin in cultured fibroblasts. J. Biol. Chem. **264:** 8729-8738.
24. Martinson, E. A., D. Goldstein & J. H. Brown. 1989. Muscarinic receptor activation of phosphatidylcholine hydrolysis: Relationship to phosphoinositide hydrolysis and diacylglycerol metabolism. J. Biol. Chem. **264:** 14748-14754.
25. Matozaki, T. & J. A. Williams. 1989. Multiple sources of 1,2-diacylglycerol in isolated rat pancreatic acini stimulated by cholecystokinin: Involvement of phosphatidylinositol bisphosphate and phosphatidylcholine hydrolysis. J. Biol. Chem. **264:** 14729-14734.
26. Wright, T. M., H. S. Shin & D. M. Raben. 1989. Sustained elevation of 1,2-diacylglycerol derived from phosphatidylcholine precedes DNA synthesis in epidermal growth factor-stimulated fibroblasts. J. Biol. Chem.: in press.
27. Go, M., K. Sekiguchi, H. Nomura, U. Kikkawa & Y. Nishizuka. 1987. Further studies on the specificity of diacylglycerol for protein kinase C activation. Biochem. Biophys. Res. Commun. **144:** 598-605.
28. Colbeau, J., J. Nachbaur & P. M. Vignais. 1971. Enzymic characterization and lipid composition of rat liver subcellular membranes. Biochim. Biophys. Acta **249:** 462-492.
29. Pandol, S. J. & M. S. Schoeffield. 1986. 1,2-Diacylglycerol, protein kinase C, and pancreatic enzyme secretion. J. Biol. Chem. **261:** 4438-4444.
30. Kester, M., M. S. Simonson, P. Mene & J. R. Sedor. 1989. Interleukin-1 generates transmembrane signals from phospholipids through novel pathways in cultured rat mesangial cells. J. Clin. Invest. **83:** 718-723.
31. Billah, M. M., J.-K. Pai, T. J. Mullmann, R. W. Egan & M. I. Siegel. 1989. Regulation of phospholipase D in HL-60 granulocytes: Activation by phorbol esters, diglyceride, and calcium ionophore via protein kinase C-independent mechanisms. J. Biol. Chem. **264:** 9069-9076.
32. Dougherty, R. W., G. R. Dubay & J. E. Niedel. 1989. Dynamics of the diacylglycerol responses of stimulated phagocytes. J. Biol. Chem. **264:** 11263-11269.
33. Anthes, J. C., S. Eckel, M. I. Siegel, R. W. Egan & M. M. Billah. 1989. Phospholipase D in homogenates from HL-60 granulocytes: Implications of calcium and G protein control. Biochem. Biophys. Res. Commun. **163:** 657-664.

DISCUSSION OF THE PAPER

A. S. Dahms (*San Diego State University, San Diego, CA*): Presumably there must be a regulated phosphatidic acid phosphatase that fits into this scheme.

Exton: Yes.

Dahms: Your observation that DAG may have alternate actions is important. Certainly there have been indications of such in the literature, but these are not well documented. I would like to call your attention to a poster this afternoon that bears upon a putative ecto-ATPase from skeletal muscle. The data show potent DAG inhibition to the ATPase, as well as inhibition by phorbol esters, including the 4-α-phorbols, which are ineffective toward protein kinase C. Additionally, the ATPase is activated by sphingosine. The data are consistent with a complete shutdown of the ecto-ATPase under conditions where protein kinase C is fully active.

EXTON: Yes, I am certain that there are undiscovered regulatory actions of DAG.

I. FRIEDBERG (*Tel Aviv University, Tel Aviv, Israel*): Did you find conditions under which extracellular ATP induces Ca^{2+} mobilization without activation of protein kinase C?

EXTON: The question has not been explored at length, that is, with different concentrations of ATP tested at different times. This is a very important point, that is, to see if Ca^{2+} mobilization can be induced without protein kinase C activation or vice versa. We are currently conducting experiments to test both possibilities.

D. SATCHELL (*University of Melbourne, Melbourne, Victoria, Australia*): Choline acts on cholinergic receptors at approximately one thousandth of the concentration of acetylcholine. Do you consider that the choline formed in your system could be acting on cholinergic receptors and having an action in this way?

EXTON: We have not measured the levels of extracellular choline that are achieved in our experiments. We do not, however, observe effects of added acetylcholine or carbamycholine in our experiments, so I do not feel that this is likely in our system.

J. S. WILEY (*Austin Hospital, Heidelberg, Australia*): Are there good data that relate increased levels of cell membrane diacylglycerol and/or phosphatidic acid to uncontrolled cell growth or oncogenesis?

EXTON: This is a complex question, and the answer is equally complex. First, many growth factors (EGF, PDGF, IL-1, IL-3) increase diacylglycerol, in part through phosphatidylcholine hydrolysis, but it is unclear what role this plays in growth stimulation. Second, addition of phosphatidic acid has been shown to induce mitogenesis in certain cells. Third, transformation of certain cells by the Ha-*ras* oncogene or Kirtsen sarcoma virus is associated with an increase in diacylglycerol that is not due to phosphoinositide breakdown. Again, it is not known what role this increase plays in phenotypic transformation.

T. D. WHITE (*Dalhousie University, Halifax, Nova Scotia, Canada*): Stimulation of sympathetic nerves stimulates adenylate cyclase-independent glycogenolysis in the liver. Also, ATP stimulates adenylate cyclase-independent glycogenolysis in hepatocytes. Do you think that this signal transduction is mediated via inositol phosphate sources or from phosphatidylcholine hydrolysis?

EXTON: The stimulation of glycogenolysis in hepatocytes by ATP is entirely attributable to the stimulation of phosphatidylinositol bisphosphate hydrolysis to yield inositol trisphosphate which mobilizes intracellular calcium. There is no evidence that phosphatidylcholine hydrolysis plays a role.

Biochemical Properties of a P_{2Y}-Purinergic Receptor

T. KENDALL HARDEN, JOSÉ LUIS BOYER, H. ALEXANDER BROWN, CHRISTY L. COOPER, ROGER A. JEFFS, AND MICHAEL W. MARTIN

Department of Pharmacology
University of North Carolina School of Medicine
Chapel Hill, North Carolina 27599

INTRODUCTION

Considerable progress has been made in the last decade in the delineation of the molecular mechanisms of signal transduction associated with cell surface receptors for a variety of hormones, neurotransmitters, and other receptor-directed signaling molecules. Association of most receptors with specific ion channel activities or second messenger pathways has been established, and in the case of a rapidly increasing set of receptors of various classes, the gene for the receptor has been cloned and a nascent understanding of the important structural features of the receptor protein has been achieved.[1-3]

As has been the case for the receptors for most extracellular signaling molecules, the cell surface receptors for adenosine (P_1-purinergic receptors) and ATP and ADP (P_2-purinergic receptors) were initially characterized by classical pharmacological studies using intact tissue preparations *in vitro*.[4-6] An amalgamation of biochemical and pharmacological studies quickly led to the subclassification of P_1-purinergic receptors into A_1-receptors and A_2-receptors.[6] The A_1-receptors inhibit adenylate cyclase and the A_2-receptors activate adenylate cyclase, although other signaling mechanisms may be responsible for the effects of adenosine in some target tissues. Both A_1- and A_2-receptors have been directly characterized with a cadre of radioligands, and inroads into purification of the receptor proteins are being made.[7-9] Much less is known about the biochemical mechanisms associated with P_2-purinergic receptor action. Burnstock and Kennedy[5] proposed in 1985 that P_2-purinergic receptors could be subclassified into P_{2X}- and P_{2Y}-purinergic receptors, and recent studies suggest that at least one additional P_2-purinergic receptor exists.[10,11] The P_{2Y}-purinergic receptor and probably a third type of P_2-purinergic receptor have been shown to stimulate inositol phosphate accumulation and elevation of cytoplasmic Ca^{2+} levels in a variety of target cells,[10-14] but the details of this potential P_2-purinergic receptor-mediated signaling mechanism have not been delineated. There have been reports that suggest, but are far from proving, that P_{2X}-purinergic receptors may have inhibitory effects on adenylate cyclase.[15] A summary of the pharmacological classification of P_2-purinergic receptor subtypes and their possible association with second messenger signaling systems is presented in TABLE 1.

TABLE 1. Classification of P_2-Purinergic Receptors[a]

Subtype	Agonist	Pharmacological Selectivity	Mechanism	Physiological Response
P_{2X}	ATP/ADP	α,β-Me-ATP = β,γ-Me-ATP > ADP > ATP > 2-Me-S-ATP	Ion Channel; (↓ cAMP ?)	Contraction
P_{2Y}	ATP/ADP	2-Me-S-ATP > ATP > ADP > α,β-Me-ATP > β,γ-Me-ATP	IP_3/DAG/Ca^{2+}	Relaxation; PGI_2 Release
P_{2Z}	ATP^{4-}	ATP^{4-} > ATP	Membrane Permeabilization	Degranulation of Mast Cells
P_{2T}	ADP	2-Me-S-ADP > ADP ATP, AMP = Antagonists	?	Aggregation, Secretion
$P_{2...}$	ATP/UTP	ATP = UTP > ADP	IP_3/DAG/Ca^{2+}	

[a] Listed are the proposed purinergic receptor subtypes; the natural agonists for these receptors; the pharmacological selectivity of the receptors in regard to ATP, ADP, and their analogues; and some proposed mechanisms of signaling associated with each putative receptor. These mechanisms, which are speculative in some cases, are almost certainly not all inclusive. The P_{2Z}-purinergic receptor refers to a phenomenon involving nonspecific permeabilization of cells in response to the ATP^{4-} form of ATP. It has not been unambiguously determined to what extent Mg-ATP versus ATP^{4-} is responsible for the effects of ATP at other purinergic receptor types. The P_{2T}-purinergic receptor has been most often described for platelet cells where its activation has been shown to result in aggregation and secretion. The $P_{2...}$ purinergic receptor refers to an unnamed putative receptor that has been observed in a variety of target tissues, including HL-60 cells, lung epithelial cells, and human fibroblasts. This receptor is effectively activated by ATP and UTP, but is not effectively activated by ADP and ADP analogues. It has been identified mainly on the basis of its capacity to stimulate inositol phosphate accumulation or to increase cytoplasmic Ca^{2+} levels. More details on this putative receptor are provided in this volume in the paper by Dubyak and Cowen.[37] DAG: diacylglycerol; IP_3: inositol trisphosphate.

Turkey erythrocytes express P_{2Y}-purinergic receptors that markedly stimulate phospholipase C (FIG. 1). This can be observed as an increase in inositol phosphate accumulation in intact erythrocytes[16] or as a guanine nucleotide-dependent stimulation of enzyme activity measured in a broken cell preparation.[17] Using the signaling response in turkey erythrocyte membranes as a model system, we have initiated investigations into the nature of P_{2Y}-purinergic receptor coupling to phospholipase C, particularly as it relates to the kinetics of receptor-promoted stimulation of the enzyme by guanine nucleotides[17] and to the potential role of G protein β- and γ-subunits.[18] Moreover, the turkey erythrocyte has a number of advantages as a means of directly identifying and purifying the components of a receptor-regulated phospholipase C signaling system. As such, we have recently purified to apparent homogeneity a phospholipase C from turkey erythrocytes[19] and have shown, by reconstitution of the purified 150-kDa protein with turkey erythrocyte acceptor membranes,[20] that this protein is regulated by P_{2Y}-purinergic receptors and G protein by properties completely analogous to the phospholipase C enzyme activity that is under receptor and G protein regulation in native membranes. In this review, we will summarize recent progress made in this laboratory directed at 1) characterization of a biochemical response (stimulation of phospholipase C) to P_{2Y}-purinergic receptor activation in turkey erythrocyte membranes; 2) direct identification by reversible radioligand binding of this membrane P_{2Y}-purinergic receptor; 3) direct identification of the P_{2Y}-purinergic receptor protein by covalent incorporation of a photoaffinity radioligand; 4) solubilization of a form of the turkey erythrocyte P_{2Y}-purinergic receptor that is in apparent association with a G protein; and 5) analysis of agonist-induced desensitization of the turkey erythrocyte P_{2Y}-purinergic receptor in a membrane preparation.

GUANINE NUCLEOTIDE-DEPENDENT REGULATION OF PHOSPHOLIPASE C BY P_{2Y}-PURINERGIC RECEPTORS

Out of a screen of a broad range of hormones, neurotransmitters, growth factors, and other receptor-active substances, we discovered that ATP and ADP were effective

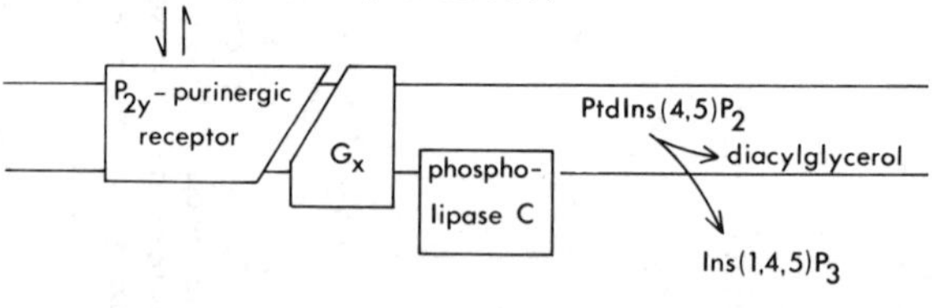

FIGURE 1. A model for the receptor-regulated phospholipase C of turkey erythrocytes. This signaling system consists of a minimum of three proteins: a cell surface P_{2Y}-purinergic receptor, a guanine nucleotide regulatory protein (G_X), and a phospholipase C. Phosphatidylinositol 4,5-bisphosphate ($PtdIns(4,5)P_2$) is the principal substrate for the activated phospholipase C, and hydrolysis of this substrate results in the release of diacylglycerol and inositol 1,4,5-trisphosphate ($Ins(1,4,5)P_3$).

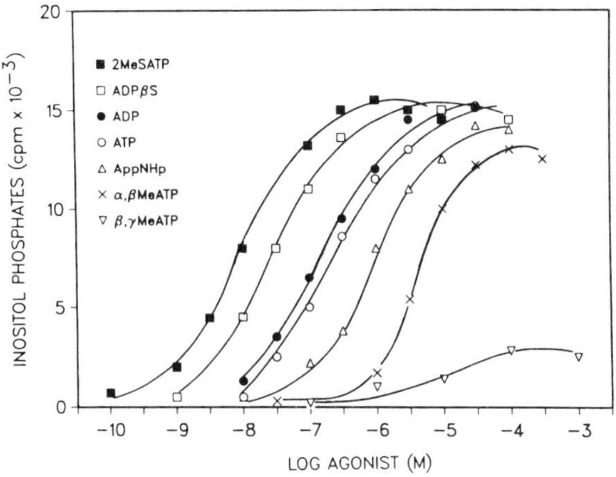

FIGURE 2. Concentration-dependent activation of turkey erythrocyte phospholipase C by P_{2Y}-purinergic receptor agonists. The capacity of a series of agonists to stimulate phosphoinositide hydrolysis was determined by a methodology described in detail in reference 17. The data are presented as inositol phosphates released from ^{3}H-labeled membrane phosphoinositides as a function of agonist concentration. UTP, AMP, and adenosine are without effect on inositol phosphate production.

stimulators of inositol phosphate accumulation in turkey erythrocytes. The response to a series of analogues of ATP and ADP[16] was in general agreement with that described by Burnstock and Kennedy[5] for a P_{2Y}-purinergic receptor, and this P_{2Y}-purinergic receptor-mediated effect was preserved in membranes prepared from turkey erythrocytes.[17]

Our initial studies[21,22] of the G protein-regulated phospholipase C of turkey erythrocyte membranes were carried out prior to realization of the presence of a P_{2Y}-purinergic receptor on these cells. Thus, high concentrations of ATP were used in early experiments to support the conversion of phosphatidylinositol to phosphatidylinositol 4-monophosphate and phosphatidylinositol 4-monophosphate to phosphatidylinositol 4,5-bisphosphate. We later demonstrated that dual effects of ATP occurred in the membrane preparation in that, in addition to maintaining membrane phosphatidylinositol 4,5-bisphosphate levels under guanine nucleotide-stimulated conditions where substrate became limiting, concentrations of ATP lower than those necessary for support of substrate levels stimulated inositol phosphate production through a purinergic receptor.[22] This effect of ATP was shown to be absolutely dependent on the presence of GTP or a stable analogue of GTP.[17] As is illustrated in FIGURE 2, the concentration-dependent effects of analogues of ATP and ADP followed an order of potency, that is, 2-Me-S-ATP > ADP-β-S > ATP-γ-S > ATP > App(NH)p = ADP > Ap(CH_2)pp > App(CH_2)p, consistent with that for a P_{2Y}-purinergic receptor. The large and relatively long-lived inositol phosphate response of the turkey erythrocyte membrane preparation has permitted an analysis of the kinetics of receptor- and G protein-mediated activation of a phospholipase C. Inositol phosphate production in the presence of the stable analogue of GTP, GTP-γ-S, occurs with a time course that exhibits a characteristic lag.[17] The P_{2Y}-purinergic receptor agonists increase the rate of GTP-γ-S-mediated activation of phospholipase C in a concentration-dependent and saturable manner. The rate of activation of phospholipase C by GTP-γ-S in the presence of a fixed concentration of P_{2Y}-purinergic receptor agonist is independent of the concentration of the guanine nucleotide. GDP-β-S competitively blocks guanine nucleotide-stimulated phospholipase C activity in turkey erythrocyte membranes.[17] In contrast, GDP-β-S has negligible effects on phospholipase C activity in membranes preactivated by the hydrolysis-resistant GTP analogues (GTP-γ-S or Gpp(NH)p); GTP-preactivated enzyme activity rapidly returns to basal levels.

These kinetic data for the turkey erythrocyte P_{2Y}-purinergic receptor and G protein-regulated phospholipase C are highly reminiscent of the properties of receptor- and G protein-regulated adenylate cyclase in this and other tissues,[23,24] and suggest a model for receptor-mediated activation of phospholipase C that is analogous to that widely accepted for adenylate cyclase. Thus, a three-protein signaling system of receptor, G protein, and phospholipase C can be envisioned (FIG. 3). In the ground state, it is predictable that the G protein exists in a GDP-liganded form, and the kinetic data with turkey erythrocyte membranes indicate that agonist-occupied P_{2Y}-purinergic receptors promote the exchange of GTP or a GTP analogue for GDP on the G protein. This rate-limiting step is followed by productive interaction of the GTP-liganded G protein with the phospholipase C catalyst resulting in the active enzymic species. The lifetime of this active complex apparently is regulated by a GTPase activity that returns the components to their ground state of GDP-liganded G protein and inactive phospholipase C. Unambiguous description of the partial reactions involved in the catalytic cycle for a receptor- and G protein-regulated phospholipase C will require purification of each component and reconstruction of the functional response in a model membrane preparation. We recently have purified the turkey erythrocyte receptor- and G protein-regulated phospholipase C,[19,20] and attempts are currently underway using this purified enzyme in an assay for purification of the yet to be identified phospholipase C-associated G protein.

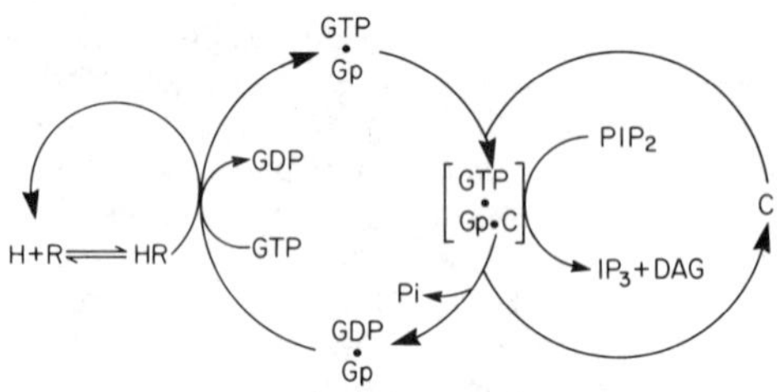

FIGURE 3. Model for receptor-promoted activation of phospholipase C. The interaction of three proteins, a cell surface receptor (R), a guanine nucleotide regulatory protein (G_p), and a phospholipase C (C), is illustrated. Hormone (H)-occupied receptors accelerate the exchange of GTP for GDP on G_p. This is the rate-limiting step in hormonal activation, with the GTP-liganded G_p now functionally activating the phospholipase C to produce an increase in the rate of formation of inositol phosphate and diacylglycerol (DAG) second messenger products. The lifetime of the active GTP-G_p-C complex is governed by a GTPase activity, which is most likely an inherent function of G_p, and which returns the system to the basal state of GDP-liganded G_p and inactive C. This model does not take into account the likelihood that G_p is a heterotrimeric G protein that undergoes dissociation and reassociation during the activation/deactivation cycle (see reference 18). It should also be pointed out that this is almost certainly an oversimplified model with the omission of a number of the partial reactions that constitute the complete regulatory cycle.

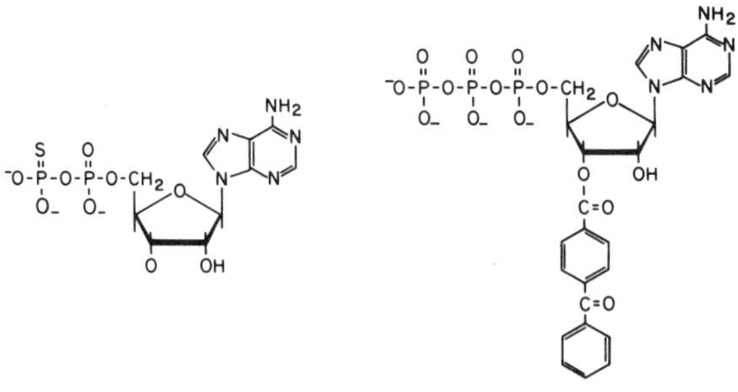

ADPβS BzATP

FIGURE 4. Structures of two ligands used to radiolabel P_{2Y}-purinergic receptors.

DIRECT IDENTIFICATION OF THE P_{2Y}-PURINERGIC RECEPTOR

Based on the marked response of the turkey erythrocyte phospholipase C to P_{2Y}-purinergic receptor agonists, we felt there was a strong possibility that this receptor could be directly identified with a high-affinity radioligand. ADP-β-S (FIG. 4) is one of the most potent P_{2Y}-purinergic receptor agonists and, relative to 2-Me-S-ATP, is stable to hydrolysis. As such, [^{35}S]ADP-β-S was synthesized as a potential radioligand for labeling P_{2Y}-purinergic receptors on purified turkey erythrocyte plasma membranes. [^{35}S]ADP-β-S bound to a single high-affinity binding site with a K_d of 5-10 nM and a B_{max} of approximately 3 pmol/mg of protein.[25] This density of binding sites is approximately 5-fold higher than the level of β-adrenergic receptors in the same tissue. Radioligand binding was inhibited with law of mass action kinetics for a single site by a series of P_{2Y}-purinergic receptor agonists. The K_i values calculated from competition curves for these compounds were essentially identical to the $K_{0.5}$ values determined for the same agonists for activation of phospholipase C in turkey erythrocyte membranes (TABLE 2). Whereas purinergic receptor agonists inhibited [^{35}S]ADP-β-S binding in a competitive fashion, guanine nucleotides inhibited radioligand binding noncompetitively and with an order of potency, that is, GTP-γ-S > Gpp(NH)p > GTP = GDP > GDP-β-S > GMP, consistent with a G protein-mediated effect. These and other results strongly suggest that a ternary complex of agonist/receptor/G protein exists in [^{35}S]ADP-β-S-labelled turkey erythrocyte plasma membranes.

We have recently initiated studies the end goal of which is the purification of the turkey erythrocyte P_{2Y}-purinergic receptor. Subsequent to solubilization of turkey erythrocyte plasma membranes with the nonionic detergent, digitonin, a soluble species can be labeled with [^{35}S]ADP-β-S that expresses pharmacological properties of a P_{2Y}-purinergic receptor.[26] A single high-affinity (K_d = 10 nM) binding site was identified in the soluble preparation, and P_{2Y}-purinergic receptor agonists competitively inhibited radioligand binding with K_i values similar to those observed in membrane binding assays. Importantly, this soluble [^{35}S]ADP-β-S binding site was still noncompetitively regulated by guanine nucleotides, suggesting that a complex of P_{2Y}-purinergic receptor and G protein is stable to solubilization. The capacity to radiolabel this soluble species presents the possibility of not only purifying the receptor, but of also identifying its associated G protein.

Another useful tool for the direct study of the turkey erythrocyte P_{2Y}-purinergic receptor has been developed. A photoaffinity analogue of ATP, 3′-*O*-(4-benzoyl)benzoyl ATP (Bz-ATP; see FIG. 4), activated turkey erythrocyte membrane P_{2Y}-purinergic receptors and phospholipase C in a guanine nucleotide-dependent fashion.[27] Although the effects of Bz-ATP were fully reversible in the absence of photolysis, after exposure of Bz-ATP-preincubated membranes to UV light, a marked increase in the inositol phosphate response to guanine nucleotides was observed. This effect was dependent on the concentration of Bz-ATP, and reversibly binding P_{2Y}-purinergic receptor agonists protected against this irreversible effect of Bz-ATP when present with Bz-ATP at the time of photolysis. Based on these results, we have concluded that Bz-ATP is a P_{2Y}-purinergic receptor agonist, which after photolysis becomes irreversibly associated with turkey erythrocyte membranes and promotes P_{2Y}-purinergic receptor-mediated guanine nucleotide-dependent activation of phospholipase C.

The results with nonradioactive Bz-ATP suggested that Bz-ATP can be photoincorporated into the turkey erythrocyte P_{2Y}-purinergic receptor. With this idea in mind, we have synthesized [^{32}P]Bz-ATP as a potential covalent photo-radiolabel for the P_{2Y}-purinergic receptor protein. In the absence of photolysis, [^{32}P]Bz-ATP reversibly interacted with a binding site on turkey erythrocyte plasma membranes with properties very similar to those observed with [^{35}S]ADP-β-S, and consistent with those for interaction of a radioligand with a P_{2Y}-purinergic receptor.[28] Upon photolysis of prelabeled plasma membranes, [^{32}P]Bz-ATP was incorporated into a 53,000-Da protein identified after autoradiography of SDS polyacrylamide gels. Photolabeling was inhibited by analogues of ATP and ADP with a potency order consistent with that for a P_{2Y}-purinergic receptor. Guanine nucleotides also inhibited photoincorporation with an order of potency consistent with effects on the G protein component of a ternary complex of [^{32}P]Bz-ATP/receptor/G protein.[28] [^{32}P]Bz-ATP was also photoincorporated into a 53,000-Da species in membranes from other tissues (for example, rat liver, brain, and astrocytes) in which P_{2Y}-purinergic receptor-mediated responses occur; [^{32}P]Bz-ATP failed to label such a species in tissues (for example, human

TABLE 2. Comparison of the Capacity of ATP and the Analogues of ATP to Activate Phospholipase C and to Inhibit [^{35}S]ADP-β-S Binding to P_{2Y}-Purinergic Receptors[a]

Agonist	Activation of Phospholipase C ($K_{0.5}$)	Inhibition of [^{35}S]ADP-β-S Binding (K_i)
2-Me-S-ATP	10 nM	10 nM
ADP-β-S	30	35
ATP	105	100
ADP	130	140
App(NH)p	400	330
α,β-Me-ATP	4	4
β,γ-Me-ATP	10	15

[a] Turkey erythrocyte membranes were used to generate concentration-effect curves for the capacity of a series of agonists to activate phospholipase C in the presence of 1 μM GTP-γ-S (see reference 17). The $K_{0.5}$ values determined in this manner are compared to the K_i values determined for the same agonists measuring their capacity to competitively inhibit [^{35}S]ADP-β-S binding (see reference 25).

platelets and human erythrocytes) where there is no evidence for the existence of P_{2Y}-purinergic receptors.

Taken together, our results with Bz-ATP and [^{32}P]Bz-ATP suggest that this photolabel can be used to irreversibly label and activate P_{2Y}-purinergic receptors. This capacity to covalently label a 53,000-Da protein exhibiting characteristics of the P_{2Y}-purinergic receptors should prove very useful during purification and characterization of this important signaling molecule.

AGONIST-INDUCED DESENSITIZATION OF P_{2Y}-PURINERGIC RECEPTOR-REGULATED PHOSPHOLIPASE C

Agonist-induced desensitization of second messenger signaling systems has been broadly described for all of the major signaling pathways. In some cases, for example, the β-adrenergic receptor-regulated adenylate cyclase, the process of desensitization now can be described in molecular terms.[29,30] In contrast, very little is known about the mechanisms responsible for agonist-induced modification of inositol phosphate/Ca^{2+} signaling, and to our knowledge there have been no reports on the modifications that may occur in P_{2Y}-purinergic receptors and their associated signaling proteins during prolonged exposure of target cells to ATP, ADP, or their analogues. In light of this lack of knowledge, we have taken advantage of the turkey erythrocyte as a model system to examine the molecular mechanisms responsible for agonist-induced desensitization of the P_{2Y}-purinergic receptor-regulated phospholipase C.

Preincubation of intact erythrocytes with ADP-β-S or other P_{2Y}-purinergic receptor agonists resulted in a rapid ($t_{1/2}$ = 2 min) loss of the capacity of P_{2Y}-purineregic receptor agonists to stimulatae phospholipase C in membranes derived from these cells.[31] The membrane response reached a quasi-steady state of reduced responsiveness (40–50% of control) within 10 min of incubation of cells with agonists, and this level of desensitization was essentially maintained over a several hour period in the continued presence of agonist. Transfer of cells to agonist-free medium resulted in a rapid ($t_{1/2}$ = 15 min) return of P_{2Y}-purinergic receptor-stimulated phospholipase C activity to control levels. The kinetic properties of the receptor- and G protein-regulated phospholipase C have been studied in membranes from agonist-preincubated cells. Desensitization occurred as a loss of maximal effect of P_{2Y}-purinergic receptor agonists on phospholipase C activity with no change occurring in the apparent affinity of agonists for stimulation of the enzyme. No change occurred in the capacity of GTP-γ-S alone to stimulate the enzyme, and the capacity of the G protein activator, $A1F_4^-$, to stimulate phospholipase C activity was also unchanged. Although there was no change in the rate of activation of phospholipase C observed with GTP-γ-S alone, the capacity of ADP-β-S to increase the rate of activation by GTP-γ-S was markedly reduced in membranes from agonist-preincubated cells.

Taken together, these results indicate that agonist-induced desensitization of the turkey erythrocyte P_{2Y}-purinergic receptor-regulated phospholipase C occurs as a consequence of a modification at the level of the P_{2Y}-purinergic receptor or at the level of receptor/G protein coupling. In preliminary experiments, we have begun to apply the recently developed P_{2Y}-purinergic receptor radioligands to the study of the desensitization process. For example, preincubation of turkey erythrocytes with P_{2Y}-purinergic receptor agonists resulted in a loss of [^{35}S]ADP-β-S binding in membranes prepared from these cells. The time course of loss of radioligand binding coincided with the time course of loss of agonist-stimulated phospholipase C activity in the same membranes. Furthermore, recovery of radioligand binding in membranes from agonist-preincubated erythrocytes that had been transferred to agonist-free medium coincided with recovery of receptor-regulated enzyme activity. The loss of [^{35}S]ADP-β-S binding in membranes from desensitized cells occurred as a decrease in B_{max} with no change occurring in the K_d of the remaining sites for radioligand. Because high-affinity [^{35}S]ADP-β-S binding apparently represents a ternary complex between radiolabeled agonist, receptor, and G protein, these results do not discriminate between an agonist-induced modification of the receptor per se and an agonist-induced alteration in some portion of either the receptor or G protein necessary for protein-protein association.

We have begun to explore the mechanisms that may be responsible for agonist-induced desensitization of the P_{2Y}-purinergic receptor signaling system of turkey erythrocytes. Second messenger-regulated protein kinases have been shown to have important regulatory effects on a variety of cell surface receptors and their associated effector proteins, and have been shown to be involved in several forms of agonist-induced desensitization.[32,33] Production of diacylglycerol and activation of protein kinase C occur as a consequence of receptor-stimulated, phospholipase C-catalyzed phosphoinositide hydrolysis. Studies with intact cells suggest that in some cases agonist-induced desensitization of receptor-regulated phospholipase C involves a feedback modification of the signaling system by diacylglycerol-activated protein kinase C.[34,35] As such, we have examined the potential role of protein kinase C in the desensitization of the P_{2Y}-purinergic receptor response of turkey erythrocytes.

Activation of protein kinase C in intact turkey erythrocytes with the phorbol ester, 4β-phorbol 12β-myristate 13α-acetate (PMA), resulted in a rapid loss of P_{2Y}-purinergic receptor-stimulated phospholipase C activity in membranes prepared from these cells.[36] The loss of responsiveness occurred with properties similar to those observed in membranes from P_{2Y}-purinergic receptor agonist-pretreated cells. Thus, there was a loss of receptor-stimulated phospholipase C activity that was observed as a decrease in the maximal effect of agonists with no change occurring in the apparent affinity for activation. The capacity of GTP-γ-S or AlF_4^- to activate phospholipase C was unchanged, but the capacity of P_{2Y}-purinergic receptor agonists to stimulate the rate of activation of the enzyme by GTP-γ-S was greatly diminished. As with agonist-induced desensitization, the effect of PMA on responsiveness of phospholipase C to receptor activation was mirrored by quantitatively similar time- and concentration (of PMA)-dependent losses of [^{35}S]ADP-β-S binding. Thus, as with agonist-induced desensitization, the PMA-induced loss of responsiveness of the turkey erythrocyte P_{2Y}-purinergic receptor/phospholipase C apparently occurs as a consequence of a lesion at the level of the P_{2Y}-purinergic receptor or at the level of receptor/G protein coupling. Does this mean that agonist-induced desensitization occurs as a *consequence* of activation of protein kinase C? Preliminary data suggest this is not the case, as conditions that apparently down-regulate protein kinase C in turkey erythrocytes did not appreciably modify the course or extent of P_{2Y}-purinergic receptor agonist-induced desensitization. Experiments are underway to more clearly define the relationship between agonist-induced desensitization and phorbol ester-induced loss of responsiveness.

SUMMARY

The turkey erythrocyte has substantial value as a model for the study of a receptor that exhibits pharmacological properties very similar to those delineated in mammalian tissues for a P_{2Y}-purinergic receptor. The G protein-dependent coupling of this receptor to phospholipase C can be studied in detail, and the availability of an abundant source of homogeneous cells from which highly purified plasma membranes can be prepared, has led to the development of a radiolabeled, reversibly binding radioligand for a P_{2Y}-purinergic receptor and a photoaffinity covalent radiolabel for this receptor. This source of plasma membranes highly enriched in P_{2Y}-purinergic receptors should also serve as a rich starting material for the eventual purification and structural characterization of this important signaling protein.

REFERENCES

1. CHANGEUX, J.-P., J. GIRAUDAT & M. DENNIS. 1987. Trends Pharmacol. Sci. **8:** 459-464.
2. STEPHENSON, E. A. 1988. Biochem. J. **249:** 21-32.
3. LEFKOWITZ, R. J. & M. G. CARON. 1988. J. Biol. Chem. **263:** 4993-4996.
4. BURNSTOCK, G. 1978. *In* Cell Membrane Receptors for Drugs and Hormones: A Multidisciplinary Approach. R. W. Straub & L. Bolis, Eds.: 107-118. Raven Press. New York, NY.
5. BURNSTOCK, G. & C. KENNEDY. 1985. Gen. Pharmacol. **16:** 433-440.
6. WILLIAMS, M. 1987. Annu. Rev. Pharmacol. Toxicol. **27:** 315-345.
7. STILES, G. L. 1986. Trends Pharmacol. Sci. **7:** 486-490.
8. BARRINGTON, W. W., K. A. JACOBSON, A. J. HUTCHINSON, M. WILLIAMS & G. L. STILES. 1989. Proc. Natl. Acad. Sci. USA **86:** 6572-6574.
9. NAKATA, H. 1989. J. Biol. Chem. **264:** 16545-16551.
10. DUBYAK, G. A., D. S. COWEN & L. M. MEULLER. 1988. J. Biol. Chem. **263:** 18108-18117.
11. OKAJIMA, F., K. SATO, M. NAZAREA, K. SHO & Y. KONDO. 1989. J. Biol. Chem. **264:** 13029-13037.
12. CHAREST, R., P. R. BLACKMORE & J. H. EXTON. 1985. J. Biol. Chem. **260:** 15789-15794.
13. PIROTTEN, S., E. RASPE, D. DEMOLLE, C. ERNEUX & J. M. BOEYNAEMS. 1987. J. Biol. Chem. **262:** 17461-17466.
14. FORSBERG, E. J., G. FEVERSTEIN, E. SHOHAMI & H. B. POLLARD. 1987. Proc. Natl. Acad. Sci. USA **84:** 5630-5634.
15. OKAJIMA, F., Y. TOKUMITSU, Y. KONDO & M. UI. 1987. J. Biol. Chem. **262:** 13483-13490.
16. BERRIE, C. P., P. T. HAWKINS, L. R. STEPHENS, T. K. HARDEN & C. P. DOWNES. 1989. Mol. Pharmacol. **35:** 526-532.
17. BOYER, J. L., C. P. DOWNES & T. K. HARDEN. 1989. J. Biol. Chem. **264:** 884-890.
18. BOYER, J. L., G. L. WALDO, T. EVANS, J. K. NORTHUP, C. P. DOWNES & T. K. HARDEN. 1989. J. Biol. Chem. **264:** 13917-13922.
19. MORRIS, A. J., G. L. WALDO, C. P. DOWNES & T. K. HARDEN. 1990. J. Biol. Chem. **265:** in press.
20. MORRIS, A. J., G. L. WALDO, C. P. DOWNES & T. K. HARDEN. 1990. J. Biol. Chem. **265:** in press.
21. HARDEN, T. K., L. STEPHENS, P. T. HAWKINS & C. P. DOWNES. 1987. J. Biol. Chem. **262:** 9057-9061.
22. HARDEN, T. K., P. T. HAWKINS, L. STEPHENS, J. L. BOYER & C. P. Downes. 1988. Biochem. J. **252:** 583-593.
23. TOLKOVSKY, A. M. & A. LEVITZKI. 1978. Biochemistry **17:** 3795-3810.
24. ROSS, E. M., M. E. MAGUIRE, T. W. STURGILL, R. L. BILTONEN & A. G. GILMAN. 1977. J. Biol. Chem. **252:** 5761-5775.
25. COOPER, C. L., A. J. MORRIS & T. K. HARDEN. 1989. J. Biol. Chem. **264:** 6202-6206.
26. JEFFS, R., C. L. COOPER, J. L. BOYER & T. K. HARDEN. 1990. Submitted for publication.
27. BOYER, J. L. & T. K. HARDEN. 1989. Mol. Pharmacol. **36:** 831-835.
28. BOYER, J. L., C. L. COOPER & T. K. HARDEN. 1990. Submitted for publication.
29. CLARK, R. B. 1986. Adv. Cyclic Nucleotide Protein Phosphoryl. Res. **20:** 151-209.

30. SIBLEY, D. R., J. L. BENOVIC, M. G. CARON & R. J. LEFKOWITZ. 1987. Cell **48:** 913-922.
31. MARTIN, M. W. & T. K. HARDEN. 1989. J. Biol. Chem. **264:** 19535-19539.
32. EDELMAN, A. M., D. K. BLUMENTHAL & E. G. KREBS. 1987. Annu. Rev. Biochem. **56:** 567-613.
33. HUNTER, T. 1987. Cell **50:** 823-829.
34. BROWN, K. D., D. M. BLAKELEY, M. H. HAMON, M. S. LAURIE & A. N. CORPS. 1987. Biochem. J. **245:** 631-639.
35. HEPLER, J. R., H. S. EARP & T. K. HARDEN. 1988. J. Biol. Chem. **263:** 7610-7619.
36. MARTIN, M. W., C. L. COOPER & T. K. HARDEN. 1990. Submitted for publication.
37. DUBYAK, G. R. & D. S. COWEN. 1990. Ann. N. Y. Acad. Sci. This volume.

Effects of Extracellular ATP on the Release of Vasoactive Mediators from Endothelium

JEREMY D. PEARSON AND THOMAS D. CARTER[a]

Section of Vascular Biology
Clinical Research Centre
Medical Research Council
Harrow, Middlesex HA1 3UJ, England

ATP ACTS AT P_{2Y}-PURINOCEPTORS ON ENDOTHELIUM

Endothelium-dependent vasodilatation in response to ATP was first described by De Mey and Vanhoutte.[1] It was subsequently demonstrated that this action, now known to be due to the synthesis and release of endothelium-derived relaxing factor (EDRF, nitric oxide),[2] occurred by stimulation of P_2-purinoceptors, that is, those recognizing ATP or ADP but not AMP or adenosine.[3] The use of ATP analogues, such as 2-methylthio-ATP (more potent than ATP) and L-ATP (almost inactive), indicated further that the P_2-purinoceptor on endothelium mediating the generation of EDRF is of the P_{2Y} subtype,[4] according to the classification of Burnstock and Kennedy.[5]

Concurrent experiments with endothelial cells in culture showed that ATP or ADP acting at P_2-purinoceptors stimulated the transient release of prostacyclin (PGI_2).[6] The receptor was again subtyped as P_{2Y} by the use of a variety of analogues.[7] Despite some apparent differences in the relative abilities of certain ATP analogues at stimulating EDRF and PGI_2 release, the most straightforward conclusion is that both of these responses are triggered by the binding of ATP to a single P_2-purinoceptor subtype.[8] Nonetheless, there are significant differences in the time course of each response, its desensitization, and its dependence on extracellular calcium, which indicate that discrete intracellular transduction pathways are likely to be involved in the generation and regulation of EDRF and PGI_2.[9]

P_{2Y}-PURINOCEPTOR ACTIVATION REGULATES CYTOSOLIC [Ca^{2+}]

Pirotton *et al.*[10] first demonstrated that in endothelial cells, as had already been shown in several other cell types, ATP acting at P_{2Y}-purinoceptors caused a rapid

[a]Present address: National Institute for Medical Research, Mill Hill, London, England.

and transient increase in inositol 1,4,5-trisphosphate concentrations, implying that release of Ca^{2+} from internal stores was likely to provide a major transduction signal. The use of endothelial cells loaded with Ca^{2+}-sensitive fluorescent dyes (indo-1, fura-2) also demonstrated directly that addition of ATP caused a rapid and transient rise in cytosolic [Ca^{2+}].[11,12] More detailed studies confirmed that the Ca^{2+} response was, as expected, mediated by action at P_{2Y}-purinoceptors, and delineated clearly two phases of the response.[13] The initial Ca^{2+} spike is due to the release of Ca^{2+} from internal stores; in the presence of extracellular Ca^{2+} there is in addition a maintained elevation of Ca^{2+}, stable for > 10 min, that is lower than the initial peak but well above resting

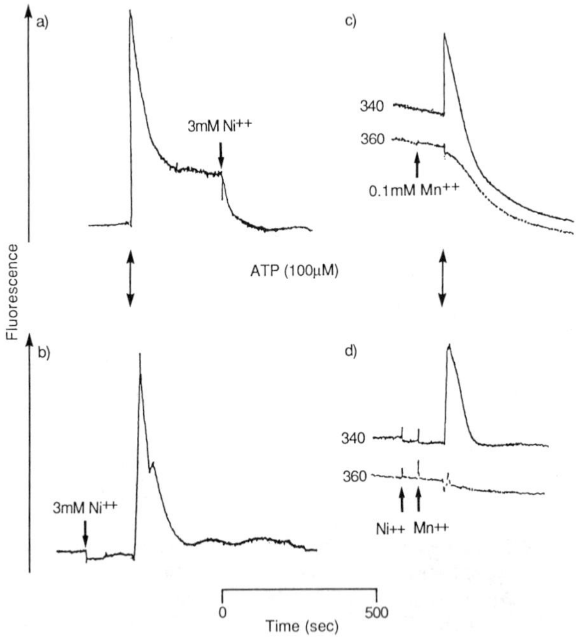

FIGURE 1. Elevations in intracellular ionized calcium in fura-2-loaded human umbilical vein cells stimulated with ATP (100 μM). Fluorescence was monitored at 500 nm to record the response to excitation at 340 nm and 380 nm (**A** & **B**—note ratioed traces) or at 340 nm and 360 nm (**C** & **D**). Elevation of Ca^{2+} has two phases: an initial transient spike due to release from internal stores, and a maintained lower level. The latter phase requires Ca^{2+} entry, which is blocked by Ni^{2+} (**A** & **B**). Direct demonstration of entry of divalent cations in response to ATP is shown by the addition of Mn^{2+}, which is translocated (unless blocked by Ni^{2+}) and quenches fluorescence at all excitation wavelengths, including 360 nm, which is insensitive to Ca^{2+} (**C** & **D**).

cytosolic [Ca^{2+}]. This second phase of the Ca^{2+} response requires influx of extracellular Ca^{2+} for its maintenance.

For other agonists that evoke similar changes in cytosolic Ca^{2+} in endothelial cells (thrombin, histamine), it has been demonstrated that the cation channel also translocates Mn^{2+} (which allows direct detection of influx because Mn^{2+} quenches the fluorescence of fura-2) and is blocked by Ni^{2+},[14] and we have recently found that ATP acts similarly (FIG. 1).[15] The channel is not gated by alterations in membrane potential and is insensitive to dihydropyridine Ca^{2+} antagonists,[12,14] and thus represents a receptor-operated channel that may be activated by direct or indirect coupling to the P_{2Y}-purinoceptor, or may be regulated by the state of the internal Ca^{2+} store.

P_{2Y}-MEDIATED TRANSIENT ELEVATIONS OF CYTOSOLIC [Ca^{2+}] ARE NECESSARY AND SUFFICIENT TO GENERATE PGI_2

The peak and steady state elevations in cytosolic [Ca^{2+}] share an identical dose-related response curve to ATP, whose minimally active concentration is ≤ 1 μM, and the same dose-response relationship is also found for ATP-stimulated PGI_2 synthesis.[13] To determine whether elevations in cytosolic Ca^{2+} were necessary for PGI_2 synthesis in response to ATP, endothelial cells were treated in the absence of extracellular Ca^{2+} with a high dose of histamine to deplete the internal stores, and then, after sufficient mepyramine to antagonize any further action of histamine completely, ATP was added. Under these conditions, elevations of cytosolic Ca^{2+} and PGI_2 synthesis induced by ATP were both blocked.[13]

ATP-stimulated PGI_2 release occurs to a similar extent, and for a similarly brief time (2-3 min after a lag of 15-30 sec), in the presence or absence of extracellular Ca^{2+}.[13] It is therefore likely that Ca^{2+} from internal stores normally provides the major or exclusive source of Ca^{2+} required for PGI_2 synthesis. Moreover, if there is no other regulatory mechanism to switch off PGI_2 production, for PGI_2 synthesis to continue cytosolic [Ca^{2+}] must be elevated above a threshold somewhat greater than the maintained steady state level. When this was tested, by concomitantly measuring PGI_2 release and peak cytosolic [Ca^{2+}] with various doses of ATP, it was found that PGI_2 synthesis was directly related to cytosolic [Ca^{2+}], but only occurred when [Ca^{2+}] ≥ 0.7 μM, compared with a resting level of $\simeq 0.1$ μM.[13] Moreover, the identical dose relationship between cytosolic [Ca^{2+}] and PGI_2 release was observed when PGI_2 synthesis was stimulated by the addition of ionomycin (FIG. 2).[13] This result implies that ATP-induced elevation of cytosolic [Ca^{2+}] is a sufficient signal to account for PGI_2 release, and that any other transduction signal generated by the agonist (such as diacylglycerol) plays no direct role. The likely site of action of Ca^{2+} is on phospholipase A_2, the enzyme at the rate-limiting step in prostanoid production, and an enzyme known to be activated by increasing [Ca^{2+}].

DESENSITIZATION OF PGI_2 SYNTHESIS

PGI_2 release in response to a variety of agonists, including ATP, is rapidly and powerfully desensitized either in the continued presence of the agonist or when the endothelial cells are sequentially exposed briefly to the agonist at short intervals.[16] Although under certain conditions (for example, at very short time intervals or in the absence of extracellular Ca^{2+}) desensitization may be due to inadequate refilling of internal Ca^{2+} stores, in general this is not the explanation because the desensitization process is agonist specific.[16]

To test whether activation of protein kinase C (PKC) (for example, via diacylglycerol generated with inositol trisphosphate[17]) played a role in this process, cells were briefly pretreated with phorbol myristate acetate (PMA). Under these conditions, PMA alone did not alter resting cytosolic Ca^{2+} or cause PGI_2 synthesis, but dose-

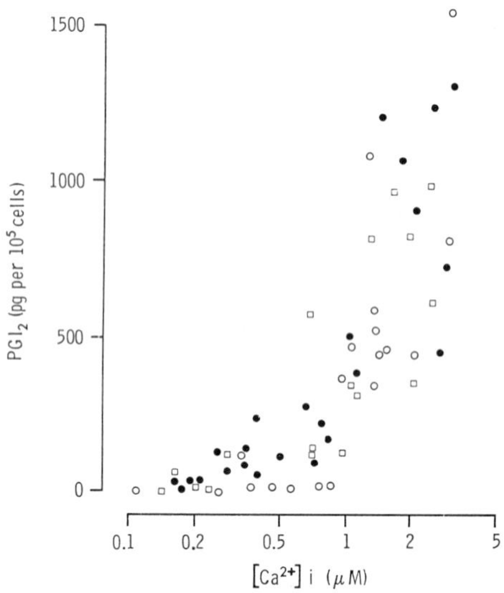

FIGURE 2. Dose-response relationship between peak cytoplasmic Ca^{2+} and prostacyclin (PGI_2) release in human umbilical vein endothelial cells in response to ionomycin (●), ATP (□), or the more potent P_{2Y} agonist 2-chloro-ATP (○). Reproduced from reference 13, with permission.

dependently inhibited both the peak and steady state elevations of cytosolic [Ca^{2+}] in response to ATP.[18] Paradoxically, however, PMA at the same time enhanced endothelial PGI_2 synthesis.[18] PKC activation thus seems unlikely to be involved in desensitization of PGI_2 release, but it does disrupt the normal relationship between cytosolic [Ca^{2+}] and PGI_2 release, by shifting the Ca^{2+} activation curve to the left and lowering the threshold level of cytosolic [Ca^{2+}] needed to induce PGI_2 synthesis.[18] The mechanism underlying this is not known, but is plausibly an effect (either direct or indirect) of PKC on phospholipase A_2 because PMA does not affect arachidonate-stimulated PGI_2 synthesis.[18] Recently, any role for PKC in homologous desensitization of PGI_2 release in response to repeated addition of ATP was ruled out, by showing that under conditions in which desensitization of ATP-induced elevation in cytosolic [Ca^{2+}] induced by pretreatment with PMA was completely reversed by the PKC inhibitor staurosporine, staurosporine failed to affect a similar degree of desensitization induced by brief pre-exposure of the cells to ATP (FIG. 3).[19]

The underlying mechanism for homologous desensitization of PGI_2 release thus has still to be determined. It may be relevant to this that P_{2Y}-purinoceptor activation causes the rapid phosphorylation of a variety of as yet undefined endothelial cell proteins, several of which are not PKC substrates.[10] Furthermore, it has been shown recently in 3T6 fibroblasts that ATP acting at P_{2Y}-purinoceptors induces inositol phosphate production without activating PKC.[20]

STEADY STATE ELEVATION OF CYTOSOLIC [Ca^{2+}]

Although the steady state elevation of cytosolic [Ca^{2+}] in the continued presence of extracellular ATP (or other agonists) does not contribute to the generation of

PGI_2, it seems likely to play an important role in the regulation of EDRF production. Direct evidence is not yet available, but several observations suggest this. The [Ca^{2+}] elevation is long-lived, resists rapid desensitization even on repeated stimulation with increasing doses of ATP,[19] requires extracellular Ca^{2+}, and is powerfully inhibited by PMA.[18] EDRF production, in contrast to PGI_2 production, is relatively long-lasting;[21] cumulative dose-response curves to ATP can be produced readily;[3] and it is blocked by activation of PKC.[22]

In platelets and smooth muscle cells, P_2-purinoceptor-mediated entry of divalent cations is extremely rapid (< 1 sec) and, from electrophysiological measurements in patch-clamp studies, apparently does not require the generation of any soluble transduction signal.[23–25] In endothelium, however, there is a delay of several seconds between addition of ATP and the detection of cation influx,[15] which suggests that coupling of P_{2Y}-purinoceptor occupation to the opening of the cation channel in endothelial cells is indirect, although more work is needed to determine the pathways involved.

CONCLUSION

Several further endothelial cell responses to P_{2Y}-purinoceptor activation have been described (reviewed in reference 26), including efflux of K^+ (dependent on Ca^{2+} influx) and efflux of choline. The functional effects of these responses, however, are not known. This review has focused particularly on the mechanisms by which ATP, acting at endothelial P_{2Y}-purinoceptors, causes elevations in cytosolic [Ca^{2+}]; has presented evidence for the central role of Ca^{2+} release from internal stores in the resultant synthesis of PGI_2; and has indicated that by contrast EDRF production may be much more dependent on the consequent sustained entry of Ca^{2+} through receptor-operated cation channels. Future studies will be directed at the definition of the molecular targets that respond to elevation of cytosolic [Ca^{2+}] by initiating the synthesis of PGI_2 and EDRF, and of the mechanism by which PGI_2 synthesis is down-regulated in an agonist-specific manner.

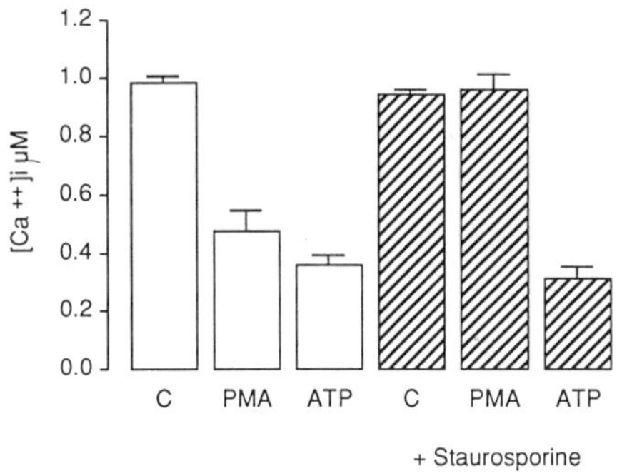

FIGURE 3. Peak cytoplasmic Ca^{2+} responses in human umbilical vein endothelial cells pretreated with PMA (20 nM) or ATP (10 μM) in the presence or absence of staurosporine (90 nM) and then rechallenged with ATP (10 μM). The experiment demonstrates that protein kinase C is not involved in homologous desensitization. Cells were exposed for 7 min to vehicle alone or phorbol ester, or for 5 min to ATP plus 2 min wash, before rechallenging. Each bar shows the S.E. of four observations.

REFERENCES

1. De Mey, J. G. & P. M. Vanhoutte. 1981. Role of the intima in cholinergic and purinergic relaxation of isolated canine femoral arteries. J. Physiol. **316:** 347-355.
2. Palmer, R. M., A. G. Ferrige & S. Moncada. 1987. Nitric oxide release accounts for the biological activity of endothelium-derived relaxing factor. Nature **327:** 524-526.
3. Gordon, J. L. & W. Martin. 1983. Endothelium-dependent relaxation of the pig aorta: Relationship to stimulation of ^{86}Rb efflux from isolated endothelial cells. Br. J. Pharmacol. **79:** 537-541.
4. Martin, W., N. J. Cusack, J. S. Carleton & J. L. Gordon. 1985. Specificity of the P_2-purinoceptor that mediates endothelium-dependent relaxation of the pig aorta. Eur. J. Pharmacol. **134:** 295-299.
5. Burnstock, G. & C. Kennedy. 1985. Is there a basis for distinguishing two types of P_2-purinoceptor? Gen. Pharmacol. **16:** 433-440.
6. Pearson, J. D., L. L. Slakey & J. L. Gordon. 1983. Stimulation of prostacyclin production through purinoceptors on cultured porcine aortic endothelial cells. Biochem. J. **214:** 273-276.
7. Needham, L., N. J. Cusack, J. D. Pearson & J. L. Gordon. 1987. Characteristics of the P_2-purinoceptor that mediates endothelial prostacyclin production. Eur. J. Pharmacol. **134:** 199-209.
8. Pearson, J. D. & T. D. Carter. 1989. Transduction of purinoceptor-mediated endothelial cell responses. *In* Vascular Endothelium: Receptors and Transduction Mechanisms. J. D. Catravas, C. N. Gillis & U. S. Ryan, Eds: 189-195. Plenum. New York, NY.
9. Pearson, J. D. & J. L. Gordon. 1989. P_2-purinoceptors in the blood vessel wall. Biochem. Pharmacol. **38:** 4157-4163.
10. Pirotton, S., E. Rapse, D. Demolle, L. Erneux & J. M. Boeynaems. 1987. Involvement of inositol 1,4,5-trisphosphate and calcium in the actions of adenine nucleotides on aortic endothelial cells. J. Biol. Chem. **262:** 17461-17467.
11. Luckhoff, A. & R. Busse. 1986. Increased free calcium in endothelial cells under stimulation with adenine nucleotides. J. Cell. Physiol. **126:** 414-420.
12. Hallam, T. J. & J. D. Pearson. 1986. Exogenous ATP raises cytoplasmic free calcium in fura-2-loaded piglet aortic endothelial cells. FEBS Lett. **176:** 139-143.
13. Carter, T. D., T. J. Hallam, N. J. Cusack & J. D. Pearson. 1988. Regulation of P_{2Y}-purinoceptor-mediated prostacyclin release from human endothelial cells by cytoplasmic calcium concentration. Br. J. Pharmacol. **94:** 1181-1190.
14. Hallam, T. J., R. Jacob & J. E. Merritt. 1988. Evidence that agonists stimulate bivalent-cation influx into human endothelial cells. Biochem. J. **255:** 179-184.
15. Carter, T. D., T. J. Hallam & J. D. Pearson. 1989. P_2-purinoceptor-stimulated mobilization of intracellular calcium precedes divalent cation influx in human endothelial cells. Br. J. Haematol. **73:** 437.
16. Toothill, V. J., L. Needham, J. L. Gordon & J. D. Pearson. 1988. Desensitization of agonist-stimulated prostacyclin release in human umbilical vein endothelial cells. Eur. J. Pharmacol. **157:** 189-196.
17. Berridge, M. J. & R. F. Irvine. 1989. Inositol phosphates and cell signalling. Nature **341:** 197-205.
18. Carter, T. D., T. J. Hallam & J. D. Pearson. 1989. Protein kinase C activation alters the sensitivity of agonist-stimulated endothelial cell prostacyclin production to intracellular ionised calcium. Biochem. J. **262:** 431-437.
19. Carter, T. D., J. S. Newton, R. Jacob & J. D. Pearson. 1989. Desensitization of ATP-mediated elevations in cytosolic [Ca^{2+}] and prostacyclin production in human endothelial cells: A protein kinase C-independent mechanism. Biochem. J.: submitted for publication.
20. Gonsalez, F. A., E. Rozengurt & L. E. Heppel. 1989. Extracellular ATP induces the release of calcium from intracellular stores without the activation of protein kinase C in Swiss 3T6 mouse fibroblasts. Proc. Natl. Acad. Sci. USA **86:** 4530-4534.
21. Kelm, M., M. Feelisch, R. Spahr, H.-M. Piper, E. Noacke & J. Schrader. 1988.

Quantitative and kinetic characterization of nitric oxide and EDRF released from cultured endothelial cells. Biochem. Biophys. Res. Commun. **154:** 236-244.
22. LEWIS, M. J. & A. H. HENDERSON. 1987. A phorbol ester inhibits the release of endothelium-derived relaxing factor. Eur. J. Pharmacol. **137:** 167-171.
23. SAGE, S. O. & T. J. RINK. 1987. The kinetics of changes in intracellular calcium concentrations in fura-2-loaded human platelets. J. Biol. Chem. **262:** 16364-16369.
24. BENHAM, C. D. & R. W. TSIEN. 1987. A novel receptor-operated, Ca^{2+}-permeable channel activated by ATP in smooth muscle. Nature **328:** 275-278.
25. SAGE, S. O., J. E. MERRITT, T. J. HALLAM & T. J. RINK. 1989. Receptor-mediated calcium entry in fura-2-loaded human platelets stimulated with ADP and thrombin. Biochem. J. **258:** 923-926.
26. BOEYNAEMS, J. M. & J. D. PEARSON. 1990. P_2-purinoceptors on vascular endothelial cells: Physiological significance and transduction mechanisms. Trends Pharmacol. Sci. **11:** 34-37.

DISCUSSION OF THE PAPER

S. P. SOLTOFF (*Tufts University, Boston, MA*): The results regarding the effects of ATP on $[Ca^{2+}]_i$ (a rapid elevation and a fall to a sustained elevated level) are similar to reports in other systems regarding the effects of bradykinin. 1) Do these cells respond to bradykinin with a similar alteration in $[Ca^{2+}]_i$? 2) If bradykinin also causes a maintained elevation, does this mean that ATP does *not* activate an ATP-gated channel?

PEARSON: 1) Yes, bradykinin produces very similar responses, as demonstrated elegantly by Schilling and colleagues (*Am. J. Physiol.* **255:** H219-H227, 1988). 2) This demonstrates that, unlike in smooth muscle cells or platelets, there is a substantial lag time before P_{2Y}-purinoceptor activation causes cation entry, suggesting that this channel is not directly receptor coupled but is gated by a second messenger. Thus the entry mechanism could be the same when bradykinin is used as an agonist, but this idea has not yet been tested.

F. BELLONI: Given the facts that 1) the same relation between the intracellular Ca^{2+} transient and prostacyclin release holds for a variety of agonists including ATP and ionomycin, and 2) protein kinase C activation can shift this relation, do you conclude that ATP does *not* activate protein kinase C in endothelial cells?

PEARSON: Our tentative conclusion is that, if ATP does lead to activation of protein kinase C, the extent or time course of activation is not sufficient to alter PGI_2 release. We are currently testing directly the ability of ATP to activate protein kinase C in these cells. Our conclusion also depends on the assumption that ionomycin does not lead to activation of protein kinase C, which we have not tested, but has been shown by Halldorsson *et al.* (*Arteriosclerosis* **8:** 147-154, 1988).

E. ROJAS (*National Institute of Diabetes and Digestive and Kidney Diseases, Bethesda, MD*): Do you have any information on the Ca^{2+} pump in endothelial cells? The reason for my question is that in the absence of extracellular Ca^{2+} you were able to induce PGI_2 secretion by using ATP activation. The presence of a Ca^{2+} pump should remove very effectively the Ca^{2+} released by the ER as a typical K_m for the pump is in the nanomolar range for Ca^{2+}.

PEARSON: PGI_2 release is essentially independent of extracellular Ca^{2+} (when cells are exposed to a single dose of agonist) because release from internal stores is sufficient to trigger PGI_2 synthesis; the transience of the response is due to the short time for which $[Ca^{2+}]_i$ remains above the threshold needed. The subsequent steady state $[Ca^{2+}]_i$ elevation presumably reflects a balance between the entry of Ca^{2+} (which as I noted continues for many minutes after addition of ATP) and the removal of Ca^{2+} by sequestration back to internal stores or pumping out of the cell. We are currently investigating whether protein kinase C activation stimulates Ca^{2+} efflux or blocks Ca^{2+} entry to achieve its effect when blocking the steady state $[Ca^{2+}]_i$ elevation.

E. W. WESTHEAD (*University of Massachusetts, Amherst, MA*): Data we have obtained with bovine chromaffin cells support your final suggestion that entry of Ca^{2+}, seen in the second phase of the Ca^{2+} transient, is the critical component for secretion. In chromaffin cells, the Ca^{2+} transient induced by ATP is almost independent of the presence of extracellular Ca^{2+}, but secretion absolutely requires external Ca^{2+}. Likewise, UTP produces a Ca^{2+} transient almost identical to that produced by ATP, but causes no secretion. So in some cells, the fura-2 signal caused by release of Ca^{2+} from internal stores is not relevant to secretion.

ATP-Gated Channels in Vascular Smooth Muscle Cells

C. D. BENHAM

Department of Pharmacology
Smith Kline & French Research, Ltd.
Welwyn, Herts AL6 9AR, England

INTRODUCTION

There is considerable evidence that ATP is an excitatory sympathetic cotransmitter (see the review by Burnstock and Kennedy[1]), and it is well established that ATP, on release, generates fast excitatory junction potentials in some arteries[2] and other smooth muscles such as the vas deferens.[3] These contractile actions of ATP in smooth muscle appear to be due to activation of the P_{2X} subtype of purinoceptor.[4] In a variety of other tissues, such as heart[5] and sensory neurons,[6] extracellular application of ATP also causes depolarization by the activation of inward currents.

In vascular smooth muscle, ATP is generally thought to stimulate contraction by activating a mainly Na^+ permeable conductance that depolarizes the cell,[7] allowing Ca^{2+} entry through voltage-gated Ca^{2+} channels.[8] Patch-clamp techniques have allowed detailed study of the ATP-activated channels in vascular smooth muscle, revealing that the conductance is cation selective,[9] and that the channels are closely coupled to the ATP receptor.[10] Results from current reversal potential measurements suggest that the channels are permeable to Ca^{2+} with a selectivity of 3:1 over Na^+.[10] This comparatively low Ca^{2+} selectivity of the channel and the much higher concentration of Na^+ present in extracellular saline gave a calculated Ca^{2+} influx of less than 10% of the total ATP-activated current. Calculations based on the modified Goldman equation depend on assumptions about the mechanism of ion permeation through ion channels that are often not valid, especially for Ca^{2+}-permeable channels.[11] This could lead to large errors in estimates of selectivity, particularly for mixed conductances such as the ATP-activated currents.

Because of this uncertainty about the relevance of this Ca^{2+} entry pathway in the Ca^{2+} economy of vascular smooth muscle cells, direct measurements of $[Ca^{2+}]_i$ have been made in smooth muscle cells during ATP application to see whether Ca^{2+} influx through ATP-gated channels was sufficient to elevate $[Ca^{2+}]_i$ and hence affect the contractile state of the cell. These results also suggest that about 10% of the current through ATP-gated channels is carried by Ca^{2+} ions at -60 mV, sufficient to significantly elevate $[Ca^{2+}]_i$ in these single cells. Thus ATP-activated channels have a dual excitatory function, depolarizing the cell leading to voltage-gated Ca^{2+} entry, and functioning as a direct Ca^{2+} entry pathway.

METHODS

Cell Preparation

Smooth muscle cells were enzymatically dissociated from the central ear arteries of adult New Zealand White rabbits (as previously described[12]) and stored on glass coverslips at 4 °C for use the same day.

Membrane Current and Fluorescence Measurements

Cells were normally bathed in an extracellular solution containing the following (constituent, mM): NaCl, 130; KCl, 5; glucose, 10; $CaCl_2$, 1.5; $MgCl_2$, 1.2; HEPES, 10 (buffered to pH 7.3 with NaOH). For patch-clamp experiments without $[Ca^{2+}]_i$ measurements, the pipette filling (intracellular) solution usually contained the following (constituent, mM): CsCl, 130; $MgCl_2$, 3; HEPES, 5; EGTA, 10 (buffered to pH 7.2 with TEA-OH). For the $[Ca^{2+}]_i$ measurements, the pipette solution was made up in MilliQ water and in most experiments contained the following (constituent, mM): KCl, 125; $MgCl_2$, 2; HEPES, 10; indo-1 (K-salt, Molecular Probes), 0.1 (buffered to pH 7.2 with NaOH). No other Ca^{2+} buffers were added to the pipette solution.

Patch-clamp experiments were performed using standard techniques.[13] Combined voltage-clamp and fluorescence measurements were performed as described previously.[14] The apparatus has also been described elsewhere.[15]

Briefly, voltage-clamp and membrane current recording were made with standard patch-clamp techniques using a List EPC-7 patch-clamp amplifier. The $[Ca^{2+}]_i$ was estimated from indo-1 fluorescence by the ratio method using single-wavelength excitation and dual emission.[16,17] Single cells loaded with indo-1 from the patch pipette were centered under a window created by a hole (diameter: 20 microns) placed in the emission beam. On breakthrough to whole-cell mode, loading could be monitored from the emitted light, which reached equilibrium 3-5 min after breakthrough with counts of 10 to 50 times background (0.1 mM indo-1). Recording was started at this point.

After background subtraction, $[Ca^{2+}]_i$ was estimated from the 405/480 ratio using a calibration for indo-1 determined within cells.[14]

RESULTS

Agonist Specificity of ATP-Activated Currents

ATP-gated current responses in the ear artery cells were subject to strong desensitization. As a consequence, it was impossible to evoke repetitive responses to ATP. Even after 10 min, wash responses had only recovered to about 20% of the first response. This prevented quantitative dose-response data from being obtained. Over a series of experiments in different cells, however, some idea of agonist potency could

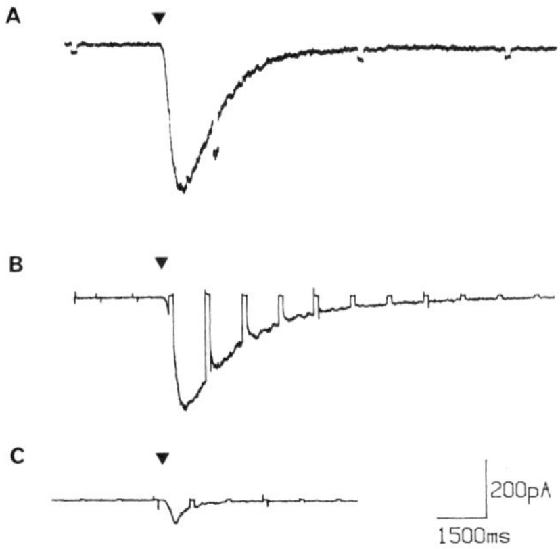

FIGURE 1. Activation of inward currents in single ear artery cells held under voltage-clamp at −60 mV by (**A**) 10^{-5} M ATP, (**B**) 10^{-5} M AMP-CPP, and (**C**) 5 × 10^{-4} M ADP. Agonists were applied for 0.5 sec. by pressure ejection from a wide-bore patch pipette positioned within 0.2 mm of the cell. Records are from three different cells. In **A,** the cell membrane potential was periodically hyperpolarized by 10 mV. In **B & C,** depolarizations to 0 mV were made for 150 msec every second.

be obtained. In whole-cell recordings, concentrations of ATP as low as 10^{-8} M activated small currents. Maximal responses were reached between 10^{-6} and 10^{-5} M ATP. α,β-Methylene ATP (AMP-CPP) was of roughly similar potency: at 10^{-5} M it was able to evoke responses similar to those evoked with ATP at 10^{-5} M (FIG. 1). In smooth muscle cells from vas deferens, where ATP activates a rather similar conductance, AMP-CPP was much less effective than ATP.[18] In the ear artery cells ATP-γ-S also had effects similar to those of ATP. ADP was much less potent, evoking only very small currents even at 0.5 mM (FIG. 1C). AMP and adenosine were ineffective at this concentration. These results are consistent with P_{2X}-purinoceptor activation.[19] As both ATP^{4-} and divalent-cation-complexed ATP are effective (FIG. 2),[10] the mechanism in these smooth muscle cells is clearly different from the non-

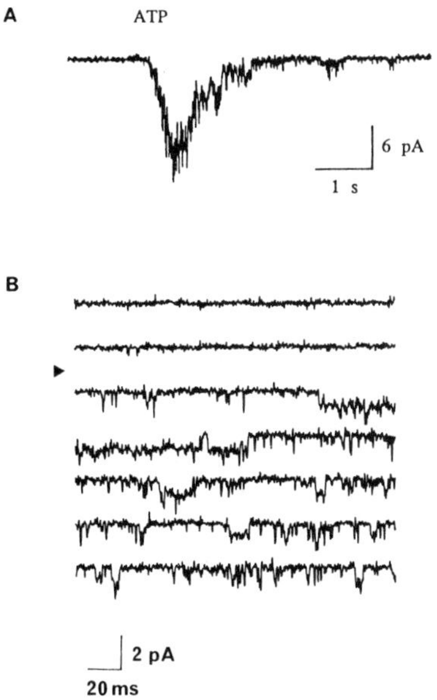

FIGURE 2. ATP activates unitary inward currents in isolated outside-out membrane patches. In **A,** 10^{-6} M ATP was applied to a patch held at −100 mV bathed in extracellular solution containing 110 mM Ca^{2+}. Many channels were activated simultaneously in this patch. In **B,** the trace shows the response of another patch held at −100 mV. This patch, however, was bathed in normal saline solution. A puffer pipette containing 10^{-7} M ATP was brought up close to the patch between the second and third traces (indicated by the arrow). The unitary current amplitude was 0.89 pA. Note the complex kinetics. (Traces the result of unpublished experiments performed with R. W. Tsien at Yale University Medical School.)

specific increase in membrane permeability seen in mast cells, which seems to be exclusively dependent on the presence of the ATP^{4-} moiety.[20]

Receptor Channel Coupling and Ca^{2+} Permeability

Application of ATP to isolated outside-out membrane patches activated inward unitary currents at negative potentials. Currents were activated with very little delay (FIG. 2A), consistent with studies on whole cells using ionophoretic application of ATP where latencies of less than 100 msec were recorded.[9] Large numbers of channels could be activated in a single patch, as shown in FIGURE 2A, where about 30 channels were simultaneously open at the peak of the current response. In some patches, and particularly if the concentration of ATP was reduced, it was often possible to activate no more than one or two channels simultaneously, revealing features of the unitary currents. FIGURE 2B shows records from a patch bathed in normal saline solution where no more than two channels were active simultaneously. The channel activity showed quite complex kinetics with openings of hundreds of milliseconds interspersed with brief, flickery events. Detailed analysis was not possible because of the rapid desensitization that occurred when channel activity completely stopped. These isolated patch experiments suggest that the receptor channel coupling must be quite tight without the involvement of readily diffusible second messengers. It is difficult, however, to exclude the possibility that a membrane-bound cofactor such as a G protein is involved without doing more experiments.

FIGURE 2A shows that the ATP-gated channels are Ca^{2+} permeable, and reversal potential measurements from whole cells suggest that the selectivity ratio for Ca^{2+} over Na^{+} was 3:1. The reliability of these estimates would depend in part on the degree of independence of ion permeation through these channels. Addition of 1.5 mM Ca^{2+} back to the 130 mM Na^{+}-containing external solution did reduce the unitary current.[10] A recording of unitary currents in 1.5 mM Ca^{2+}-containing external solution is shown in FIGURE 2B. Currents are about 20% smaller than in the absence of any divalent cations. This reduction can be interpreted in terms of Ca^{2+} binding sites within the pore.[11] Doing so brings into question the interpretation of the reversal potential measurements. Thus it was of interest to use a more direct method of ascertaining the importance of this channel as a Ca^{2+} entry pathway.

ATP-Activated Inward Current is Associated with a Rise in $[Ca^{2+}]_i$

FIGURE 3A shows simultaneous recordings of membrane current and $[Ca^{2+}]_i$ from a cell bathed in normal saline (1.5 mM Ca^{2+}) and loaded with pipette solution containing 10^{-4} M indo-1. Application of 10^{-6} M ATP to this cell held under voltage-clamp evoked an inward current that reached a peak of about 500 pA and then declined in the continued presence of ATP. The apparent delay in onset of response to ATP reflects the dead space in the solution exchange system. Hyperpolarizing voltage jumps to -80 mV were used to monitor the cell conductance and the access resistance. During the ATP response the voltage jumps evoked larger inward current steps, showing that there was an increase in membrane conductance. A rise in $[Ca^{2+}]_i$

from 140 nM to 300 nM was associated with this inward current. The $[Ca^{2+}]_i$, was also transient, but declined more slowly than the current. The maximum rate of rise of $[Ca^{2+}]_i$ coincided with the peak of the inward current—as would be expected if this were the cause of the rise.

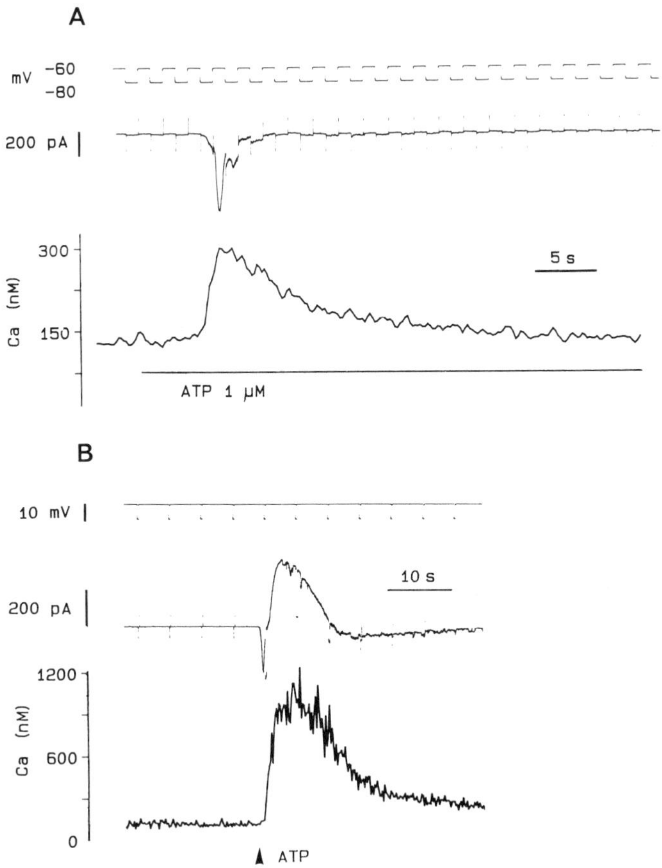

FIGURE 3. Bath application of ATP elevates $[Ca^{2+}]_i$ in vascular smooth muscle cells under voltage-clamp. (**A**) Membrane potential (*top*), membrane current (*middle*), and estimated $[Ca^{2+}]_i$ (*bottom*) responses of a single rabbit ear artery cell to bath application of 10^{-6} M ATP (horizontal bar) in the presence of 1.5 mM Ca^{2+}. Brief hyperpolarizing voltage steps were made throughout the recording to monitor cell input resistance and pipette series resistance. (Reproduced from Benham[24] with permission.) (**B**) Another cell exposed to a brief puffer pipette application of 10^{-5} M ATP. Note the biphasic current response.

The current response was biphasic in cells that showed a large rise in $[Ca^{2+}]_i$, the usual inward current being followed by a large outward current as Ca^{2+} rose within the cell (FIG. 3B). Net outward currents were seen when the rise in $[Ca^{2+}]_i$ exceeded 350 nM, close to the $[Ca^{2+}]_i$ expected to open significant numbers of Ca^{2+}-activated K^+ channels at this membrane potential of -60 mV.[21]

Effect of Extracellular Ca^{2+} Removal

Removal of extracellular Ca^{2+} did not inhibit the ATP-activated inward current. Inward currents of similar amplitude were seen (peak amplitude: 256 ± 60 pA, $N = 5$), presumably due to monovalent cation movement through the ATP-activated channels (FIG. 4).[10] Biphasic responses with secondary outward currents, however, were abolished, and there was very little change in $[Ca^{2+}]_i$. The very small rise in $[Ca^{2+}]_i$ seen in two of these five cells (one of which is shown in FIG. 4) could be due to residual Ca^{2+} in the bathing medium, as no Ca^{2+} chelator was added. So, the rise in $[Ca^{2+}]_i$ seemed to be dependent on extracellular Ca^{2+} and was thus unlikely to be due to intracellular Ca^{2+} store release stimulated by generation of a second messenger such as inositol trisphosphate. Effective voltage-clamp of these cells under these conditions[9] means that it is unlikely that any depolarization occurs during the ATP response that might open voltage-gated Ca^{2+} channels leading to Ca^{2+} influx through this route. However, as release of Ca^{2+} stores may also be triggered by a rise in $[Ca^{2+}]_i$ in some cells,[22] some of the increase in $[Ca^{2+}]_i$ seen in response to ATP could still be due to Ca^{2+}-induced Ca^{2+} release.

Prerelease of Internal Ca^{2+} Stores Does Not Affect Response to ATP

Cells were pretreated with caffeine or noradrenaline, which both cause store release in arterial smooth muscle cells.[23] ATP was then applied to these store-depleted cells. Both noradrenaline and caffeine stimulated a rise in $[Ca^{2+}]_i$ and an associated outward current that declined back to basal levels in the continued presence of these compounds.[24] After $[Ca^{2+}]_i$ returned to basal levels, the application of ATP evoked an inward current and a rise in $[Ca^{2+}]_i$, as in untreated cells. The increase after nora-

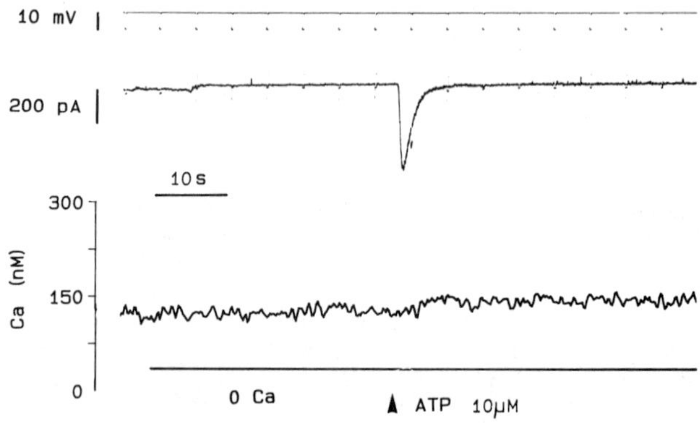

FIGURE 4. Response of a cell bathed in nominally Ca^{2+}-free saline to application of ATP by pressure ejection from a micropipette. Note how the magnitude and time course of the ATP-activated current are similar to those in FIGURE 3. Also note, however, the complete absence of a $Ca^{2+}{}_i$ signal. The traces are arranged as in FIGURE 3A. (Reproduced from Benham[24] with permission.)

TABLE 1. Responses of $[Ca^{2+}]_i$ to ATP Application[a]

Bathing Solution	N	Basal $[Ca^{2+}]_i$ (nM)	Peak $[Ca^{2+}]_i$ (nM)
Control (1.5 mM $Ca^{2+}{}_o$)	6	149 ± 9	495 ± 105
+ Noradrenaline (10 M)	5	174 ± 20	348 ± 80
+ Caffeine (10 mM)	4	176 ± 17	572 ± 72
$Ca^{2+}{}_o$-free solution	5	140 ± 14	143 ± 13

[a] Basal $[Ca^{2+}]_i$ values were measured over the 10 sec preceding ATP application. Peak $[Ca^{2+}]_i$ values were taken as the mean over 1 sec (five points) spanning the highest point. Experiments were started in control solution, and then the bathing solution was changed to the test solution 40-60 sec before applying a maximal concentration of ATP (10^{-6}-10^{-5} M in the bathing solution or 10^{-5} M pressure ejected from a wide-bore patch pipette). No Ca^{2+} chelator was added to the $Ca^{2+}{}_o$-free solution. Each value is a mean ± SEM. (Adapted from Benham.[24])

drenaline pretreatment was smaller, and after caffeine, larger—but not significantly so (TABLE 1), indicating that the ATP response was not dependent on the noradrenaline- or caffeine-releasable store. These results are consistent with most of the response to ATP being attributed solely to Ca^{2+} entry with little contribution from stores. At least those stores sensitive to noradrenaline or caffeine did not appear to be involved.

The ATP-Evoked Rise in $[Ca^{2+}]_i$ Is Voltage Dependent

If the Ca^{2+} influx that was responsible for the rise in $[Ca^{2+}]_i$ was through the ATP-gated channels and not through the release of some unidentified store, then the $[Ca^{2+}]_i$ signal should be voltage dependent such that at strongly depolarized potentials (where the driving force for Ca^{2+} is much reduced) there should be negligible Ca^{2+} influx. This was tested by applying ATP to cells held at depolarized potentials. Cells were perfused with pipette solution containing CsCl rather than KCl to eliminate outward voltage-gated K^+ currents that might obscure responses at depolarized potentials, and nifedipine was added to the bathing solution to block voltage-gated Ca^{2+} entry at depolarized potentials. In these conditions, ATP evoked inward currents at −60 mV with rises in $[Ca^{2+}]_i$ comparable to responses in KCl-filled cells (FIG. 5). The voltage dependence of voltage-gated Ca^{2+} channels has been reported to be shifted in a negative direction by noradrenaline in some vascular smooth muscle cells,[25] but not in these ear artery cells.[26] As nifedipine (an L-type Ca^{2+} channel antagonist) did not block the rise in $[Ca^{2+}]_i$ evoked by ATP, it was unlikely that ATP was raising $[Ca^{2+}]_i$ by allowing voltage-gated Ca^{2+} channels to open in the absence of any change in membrane potential.

For a holding potential of +50 mV, calculations from constant field theory suggest that very little inward Ca^{2+} flux should occur through the ATP-gated channels.[27] ATP evoked outward currents but with insignificant effects on $[Ca^{2+}]_i$ in all four cells tested at this potential (even though the magnitude of the outward currents suggested that the degree of ATP receptor activation was roughly comparable).

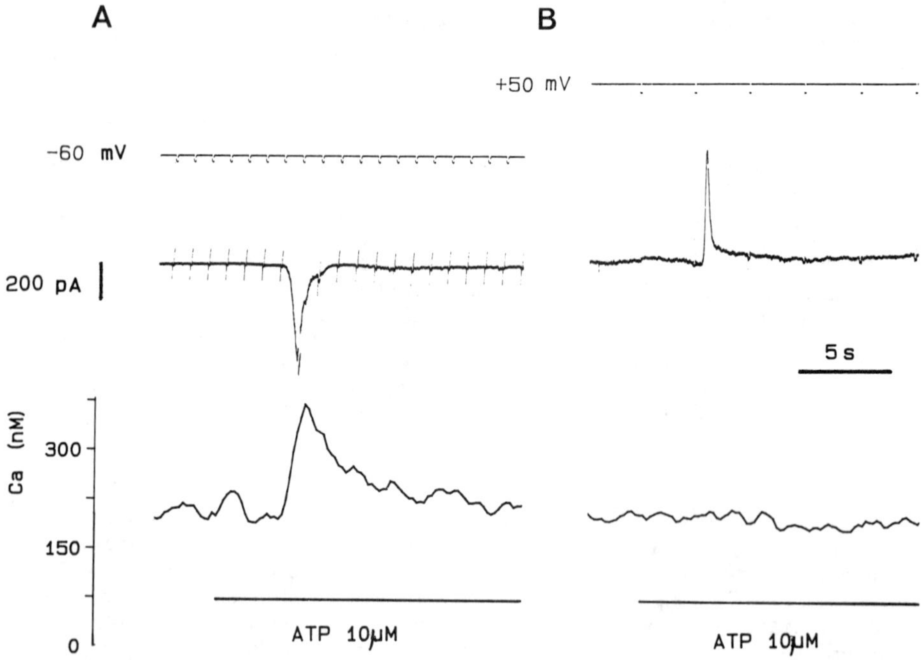

FIGURE 5. Voltage dependence of $[Ca^{2+}]_i$ changes. Responses of two cells bathed in saline containing 10^{-6} M nifedipine to application of 10^{-5} M ATP. (**A**) Response of a cell at a holding potential of −60 mV. (**B**) Response of another cell at a holding potential of +50 mV. Note that the outward current is not associated with any rise in $[Ca^{2+}]_i$. For these experiments, KCl in external and pipette solutions was replaced with NaCl and KCl, respectively, so that outward K^+ currents did not obscure the ATP-gated current at depolarized potentials. (Reproduced from Benham[24] with permission.)

Physiological Role

ATP is thought to be an excitatory cotransmitter with noradrenaline at sympathetic nerve terminals on vascular smooth muscle cells.[1] It is probable that the ATP-gated conductance generates excitatory junction potentials in these tissues. The properties of the ATP-activated conductance described here are consistent with the conductance responsible for generating the fast excitatory junction potentials in vascular smooth muscle. The most detailed information on the transmitter-activated current is provided by the work of Finkel *et al.*[28] This study, using short segments of arteriole held under voltage-clamp, shows that the current has a reversal potential close to 0 mV with a linear current-voltage relationship—until the membrane potential is made more negative than −60 mV, when inward rectification is observed. Scaling of the current-voltage data from Finkel *et al.* and plotting on the same axes as the ATP-gated current in ear artery cells shows the close resemblance in their current-voltage characteristics (FIG. 6). Clearly this is not a very rigorous test, but it does illustrate that the ATP-gated conductance (the transmitter could be another agonist opening the channels) is a good candidate to explain this neurally mediated fast depolarization.

ATP activates direct receptor-gated Na^+ and Ca^{2+} entry at resting membrane potentials. In cells that are not held under voltage-clamp, activation of ATP-gated channels will depolarize the cells because of the predominant Na^+ influx. This will tend to limit the rise in $[Ca^{2+}]_i$ through the ATP-gated channels by reducing the driving force for Ca^{2+}. Significant Ca^{2+} entry, however, could occur over a period of repeated tonic sympathetic stimulation. Also, the Ca^{2+} influx through these channels would be maximized if other simultaneous hyperpolarizing conductance changes occurred.

A further factor to consider is the anatomical arrangement of the tissue as the degree of innervation varies. During sympathetic nerve stimulation, activation of the ATP-gated channels allows discretely graded entry of Na^+ and Ca^{2+} into cells exposed to transmitter. This will be most important in the smaller arterioles where a high proportion of the smooth muscle cells are in direct contact with synaptic varicosities, and hence directly exposed to transmitter.[29] The depolarization, however, will spread electrotonically, and on strong stimulation a propagating wave of voltage-gated Ca^{2+} entry and contraction will occur in association with action potential generation.[30] The parallel increase in $[Na^+]_i$ might also have implications for Ca^{2+} homeostasis through an effect on the Na^+/Ca^{2+} exchanger. An increase in $[Na^+]_i$ might lead to an increase in $[Ca^{2+}]_i$ through this mechanism.[31]

SUMMARY AND CONCLUSION

ATP acting through P_{2X}-purinoceptors activates cation channels with some similarities to the activation of channels gated by acetylcholine and glutamate (channels that can also act as fast excitatory transmitters). These experiments clearly demonstrate an ATP-mediated Ca^{2+} influx through agonist-gated channels and a consequent elevation of $[Ca^{2+}]_i$ in these single vascular smooth muscle cells. The combination of the ability to hold these cells under voltage-clamp and to measure $[Ca^{2+}]_i$ simultaneously has allowed us to exclude other possible explanations for the rise in $[Ca^{2+}]_i$

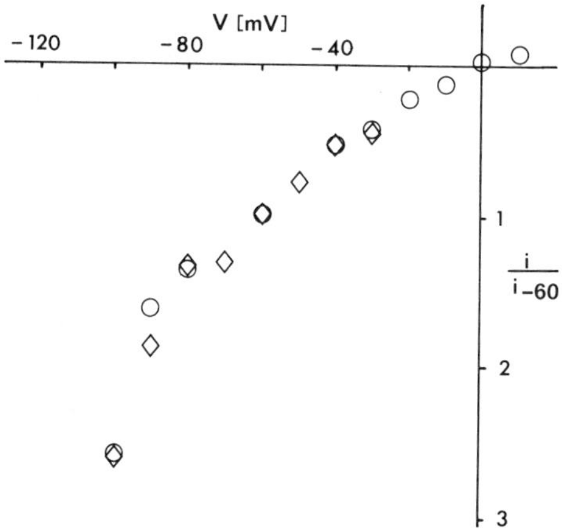

FIGURE 6. Comparison of current-voltage relationship for ATP-activated current and nerve-evoked excitatory junction currents. One set of data (◇) was derived from excitatory junction current data obtained in small segments of submucosal arterioles (Finkel *et al.*[28]). Another set of data (○) represents ATP-gated currents in ear artery cells (Benham and Tsien[10]). Both sets of data were normalized to the current observed at −60 mV.

under these conditions. Thus, although the major cation entering through the channels is Na^+, ATP receptor activation will also generate subtle, localized increases in $[Ca^{2+}]_i$. These increases might directly activate contractile proteins or, if insufficient to do this, might upregulate other Ca^{2+}-dependent enzymes modulating the contractile process and provide an enhanced source of Ca^{2+} for uptake into internal Ca^{2+} stores. Further understanding of the physiological role of this conductance pathway may require the development of specific receptor antagonists or channel blockers.

ACKNOWLEDGMENTS

Some of this work has recently been published in the *Journal of Physiology.*[24]

REFERENCES

1. Burnstock, G. & C. Kennedy. 1986. A dual function for adenosine 5′-triphosphate in the regulation of vascular tone. Circ. Res. **58:** 319-330.
2. Stjarne, L. 1986. New paradigm: Sympathetic transmission by multiple messengers and lateral interaction between monoquantal release sites? Trends Neurosci. **9:** 547-548.
3. Sneddon, P., D. P. Westfall & J. S. Fedan. 1982. Co-transmitters in the motor nerves of the guinea-pig vas deferens: Electrophysiological evidence. Science **218:** 693-695.
4. Burnstock, G. & C. Kennedy. 1985. Is there a basis for distinguishing two types of P_2-purinoceptor? Gen. Pharmacol. **16:** 433-440.
5. Friel, D. D. & B. P. Bean. 1987. Two ATP-activated conductances in bullfrog atrial cells. J. Gen. Physiol. **91:** 1-27.
6. Krishtal, O. A., S. M. Marchenko & V. I. Pidoplichko. 1983. Receptor for ATP in the membrane of mammalian sensory neurones. Neurosci. Lett. **35:** 41-45.
7. Suzuki, H. 1985. Electrical responses of smooth muscle cells of the rabbit ear artery to adenosine triphosphate. J. Physiol. **359:** 401-415.
8. Burnstock, G. 1988. Sympathetic purinergic transmission in small blood vessels. Trends Pharm. Sci. **9:** 116-117.
9. Benham, C. D., T. B. Bolton, N. G. Byrne & W. A. Large. 1987. Action of extracellular adenosine triphosphate in single smooth muscle cells dispersed from the rabbit ear artery. J. Physiol. **387:** 473-488.
10. Benham, C. D. & R. W. Tsien. 1987. Receptor-operated, Ca-permeable channels activated by ATP in arterial smooth muscle. Nature **328:** 275-278.
11. Tsien, R. W., P. Hess, E. W. McCleskey & R. L. Rosenberg. 1987. Calcium channels: Mechanisms of selectivity, permeation and block. Annu. Rev. Biophys. Biophys. Chem. **16:** 265-290.
12. Benham, C. D. & T. B. Bolton. 1986. Spontaneous transient outward currents in single visceral and vascular smooth muscle cells of rabbit. J. Physiol. **381:** 385-406.
13. Hamill, O. P., A. Marty, E. Neher, B. Sakmann & F. J. Sigworth. 1981. Improved patch-clamp techniques for high-resolution current recording from cells and cell-free membrane patches. Pfluegers Arch. Gesamte Physiol. Menschen Tiere **391:** 85-100.
14. Benham, C. D. 1989. Voltage-gated and agonist-mediated rises in intracellular Ca in rat clonal pituitary cells (GH_3) held under voltage clamp. J. Physiol. **415:** 143-158.
15. Jacob, R. & C. D. Benham. 1989. Measuring cytoplasmic calcium in single living cells using fluorescent probes. *In* New Techniques of Optical Microscopy and Microspectrophotometry. R. J. Cherry, Ed.: in press. Macmillan. New York, NY.
16. Grynkiewicsz, G., M. Poenie & R. Y. Tsien. 1985. A new generation of Ca indicators with greatly improved fluorescence properties. J. Biol. Chem. **260:** 3440-3450.

17. COBBOLD, P. & T. J. RINK. 1987. Fluorescence and bioluminescence measurement of cytoplasmic free calcium. Biochem. J. **248:** 313-328.
18. FRIEL, D. D. 1988. An ATP-sensitive conductance in single smooth muscle cells from the rat vas deferens. J. Physiol. **401:** 361-380.
19. GORDON, J. L. 1986. Extracellular ATP: Effects, sources and fate. Biochem. J. **233:** 309-319.
20. COCKROFT, S. & B. D. GOMPERTS. 1980. The ATP^{4-} receptor of rat mast cells. Biochem. J. **188:** 789-798.
21. BENHAM, C. D., T. B. BOLTON, R. J. LANG & T. TAKEWAKI. 1986. Calcium-activated potassium channels in single dispersed smooth muscle cells of rabbit jejunum and guinea-pig mesenteric artery. J. Physiol. **371:** 45-67.
22. ENDO, M. 1977. Calcium release from the sarcoplasmic reticulum. Physiol. Rev. **57:** 71-108.
23. ITOH, T., M. KAJIWARA, K. KITAMURA & H. KURIYAMA. 1982. Roles of stored calcium on the mechanical response evoked in smooth muscle cells of the porcine coronary artery. J. Physiol. **322:** 107-125.
24. BENHAM, C. D. 1989. ATP-activated channels gate calcium entry in single smooth muscle cells dissociated from rabbit ear artery. J. Physiol. **419:** 689-701.
25. NELSON, M. T., N. B. STANDEN, J. E. BRAYDEN & J. F. WORLEY. 1988. Noradrenaline contracts arteries by activating voltage-dependent calcium channels. Nature **336:** 382-385.
26. BENHAM, C. D. & R. W. TSIEN. 1988. Noradrenaline modulation of calcium channels in single smooth muscle cells from rabbit ear artery. J. Physiol. **404:** 767-784.
27. MAYER, M. L. & G. L. WESTBROOK. 1987. Permeation and block of *N*-methyl-D-aspartic acid receptor channels by divalent cations in mouse cultured central neurones. J. Physiol. **394:** 501-527.
28. FINKEL, A. S., G. D. S. HIRST & D. P. VAN HELDEN. 1984. Some properties of excitatory junction currents recorded from submucosal arterioles of guinea-pig ileum. J. Physiol. **351:** 87-98.
29. LUFF, S. E., E. M. MCLACHLAN & G. D. S. HIRST. 1987. An ultrastructural analysis of the sympathetic neuromuscular junctions on arterioles of the submucosa of the guinea-pig ileum. J. Comp. Neurol. **257:** 578-594.
30. KAJIWARA, M. & R. CASTEELS. 1983. Effects of Ca antagonists on neuromuscular transmission in the rabbit ear artery. Pfluegers Arch. Gesamte Physiol. Menschen Tiere **396:** 1-7.
31. BLAUSTEIN, M. P. 1977. Sodium ions, calcium ions, blood pressure regulation, and hypertension: A reassessment and a hypothesis. Am. J. Physiol. **232**(3): C165-C173.

DISCUSSION OF THE PAPER

E. M. SILINSKY (*Northwestern Medical School, Chicago, IL*): Given that it is possible to consider that P_2-purinoceptors are members of a family of receptors linked to rapid excitation via ion channels (this family includes nicotinic, 5HT3, and glutamate receptors), is the rate of recovery from desensitization produced by P_{2X} agonists dependent upon the nature of the agonist (as has been shown for nicotinic receptors)? An independence would suggest that the rate-limiting step in recovery would be due to the unliganded desensitized state isomerizing to an undesensitized, activatable receptor. It might be possible to determine whether a G protein is involved by lowering the temperature to levels at which G protein-linked effects would be impaired (by impairing the dissociation of the α subunit from the β complex) with minimal effects on channels directly linked to agonist binding and unlinked to G proteins.

BENHAM: I agree that examining the temperature dependence of the response would be simple to do. My feeling is that the channels are probably directly coupled to the ATP receptor—directly analogous to the acetylcholine receptor channel. The kinetic experiments of Bruce Bean and David Friel support this interpretation.

E. ROJAS (*National Institute of Diabetes and Digestive and Kidney Diseases, Bethesda, MD*): What can you say about the minimal requirements for the maintenance of the Ca^{2+} conductance? You showed indo-1 records from cells in which the cytoplasm was exposed to the solution in the pipette. Did you get these records with physiological extracellular medium?

BENHAM: All the experiments in which cytoplasmic Ca^{2+} was measured with indo-1 were performed with a normal physiological external solution containing 1.5 mM Ca^{2+}. The pipette solution contained no Ca^{2+} buffers other than the indo-1 itself (100 μM) and for most experiments was potassium chloride based. Thus, the cation gradients were as close as possible to a physiological situation, as was the membrane potential at which the cells were held (-60 mV). So, I think that we can be quite confident that a significant amount of Ca^{2+} will enter the cells when this pathway is activated.

R. I. HUME (*University of Michigan, Ann Arbor, MI*): The responses you described desensitize quite profoundly. How can this be reconciled with the idea that this is the fast excitatory neurotransmitter?

BENHAM: Desensitization may be accentuated in these dialyzed cells, or rather, recovery from it may be slowed. The concentrations of ATP used were maximal doses. At lower concentrations, say 10 nM, smaller currents can be recorded showing much less desensitization, and this may be a more physiological stimulus. It may also be relevant that transmitter release is intermittent so that individual cells are not exposed to transmitter following every stimulus.

Neural Release of ATP and Adenosine[a]

THOMAS D. WHITE AND WENDA F. MACDONALD

Department of Pharmacology
Dalhousie University
Halifax, Nova Scotia, Canada B3H 4H7

INTRODUCTION

There is mounting evidence that extracellular purines such as adenosine and ATP perform important functions in the central and peripheral nervous systems, where their actions are mediated by occupation of P_1- and P_2-purinoceptors, respectively. In order to elicit these responses, the purines must first be released from nerves. Since 1977, we have been characterizing the release of ATP and adenosine in several preparations from the central and autonomic nervous systems. The release and actions of adenosine and ATP have been reviewed extensively elsewhere.[1–10] Here we present some of our findings on the potential sources of neuronally released ATP and adenosine. In the first section of this paper, evidence is presented that ATP is coreleased with noradrenaline from sympathetic nerves. In certain cases, however, it does not appear that ATP is cosecreted with noradrenaline from noradrenergic vesicles. In the second section of this paper, we present evidence that depolarization-evoked adenosine release from rat brain synaptosomes occurs via the nucleoside transporter, and that the synaptosomal release of adenosine varies from region to region in the brain.

SOURCE OF ATP RELEASE IN THE AUTONOMIC NERVOUS SYSTEM

Release of ATP in the Autonomic Nervous System

In the early 1970s, Burnstock *et al.*[11] first proposed that purinergic nerves, which release ATP as their primary neurotransmitter, might mediate nonadrenergic, noncholinergic (NANC) relaxations of gastrointestinal smooth muscle. He and his colleagues showed that [^{3}H]adenosine was incorporated into [^{3}H]ATP in the guinea pig taenia coli and that [^{3}H]purines were released during transmural electrical stimulation, about 33% apparently arising from the nerves and the remainder from the smooth

[a]This research was supported by a grant to T.D.W. from the Medical Research Council of Canada.

muscle.[12,13] Release of endogenous, unlabeled ATP from the taenia coli following transmural electrical stimulation has been reported,[14] although subsequent experiments could only demonstrate ATP release at stimulation parameters that might have directly depolarized the smooth muscle and nerves of the taenia coli.[15]

Unlike the situation in the guinea pig taenia coli, there is considerable evidence against the concept that ATP mediates NANC relaxations in the guinea pig ileum.[10,16,17] Nevertheless, Ca^{2+}-dependent release of endogenous ATP has been observed during exposure of isolated myenteric nerve varicosities to elevated extracellular K^+, the Na^+-channel activator veratridine, acetylcholine, and 5-hydroxytryptamine (5-HT).[18–22] The released ATP could, following extracellular degradation to adenosine by ecto-ATPase, ecto-ADPase, and ecto-5′-nucleotidase, provide a source of adenosine that could act at presynaptic P_1-purinoceptors to diminish the release of acetylcholine.[23–30] Much of the released ATP appears to arise from noradrenergic varicosities insofar as release is reduced substantially following chemical sympathectomy with 6-hydroxydopamine (FIG. 1). This treatment appears to be selective for noradrenergic nerves because it does not destroy serotonergic varicosities.[22]

By analogy with adrenal chromaffin cells, where there is evidence that ATP is costored with catecholamines within the chromaffin granules in the ratio of 4 moles catecholamine:1 mole ATP,[31] it has generally been assumed that substantial amounts of ATP are similarly costored with noradrenaline within sympathetic vesicles. Some studies, however, have suggested that the molar ratio of noradrenaline to ATP may only be 12:1 in purified sympathetic vesicles,[32] raising the question of whether noradrenergic vesicles can secrete significant amounts of ATP. In this regard, we have found that the α_2 agonist, clonidine, diminishes the release of [^{3}H]noradrenaline

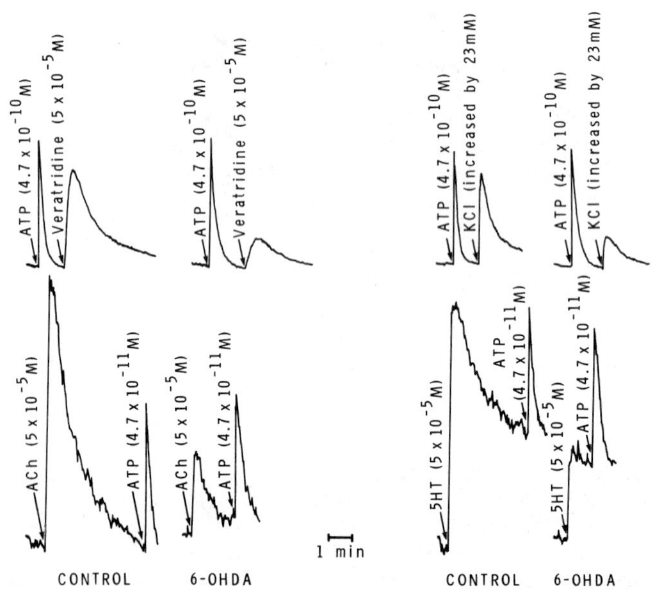

FIGURE 1. Effect of pretreatment of guinea pigs with 6-hydroxydopamine (250 mg/kg, i.p.) on evoked release of ATP from guinea pig ileal myenteric synaptosomes. ATP release was directly and continuously monitored in the presence of firefly luciferin-luciferase.[18] Results shown (derived from references 20-22) are typical of four to six separate experiments.

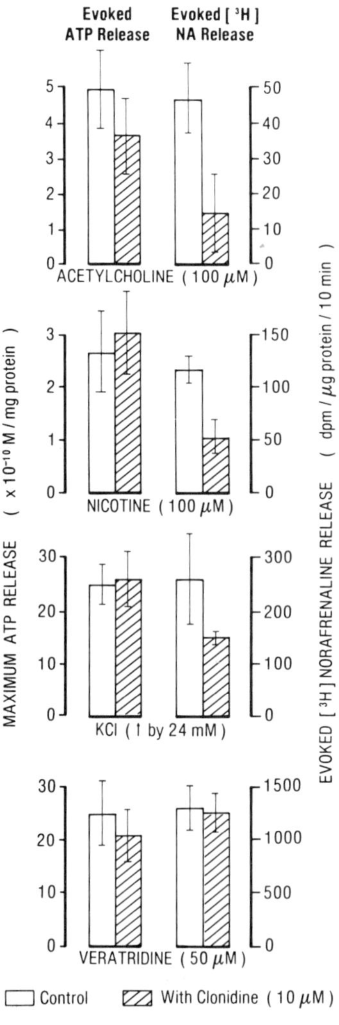

FIGURE 2. Effect of clonidine (10 μM) on evoked release of ATP and [^{3}H]noradrenaline from myenteric synaptosomes. Each bar represents the mean ± SEM for five experiments. Reproduced from Hammond *et al.*[33] with the permission of the National Research Council of Canada.

evoked by K^+, acetylcholine, and nicotine from isolated myenteric varicosities, but has no effect on the release of ATP evoked by these agents (FIG. 2).[33] This observation is difficult, if not impossible, to reconcile with the idea that ATP is coreleased with noradrenaline from the same synaptic vesicles. The implications of this observation are discussed below.

There is excellent evidence that ATP is an excitatory cotransmitter with noradrenaline in the vas deferens, where it produces excitatory junction potentials and mediates the first, rapid phase of contraction in the smooth muscle; noradrenaline mediates a slow depolarization and the second phase of contraction.[10] Stimulation of the hypogastric nerve innervating the guinea pig vas deferens releases endogenous ATP, which can be detected using a firefly luciferin-luciferase method (FIG. 3).[34] Release is not due to muscle contraction because contraction of the muscle with phenylephrine does not release ATP.[34] Moreover, block of the slow contractile response with prazosin, and of the initial twitch response following desensitization of the P_{2X}-purinoceptors with α,β-methylene-ATP, does not diminish the evoked release of ATP (FIG. 3), indicating that release occurred from the sympathetic nerves and not post-synaptically from the muscle.

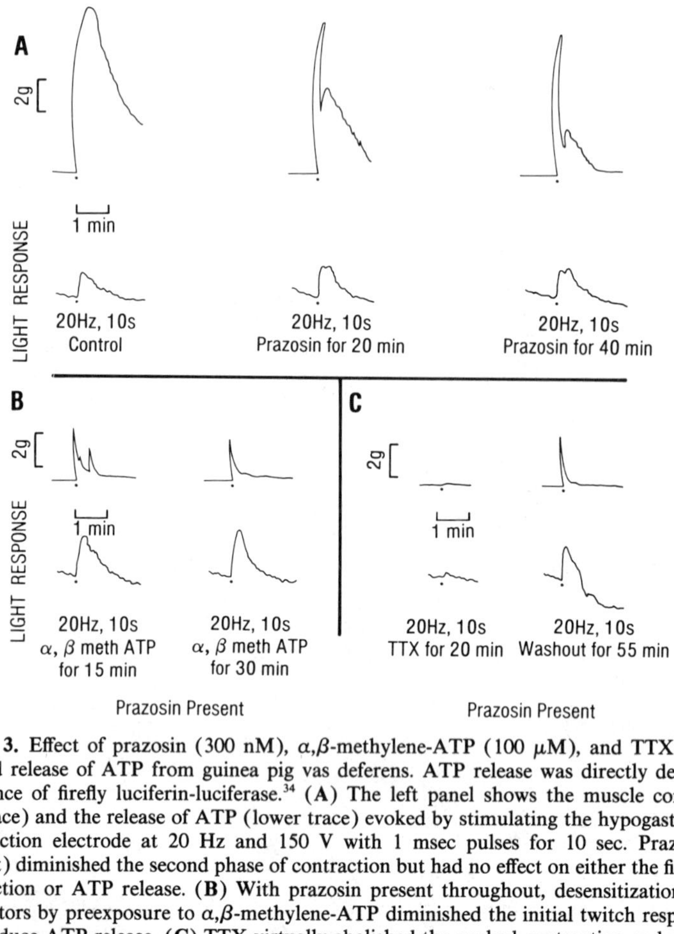

FIGURE 3. Effect of prazosin (300 nM), α,β-methylene-ATP (100 μM), and TTX (1 μM) on evoked release of ATP from guinea pig vas deferens. ATP release was directly detected in the presence of firefly luciferin-luciferase.[34] (**A**) The left panel shows the muscle contraction (upper trace) and the release of ATP (lower trace) evoked by stimulating the hypogastric nerve with a suction electrode at 20 Hz and 150 V with 1 msec pulses for 10 sec. Prazosin (α_1 antagonist) diminished the second phase of contraction but had no effect on either the first phase of contraction or ATP release. (**B**) With prazosin present throughout, desensitization of P_{2X}-purinoceptors by preexposure to α,β-methylene-ATP diminished the initial twitch response but did not reduce ATP release. (**C**) TTX virtually abolished the evoked contraction and release of ATP in a reversible manner. Reproduced from Lew and White[34] with the permission of the Macmillan Press.

Is ATP Secreted from Noradrenergic Synaptic Vesicles?

As was discussed above, clonidine modulates the evoked release of [^{3}H]noradrenaline without affecting the release of ATP from the same preparation of ileal myenteric varicosities.[33] Yet much of the ATP release appears to arise from noradrenergic varicosities, insofar as release is greatly diminished following pretreatment of the animals with 6-hydroxydopamine, which selectively destroys the noradrenergic varicosities.[20-22] These findings are not consistent with the proposition that ATP is cosecreted with noradrenaline from identical synaptic vesicles because, in this case, one would expect the release of both noradrenaline and ATP to be modulated in an exactly parallel fashion. Similar discrepancies can be derived from recent studies with rabbit vas deferens where prostaglandin E_2 inhibits nonadrenergic but enhances adrenergic responses to electrically mediated (or nerve-mediated) contractions.[35] Moreover, acute treatments with the adrenergic neuron blocker guanethidine, with the α_2 agonist guanabenz, and with 6-hydroxydopamine, all have more pronounced effects on adrenergic responses than on nonadrenergic responses to nerve stimulation. The authors concluded from these indirect measurements that the ratio of noradrenaline and ATP released by electrical stimulation differs depending on the pharmacological manipulations applied, and that these results argue against cosecretion of noradrenaline and ATP from the same vesicles. Direct studies of the effects of presynaptic modulators on the evoked release of ATP and noradrenaline from vas deferens are required before firm conclusions can be reached, but the results of this study, together with our findings on ATP and [^{3}H]noradrenaline release from ileal myenteric varicosities, certainly do not support the concept of cosecretion of ATP and noradrenaline from the same synaptic vesicles.

What then is the intracellular source of ATP that is released from sympathetic nerves? ATP could be released from autonomic nerves, which are nonadrenergic but susceptible to chemical sympathectomy with 6-hydroxydopamine. The actions of 6-hydroxydopamine in the guinea pig myenteric plexus, however, appear to be selective to noradrenergic fibers because serotonergic varicosities are not destroyed.[22] The most plausible explanation is that ATP and noradrenaline are released from different intracellular sites in noradrenergic sympathetic nerves. The Ca^{2+} dependency of evoked ATP release from myenteric varicosities suggests that this release occurs from synaptic vesicles. Given that noradrenergic nerve terminals contain a variety of morphologically distinguishable synaptic vesicles,[18] it is possible that ATP is stored separately from noradrenaline in another population of synaptic vesicles located within noradrenergic varicosities. It is also possible that ATP is released directly from the cytoplasm, rather than from synaptic vesicles, by some process that is Ca^{2+} dependent. Regardless of the source of ATP that is released, the observation that noradrenaline release is modulated under circumstances where ATP release is not raises important issues concerning the consequences that this might have on synaptic functions in those cases where both noradrenaline and ATP act as cotransmitters.

ADENOSINE RELEASE FROM SYNAPTOSOMES

Evoked Adenosine Release Occurs on the Nucleoside Transporter

Previously, we have demonstrated that depolarization of rat brain synaptosomes with K^+ or veratridine releases endogenous adenosine.[36] Most of the basal release of adenosine appeared to arise from the extracellular degradation of a released nucleotide, insofar as inhibition of ecto-5′-nucleotidase by 90% with α,β-methylene-ADP and GMP diminished basal extrasynaptosomal adenosine levels by 74%. In contrast, about half of the veratridine-evoked and most of the K^+-evoked release of adenosine arose from adenosine released in its own right rather than from the extracellular metabolism of released nucleotide.

Whereas veratridine-evoked release of adenosine was not diminished in the absence of extracellular Ca^{2+}, K^+-evoked release required extracellular Ca^{2+} and consequently resembled the secretion of classical neurotransmitters. Nevertheless, evidence that adenosine is stored in synaptic vesicles is lacking. Carrier-mediated, facilitated diffusion of nucleosides into various tissues and cells including neuronal preparations has been described previously.[37–45] This facilitated diffusion system transports adenosine down its concentration gradient and is bidirectional. Therefore, not only would the nucleoside transporter facilitate the movement of adenosine into cells when the extracellular concentration of adenosine exceeds the intracellular concentration, but it should also facilitate the efflux of adenosine from cells when the intracellular concentration of adenosine exceeds the extracellular concentration. Indeed, previous studies have shown that the evoked release of radiolabeled purines is diminished when the nucleoside transporter is inhibited by dipyridamole.[46–49]

Both dipyridamole and nitrobenzylthioinosine (NBI) have been used extensively to inhibit nucleoside transport. Because there is evidence that there may be NBI-insensitive nucleoside transporters in certain tissues including the CNS,[50–53] we studied the effect of 20 μM dipyridamole on the release of adenosine from rat brain synaptosomes. This concentration of dipyridamole maximally inhibits the uptake of [^{3}H]adenosine by rat synaptoneurosomes.[52]

In the presence of 20 μM dipyridamole, basal release of adenosine was increased 3-fold (TABLE 1). Because most of the basal extrasynaptosomal adenosine is due to release of a nucleotide followed by its extrasynaptosomal metabolism to adenosine,[36] block of the nucleoside transporter prevents this adenosine from being taken up into the synaptosomes and elevates the levels of adenosine accumulating in the medium.

In contrast, K^+-evoked release of adenosine was diminished by 25% when the nucleoside transporter was inhibited with dipyridamole, a result consistent with the notion that part of the adenosine released by K^+ occurs on the nucleoside transporter. Similarly, veratridine-evoked release of adenosine was diminished by 22% when the nucleoside transporter was inhibited by dipyridamole. The reason that inhibition of the nucleoside transporter does not produce a more substantial block of evoked adenosine release probably relates to the fact that depolarization also releases a nucleotide (possibly ATP[54,55]) in addition to adenosine. Adenosine formed extrasynaptosomally from released nucleotide would accumulate in the medium when the nucleoside transporter is inhibited, and this would tend to offset the inhibitory effect that dipyridamole has on release of adenosine. It is also possible, if the intracellular levels are increased sufficiently, that some of the adenosine exits the synaptosomes via passive diffusion, which would not occur on the nucleoside transporter.[48] These effects of dipyridamole on evoked adenosine release do not appear to be mediated by inhibition of cyclic AMP-phosphodiesterase because 100 μM rolipram, a concentration three times greater than that which maximally elevates cyclic AMP levels in rat brain slices,[56] had no effect on basal or evoked release of adenosine (data not shown).

Recently, a high-affinity, Na^+-dependent uptake system for adenosine into dissociated rat brain cells was described.[45] This system, which is capable of actively accumulating adenosine and is inhibited by 10 μM dipyridamole, could be distinguished from the lower affinity, facilitated diffusion system for nucleosides. If such a high-affinity, Na^+-dependent transport system exists in synaptosomes, as the work of Bender *et al.*[39] suggests, then depolarization with K^+ would diminish the inward-directed electrical gradient for Na^+ and could reverse the Na^+-coupled transport of adenosine from an inward to an outward direction. Similarly, veratridine would open Na^+ channels and diminish the inward electrical and chemical gradients for Na^+, so that the Na^+-coupled outward transport of adenosine would be favored.

It would appear from the above discussion that carrier-mediated transport systems for adenosine could function not only to remove extracellular adenosine, thereby terminating its extracellular effects and salvaging the purine, but also to provide a means of releasing adenosine into the extracellular space, where it can elicit its neuromodulatory effects at P_1-purinoceptors. Inhibition of the nucleoside transporter(s) might potentiate the actions of extracellular adenosine when the adenosine is derived from a released nucleotide, but it might, in some cases, diminish the extracellular actions of adenosine if that adenosine is released on the nucleoside transporter.

Regional Distribution of Adenosine Release in Rat Brain

If depolarization-evoked release of adenosine from synaptosomes is distributed uniformly throughout the brain, this may indicate that purine release occurs as a consequence of depolarization of all nerve endings. Conversely, regional variations

TABLE 1. Effect of Dipyridamole (DPR) on Synaptosomal Adenosine Release[a]

		Adenosine Release (pmol/mg protein/10 min)	
		Total (% control)	Evoked (% control)
Basal	Control	68.37 ± 18.15 (100)	
	+DPR (20 μM)	220.36 ± 17.34 (322)[b]	
K^+	Control	206.28 ± 17.58	137.91 ± 14.10 (100)
	+DPR (20 μM)	323.14 ± 26.66	102.78 ± 29.85 (75)[b]
Ver	Control	766.69 ± 60.05	698.32 ± 68.84 (100)
	+DPR (20 μM)	762.43 ± 66.47	542.07 ± 81.88 (78)[b]

[a] Synaptosomes (P_2) were incubated in the absence and presence of 20 μM dipyridamole to inhibit adenosine transport. Then, K^+ (increased by 24 mM) or veratridine (50 μM) was added to depolarize synaptosomes. Evoked release of adenosine is the value obtained with the depolarizing agent minus the corresponding basal release for 10 min. Adenosine was determined as described previously.[36] Each value is a mean ± SEM from six experiments.

[b] Significantly different from the respective control ($p < .05$, paired t test).

would perhaps suggest more discrete events restricted to certain types of nerve terminals or certain areas of the brain.

In order to investigate possible regional variations in adenosine release, rat brains were dissected according to the method of Glowinski and Iversen[57] and P_2 synaptosomes were prepared.[36] Occluded lactate dehydrogenase (LDH) was determined on each preparation to standardize the amount of adenosine release in terms of synaptosomal contents.[58] The results are expressed as pmol adenosine released/1000 occluded LDH units, as shown in TABLE 2.

Release of endogenous adenosine is not uniformly distributed for synaptosomes prepared from different regions of rat brain. Basal adenosine release exhibited the following distribution: cerebellum > hippocampus ≥ amygdala, striatum, cortex, hypothalamus > pons/medulla > thalamus. The K^+-evoked release of adenosine exhibited this distribution: amydala > cortex > striatum > hippocampus > thalamus > hypothalamus > cerebellum > pons/medulla. The veratridine-evoked release was as follows: amygdala > cortex > thalamus, cerebellum > hippocampus, striatum >

hypothalamus > pons/medulla. The most striking observation is that both K^+- and veratridine-evoked release of adenosine was relatively high from amygdala synaptosomes and low from hypothalamic synaptosomes. On the other hand, there appeared to be regional differences in K^+- and veratridine-evoked adenosine release. Thus, K^+ released little adenosine from the thalamus and cerebellum, whereas veratridine released considerable amounts of adenosine from both of these regions. The significance of these observations is not yet apparent.

There are no apparent correlations between the distributions of evoked adenosine release and the distributions of evoked ATP release,[58] A_1 receptors,[59–61] A_2 receptors,[62] adenosine uptake sites,[60,63] adenosine-like immunoreactivity,[64] adenosine deaminase,[65] or 5′-nucleotidase activity.[66] It should be noted that there is also little apparent correlation among the above markers of purinergic function. The variations in the distributions of these potential markers probably reflect the complexity of the interrelations of the various processes involved. The interplay between some or all of these processes may determine the levels of extraneuronal adenosine that can occur in the various regions.

TABLE 2. Synaptosomal Adenosine Release from Various Brain Regions[a]

Brain Region	Adenosine Release (pmol/1000 occluded LDH units)		
		Evoked	
	Basal	K^+	Veratridine
Striatum	92.7 ± 10.2	64.6 ± 9.2	272.9 ± 36.2
Amygdala	95.0 ± 9.4	120.8 ± 18.4	401.2 ± 45.7
Hypothalamus	86.4 ± 4.9	22.7 ± 6.0	229.3 ± 18.0
Hippocampus	106.8 ± 12.0	49.3 ± 6.6	270.0 ± 18.6
Thalamus	51.3 ± 8.1	30.4 ± 9.6	306.9 ± 23.6
Cortex	92.2 ± 15.1	75.3 ± 9.2	341.3 ± 24.0
Cerebellum	225.0 ± 27.3	10.8 ± 11.3	292.2 ± 2.0
Pons/Medulla	71.9 ± 7.3	0.9 ± 3.9	135.4 ± 13.4

[a] Synaptosomes (P_2) were prepared from eight brain regions, as described in the text. Then, K^+ (increased by 24 mM) or veratridine (50 μM) was added to depolarize synaptosomes. Adenosine release was determined as described previously.[36] Each value is a mean ± SEM from five experiments.

SUMMARY

Release of ATP can be evoked from noradrenergic nerve varicosities isolated from guinea pig ileal myenteric plexus by depolarization with K^+ and veratridine and during exposure to acetylcholine or 5-HT. Clonidine, however, modulates the release of [^{3}H]noradrenaline without affecting the release of ATP. ATP is also released from noradrenergic sympathetic nerves in the vas deferens, where it mediates the initial depolarization and contraction in the smooth muscle. Factors that apparently modulate the release of noradrenaline do not produce corresponding effects on ATP release. The above results are best explained by the hypothesis that ATP and noradrenaline are stored in separate populations of vesicles within sympathetic nerves and that these pools are subject to differential presynaptic modulation.

Depolarization of rat brain synaptosomes releases adenosine by a process that is mediated, at least in part, by efflux on the nucleoside transporter. Drugs that block the nucleoside transporter (such as dipyridamole) reduce evoked adenosine release and may thereby diminish, rather than augment, the actions of adenosine at its receptors. Release of adenosine does not appear to be uniformly distributed throughout the brain insofar as release varies from synaptosomes prepared from different regions. Although the distribution of several markers for possible adenosine pathways in the brain, including adenosine release, do not show any consistent correlations, the non-uniform distribution for these markers suggests that adenosine may have differential functions in various brain regions.

REFERENCES

1. BAER, H. P. & G. I. DRUMMOND, EDS. 1979. Physiological and Regulatory Functions of Adenosine and Adenine Nucleotides. Raven Press. New York, NY.
2. STONE, T. W. 1981. Physiological roles for adenosine and adenosine 5′-triphosphate in the nervous system. Neuroscience **6:** 523-555.
3. PHILLIS, J. H. & P. H. WU. 1981. The role of adenosine and its nucleotides in central synaptic transmission. Prog. Neurobiol. **16:** 187-239.
4. PATON, D. M., ED. 1985. Methods Used in Adenosine Research. Methods in Pharmacology. Vol. 6. Plenum. New York, NY.
5. GORDON, J. L. 1986. Extracellular ATP: Effects, sources and fate. Biochem. J. **233:** 309-319.
6. WILLIAMS, M. 1987. Purine receptors in mammalian tissues: Pharmacology and functional significance. Annu. Rev. Pharmacol. Toxicol. **27:** 315-345.
7. WHITE, T. D. 1984. Characteristics of neuronal release of ATP. Prog. Neuro-Psychopharmacol. Biol. Psychiatry **8:** 487-493.
8. WHITE, T. D. 1985. The demonstration and measurement of adenosine triphosphate release from nerves. *In* Methods in Pharmacology. D. M. Paton, Ed. Vol. 6: 43-46. Plenum. New York, NY.
9. WHITE, T. D. 1985. Release of ATP from central and peripheral nerve terminals. *In* Purines: Pharmacology and Physiological Roles. T. W. Stone, Ed.: 95-105. Macmillan. London.
10. WHITE, T. D. 1988. Role of adenine compounds in autonomic neurotransmission. Pharmacol. Ther. **38:** 129-168.
11. BURNSTOCK, G., G. CAMPBELL, D. G. SATCHELL & A. SMYTHE. 1970. Evidence that adenosine triphosphate or a related nucleotide is the transmitter substance released by nonadrenergic inhibitory nerves in the gut. Br. J. Pharmacol. **40:** 668-688.
12. SU, C., J. A. BEVAN & G. BURNSTOCK. 1971. [^{3}H]Adenosine triphosphate: Release during stimulation of enteric nerves. Science **173:** 336-338.
13. RUTHERFORD, A. & G. BURNSTOCK. 1978. Neuronal and nonneuronal components in the overflow of labelled adenyl compounds from guinea-pig taenia coli. Eur. J. Pharmacol. **48:** 192-202.
14. BURNSTOCK, G., T. COCKS, L. KASAKOV & H. WONG. 1978. Direct evidence for ATP release from non-adrenergic, non-cholinergic ('purinergic') nerves in the guinea-pig taenia coli and bladder. Eur. J. Pharmacol. **49:** 145-149.
15. WHITE, T., P. POTTER, C. MOODY & G. BURNSTOCK. 1981. Tetrodotoxin-resistant release

of ATP from guinea-pig taenia coli and vas deferens during electrical field stimulation in the presence of luciferin-luciferase. Can. J. Physiol. Pharmacol. **59:** 1094-1100.
16. BAUER, V. & H. KURIYAMA. 1982. The nature of non-cholinergic, non-adrenergic transmission in longitudinal and circular muscles of the guinea-pig ileum. J. Physiol. (London) **332:** 375-391.
17. MATUSAK, O. & V. BAUER. 1986. Effect of desensitization induced by adenosine 5′-triphosphate, substance P, bradykinin, τ-aminobutyric acid and endogenous noncholinergic, nonadrenergic transmitter in the guinea-pig ileum. Eur. J. Pharmacol. **126:** 199-209.
18. WHITE, T. D. & R. A. LESLIE. 1982. Depolarization-induced release of adenosine 5′-triphosphate from isolated varicosities derived from the myenteric plexus of the guinea-pig small intestine. J. Neurosci. **2:** 205-215.
19. WHITE, T. D. 1982. Release of ATP from isolated myenteric varicosities by nicotinic agonists. Eur. J. Pharmacol. **79:** 333-334.
20. WHITE, T. D. & M. AL-HUMAYYD. 1983. Acetylcholine releases ATP from varicosities isolated from the guinea-pig myenteric plexus. J. Neurochem. **40:** 1069-1075.
21. AL-HUMAYYD, M. & T. D. WHITE. 1985. 5-Hydroxytryptamine releases adenosine 5′-triphosphate from nerve varicosities isolated from the myenteric plexus of guinea-pig ileum. Br. J. Pharmacol. **84:** 27-34.
22. AL-HUMAYYD, M. & T. D. WHITE. 1985. Adrenergic and possible nonadrenergic sources of adenosine 5′-triphosphate release from nerve varicosities isolated from ileal myenteric plexus. J. Pharmac. Exp. Ther. **233:** 796-800.
23. SAWYNOK, J. & K. H. JHAMANDAS. 1976. Inhibition of acetylcholine release from cholinergic nerves by adenosine, adenine nucleotides and morphine: Antagonism by theophylline. J. Pharmac. Exp. Ther. **197:** 379-390.
24. GUSTAFSSON, L., P. HEDQVIST, B. B. FREDHOLM & G. LUNDGREN. 1978. Inhibition of acetylcholine release in guinea-pig ileum by adenosine. Acta Physiol. Scand. **104:** 469-478.
25. DOWDLE, E. B. & R. MASKE. 1980. The effects of calcium concentration on the inhibition of cholinergic neurotransmission in the myenteric plexus of guinea-pig ileum by adenine nucleotides. Br. J. Pharmacol. **71:** 245-252.
26. WATT, A. J. 1982. Direct and indirect effects of adenosine 5′-triphosphate on guinea-pig ileum. Br. J. Pharmacol. **77:** 725-730.
27. PAUL, M. L., D. L. MILES & M. A. COOK. 1982. The influence of glycosidic conformation and charge distribution on activity of adenine nucleotides as presynaptic inhibitors of acetylcholine release. J. Pharmac. Exp. Ther. **222:** 241-245.
28. MOODY, C. J. & G. BURNSTOCK. 1982. Evidence for the presence of P_1-purinoceptors on cholinergic nerve terminals in the guinea-pig ileum. Eur. J. Pharmacol. **77:** 1-9.
29. REESE, J. H. & J. R. COOPER. 1982. Modulation of the release of acetylcholine from ileal synaptosomes by adenosine and adenosine 5′-triphosphate. J. Pharmac. Exp. Ther. **223:** 612-616.
30. SHINOZUKA, K., T. MAEDA & E. HAYASHI. 1985. Effects of adenosine on ^{45}Ca uptake and [^{3}H]acetylcholine release in synaptosomal preparation from guinea-pig ileum myenteric plexus. Eur. J. Pharmacol. **113:** 417-424.
31. WINKLER, H. & S. W. CARMICHAEL. 1982. The chromaffin granule. *In* The Secretory Granule. A. M. Poisner & J. M. Trifaro, Eds.: 3-79. Elsevier. Amsterdam.
32. FRIED, G., H. LAGERCRANTZ & T. HOKFELT. 1978. Improved isolation of small noradrenergic vesicles from rat seminal ducts following castration: A density gradient centrifugation and morphological study. Neuroscience **3:** 1271-1291.
33. HAMMOND, J. R., W. F. MACDONALD & T. D. WHITE. 1988. Evoked secretion of [^{3}H]noradrenaline and ATP from nerve varicosities isolated from the myenteric plexus of guinea-pig ileum. Can. J. Physiol. Pharmacol. **66:** 369-375.
34. LEW, M. J. & T. D. WHITE. 1987. Release of endogenous ATP during sympathetic nerve stimulation. Br. J. Pharmacol. **92:** 349-355.
35. TRACHTE, G. S., S. B. BINDER & M. J. PEACH. 1989. Indirect evidence for separate vesicular neuronal origins of norepinephrine and ATP in the rabbit vas deferens. Eur. J. Pharmacol. **164:** 425-433.

36. MacDonald, W. F. & T. D. White. 1985. Nature of extrasynaptosomal accumulation of endogenous adenosine evoked by K^+ and veratridine. J. Neurochem. **45:** 791-797.
37. Shimuzu, H., C. R. Creveling & J. Daly. 1970. Stimulated formation of adenosine 3′,5′-cyclic phosphate in cerebral cortex: Synergism between electrical activity and biogenic amines. Proc. Natl. Acad. Sci. USA **65:** 1033-1040.
38. Hertz, L. 1978. Kinetics of adenosine uptake into astrocytes. J. Neurochem. **31:** 55-96.
39. Bender, A. S., P. H. Wu & J. W. Phillis. 1980. The characterization of [^{3}H]adenosine uptake into rat cerebral cortical synaptosomes. J. Neurochem. **35:** 629-640.
40. Bender, A. S., P. H. Wu & J. W. Phillis. 1981. The rapid uptake and release of [^{3}H]adenosine by rat cerebral cortical synaptosomes. J. Neurochem. **36:** 651-660.
41. Bender, A. S., P. H. Wu & J. W. Phillis. 1981. Some biochemical properties of the rapid adenosine uptake system in rat brain synaptosomes. J. Neurochem. **37:** 1282-1290.
42. Barberis, C., A. Minn & H. McIlwain. 1981. Adenosine transport into guinea-pig synaptosomes. J. Neurochem. **36:** 347-354.
43. Thampy, K. G. & E. M. Barnes. 1983. Adenosine transport by primary cultures of neurons from chick embryo brain. J. Neurochem. **40:** 874-879.
44. Geiger, J. D., M. E. Johnston & V. Yago. 1988. Pharmacological characterization of rapidly accumulating adenosine by dissociated brain cells from adult rat. J. Neurochem. **51:** 283-291.
45. Johnston, M. E. & J. D. Geiger. 1989. Sodium-dependent uptake of nucleosides by dissociated brain cells from the rat. J. Neurochem. **52:** 75-81.
46. Fredholm, B. B., A. Sollevi, L. Vernet & P. Hedqvist. 1980. Inhibition by dipyridamole of stimulated purine release. Naunyn-Schmiedeberg's Arch. Pharmacol. **313:** R18.
47. Fredholm, B. B., E. Lindgren, K. Lindstrom & L. Vernet. 1983. The effect of some drugs with purported antianoxic effect on veratridine-induced purine release from isolated rat hypothalamic synaptosomes. Acta Pharmacol. Toxicol. **53:** 236-244.
48. Fredholm, B. B., M. Duner-Engstrom, J. Fastbom, B. Jonzon, E. Lindgren & C. Nordstedt. 1988. Formation and actions of adenosine in the rat hippocampus, with special reference to the interactions with classical transmitters. *In* Neurotransmitters and Cortical Function: From Molecules to Mind. M. Avoli, T. A. Reader, R. W. Dykes & P. Gloor, Eds.: 437-451. Plenum. New York, NY.
49. Jonzon, B. & B. B. Fredholm. 1985. Release of purines, noradrenaline, and GABA from rat hippocampal slices by field stimulation. J. Neurochem. **44:** 217-224.
50. Belt, J. A. 1983. Heterogeneity of nucleoside transport in mammalian cells: Two types of transport activity in L1210 and other cultured neoplastic cells. Mol. Pharmacol. **24:** 479-484.
51. Davies, L. P. & J. W. Hambley. 1986. Regional distribution of adenosine uptake in guinea-pig brain slices and the effects of some inhibitors: Evidence for nitrobenzylthioinosine-sensitive and insensitive sites? Neurochem. Int. **8:** 103-108.
52. Morgan, P. F. & P. J. Marangos. 1987. Comparative aspects of nitrobenzylthioinosine and dipyridamole inhibition of adenosine accumulation in rat and guinea pig synaptoneurosomes. Neurochem. Int. **11:** 339-346.
53. Marangos, P. J. & J. Deckert. 1987. [^{3}H]Dipyridamole binding to guinea-pig brain membranes: Possible heterogeneity of central adenosine uptake sites. J. Neurochem. **48:** 1231-1236.
54. White, T. D. 1977. Direct detection of depolarization-induced release of ATP from a synaptosomal preparation. Nature **267:** 67-68.
55. White, T. D. 1978. Release of ATP from a synaptosomal preparation by elevated extracellular K^+ and veratridine. J. Neurochem. **30:** 329-336.
56. Schwabe, U., M. Miyake, Y. Ohga & J. W. Daly. 1976. 4-(3-Cyclopentyloxy-4-methoxyphenyl)-2-pyrrolidine (ZK 62711): A potent inhibitor of adenosine cyclic 3′,5′-monophosphate phosphodiesterases in homogenates and tissue slices from rat brain. Mol. Pharmacol. **12:** 900-910.
57. Glowinski, J. & L. L. Iversen. 1966. Regional studies of catecholamines in the rat brain.

I. The disposition of [^{3}H]norepinephrine, [^{3}H]dopamine and [^{3}H]dopa in various regions of the brain. J. Neurochem. **13:** 655-669.

58. POTTER, P. E. & T. D. WHITE. 1980. Release of adenosine 5′-triphosphate from synaptosomes from different regions of rat brain. Neuroscience **5:** 1351-1356.
59. PATEL, J., P. J. MARANGOS, J. STIVERS & F. K. GOODWIN. 1982. Characterization of adenosine receptors in brain using N^6-cyclohexyl [^{3}H]adenosine. Brain Res. **237:** 203-214.
60. GEIGER, J. D. & J. I. NAGY. 1984. Heterogeneous distribution of adenosine transport sites labelled by [^{3}H]nitrobenzylthioinosine in rat brain: An autoradiographic and membrane-binding study. Brain Res. Bull. **13:** 657-666.
61. GOODMAN, R. R. & S. H. SNYDER. 1982. Autoradiographic localization of adenosine receptors in rat brain using [^{3}H]cyclohexyladenosine. J. Neurosci. **2:** 1230-1241.
62. BRUNS, R. F., G. H. LU & T. A. PUGSLEY. 1986. Characterization of the adenosine A_2 receptor labelled by [^{3}H]NECA in rat striatal membranes. Mol. Pharmacol. **29:** 331-346.
63. BISSERBE, J. C., J. PATEL & P. J. MARANGOS. 1985. Autoradiographic localization of adenosine uptake sites in rat brain using [^{3}H]nitrobenzylthioinosine. J. Neurosci. **5:** 544-550.
64. BRAAS, K. M., A. C. NEWBY, V. S. WILSON & S. H. SNYDER. 1986. Adenosine-containing neurons in the brain localized by immunocytochemistry. J. Neurosci. **6:** 1962-1961.
65. GEIGER, J. D. & J. I. NAGY. 1986. Distribution of adenosine deaminase activity in rat brain and spinal cord. J. Neurosci. **6:** 2707-2714.
66. NAGATA, H., Y. MIMORI, S. NAKAMURA & M. KAMEYAMA. 1984. Regional and subcellular distribution in mammalian brain of the enzyme producing adenosine. J. Neurochem. **42:** 1001-1007.

DISCUSSION OF THE PAPER

A. WIERASZKO (*Boston College, Chestnut Hill, MA*): 1) Why is the basal release of adenosine from cerebellum higher than K^+-stimulated release? 2) Was this release Ca^{2+} dependent? 3) What is the mechanism *in vivo* by which the release of ATP and noradrenaline from different pools of vesicles is spontaneously regulated?

WHITE: 1) The K^+-stimulated release represents the real increase in adenosine release. 2) The K^+-stimulated release was Ca^{2+} dependent, although veratridine-stimulated release was not. 3) It is known in other systems too that ATP may be stored in a pool separate from that for the neurotransmitter. For instance, as we have shown, clonidine can modulate the release of noradrenaline from myenteric synaptosomes without affecting the release of endogenous ATP. This indicates that these two substances are not likely released from the same synaptic vesicle.

Y. H. EHRLICH (*College of Staten Island, New York, NY*): We have reported (*Brain Res. Bull.,* 1988) that cultured neurostriatal neurons that have undergone synaptogenesis release ATP in response to veratridine, and that 80% of this release is blocked by TTX. High K^+ depolarization also triggers ATP release from these neurons, 15% of which depend on the presence of extracellular Ca^{2+}. It thus seems that 20-25% of the ATP released from CNS neurons is not vesicular but cytoplasmic. Could you please comment on the possible significance of cytoplasmic ATP release in brain function?

WHITE: I am open to the possibility of nonvesicular release of ATP from tissues. On the other hand, your inability to entirely block veratridine-evoked release with TTX suggests that part of the effect of veratridine was not mediated by the opening of voltage-sensitive Na^+ channels. We see complete block of veratridine-evoked release of both ATP and adenosine by TTX.

E. ROJAS (*National Institute of Diabetes and Digestive and Kidney Diseases, Bethesda MD*): We have detected two components of the K^+-evoked ATP release. One that is extracellular Ca^{2+} dependent and the other that proceeds in the complete absence of Ca^{2+}. The second component is rather slow. I wonder if your inability to detect this second component is related to the fast degradation of the secreted ATP, which would be indicative of the presence of ecto-ATPases.

WHITE: I do not know. We have never observed a Ca^{2+}-independent, K^+-evoked release of either ATP or adenosine from rat brain synaptosomes.

ATP as a Cotransmitter[a]

DAVID P. WESTFALL, KHALED O. SEDAA,
KAZUMASA SHINOZUKA, RICHARD A. BJUR, AND
IAIN L. O. BUXTON

Department of Pharmacology
University of Nevada School of Medicine
Reno, Nevada 89557

INTRODUCTION

The notion that one neuron synthesizes, stores, and releases only one transmitter has been seriously questioned recently.[1] The possible existence of cotransmitters in postganglionic sympathetic and parasympathetic neurons has been of particular research interest of late, and much credit for focusing attention on this issue goes to G. Burnstock, who published a commentary in *Neuroscience* in 1976 entitled "Do Some Nerve Cells Release More Than One Transmitter?"[2]

Although a number of substances have been proposed to act as cotransmitters with the "classical" monoamine or amino acid transmitters,[3] the evidence to date is strongest for a nucleotide, most probably ATP. There has been, naturally enough, a general reluctance in accepting the notion that ATP, a compound that plays an essential role in intracellular function, could be released extracellularly. Nevertheless, the evidence is now abundant that adenine nucleotides and nucleosides do appear extracellularly upon stimulation of nerves, and this, together with the knowledge that these compounds exert a variety of pharmacological actions, makes it reasonable to postulate roles as neuromodulators and neurotransmitters.

The tissue for which the data is probably the most extensive in support of a cotransmitter role for ATP is the sympathetically innervated vas deferens. In this paper, we will begin by reviewing the evidence from our laboratory and the laboratories of others that norepinephrine (NE) and ATP are cotransmitters in the vas deferens. We will then discuss other neuroeffector junctions, especially vascular smooth muscle, from the standpoint of cotransmission. We will also consider the potential sites of release of endogenous ATP upon transmural nerve stimulation of blood vessels.

COTRANSMISSION IN VAS DEFERENS

For years there has been confusion about the nature of the motor transmission of the vas deferens.[4,5] Individual stimuli or low-frequency stimulation of nerves for short

[a]This work was supported in part by Grant HL 38126 from the National Institutes of Health and by grants from the American Heart Association.

periods produces a phasic, short-lasting contraction of the vas deferens. Longer periods of stimulation produce, following the initial phasic contraction, a later contraction, which is more tonic in character (see FIG. 1). The phasic contraction in response to nerve stimulation is considerably less sensitive to α-adrenoceptor blocking agents than the tonic contraction. After depletion of endogenous NE by pretreatment with reserpine, the tonic response to transmural nerve stimulation is greatly reduced, whereas the phasic contraction remains largely unaffected.[5,6] Among the many possible explanations for these results is the possibility that there is another transmitter.[4] Because ATP was known to be present as part of the adrenergic neurotransmitter-storage complex,[7] we sought evidence for ATP's participation in the neurotransmission process in the vas deferens.[8] ATP was found by us and by others[9] to be an effective contraction-inducing agent in the vas deferens. The response was transient in character and resembled the phasic contraction of the neurogenic response. ADP and AMP also produced contraction but were less potent than ATP. Adenosine failed to produce contraction. We also sought evidence for the release of ATP upon transmural nerve stimulation. The basic approach was to stimulate tissues that were preincubated with [^{3}H]adenosine. Preincubation in this manner was known to be effective for labeling endogenous ATP in smooth muscle.[10,11] Tissues so labeled released significant amounts of tritium upon transmural stimulation, supporting the idea of the neuronal release of ATP. (More recent studies using fluorescence and chemiluminescence techniques have confirmed the neuronal release of endogenous ATP in the vas deferens.[12,13])

Although these results helped to establish ATP as a cotransmitter in the vas deferens there was a major impediment to fully embracing the hypothesis, the impediment being the unavailability of a pharmacological antagonist that was specific for responses mediated by receptors for ATP. This problem was overcome by the introduction by Hogaboom and collaborators in 1980 of 3′-*O*-(3[*N*-(4-azido-2-nitrophe nyl)amino]propionyl)adenosine 5′-triphosphate (referred to as $ANAPP_3$) as an ATP antagonist.[14] $ANAPP_3$ is a photoaffinity analogue of ATP. Without photoactivation, $ANAPP_3$ produces an antagonism of ATP that is reversible. With photoactivation by visible light, the antagonism is quite persistent. Presumably, photolysis activates the compound to a reactive nitrene intermediate that is capable of forming covalent bonds with sites at or near the P_2-purinoceptor. Hogaboom *et al.*[14] showed that $ANAPP_3$ antagonized responses of the quinea pig vas deferens to adenine nucleotides but not to NE, acetylcholine (ACh), histamine, or KCl. Thus the compound possessed the requisite specificity to test the cotransmitter hypothesis. We used $ANAPP_3$ to determine whether adenine nucleotides released from nerves of the guinea pig vas deferens contributed to the neurogenic contraction.[16] $ANAPP_3$ preferentially antagonized the initial phasic component of the neurogenic response, whereas the secondary tonic component of the response was preferentially antagonized by the α-adrenoceptor antagonist prazosin (FIG. 1). Our studies also showed that the initial phasic contraction was maintained after pretreatment with reserpine but that both phases of the response were nearly abolished by destruction of the sympathetic nerves by 6-hydroxydopamine. This evidence added strong support to the idea that ATP, originating from sympathetic nerves, acts as a cotransmitter with NE in this tissue.

Mimicry by ATP and antagonism by $ANAPP_3$ was subsequently shown to occur at the electrophysiological level.[17,18] Intracellular microelectrodes were used to record the electrical response of the smooth muscle cells of the vas deferens to nerve stimulation and to exogenously applied ATP and NE. In the vas deferens, the initial phasic contraction is dependent upon the summation of excitatory junction potentials (EJPs) to threshold and the firing of action potentials producing the contraction. $ANAPP_3$ antagonizes both the initial phasic contraction and the EJPs. Prazosin does not (FIG. 2). Local application of ATP by pressure ejection from a micropipette

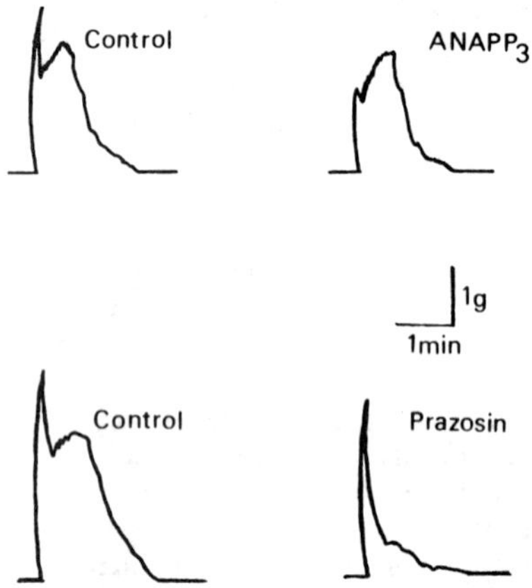

FIGURE 1. Polygraph tracings of the contractile response of the guinea pig vas deferens stimulated at 16 Hz (0.5 msec pulses at supramaximal voltage) for 30 sec. Tension was recorded isometrically in a continuously suffused organ bath, and field stimulation was applied with two platinum ring electrodes. Note that treatment with $ANAPP_3$ (10^{-4} M followed by photolysis) reduced the initial phasic component of the contraction and that prazosin (10^{-6} M) reduced the secondary tonic component of the contraction. For each drug treatment, the control response of the tissue is shown on the left. Reproduced from reference 17 with permission.

produced a depolarization similar in magnitude and time course to the EJP. NE produced no such response.

Although $ANAPP_3$ is an important tool, the compound is not available commercially and thus is not widely available to investigators wishing to study the potential role of ATP as a cotransmitter. In this context, the work of Kasakov and Burnstock in 1982 reporting the use of the slowly degradable analogue of ATP, α,β-methylene ATP (APCPP) was an important advance.[19] APCPP, which is available commercially, acts as an agonist at P_2-purinoceptors. Upon repeated exposure to or in the continual presence of APCPP there is a desensitization of P_2-purinoceptors such that responses to ATP and other agonists at P_2-purinoceptors are reduced. Used in this manner, APCPP can be viewed as an "antagonist."

Meldrum and Burnstock[20] subsequently showed that APCPP acted much the same way in the guinea pig vas deferens as did $ANAPP_3$; that is, after an initial contraction, APCPP reduced the initial phasic neurogenic contraction and the response to exogenous ATP while leaving the secondary tonic contraction essentially unchanged. APCPP was also shown to decrease the size of the EJP in the guinea pig vas deferens as well as the depolarization in response to the local application of ATP.[21]

Based on the evidence reviewed above, as well as a considerable number of other studies that have employed similar pharmacological and electrophysiological approaches (TABLE 1), the conclusion seems inescapable that neurotransmission in the vas deferens involves ATP and NE as cotransmitters.

COTRANSMISSION IN URINARY BLADDER

The smooth muscle of the body of the mammalian urinary bladder is another tissue that has confounded investigators, the difficulties having to do with the excitatory innervation of this tissue. The classical view has been that the bladder received excitatory cholinergic innervation exclusively, and indeed exogenous ACh mimics the motor response evoked by transmural nerve stimulation. Nevertheless, in spite of profound antagonism of the response of the bladder to exogenous ACh by the muscarinic cholinoceptor antagonist atropine, the neurogenic response of the bladder exhibits considerable resistance to antagonism by atropine. Consequently, there have been numerous suggestions over the years that excitatory neurotransmission to the bladder is partly noncholinergic.[30–38] The suggestion was made in 1972 by Burnstock and colleagues[32] and later by others[37] that ATP may be the noncholinergic excitatory transmitter in the urinary bladder. Originally the concept seemed to be that there may be a specific population of neurons that utilized ATP exclusively as a transmitter (that is, purinergic nerves), but this notion has been expanded to include the possibility that ATP is a cotransmitter released in conjunction with ACh and/or NE.[19,39] The experimental approaches that have supported the idea that ATP is a cotransmitter in bladder are similar to those discussed for the vas deferens, especially the use of $ANAPP_3$, the use of APCPP, and analyses of EJPs. Some of the papers that provide such information are listed in TABLE 2. Other smooth muscles for which there is

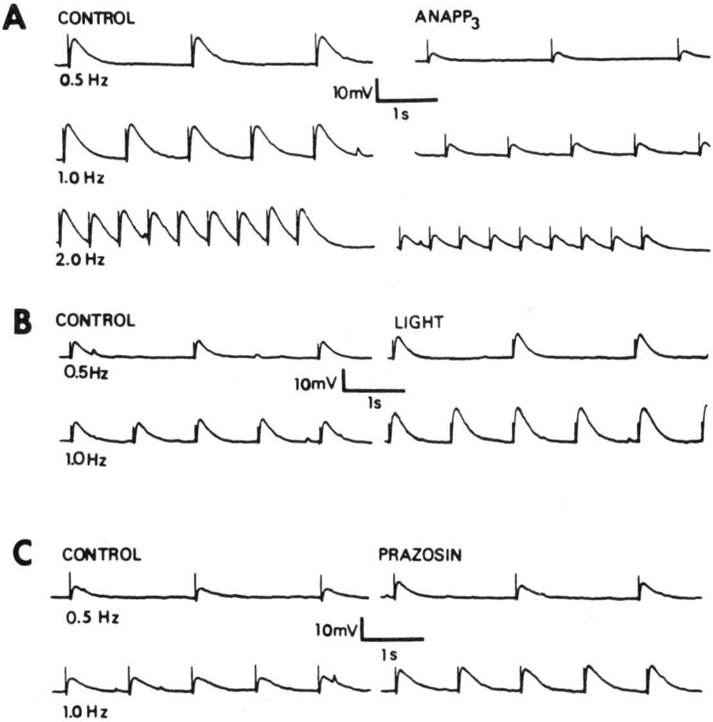

FIGURE 2. (**A**) Excitatory junction potentials (EJPs) recorded from guinea pig vas deferens at stimulation frequencies of 0.5, 1.0, and 2.0 Hz. Responses are from cells in the same muscle recorded before and after treatment with the P_2-antagonist $ANAPP_3$. The vertical line preceding each EJP is the stimulus artifact. (**B**) EJPs recorded from cells in the same muscle before and after treatment with light only (as a control for the photolysis of $ANAPP_3$). Note that the EJPs are substantially reduced after $ANAPP_3$ treatment (but not after light alone). (**C**) Continuous recording from the same cell of EJPs at 0.5 and 1.0 Hz before and 15 min after treatment with 10^{-6} M prazosin. Note that the EJPs are not reduced in size. Reproduced from reference 17 with permission.

TABLE 1. Examples of Papers Providing Evidence for ATP and NE as Cotransmitters in the Vas Deferens.

Species	Investigators	References
Guinea pig	Westfall *et al.*	8
	Fedan *et al.*	16
	Sneddon *et al.*	17
	Meldrum & Burnstock	20
	Sneddon & Westfall	18
	Sneddon & Burnstock	21
	Stjärne & Åstrand	22, 23
	Allcorn *et al.*	24
	Cunnane & Manchanda	25
	Suzuki & Gomi	26
Rat	French & Scott	27
	Sneddon *et al.*	28
	Satchell	29
Mouse	Stjärne & Åstrand	22, 23
	Allcorn *et al.*	24
Rabbit	Sneddon *et al.*	28

some evidence for ATP as a cotransmitter are the seminal vesicle of the guinea pig[46] and rat[47] and the nictitating membrane of the cat.[48,49]

COTRANSMISSION IN BLOOD VESSELS

Still another sympathetically innervated system for which there is a rapidly accumulating body of information suggesting a cotransmitter role for ATP is the vasculature. One of the earliest observations that focused attention on this possibility was made by the late Che Su.[50] In 1975, Su showed that stimulation of sympathetic nerves of the thoracic aorta and portal vein of the rabbit is associated with the release of purines. This knowledge—along with observations that in several blood vessels vasoconstriction and EJPs are resistant, or partially so, to α-adrenoceptor

TABLE 2. Examples of Papers Based on Studies with P_2-Antagonists Supporting a Role of ATP as a Cotransmitter in the Urinary Bladder

Species	Investigators	References
Human	Belis *et al.*	40
Cat	Theobald	41, 42
Guinea pig	Kasakov & Burnstock	19
	Westfall *et al.*	39
	Moss & Burnstock	43
Rabbit	Longhurst *et al.*	44
	Hoyle & Burnstock	45
Ferret and marmoset	Moss & Burnstock	43

antagonism[51–53]—has lead to a consideration that perhaps, in a manner analogous to that in the vas deferens, ATP and NE act as cotransmitters in some blood vessels.

Listed in TABLE 3 are some of the blood vessels for which there is compelling evidence for a cotransmitter role of ATP. The references that are provided are restricted to those that have used the types of approaches already discussed for the vas deferens and urinary bladder, that is, an analysis of the mechanical (contraction and/or vasoconstriction) and electrophysiological (EJPs and/or membrane potential) responses with and without α-adrenoceptor and P_2-purinoceptor perturbations with compounds such as prazosin, $ANAPP_3$, and APCPP. An analysis of these studies indicates that in a number of blood vessels the neurogenically evoked EJPs and the α-blocker-resistant contractions and vasoconstriction are likely to be mediated by ATP rather than NE. The extent of the purinergic component may vary among blood vessels and even within blood vessels.

TABLE 3. Examples of Papers Providing Evidence for ATP and NE as Cotransmitters in Blood Vessels

Tissue	Investigators	References
Rat tail artery	Sneddon & Burnstock	54
	Neild & Kotecha	55
	Vidal *et al.*	56
	Dalziel *et al.*	57
	Bao *et al.*	58
Guinea pig mesentery	Ishikawa	59
Rabbit mesentery	Ishikawa	59
	Kügelgen & Starke	60
	Ramme *et al.*	61
Dog mesentery	Muramatsu	62
	Machaly *et al.*	63
	Muramatsu *et al.*	64
Rabbit ear artery	Suzuki	65
	Kennedy *et al.*	66
	Saville & Burnstock	67
Guinea pig saphenous vein	Cheung & Fujioka	68
Rabbit saphenous vein	Burnstock & Warland	69
	Warland & Burnstock	70
Dog basilar artery	Muramatsu *et al.*	71
	Muramatsu & Kigoshi	72

RELEASE OF ATP FROM THE VASCULAR NEUROEFFECTOR JUNCTION

In spite of such evidence, there has been reluctance to accept the notion that ATP serves as a sympathetic cotransmitter in blood vessels, primarily, it seems, because of a lack of an unequivocal demonstration of the release of ATP.[73] There is a good deal of indirect evidence for the release of ATP, however. For example, the studies of Su and colleagues[50,74] showed the release of radiolabeled purines in thoracic aorta, portal vein, main pulmonary artery, and several other tissues. Other investigators have employed similar techniques.[71,75,76] The method of analysis is to quantify the release of tritium from tissues that are prelabeled with [^{3}H]adenosine. The strategy is that because tissues take up [^{3}H]adenosine and incorporate it into nucleotides, the release of tritium reflects the release of [^{3}H]nucleotides as well as the [^{3}H]nucleosides. This could be incorrect, and the release of tritium might originate mostly or solely from [^{3}H]adenosine.

As an approach to the question of whether ATP itself is released and at what concentration, we have used high-performance liquid chromatography (HPLC) coupled with fluorescence detection[12] to quantify the release of endogenous ATP and its 6-aminopurine analogues from blood vessels.[77,78] One of the advantages of this technique over other approaches, such as the firefly chemiluminescence assay,[79] is that one can measure simultaneously ATP, ADP, AMP, and adenosine. If, as is generally believed, the nucleotides are rapidly metabolized to adenosine by ectoenzymes,[80] it would be useful to be able to quantify all four of these adenine-containing compounds. Specificity in the HPLC-fluorescence assay resides with the development of fluorescent derivatives. This formation is dependent upon the attachment of chloroacetaldehyde to the NH_2 group on the 6-position of the purine ring. Because of this, only the 6-amino purine compounds can be detected.

Using this method, we have demonstrated the release by transmural nerve stimulation of ATP, ADP, AMP, and adenosine from the rat tail artery and the rabbit thoracic aorta.[77,78] A remarkable finding was that the release of ATP far exceeds the release of NE. In the aorta, for example, there was 30-fold more ATP than NE released by stimulation at 16 Hz for 3 min. If one considers not only the release of ATP but of the entire pool of adenine nucleosides and nucleotides, the release exceeds that of NE by over 300-fold.

In view of the disparity in the amounts of release of NE and ATP, and particularly because the suspected ratio of NE to ATP in adrenergic neuron storage vesicles is approximately four,[3] it seemed reasonable to us that the release of the adenine-containing compounds is not exclusively from neuronal sources. Indeed, we, as well as others,[8,76,81] have suggested that nucleotide and nucleoside release occurs from extraneuronal as well as neuronal sites, perhaps as a result of NE exerting a postjunctional action on α-adrenoceptors.

Evidence supporting the idea that α-adrenoceptor stimulation leads to the release of adenine compounds comes from two approaches: 1) Methoxamine, an α_1-adrenoceptor agonist whose site of action is considered to be postjunctional, causes the release of nucleotides and nucleosides. 2) Prazosin, an α_1-adrenoceptor antagonist, reduces both the transmural nerve stimulation-induced release and the methoxamine-induced release.[78]

Although it seems natural to consider the smooth muscle cells as a likely postjunctional source of release of purines, there are other sources as well. We focused attention on the endothelial cells of the aorta. There were several reasons to consider the endothelium. One reason is that endothelial cells grown in culture have a high concentration of ATP. Furthermore, these cells can release ATP upon challenge with agents such as thrombin.[82,83] We wondered whether α-agonists such as NE and methoxamine could do the same.

Removing the endothelium from the aorta (FIG. 3) greatly diminished the release of adenine nucleotides and nucleosides produced by transmural nerve stimulation. The release of these compounds was only 10% of that from the intact vessels. This suggests that approximately 90% of the transmural nerve stimulation-induced overflow of purines is dependent upon the endothelium. The removal of the endothelium also greatly (by 93%) diminished the release of adenine nucleotides and nucleosides by methoxamine.

These results with endothelium-denuded blood vessels suggests that these cells may be a source of purines. To test this idea directly, we determined whether exposure of cultured endothelial cells to NE brought about the liberation of ATP. Preliminary results are shown in FIGURE 4. NE (10 μM) was able to cause the release of large amounts of ATP from endothelial cells in primary culture. The effect of NE was antagonized by prazosin, suggesting that it was mediated by α-adrenoceptors.

Although the endothelium appears to be a major source of transmural nerve stimulation-induced overflow of adenosine and the adenine nucleotides, it is not the

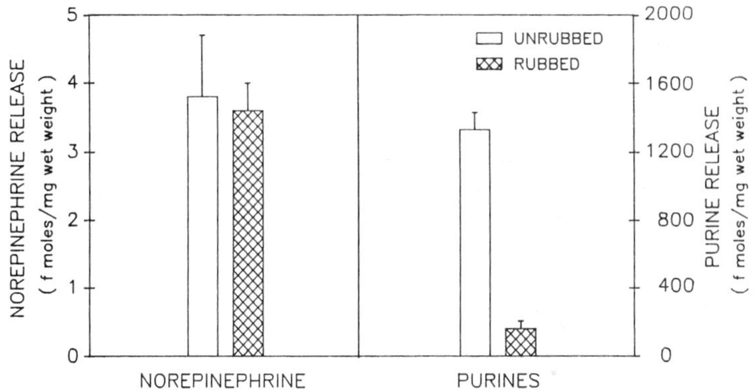

FIGURE 3. Transmural nerve stimulation (16 Hz) induced release of NE and adenine nucleotides and nucleosides (the sum of ATP, ADP, AMP, and adenosine) from unrubbed and endothelium-denuded (by rubbing) segments of rabbit thoracic aorta. Removing the endothelium did not significantly affect the release of NE but reduced the release of the purines by approximately 90%. With both the intact and rubbed preparations, ATP constituted about 10% of the pool of adenine compounds. Note the difference in vertical scales for NE and the purines. Reproduced from reference 78 with permission.

exclusive source. There is about 10% of the total that is independent of the endothelium, and, of this endothelium-independent release, 70% is blocked by prazosin. This second site of release then, which we believe to be the smooth muscle, accounts for 7% of the total.

Now we can consider the question of neuronal release. If 90% of the total overflow is from the endothelium and 7% is from the smooth muscle, only 3% of the total release of adenine nucleotides and nucleosides could arise from the nerves. Is this enough to support a cotransmitter function of ATP? We believe so. Transmural nerve stimulation released a total of 125 fmoles ATP/mg wet weight. If we may assume that 3% came from the nerves, 3.7 fmoles ATP/mg wet weight must have been released, compared to a release of 3.8 fmoles NE/mg wet weight; that is, the ratio of NE to ATP is 1. The amount of ATP may actually be higher if indeed some of

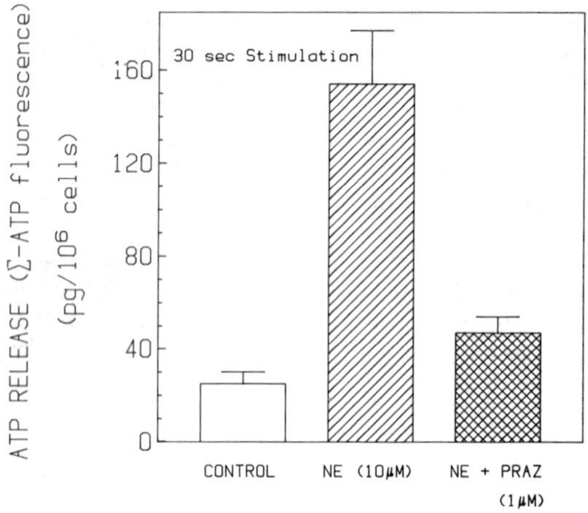

FIGURE 4. The α_1-receptor-stimulated ATP release from cardiac endothelial cells. Endothelial cells were grown for 7 days in plastic 24-well culture dishes. Cells were stimulated following removal of growth medium and by addition of physiological salt solution without NE (control), with NE (10 μM), or with NE plus prazosin (1 μM). Data are the mean of three experiments expressed as pg/10^6 cells.

the released ATP was metabolized to other nucleotides or to adenosine. In any case, the synaptic concentrations of ATP and NE, each arising from neuronal sources, are comparable. Thus there does seem to be direct evidence for the neuronal release of ATP, and neuronally released ATP seems to appear in sufficient quantities to act as a cotransmitter in vascular smooth muscle.

REFERENCES

1. CUELLO, A. C., ED. 1982. Co-transmission. Macmillian. London.
2. BURNSTOCK, G. 1976. Neuroscience **1:** 239-248.
3. STJÄRNE, L. 1989. Rev. Physiol. Biochem. Pharmacol. **112:** 1-137.
4. AMBACHE, N. & M. A. ZAR. 1971. J. Physiol. **216:** 359-389.
5. SWEDIN, A. 1971. Acta Physiol. Scand. Suppl. **83:** 369.
6. WAKADE, A. B. & J. KRUSZ. 1972. J. Pharmacol. Exp. Ther. **181:** 310-317.
7. SMITH, A. D. 1972. Pharmacol. Rev. **24:** 435-457.
8. WESTFALL, D. P., R. E. STITZEL & J. N. ROWE. 1978. Eur. J. Pharmacol. **50:** 27-38.
9. HOLCK, M. I. & B. H. MARKS. 1978. J. Pharmacol. Exp. Ther. **205:** 104-117.
10. SU, C. 1975. J. Pharmacol. Exp. Ther. **195:** 159-166.
11. ROWE, J. N., K. VAN DYKE, D. P. WESTFALL & R. E. STITZEL. 1976. Pharmacology **14:** 193-204.
12. LEVITT, B., R. J. HEAD & D. P. WESTFALL. 1984. Anal. Biochem. **137:** 93-100.
13. LEW, M. J. & T. D. WHITE. 1987. Br. J. Pharmacol. **92:** 349-355.
14. HOGABOOM, G. K., J. P. O'DONNELL & J. S. FEDAN. 1980. Science **208:** 1273-1274.
15. FEDAN, J. S., G. K. HOGABOOM, J. P. O'DONNELL & D. P. WESTFALL. 1985. *In* Methods in Pharmacology, D. M. Paton, Ed. Vol. 6: 279-292. Plenum. New York, NY.
16. FEDAN, J. S., G. K. HOGABOOM, J. P. O'DONNELL, J. COLBY & D. P. WESTFALL. 1981. Eur. J. Pharmacol. **69:** 41-53.
17. SNEDDON, P., D. P. WESTFALL & J. S. FEDAN. 1982. Science **218:** 293-695.
18. SNEDDON, P. & D. P. WESTFALL. 1984. J. Physiol. **347:** 561-580.
19. KASAKOV, L. & G. BURNSTOCK. 1983. Eur. J. Pharmacol. **86:** 291-294.
20. MELDRUM, L. & G. BURNSTOCK. 1983. Eur. J. Pharmacol. **92:** 161-163.
21. SNEDDON, P. & G. BURNSTOCK. 1984. Eur. J. Pharmacol. **100:** 85-90.

22. STJÄRNE, L. & P. ÅSTRAND. 1984. Neurosci. **13:** 21-28.
23. STJÄRNE, L. & P. ÅSTRAND. 1985. Neurosci. **14:** 929-946.
24. ALLCORN, R. J., T. C. CUNNANE & K. KIRKPATRICK. 1986. Br. J. Pharmacol. **89:** 647-659.
25. CUNNANE, T. C. & R. MANCHANDA. 1988. J. Physiol. **404:** 349-364.
26. SUZUKI, N. & Y. GOMI. 1989. Jpn. J. Pharmacol. **49:** 59-65.
27. FRENCH, A. M. & N. C. SCOTT. 1983. Experientia **39:** 264-266.
28. SNEDDON, P., D. P. WESTFALL, J. COLBY & J. S. FEDAN. 1984. Life Sci. **35:** 1903-1912.
29. SATCHELL, D. 1986. Eur. J. Pharmacol. **132:** 305-308.
30. AMBACHE, N. & M. A. ZAR. 1970. J. Physiol. **210:** 761-783.
31. DUMSDAY, B. H. 1971. J. Pharm. Pharmacol. **23:** 222-225.
32. BURNSTOCK, G., B. DUMSDAY & A. SMYTHE. 1972. Br. J. Pharmacol. **44:** 451-461.
33. BURNSTOCK, G., T. COCKS, R. CROWE & L. KASAKOV. 1978. Br. J. Pharmacol. **63:** 125-138.
34. RAEZER, D. M., A. J. WEIN, D. J. JACOBOWITZ & J. N. CORRIERE, JR. 1973. Urology **2:** 211-221.
35. DEGROAT, W. C. & W. R. SAUM. 1972. J. Physiol. **220:** 297-314.
36. DOWNIE, J. W. & D. M. DEAN. 1977. J. Pharmacol. Exp. Ther. **203:** 417-425.
37. DEAN, D. M. & J. W. DOWNIE. 1978. J. Pharmacol. Exp. Ther. **207:** 431-445.
38. KRELL, R. D., J. L. MCCOY & P. T. RIDLEY. 1981. Br. J. Pharmacol. **74:** 15-22.
39. WESTFALL, D. P., J. S. FEDAN, J. COLBY, G. K. HOGABOOM & J. P. O'DONNELL. 1983. Eur. J. Pharmacol. **87:** 415-422.
40. BELIS, J. A., J. E. COLBY, L. B. ADLESTEIN & D. P. WESTFALL. 1981. Surg. Forum **32:** 625-627.
41. THEOBALD, R. J. 1982. J. Auton. Pharmacol. **3:** 175-179.
42. THEOBALD, R. J. 1983. Life Sci. **32:** 2479-2484.
43. MOSS, H. E. & G. BURNSTOCK. 1985. Eur. J. Pharmacol. **114:** 311-316.
44. LONGHURST, P. E., J. A. BELIS, J. P. O'DONNELL, J. R. GALIE & D. P. WESTFALL. 1984. Eur. J. Pharmacol. **99:** 295-302.
45. HOYLE, C. H. V. & G. BURNSTOCK. 1985. Eur. J. Pharmacol. **114:** 239-240.
46. NAKANISHI, H. & H. TAKEDA. 1973. Jpn. J. Pharmacol. **23:** 479-485.
47. WALI, F. A. & E. GREENIDGE. 1989. Pharmacol. Res. **21:** 397-404.
48. LANGER, S. Z. & J. E. B. PINTO. 1976. J. Pharmacol. Exp. Ther. **196:** 697-713.
49. DUVAL, N., P. E. HICKS & S. Z. LANGER. 1985. Eur. J. Pharmacol. **110:** 373-377.
50. SU, C. 1977. J. Pharmacol. Exp. Ther. **204:** 351-361.
51. BEVAN, J. A. & C. SU. 1971. Circ. Res. **28:** 179-187.
52. HIRST, G. D. S. & T. O. NEILD. 1980. Nature **283:** 767-768.
53. HOLMAN, M. E. & A. M. SURPRENTANT. 1980. Br. J. Pharmacol. **71:** 651-661.
54. SNEDDON, P. & G. BURNSTOCK. 1984. Eur. J. Pharmacol. **106:** 149-152.
55. NEILD, T. O. & N. KOTECHA. 1986. Gen. Pharmacol. **17:** 461-464.
56. VIDAL, M., P. E. HICKS & S. Z. LANGER. 1986. Naunyn-Schmiedeberg's Arch. Pharmacol. **332:** 384-390.
57. DALZIEL, H. H., M. MACHALY & P. SNEDDON. 1988. Br. J. Pharmacol. **95:** 187P.
58. BAO, J. X., I. E. ERIKSSON & L. STJÄRNE. 1989. Acta Physiol. Scand. **136:** 139-140.
59. ISHIKAWA, S. 1985. Br. J. Pharmacol. **86:** 777-787.
60. KÜGELGEN, I. V. & K. STARKE. 1985. J. Physiol. **367:** 435-455.
61. RAMME, D., J. T. REGENOLD, K. STARKE, R. BUSSE & P. ILLES. 1987. Naunyn-Schmiedeberg's Arch. Pharmacol. **336:** 267-273.
62. MURAMATSU, I. 1986. Br. J. Pharmacol. **87:** 478-480.
63. MACHALY, M., H. H. DALZIEL & P. SNEDDON. 1988. Eur. J. Pharmacol. **147:** 83-91.
64. MURAMATSU, I., T. OHMURA & M. OSHITA. 1989. J. Physiol. **411:** 227-243.
65. SUZUKI, H. 1985. J. Physiol. **359:** 401-415.
66. KENNEDY, C., V. L. SAVILLE & G. BURNSTOCK. 1986. Eur. J. Pharmacol. **122:** 291-300.
67. SAVILLE, V. L. & G. BURNSTOCK. 1988. Eur. J. Pharmacol. **155:** 271-277.
68. CHEUNG, D. W. & M. FUJIOKA. 1986. Br. J. Pharmacol. **89:** 3-5.
69. BURNSTOCK, G. & J. J. I. WARLAND. 1987. Br. J. Pharmacol. **90:** 111-120.
70. WARLAND, J. J. I. & G. BURNSTOCK. 1987. Br. J. Pharmacol. **92:** 871-880.
71. MURAMATSU, I., M. FUJIWARA, A. MIURA & Y. SAKAKIBARA. 1981. J. Pharmacol. Exp. Ther. **216:** 401-409.

72. MURAMATSU, I. & S. KIGOSHI. 1987. Br. J. Pharmacol. **92:** 901-908.
73. WHITE, T. D. 1988. Pharmacol. Ther. Rev. **38:** 129-168.
74. KATSURAGI, T. & C. SU. 1980. J. Pharmacol. Exp. Ther. **215:** 685-690.
75. BURNSTOCK, G., R. CROWE & H. K. WONG. 1979. Br. J. Pharmacol. **65:** 377-388.
76. LEVITT, B. & D. P. WESTFALL. 1982. Blood Vessels **19:** 30-40.
77. WESTFALL, D. P., K. SEDAA & R. A. BJUR. 1987. Blood Vessels **24:** 125-127.
78. SEDAA, K. O., R. A. BJUR & D. P. WESTFALL. 1989. J. Pharmacol. Exp. Ther. **252:** 1060-1067.
79. WHITE, T. D. 1985. *In* Methods in Pharmacology. D. M. Paton, Ed. Vol. 6: 43-63. Plenum. New York, NY.
80. PEARSON, J. D. 1985. *In* Methods in Pharmacology. D. M. Paton, Ed. Vol. 6: 83-107. Plenum. New York, NY.
81. FREDHOLM, B. B., P. HEDQVIST & L. VERNET. 1979. Acta Physiol. Scand. **106:** 381-382.
82. PEARSON, J. D. & E. L. GORDON. 1979. Nature **281:** 384-386.
83. NEES, S. & E. GERLACH. 1983. *In* Regulatory Function of Adenosine. R. M. Berne, T. N. Rall & R. Rubio, Eds.: 347-355. Martinus Nijhoff. Boston, MA.

Characteristics of Receptor-Operated and Membrane Potential-Dependent ATP Secretion from Adrenal Medullary Chromaffin Cells

EDUARDO ROJAS,[a] VALENTÍN CEÑA,[b]
ANDRES STUTZIN,[c] ERIK FORSBERG, AND
HARVEY B. POLLARD

Laboratory of Cell Biology and Genetics
National Institute of Diabetes and
Digestive and Kidney Diseases
National Institutes of Health
Bethesda, Maryland 20892

INTRODUCTION

Catecholamine secretion from medullary chromaffin cells is widely appreciated to occur by exocytosis, and the system has been closely studied over the past decades as a biological prototype for the process.[1] Nonetheless, the mechanisms by which chromaffin granules eventually fuse with plasma membranes and release their content outside the cell remain poorly understood.[2] It is known that a rapid elevation in the free calcium concentration, in the cytosol, in response to acetylcholine (ACh), precedes exocytosis, and that the calcium increase is dependent on the function of receptor-operated (cholinergic) and voltage-sensitive calcium channel in the plasma membrane. The process by which calcium subsequently induces granule and plasma membrane contact, however, is not known. Indeed, protein mediators such as synexin, and/or other members of the synexin (annexin) gene family, have been implicated.[3]

Furthermore, once calcium acts, at least two types of membrane fusion processes occur. Simple exocytosis involves fusion of a granule membrane with the plasma membrane. Chromaffin granules situated more deeply within the cell, however, also fuse with granule membranes that have already secreted by the simple process, and that are still attached to the plasma membrane. Such secretion is termed compound exocytosis. In addition to such heterogeneity, one may note that the typical chromaffin cell preparation contains cells specific for either epinephrine or norepinephrine.

[a]To whom correspondence should be addressed.

[b]Present address: Department of Neurochemistry, University of Alicante, Alicante, Spain.

[c]Present address: Department of Experimental Medicine, Faculty of Medicine, University of Chile, Santiago, Chile.

In the face of such mechanistic complexity, one might anticipate that the kinetics of exocytosis might also be extraordinarily complex. Surprisingly, however, the kinetics of this process in chromaffin cells has been barely studied.[4,5] The reason is that until very recently secretion could only be studied by a quantitative but cumbersome endpoint method for catecholamines, with low time resolution. Low time resolution was inescapable and implicit in the older approach because detection depended not only on catecholamines being released from the cell, but also on their being diffused into the bulk phase for collection and analysis. By contrast, the new methodology involves a quantitative, on-line method using highly purified luciferase for detecting the ATP molecules that are cosecreted with the catecholamines at the anatomic point of exocytosis.[6] The time resolution is about 1 msec, and noise analysis on small populations of chromaffin cells (10^5) has even allowed us to detect "unitary events" of about 25 msec in duration, using a sensitive photometer of our own design.[7]

EARLY TIME COURSE OF ACh-STIMULATED ATP RELEASE

FIGURE 1B depicts a typical record of the time course of the light emitted by the luciferin-luciferase mixture after the application of ACh (1 μM) to the cells at 23 °C. Application of the physiological secretagogue caused an immediate increase in the light signal, which reached a steady state level after a few minutes.

Given the assumption that ATP is released by two parallel processes, the time course of the secretory response can be represented by the empirical function

$$\mathrm{ATP}(t) = B + A_f\{1-\exp(-t/\tau_f)\}^3 + A_s(1-\exp(-t/\tau_s) \qquad (1)$$

where B represents the basal level of ATP, A_f and A_s represent the maximum amount released from the fast and slow compartments, and τ_f and τ_s represent the corresponding time constants.[8]

The luciferin-luciferase kit used in the present series of experiments contained luciferin in excess.[6] Therefore, the rate of reaction should just be proportional to the concentration of ATP. To verify this assumption, we measured the rate of decay of the light signal as a function of the amount of ATP added to the reaction chamber (10-40 pmol). A single time constant was required to fit the decay in the light signal (FIG. 1A), and the time constant of decay was independent of the amount of ATP added. As expected, lowering the temperature from 37 to 23 °C increased the time constant of decay from 549 $\pm$ 11.6 to 1683.4 $\pm$ 62.7 sec (FIG. 1A).

FIGURE 1B illustrates the effect of the decay of the light signal caused by the ATP hydrolysis on the evaluation of the kinetic parameters of the early phase of the secretory response at 23 °C. The best fit of the noisy record gave values for τ_f and τ_s of 6.8 and 32.4 sec, respectively. To correct for inactivation, we divided the experimental curve by an exponential function with the appropriate time constant of inactivation

$$\mathrm{ATP}_{\mathrm{corrected}}(t) = \mathrm{ATP}(t)/\exp(-t/\tau) \qquad (2)$$

where τ represents the time constant of inactivation. ATP release after correction for ATP hydrolysis was then fitted using equation 2. At 23 °C, τ was about 1700 sec and the time constants τ_f and τ_s obtained from the fit of the experimental record using equation 1 before and after correction for inactivation decreased from 6.8 and 32.4 sec to 6.8 and 31.8 sec, respectively. This result shows that the uncorrected time constants are very similar to the true time constants. For this reason, in the analysis presented here, we considered only the original uncorrected records (FIG. 1B). It

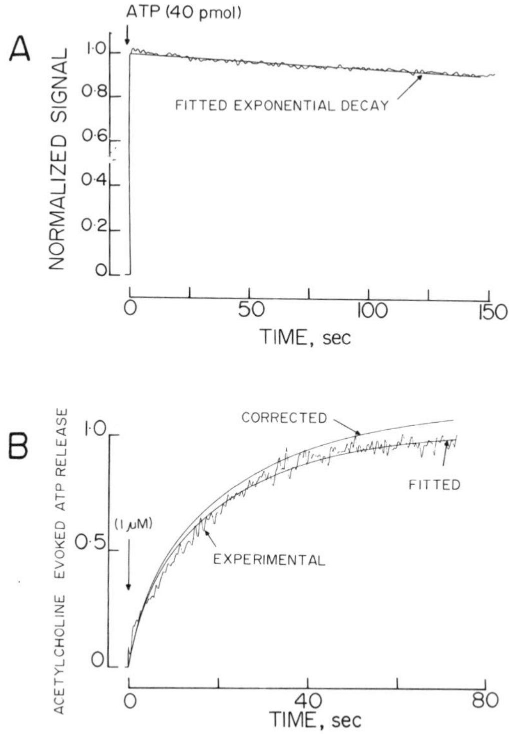

FIGURE 1. (**A**) Calibration of the reaction mixture and luminescent detection. Record of the photomultiplier output at 23 °C. At time zero, 10 μl of Krebs solution containing 40 pmol of ATP (arrow) were added to the reaction mixture. Fitted curve is a least-squares fit of the following equation: ATP(t) = 40 exp ($-t/\tau$) pmol, where τ represents a time constant of decay equal to 1688 ± 62.7 sec. (**B**) ACh-evoked ATP release from freely suspended chromaffin cells. Record of the photomultiplier output representing the time integral of the ATP released in the reaction chamber ("experimental"). About 10,000 cells were present in the reaction mixture. The arrow indicates the addition of 10 μl of Krebs solution containing 20 μM acetylcholine to 190 μl of reaction medium. The curve labeled "fitted" represents the best possible fitting to the experimental record using equation 1. The curve labeled "corrected" indicates the data corrected by inactivation of the luminiscence signal. Temperature: 23 °C.

should be emphasized that the extent of ATP release measured 80 sec after the addition of the secretagogue represents about 90% of the ATP level that would have been reached if the reaction had occurred in the absence of ATP hydrolysis (FIG. 1B).

FIGURE 2 depicts the average time course of the ATP secretion evoked by nicotine (22.5 μM) from three experiments carried out at 23 °C. The curve was calculated with equation 1. The values of A_s and A_f that gave the best fit were 12.2 pmol and 2.4 pmol, respectively, and the corresponding time constants were 16 and 1.5 sec, respectively. With these values, we can calculate the time course of the rates of ATP secretion from each process. The reason for this is that the ATP released by the cells is confined to the reaction chamber, and, therefore, the light output from the reaction represents accumulative ATP released. Thus, the time derivative of this signal represents the time course of the rate of ATP release (FIG. 2, insert). The rates of secretion start decaying from their initial values A_f/τ_f and A_s/τ_s (equal to 0.76 and 1.6 pmol/sec, respectively) toward the basal rate with time constants τ_f and τ_s. Thus, in the maintained presence of the agonist, ATP secretion rate returned to basal levels in a few minutes, consistent with data for the rate of secretion of catecholamines.[9]

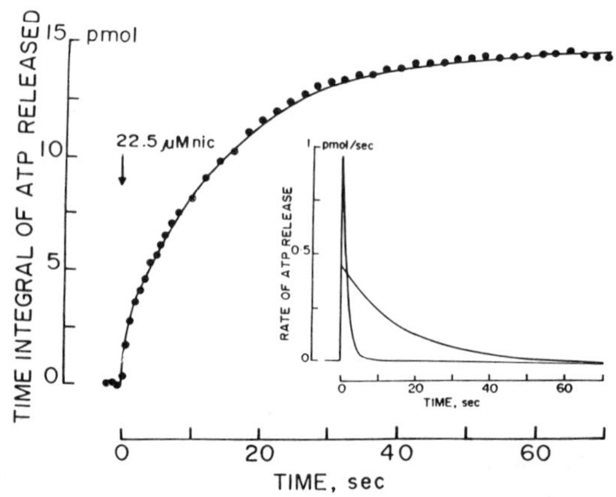

FIGURE 2. Kinetic analysis of ATP release. Time course of ATP release in response to the addition of nicotine (22.5 μM). The filled circles represent the average data from three experiments. The solid curve, which represents the best fit to the experimental data, was calculated as explained in the text using equation 1. The inset shows the calculated time course of the rates of release from two different ATP-releasable pools. Single exponentials were calculated using A_s/τ_s (0.76 pmol/sec) and A_f/τ_f (1.6 pmol/sec) from the least-squares fit to the data.

EFFECT OF Ca^{2+} ON NICOTINE-EVOKED ATP RELEASE

In chromaffin cells from the adrenal gland, the secretion of catecholamines is evoked by stimulation of cholinergic receptors and depends on calcium in the medium. As shown in FIGURE 3, this is also the case for nicotine-stimulated ATP release. Nicotine (30 μM) was applied to cells in the luciferin-luciferase medium in presence of different $[Ca^{2+}]_o$. A few minutes prior to the stimulation of the cells, EGTA was added to adjust $[Ca^{2+}]_o$ at the desired level. In presence of 1.9 mM Ca^{2+}, nearly 20% of the cellular ATP was released as a result of nicotinic stimulation (FIG. 3C). The secretagogues used in this study, namely, ACh, nicotine, and high K^+, were

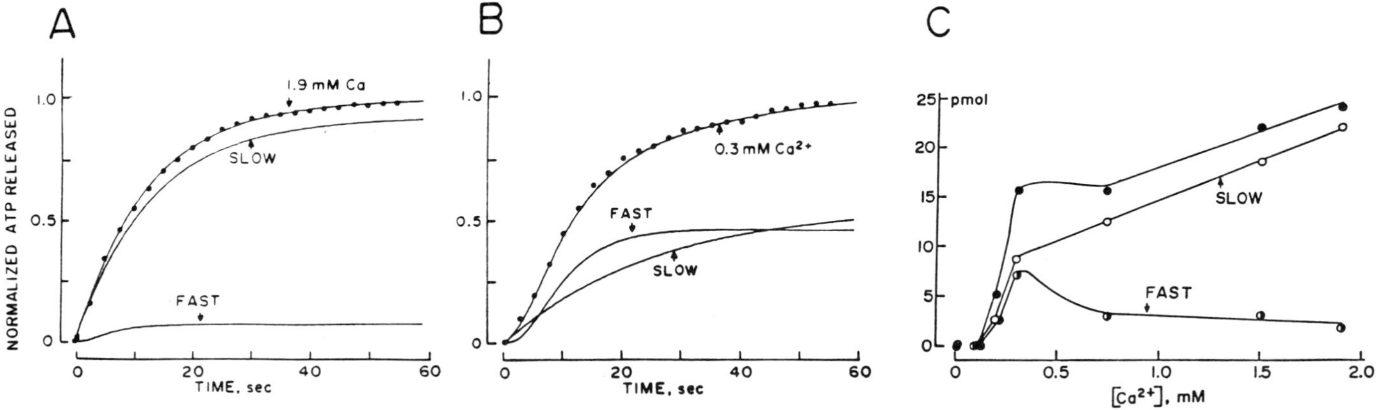

FIGURE 3. Calcium dependence of nicotine-induced ATP release. (**A**) Time integral of the ATP released by about 10,000 cells in the reaction mixture in response to nicotine (30 μM). The low-pass filter was set at 5 Hz. Free-calcium concentration (measured with a calcium electrode) was 1.9 mM. Curves labeled "fast" and "slow" represent the hypothetical time courses of the cumulative ATP released by the fast and slow processes, respectively. They represent the two components of the double exponential required to fit the original data. (**B**) Cumulative ATP released in response to nicotine (30 μM) in the presence of 0.3 mM Ca^{2+}. (**C**) Relative contribution of the components (fast and slow) to the extent of nicotine-evoked release as a function of free-calcium concentration in the reaction medium. Temperature: 23 °C.

ineffective in eliciting ATP secretion from cells in Ca^{2+}-free media. The lowest $[Ca^{2+}]_o$ required to obtain a measurable secretory response to nicotine was 0.2 mM.

In contrast to the marked effect of $[Ca^{2+}]_o$ on the extent of ATP release, the time constants required to fit the kinetics of ATP release were less affected by $[Ca^{2+}]_o$. This property of ATP secretion is illustrated in FIGURES 3A & 3B. Two examples of our analysis of the effects of $[Ca^{2+}]_o$ on the time course of the two components of the release process are shown. The components of a record made in the presence of 1.9 mM $[Ca^{2+}]_o$ (FIG. 3A) can be compared to those from another record obtained at 0.3 mM (FIG. 3B). To facilitate the comparison, the records were normalized by dividing all the values by the value at 70 sec. At 1.9 mM $[Ca^{2+}]_o$, τ_f and τ_s were equal to 3.7 and 12.5 sec, respectively. At 0.3 mM $[Ca^{2+}]_o$, the corresponding values were 5.6 and 25.6 sec, respectively, the differences being significant. The extent of the ATP secreted by the fast and the slow processes have been plotted in FIGURE 3C, together with the total.

MEMBRANE POTENTIAL DEPENDENCE OF ATP RELEASE

To determine the relationship between membrane potential and nicotine-evoked ATP release, the $[K^+]_o$ was increased from 5 mM to different levels. After the K^+-evoked ATP secretion reached a steady state level, nicotine was applied. The time course of the responses obtained with this sequence of stimuli are shown in FIGURE 4 (upper panel). Application of K^+ (68 mM) induced ATP release (18% of the total ATP in the cells), and the signal leveled after 65 sec. Application of nicotine (30 μM) also resulted in further release of ATP (5% of the total ATP). The nicotine-induced secretory responses obtained at different $[K^+]_o$ values ranging from 5 to 120 mM are shown in FIGURE 4 (lower panel). Each response has been normalized; that is, the vertical axis represents the extent of the nicotine-evoked ATP release at a given $[K^+]_o$ divided by the extent of the nicotine-evoked ATP release at a $[K^+]_o$ of 5 mM. Each point represents the mean value from three determinations. Each determination was corrected to take into account the inhibition of the luciferin-luciferase system by $[K^+]_o$. Corrected values are plotted in FIGURE 4 (lower panel) as a function of the membrane potential at each $[K^+]_o$. Membrane potentials were calculated assuming that $[Na^+]_i$ was 38 mM, that $[K^+]_o$ plus $[Na^+]_o$ equaled 150 mM, and that P_{Na}/P_K was 0.025.[10] It may be seen that even when the chromaffin cell membrane is completely depolarized, further ATP secretion is evoked by the activation of nicotinic receptor channels.

PHENOTHIAZINE DRUGS BLOCK NICOTINE- AND K^+-EVOKED ATP RELEASE

Phenothiazine drugs such as trifluoperazine (TFP) and promethazine (PMTHZ) block secretion at quite low concentrations.[11–13] TFP also blocks ACh-induced ATP release from bovine adrenal chromaffin cells.[14] We confirmed this result by showing that TFP is also effective in blocking nicotine-evoked ATP release (FIG. 5A, a control experiment testing whether TFP (5 μM) induces ATP secretion; FIG. 5B, the results after a 2-min preincubation period in the presence of 1 μM TFP). The responses to nicotine (3.2 μM) and to elevated $[K^+]_o$ from cells in the presence and absence of TFP can be compared. The blockade of the nicotinic receptor-mediated secretory response was rapid (FIG. 5A, lower record) and complete in less than 2 min (FIG.

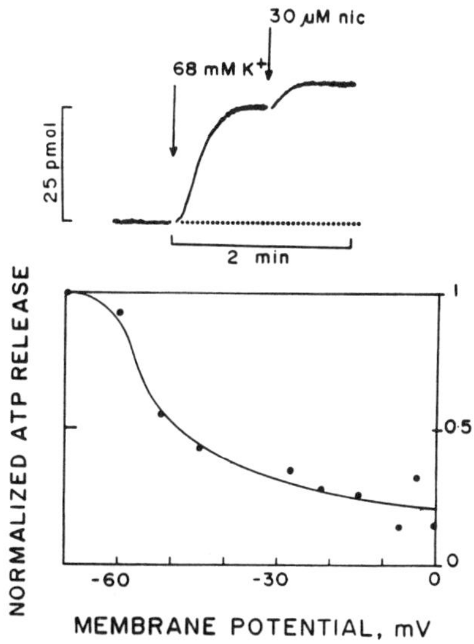

FIGURE 4. Dependence on membrane potential of nicotine-evoked ATP release. Upper trace: Time course of the ATP release induced by a sudden elevation of $[K^+]_o$ from 5 to 68 mM. After 68 sec, nicotine (30 μM) was added. Lower trace: Normalized and corrected ATP release induced by nicotinic receptor stimulation (nicotine: 30 μM) at different membrane potentials. Vertical axis: (steady state ATP released by nicotine at the indicated $[K^+]_o$)/(steady state ATP released by nicotine at resting membrane potential). Each point represents the mean value of three experiments. The abscissa was calculated as $(RT/F)\ \ln\ \{(P_{Na}/P_K\ [Na^+]_o + [K^+]_o)/(P_{Na}/P_K[Na^+]_i + [K^+]_i)\}$, assuming that $[Na^+]_i + [K^+]_i = 150$ mM and that $[K^+]_i = 120$ mM. Note that R, T, and F have their usual meanings, and that P_{Na}/P_K was taken as 0.025. Temperature: 20-23 °C.

5B). As for the ATP secretion induced by high $[K^+]_o$, TFP was less effective and the response in the presence of TFP was slower than in the control experiment. PMTHZ was 10-fold less effective than TFP in blocking nicotine-induced ATP secretion (data not shown). Because TFP and PMTHZ also block chromaffin granule aggregation induced by the cytosolic protein synexin,[15] we interpret the results shown in FIGURE 5 to be consistent with the hypothesis that synexin molecules are the receptor/target molecules for the phenothiazine drugs.

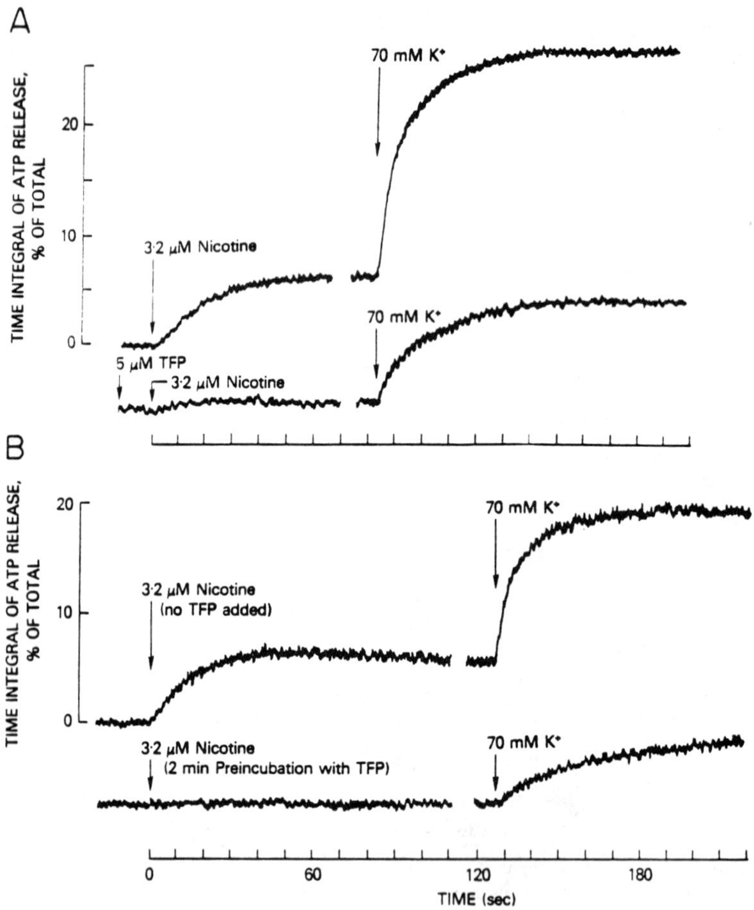

FIGURE 5. TFP blockade of nicotine-induced ATP secretion. (**A**) Sequential stimulation of bovine chromaffin cells first with nicotine (3.2 μM) and then high K^+ (70 mM) of the cells in the absence of TFP (upper trace) and 15 sec after the application of TFP (5 μM, lower trace). (**B**) Similar sequence of stimuli in the absence of TFP (upper trace) and after 2 min of incubation in the presence of TFP (1 μM, lower trace). (**C**) The extent of the ATP secreted by the fast (A_f, ○) and the slow (A_s, ○) processes, and the total ATP secretion ($A_f + A_s$, ●). Temperature: 37 °C.

INTRACELLULAR DISTRIBUTION OF ATP AND GRANULAR ORIGIN OF THE SECRETED ATP

ATP is distributed both in the secretory granules and in the cytosol of the adrenal medullary cells.[16,17] To establish the origin of the secreted ATP, we permeabilized the cells by application of digitonin.[18,19]

Chromaffin cells, which were suspended in a medium with low Ca^{2+}, low Cl^- (isethionate in place of chloride), and luciferin-luciferase, were subjected to digitonin treatment. FIGURE 6 shows that about 15% of the ATP in the cells was rapidly released to the medium. Under these conditions, application of ACh (100 μM) failed to evoke further release. This ATP may represent the sum of the ATP in the cytosol plus a small fraction of the ATP from the granular pool, releasable by Ca^{2+} at 1 μM.[18,19] Finally, treatment of the reaction mixture with 1% triton X-100 released the remainder of the ATP in the cells. The ATP released had probably been contained within intact chromaffin granules. The light level reached with the triton X-100 treatment gave the total ATP as 2,000 pmol (FIG. 6). Because the reaction mixture for this particular experiment contained about 250,000 cells, the measured ATP content per chromaffin cell was 8 fmol. Under these experimental conditions, namely isethionate in place of chloride and low $[Ca^{2+}]_o$ (1 μM) in the external medium, 10 to 16% of the ATP in the cells is liberated by digitonin (N = 40).

ATP RELEASE AS A MARKER OF CATECHOLAMINE SECRETION

In previous preliminary communications[6,7] and, in more detail, in the present work, we have shown that Ca^{2+}-dependent secretion from chromaffin cells can be quantitatively measured using a highly purified luciferin-luciferase preparation to assay released ATP. With this new technique it is possible to determine the kinetic parameters of secretion with a time resolution in the millisecond range.[8]

ACh-induced ATP release shares many properties with catecholamine secretion. Our studies of the $[Ca^{2+}]_o$ dependence of the nicotinic receptor-activated ATP release

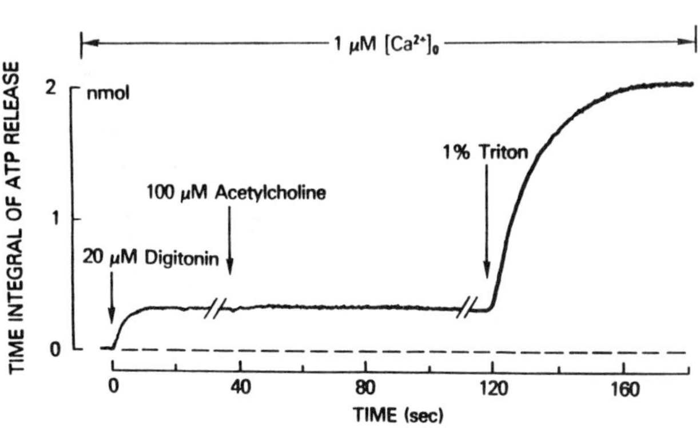

FIGURE 6. Identification of the origin of the secreted ATP. Record represents the time integral of the ATP released from about 250,000 chromaffin cells in response to the following: an application of 20 μM digitonin (arrow on the left), a subsequent application of 100 μM acetylcholine (middle arrow), and a subsequent application of 1% triton X-100 (arrow on the right). The low-$[Ca^{2+}]_o$ reaction medium was made with isethionate in place of chloride ions. The $[Ca^{2+}]_o$ levels were maintained at 1 μM.

and our studies of the blockade by Ca^{2+} channel antagonists show that Ca^{2+} entry is mandatory for ATP release. Furthermore, phenothiazine drugs such as TFP and PMTHZ were effective inhibitors of ATP release. Both drugs are known to block catecholamine secretion.[15,20] TFP also blocks ATP release at concentrations below TFP concentrations reported to interfere with either Ca^{2+} influx[20] or catecholamine transport.[21] Another important similarity between the catecholamine and ATP secretion processes is the extent of the release. A maximum ATP release of 26% of the cellular ATP was obtained by application of 100 μM ACh. Similar catecholamine responses have been reported for bovine chromaffin cells.

ORIGIN OF MULTICOMPONENT KINETICS OF ATP RELEASE

The data clearly showed at least two components to the secretory process, which we have termed "fast" and "slow." There are a number of possible alternative origins for these two components. The initial concept,[7] for example, attributed the fast component to simple exocytosis and the slow component to compound exocytosis. The two components could also correspond to secretion of epinephrine and norepinephrine because these catecholamines are in separate cells. There is no other evidence, however, to suggest that these two processes are kinetically distinct. Because the relative contributions of both components to secretion are nearly equal and independent of the concentration of the cholinergic agonists, the existence of these components appears to be a specific property of the secretion event per se. Perhaps a vesicle, upon fusion, secretes a readily available fraction of its ATP, and perhaps this ATP, which is possibly deeper in the core, is released at a slower rate. Evidence that this may be a valid possibility rests in unpublished observations from our laboratory that ATP released from isolated chromaffin granules during osmotic lysis follows a process defined by two time constants.

ORIGIN OF THE ATP RELEASED BY PERMEABILIZATION

ATP is distributed both in secretory granules and cytosol. We have seen that in low $[Ca^{2+}]_o$ (1 μM for FIG. 6) close to 16% of the cellular ATP is liberated after treatment of the cells with digitonin. Digitonin solubilizes the plasma membrane of the chromaffin cell, leaving the organelles such as chromaffin granules, mitochondria, and endoplasmic reticulum intact.[18,19] The permeabilized cell loses large molecules such as lactate dehydrogenase (LDH), but not dopamine β-hydroxylase (DBH) or catecholamines, which are found in the granules. On the other hand, previous reports show that the high electric field permeabilization technique induced the release of about 7% of the cellular ATP.[14,22] Application of high electric field pulses, a few microseconds in duration, makes holes in the plasma membrane estimated to be 4-5 Å in diameter. These holes do not allow LDH to escape.[14] Thus, it can be assumed that luciferase cannot enter the permeabilized cells, and, therefore, the detection of ATP in our early studies[6] took place outside the cell boundary, unlike the situation with digitonin-permeabilized cells. It seems logical to assume that the greater percentages obtained with digitonin, namely 16% compared with 7%[6,7] obtained with the high-voltage pulses, may be attributable to the more efficient membrane removal by digitonin. We have seen that the extent of the ATP release evoked by application of ACh may reach levels as high as 30% of the total cellular ATP (FIG. 4). We have also shown the ATP release from cells permeabilized using digitonin was only 16% of the ATP in the cells. Thus, there can be no doubt that at least this additional ATP release after ACh stimulation must originate in a compartment other than the cytosol.

The total ATP that can be extracted from the cell is close to 8 fmol/cell (FIG. 6). Because 16% of the total ATP (1.3 fmol) is distributed in about 50% of the cell volume (2.5×10^{-10} cm^3), the overall concentration of ATP in the cytosolic compartment is calculated to be 10 μmol/cm^3. Taking the mean volume of the granule as 4.2×10^{-15} cm^3, and estimating that there are 10^4 granules per cell[16], we calculate that the ATP concentration in the granular pool is 160 μmol/cm^3.

ACKNOWLEDGMENTS

The authors wish to thank Diane Seaton for preparation of the chromaffin cells.

REFERENCES

1. POLLARD, H. B., R. ORNBERG, M. LEVINE, E. HELDMAN, K. MORITA, K. KELNER, P. I. LELKES, K. BROCKLEHURST, E. FORSBERG, L. DUONG, R. LEVINE & M. B. H. YOUDIM. 1985. Hormone packaging and secretion by exocytosis: A view from the chromaffin cell. Vitam. Horm. **42:** 109-196.
2. POLLARD, H. B., C. E. CREUTZ, V. M. FOWLER, J. H. SCOTT & C. J. PAZOLES. 1982. Calcium-dependent regulation of chromaffin granule movement, membrane contact and fusion during exocytosis. Cold Spring Harbor Symp. Quant. Biol. **46:** 819-833.
3. POLLARD, H. B., E. ROJAS & L. BURNS. 1987. Synexin and chromaffin granule membrane fusion. Ann. N.Y. Acad. Sci. **493:** 524-541.
4. ITO, S. 1983. Time course of release of catecholamines and other granular contents from perfused adrenal chromaffin cells of guinea-pig. J. Physiol. **341:** 153-167.
5. KAO, L. S. & E. W. WESTHEAD. 1984. Temperature dependence of catecholamine secretion from cultured bovine chromaffin cells. J. Neurochem. **43:** 590-592.
6. ROJAS, E., H. B. POLLARD & E. HELDMAN. 1985. Real-time measurements of acetylcholine-induced release of ATP from bovine medullary chromaffin cells. FEBS Lett. **185:** 323-327.
7. ROJAS, E., E. FORSBERG & H. B. POLLARD. 1986. Optical detection of calcium-dependent ATP release from stimulated medullary chromaffin cells. Adv. Exp. Med. Biol. **211:** 7-29.
8. CEÑA, V. & E. ROJAS. 1989. Kinetic characteristics of calcium-dependent, cholinergic receptor-controlled ATP secretion from adrenal medullary chromaffin cells. Biochim. Biophys. Acta: in press.
9. OKA, M., M. ISOSAKI & J. WATANABE. 1980. Calcium flux and catecholamine release in isolated bovine adrenal medullary cells: Effects of nicotinic and muscarinic stimulation. *In* Synthesis, Storage and Secretion of Adrenal Catecholamines. Adv. Biosci. **36:** 29-36.
10. NASSAR-GENTINA, V., H. B. POLLARD & E. ROJAS. 1988. Electrical activity in chromaffin cells of intact mouse adrenal gland. Am. J. Physiol. **254:** C675-C683.
11. SCHUBART, U. K., N. LEISCHER & J. ERLICHMAN. 1980. Calcium-dependent protein phosphorylation and insulin release in intact hamster insulinoma cells: Inhibition by trifluoperazine. J. Biol. Chem. **255:** 11063-11066.
12. FEINSTEIN, M. B. & R. A. HADJIAN. 1982. Effects of the calmodulin antagonist trifluoperazine on stimulus-induced calcium mobilization, aggregation, secretion and protein phosphorylation in platelets. Mol. Pharmacol. **21:** 422-431.

13. SUSSMAN, K. E. & J. W. LEITNER. 1977. Conversion of ATP into other adenine nucleotides within isolated islet secretory vesicles: Effect of cyclic AMP on phosphorous translocation. Endocrinology **101:** 694-701.
14. KNIGHT, D. E. & P. F. BAKER. 1982. Calcium dependence of catecholamine release from bovine adrenal medullary cells after exposure to intense electric fields. J. Membr. Biol. **68:** 107-140.
15. POLLARD, H. B., J. H. SCOTT & C. E. CREUTZ. 1983. Inhibition of synexin activity and exocytosis from chromaffin cells by phenothiazine drugs. Biochem. Biophys. Res. Commun. **113:** 908-915.
16. WINKLER, H. & E. WESTHEAD. 1980. The molecular organization of adrenal chromaffin granules. Neuroscience **5:** 1803-1823.
17. UNGAR, A. & J. H. PHILLIPS. 1983. Regulation of the adrenal medulla. Physiol. Rev. **63:** 787-843.
18. WILSON, S. P. & N. KIRSHNER. 1983. Calcium-evoked secretion from digitonin-permeabilized adrenal medullary chromaffin cells. J. Biol. Chem. **258:** 4994-5000.
19. DUNN, L. A. & R. W. HOLZ. 1983. Catecholamine secretion from digitonin-treated adrenal medullary chromaffin cells. J. Biol. Chem. **258:** 4989-4993.
20. BAKER, P. F. & D. E. KNIGHT. 1981. Calcium control of exocytosis and endocytosis in bovine adrenomedullary cells. Philos. Trans. R. Soc. London, Ser. B **296:** 83-101.
21. DRAKE, R. A. L., S. A. HARVEY, D. NJUS & G. K. RADDA. 1979. The effect of chlorpromazine on bioenergetic processes in chromaffin granule membrane. Neuroscience **4:** 853-861.
22. ZIMMERMANN, U., G. PILWAT & F. RIEMANN. 1974. Dielectric breakdown of cell membranes. Biophys. J. **14:** 881-899.

DISCUSSION OF THE PAPER

E. M. SILINSKI (*Northwestern University Medical School, Chicago, IL*): Have you tried to detect quantal ATP release electrophysiologically by taking patches of membranes with fast ATP-gated K^+ channels, placing the electrode with ATP receptor channel near your chromaffin cells and then measuring the ionic currents in your patch produced by quantal ATP release?

ROJAS: We have not tried this experiment. We have determined instead the quantal nature of the cholinergic receptor-activated ATP release by cross-correlation analysis of the signals from two identical light detectors receiving the light from the reaction chamber containing about 10,000 bovine medullary chromaffin cells (E. Rojas, E. Forsberg & E. B. Pollard, *Adv. Exp. Med. Biol.* **211:** 7-29, 1986). The mean duration of each quantal event changes in time after the application of the cholinergic agonist, from about 50 msec (measured immediately after the application of acetylcholine) to 120 msec (about 100 sec of exposure to the agonist). The size of each individual event (estimated by us to be of the order of 5×10^{-15} mol) could not be obtained using cross-correlation analysis because the ATP concentration at the release site is presumably 130 mM and because, at this extremely high concentration, luciferine is rapidly consumed, which makes it impossible to calibrate the system.

G. BURNSTOCK (*University College, London, England*): Would you care to speculate about the role(s) of ATP released from adrenal medullary chromaffin cells?

ROJAS: Perhaps this question should be answered by Dr. E. Forsberg who is attending this meeting and who has been working in our laboratory for the past three years on the action(s) of ATP using endothelial cells from the adrenal medulla in culture. In summary, endothelial cells respond to ATP in many different ways. For example, the phosphoinositide metabolism is rapidly activated by externally applied ATP, and the subsequent generation of inositol trisphosphate and diacylglycerol from phosphatidylinositol 4,5-bisphosphate is accompanied by Ca^{2+} release from internal stores. Thus, we propose that one action of ATP is to reduce the capillary permeability barriers by acting directly on the endothelial cell and thus to facilitate the export of the catecholamines into the main blood stream.

T. D. WHITE (*Dalhousie University, Halifax, Nova Scotia, Canada*): Do you think that your fast phase of ATP release might correlate with adrenaline secretion and your slow phase of ATP secretion with noradrenaline release, as we showed previously?

ROJAS: Because the time constant involved in the fast phase of ATP release is close to the sampling time in your catecholamine experiment (T. D. White, J. E. Bourke & B. G. Livett, *Neurochemistry* **49:** 1266-1273, 1987), it is difficult to compare both sets of measurements. The simplest interpretation of our results is to consider that secreted ATP originates in two different pools. The fastest release process (time constant: 2-5 sec at 23 °C) may correspond to nucleotide released from a pool in close proximity to the cell membrane. The slow secretory process (time constant: 20-50 sec) may correspond to ATP release from another pool, further away from the membrane. Another explanation for the two components of release could be the concurrence of both direct and compound exocytosis. Regardless of the origin of the secreted ATP, however, our results clearly establish the biphasic nature of ATP secretion induced by cholinergic agonists and by membrane depolarization.

WHITE: We also observed that the initial rate of ATP secretion evoked by acetylcholine (or nicotine) was dose dependent, whereas you see no difference. Would you care to comment on this?

ROJAS: We found that the kinetics of ATP release induced by acetylcholine (or nicotine) was independent of the dose used to stimulate chromaffin cells. We also found that, in contrast, the kinetics of ATP secretion was markedly dependent on $[K^+]_o$. Although it is generally accepted that many of the steps in the sequence of events leading to exocytosis may be common to both modes of activation of the secretory response, some profound differences must exist between the two modes of activation. Indeed, we and others have shown that the extent of membrane depolarization evoked by a maximal dose of acetylcholine (100 μM) is comparable to that obtained with a minimal dose of K^+ (12.5 mM). Thus, K^+ at all concentrations tested (30-100 mM) was more effective than acetylcholine (or nicotine) at maximal dose in depolarizing the chromaffin cell membrane. Therefore, we may expect that a larger number of non-inactivating Ca^{2+} channels might be activated with K^+ than with acetylcholine (or nicotine).

Prejunctional Adenosine and ATP Receptors[a]

E. M. SILINSKY, J. M. HUNT, C. S. SOLSONA,
AND J. K. HIRSH

Department of Pharmacology
Northwestern University Medical School
Chicago, Illinois 60611

INTRODUCTION

ATP and adenosine are potent inhibitors of neurotransmitter secretion in the vertebrate nervous system.[1-6] At the majority of synapses investigated, the prejunctional inhibitory effects of ATP are mediated by the hydrolysis product adenosine acting on adenosine receptors.[7] At some synaptic loci, however, unhydrolyzed ATP exerts the presynaptic inhibitory effect.[3,8]

The first purpose of this paper is to discuss the likely mechanism by which adenosine, via adenosine receptors, inhibits the release of acetylcholine (ACh) from peripheral nerve endings in frog and electric fish. The second purpose is to provide a discussion of the direct inhibitory action of ATP on certain prejunctional purinoceptors.

The processes by which adenosine inhibits transmitter release are controversial at present and may involve different cellular mechanisms at different presynaptic membranes.[6] We will focus on the effects of adenosine at presynaptic motor nerve endings in frog and electric fish, where a more cohesive picture of the action of adenosine is emerging.

The presynaptic receptor responsible for the inhibitory effects of adenosine is a P_1-purinoceptor.[9,10] The potential mechanisms underlying this inhibition are depicted in FIGURE 1, sites 1-4. First, adenosine could inhibit transmitter release by effects on the Na^+ and K^+ currents underlying the nerve terminal action potential (ntp) (site 1). Next, adenosine could inhibit transmitter release by reducing Ca^{2+} entry through voltage-gated Ca^{2+} channels, thus impairing the delivery of Ca^{2+} to strategic regions associated with neurotransmitter secretion (site 2). Based upon studies of ionic currents in cell soma, two particularly favored explanations for the inhibitory action of adenosine are 1) that adenosine decreases Ca^{2+} entry and/or 2) that adenosine increases

[a]This research was supported by Grant NS 12782 from the National Institutes of Health. J. K. H. was supported by a predoctoral fellowship from the Lucille P. Markey Charitable Trust Foundation (and by NS 12782 for postdoctoral work); J. M. H. was supported by a predoctoral training grant from the National Institute of Neurological and Communicative Disorders and Stroke.

K^+ conductance, thus hyperpolarizing the membrane and preventing activation of Ca^{2+} channels (for a discussion, see Silinsky[11]).

In contrast to the potential regulation of Ca^{2+} availability by effects on membrane ionic channels, adenosine could also regulate Ca^{2+} availability via intracellular processes. For example, adenosine could inhibit ACh release by increasing the uptake of Ca^{2+} into storage sites, thereby reducing free intracellular Ca^{2+} concentrations (site 3). Finally, adenosine could act directly on the secretory process by decreasing the ability of Ca^{2+} to promote ACh release, for example, by decreasing the apparent affinity for Ca^{2+} or decreasing the number of activatable release sites (site 4).

Earlier evidence from this laboratory suggested that the inhibitory action of adenosine was at the level of the secretory apparatus (site 4) in frog motor nerve.[4,12,13] Further evidence in support of this viewpoint from our laboratory and from others[14] will be presented in this paper.

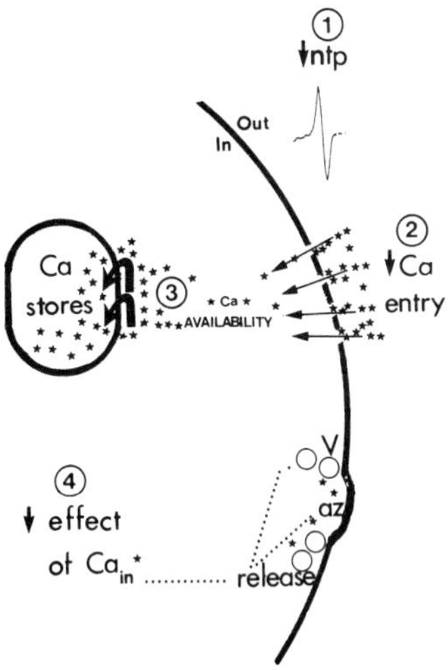

FIGURE 1. Possible target sites for the inhibitory effects of adenosine on neurotransmitter secretion. Asterisks: Ca^{2+}; ntp: nerve terminal action potential; V: synaptic vesicle; az: active zone of secretion.

METHODS

Conventional electrophysiological methods for intracellular and focal (loose patch) recording were employed in this study on frog cutaneous pectoris nerve-muscle preparations.[15,16] Effects of adenosine derivatives on quantal ACh release were assessed by changes in 1) the mean number of ACh quanta released synchronously ($\overline{m}$) as determined from the ratio of the mean end-plate potential (epp) to miniature end-plate potential (mepp) amplitudes, 2) the epp or end-plate current (epc) amplitude (as adenosine does not affect the amplitude of the mepp), or 3) the frequency of occurrence of mepps.

RESULTS

Adenosine Derivatives Do Not Alter Na^+ and K^+ Currents Associated with the Action Potential in Frog Motor Nerve Endings

Using focal (loose patch) recording, which measures both the prejunctional action currents and the epc with the same recording electrode, it was found that neither Na^+ nor K^+ currents were impaired under conditions in which adenosine derivatives inhibits ACh release[4] (FIGS. 2a & 2b). In recent experiments, we found that concentrations of tetraethylammonium (TEA, 1 mM) and 3,4-diaminopyridine (DAP, 100 μM) that eliminate all K^+ currents in motor nerve endings did not prevent the inhibitory effects of adenosine (FIG. 2c). It thus appears that adenosine does not affect the Na^+ or K^+ currents associated with the presecretory nerve terminal action potential.

Blockade of Ca^{2+} Entry Is Unlikely to Be Responsible for the Inhibitory Effects of Adenosine in Frog and Electric Fish

Measurements of Ca^{2+} Entry

Direct measurements of the entry of radiolabeled Ca^{2+} ($^{45}Ca^{2+}$) into *Torpedo* nerve terminals and the concomitant release of ACh were made by Muller *et al.*[14] These workers found that under conditions in which adenosine reduced ACh release (FIG. 3a), this nucleoside did not reduce Ca^{2+} entry. In contrast, Ca^{2+} entry and evoked ACh release were blocked competitively by Ca^{2+} channel blockers such as Cd^{2+} (FIG. 3b). In preliminary experiments from our laboratory, Ca^{2+} currents were measured directly using loose patch recordings from the heminode or from the nerve ending of frog motor nerve after complete blockade of K^+ channels with TEA and DAP.[15] Neither adenosine (50 μM) nor 2-chloroadenosine (2.5 μM) inhibited the Ca^{2+} currents under conditions in which the Ca^{2+} channel blocker Co^{2+} (2.5-5.0 mM) blocked such currents.

Indirect measurements of Ca^{2+} currents were made using the reverse gradient approach. By this method, a depolarizing stimulus applied in Ca^{2+}-free EGTA Ringer decreases rather than increases ACh release. This is because Ca^{2+}-free EGTA Ringer generates a reverse concentration gradient whereby Ca^{2+} is in higher concentration inside the cell than outside. Depolarization under reverse gradient conditions thus elicits the efflux of Ca^{2+} from the cell through voltage-gated Ca^{2+} channels, reduces the Ca^{2+} concentrations at release sites, and reduces mepp frequency.[17] If adenosine blocks Ca^{2+} channels, then adenosine should increase ACh release under reverse gradient conditions by blocking the egress of Ca^{2+}. This is not the case, however. Adenosine derivatives continue to inhibit ACh secretion under reverse gradient conditions—conditions in which the Ca^{2+} channel blocker Co^{2+} increases release (N = 8). The results of Muller *et al.*[14] and our own data thus suggest that Ca^{2+} channel blockade is unlikely to be responsible for the inhibitory effect of adenosine.

Effects of Adenosine on ACh Release Evoked Independently of Ca^{2+} Channels

Two methods that evoke Ca^{2+}-dependent ACh release by mechanisms bypassing active Ca^{2+} channels were employed: Ca^{2+}-containing liposomes[4] and the Ca^{2+} ionophore ionomycin.[18] It was found previously that adenosine derivatives inhibit ACh

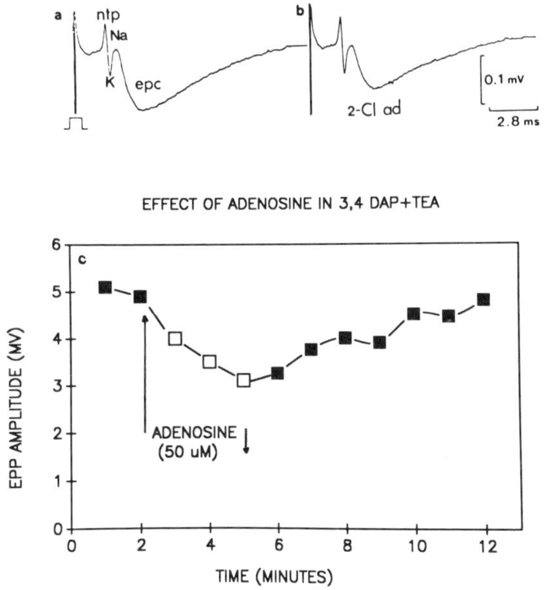

FIGURE 2. Evidence that an increased K^+ conductance is not responsible for the inhibitory effects of adenosine. (**a**) Control response recorded with a focal microelectrode. (**b**) Responses after an 8-min superfusion with 10 μM 2-chloroadenosine. Note the decline in the end-plate current (epc) without an effect on the currents underlying the nerve terminal action potential (ntp). The upward-going phase of the ntp reflects the Na^+ component of the action potential whereas the downward-going phase represents the K^+ currents recorded at the frog nerve ending (for details, see reference 15). Ringer contained low Ca^{2+} and high Mg^{2+}. Traces in **a** and **b** are the averaged responses to 128 stimuli (1 Hz). Reproduced with permission from reference 4. (**c**) Adenosine still inhibits ACh release after treatment with sufficient concentrations of K^+ channel blockers (TEA, 1 mM; 3,4-diaminopyridine, 100 μM) to block completely K^+ channels at frog motor nerve endings. Each record is the response to a single stimulus. As these K^+ channel blockers produce large increases in evoked ACh release, the Ringer contained reduced Ca^{2+}, elevated Mg^{2+}, and tubocurarine to reduce the ACh release to levels at which the immediately available store of quanta would not be depleted and to maintain subthreshold epps.

release produced by Ca^{2+}-containing liposomes.[4] In two preliminary experiments using ionomycin (1.3-50 μM), increases in mepp frequency produced by this ionophore were reversibly inhibited by adenosine. It thus appears that active Ca^{2+} channels are not required for inhibitory effects of adenosine to be observed on Ca^{2+}-evoked ACh secretion.

The Cellular Ca^{2+} Buffer BAPTA neither Mimics the Effect of Adenosine nor Occludes the Inhibitory Effects of Adenosine

At frog motor nerve endings, 1,2-bis(2-aminophenoxy)ethane-*N,N,N',N'*-tetra-acetic acid (BAPTA, delivered to the cytoplasm by using extracellular BAPTA-AM) was found to reduce mepp frequency and to inhibit facilitation; yet it did not impair synchronous evoked ACh release in response to one nerve impulse ($\overline{m}$).[19] Apart from the similar decrease in mepp frequency, the other effects of BAPTA differ from those of adenosine. Specifically, adenosine produces a decrease in both mepp frequency and $\overline{m}$ without inhibiting facilitation. It is thus unlikely that evoked ACh release is inhibited by a decrease in the free intracellular Ca^{2+} concentration. If adenosine inhibited spontaneous ACh release by decreasing free Ca^{2+} levels in the nerve ending, then the similar reductions in mepp frequency reported for BAPTA[19] and for adenosine[1,2,4]

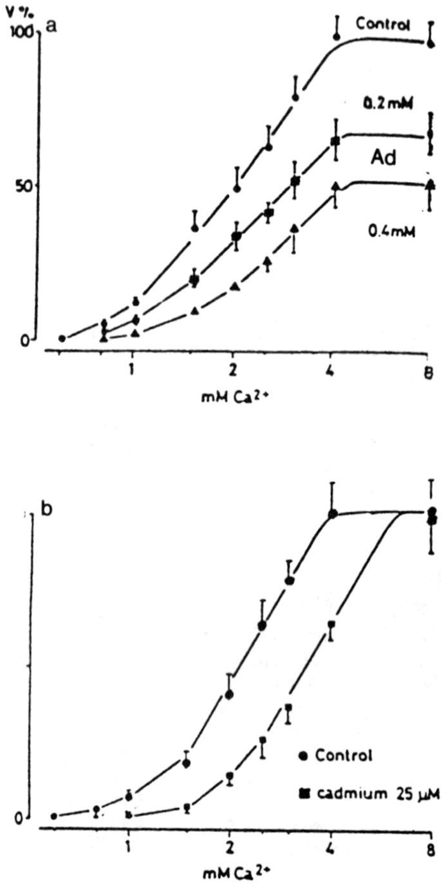

FIGURE 3. Noncompetitive blockade of evoked ACh release by adenosine (**a**) and competitive blockade of ACh release by the Ca^{2+} channel blocker Cd^{2+} (25 μM) (**b**). Under these conditions, adenosine (0.5 mM) had no effect on Ca^{2+} entry into nerve endings whereas Cd^{2+} inhibited both evoked Ca^{2+} entry and ACh release. Reprinted with permission from reference 14.

suggest that BAPTA has a buffering capacity near that of the putative adenosine-sensitive Ca^{2+} translocation mechanism. It would thus be predicted that BAPTA should occlude the inhibitory effects of adenosine on ACh release. In nine experiments on BAPTized preparations, we found[20] that BAPTA did not alter the inhibitory effects of adenosine on mepp frequency (mean of 41% inhibition ±5% standard error of the mean, which is in the range of control values found in other studies in our laboratory). Evidence that BAPTA was actually buffering Ca^{2+} is based upon the observations that 1) BAPTA blocked the excitatory effects of caffeine and that 2) BAPTA delayed that excitatory effects of high K^+ and ionomycin on ACh release (the delay being due to the time required to saturate the intracellular BAPTA with Ca^{2+}). These results,[20] when taken in conjunction with earlier results,[4] suggest that adenosine is unlikely to inhibit ACh release by decreasing free cytoplasmic Ca^{2+} concentrations.

Adenosine Decreases the Ability of Ca^{2+} to Promote ACh Secretion

Previous results from this laboratory have found that dose-response relationships for extracellular Ca^{2+} and Sr^{2+} in the process of evoked ACh release are incompatible with simple competitive block of alkaline earth cation entry by adenosine.[4,12] Computer modeling demonstrated that a reduction in the apparent affinity of Ca^{2+} for a strategic component of the secretory apparatus accurately simulates the actual effect of adenosine analogues on evoked ACh release.[13,21] FIGURE 3 shows results from the work of Muller *et al.*[14] demonstrating that adenosine acts very differently from Ca^{2+} channel blockers in electric fish. The effects observed in FIGURE 3 are thus incompatible with competitive inhibition of alkaline earth cation entry by adenosine derivatives. Muller *et al.*[14] suggested that adenosine reduces the number of active release sites. These results from frog and electric fish suggest that the inhibition of ACh release by adenosine is primarily due to the decreased ability of a fixed cellular concentration of Ca^{2+} to promote the secretory process.

What Type of Adenosine Receptor Is Responsible for the Inhibition of ACh Release?

8-Cyclopentyltheophylline is a potent inhibitor of ACh release.[6,22] This observation, when coupled with the results to be described below, suggests that the receptor responsible for the inhibition of neurotransmitter release is a P_1-purinoceptor of the A_1 flavor. (Considerable controversy does exist, however, and the reader is referred to references 5, 11, and 21 for discussion.) At rat[23] and mouse[24] motor nerve endings, pertussis toxin (PTX) consistently blocked the effects of adenosine in inhibiting ACh release. In some experiments on frog, PTX blocked the inhibitory effects of adenosine on evoked ACh release.[23] Unfortunately, the effects of PTX in frog were inconsistent.[23] The successful experiments from both species suggest, however, that a PTX-sensitive guanine nucleotide-binding protein (G protein) is responsible for the inhibitory effect of adenosine.

Specific PTX-sensitive G proteins include G_i, which links adenosine to the inhibition of adenylyl cyclase, and G_o, which may directly link adenosine to a cellular effector without the need for phosphorylation.[25] To determine whether phosphorylation via the cyclic AMP system is involved in the inhibitory effects of adenosine, we treated preparations with a nonspecific isoquinoline sulfonamide inhibitor of protein kinases, H-7.[26] In the presence of concentrations of H-7 sufficient to block the effects of the cyclic AMP analogue 8-(4-chlorophenylthio)-cAMP[27] and of phorbol esters,[28] adenosine still exerted its usual inhibitory effects in frog or rat.[27] Adenosine thus appears to inhibit ACh release independently of phosphorylation via cAMP-dependent protein kinase or protein kinase C.

ATP and the Inhibition of Transmitter Release

As discussed above, the prejunctional effects of ATP at most synapses are due to the hydrolysis product, adenosine. There are a number of instances, however, where ATP directly inhibits ACh release. FIGURE 4 shows that in frog sympathetic chain ganglia, ATP but not adenosine inhibits ACh release.[3] 2-Chloroadenosine did not significantly inhibit ACh release from these synapses at concentrations two orders of magnitude greater than that necessary to produce maximal inhibition of ACh release from motor nerve endings of the same species.[3] FIGURE 5 shows that α,β-methylene-ATP was without effect (suggesting that a P_{2X}-purinoceptor was not involved in inhibition at preganglionic nerve endings), but that theophylline did block the effect of ATP. Because of such results, it has been suggested that a new purinoceptor subtype (P_3) may be necessary to explain some of these effects of ATP.[31] In bullfrog sympathetic ganglia, the potency order of purines was ATP > ADP > AMP > adenosine for the inhibition of ACh release. In bullfrog, the effect has been attributed to a depolarization of the nerve ending.[32] Interestingly, in guinea pig ileum, ATP, in a theophylline-sensitive manner, exerted a direct action to inhibit ACh release.[8] This effect of ATP was not due to degradation to adenosine, and was not mimicked by α,β-methylene-ATP. The effects of ATP derivatives at cholinergic nerve endings in frog sympathetic ganglia and in guinea pig ileum are thus remarkably similar. Adenosine was also an effective inhibitor of ileal ACh release, however, through the same purinoceptor as ATP.

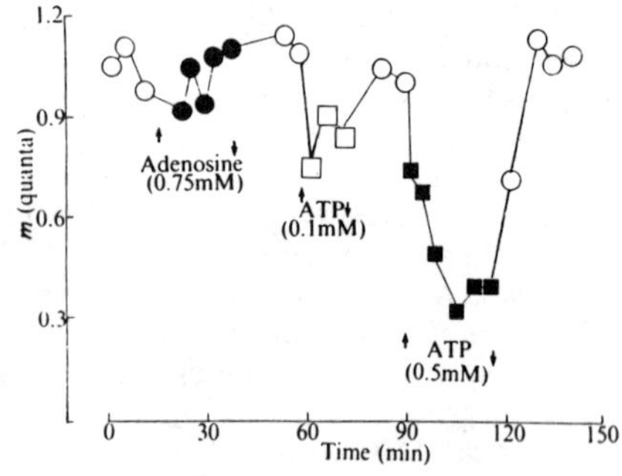

FIGURE 4. Inhibition of ACh release from frog sympathetic chain ganglion by ATP but not adenosine. The electrophysiological methods were similar to those described in the text for frog neuromuscular junctions. Ringer contained 1 mM Ca^{2+} and 18 mM Mg^{2+}. Each point represents $\overline{m}$ calculated by the method of failures[16] in response to 128 preganglionic nerve stimuli delivered at a frequency of 1 Hz. Open circles: control $\overline{m}$. Reprinted with permission from reference 3.

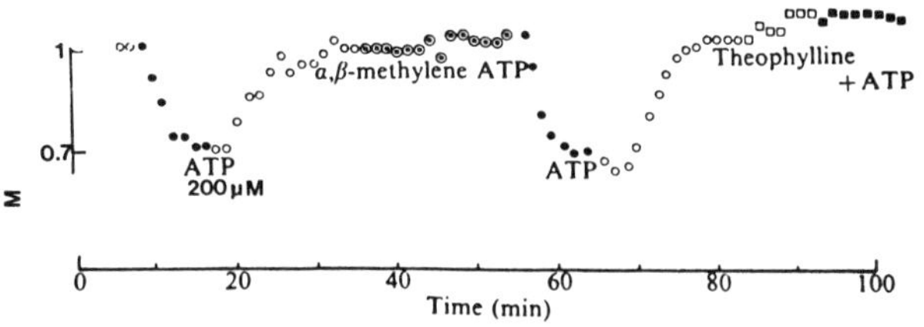

FIGURE 5. Effects of ATP (200 μM), α,β-methylene-ATP (200 μM), and theophylline (2 mM) on evoked ACh release ($\overline{m}$) in frog sympathetic chain ganglia. Evoked ACh release is expressed relative to the control level, which is taken as 1; values below 1 indicate inhibition. Ringer solution contained 1 mM Ca^{2+} and 17 mM Mg^{2+}. Modified with permission from reference 3.

DISCUSSION AND CONCLUSIONS

The results suggest that activation of P_1-purinoceptors by adenosine inhibits ACh release by impairing the ability of Ca^{2+} to promote secretion at frog and electric fish motor nerve endings. Inhibition of ACh secretion by adenosine is unlikely to be mediated by A or C kinases. It is possible that adenosine receptor sites may be directly coupled to the secretory apparatus via G proteins. The independent suggestions of a decreased Ca^{2+} affinity,[4,12] and of a decreased number of activatable release sites[14] are compatible as it has been found that a release site is likely to require a certain concentration of bound Ca^{2+} to be in an activatable form.[29] The generality of these results to other synapses is unknown at present, although a case may be made for a similar mechanism for the inhibition of transmitter release by adenosine from the CA1 region of the hippocampus (as argued in reference 11, but see references 6 and 30 for other possibilities).

It also appears that in some systems ATP is capable of inhibiting ACh release in a theophylline-sensitive manner, and that it does not do so by degradation to adenosine. Although such results may be due to a P_3 receptor, it is plausible to suggest that P_1-purinoceptors be further divided into a P_{1X} subtype—the conventional adenosine receptor—and a P_{1Y} subtype—the theophylline-sensitive site at which ATP is more potent than the other naturally occurring adenosine derivatives. Such a subdivision would be consistent with the traditional approach of defining receptors by selective competitive antagonists.

REFERENCES

1. GINSBORG, B. L. & G. D. S. HIRST. 1972. The effect of adenosine on the release of the transmitter from the phrenic nerve of the rat. J. Physiol. **224:** 629-645.
2. SILINSKY, E. M. 1980. Evidence for specific adenosine receptors at cholinergic nerve endings. Br. J. Pharmacol. **71:** 191-194.
3. SILINSKY, E. M. & B. L. GINSBORG. 1983. Inhibition of acetylcholine release from preganglionic frog nerves by ATP but not adenosine. Nature **305:** 327-328.

4. SILINSKY, E. M. 1984. On the mechanism by which adenosine receptor activation inhibits the release of acetylcholine from motor nerve endings. J. Physiol. **346:** 243-256.
5. RIBEIRO, J. A. & A. M. SEBASTIAO. 1986. Adenosine receptors and calcium: Basis for proposing a third (A_3) adenosine receptor. Prog. Neurobiol. **26:** 279-309.
6. FREDHOLM, B. B. & T. V. DUNWIDDIE. 1988. How does adenosine inhibit transmitter release? Trends Pharmacol. Sci. **9:** 130-134.
7. RIBEIRO, J. A. & A. M. SEBASTIAO. 1987. On the role, inactivation and origin of endogenous adenosine at the frog neuromuscular junction. J. Physiol. **385:** 571-585.
8. WIKLUND, N. P., L. E. GUSTAFSSON & J. LUNDIN. 1985. Pre- and postjunctional modulaton of cholinergic neuroeffector transmission by adenine nucleotides: Experiments with agonist and antagonist. Acta Physiol. Scand. **125:** 681-691.
9. BURNSTOCK, G. 1990. Overview: Purinergic mechanisms. Ann. N.Y. Acad. Sci. This volume.
10. BURNSTOCK, G. 1990. Dual control of local blood flow by purines. Ann. N.Y. Acad. Sci. This volume.
11. SILINSKY, E. M. 1989. Adenosine derivatives and neuronal function. Semin. Neurosci.: **1:** in press.
12. SILINSKY, E. M. 1981. On the calcium receptor that mediates depolarization-secretion coupling at cholinergic motor nerve terminals. Br. J. Pharmacol. **73:** 413-429.
13. SILINSKY, E. M. 1986. Inhibition of transmitter release by adenosine: Are calcium currents depressed or are the intracellular effects of calcium impaired? Trends Pharmacol. Sci. **7:** 180-185.
14. MULLER, D., F. LOCTIN & Y. DUNANT. 1987. Inhibition of evoked acetylcholine release: Two different mechanisms in the *Torpedo* electric organ. Eur. J. Pharmacol. **133:** 225-234.
15. MALLART, A. 1984. Presynaptic currents in frog motor endings. Pfluegers Arch. (Eur. J. Physiol.) **400:** 8-13.
16. SILINSKY, E. M. 1987. Electrophysiological methods for studying acetylcholine secretion. *In* The Secretory Process. Vol. 3: *In Vitro* Methods for Studying Secretion. A. M. Poisner & J. T. Trifaro, Eds.: 255-271. Elsevier. Amsterdam.
17. SHIMONI, Y., E. ALNAES & R. RAHAMIMOFF. 1977. Is hyperosmotic neurosecretion from motor nerve endings a calcium-dependent process? Nature **267:** 170-172.
18. BEELER, T. J., I. JONA & A. MARTONOSI. 1979. The effect of ionomycin on calcium fluxes in sarcoplasmic reticulum vesicles and liposomes. J. Biol. Chem. **254:** 6229-6231.
19. KIJIMA, H. & N. TANABE. 1988. Calcium-independent increase of transmitter release at frog end-plate by trinitrobenzene sulphonic acid. J. Physiol. **403:** 135-149.
20. HUNT J. M. & E. M. SILINSKY. 1989. BAPTism of frog motor nerve terminals does not impair the inhibitory actions of adenosine on acetylcholine release. Soc. Neurosci. Abstr. **15:** 484.
21. SILINSKY, E. M., C. S. SOLSONA, J. K. HIRSH & J. M. HUNT. 1989. Calcium-dependent acetylcholine secretion: Influence of adenosine. *In* Adenosine Receptors in the Nervous System. J. A. Ribeiro, Ed.: 141-149. Taylor & Francis. London.
22. SEBASTIAO, A. M. & J. A. RIBEIRO. 1989. 1,3,8- and 1,3,7-Substituted xanthines: Relative potency as adenosine receptor antagonists at the frog neuromuscular junction. Br. J. Pharmacol. **96:** 211-219.
23. SILINSKY, E. M. C. S. SOLSONA & J. K. HIRSH. 1989. Pertussis toxin prevents the inhibitory effect of adenosine and unmasks adenosine-induced excitation at mammalian motor nerve endings. Br. J. Pharmacol. **97:** 16-18.
24. CHEN, H., Y. N. SINGH & W. F. DRYDEN. 1989. Transduction mechanism involving the presynaptic adenosine receptor at mouse motor nerve terminals. Neurosci. Lett. **96:** 318-322.
25. KURACHI, Y., T. NAKAJIMA & T. SUGIMOTO. 1986. On the mechanism of activation of muscarinic K^+ channels by adenosine in isolated atrial cells: Involvement of GTP-binding proteins. Pfluegers Arch. (Eur. J. Physiol.) **407:** 264-274.
26. HIDAKA, H., M. INAGAKI, S. KAWAMOTO & Y. SASAKI. 1984. Isoquinolinesulfonamides: Novel and potent inhibitors of cyclic nucleotide-dependent protein kinase and protein kinase C. Biochemistry **23:** 5036-5041.
27. HIRSH, J. K. & E. M. SILINSKY. 1989. Signal transduction and the adenosine receptor

inhibitory to acetylcholine release in frog motor nerve endings. Soc. Neurosci. Abstr. **15:** 484.
28. CARATSCH, C. G., S. SCHUMACHER, F. GRASSI & F. EUSEBI. 1988. Influence of protein kinase C stimulation by a phorbol ester on neurotransmitter release at frog end-plates. Naunyn-Schmeideberg's Arch. Pharmacol. **337:** 9-12.
29. SILINSKY, E. M. 1985. The biophysical pharmacology of calcium-dependent acetylcholine secretion. Pharmacol. Rev. **37:** 81-132.
30. HAMILTON, B. R., Z. LU & D. O. SMITH. 1988. Modulation of calcium currents in mammalian motor nerve terminals. Soc. Neurosci. Abstr. **14:** 69.
31. STONE, T. 1985. Some unresolved problems. *In* Purines: Pharmacological and Physiological Roles. T. W. Stone, Ed.: 245-251. Macmillan. London.
32. AKASU, T., P. SHINNICK-GALLAGHER & P. GALLAGHER. 1985. Actions of purines in autonomic ganglia. *In* Purines: Pharmacological and Physiological Roles. T. W. Stone, Ed.: 57-66. Macmillan. London.

DISCUSSION OF THE PAPER

E. W. WESTHEAD (*University of Massachusetts, Amherst, MA*): Have you any way of determining whether adenosine or ATP reduces the Ca^{2+} transient produced by stimulation? In chromaffin cells, we were surprised to find that even ionomycin-stimulated Ca^{2+} transients are reduced by adenosine, suggesting that adenosine increases the rate of sequestration or extrusion of Ca^{2+}.

SILINSKY: This would indeed be an ideal experiment, but at this stage fura-2 injection into the very small presynaptic motor neuron endings in frog and subsequent quantitative measurements of Ca^{2+} transients is technically impossible for mere mortals. Your result with chromaffin cells is most interesting. It suggests that active membrane ionic channels are not necessary for the inhibitory effects of adenosine in chromaffin cells. We need to do another series of experiments with BAPTA-AM and ionomycin before direct comparison between your results and ours may be made, especially if adenosine is capable of reducing $[Ca^{2+}]_i$ below that buffered by BAPTA. For example, if adenosine is decreasing mepp frequency by decreasing $[Ca^{2+}]_i$, then in the presence of BAPTA, after adenosine has produced its inhibitory effect, ionomycin should produce rapid restoration of mepp frequency to the pre-adenosine level (as this Ca^{2+} would not be buffered by BAPTA). If adenosine works by another mechanism, then ionomycin would have its usual delay in BAPTA before producing any increase in mepp frequency. These experiments are currently in progress by JimBob Hunt.

N. P. WIKLUND (*Karolinska Institutet, Stockholm, Sweden*): Have you tried determining whether inhibitors of ATP degradation have any effect on the proposed prejunctional ATP effects? In the guinea pig ileum, inhibition of 5′-nucleotidases with such inhibitors as α,β-Me-ADP and IMP enhances the inhibitory effects of ATP or ADP on contractile responses to nerve stimulation, suggesting that ATP and ADP can act per se on prejunctional P_1-purinoceptors (Wiklund *et. al., Acta Physiol. Scand.* **126:** 217-223, 1986). Furthermore Schild analysis on inhibition of contractile responses to nerve stimulation by adenine nucleotides in guinea pig ileum, using the selective adenosine receptor antagonist 8-*p*-sulfophenyltheophylline, showed slopes of unity for several nucleotides, including ATP, ADP, AMPPNP, and β,γ-Me-ATP. The nucleotides had pA2-values not significantly different from the pA2-value of 2-chloroadenosine, strongly suggesting action by adenine nucleotides per se on prejunctional adenosine receptors (*cf. Acta Physiol. Scand.* **125:** 681-691, 1985).

SILINSKY: We have not tried the experiment you described. Akasu and colleagues have found in bullfrog that ADP and AMP have slight effects, with the following sequence: ATP > ADP > AMP > adenosine. We definitely need to pursue a more extensive series of analogues, including adenosine deaminase inhibitors and 5′-nucleotidases. We should also explore the possible antagonism of ATP effects by α,β-Me-ATP and other nonhydrolyzable analogues. Based upon 1) the Wiklund *et al.* reference cited by you and in our paper, 2) the work of Collis and Pettinger, 3) the poster from Dr. Westfall's laboratory at this meeting, and 4) our results, it is predicted that the putative P_{2Y}-purinoceptor would not be activated by α,β-Me-ATP (which might be antagonistic), β,γ-Me-ATP (which should be an agonist), and blockade by 8-phenyltheophylline. The stimulating presentations and discussions at this meeting have provided an impetus for us to resume work on the inhibitory purine receptor at frog preganglionic nerve endings.

Y. H. EHRLICH (*College of Staten Island, New York, NY*): Could inhibition of the ATP effect by theophylline be mediated by a mechanism other than blockade of a P_1-type receptor?

SILINSKY: Theoretically, many of the secondary effects reported for theophylline would certainly suggest that your question is a valid one. My feeling, however, is that theophylline is indeed working as a blocker of P_1 (P_{1Y}?) sites at frog preganglionic nerves. At P_1 sites in peripheral cholinergic nerves of the same species, phosphodiesterase inhibitors (including caffeine) increase the level of inhibition by P_1 agonists, rather than blocking their effects. Our now rather dated 1983 experiments on ganglion definitely need to be repeated with newer, more selective alkyl xanthines.

C. LONDOS (*National Institutes of Health, Bethesda, MD*): Perhaps another explanation for theophylline reversal of an ATP effect is that the ATP effect is being facilitated by adenosine arising from ATP metabolism. One example of facilitation is found in PC12 pheochromicytoma cells in which adenosine receptor agonists, which have no effect of neurite outgrowth, greatly enhance the stimulation of neurite outgrowth by nerve factor.

SILINSKY: Your suggestion is intriguing. My belief is that adenosine deaminase would not alter the effects of ATP, but we have not done these experiments. From work we have done, 1 mM adenosine does not influence the action of concurrently superfused ATP, so at least exogenous adenosine is not influencing the action of ATP.

Release of ATP from Heart

Presentation of a Release Model Using Human Erythrocyte[a]

T. FORRESTER

Department of Physiology
St. Louis University Medical Center
St. Louis, Missouri 63104

HISTORICAL INTRODUCTION

The mammalian heart is dependent upon a supply of oxygen that must be immediate and sufficient for myocardial survival. It is generally accepted that coronary blood vessels are profoundly affected by products of metabolism that act as vasodilators; many substances have been proposed to act as so-called vasodilator metabolites. The term purine bodies was used by Barcroft and Dixon[1] in reference to such a vasodilator action, but how the concept of purine release arose remains unclear. Zipf[2–4] in the early 1930s suggested that vasodilator substances in the effluent blood from ischemic tissues might be adenylic acid or adenosine. Also at this time the extracellular actions upon the heart of ATP and its breakdown products were being closely studied,[6–10] and the great potency of these agents as vasodilators placed them as possible regulators of coronary blood flow. Two theoretical obstacles arose, however, to discourage the view that ATP could traverse cell membranes. First, Lipmann[11] had established ATP as the predominant energy carrier for the many recently discovered biochemical pathways, and it seemed very unlikely that this valuable molecule should be lost to the external milieu in any physiological process. Second, investigations into the properties of cell membranes with regard to the selective permeability to charged particles[12] discouraged the view that ATP could cross cell membranes by pointing out that ATP carried such a large negative charge. Nevertheless, attempts were made to detect the release of ATP from the myocardium, for example, by testing the venous effluent from the coronary sinus,[13] and later from skeletal muscle,[5] but inadequate technology coupled with the aforementioned theoretical objections led to rejection of ATP as a local vasodilator substance in the coronary circulation.

In 1950, Green and Stoner[14] strongly suspected that circulating nucleotides contributed significantly to the condition of hemorrhagic shock seen in battlefield injuries but at that time had no way of assessing the concentrations of these substances in the blood plasma with the methodology then available.

[a]This work was supported by the Department of Physiology, St. Louis University Medical Center.

The physiological significance of extracellular nucleotides could not be appreciated until techniques were developed that could measure with extreme precision very small quantities of ATP within a very short space of time. In a search to define the energy source for bioluminescence, McElroy[15] showed that when Lampyrid beetles (*Photinus pyralis*) are ground up with sand and water a crude extract is obtained that responds with a flash of light when microgram quantities of ATP are added. It was found that the extract could also respond to other triphosphates, but this was shown to be due to the presence of transphosphorylases[16] that could catalyze the following reaction: xTP + ADP = xDP + ATP, where x is any nucleotide base. The factors required for the light reaction in response to ATP have now been prepared in crystalline form,[17] however, and the specificity of the firefly reaction for ATP using these reagents is now thought to be absolute. It is now known that one quantum of light is emitted for each luciferin molecule excited by ATP.[17] With refinements in photodetection and the development of more powerful photomultiplier tubes, the firefly technique was soon adapted to measure extremely small amounts of ATP in biological fluids.

The first direct demonstration of ATP release from tissues using the firefly technique was published by Pamela Holton,[18] who tested the effluent from rabbit ears perfused with Locke's solution. When the auricular nerve was stimulated, ATP appeared in the perfusate; no ATP was detected in the control perfusions. As another control, ATP was measured in samples of rabbit blood obtained by heart puncture. These plasma levels were found to vary little from the mean value of 1.0 $\mu mol \cdot ml^{-1}$ blood. This was probably the first published demonstration of normal levels of ATP in circulating plasma.

Holton's work ushered in a new era of investigation with the bioluminescence technique, and all subsequent experiments reviewed here have assessed effluent ATP in this way. *Ex scintilla incendium*—from a spark a conflagration (proverb).

RELEASE OF ATP FROM LANGENDORFF-PERFUSED HEARTS

Evidence that ATP was released from *in vitro* stimulation of frog skeletal muscle[19] and that it increased in concentration in the venous effluent from active human forearm muscle[20,21] prompted Paddle and Burnstock[22] to investigate whether ATP was released from hypoxic myocardium. Hypoxia was induced in guinea pig hearts by switching from a perfusion medium of 95% O_2 + 5% CO_2 to a medium gassed with 95% N_2 + 5% CO_2 for a period of 90 sec. This time of exposure to the hypoxic gas mixture was found to produce maximum vasodilatation of the coronary circulation. FIGURE 1 shows the result of one such experiment (see experiment 2 in TABLE 1). Perfusion pressure diminished rapidly, indicating sudden vasodilatation of the coronary circulation in this constant flow arrangement. TABLE 1 gives the results from all experiments performed. ATP concentration ($pmol \cdot ml^{-1}$) in the control perfusate was 0.21 ± 0.05 (mean ± SEM) compared to 0.64 ± 0.009 at the end of the hypoxic period (N = 10). This result was verified by Stowe *et al.*[23] A similar result was obtained using rabbit hearts[24] where ATP and myoglobin were measured. FIGURE 2 indicates that while ATP was released into the coronary effluent immediately following the induction of hypoxia, there was a delay of approximately 7 min before a peak of myoglobin concentration was achieved.[24]

It was soon realized that the ATP detected in all of these experiments was a residual amount that had escaped degradation by outward-facing membrane ATPases.

The perfused heart preparation was well known to have a powerful extracellular ATPase activity,[25] and this was further explored by Paddle and Burnstock.[22] It was found that when ATP at 100-200 pmol·ml^{-1} was perfused through the coronary circulation, about 99% was degraded in a single transit, leaving only 1-2 pmol·ml^{-1} in the perfusate, almost equal to that measured as a result of applying hypoxia. This clearance pattern has been confirmed.[26,27] A simple conclusion is that the *source* concentration of ATP in hypoxia must approach one hundred times that measured in the coronary sinus effluent. When hypoxia was imposed after 30-40 min of perfusion with ouabain (0.25 μg·ml^{-1}), an Na^+- and K^+-activated ATPase inhibitor, the mean concentration of ATP in the perfusate rose from 0.39 to 2.17 pmol·ml^{-1} (see Table III in reference 22). Because the mean concentration of ATP in the perfusate from hypoxic hearts without ouabain was 0.64 pmol·ml^{-1}, it was concluded that the

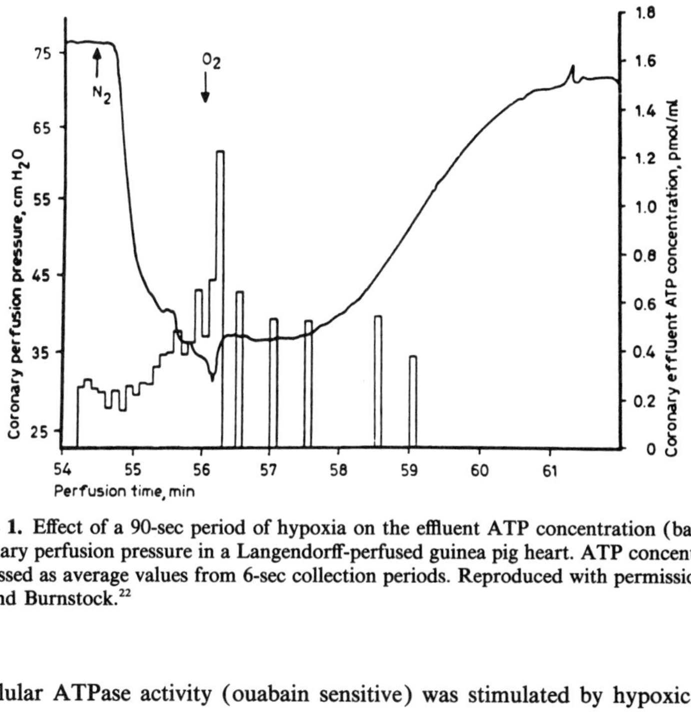

FIGURE 1. Effect of a 90-sec period of hypoxia on the effluent ATP concentration (bars) and the coronary perfusion pressure in a Langendorff-perfused guinea pig heart. ATP concentrations are expressed as average values from 6-sec collection periods. Reproduced with permission from Paddle and Burnstock.[22]

extracellular ATPase activity (ouabain sensitive) was stimulated by hypoxic conditions.[22]

The source of the released ATP in these experiments could have been endothelial cells, smooth muscle, cardiac muscle, or blood elements. Vascular endothelial and smooth muscle cells in culture have been shown to release ATP selectively in response to potentially damaging stimuli[28]; of course, damaged blood elements, especially platelets, are always capable of releasing nucleotides if washout procedures have not been thorough. It is unlikely, however, that this latter situation arose, as hearts were perfused for about 1 hr before the application of hypoxia.[22]

Pharmacological modulation of ATP release from the perfused heart has been clearly demonstrated. Stimulation of cardiac muscarinic receptors by acetylcholine induces a large release of ATP, as well as AMP and adenosine, into the coronary circulation of guinea pig hearts.[29] This event, being independent of oxygen consumption, was interpreted as indicating that it was very likely that the coronary vasodilatation following infusion with acetylcholine was in part or even fully mediated by a secondary release of vasoactive purines. This interpretation was supported by the finding that theophylline diminished by about half the coronary flow response to a 10^{-6} M infusion of acetylcholine. The potent coronary vasoconstrictor substance leukotriene D_4 (LTD_4) invoked a release of ATP that was blocked by a specific LTD_4 receptor blocker.[27] It was also found that infusion of norepinephrine released ATP in comparable amounts; this effect was blocked by an α-adrenergic blocking agent (phentolamine). A vasoconstriction control performed with arginine vasopressin did not result in ATP release, indicating that vasoconstriction (or activation of a myogenic stretch response) itself was not the mechanism of release. The possibility that muscarinic cholinergic receptors were involved was ruled out because pretreatment with scopolamine and guanethidine did not influence ATP release, supporting the findings of Paddle and Burnstock.[22] Darius *et al.*[27] proposed that ATP release specifically induced by LTD_4 and norepinephrine could be an acute response of the heart to prolonged coronary vasoconstriction and myocardial ischemia, since blood levels of LTD_4 and norepinephrine are elevated in acute cardiovascular stress.

TABLE 1. Effects of Hypoxia on Perfusion Pressure and Coronary Effluent ATP Concentration in the Isolated Guinea Pig Heart[a]

		Control (aerobic)		Hypoxia[b]	
Experiment	Perfusion Rate (ml/min)	Perfusion Pressure (cm H_2O)	Coronary Effluent ATP Concentration (pmol/ml)	Perfusion Pressure (cm H_2O)	Coronary Effluent ATP Concentration (pmol/ml)
1	6.8	70	0.11	36	0.46
2	4.8	77	0.26	35	0.66
3	5.1	74	0.18	38	0.48
4		78	0.21	37	0.41
5	4.1	81	0.13	41	0.43
6		72	0.13	39	0.64
7	4.7	74	0.60	25	1.33
8	8.0	74	0.08	34	0.56
9	5.7	79	0.22	28	0.91
10	7.2	78	0.22	34	0.56
			0.21 ± 0.05[c]		0.64 ± 0.09[c]

[a] Adapted from Paddle and Burnstock.[22]
[b] Duration of hypoxia: 1.5 min.
[c] Mean ± SEM.

Valuable information was obtained from a series of elegant experiments by Vial, Owen, Opie, and Posel.[30] The release of ATP, adenosine, and the enzyme lactic dehydrogenase (LDH) was measured in the perfusate from rat hearts in response to sustained hypoxia. In agreement with previous findings, the onset of hypoxia induced a rapid increase of ATP concentration in the coronary effluent from 0.8 ± 0.2 nM to an average peak value of 1.3 ± 0.2 nM occurring at 2 ± 0.5 min after the onset

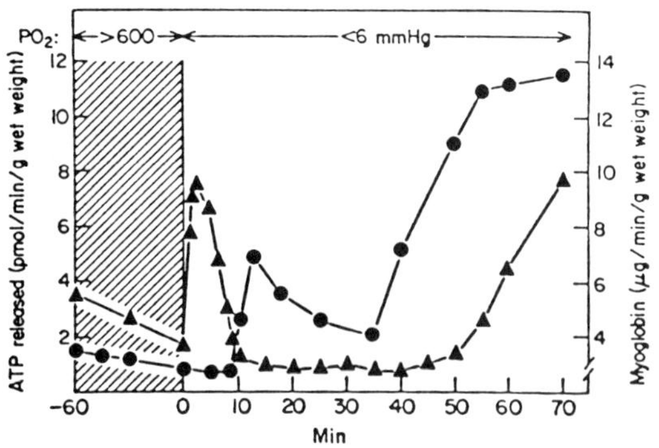

FIGURE 2. Release of ATP (▲) and myoglobin (●) into the coronary effluent of a Langendorff-perfused rabbit heart after introduction of hypoxia. Note the rapid release of ATP in response to a change of Po_2 from 600 mmHg to <6 mmHg. Reproduced with permission from Nayler *et al.*[24]

of hypoxia. FIGURE 3 shows the pattern of release of ATP, adenosine, and LDH in relation to the coronary flow in response to sustained hypoxia. There is a close correspondence between the pattern of release of ATP and the increase in coronary flow. The rate of flow decline also matched the decline in ATP concentrations. Although peak rates of release of adenosine were high, and amounted to vasodilator quantities, the peak concentration of adenosine release occurred some 7 min after the onset of hypoxia. Peak release of LDH occurred after that of adenosine. A two-stage approach was used to dissociate work of the heart from the effects of hypoxia. Hypoxia was imposed without work by arresting the hearts with a solution of 24 mM KCl and then exposing them to hypoxia. FIGURE 4 shows that the relationship of coronary flow increase to the rate of ATP release is maintained, whereas the release of adenosine becomes even further dissociated than that depicted in FIGURE 3. Work was increased without imposed hypoxia by perfusing the hearts with adrenaline. There was a rapid release of ATP, similar to that observed during hypoxia. Peak release of adenosine once again occurred after peak release of ATP. The increase in coronary flow was less than that observed in sustained hypoxia. The hypoxia-induced release of ATP was not inhibited by 10^{-6} M propranolol, a β-adrenergic receptor blocking drug.

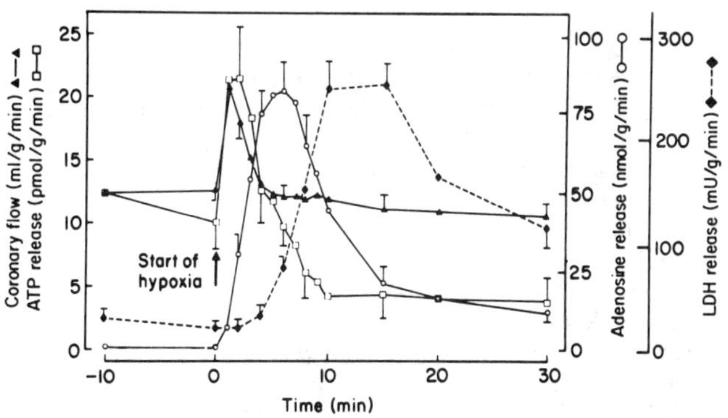

FIGURE 3. Release of ATP (□), adenosine (○), and LDH (◆) after the onset of hypoxia in a Langendorff-perfused rat heart. Note that coronary flow (▲) and ATP release are immediately increased at the onset of hypoxic perfusion. Reproduced with permission from Vial *et al.*[30]

RELEASE OF ATP FROM WORKING HEARTS

Hearts perfused via a retrograde aortic cannula in the Langendorff mode perform a low and variable amount of work that is very difficult to assess. A working perfused heart preparation developed by Neely *et al.*[31] arranged the left ventricle to perform work against a controlled afterload, allowing the work of the chamber to be calculated. Mechanical work was determined from the cardiac output and aortic pressure. Left ventricular power was also calculated.[32] Oxygen uptake of the left ventricle was monitored by continuous recording of the oxygen tension difference between the buffer entering the heart (aortic line) and that of the coronary sinus effluent.[32] FIGURE 5 shows the ATP concentrations measured in the coronary sinus effluent before, during, and after exposure of the heart to 90 sec of hypoxia (95% N_2 + 5% CO_2). In four hearts, the total amounts of ATP before, during, and after the hypoxic period were 5.9 ± 0.9, 46.1 ± 6.0, and 5.5 ± 1.2 pmol·min^{-1}, respectively (each value a mean

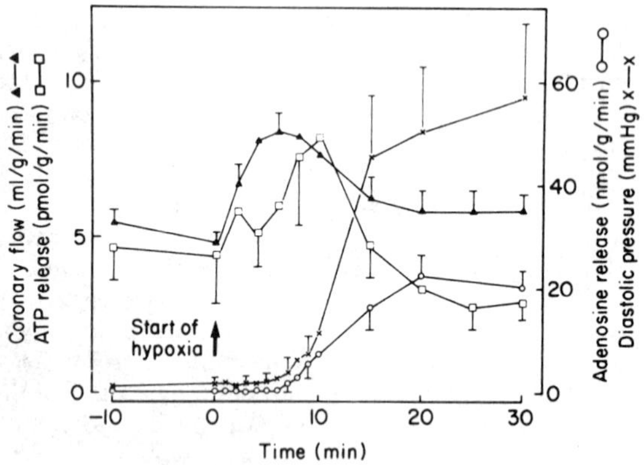

FIGURE 4. Time course of ATP (□) and adenosine (○) release into the coronary perfusate from hearts arrested with 24 mM K^+ and exposed to sustained hypoxia. Coronary flow (▲) increase remains coincident with ATP release. Reproduced with permission from Vial *et al.*[30]

± SEM). The critical point is that the working preparation produced about eight times the amount of ATP in the sinus effluent compared with the nonworking preparation.[22,32]

TABLE 2 shows a comparison of the concentrations found in the coronary sinus effluent from working and nonworking hearts. No account has been taken of the difference in size of the hearts; presumably the guinea pig hearts are slightly larger than the rat hearts. It is evident that an increase in time of hypoxia or the work of the heart will increase the concentrations of ATP released.

It seemed likely that release of ATP was proportional to the myocardial workload in the perfused mammalian hearts, yet the possibility remained that these buffer-perfused preparations were subjected to unusual degrees of hypoxia and that sudden imposition of a heavy afterload could produce some myocardial damage, resulting in the pathological appearance of ATP in the effluent. A much more accurately controlled preparation was adopted for the same type of experiment. Frog hearts from *Rana pipiens* were perfused *in situ* through the posterior vena cava with frog Ringer's

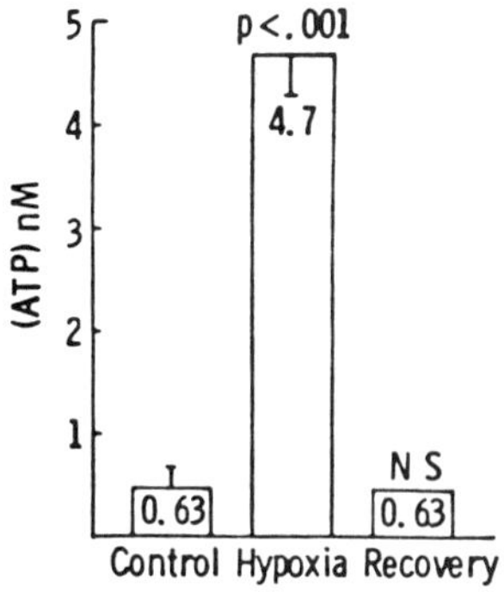

FIGURE 5. Concentration of ATP in the coronary effluent from isolated working hearts perfused with oxygenated buffer (control) during the last 30 sec of hypoxia (hypoxia) and after a 5-min perfusion with oxygenated buffer (recovery). Each value is a mean ± SEM ($N = 4$). NS: not significantly different from control. Reproduced from Clemens and Forrester.[32]

solution.[33] The single ventricle contracted against an artificial resistance of constant value, and an increase in workload was achieved simply by elevating the perfusion pressure. No complications with coronary artery perfusion occurred because frog hearts do not possess such a circulation. The myocardium is bathed in perfusate, and oxygen reaches the sponge-like myocardium by diffusion from the perfusate. ATP in the perfusate was assayed by firefly technique in the usual way except that the sensitivity was increased by nearly 10-fold (threshold $<10^{-9}$ M) with the addition of synthetic D-luciferin to the crude extract.[33] FIGURE 6 indicates the relationship obtained between workload and rate of release of ATP into the perfusate. The work of the heart was calculated using the pressure-power equation derived by Kannengiesser, Opie, and Van der Werff.[34] Kinetic energy factors were not included in the assessment of workload because momentum in this perfusion system was negligible.

A question may be raised about the function of ATP release in frog myocardium if there are no coronary arteries to be dilated. It has long been known that frog myocardium is very sensitive to extracellularly applied ATP, and indeed this was the original observation that helped to define the release of ATP from frog skeletal muscle.[19] ATP produces an increase in myocardial contractility, and its release from hypoxic cells could produce a stronger pumping action, encouraging a greater flow of blood through the ventricle and enhancing oxygen delivery. Niedergerke and Page[35] proposed that ATP may participate in circulatory control of the frog by augmenting atrial pumping action. Whatever the precise function may be, the release of ATP from this tissue in response to an increasing workload fortifies the proposal that an increase in work of the mammalian heart increases ATP release from the myocardial cells, with perhaps hypoxia being the intermediate stimulus.

TABLE 2. Release of ATP from Working and Nonworking Hearts

Reference	Period of Hypoxia (sec)	Coronary Effluent ATP (nM)	Heart Preparation
Paddle and Burnstock[22]	90	0.63	Langendorff guinea pig
Stowe *et al.*[23]	180	2.00	Langendorff guinea pig
Clemens and Forrester[32]	90	4.70	Working rat
Vial *et al.*[30]	120	1.30	Langendorff rat

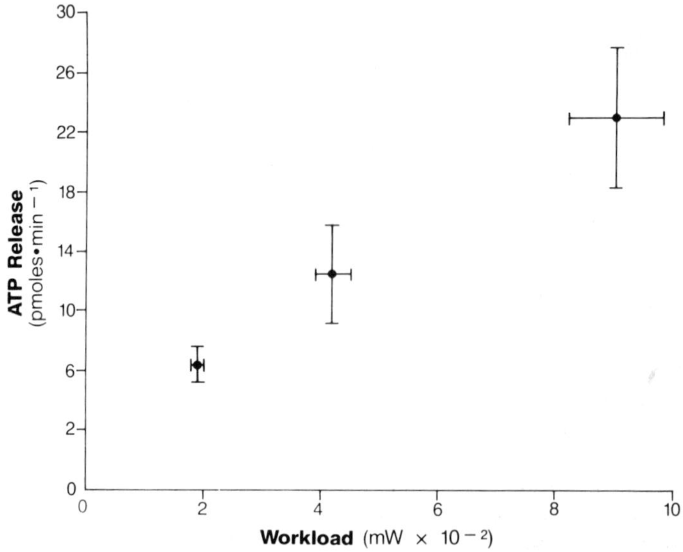

FIGURE 6. Relationship between workload (milliwatts $\times$ 10^{-2}) and rate of release of ATP (pmoles·min^{-1}) into the perfusate from isolated frog heart. Samples were collected 30 sec after commencement of increased workload. Horizontal and vertical bars at each point represent one SEM.

RELEASE OF ATP FROM ISOLATED CARDIOCYTES IN RESPONSE TO HYPOXIA

The isolated adult myocardial cell has been shown to release ATP in response to a brief period of hypoxia.[36] Adult rat heart cells were isolated enzymatically, and ATP was identified in the cell suspension using the firefly technique. It was of interest to note that ATP was not detected in cell suspensions obtained from hearts that had been left asystolic for 10 min before the harvesting procedure. This was taken to indicate that dead cells in suspension could not contribute to the light signal evoked from firefly extract in response to hypoxia. It was found that 0.34 ± 0.22 μmol ATP·mg^{-1} protein was released by cells kept in an oxygenated environment, whereas 1.28 ± 0.41 μmol·mg^{-1} protein was released initially by cells exposed briefly to hypoxia (FIG. 7). The oxygenated cells continued to release ATP, leading to a constant amount in the suspension fluid equivalent to 0.28 μmol·mg^{-1} protein for at least 60 min of incubation at 37 °C. In contrast, the amount of ATP released by the hypoxic cell population, while falling significantly over 10 min following the hypoxic period, remained higher at 0.45 μmol·mg^{-1} protein.

It was calculated that over a period of 1 min the release of ATP from cells exposed to hypoxia amounted to 0.5% of the total intracellular ATP. In the oxygenated state the proportion released was 0.05%.[36]

It was important to exclude the possibility that the isolated cells simply extruded ATP as a function of time, perhaps indicating steady cell deterioration. FIGURE 8 shows the response of firefly extract to a cell suspension exposed alternately to hypoxic and oxygenated buffer solutions. Each time the hypoxic cells were introduced into the oxygenated environment, the levels of ATP released approximately equaled the amounts of ATP released by the cells in a normoxic environment. FIGURE 9 summarizes results from six experiments of this kind.

Some idea of the dynamism of the ATP release process was gained from the great reduction in light signal following the rapid removal of cells from the fluid suspension. When cells were present a signal equivalent to 0.39 μM ATP was produced, but following removal of cells by a 10-sec centrifugation interval the cell-free fluid sample gave a signal of only 0.11 μM equivalent. In fact, this meant that 0.027 $\mu M \cdot sec^{-1}$ was lost after removal of cells. According to the estimated rate of hydrolysis of ATP added to suspension fluid from which the cells had been removed, measured at substrate concentrations between 0.5 and 1.0 μM ATP (TABLE 3), between 0.009 and 0.040 $\mu M \cdot sec^{-1}$ can be hydrolyzed; therefore the disappearance of 0.274 μM ATP after the 10-sec centrifugation seemed to be accounted for by this process. A preliminary characterization of this extracellular ATPase activity was undertaken, and the results are shown in TABLE 3. The K_m value was 13 μM and the $V/2$ of hydrolyzed ATP was 18.3 $\mu M \cdot min^{-1}$ at 37 °C. The Q_{10} was found to be between 25 and 37 °C. The enzyme activity remained unaffected by hypoxic conditions or ouabain.[36]

A constant concern throughout these experiments with isolated adult cells was the possibility that the enzyme dissection technique damaged the cell membranes. The cells were routinely isolated in bicarbonate-phosphate buffer solution with no added Ca^{2+}. Direct measurement of Ca^{2+} in the enzyme solution amounted to only 14.4 μM.[35] In a further series,[37] cells maintained in oxygenated buffer with no added Ca^{2+} released 88.8 ± 49 pmol $ATP \cdot mg^{-1}$ protein (mean ± SD) after 5 min of incubation at 37 °C. Addition of up to 2 mM Ca^{2+} to the oxygenated buffer did not significantly alter the amounts of ATP found in the suspension medium.[37] Thus it was not felt that these cells were calcium sensitive as far as ATP extrusion was concerned.

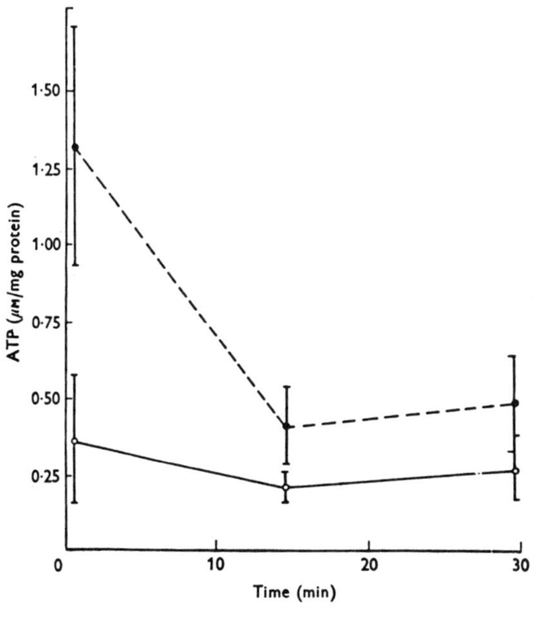

FIGURE 7. Amounts of ATP measured in fluid suspending isolated adult cardiocytes from rat heart. Cell suspensions were exposed to hypoxia (●) and tested for ATP at 1, 15, and 30 min. Oxygenated controls (○) were tested for ATP at the same times. Each point is mean of six determinations. Vertical bars indicate one SD. Temperature: 37 °C. Reproduced from Forrester and Williams.[36]

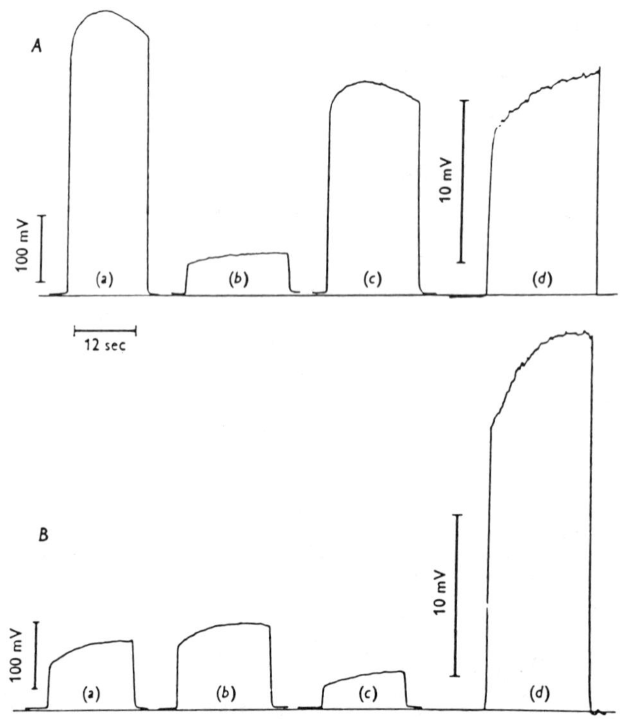

FIGURE 8. Emission of light from firefly extract in response to a cell suspension exposed alternately to hypoxic and oxygenated buffer solution. **A**: (a), signal after exposure of cells to hypoxia; (b), response after cells had been returned to oxygenated medium; (c), response when returned to hypoxic buffer; (d), response when finally returned to oxygenated medium. Note alteration in pen-recorder amplification. **B**: (a)-(d), paired oxygenated controls matching the solutions tested in **A**.

Is the enzymatic treatment for cell separation responsible for hypoxia-induced release of ATP? To test for this possibility, the cells were harvested with three different types of proteolytic enzyme mixtures: trypsin + collagenase[36]; hyaluronidase + collagenase[37]; and dispase + collagenase.[37] TABLE 4 shows ATP release in response to normoxic and hypoxic conditions after incubation at 37 °C for 20 min. From 3.5 to 5 times the ATP concentration of the normoxic controls was produced in the suspension fluid by hypoxia. The very small absolute levels produced by the hyaluronidase + collagenase mixture are not understood at present, but recent work with erythrocyte membranes[38] would suggest that membrane protein associated with hypoxia-induced ATP efflux might be impaired specifically by this proteolytic enzyme mixture.

Another approach to assessing cellular damage was to measure the oxygen consumption of the isolated cell population.[37] TABLE 5 shows the oxygen consumption and intracellular concentrations of ATP and pyruvate during normoxic and hypoxic conditions in cells where the isolation procedure was with trypsin + collagenase. Cells maintained in an oxygenated buffer at pH 7.4 and 37 °C used 15.5 ± 1.64 mmol $O_2 \cdot min^{-1} \cdot 10^{-5}$ cells. After incubation for 30 min, the intracellular concentration of

ATP declined slightly from 5.36 ± 1.53 to 4.41 ± 1.47 mM (each value a mean ± SD). The intracellular concentration of ATP after 1 min in hypoxia was 4.67 ± 0.70 mM (mean ± SD). Pyruvate levels remained constant during the first 10 min of incubation, averaging 4.5 to 4.9 mmol·mg^{-1} protein. The same general pattern was seen in hypoxia. All levels of pyruvate were elevated over normal in hypoxic conditions.

Membrane integrity was also assessed by measuring cellular and extracellular LDH levels.[37] It was found that the activity of cellular LDH remained the same during a 30-min period of incubation in normoxic conditions (324-395 IU·mg^{-1} protein) and that LDH levels in the cell-free supernatant also remained at a low, but constant, level (59.6 IU·mg^{-1} protein). Addition of 2 mM Ca^{2+} to oxygenated cells increased the intracellular LDH activity, but it remained constant over a 30-min period, averaging 891 IU·mg^{-1} protein. Addition of Ca^{2+} did not significantly raise extracellular LDH activity found without added Ca^{2+}. More significantly, though, the level of

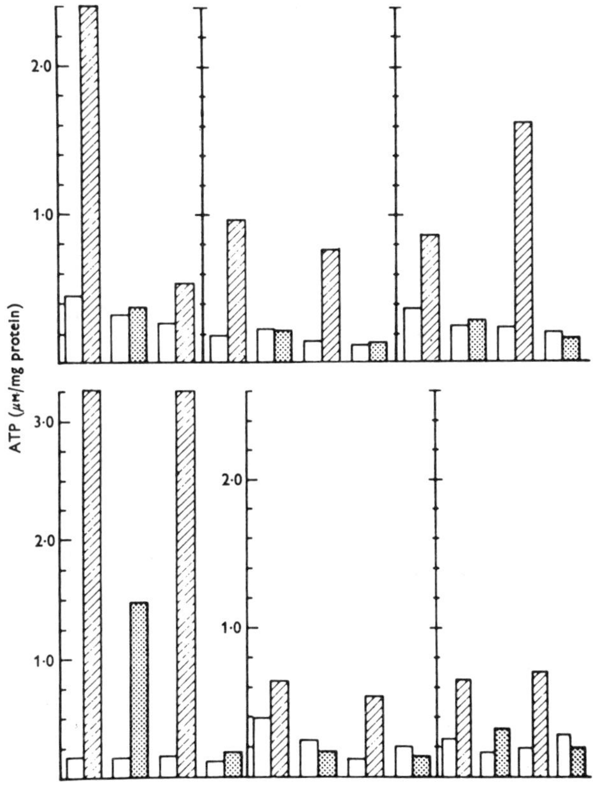

FIGURE 9. Levels of ATP release from cells alternately exposed to hypoxic and oxygenated buffer solution. Results from six experiments. Open columns, cells continuously oxygenated. Hatched columns, cells first exposed to hypoxic buffer. Stippled columns, cells previously suspended in hypoxic buffer restored to oxygenated medium. Reproduced from Forrester and Williams.[36]

LDH activity found extracellularly increased only slightly during hypoxia and was not significantly higher than those levels measured under normoxic conditions.[37]

Further evidence to support the innocuousness of the enzymatic separation was provided by insulin receptor studies.[39,40] The cells were shown to retain insulin sensitivity. Both glucose incorporation into glycogen and the activity of the I form of glycogen synthetase (EC 2.4.1.11) were stimulated by application of insulin.

Effect of Extracellular pH on Release of ATP from Isolated Myocytes

A decrease in the pH to 6.8 diminished the release of ATP in both normoxic and hypoxic conditions.[37] This pH value was similar to pH values recorded from ventricles during ischemia.[41] Although the levels of ATP released were diminished by acidic conditions, it was calculated that significant amounts of ATP were released to cause dilatation of the coronary arteries in the intact myocardium.[36]

TABLE 3. Characterization of Extracellular ATPase Activity of Suspended Cardiocytes Isolated from Adult Rat Heart[a]

Initial ATP Concentration (μM)	Rate of Hydrolysis (μM · hr^{-1})		
	0 °C/ $O_2 : CO_2$	37 °C/ $O_2 : CO_2$	37 °C/N_2
0.5	4	33	77
1.0	10	147	88
2.0	18	598	185
5.0	—	—	540
10.0	111	690	1050
15.0	147	900	1230
25.0	132	2220	2520
50.0	180	2160	1257
	K_m (μM)		
	10.5-11	12-13	12.5
	$V/2$ (μM/min)		
	1.5	18.3	20.8

[a] Adapted from Forrester and Williams.[36]

In the context of acidosis it is of interest to note that Pentilla and Trump[42] found that acidosis protected the renal cortex of rat and Ehrlich ascites tumor cells against loss of cellular ATP during hypoxia. They suggested that a lowered pH would modulate membrane permeability through interactions of H^+ with membrane proteins and/or lipids. Also, Bing *et al.*[41] found that papillary and trabecular myocardium recovered greater mechanical activity during reoxygenation if the muscles were kept at an acid

TABLE 4. Effect of Different Enzyme Separation Techniques on ATP Release from Isolated Cardiocytes in Response to Hypoxia[a]

	ATP Release (pmol · mg protein^{-1})		
	Trypsin : Collagenase	Hyaluronidase : Collagenase	Dispase : Collagenase
Oxygenated	340	17.1	361.3
Hypoxic	1280	82.2	826.7

[a] Cells were incubated for 20 min at 37 °C.

pH (6.8) during the 15-min period of hypoxia. Nayler *et al.*[43] also demonstrated a protective effect of acidosis on hypoxic heart muscle.

It has been proposed that ATP produced by cellular glycolysis may function in the maintenance of membrane integrity.[44,45] Bing *et al.*[41] felt that a pH fall inhibited the glycolysis cascade, thereby conserving carbohydrates; depression of mechanical activity also aided in energy conservation. Thus glycolysis may function as the metabolic source of ATP, a small portion of which is released from the cell in response to hypoxia.

A Model for the Mechanism of ATP Release from Cardiocytes

The existence of a nucleotide-Ca-protein complex was first proposed by R. J. P. Williams,[46] and a similar model has been proposed for excitable membranes.[47–50] Some investigators[47,48] noted that conditions that affected nucleotide release also increased

TABLE 5. Metabolic Indices of Suspended Cardiocytes Isolated from Adult Rat Heart

	Oxygen Consumption (nmol · min^{-1} · 10^{-5} cells)	ATP[a] (min, mol · liter^{-1})	Pyruvate[b] (min, nmol · mg protein^{-1})
Oxygenated cells (pH 7.4)	15.5 ± 1.64	1, 5.36 ± 1.53	1, 4.5 ± 1.3
		10, 2.35 ± 1.11	10, 4.9 ± 1.2
		20, 4.24 ± 0.78	20, 6.2 ± 1.5
		30, 4.41 ± 1.47	
Anoxic cells (pH 7.4)		1, 4.67 ± 0.70	1, 5.7 ± 2.3
		10, 3.47 ± 0.09	10, 5.4 ± 1.9
		20, 1.65 ± 0.73	20, 9.7 ± 2.4
		30, 2.92 ± 1.39	

[a] Intracellular concentration calculated from water content and mean cell volume.
[b] Intracellular content corrected for cellular protein concentration.

the rate of calcium efflux. Incubation of oxygenated cardiocytes in the presence of EDTA in concentrations of 50 μM had no effect on the release of ATP.[37] In concentrations above 50 μM EDTA, ATP release was increased 7-fold during the first 5 min of incubation.[37] It was determined by radioactive labeling of EDTA that no more than 0.6% applied EDTA was associated with any cell fraction and that up to 98% EDTA remained in the cell supernatant.[37] Thus it seemed likely that extracellular EDTA (>50 μM) was chelating Ca^{2+} *off* the membrane that was accompanied by the anionic ATP molecule.

RELEASE OF ATP FROM HUMAN ERYTHROCYTES IN RESPONSE TO HYPOXIA

Release of ATP from myocardial tissues in response to an hypoxic situation has now been demonstrated by many workers, yet a mechanism for the release process has not been defined. A theoretical model has been proposed[37,39] based on what could be termed circumstantial evidence. Tissue damage remains a critical point with both perfused myocardial preparations and with cardiocytes in culture. Because ATP is released from a great diversity of tissues, it seemed prudent to test cells that were easy to handle, that were difficult to damage, and that showed a functional response to hypoxia. Erythrocytes come close to satisfying these criteria.

Initial observations in this laboratory have shown that the release of ATP from frog erythrocytes could be blocked with the anion channel blocking drugs probenecid and furosemide.[51,52] It was postulated that ATP, being strongly anionic, might traverse the membrane via anion channels.[52] It was further speculated that in order to preserve electrical balance extracellular chloride would need to enter the cell.[51] These findings led to similar experiments using the human erythrocyte *in vitro.* It was found that an oxygenated Krebs-Henseleit solution (pH 7.4) containing approximately 5000 cells$\cdot\mu l^{-1}$ in suspension had a concentration of 500,000 molecules ATP$\cdot$cell^{-1}, as measured by the firefly technique.[38] This was correlated with a 1% hemolysis, verified by measurement of free hemoglobin in the suspension fluid. Cell suspensions were injected into an "anerobic" chamber (Po_2: <3 mm Hg; Pco_2: 60 mm Hg; 37 °C), gently mixed, and then withdrawn after a 50-sec exposure to the hypoxic environment. The suspension fluid was then assayed for ATP using firefly extract + purified luciferin.[38] An ATP concentration of 2.65 ($\pm$0.15 SEM) $\times$ 10^6 molecules$\cdot$cell^{-1} was measured. This hypoxia-induced ATP release could be blocked in three different ways. First, by application of anion channel blockers niflumic acid (50 μM), DIDS (50 μM), and dipyridamole (50 M); second, by substitution of extracellular permeant anions chloride and bicarbonate with the impermeant anion methanesulfonate; third, by incubating the cells with nitrobenzylthioinosine, a *nucleoside* transport blocking agent,[53] in a concentration of only 5 nM.[38] The background levels of ATP produced by spontaneous hemolysis remained unaffected by these blocking techniques.

These findings allow the following model to be proposed for the passage of ATP across the hypoxic erythrocyte membrane. Hypoxia has the effect of opening a channel in the nucleoside transporter (associated with band 4.5 protein) and allowing intracellular ATP to travel down a steep electrochemical gradient (transmembrane potential measurement of the human erythrocyte has been measured at -11 mV inside[54]). In order to preserve electrical silence across the membrane, the efflux of ATP^{4-} is accompanied by simultaneous influx of Cl^- and/or HCO_3^-, which occurs through band 3 protein (anion channels) in the erythrocytic membrane.

COMMENTARY AND CONCLUSIONS

There seems little doubt that ATP is released into the coronary circulation as a result of hypoxia. Release from endothelial cells, vascular smooth muscle, isolated adult cardiocytes, and human erythrocytes *in vitro* suggests multiple sources when hypoxia is imposed upon the whole-heart preparation.

Release of ATP in proportion to work performed[33] could mean that the greater the workload the greater the degree of transient hypoxia that occurs, leading to an increased rate of ATP release. It is not yet known how the activity of extracellular ATPases is affected by oxygen tension; however, one study[22] did suggest that hypoxia could increase ouabain-sensitive ATPase activity. Extracellular dephosphorylation is evident by the appearance of a peak amount of adenosine occurring a few minutes after the peak for ATP.[30] No doubt some of the adenosine could have been released directly from hypoxic myocardium, but data concerning ATP clearance across the coronary vascular bed indicates that 97-99% is degraded.[22,26,27]

It is interesting to note that the amount of ATP released from cardiocytes was maximal at the *onset* of hypoxia[36] and that the ATP release from perfused hearts diminished with successive applications of hypoxia.[22] Also, in sustained hypoxia the ATP release was maximal immediately after hypoxia was introduced, but after a very short time (1 min) fell to steady levels.[24,30] It may be highly significant that the coronary flow rate also fell to a steady value in spite of the continued application of hypoxia.[30] Teleologically it would be highly advantageous for a powerful vasodilator to be mobilized in the coronary circulation at the very onset of hypoxia.

Hypoxia-induced release of ATP from human erythrocytes *in vitro* raises a new speculation concerning the control of blood vessel caliber in conditions of local hypoxia. Intravascular ATP has been shown to act as a vasodilator through stimulating the release of endothelial cell derived relaxing factor (EDRF) from vascular endothelium.[55] If a portion of myocardium becomes acutely hypoxic, then erythrocytes circulating into that region may release vasodilator amounts of ATP, acting via EDRF release.

All these findings point to something of far broader significance concerning cellular regulation of the "milieu intérieur." Considering the great variety of cell types that exhibit nucleotide release, hypoxia-induced release of ATP from erythrocytes and myocardium may be only one example of a more widespread extracellular nucleotide system providing *inter*cellular routes of metabolic communication.

REFERENCES

1. BARCROFT, J. & W. E. DIXON. 1907. The gaseous metabolism of the mammalian heart (part I). J. Physiol. (London) **35:** 182-204.
2. ZIPF, K. 1930. Die chemische Natur des Fruhgiftes. Arch. Exp. Pathol. Pharmakol. **157:** 97-100.
3. ZIPF, K. 1931. Die chemische Natur der "depressorischen Substanz" des Blutes. Arch. Exp. Pathol. Pharmakol. **160:** 579-598.
4. ZIPF, K. 1932. Zur Kreislaufwirkung des frisch defibrinierten Blutes. Arch. Exp. Pathol. Pharmakol. **167:** 621-637.
5. BOCKMAN, E. L., R. M. BERNE & R. RUBIO. 1975. Release of adenosine and lack of release of ATP from contracting skeletal muscle. Pfluegers Arch. **355:** 229-241.
6. FLEISCH, A. & P. WEGER. 1938. Die gefasserweiternde Wirkung der phosphorylierten Stoffwechselproduckte. Pfluegers Arch. **239:** 362-369.
7. GILLESPIE, J. H. 1934. The biological significance of the linkages in adenosine triphosphoric acid. J. Physiol. (London) **80:** 345-359.

8. RIGLER, R. 1932. Uber die Ursache der vermehrten Durchblutung des Muskels wahrend der Arbeit. Arch. Exp. Pathol. Pharmakol. **167:** 54-56.
9. ROTHMANN, H. 1930. Der Einfluss der Adenosinphosphorsaure auf die Herztatigkeit. Arch. Exp. Pathol. Pharmakol. **155:** 129-138.
10. WEDD, A. M. 1931. The action of adenosine and certain related compounds on the coronary flow of the perfused heart of the rabbit. J. Pharmacol. Exp. Ther. **41:** 335-366.
11. LIPMANN, F. A. 1950. Biosynthetic mechanisms. Harvey Lect. **44:** 99-123.
12. HODGKIN, A. L. & A. F. HUXLEY. 1952. Currents carried by sodium and potassium ions through the membrane of the giant axon of *Loligo.* J. Physiol. (London) **116:** 449-472.
13. BERNE, R. M. 1963. Cardiac nucleotides in hypoxia: Possible role in regulation of coronary flow. Am. J. Physiol. **204:** 317-322.
14. GREEN, H. N. & H. B. STONER. 1950. Biological Actions of the Adenine Nucleotides: 65-103. H. K. Lewis. London.
15. McELROY, W. D. 1947. The energy source for bioluminescence in an isolated system. Proc. Natl. Acad. Sci. USA **33:** 342-345.
16. BALFOUR, W. M. & F. E. SAMSON. 1959. Transphosphorylases in the firefly lantern. Arch. Biochem. Biophys. **84:** 140-142.
17. SELIGER, H. H. & W. D. McELROY. 1960. Spectral emission and quantum yield of firefly luminescence. Arch. Biochem. Biophys. **88:** 136-141.
18. HOLTON, P. 1959. The liberation of adenosine triphosphate on antidromic stimulation of sensory nerves. J. Physiol. (London) **145:** 494-504.
19. BOYD, I. A. & T. FORRESTER. 1968. The release of adenosine triphosphate from frog skeletal muscle *in vitro.* J. Physiol. (London) **199:** 115-135.
20. FORRESTER, T. & A. R. LIND. 1969. Identification of adenosine triphosphate in human plasma and the concentration in the venous effluent of forearm muscles before, during and after sustained contractions. J. Physiol. (London) **204:** 347-364.
21. FORRESTER, T. 1972. An estimate of adenosine triphosphate release into the venous effluent from exercising human forearm muscle. J. Physiol. (London) **224:** 611-628.
22. PADDLE, B. M. & G. BURNSTOCK. 1974. Release of ATP from perfused heart during coronary vasodilatation. Blood Vessels **11:** 110-119.
23. STOWE, D. F., T. E. SULLIVAN, J. M. DABNEY, J. B. SCOTT & F. J. HADDY. 1974. Role of ATP in coronary flow regulation in isolated perfused guinea pig heart. Physiologist **17:** 339.
24. NAYLER, W. G., P. A. POOLE-WILSON & A. WILLIAMS. 1979. Hypoxia and calcium. J. Mol. Cell. Cardiol. **11:** 683-706.
25. WILLIAMSON, J. R. & D. L. DIPIETRO. 1965. Evidence for extracellular enzymatic activity of the isolated perfused rat heart. Biochem. J. **95:** 226-232.
26. BELARDINELLI, L., J. SHRYOCK, G. A. WEST, H. F. CLEMO, J. P. DIMARCO & R. M. BERNE. 1984. Effects of adenosine and adenine nucleotides on the atrioventricular node of isolated guinea pig hearts. Circulation **70:** 1083-1091.
27. DARIUS, H., G. L. STAHL & A. M. LEFER. 1987. Pharmacologic modulation of ATP release from isolated rat hearts in response to vasoconstrictor stimuli using a continuous flow technique. J. Pharmacol. Exp. Ther. **240:** 542-547.
28. PEARSON, J. D. & J. L. GORDON. 1979. Vascular endothelial and smooth muscle cells in culture selectively release adenine nucleotides. Nature (London) **281:** 384-386.
29. SCHRADER, J., C. J. THOMSON, G. HIENDLMAYER & E. GERLACH. 1982. Role of purines in acetylcholine-induced vasodilation. J. Mol. Cell. Cardiol. **14:** 427-430.
30. VIAL, C., P. OWEN, L. H. OPIE & D. POSEL. 1987. Significance of release of adenosine triphosphate and adenosine induced by hypoxia and adrenaline in perfused rat heart. J. Mol. Cell. Cardiol. **19:** 187-197.
31. NEELY, J. R., H. LIEBERMEISTER, E. J. BATTERSBY & H. E. MORGAN. 1967. Effect of pressure development on oxygen consumption by isolated rat heart. Am. J. Physiol. **212:** 804-814.
32. CLEMENS, M. G. & T. FORRESTER. 1981. Appearance of adenosine triphosphate in the coronary sinus effluent from isolated working rat heart in response to hypoxia. J. Physiol. (London) **312:** 143-158.
33. DOYLE, T. B. & T. FORRESTER. 1985. Appearance of adenosine triphosphate in the perfusate from working frog heart. Pfluegers Arch. **405:** 80-82.
34. KANNENGIESSER, G. J., L. H. OPIE & J. J. VAN DER WERFF. 1979. Impaired cardiac

work and oxygen uptake after reperfusion of regionally ischemic myocardium. J. Mol. Cell. Cardiol. **11:** 197-207.
35. NIEDERGERKE, R. & S. PAGE. 1981. Two physiological agents that appear to facilitate calcium discharge from the sarcoplasmic reticulum in frog heart cells: Adrenalin and ATP. Proc. R. Soc. London, Ser. B **213:** 325-344.
36. FORRESTER, T. & C. A. WILLIAMS. 1977. Release of adenosine triphosphate from isolated adult heart cells in response to hypoxia. J. Physiol. (London) **268:** 371-390.
37. WILLIAMS, C. A. & T. FORRESTER. 1983. Possible source of adenosine triphosphate released from rat myocytes in response to hypoxia and acidosis. Cardiovasc. Res. **17:** 301-312.
38. BERGFELD, G. R. & T. FORRESTER. 1989. Efflux of adenosine triphosphate from human erythrocytes in response to a brief pulse of hypoxia. J. Physiol. (London) **418:** 88P.
39. WILLIAMS, C. A. 1977. Responses of Isolated Heart Cells to Hypoxia. Ph.D. Thesis. St. Louis University.
40. WILLIAMS, C. A. & A. H. GOLD. 1975. The effect of insulin on glycogen metabolism of rat heart cells in suspension culture. Fed. Proc. **34:** 262.
41. BING, O. H. L., W. W. BROOKS & J. V. MESSER. 1973. Heart muscle viability following hypoxia: Protective effect of acidosis. Science **180:** 1297-1298.
42. PENTILLA, A. & B. F. TRUMP. 1974. Extracellular acidosis protects Ehrlich ascites tumor cells and rat renal cortex against anoxic injury. Science **185:** 277-278.
43. NAYLER, W. G., R. FERRARI, P. A. POOLE-WILSON & C. E. YEPEZ. 1979. A protective effect of a mild acidosis on hypoxic heart muscle. J. Mol. Cell. Cardiol. **11:** 1053-1071.
44. BRICKNELL, O. L. & L. H. OPIE. 1978. Effects of substrates on tissue metabolite changes in the isolated rat heart during underperfusion and on release of LDH and arrhythmias during reperfusion. Circ. Res. **43:** 102-115.
45. MCDONALD, T. F., E. G. HUNTER & D. P. MACLEOD. 1971. ATP partition in cardiac muscle with respect to transmembrane electrical activity. Pfluegers Arch. **322:** 95-108.
46. WILLIAMS, R. J. P. 1959. Coordination, chelation and catalysis. *In* The Enzymes. P. D. Boyer, H. Lardy, K. Myrback, Eds. Vol. 1: 391-441. Academic Press. New York, NY.
47. ABOOD, L. G., K. KOKETSU & S. MIYAMOTO. 1962. Outflux of various phosphates during membrane depolarization of excitable tissues. Am. J. Physiol. **202:** 469-474.
48. KUPERMAN, A. S., W. A. VOLPERT & M. OKAMOTO. 1964. Release of adenine nucleotide from nerve axons. Nature (London) **204:** 1000-1001.
49. SINGER, S. J. & G. L. NICHOLSON. 1972. The fluid mosaic model of the structure of cell membranes. Science **175:** 720-731.
50. LING, G. N. 1962. A physical theory of the living state: The association-induction hypothesis. Blaisdell. New York, NY.
51. DIXON, J. P. 1986. Extracellular Adenosine Triphosphate Associated with Amphibian Erythrocytes: Inhibition of ATP Release by Anion Channel Blockers. Ph.D. Thesis. St. Louis University.
52. DIXON, J. P. & T. FORRESTER. 1986. Inhibition of ATP release from frog erythrocytes by probenecid and furosemide. Fed. Proc. **45:** 165.
53. JARVIS, S. M. & J. D. YOUNG. 1981. Extraction and partial purification of the nucleoside-transport system from human erythrocytes based on the assay of nitrobenzylthioinosine-binding activity. Biochem. J. **194:** 331-339.
54. LASSEN, U. V. & O. STEN-KNUDSEN. 1968. Direct measurements of membrane potential and membrane resistance of human red cells. J. Physiol. (London) **195:** 681-696.
55. DEMEY, J. G. & P. M. VANHOUTTE. 1981. Role of the intima in cholinergic and purinergic relaxation of isolated canine femoral arteries. J. Physiol. (London) **316:** 347-355.

DISCUSSION OF THE PAPER

J. BARANKIEWICZ (*Gensia Pharmaceuticals, San Diego, CA*): I agree that red blood cells (RBCs) represent good material for the study of ATP release from cells because RBCs have relatively low activity of ectonucleotidases. We recently did several experiments of ATP release from RBCs, endothelial cells, and lymphocytes under ischemic conditions *in situ.* Whereas lymphocytes and endothelial cells release significant amounts of adenine nucleotides, RBCs were the worst source of ATP where the results were calculated per the same number of cells (10^6 cells).

T. D. WHITE (*Dalhousie University, Halifax, Nova Scotia, Canada*): Just a comment that K^+- and veratridine-evoked release of ATP from rat brain synaptosomes is *not* blocked by dipyridamole (20 μM), suggesting that this release does not involve a nucleotide transporter.

L. L. SLAKEY (*University of Massachusetts, Amherst, MA*): Your work with RBC ATP efflux is very elegant, and I think ushers in a period of serious attention to the problem of the transport of phosphorylated compounds at the plasma membrane. In response to a previous question, I think it is likely that efflux of ATP and cAMP are different processes. We have studied hormone-induced efflux of cAMP in smooth muscle cells, and find that ATP does not come out under conditions in which there is substantial efflux of cAMP.

D. SATCHELL (*University of Melbourne, Melbourne Victoria, Australia*): Hypoxia releases catecholamines from tissues. I was wondering to what extent this might contribute to the efforts of hypoxia on ATP release in heart.

FORRESTER: An interesting question that gives the opportunity to bring up some general points. First, it is highly probable that hypoxia-induced ATP release from a whole heart preparation has neural tissue as one source, and that would include the sympathetic nerve supply. There is now much evidence for the release of ATP from depolarized neural tissue (for example, the synaptosome preparation of Tom White). Second, there would seem to be a specific release effect of hypoxia on both isolated cardiocytes and human erythrocytes in suspension. I have implicated the nucleoside transporter in this response (RBCs), but more work has to be done to explore the effect of hypoxia on the permeability of membranes to ATP.

ATP Compartmentation in Neuroendocrine Secretory Vesicles

CHRISTOPHER D. UNSWORTH AND
ROBERT G. JOHNSON, JR.[a]

University of Pennsylvania Medical Center
Philadelphia, Pennsylvania 19104

INTRODUCTION

One of the exciting discoveries made in neurobiology during the past decade has been the observation that ATP appears in the extracellular space surrounding a variety of neurones under physiological conditions. This has led to the inclusion of ATP and its metabolites, particularly adenosine, on a growing list of neurotransmitter or neuromodulator candidates. Many potential actions for this extracellular ATP have been proposed; these effects may be mediated through several classes of the ATP receptor, which are in turn coupled to a variety of intracellular signaling mechanisms.[1,2] Many of the mechanisms involved in the regulation of extracellular ATP levels are poorly characterized, and this is true for the immediate intracellular origin of the ATP found extracellularly, and the mechanisms involved in its release. As for other neurotransmitters or neuromodulators, characterization of these processes involved in ATP release is essential for a complete understanding of the biological role of extracellular ATP.

Potential sources of extracellular ATP within neuronal systems include 1) cytosol of the presynaptic and/or postsynaptic cell; 2) endothelial cells of nearby blood vessels; 3) adjacent supporting cells; and 4) synaptic vesicles of the nerve terminal (FIG. 1). In this discussion, the role of the ATP localized within neuroendocrine secretory vesicles will be reviewed.

DISTRIBUTION OF SECRETORY VESICLES CONTAINING ATP

A variety of hormone- and neurotransmitter-containing secretory vesicles have been purified in sufficient quantities for biochemical analysis of their contents. Many of these isolated vesicles—including chromaffin granules from the adrenal medulla, peripheral and central adrenergic vesicles, cholinergic synaptic vesicles, and platelet-dense granules—contain a large quantity of nucleotides (TABLE 1).

[a] Address for correspondence: Howard Hughes Medical Institute Research Laboratories, University of Pennsylvania School of Medicine, Room 360, Johnson Pavilion, 36th and Hamilton Walk, Philadelphia, Pennsylvania 19104.

TABLE 1. Hormone- and Neurotransmitter-Containing Secretory Vesicles

Cell Type	Organelle	Composition	Reference
	Hormone-Containing Secretory Vesicles		
Chromaffin cell	Chromaffin granule	Catecholamines, ATP, enkephalin	3
Platelet	Dense granule	Serotonin, ADP, ATP	4
	Neurotransmitter-Containing Secretory Vesicles		
Peripheral adrenergic nerve ending	Synaptic vesicle	Catecholamines, ATP, enkephalin	5
Central adrenergeic nerve ending	Synaptic vesicle	Catecholamines, ATP	6
Peripheral cholinergic nerve ending	Synaptic vesicle	Acetylcholine, ATP	7
Central cholinergic nerve ending	Synaptic vesicle	Acetylcholine, ATP	6,7

The chromaffin granule, the subcellular organelle within the chromaffin cells of the adrenal medulla that accumulates and stores catecholamines, is one of the most studied secretory granules. Because the equilibrium sedimentation density of the chromaffin granule is equivalent to 2 M sucrose, and because large amounts of starting material can be obtained from slaughterhouses, the granules can be isolated in extremely high purity and yield. The measured ATP content of bovine adrenal chromaffin granules is approximately 560 nmol/mg of protein.[8,9] Based upon the calculated intravesicular water space, this would correspond to an apparent concentration of 125 mM if all of the ATP was free in solution. The ATP content is often expressed as the catecholamine-to-ATP ratio, and reported values range from 3:1 to 20:1.[8,9] These disparate numbers may represent differing degrees of granule maturity, or may reflect damage to the granules during the isolation procedure.

Although ATP remains the predominant nucleotide within the chromaffin granule, smaller amounts of GTP, UTP, and CTP as well as ADP and AMP have also been measured.[10] It is not known whether the granular content of these other nucleotides is physiologically important, or merely reflects their accumulation into the chromaffin granule in a ratio similar to that of the cytosol.

In addition to catecholamines and ATP, the chromaffin granule contains other soluble proteins including large amounts of enkephalin-related peptides as well as other neuropeptides.[3] The coexistence of catecholamines, ATP, and enkephalins within the chromaffin granule has proved valuable for the study of the synthesis, copackaging, and cosecretion of these putative messengers. Peripheral adrenergic nerve vesicles also contain a high content of catecholamines and ATP.[5] Cholinergic secretory vesicles have been isolated from a variety of tissues innervated by cholinergic nerves including brain, cervical ganglion, myenteric plexus, and diaphragm.[6,7,11–13] Dense granules from platelets containing the biogenic amine serotonin also contain ATP (and sometimes ADP in certain species), with an apparent nucleotide concentration of 150 mM.[4,17]

PROPERTIES OF ATP TRANSPORT INTO SECRETORY VESICLES

As previously noted, based upon the measured ATP and catecholamine content within the chromaffin granule and the calculated internal water space, an apparent intragranular concentration of 125 mM for ATP and 0.5 M for catecholamines can be calculated. A portion of this ATP is certainly bound to catecholamines in order to maintain the internal osmolality.[18] Given a cytosolic ATP concentration of 3-5 mM, however, a concentration gradient of ATP of approximately 30 to 1 probably exists across the chromaffin granule membrane. Despite the potential advantages of the chromaffin granule system and the ability to form and isolate chromaffin ghosts devoid of endogenous catecholamines or ATP, no *net* uptake of nucleotides has yet been demonstrated. The majority of experiments have been performed using radio-labeled nucleotides and intact chromaffin granules where the exchange reactions, leakage of endogenous nucleotides into the incubation medium, and competition of endogenous substrate may all contribute to the flux measurements. Many of the properties of ATP transport into chromaffin granules have been elucidated in the laboratories of Winkler and Apps.[19–23]

A summary of the measured properties of ATP transport are listed in TABLE 2. Uptake has been found to be a saturable process with a K_m within the concentration range of the cytosolic ATP (3-5 mM) and a V_{max} an order of magnitude faster than that for biogenic amine accumulation (300 nmol/mg protein/min for ATP uptake versus 15 nmol/mg protein/min for biogenic amine uptake).[24] The kinetic parameters for biogenic amine uptake were obtained under conditions where *net* uptake is measured.

The precise driving force for ATP transport has not been firmly established. Because of the presence of a H^+-translocating ATPase within the membrane of the chromaffin granule responsible for the generation of a ΔpH (acidic inside, pH 5.5) and a ΔΨ (positive inside), it has been hypothesized that the electrochemical proton gradient is the driving force for ATP accumulation. The electrochemical proton gradient is expressed as

$$\Delta \tilde{\mu}_{H^+} = \Delta\Psi - 2.3\, Z\, \Delta pH \text{ (in mV)}$$

where $Z = RT/F$, and where R is a constant, T is the temperature, and F is the Faraday constant.[25]

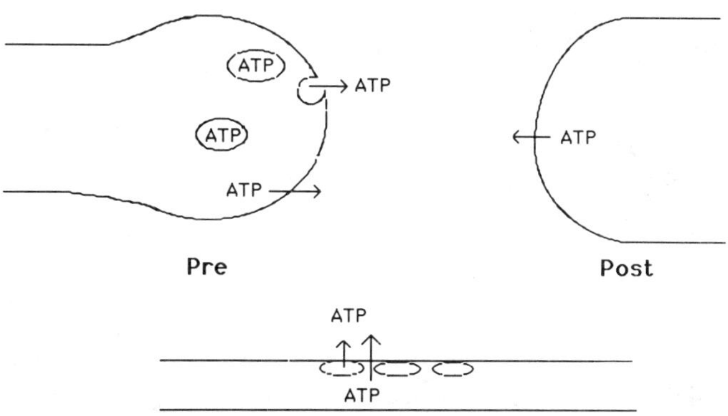

FIGURE 1. Schematic of a neuronal system outlining the potential sources of the ATP appearing in the extracellular space. These include 1) cytoplasm of the presynaptic cell; 2) cytoplasm of the postsynaptic cell; 3) release of ATP from neurotransmitter-containing secretory vesicles by exocytosis; 4) endothelial cells; and 5) adjacent stromal cells.

TABLE 2. Properties of Nucleotide Transport into Isolated Chromaffin Granules

K_m	0.9-1.4 mM
V_{max}	476 nmol·min^{-1}·mg protein^{-1} (at 37 °C)
pH dependence	None
Driving force	$\Delta\Psi$
Specificity	ATP-GTP-UTP > ADP > AMP
Analogues	SO_4^{2-}, phosphoenolpyruvate
Inhibitors/K_i (50%)	SO_4^{2-}, PO_4^{2-}, phosphoenolpyruvate/1 mM, atractyloside/0.1 mM, carboxyatractyloside/0.4 mM, phenylglyoxal/5 mM

For ATP, steady state conditions are achieved whenever the electrochemical potential for ATP distribution is equal to the electrochemical proton gradient, which is defined as

$$\Delta\ \tilde{\mu}_{ATP} = n\ \Delta\ \tilde{\mu}_{H^+}$$

where n is the coupling stoichiometry, the number of protons translocated either in the same direction as the ATP (symport) or in the opposite direction (antiport).[26] Because of the ΔpH (acidic inside) and the $\Delta\Psi$ (positive inside) generated by the H^+-translocating ATP, the diffusion gradient for protons is in the outward direction. Therefore, it would be predicted that the ATP influx would be coupled to the H^+ efflux (FIG. 2).

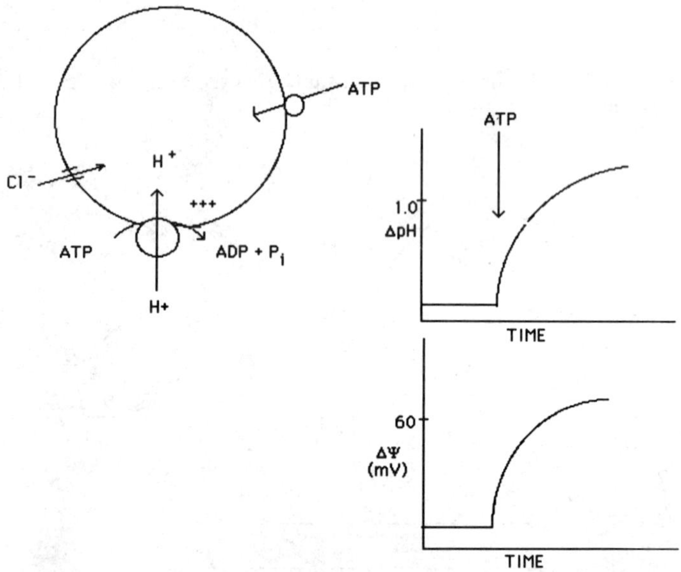

FIGURE 2. Schematic of a secretory vesicle demonstrating the H^+-translocating ATPase and the generation of a transmembrane electrochemical proton gradient. This gradient is the driving force for ATP accumulation.

The effect of the ΔpH and the ΔΨ upon ATP uptake has been investigated (FIG. 3).[19] It was concluded that the major driving force for ATP influx was the transmembrane electrical potential because the ATP uptake was found to be independent of the external pH. Because only the external pH was varied, this experiment does not exclude a possible contribution of the ΔpH to the thermodynamics of ATP accumulation. More detailed experimentation will be required before the precise coupling of this electrochemical proton gradient to the ATP accumulation is firmly established.

The uptake process is extremely nonselective and transports not only the nucleotides ATP, GTP, and UTP,[23] but also phosphoenolpyruvate, sulfate, and phosphate,[22]

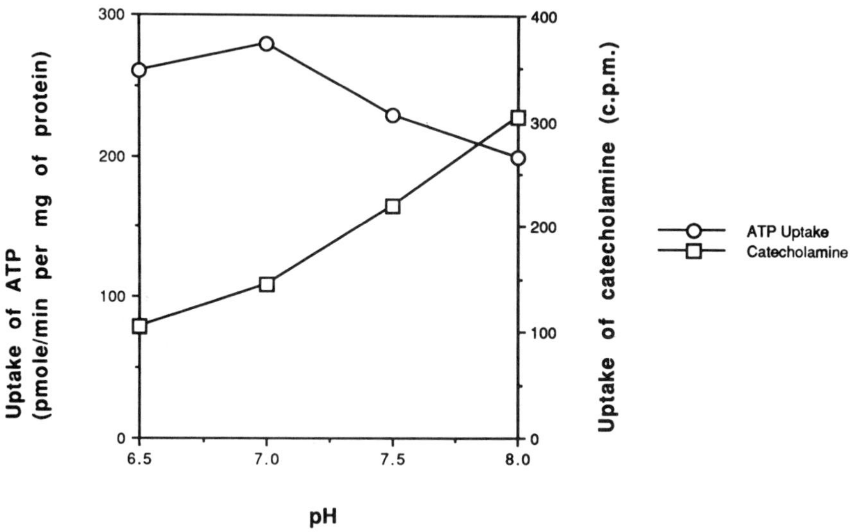

FIGURE 3. Uptake of [^{3}H]ATP and [^{14}C]epinephrine into isolated chromaffin granules as a function of the external pH, from data of Aberer *et al.*[19] Isolated chromaffin granules were incubated with [^{3}H]ATP and [^{14}C]epinephrine in a medium containing 0.3 M sucrose-10 mM Tris/HCl at the indicated pH for 5 min at 37 °C. After the 5-min incubation, the granules were diluted into cold sucrose, centrifuged, washed in cold sucrose, and counted for radioactivity. No corrections were made for endogenous nucleotides or amines. The traces, which have been used with permission, have been redrawn from reference 19.

the only common property of these compounds being that they each possess two negative charges. ATP accumulation can be inhibited by two classes of inhibitors: 1) those that dissipate the ΔpH and/or ΔΨ and therefore decrease the driving force and 2) those that inhibit transporter function. The latter group includes phenylglyoxal, an arginine-specific reagent, and atractyloside, an inhibitor used for the ATP/ADP translocator in mitochondria.[21–23] The IC_{50} for atractyloside inhibition, however, is several orders of magnitude higher than in the mitochondrion. One possible interpretation is that the atractyloside binding site is on the inside of the vesicle (and therefore not freely available to the externally added compound); another is that the two proteins are structurally dissimilar.

In chromaffin granules it is hypothesized that an electrostatic interaction occurs between the cationic biogenic amines and the anionic ATP. The most elegant data arises from the measurement of the colligative property of ATP using vapor pressure osmolality.[27] These results indicate that the ATP-catecholamine mixtures form highly ideal solutions with a mixture of 0.6 M epinephrine and 0.15 M ATP, showing an effective osmotic pressure of only 0.25 osmol at 36 °C. This catecholamine-to-ATP ratio closely matches that generally accepted for the chromaffin granule contents. Therefore, ATP may have a physiological role within the secretory vesicle facilitating the storage of catecholamines, and an additional biological role following exocytotic release.

Given the accumulating information that extracellular purines mediate a variety of important physiological and pharmacological effects, it is important that the mechanism and regulation of nucleotide (predominantly ATP) accumulation into secretory vesicles is understood.

SOURCE OF ATP RELEASED FROM NERVE TERMINALS

Although it is accepted that ATP is coreleased with catecholamines and other intragranular components from the adrenal medulla chromaffin cell by an exocytotic process, the mechanism of release from nerve terminals is less well characterized. This is particularly true for ATP release from cholinergic nerve terminals, where several reports suggest separate origins for the released acetylcholine (ACh) and ATP, even though these substances are colocalized in the synaptic vesicles (see reference 28 for review). Clearly, characterization of the immediate intracellular origin of released ATP (and the mechanism by which it is released) is essential for a full understanding of the role ATP may serve as a neurotransmitter or neuromodulator in this cholinergic system.

We utilized the electric organ of *Narcine brasiliensis* as a source of purely cholinergic synaptosomes, which allowed us to study the corelease of ACh and ATP.[28] Established methods of tissue disruption, in combination with density gradient centrifugation, allowed the isolation of synaptosomes containing ACh and ATP with specific activities of 248.5 ± 21.6 ($N = 8$) and 27.9 ± 3.4 ($N = 4$) nmol/mg protein, respectively. These values compare favorably with previous reports and indicate a high degree of synaptosomal purity.[28]

When these synaptosomes were depolarized by elevating the extracellular K^+ concentration, ACh release could be detected by an increased content in the incubation medium of either choline, or ACh if any acetylcholinesterase (AChE) inhibitor (phospholine iodide) had been used. Under these conditions, little or no ATP release into the medium could be detected. In light of previous reports of an ATPase associated with cholinergic synaptosomes isolated from elasmobranch electric organ and the mammalian central nervous system,[29–31] the ATPase activity in our synaptosome preparation was evaluated. As shown in FIGURE 4, there was an ATPase activity associated with this synaptosomal preparation with a K_m of 0.88 μM and a V_{max} of 3.00 nmol/min/mg protein. Because of this low K_m value, it appeared that this activity could be responsible for hydrolyzing ATP released from the synaptosomes, and three analogues of ATP were tested as inhibitors of this ATPase. All these analogues inhibited the ATPase, but α,β-methylene ATP was the most effective with an IC_{50} of 25 μM (FIG. 5). In contrast, both α,β-methylene ADP and β,γ-methylene ATP had IC_{50}

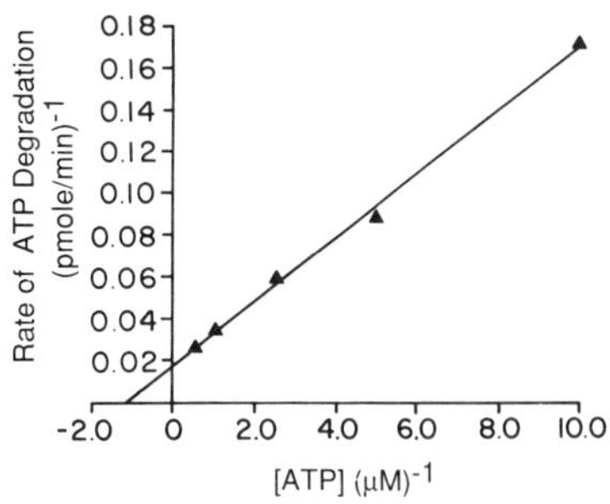

FIGURE 4. Ecto-ATPase activity of isolated cholinergic synaptosomes. First, 50-100 μl of the synaptosomal suspension (50-100 μg of protein) was added to 350 μl of physiological medium containing 50 μl of ATP-monitoring reagent (LKB) in a cuvette. The cuvette was then placed in a LKB luminometer (model 1250) attached to a strip chart recorder. The rate of ATP hydrolysis was calculated from the decrease in the chemoluminescence. The temperature was 24 °C.

values greater than 500 μM. As shown in FIGURE 5, α,β-methylene ATP inhibited ATP degradation by 85% at a concentration of 500 μM. Using a combination of phospholine iodide and α,β-methylene ATP (to inhibit AChE and ATPase, respectively), it was possible to quantitate the release of ACh and ATP from synaptosomes. Values were corrected by assessing the recovery of exogenous ACh and ATP from a parallel synaptosome incubation.

Under these conditions, the corelease of ACh and ATP could be readily demonstrated (FIG. 6). The amount of ACh and ATP that was released rose with increasing degrees of K^+ depolarization; however, the ACh:ATP ratio remained constant at approximately 6:1. Other secretogogues, such as veratridine (100 μM) and ionomycin (2.5 μM) produced different degrees of ACh and ATP release, but the ACh:ATP ratio again remained constant. This 6:1 (ACh:ATP) ratio is consistent with the generally accepted stoichiometry of ACh:ATP within cholinergic vesicles isolated from electric organ. The fact that the amount of released ACh and ATP varied under different conditions of stimulation while the stoichiometry remained constant is consistent with the notion that the synaptic vesicle is the immediate source of ACh and ATP release. A second calcium ionophore, A23187, consistently induced the corelease of ACh and ATP in a higher ratio than that found for other secretogogues; at present the reason for this is unclear.

The demonstration of stoichiometric corelease of ACh and ATP from synaptosomes did not exclude the possibility that all or a portion of the ACh and ATP was released in this ratio from the cytosol by a mechanism other than exocytosis. Because of this, a series of radiolabeling experiments was undertaken, making use of the ability of the synaptosomes to rapidly accumulate and acetylate exogenous choline. The compound

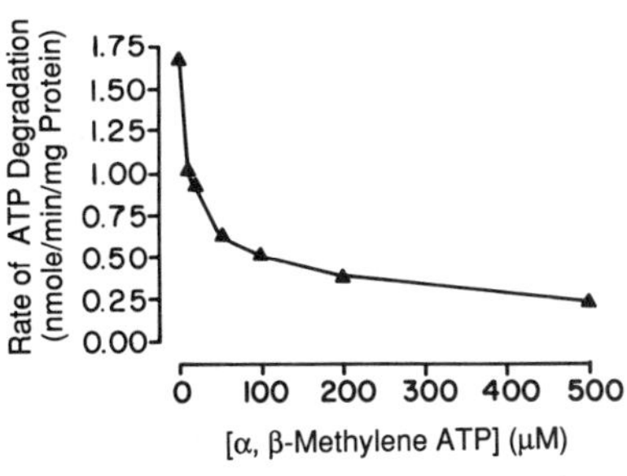

FIGURE 5. Dose dependence of α,β-methylene ATP upon the ecto-ATPase activity of isolated synaptosomes. Conditions were identical to those described in FIGURE 4, except α,β-methylene ATP was added to the incubation mixture at the indicated concentration.

2-(4-phenylpiperidino)cyclohexanol (AH5183) was used to inhibit the accumulation of ACh into the synaptic vesicle, thereby allowing us to radiolabel either the entire ACh pool (cytosolic and vesicular) or preferentially the cytosolic pool (FIG. 7).

When the synaptosomes that had been radiolabeled in this manner were depolarized and the extracellular level of ACh and radiolabeled ACh determined, it was clear that AH5183 did not affect the total amount of ACh released but greatly inhibited the release of newly synthesized (radiolabeled) ACh (FIG. 8). The conclusion from this and similar studies is that by inhibiting the entry of ACh into the synaptic vesicle, the release of ACh is also inhibited. The fact that the radiolabeled ACh content of the cytosol is unaffected by AH5183 indicates that this response is not a consequence of disrupted ACh synthesis. Also, addition of AH5183 after the labeling period did

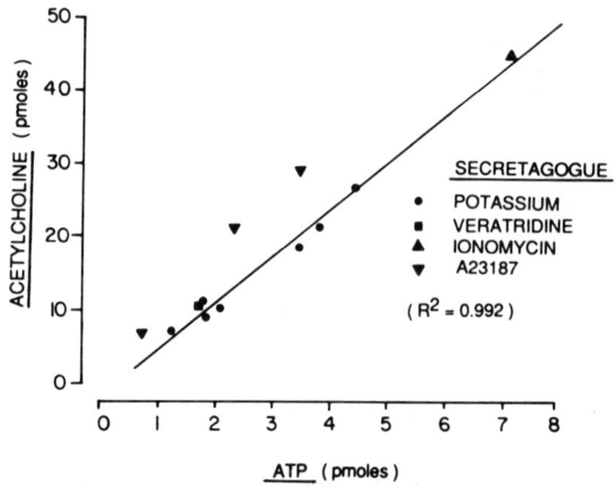

FIGURE 6. Stoichiometry of acetylcholine and ATP release from isolated synaptosomes. Washed synaptosomes incubated with phospholine (to inhibit acetylcholinesterase activity) and α,β-methylene ATP (to inhibit ecto-ATPase activity)—at concentrations of 50 μM and 500 μM, respectively—were added to physiologic medium containing either different amounts of K^+, veratridine, ionomycin, or A23187. In the samples containing K^+, the urea was decreased proportionately to maintain osmolality. After 4 min of incubation, the samples were centrifuged for 4 min at room temperature and placed on ice, and aliquots of the supernatant were assayed for acetylcholine and ATP. Recovery of acetylcholine and ATP was calculated from the synaptosomal fraction to which known amounts of acetylcholine and ATP standard had been added. The data have been redrawn from reference 28.

not affect release of radiolabeled ACh, indicating that this inhibition of ACh release was not mediated upon some AH5183-sensitive transporter in the plasma membrane. Similar experiments were performed using electric organ tissue slices, in which the cholinergic nerve terminals are much less disrupted compared to synaptosomes, as evidenced by their greater capacity to accumulate and acetylate radiolabeled choline, and their ability to release up to 50% of their ACh content following K^+ depolarization. The effect of AH5183 in this preparation was qualitatively similar to that established with synaptosomes; that is, inhibiting ACh entry into the synaptic vesicle inhibits its release. Assuming that AH5183 is 100% effective at inhibiting vesicular ACh entry, and that the residual radiolabeled ACh release originated directly from the cytosol, it can be calculated that a maximum of 5% of the ACh release can be attributed to such a process. This series of radiolabeling experiments clearly indicates

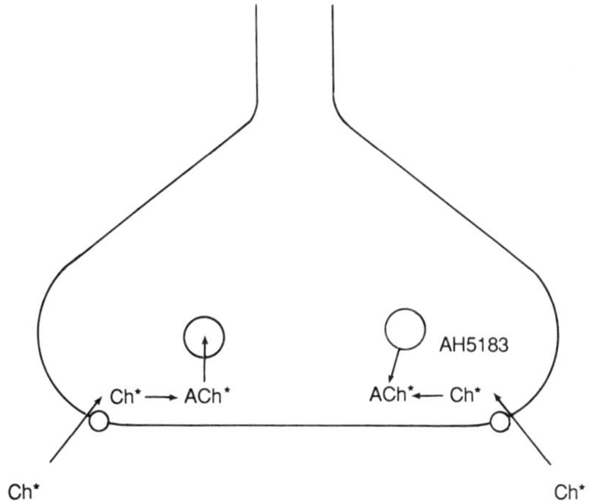

FIGURE 7. Schematic of the uptake of radiolabeled choline into a presynaptic cholinergic nerve ending, the conversion of choline to acetylcholine, and the effect of AH5183.

that greater than 95% of the released ACh originates from the vesicular pool. This finding confirms the conclusion from the first part of this study, where the ACh:ATP ratio was consistent with a vesicular origin. Together, these data clearly identify the synaptic vesicle as the immediate origin of ACh and ATP corelease upon depolarization.

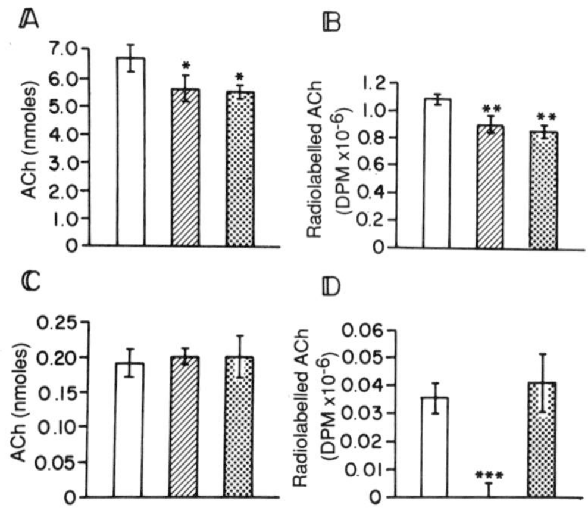

FIGURE 8. The effect of AH5183 upon synaptosomal content of endogenous acetylcholine (**A**) and radiolabeled acetylcholine (**B**), and upon the release of endogenous acetylcholine (**C**) and radiolabeled acetylcholine (**D**) during depolarization with 75 mM potassium. Synaptosomes were labeled for 5 hr at room temperature with [*methyl*-^{3}H]choline. The three groups shown included 1) no additions (open bars); 2) 10 μM AH5183 present during the radiolabeling period and release experiments (hatched bar); and 3) 10 μM AH5183 added for 15 min after the labeling period and during the release experiments (stippled bar). Each bar represents a mean ± SD ($N = 4$). The values indicated have been compared to their respective controls (one-way analysis of variance): *: $p < .05$; **: $p < .01$; ***: $p < .01$. Reproduced from reference 28 with permission.

CONCLUSION

Although the fundamental importance of intracellular ATP levels in cellular metabolism has been recognized for many years, the observation that extracellular ATP can influence a variety of biological processes, including platelet aggregation, vascular tone, neurotransmission, cardiac function, and muscle contraction, is more recent. Currently, the mechanisms involved in regulating extracellular ATP levels are only partially understood. In addition to identifying the receptors for ATP and the cellular processes to which they are coupled, characterizing the actions of ATP and its influence on the effects mediated by other colocalized transmitters, and investigating the metabolism of extracellular ATP by ectonucleotidases into inactive forms or perhaps alternative modulators (that is, adenosine), the intracellular origin of the extracellular ATP and the mechanisms involved in its release need to be established. One potential source of the ATP is the secretory vesicle, which is capable of accumulating ATP against a large concentration gradient via a membrane transporter. All the evidence, from the studies on the cholinergic system reported here, indicates that the ATP released upon stimulation originates directly from the cholinergic synaptic vesicle. This identification of the ATP in the synaptic vesicle as the physiologically relevant pool that is released upon depolarization, indicates that further study of the regulation of vesicular ATP accumulation and the effects of this ATP once it is released, either as a neurotransmitter in its own right or as a neuromodulator of ACh neurotransmission, will be pertinent to a complete understanding of the role ATP serves in this cholinergic transmitter system.

REFERENCES

1. Burnstock, G. 1990. Overview: Purinergic mechanisms. Ann. N.Y. Acad. Sci. This volume.
2. Burnstock, G. 1990. Dual control of local blood flow by purines. Ann. N.Y. Acad. Sci. This volume.
3. Winkler, H., M. Sietzen & M. Schober. 1987. The life cycle of catecholamine-storing vesicles. Ann. N.Y. Acad. Sci. **493:** 3-19.
4. Da Prada, M., J. G. Richards & H. P. Lorez. 1978. Blood platelets and biogenic monoamines: Biochemical, pharmacological and morphological studies. *In* Platelets: A Multidisciplinary Approach. G. Gaetano & S. Garattini, Eds.: 331-353. Raven Press. New York, NY.
5. Klein, R. L. & A. K. Thureson-Klein. 1982. Noradrenergic vesicles. Handb. Neurochem. **7:** 71-109.
6. Whittaker, V. P. 1982. The synaptic vesicle. Handb. Neurochem. **7:** 41-69.
7. Zimmerman, H. 1988. Cholinergic synaptic vesicles. Handb. Exp. Pharmacol. **86:** 349-375.
8. Van Kyke, K., R. Robinson, P. Urquilla, D. Smith, M. Taylor, M. Trush & M. Wilson. 1977. An analysis of nucleotides and catecholamines in bovine medullary granules by anion exchange high-pressure liquid chromatography and fluorescence: Evidence that most of the catecholamines in chromaffin granules are stored without associated ATP. Pharmacology **15:** 377-391.
9. Hillarp, N. A. & B. Nilson. 1954. The structure of the adrenaline- and noradrenaline-containing granules in the adrenal medullary cells with reference to the storage and release of the sympathomimetic amines. Acta Physiol. Scand. Suppl. **45:** 328-338.

10. HILLARP, N. A. & G. THIEME. 1959. Nucleotides in the catecholamine granules of the adrenal medulla. Acta Physiol. Scand. **45:** 328-338.
11. DOWDALL, M. J., A. F. BOYNE & V. P. WHITTAKER. 1974. Adenosine triphosphate: A constituent of cholinergic synaptic vesicles. Biochem. J. **140:** 1-12.
12. CARLSON, S. S., J. A. WAGNER & R. G. KELLY. 1978. Purification of synaptic vesicles from elasmobranch electric organ and the use of biophysical criteria to demonstrate purity. Biochemistry **17:** 1199-1206.
13. TASHIRO, T. & H. STADLER. 1978. Chemical composition of cholinergic synaptic vesicles from *Torpedo marmorata* based on improved purification. Eur. J. Biochem. **90:** 479-487.
14. WAGNER, J. A., S. S. CARLSON & R. B. KELLY. 1978. Chemical and physical characterization of cholinergic synaptic vesicles. Biochemistry **17:** 1199-1206.
15. MAYCOX, P. R., T. DECKWERTH, J. W. HELL & R. JAHN. 1988. Glutamate uptake by brain synaptic vesicles: Energy dependence of transport and functional reconstitution in proteoliposomes. J. Biol. Chem. **263:** 15,423-15,428.
16. HELL, J. W., P. R. MAYCOX, H. STADLER & R. JAHN. 1988. Uptake of GABA by rat brain synaptic vesicles isolated by a new procedure. EMBO J. **7:** 3023-3029.
17. CARTY, S. E., R. G. JOHNSON & A. SCARPA. 1981. Serotonin transport in isolated platelet granules: Coupling to the electrochemical proton gradient. J. Biol. Chem. **256:** 11,244-11,250.
18. BERNEIS, K. H., M. DA PRADA & A. A. PLETSCHER. 1971. A possible mechanism for uptake of biogenic amines by storage organelles: Incorporation into nucleotide metal aggregates. Experientia (Basel) **27:** 917-918.
19. ABERER, W., H. KOSTRON, E. HUBER & H. WINKLER. 1978. A characterization of the nucleotide uptake of chromaffin granules of bovine adrenal medulla. Biochem. J. **172:** 353-360.
20. CARMICHAEL, S. W., A. WEBER & H. WINKLER. 1980. Uptake of nucleotides and catecholamines by chromaffin granules from pig and horse adrenal medulla. J. Neurochem. **35:** 270-272.
21. KOSTRON, H., H. WINKLER, L. J. PEER & P. KONIG. 1977. Uptake of adenosine triphosphate by isolated adrenal chromaffin granules: A carrier-mediated transport. Neuroscience **2:** 159-166.
22. WEBER, A., E. W. WESTHEAD & H. WINKLER. 1983. Specificity and properties of the nucleotide carrier in chromaffin granules from bovine adrenal medulla. Biochem. J. **210:** 789-794.
23. WEBER, A. & H. WINKLER. 1981. Specificity and mechanism of nucleotide uptake by adrenal chromaffin granules. Neuroscience **6:** 2269-2276.
24. CARTY, S. E., R. G. JOHNSON, T. VAUGHAN, A. PALLANT & A. SCARPA. 1985. Amine transport into chromaffin ghosts: Kinetic measurements of net uptake of biologically and pharmacologically relevant amines using an on-line amperometric technique. Eur. J. Biochem. **147:** 447-452.
25. MITCHELL, P. 1968. Chemiosmotic Coupling and Energy Transduction. Glynn Research. Bodmin.
26. PADAN, E. & H. ROTTENBERG. 1973. Respiratory control and the proton electrochemical gradient in mitochondria. Eur. J. Biochem. **40:** 431-437.
27. KOPELL, W. N. & E. W. WESTHEAD. 1982. Osmotic pressures of solutions of ATP and catecholamines relating to storage in chromaffin granules. J. Biol. Chem. **257:** 5707-5710.
28. UNSWORTH, C. D. & R. G. JOHNSON. 1990. Acetylcholine and ATP are coreleased from the electromotor nerve terminals of *Narcine brasiliensis* by an exocytotic mechanism. Proc. Natl. Acad. Sci. USA **87:** 553-557.
29. KELLER, F. & H. ZIMMERMANN. 1983. Ectoadenosine triphosphatase activity at the cholinergic nerve endings of the *Torpedo* electric organ. Life Sci. **33:** 2635-2641.
30. GRONDAL, E. J. M. & H. ZIMMERMANN. 1986. Ectonucleotidase activities associated with cholinergic synaptosomes isolated from *Torpedo* electric organ. J. Neurochem. **47:** 871-881.
31. NAGY, A. K., T. A. SHUSTER & A. V. DELGADO-ESCUETA. 1986. Ecto-ATPase of mammalian synaptosomes: Identification and enzyme characterization. J. Neurochem. **47:** 871-881.

DISCUSSION OF THE PAPER

E. RAPAPORT (*Boston University School of Medicine, Boston, MA*): My question is with regard to the platelet-dense granules. The ATP and ADP in the platelet-dense granule are metabolically stable, and they cannot be radioactively labeled after [^{3}H]adenosine labeling of the cytoplasmic ATP in platelets (which are 55-60% of the total ATP). What would be the differences between the uptake of cytoplasmic ATP in the precursor megakaryocyte cell and the platelets?

JOHNSON: At this time, we can only speculate as to the difference between the uptake of cytoplasmic ATP and ADP into the dense granules of megakaryocytes and platelets. However, given that the dense granules contained in platelets arise from the megakaryocyte, one would predict that they are identical (or at least highly similar).

Your question also touches upon another unanswered question: How do platelet-dense granules of certain species accumulate large amounts of ADP in addition to the ATP? Because the cytosolic concentration is quite low compared to ATP, and because the specificity of the translocator, at least in chromaffin cells, is higher for ATP, the transporter on platelet-dense granules differs from the chromaffin granule carrier (the biogenic amine transporter does not, by the way), or the specificity of the dense granule carrier differs, or a second ADP transporter exists. Your question points out the need for more study in this area.

E. M. SILINSKY (*Northwestern University Medical School, Chicago, IL*): 1. A number of workers have found that the intimacy between ATP release and ACh release is so tight, that the firefly luciferase method for ATP may be used as a quantitative measure for coreleased ACh. The one disparity in the literature is the dissociation of ATP release from ACh release by botulinum toxin. Have you investigated this phenomenon and/or do you have any explanation for this troublesome effect of botulinum toxin?

2. Do you have any evidence for or against the notion that nonquantal ACh release at the neuromuscular junction is blocked by vesamicol because of the incorporation of vesamicol-sensitive ACh transporters into the plasma membrane by exocytosis?

JOHNSON: 1. My colleague, Dr. Unsworth, speculates that botulinum toxin inhibits the release of both ACh and ATP, but the methods used to quantitate these released substances may be misleading. ACh released by synaptosomes is rapidly hydrolyzed to choline, unless the AChE associated with the preparation is inhibited. Accurate quantitation of released ACh is therefore possible by either measuring choline, or ACh, if an AChE inhibitor is used. Similarly, we have shown the necessity of employing an ATPase inhibitor for the accurate quantitation of released ATP. In the absence of such inhibitors, ATP released into the medium will be hydrolyzed. Depending on the level of basal ATP release, and the kinetic parameters of any ATPase present, a higher level of ATP may be detected in the release medium of K^+-stimulated synaptosomes compared to unstimulated synaptosomes. However, this is an underestimation of the absolute amount of ATP released. After treatment with botulinum toxin the total amount of ATP released may be reduced, but in the absence of ATPase inhibitors the level detected in the medium of K^+-stimulated synaptosomes is likely to be similar to non-toxin treated synaptosomes, as the amount of ATP recovered will depend on the kinetic parameters of the ATPase. In both cases (with or without toxin) the ATP content of the medium will by hydrolyzed to a similar final concentration related to the K_m and V_{max} of the ATPase present. Thus it may appear that ATP release is unaffected while ACh release is inhibited. Similar experiments are needed with a more thorough evaluation of the recovery of ATP coreleased with ACh, before the actions of botulinum toxin can be fully understood.

2. We have no evidence for this, and would question the contribution of such a process if a transmembrane ΔpH is necessary for this transporter's activity and if the binding characteristic of the transporter are asymmetric.

E. M. WESTHEAD (*University of Massachusetts, Amherst, MA*): I would just like to point out that secretory vesicles of chromaffin cells are quite heterogeneous in their ratio of catecholamines to ATP. If you sediment chromaffin granules through a sucrose density gradient you find that in the most dense vesicles the ratio of ATP to catecholamines can be as high as 10 or 12 to 1. Furthermore, Karen Helle has shown that in Norwegian cows (which are presumably under considerable stress in winter) the ratio of ATP to catecholamines varies considerably through the course of the year.

JOHNSON: Thank you for your comments. In our own work, we also see a wide range of catecholamine/ATP ratios within the gradient. I am not surprised that the catecholamine/ATP ratio varies in Norwegian cows—we previously found that the composition of the membrane lipids also varies throughout the year. Your comments point out again how phenomenologic our results are, and how little we know about the cellular and molecular basis of secretory vesicle content.

J. S. FEDAN (*National Institute for Occupational Safety and Health, Morgantown, WV*), Can you comment on the relative proportions of free and metal-complexed ATP in adrenergic granules, and how rapidly ATP might dissociate from other constituents in the granule to which it is bound?

JOHNSON: This is an excellent question, and one to which I cannot give a definite answer. The concentration of metal ions (particularly Mg^{2+}) within a chromaffin granule (the best studied example) is approximately 40 mM, with the ATP concentration approaching 125 mM. Given an acidic pH (pH 5.5), which would decrease the metal complexed to ATP, most of the Mg^{2+}, due to its affinity for ATP, should be complexed to ATP, but the ATP will not be saturated with metal ions. Although there is excellent *thermodynamic* evidence that an intragranular complex is formed within the chromaffin granule, there are no data on the kinetics of the process, which could be an interesting physiologic question. It is conceivable, for example, that after release this intragranular complex could diffuse away from the plasma membrane as a complex protecting one or several of the constituents from membrane-bound degradative enzymes. In nerve or neuromuscular junction, because of the speed of chemical transmission, a delay of even 1 msec in the delivery of an unknown constituent may have important physiologic implications.

A Comparison of Ectonucleotidase Activities on Vascular Endothelial and Smooth Muscle Cells[a]

LINDA L. SLAKEY AND ELLEN L. GORDON[b]

Department of Biochemistry
University of Massachusetts
Amherst, Massachusetts 01003

JEREMY D. PEARSON

Medical Research Council
Clinical Research Centre
Harrow, Middlesex HA1 3UJ, England

INTRODUCTION

Extracellular adenine nucleotides can be sequentially hydrolyzed to yield adenosine by ectoenzymes found on many cell types.[1] Because the extent of phosphorylation can profoundly modify the physiologic effects of extracellular adenine nucleotides, we thought it highly likely that the regulation of the time course of their hydrolysis would be complex and would be tuned to serve different purposes in different locations. Regulation of the rates of nucleotide hydrolysis, together with an array of cell-specific responses to adenine nucleotides and adenosine, provides a rich network of regulatory elements with which to integrate the time course of cell and tissue response during crisis or signaling. We have investigated the whole time course of hydrolysis of ATP, ADP, and AMP by cultured endothelial cells and smooth muscle cells from pig aorta.[2,3] Cells were grown attached to polystyrene beads. Cell-coated beads were loaded into chromatography columns, and substrates were perfused over them (FIG. 1). In single-pass perfusion, this gives an incubation volume-to-cell-surface ratio close to that observed in intact issue. When substrates are recirculated over the cells, the volume-to-surface ratio is comparable to that found at the surface of large blood vessels. We found that endothelial cells and smooth muscle cells differ strikingly in their means of regulation of the rate of adenosine appearance from ATP or ADP.

[a]This work was supported by Grants HL 31854 and HL 38130 from the National Institutes of Health.

[b]Present address: Department of Neurological Surgery, Harborview Medical Center, Seattle, Washington 98104.

ENDOTHELIAL ECTONUCLEOTIDASE ACTIVITIES

FIGURE 2 shows a characteristic time course of change in nucleotide and adenosine concentrations observed during hydrolysis of a bolus of ATP by endothelial cells. Substrate (0.68 ml of 250 μM ATP) was recirculated over columns and aliquots withdrawn from the reservoir at different times for analysis. Columns containing different numbers of cells were tested. The figure shows that the time course of hydrolysis was proportional to cell number, as would be expected if the extent of reaction at any time were simply the sum of the contributions of all the cells present. This control establishes that each reaction is at apparent steady state throughout the time course observed. It also shows that the column is functioning as a differential reactor; that is, the extent of reaction during each pass of substrate across the column is small enough that reaction rate is constant within experimental error across the column. Under these conditions, the time course of events in the reservoir reflects the effect of enzymic reaction on the whole bulk phase.

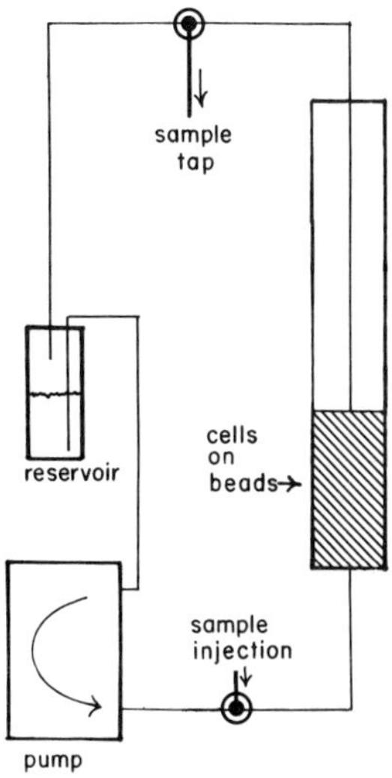

FIGURE 1. Schematic of experimental arrangement for using cells on beads for kinetics experiments. Cells are grown attached to polystyrene beads as described in references 2 and 3. Reproduced from reference 18 with permission.

It was known from work with the purified enzyme,[4,5] and from initial velocity studies with cultured vascular cells,[6] that 5′-nucleotidase is inhibited by ADP and, less potently, by ATP. We predicted that feed-forward inhibition of AMP hydrolysis by ADP and ATP would play a major role in regulating the supply of adenosine from ATP at cell surfaces. This prediction was borne out in work with endothelial cells. Nearly all the initial bolus of nucleotide accumulates as AMP, and significant hydrolysis of AMP begins only after removal of ATP and ADP is nearly complete (FIG. 2).

Feed-forward inhibition also predicts that under some conditions the rate of appearance of end product will depend inversely on the concentration of starting material. FIGURE 3 shows the rate of adenosine appearance from ATP, ADP, or AMP as initial

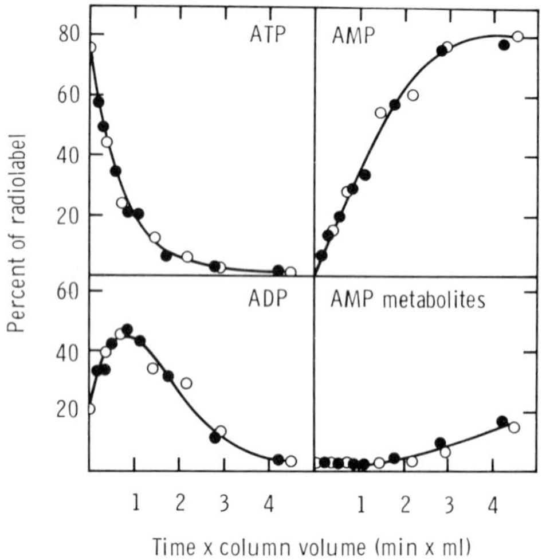

FIGURE 2. Time course of ATP hydrolysis by endothelial cells. ATP at 250 μM was recirculated over columns containing 0.07 ml (●) or 0.18 ml (○) of cell-covered beads. Cell density: 10^6 cells/0.1 ml. Column volume is proportional to cell number. AMP metabolites include adenosine, inosine, and hypoxanthine. Adenosine was greater than 70% of AMP metabolites at all times shown. Reproduced from reference 2 with permission.

substrate, with each substrate tested at 100 or 500 μM, concentrations that may be easily reached in the immediate environment of a forming thrombus.[2] The rate of adenosine production is slightly greater from 500 μM AMP as initial substrate than from 100 μM AMP, as would be predicted from the apparent K_m of endothelial 5′-nucleotidase for AMP (28 ± 13 μM).[2] When either ADP or ATP is used as initial substrate, the rate of appearance of adenosine is substantially less at a starting concentration of 500 μM than it is at 100 μM. This relationship (faster initial production of adenosine at lower initial ATP or ADP concentrations) held for seven experiments done with four different cell lines, within a concentration range from 100 to 1000 μM.

Simulation of nucleotide hydrolysis and fitting simulated to observed time courses were undertaken to estimate the values of the kinetic parameters governing the progress

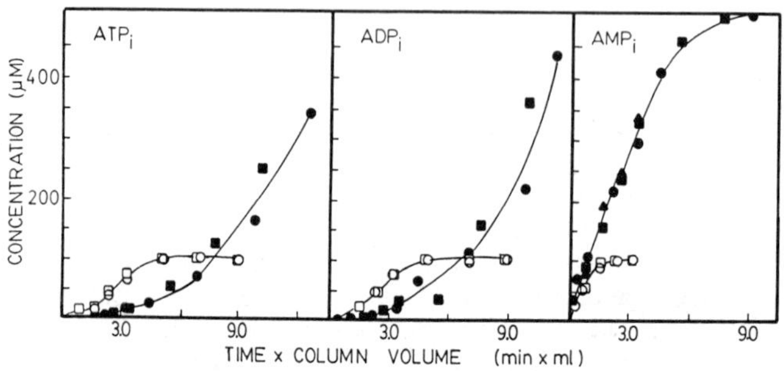

FIGURE 3. Appearance of adenosine as a function of initial substrate concentration on endothelial cells. Each panel shows the results obtained with a different initial substrate. Open symbols: initial concentration 100 μM. Closed symbols: initial concentration 500 μM. Column volumes used: ○ and □, 0.20; ■, 0.22; and ●, 0.28. Reproduced from reference 2 with permission.

of the overall conversion of ATP to adenosine. Simulation and fitting were also used to assess the impact of feed-forward inhibition in governing the rate of adenosine appearance. Each of the time courses of disappearance of ATP, ADP, or AMP as initial substrate fitted well to the assumption of Michaelis-Menten kinetics, and the K_m values observed agreed well with those previously estimated by initial velocity measurements.[2,7] FIGURE 4 shows a comparison of simulated and observed time courses for the complete hydrolysis of 500 μM ATP by endothelial cells. The time course of appearance and disappearance of ADP was well predicted by kinetic parameters found for the ATPase and ADPase individually. Fitting the time courses of appearance and disappearance of AMP, and of appearance of adenosine, required the assumption of inhibition of 5′-nucleotidase by ADP to yield values of K_m for AMP like those observed when AMP was the initial substrate. In addition, simulation of the behavior shown in FIGURE 3 required the assumption of inhibition of AMP hydrolysis by ADP. ATP is also known to inhibit 5′-nucleotidase.[8] The assumption that ATP and ADP inhibited

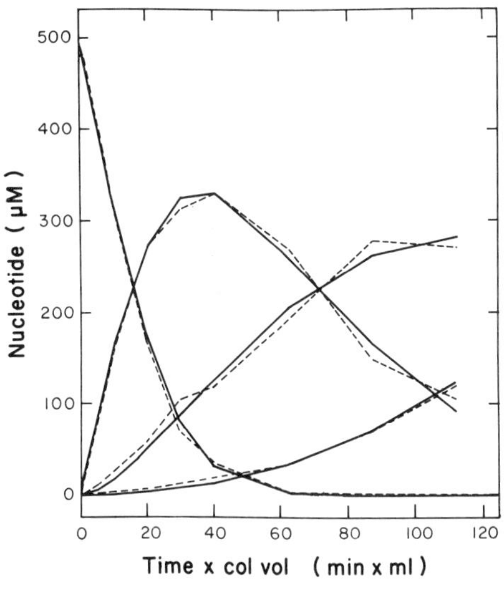

FIGURE 4. Comparison of simulated (solid line) and observed (dashed line) time courses of hydrolysis of 500 μM ATP by endothelial cells. The kinetic model assumed competitive inhibition of 5′-nucleotidase by ADP. The plot shows point-to-point connections of the times for which there are observed values. Reproduced from reference 17 with permission.

AMP hydrolysis also led to good agreement between simulated and observed time courses, but did not improve on the goodness of fit obtained with ADP alone as an inhibitor.

Simulation and data fitting also permitted us to test whether any of the reaction rates were affected by preferential delivery of products to be substrates at the cell surface. When enzymes are immobilized, the possibility arises that local concentrations may be different from those in the bulk phase that we observe (FIG. 5). If delivery of products of upstream enzymes to be substrates for downstream ones is fast compared to the rate of equilibration of the surface and bulk phases, the effective concentrations of intermediate metabolites will be higher than the bulk-phase concentrations as long as the previous reaction is the principal source of any given metabolite. We can predict that the apparent K_m for an intermediate such as ADP obtained by fitting the time course of appearance and disappearance of that intermediate will be less than that

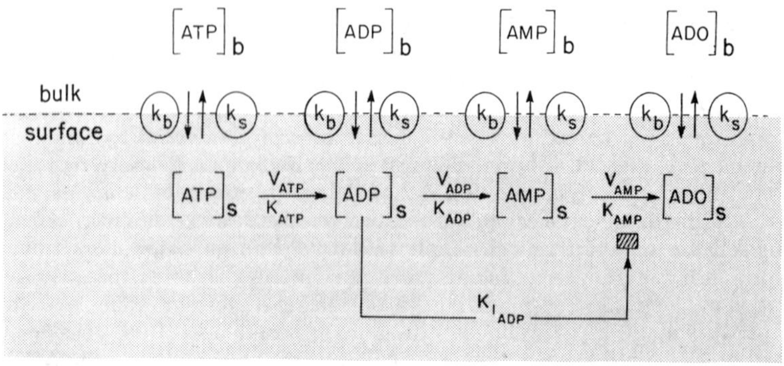

FIGURE 5. Two-compartment model for ATP hydrolysis at cell surfaces. Abbreviations: b, bulk; s, surface; k_b and k_s, the first-order rate constants for transfer from the bulk and the surface phases, respectively. Reproduced from reference 10 with permission.

observed when the compound is used as initial substrate. TABLE 1 shows that the K_m values observed for endothelial ADPase and 5′-nucleotidase when their substrates were produced from preceding reactions were not significantly different from those observed when each reactant was used as initial substrate.

Kinetic barriers to equilibration of the bulk and surface phases can also lead to inability to fit the same kinetic parameters to the rising and falling segments of the time course of change in concentration of an intermediate metabolite. While the change in concentration is driven by production, the surface concentration will be higher than the bulk. When the precursor pool is exhausted, the surface is fed by diffusion back from the bulk phase, and surface concentration will be less than bulk. In fitting data observed with endothelial cells, it was always possible to obtain kinetic parameters that predicted the rising and falling segments of overall time courses equally well.

Thus, there is no evidence that surface effects have any impact on the regulation of adenosine production on cultured aortic endothelial cells. Feed-forward inhibition of 5′-nucleotidase plays the major role in determining the rate of appearance of adenosine from ATP or ADP. This is evident both from the accumulation of AMP during ADP or ATP hydrolysis, even though the efficiency of 5′-nucleotidase is greater than that of ADPase, and from the inverse dependence of the rate of adenosine appearance from ADP or ATP upon the initial concentration of substrate.

TABLE 1. Ectonucleotidase Kinetic Parameters Observed on Endothelial Cells[a]

	Parameter			
Initial Substrate	K_{ADP} (μM)	V_{ADP} (fmol/cell)	K_{AMP} (μM)	V_{AMP} (fmol/cell)
AMP			21	6.1
ADP	342	10	31	5.6
ATP	354	11	27	5.2

[a] Data from experiments done on the same day with a single cell batch are compared. For each nucleotide, 500 μM was used as initial substrate. Reprinted from reference 2, which also contains additional experiments. Strategies for fitting parameters to data are described in detail in references 10 and 17.

SMOOTH MUSCLE ECTONUCLEOTIDASE ACTIVITIES

The time course of hydrolysis of 500 μM ATP by arterial smooth muscle cells is compared in FIGURE 6 to a typical endothelial pattern. The two cell types have similar capacities to hydrolyze ATP but differ strikingly in their rates of production of adenosine. The immediate and rapid production of adenosine from ATP by smooth muscle cells was unexpected because 5'-nucleotidase on smooth muscle is known to be inhibited by ADP, as it is on other cell types.[6,9] FIGURE 7 shows a test of the relationship between rate of adenosine production from ATP and initial ATP concentration. Production of adenosine was lower at 1000 μM initial ATP than at 500 μM but higher at 500 μM than at 100 μM. In two other experiments, the rate of adenosine production from ATP was the same for starting concentrations from 250 μM to 750 μM. Four out of seven experiments done with different cell batches gave patterns in which production of adenosine was slower at higher initial ATP concentrations, over the range 100-1000 μM initial ATP. We separately determined that all

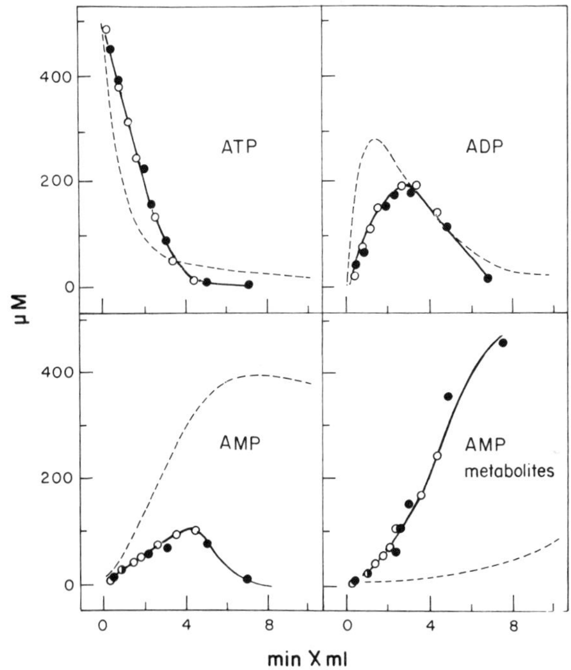

FIGURE 6. Comparison of time course of hydrolysis of 500 μM ATP by endothelial cells (dashed line) and smooth muscle cells (symbols and solid line). The endothelial cell data were taken in experiments with two columns of packed beads. One column had 0.11-ml beads, and the other had 0.30-ml beads. The smooth muscle cell data were also taken from experiments with two different columns: ○, 0.10-ml beads; ●, 0.20-ml beads. Reproduced from reference 3 with permission.

of the activity for hydrolysis of AMP on smooth muscle cells is inhibited by the α,β-methylene analogue of ADP (not shown). This suggests that feed-forward inhibition plays a role on smooth muscle cells in regulating the rate of adenosine production from ATP, but it has less impact, and the effect is more variable from cell line to cell line, than on endothelial cells.

To help consider possible explanations for the variability from cell line to cell line, the kinetic equation for competitive inhibition is reviewed in FIGURE 8. It is convenient to define the term M, the modulator of observed velocity, containing the substrate and inhibitor concentrations and their respective kinetic constants. At constant M, the velocity will be constant. If $[S]$ and $[I]$ both change so that $\Delta[S] = (1/M)(K_m/K_i)\Delta[I]$, velocity will not change. Greater changes in $[S]$ relative to $[I]$ will lead to increased velocities, and greater changes in $[I]$ will lead to decreased velocities. Smooth muscle cells, unlike endothelial cells, appear to be poised so that small changes in the relative rates of production of AMP and ADP alter whether or not feed-forward inhibition is manifested by an inverse dependence of rate of adenosine appearance on initial ATP. Smooth muscle cells do manifest greater variability from cell line to cell line in relative activities of each nucleotidase than do endothelial cells.

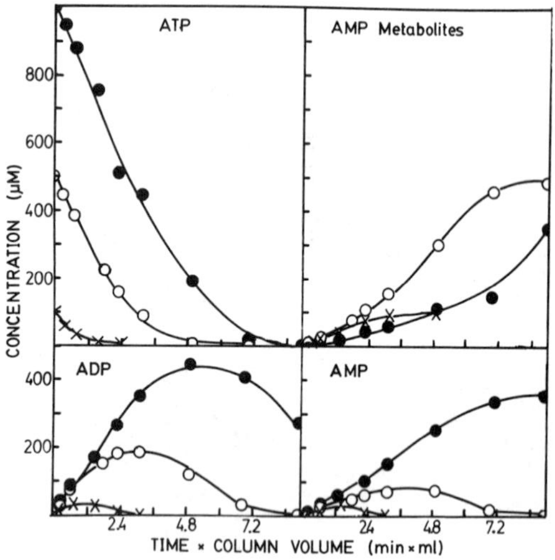

FIGURE 7. Hydrolysis of different initial concentrations of ATP by smooth muscle cells (0.20-ml column). Initial ATP concentrations: ×, 100 µM; ○, 500 µM; ●, 1000 µM.

With smooth muscle cells, again unlike endothelial cells, the kinetic parameters for ADPase and 5′-nucleotidase estimated from progress-of-reaction curves in which ADP or AMP were intermediates were very different from those found when either compound was the initial substrate (TABLE 2). The apparent K_m for ADP when it is the product of ATP hydrolysis is very much less than when it is the initial substrate, suggesting that when ADP is produced at the cell surface, the ADPase is operating at or near its maximum velocity over the whole time course observed. The apparent K_m for AMP is less when it is produced from ADP than when it is the initial substrate, and lower still when it is produced from substrate supplied one step further upstream, as ATP. Similarly, ADP appears to be more effective as an inhibitor when it is supplied at the surface from ATP hydrolysis than when it is supplied from the bulk phase.

$$v = \frac{V_{max}}{1 + \frac{K_m}{[S]}\left(1 + \frac{[I]}{K_i}\right)}$$

$$M = \frac{K_m}{[S]}\left(1 + \frac{[I]}{K_i}\right)$$

For constant M :

$$\Delta[S] = \frac{1}{M} \cdot \frac{K_m}{K_i} \cdot \Delta[I]$$

FIGURE 8. Relationship of changing substrate and inhibitor concentrations to velocity of competitively inhibited enzyme.

Changes in bulk and surface concentrations of a two-compartment model like that shown in FIGURE 5 were simulated assuming 300 μM ATP initially in the bulk phase, and a surface phase volume 5% of the total volume. Kinetic parameters for the reactions in the surface compartment were chosen to yield a pattern in the bulk phase similar to those observed with smooth muscle cells. The first-order constants for transfer of each compound to and from the surface were assumed to be equal to each other. The time course of change in nucleotide and adenosine concentrations predicted by this model are shown in FIGURE 9 for ATP as initial substrate (9a-d), ADP as initial substrate (9e-g), and AMP as initial substrate (9f). In each case, the surface concentration of the initial substrate is consistently less than the bulk concentration, and the converse is true of the final product. For intermediate compounds, the surface concentration is higher than the bulk concentration initially, and the situation reverses as precursor diminishes and back transfer from the bulk phase becomes the dominant source of supply.

Kinetic parameters were estimated from the changing bulk phase concentrations observed in the simulation shown in FIGURE 9, fitted to a one-compartment model. The apparent kinetic parameters are shown in TABLE 3, together with the parameters used for the two-compartment simulation. The relationships among the estimated parameters are like those observed with smooth muscle cells. (The actual values are not identical to any observed set because we have not yet estimated the real transfer constants or pool sizes, and because the pattern to which the bulk phase was matched was based on average properties from many experiments.) Two properties of a partially mass-transfer-limited system emerge from this exercise that are helpful in interpreting

TABLE 2. Ectonucleotidase Kinetic Parameters Observed on Smooth Muscle Cells[a]

	Parameter				
Initial Substrate	K_{ADP} (μM)	V_{ADP} (fmol/cell)	K_{AMP} (μM)	V_{AMP} (fmol/cell)	K_I (μM)
AMP			66	39	
ADP	151	13	15	12	50
ATP	0.03	6	0.8	8	3

[a] Data from experiments done on the same day with a single cell batch are compared. For each nucleotide, 500 μM was used as initial substrate. Reprinted from reference 3, which also contains additional experiments.

the smooth muscle data. First, the apparent K_m values estimated from hydrolysis of initial substrates are higher than the ones used for the simulation, and the discrepancy increases with the velocity of the reaction. Second, the kinetic advantage of supply at the cell surface is greater the farther upstream the metabolite used as initial substrate. As was pointed out above, we also anticipated difficultly in fitting the same kinetic parameters across the whole time course when there are significant surface effects, and this was observed both for fitting the bulk-phase data in FIGURE 9 and many of the smooth muscle data sets.[3,10] Although the sets could be fitted within experimental error, the distribution of error was not random along the time course, and the quality of fit could be improved by examining shorter segments of the time course.

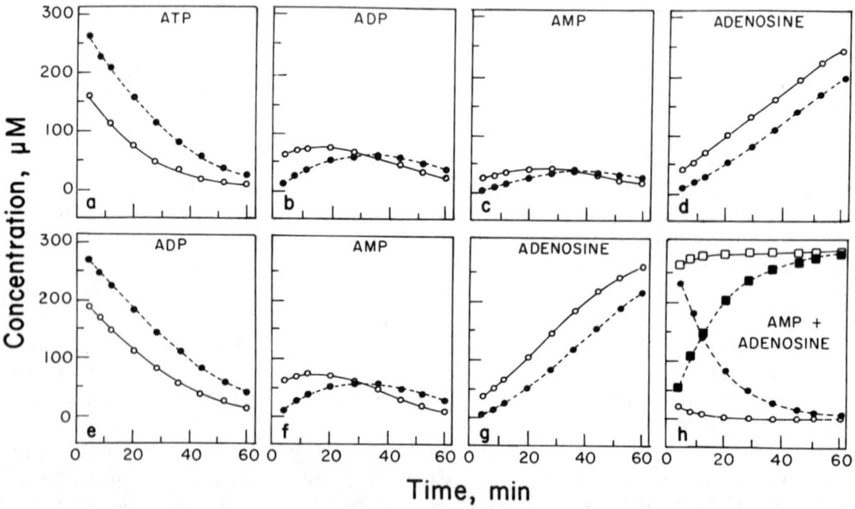

FIGURE 9. Comparison of bulk-phase and surface-phase concentrations in the two-compartment model. The kinetic parameters used for simulation were as follows: $k_b = k_s = 0.7$/min; $K_{ATP} = 50$, $V_{ATP} = 10$; $K_{ADP} = 50$, $V_{ADP} = 7.5$; $K_{AMP} = 20$, $V_{AMP} = 30$; $K_{IADP} = 5$. Each K value is in μM. Each V value is in nmol/min. ADP inhibited AMP hydrolysis competitively. The volume of the surface phase was 5% of the total volume. Total volume was 0.68 ml. The initial concentration of substrate was 300 μM in the bulk phase. Data are shown from 4 to 60 min, after equilibration of the net nucleotide concentration between the phases. Panels **a-d:** ATP was the initial substrate. Panels **e-g:** ADP was the initial substrate. Panel **h:** AMP was the initial substrate. Open symbols: surface concentrations. Closed symbols: bulk-phase concentrations. Reproduced from reference 10 with permission.

Because the kinetic parameters of the individual nucleotidases fail to predict the time course of ATP hydrolysis on smooth muscle cells, we can unequivocally rule out a simple sequential pathway with each step governed independently by Michaelis-Menten kinetics, with or without inhibition of 5′-nucleotidase. Rather, the observed efficiencies of the smooth muscle ADPase and 5′-nucleotidase are very much greater when their substrates are supplied from upstream reactions than when they are supplied initially from the bulk phase (TABLE 4). Thus, although smooth muscle 5′-nucleotidase is inhibited by ADP, the predominant factor governing the rate of production of adenosine from ATP or ADP is preferential delivery of intermediates at the cell surface, leading to very efficient production of adenosine. The opposite situation prevails on endothelial cells, on which there is no evidence of surface effects, and feed-

TABLE 3. Apparent Kinetic Parameters Estimated from the Bulk-Phase Concentrations of the Two-Compartment Model[a]

Initial Substrate	Parameter				
	K_{ADP}	V_{ADP}	K_{AMP}	V_{AMP}	K_I
	Values Used for Simulation				
	50	7.5	20	30	5
	Values Found by Fitting to the Bulk Phase				
AMP			1550	75	
ADP	158	6.2	13	4.2	133
ATP	.0001	2.6	2	6.3	3

[a] Fitted values were obtained by optimizing the fit of the one-compartment model to the bulk-phase data shown in FIGURE 9. Taken from reference 10.

forward inhibition leads to a pronounced lag in production of adenosine from ATP or ADP.

Endothelial and smooth muscle cells also differ in the efficiency of their further metabolism of extracellular adenosine. Although in endothelial experiments (like those shown in FIGS. 2 & 3) most of the product of AMP hydrolysis appears as adenosine, by late time points up to 20-30% appears as inosine plus hypoxanthine. This is a consequence of uptake and intracellular metabolism, and it can be completely blocked by dipyridamole.[2] Adenosine is also cleared from the bloodstream rapidly by blood cells.[11,12] Smooth muscle cells lack the high-affinity adenosine transport system[13] and convert very little of the adenosine produced extracellularly to downstream metabolites (FIG. 10). Thus, smooth muscle cells exposed to ADP or ATP are poised to produce adenosine rapidly and consume it slowly. Endothelial cells, on the other hand, are poised to produce adenosine from ATP or ADP only after a substantial lag time, and both they and blood cells will tend to clear it from the circulation rapidly.

DISCUSSION

The differences in regulation of nucleotide hydrolysis on endothelial and smooth muscle cells prompt a consideration of the different ways the two cell types may use

TABLE 4. Apparent Efficiencies of Ectonucleotidases (V/K)

Enzyme	Efficiency (V/K)	
	Endothelial Cells	Smooth Muscle Cells
	By Initial Velocity	
ADPase	0.03	0.09
AMPase	0.30	0.60
	At 500 μM ATP as Initial Substrate	
ADPase	0.03	200
AMPase	0.20	10

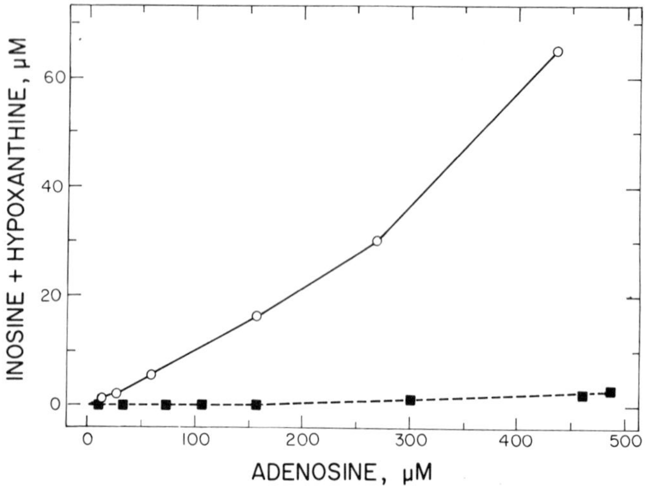

FIGURE 10. Comparison of the extent of conversion of adenosine to downstream metabolites during adenine nucleotide hydrolysis. The initial substrate in each case was 500 μM ATP. The plot shows the amount of inosine and hypoxanthine formed as a function of the amount of adenosine present in the extracellular space. ○, endothelial cells; ■, smooth muscle cells.

this pathway to integrate the responses of neighboring cells. Smooth muscle cells are exposed to extracellular adenine nucleotides principally during neurotransmission (FIG. 11). The multiple roles of the nucleotides and adenosine in this context are discussed in detail in other papers in this volume.

One context in which the endothelial cell surface is exposed transiently to high concentrations of ADP and ATP is during platelet thrombus formation (FIG. 12). The combined ATP and ADP concentration in the platelet dense granule is 1 molar.

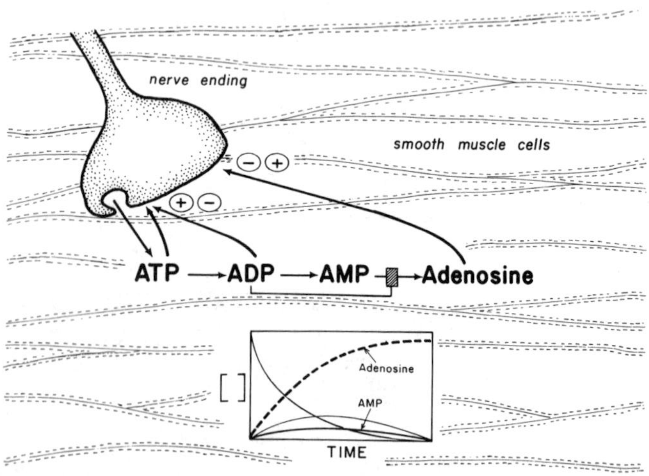

FIGURE 11. The place of smooth muscle ectonucleotidases in integration of signaling.

Thus, even with substantial dilution of the granule contents upon release, nucleotide concentrations in the high micromolar range should prevail in the unstirred region created by a forming thrombus.[14] ADP is proaggregatory and is one of several locally released mediators that make platelet aggregation autocatalytic. Adenosine is antiaggregatory and opposes platelet aggregation regardless of the initial stimulus by causing a rise in platelet cAMP.[15] The endothelial ectonucleotidases provide a mechanism that maximizes the time separation between pools of ADP and adenosine. By creating a time gap proportional to the initial release of ATP and ADP between release at the site of a forming thrombus and the generation of the antiaggregatory adenosine, this mechanism could help to insure that the extracellular milieu in the immediate neighborhood of the thrombus remains proaggregatory for a sufficient time to consolidate the forming thrombus, while minimizing the likelihood it will spread beyond the site of damage. Marcus *et al.* have recently shown that hydrolysis of ADP accounts for a significant fraction of the total potential of endothelial cells to oppose platelet aggregation.[16]

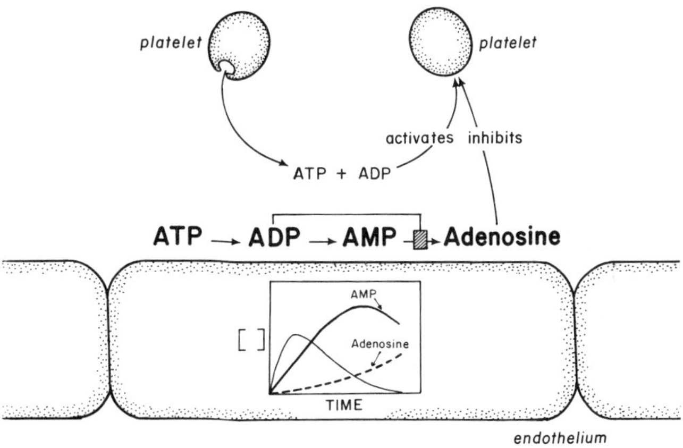

FIGURE 12. The role of endothelial ectonucleotidases during platelet activation.

ACKNOWLEDGMENTS

We are grateful for the technical assistance of Ellen Dickinson and Deborah Moreau, and for the editorial assistance of Christine Decker.

REFERENCES

1. PEARSON, J. D. 1985. Ectonucleotidases: Measurement of activities and use of inhibitors. Methods Pharmacol. **6:** 83.

2. GORDON, E. L., J. D. PEARSON & L. L. SLAKEY. 1986. The hydrolysis of extracellular adenine nucleotides by cultured endothelial cells from pig aorta: Feed-forward inhibition of adenosine production at the cell surface. J. Biol. Chem. **261:** 15496.
3. GORDON, E. L., J. D. PEARSON, E. S. DICKINSON, D. MOREAU & L. L. SLAKEY. 1989. The hydrolysis of extracellular adenine nucleotides by arterial smooth muscle cells: Regulation of adenosine production at the cell surface. J. Biol. Chem. **264:** 18986.
4. SULLIVAN, J. M. & J. B. ALPERS. 1971. *In vitro* regulation of rat heart 5′-nucleotidase by adenine nucleotides and magnesium. J. Biol. Chem. **246:** 3057.
5. NAITO, Y. & J. M. LOWENSTEIN. 1985. 5′-Nucleotidase from rat heart membranes. Biochem. J. **226:** 645.
6. PEARSON, J. D., J. CARLETON & J. L. GORDON. 1980. Metabolism of adenine nucleotides by ectoenzymes of vascular endothelial and smooth muscle cells in culture. Biochem. J. **190:** 421.
7. CUSACK, N. J., J. D. PEARSON & J. L. GORDON. 1983. Stereoselectivity of ectonucleotidases on vascular endothelial cells. Biochem. J. **214:** 975.
8. BURGER, R. M. & J. M. LOWENSTEIN. 1975. 5′-Nucleotidase from smooth muscle of small intestine and from brain: Inhibition by nucleotides. Biochemistry **14: 2362.**
9. BURGER, R. M. & J. M. LOWENSTEIN. 1970. Preparation and properties of 5′-nucleotidase from smooth muscle of small intestine. J. Biol. Chem. **245:** 6274.
10. SLAKEY, L. L., J. EIKENBERRY, D. R. HAYES & E. L. GORDON. 1989. Simulation of extracellular nucleotide hydrolysis on pig aortic smooth muscle cells: Comparison of one-compartment and two-compartment models (appendix). J. Biol. Chem. **264:** 18993.
11. PEARSON, J. D. & J. L. GORDON. 1985. Nucleotide metabolism by endothelium. Annu. Rev. Physiol. **47: 617.**
12. CATRAVAS, J. D. 1984. Removal of adenosine from the rabbit pulmonary circulation, *in vivo* and *in vitro*. Circ. Res. **54:** 603.
13. PEARSON, J. D., S. J. CARLETON, A. HUTCHINGS & J. L. GORDON. 1978. Uptake and metabolism of adenosine by pig aortic endothelial and smooth muscle cells in culture. Biochem. J. **170: 265.**
14. UGURBIL, K. & H. HOLMSEN. 1981. Nucleotide compartmentalization: Radioisotopic and nuclear magnetic resonance studies. *In* Research Monographs in Cell and Tissue Physiology. Vol. 5: Platelets in Biology and Pathology—2. J. L. Gordon Ed.: 147. Elsevier/North-Holland. Amsterdam.
15. HASLAM, R. J. & N. J. CUSACK. 1981. Blood platelet receptors for ADP and for adenosine. *In* Purinergic Receptors. G. Burnstock, Ed.: 221. Chapman & Hall. London.
16. MARCUS, A. J., K. A. HAJJAR, L. B. SAFIER, H. L. ULLMAN, N. ISLAM & M. J. BROEKMAN. 1989. Inhibition of platelet function by human endothelial cells: Aspirin-sensitive and -insensitive mechanisms. Clin. Res. **37: 547a.**
17. SLAKEY, L. L., K. COSIMI, J. P. EARLS, C. THOMAS & E. L. GORDON. 1986. Simulation of extracellular nucleotide hydrolysis and determination of kinetic constants for the ectonucleotidases (appendix). J. Biol. Chem. **261:** 15505.
18. SLAKEY, L. L. & E. L. GORDON. 1987. Extracellular nucleotide hydrolysis and the integration of signaling. *In* Cardiovascular Disease: Molecular and Cellular Mechanisms, Prevention, and Treatment. L. L. Gallo, Ed.: 323. Plenum. New York, NY.

DISCUSSION OF THE PAPER

W. R. RICE (*University of Cincinnati Medical Center, Cincinnati, OH*): Linda, that was very nice work. I do have a question about the model presented for smooth muscle cells in which ATP potentiates cAMP generation in response to adenosine.

Do you know what second messenger system is involved in the ATP-induced potentiation?

SLAKEY: No, I do not. I doubt it is Ca^{2+}. To do the experiments, we used AOPCP to block conversion of ATP to adenosine, and the controls showed that AOPCP did not potentiate the response to adenosine. We have also been looking at Ca^{2+} transients in these cells, and ATP and AOPCP cause very similar responses, so I think it is not the Ca^{2+} rise that causes the potentiation.

B. WALKER: I wonder if you have considered the effect of flow across the surface of these cells. Endothelial cells will presumably have a higher relative flow rate across their surface.

SLAKEY: The flow rates we use are slow compared to those expected in large blood vessels. I think the physiologic context in which the kinetic pattern I presented (initial substrate concentrations in the high micromolar range) is probably most relevant is the environment of a forming thrombus, which will create a relatively stagnant pool around itself.

J. BARANKIEWICZ (*Gensia Pharmaceuticals, San Diego, CA*): In your systems, adenosine is the end product of extracellular ATP degradation. Did you find any inosine and hypoxanthine produced from extracellular ATP? If yes, what was the proportion between adenosine, inosine, and hypoxanthine? Does this ratio depend on the initial concentrations of exogenous ATP?

SLAKEY: Inosine and hypoxanthine are produced from extracellular adenosine by both endothelial cells and smooth muscle cells, but with very different efficiencies. In the endothelial time course shown, about 20% of the AMP metabolites were inosine and hypoxanthine at the end of the experiment. In the smooth muscle experiment shown, they were less than 2%. Endothelial conversion of adenosine to inosine and hypoxanthine is totally blocked by dipyridamol. Smooth muscle cells lack the high-affinity, dipyridamol-sensitive adenosine transport systems. We have not asked how conversion to adenosine metabolites depends on initial ATP concentration.

Extracellular ATP Metabolism in B and T Lymphocytes

JERZY BARANKIEWICZ

Gensia Pharmaceuticals, Inc.
San Diego, California 92130

AMOS COHEN

Division of Immunology and Rheumatology
Hospital for Sick Children
Toronto, Ontario M5G1X8, Canada

INTRODUCTION

Although nucleotide degradation ectoenzymes were first described in 1948,[1] these enzymes received relatively little attention because of the difficulty in distinguishing between ectoenzyme activities per se and enzyme activities derived from the interior of broken cells in an experimental preparation. In addition, nucleotides were attributed only to the intracellular space, where they are biosynthesized and accumulate, bordered by the plasma membrane, which, in general, prevents nucleotides from escaping into the extracellular environment. Over the last decade, however, the occurrence of ectonucleotidases, and extracellular ATP, especially in endothelial cells, smooth muscle cells, platelets, and mast cells, has been well established and has been met with great interest (for recent reviews see references 2 and 3). De Pierre and Karnovsky were the first who undertook the detailed study of ectonucleotidases in immune system cells (guinea pig polymorphonuclear leukocytes).[4,5] Although some cells in the immune system readily metabolize ATP extracellularly (FIG. 1), other immune cells have little or no ectonucleotidase activity.[4–8] Later it was found that nucleotides can be released from lymphocytes and that the extracellular action of these nucleotides can result in immunomodulation.[2]

EXPERIMENTAL APPROACHES

Cells of the immune system, because they represent free-floating cells, may be obtained for *in vitro* study without impairing the cell surface—as compared with cells from solid tissues, which must be isolated by collagenase or trypsin treatment. If prepared gently, rapidly, and in optimal metabolic conditions (with glucose and oxygen), immune cells have minimal plasma membrane damage. The experimental problem that occurs is related to the use of serum that is added to the growth medium. Fetal calf serum, even if heat inactivated (56 °C, 60 min), contains significant activities of several purine enzymes, including adenosine deaminase and purine nucleoside phosphorylase. Although activities that are able to degrade ATP are relatively low in growth medium (RPMI) containing 10% heat-inactivated fetal calf serum, they can still contribute to degradation of ATP (500 μM) up to 10% in a 60-min incubation (J. Barankiewicz, unpublished observations). On the other hand, lymphocytes or lymphoblasts tend to become leaky or to lyse entirely when incubated or even washed in the absence of protein.[9] To overcome these problems, most ectometabolic studies with lymphocytes should be done with short incubation periods, no longer than 60 min, and should be carried out in media without fetal calf serum or in the presence of 5% highly purified bovine serum albumin.

Common methods for evaluating lymphocyte integrity include the determination of lactate dehydrogenase or pyruvate kinase activities (Sigma KITS) in extracellular medium and the trypan blue dye exclusion method.[10] If approached carefully, lym-

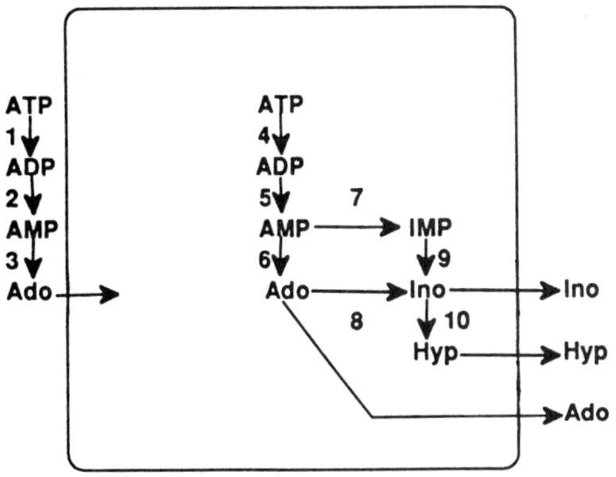

FIGURE 1. Extracellular and intracellular ATP degradation in B lymphocytes. Extracellular enzymes: 1: ecto-ATPase; 2: ecto-ADPase; 3: ecto-AMPase (ecto-5′-nucleotidase). Intracellular enzymes: 4: ATPase; 5: ADPase; 6: cytosolic AMP-5′-nucleotidase; 7: AMP deaminase; 8: adenosine deaminase (ADA); 9: cytosolic IMP-5′-nucleotidase; 10: purine nucleoside phosphorylase (PNP).

phocytes and lymphoblasts sustain cell breakage of only a few percent under experimental manipulation.

The most convenient method for analyzing radioactive products of extracellular labeled ATP metabolism is thin-layer chromatography (TLC). All purine nucleotides present in extracellular medium or cell extracts can be separated using polyethyleneimine (PEI)-cellulose TLC.[11] Nucleosides and bases formed extracellularly or released into the medium can be separated using Kodak cellulose TLC.[11] Alternatively, HPLC analysis of purines can also be used.

EXTRACELLULAR ATP METABOLISM

Many immune system cells contain active ecto-ATPase, ecto-AMPase, and ecto-*p*-nitrophenylphosphatase. Activities of these ectoenzymes were first demonstrated in guinea pig polymorphonuclear leukocytes.[4,5] Later, it was found that the highly active ectonucleotidases were present in other immune cells, including B lymphocytes and other cells of B lineage, as well as macrophages, eosinophils, and granulocytes. Ecto-ATPase has been differentiated from Na^+, K^+-stimulated ATPase present in the plasma membrane,[5] and marked differences in the activities of ecto-ATPase and ecto-AMPase were observed in many cells.[6]

The high ability of B cells to degrade extracellular ATP has been reported[6,7] for peripheral blood B lymphocytes, 34L83C B lymphoblasts, tonsil B lymphoblasts, HSC-3 B lymphoblasts, M_4 B lymphoblasts, and B lymphoma cells. Recent studies using WI-L2 B lymphoblasts (FIG. 2) showed that exogenous ATP is rapidly degraded by these cells via ADP to AMP, which accumulates extracellularly. Other products of extracellular ATP degradation that also accumulate in the medium include hypoxanthine, inosine, and adenosine (FIG. 2). Inosine and hypoxanthine seem to be produced by B cells intracellularly because their formation is reduced in the presence of dipyridamole (50 μM), a nucleoside transport inhibitor.[6] Ectonucleotidases of tonsilar B lymphocytes actively degrade not only ATP, but also GTP and deoxyribonucleoside triphosphates (dATP, dGTP, dCTP).[6]

Catabolism of extracellular ATP by rat granulocytes was similar to that in rat peripheral blood lymphocytes (FIG. 3); during ATP degradation, transient extracellular AMP accumulation was also observed. Hypoxanthine, inosine, and adenosine were also found as the products of extracellular ATP degradation, and, as in the B lymphocytes, inosine and hypoxanthine are probably formed intracellularly.

Macrophages of the J774 cell line[8] and of guinea pig peritoneum[12] have also been reported to be able to degrade extracellular ATP. Ecto-AMPase activity has been studied in different macrophage states and found to be highest in resident cells, lower in activated cells, and very low in elicited cells.[12]

Among ectonucleotidases in the immune system, the most extensively studied has been ecto-5′-nucleotidase (ecto-5′-NT)(ecto-AMPase). Ecto-5′-NT,[13,14] which exists beside cytoplasmic 5′-NT[15] in many human cells, has been cytochemically localized on the lymphocyte surface and has been found to be capable of dephosphorylating both ribonucleotides and deoxyribonucleotides outside of the cells.[16–18] It has been clearly shown that plasma membrane 5′-NT hydrolyzes only extracellular substrate and does not participate in intracellular degradation of AMP.[19,20] All B cells, except B lymphoma cells, express active ecto-5′-NT; however, its activity in human lymphoblasts is markedly lower in comparison with the activities of ecto-ATPase or ecto-ADPase.[6] Activity of ecto-5′-NT showed much higher activity in human B lymphocytes than in T lymphocytes,[21] although cytosolic 5′-NT has been similar in both T and B lymphocytes. Reduced activity of ecto-5′-NT was observed in lymphocytes from patients with common variable immunodeficiency, congenital X-linked agammaglobulinemia, and severe combined immunodeficiency.[22–26]

Whereas the existence of ectonucleotidases in B cells is well established, the presence or absence of ectonucleotidases in T cells is still controversial. It has been reported that human T cells, in contrast to human B cells, have no or very low activity of ectonucleotidases.[6,7] Human thymocytes, human T peripheral blood lymphocytes, Molt-3 T lymphoblasts, Jurkat T lymphoblasts, HSB-2 T lymphoblasts, CEM T lymphoblasts, T_4 lymphoblasts, and Sup T_1 lymphoblasts, as well as non-B non-T lymphocytes from ALL and AML patients, have low or no abilities to degrade extracellular ATP. Exogenous ATP degradation in T lymphoblasts is very slow, and practically no inosine, hypoxanthine, or adenosine can be detected. The slow degradation of exogenous ATP may result from endonucleotidases derived from broken cells as well as from ectonucleotidases on cell surfaces, but more detailed studies are needed to resolve this issue.

Although our studies showed practically no ectonucleotidase in different T cell populations,[6] there are reports of ecto-ATPase and ecto-AMPase activities found in rat thymocytes[27] and mouse thymocytes.[28] Ecto-ATPase activity in rats was significantly (3-fold) higher than that in mice.[28] A comparison of T and B lymphocytes from human blood, separated by two different methods, showed significantly higher ecto-ATPase activity in B lymphocytes than in T lymphocytes.[29] T cells, however, showed measurable activity of this enzyme.[29]

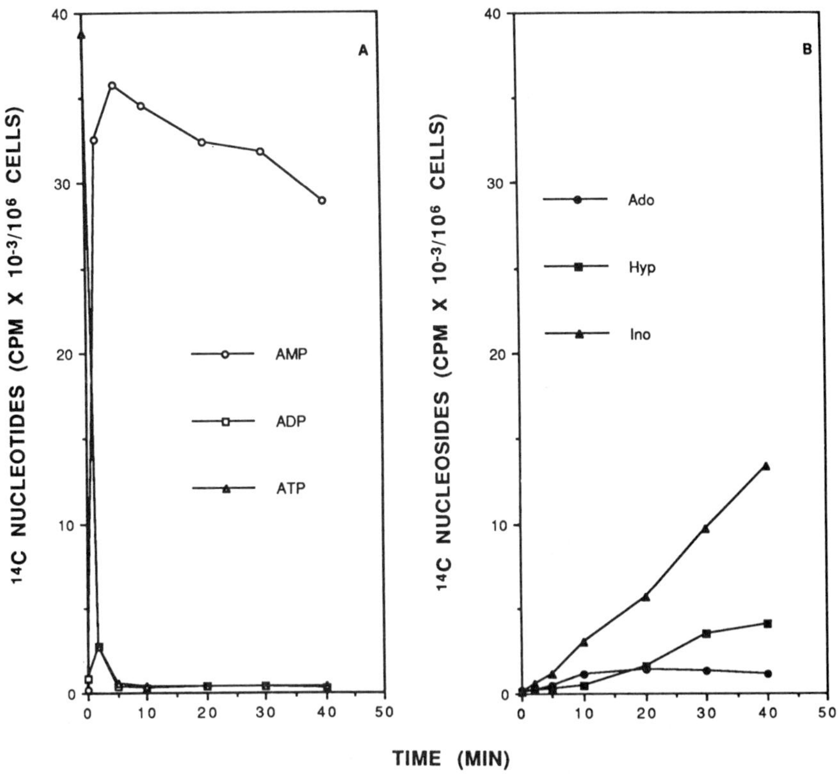

FIGURE 2. Extracellular ATP degradation by human B lymphoblasts. Human B lymphoblasts (WI-L2) (1×10^6 cells/0.1 ml RPMI without fetal calf serum) were incubated with 2 μCi of [8-^{14}C]ATP (500 μM initial concentration) for 60 min. After the cells were incubated and centrifuged, thin-layer chromatography was used to analyze nucleotides (**A**) and nucleosides and bases (**B**) according to the method of Henderson *et al.*[11]

Activities of ecto-ATPase in unseparated human thymocytes are very low or nonexistent (FIG. 4). When human thymocytes are separated into large and small thymocyte subpopulations, however, large thymocytes—in contrast to nonactive small thymocytes—express low activity of ecto-ATPase (FIG. 5).

It has also been reported that isolated mouse spleen lymphocytes hydrolyze UDP-galactose by ectonucleotide pyrophosphatase.[30] Ability to degrade ATP was not found on rat (FIG. 3) or human (J. Barankiewicz, unpublished observations) red blood cells.

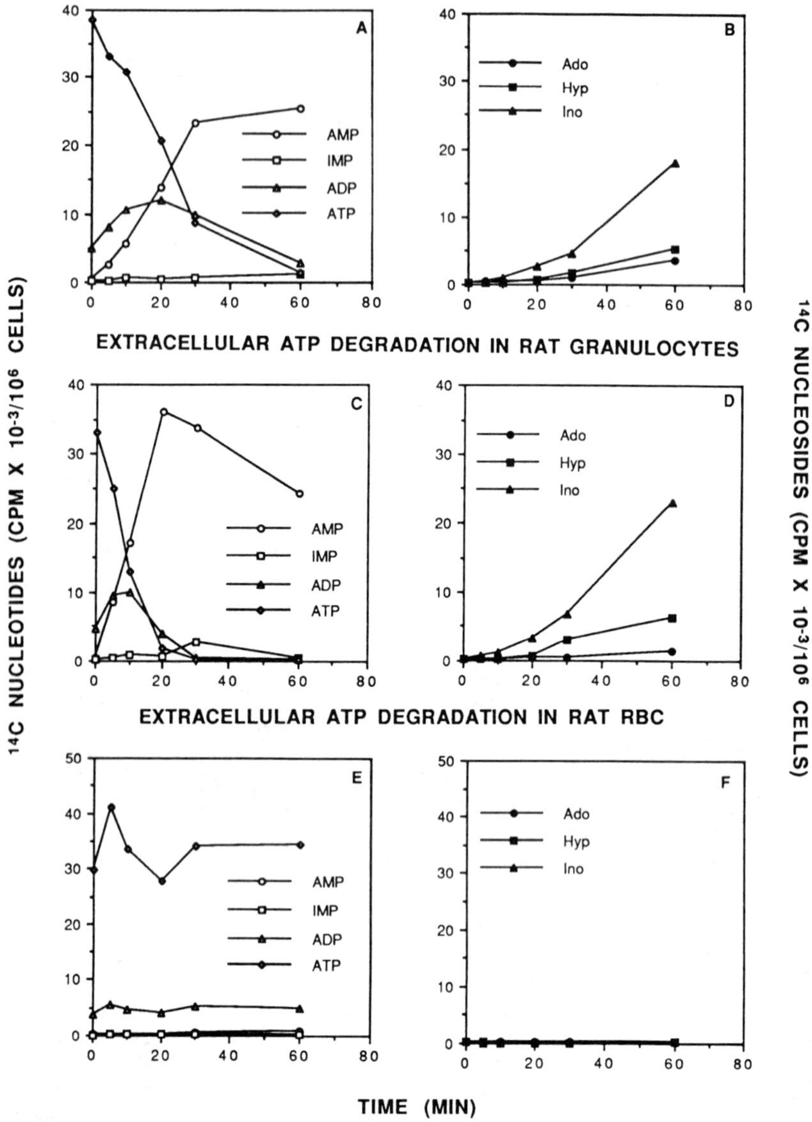

FIGURE 3. Extracellular ATP degradation by rat peripheral blood lymphocytes (**A** & **B**), granulocytes (**C** & **D**), and red blood cells (**E** & **F**). Cells (1×10^6 cells/0.1 ml of RPMI without fetal calf serum) were incubated with 2 μCi of [8-^{14}C]ATP for 60 min. After the cells were incubated and centrifuged, thin-layer chromatography was used to analyze nucleotides (**A**, **C** & **E**) and nucleosides and bases (**B**, **D** & **F**) according to method of Henderson *et al.*[11]

ECTONUCLEOTIDASE ACTIVITY IN LYMPHOCYTE DIFFERENTIATION

The activities of ecto-ATPase, ecto-ADPase, and ecto-AMPase (ecto-5′-NT) change during B cell ontogeny.[6,31,32] Activities of ecto-ATPase and ecto-ADPase in these cells increase continuously during maturation and reach maximum activities in the final stage.[6] Conversely, the activity of ecto-AMPase has been reported to increase

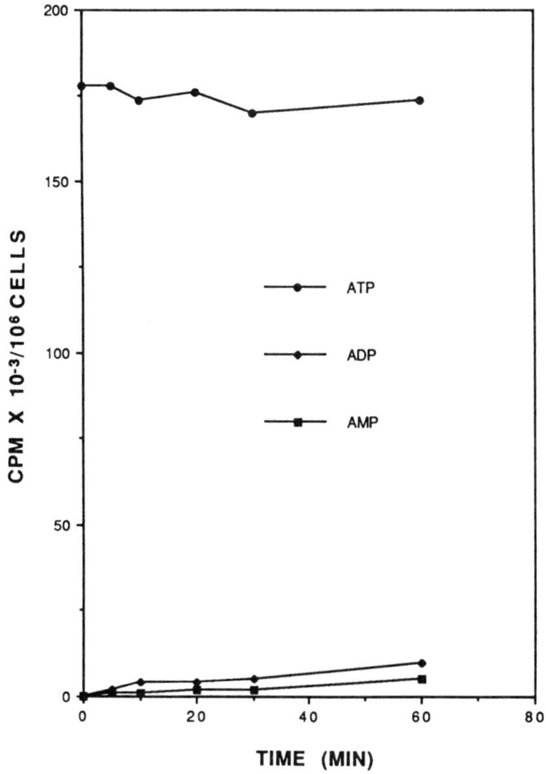

FIGURE 4. Extracellular ATP degradation by human thymocytes. For details see FIGURE 2.

from early B stage to reach maximum in late, pre-B cells, and then to decline in mature B cells. Other studies have shown that the activity of ecto-5′-NT in B cell preparations from adult peripheral blood lymphocytes is significantly higher than that from cord blood.[31]

Ecto-ATPase[33] and the ecto-5′-NT[34–37] activities have also been reported to change during the maturation of T cells. The activity of ecto-5′-NT has been reported to be low in immature T cells, thymocytes, and cord blood mononuclear cells, but significantly higher in mature T and B cells. Thymic derived factors, which can stimulate

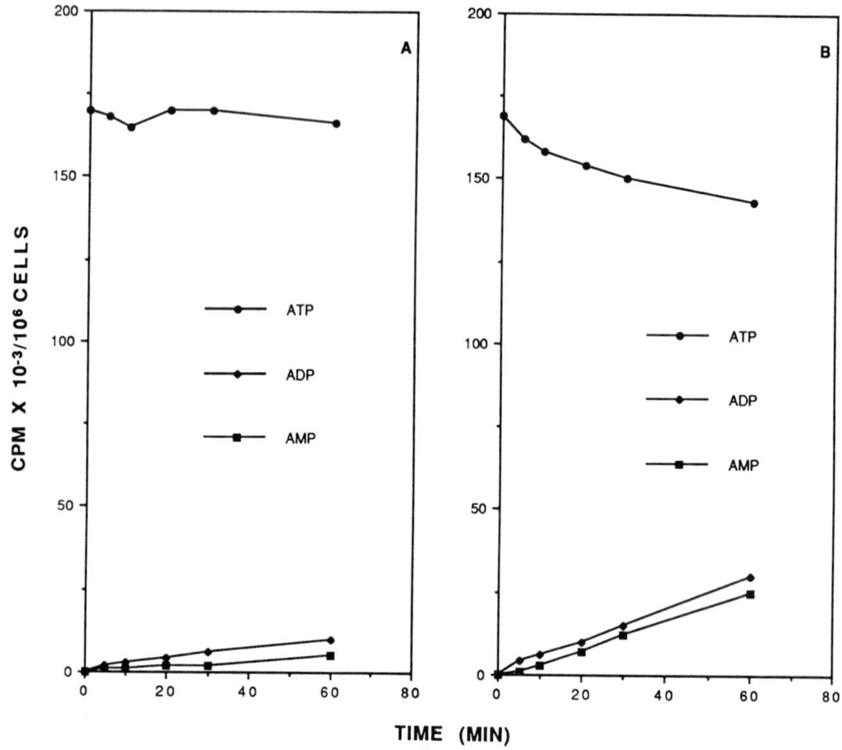

FIGURE 5. Extracellular ATP degradation by human subpopulations of thymocytes. Small intrathymic (**A**) and large (**B**) thymocyte subpopulations (1×10^6 cells/0.1 ml) were incubated with 1 μCi of [2.8-^{14}C]ATP (500 μM initial concentration). After the cells were incubated, thin-layer chromatography was used to analyze nucleotides in the medium according to the method of Henderson *et al.*[11]

T cell differentiation, increase the activity of ecto-5′-nucleotidase on cultured human thymocytes.[37] The activity of ecto-ATPase in mouse thymocytes has been reported to be lower in immature than in mature mouse thymocytes.[33] As mentioned before, the activity of ecto-5′-NT on macrophages is lower in some cell states than in others; that is, the activity on resident cells (macrophages) is greater than that on activated cells, which in turn is greater than that on elicited cells.[12]

MODULATION OF ECTONUCLEOTIDASE ACTIVITY

Very little is known regarding what factors modulate the activity of ectonucleotidases in lymphocytes. Ecto-ATPase on B lymphocytes was found to be stimulated by Mg^{2+} (FIG. 6). Stimulation of ecto-ATPase by Mg^{2+} and Ca^{2+} was also reported for human peripheral blood lymphocytes and several hematopoietic cell lines.[38] Concanavalin A has been found to stimulate rat lymphocyte ATPase,[39] but also to inhibit ecto-5′-NT from pig lymphocytes.[40,41] Attempts to find whether ecto-5′-nucleotidase is essential for the stimulation of human lymphocytes showed that this enzyme does not act as a mitogenic receptor for lectins.[42] On the other hand, extracellular ATP, adenosine, and 2-chloroadenosine have been found to block T cell mitogenesis induced by concanavalin A in a dose-dependent fashion.[43]

RELEASE OF ATP

Release of ATP or other nucleotides from the intracellular environment to extracellular space by immune system cells has not been reported up to now. Although the plasma membrane maintains intracellular nucleotide levels, it has been reported

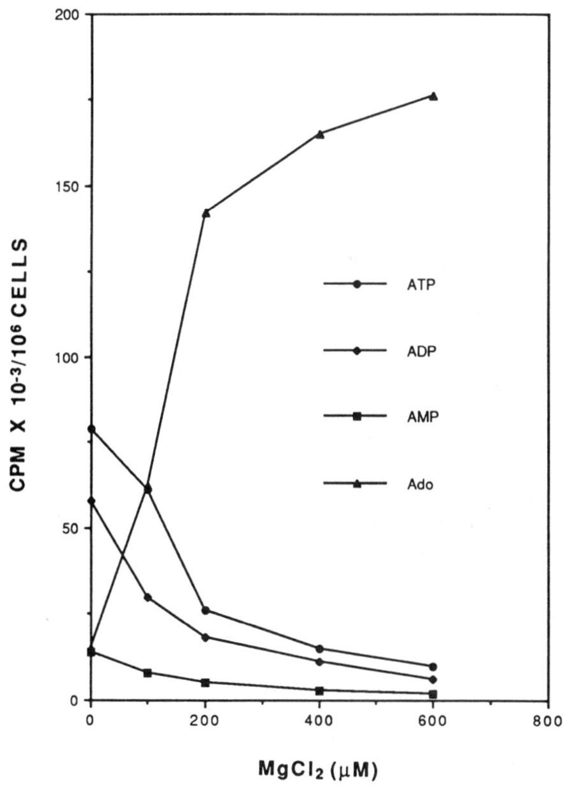

FIGURE 6. Effect of Mg^{2+} on extracellular ATP degradation by human tonsilar B lymphocytes. B lymphocytes (1×10^6 cells/0.1 ml) were incubated with [2.8-^{14}C]ATP (500 μM initial concentration) and different concentrations of $MgCl_2$. After the cells were incubated, thin-layer chromatography was used to analyze nucleotides in medium according to the method of Henderson *et al.*[11]

that ATP was released from different cells (including platelets, neurons, and adrenal medulla) and that the cells remained viable.[2] Human B lymphoblasts (WI-L2) that were preloaded with radioactive ATP and subsequently exposed to ischemic conditions in the presence of α,β-methylene ADP, an ecto-5′-NT inhibitor, accumulated radioactive AMP in the culture medium (FIG. 7). Simultaneously, no cell impairment for the first 2 hr of ischemia was observed. Since ectonucleotidases are very active on the surface of B cells, it is not clear whether 1) ATP was released from these cells and

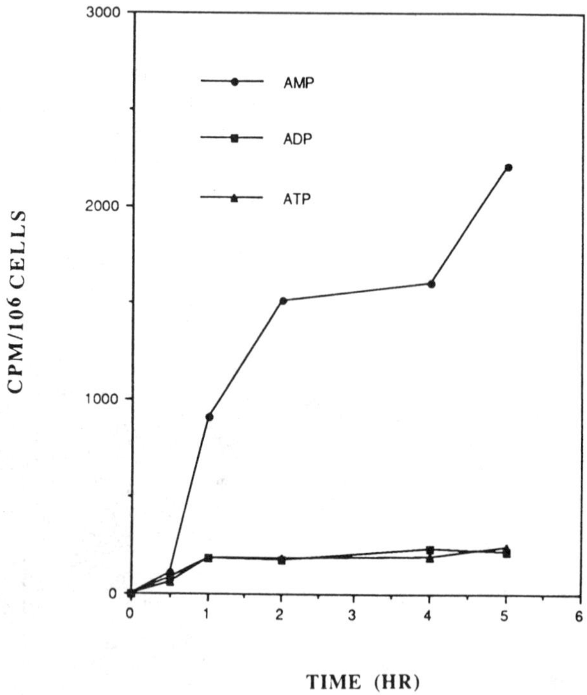

FIGURE 7. Release of adenine nucleotides by human B lymphoblasts. Human B lymphoblasts (WI-L2) (20 × 10^6/2 ml) were first incubated for 1 hr with 5 μCi of [8-^{14}C]adenine to label intracellular ATP. After incubation, unused radioactive adenine was washed out and cells were resuspended in glucose-free RPMI medium and incubated under argon atmosphere with α,β-methylene ADP. After the cells were incubated, thin-layer chromatography was used to analyze the medium for radioactive nucleotides. The medium was also analyzed according to the trypan blue exclusion method.[10] There was no measurable activity of lactate dehydrogenase in the medium for up to 5 hr, and no cell death was observed for up to 3 hr.

degraded to AMP or 2) AMP was accumulated intracellularly during the ischemic condition and directly released. No information is available at present regarding the release of nucleotides from other immune cells. (An important source of extracellular nucleotides can be the thymus, where nucleotides can be released in large amounts after thymocyte death.) Also, there are currently no data available regarding the possibility that extracellular ATP might be incorporated by lymphocytes, although this phenomenon has been reported for several other cells.[44]

EFFECTS OF EXTRACELLULAR ATP ON CELLS OF THE IMMUNE SYSTEM

High concentrations of ATP stimulate *in vivo* DNA synthesis in bone marrow and in thymocytes.[37,38] In contrast, ATP inhibits DNA synthesis in spleen, lymph nodes, and peripheral blood lymphocytes.[45,46] Extracellular ATP inhibits the activities of natural killer cells.[47,48] At micromolar concentrations (10-100 μM), extracellular ATP, but not GTP, UTP, or CTP, markedly inhibits neutrophil-mediated cytotoxicity.[49] Other adenosine nucleotides, AMP and ADP, express inhibitory activity similar to that of ATP.

Exogenous ATP has been found to cause depolarization of polymorphonuclear phagocytes.[50] In the mouse macrophage cell line, J774, extracellular ATP induces plasma membrane depolarization, Na^+ influx, K^+ efflux, and an increase in cytosolic free-Ca^{2+}.[51] Ecto-ATPase has been shown not to mediate the effect of ATP on J774 cells, and ATP^{4-} rather than $MgATP^{2-}$ has been shown to bind to the receptor on these cells.[8] ATP^{4-} also permeabilized the membrane of mouse macrophages.[52] Exogenous ATP added to mice thymocytes increased Ca^{2+} uptake up to 20-fold without any effect on cell integrity.[53] Since Mg^{2+} is essential for the stimulation, ATP probably works in complex with Mg^{2+}. Other nucleotides including deoxy-ATP did not significantly increase Ca^{2+} uptake, with the exception of GTP, which showed moderate stimulation.

In Friend erythroleukemia cells, extracellular ATP (1 mM) caused changes in the permeability to ions as well as in the inhibition of cell growth.[54] Moreover, in chronic lymphocytic leukemic lymphocytes, extracellular ATP increases cation permeability.[55] Although ecto-ATP produced a 12-fold increase in Na^+ influx in these lymphocytes, only a 3.5-fold increase in this flux has been observed in peripheral blood lymphocytes. In recent studies, it has been shown that extracellular ATP (0.3 mM), and to a lesser extent ADP were highly mitogenic to mature medullary thymocytes in combination with phorbol myristate acetate.[56] Since adenosine or other nucleotides were not mitogenic, it was suggested that thymocytes express receptors on their surface specific for ATP. Such receptors, possibly P_2 receptors, could be involved in the lymphocyte response through the regulation of intracellular free Ca^{2+} levels.

Certain untransformed cells with a secretory function respond to extracellular ATP. The most extensive studies have been done with mast cells.[57] In experimenting with B cells, we have not been able to show any changes in antibody production in the presence of external ATP at concentrations up to 500 μM (J. Barankiewicz, unpublished observations).

RELATIONSHIP BETWEEN EXTRACELLULAR AND INTRACELLULAR ATP METABOLISM

Since intracellular ATP degradation proceeds exclusively or mainly via AMP deamination in lymphocytes,[58] intracellular ATP catabolism seems to be a poor source of adenosine. In contrast, extracellular ATP degradation proceeds exclusively via AMP dephosphorylation resulting in the formation of significant amounts of adenosine as the end product. Amounts of extracellularly formed adenosine depend on the concentration of exogenous ATP, the activities of ectonucleotidases, and the ability of cells to utilize adenosine intracellularly.

Adenosine can act via both intracellular and extracellular adenosine receptors[59] and can regulate various lymphocyte functions through the activation of adenylate cyclase.[60–62] On the other hand, adenosine can easily enter B and T cells and function as a substrate for nucleotide synthesis.[63] We have found that extracellular ATP can induce degradation of prelabeled intracellular ATP (FIG. 8). Adenosine produced

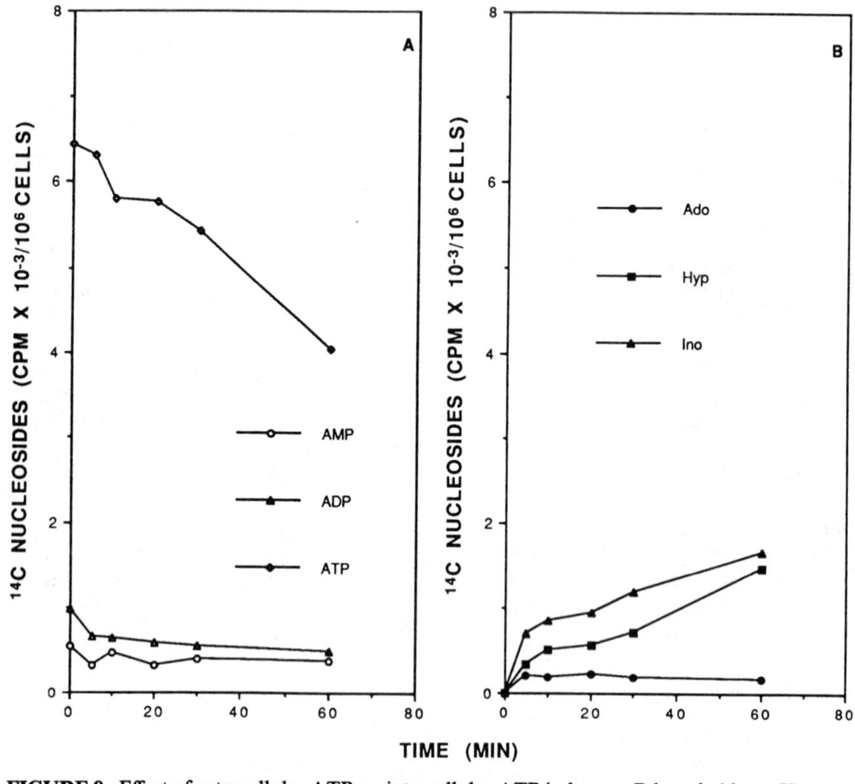

FIGURE 8. Effect of extracellular ATP on intracellular ATP in human B lymphoblasts. Human B lymphoblasts (WI-L2) (20×10^6/1 ml) were incubated for 1 hr with 5 μCi of [8-^{14}C]adenine. After incubation, unused radioactive adenine was washed out and cells containing intracellular radioactive ATP were incubated with 500 μM extracellular ATP. Thin-layer chromatography was used to analyze radioactive adenine nucleotides in cell extracts and nucleosides and bases in medium.[11]

extracellularly may also affect the biosynthesis of purine *de novo,* as well as the salvage of adenosine, hypoxanthine, guanosine, and thymidine in both B and T lymphocytes.[63] The activities of ectoenzymes in human lymphoid cells (ecto-5′-NT) are sufficient to produce significant amounts of purines required for rapidly dividing human B lymphoblasts or mitogen-stimulated T cells.[64,65] It is still unknown whether a regulatory relationship exists between B cells, which produce adenosine, and T cells, which express receptors for adenosine.

ACKNOWLEDGMENTS

We thank Dr. J. Wiesner for his critical review of this manuscript and R. Jimenez for expert technical assistance.

REFERENCES

1. Rothstein, A. & R. Meier. 1948. J. Cell Comp. Physiol. **32:** 77-95.
2. Gordon, J. L. 1986. Biochem. J. **233:** 309-318.
3. Luthje, J. 1989. Klin. Wochenschr. **67:** 317-327.
4. De Pierre, J. W. & M. L. Karnovsky. 1974. J. Biol. Chem. **249:** 7111-7120.
5. De Pierre, J. W. & M. L. Karnovsky. 1974. J. Biol. Chem. **249:** 7121-7129.
6. Barankiewicz, J., H.-M. Dosch & A. Cohen. 1988. J. Biol. Chem. **263:** 7094-7098.
7. Barankiewicz, J., M. Hui, A. Cohen, & H.-M. Dosch. 1989. *In* Purine Metabolism in Man VI. K. Mikanagi, K. Nishioka & W. N. Kelly, Eds.: 455-461. Plenum. New York, NY.
8. Steinberg, T. H. & S. C. Silverstein. 1987. J. Biol. Chem. **262:** 3118-3122.
9. Henderson, J. F. 1985. *In* Methods in Pharmacology, Vol. 6. Methods Used in Adenosine Research. D. M. Paton, Ed.: 67-82. Plenum. New York, NY.
10. Hanks, J. H. & J. H. Wallace. 1958. Proc. Soc. Exp. Biol. Med. **98:** 188-192.
11. Henderson, J. F., J. H. Fraser & E. E. McCoy. 1974. Clin. Biochem. **7:** 339-358.
12. Karnovsky, M. L. & M.-M. Philippeaux. 1985. Ann. N.Y. Acad. Sci. **451:** 250-255.
13. Manery, J. F. & E. E. Dryden. 1979. *In* Physiological and Regulatory Functions of Adenosine and Adenosine Nucleotides. H P. Baer & G. I. Drummond, Eds.: 323-339. Raven Press. New York, NY.
14. Pearson, J. D. 1985. *In* Methods in Pharmacology. Vol. 6. Methods Used in Adenosine Research. D. M. Paton, Ed.: 83-107. Plenum. New York, NY.
15. Carson, D. A., J. Kaye & D. B. Wasson. 1981. J. Immunol. **126:** 348-352.
16. De Pierre, J. W. & M. L. Karnovsky. 1974. Science **183:** 1096-1098.
17. Uusitalo, R. J. & M. J. Karnovsky. 1977. J. Histochem. Cytochem. **25:** 87-96.
18. Fleit, H., M. Conklyn, R. D. Stebbins & R. Silber. 1975. J. Biol. Chem. **250:** 8889-8892.
19. Edwards, N. L., D. Recker, J. Manfredi, R. Remback & I. H. Fox. 1982. Am. J. Physiol. **243:** C270-C277.
20. Newby, A. C., Y. Worku & P. Meghji. 1987. *In* Topics and Perspectives in Adenosine Research. E. Gerlach & B. F. Becker, Eds.: 155-163. Springer-Verlag. Berlin.
21. Burgess, F. W., M. H. El Kouni & R. E. Parks, Jr. 1985. Biochem. Pharmacol. **34:** 3061-3070.
22. Johnson, S. M., M. E. North, G. L. Asherson, J. Allsop, R. W. E. Watts & A. D. B. Webster. 1977. Lancet **1:** 168-170.
23. Edwards, N. L., D. B. Magilavy, J. T. Cassidy & I. H. Fox. 1978. Science **201:** 628-630.
24. Thompson, L. F., G. R. Boss, H. L. Spiegelberg, I. V. Jansen, R. D. O'Connor, T. A. Waldmann, R. N. Hammburger & J. E. Seegmiller. 1979. J. Immunol. **123:** 2475-2478.
25. Cohen, A., A. Mansour, H.-M. Dosch & E. W. Gelfand. 1980. Clin. Immunol. Immunopathol. **15:** 245.

26. Gelfand, E. W., D. McCurdy & A. Cohen. 1984. Blood **63:** 1475-1480.
27. Dmitrenko, N. P. & G. V. Goroshnikova. 1987. Ukr. Biokhim. Zh. **59:** 61-65.
28. Hennighausen, G., E.-M. Hehl & J. Rychly. 1985. Arch. Toxicol. Suppl. **8:** 500-503.
29. Kragballe, K. & J. Ellegaard. 1978. Scand. J. Haematol. **20:** 271-279.
30. Abney, E. R., W. H. Evans & M. E. Parkhouse. 1976. Biochem. J. **159:** 293-299.
31. Thompson, L. F., J. M. Ruedi, R. D. O'Connor & J. F. Bastian. 1986. J. Immunol. **137:** 2496-2500.
32. Bastian, J. F., J. M. Ruedi, G. A. Macpherson, H. E. Golembesky, R. D. O'Connor & L. F. Thompson. 1984. J. Immunol. **132:** 1767-1772.
33. Hennighausen, G. & P. Lange. 1985. Biomed. Biochem. Acta **44:** 1269-1272.
34. Freire-Moar, J. M., D. Rodrigues, S. Rodriguez-Segade & M. Freire. 1984. Int. J. Biochem. **16:** 225-229.
35. Edwards, N. L., E. W. Gelfand, L. Burk, H.-M. Dosch & I. H. Fox. 1979. Proc. Natl. Acad. Sci. USA **76:** 3474-3476.
36. Cohen, A., J. W. W. Lee, H.-M. Dosch & E. W. Gelfand. 1980. J. Immunol. **125:** 1578-1582.
37. Cohen, A., H.-M. Dosch & E. W. Gelfand. 1981. Clin. Immunol. Immunopathol. **18:** 287-290.
38. Agren, G., K. Nilsson & G. Ronquist. 1976. Acta Physiol. Scand. **98:** 67.
39. Novogrodsky, A. 1972. Biochem. Biophys. Acta **266:** 343-349.
40. Dornand, J., C. Reminiac & J.-C. Mani. 1977. Biochimie **59:** 425-432.
41. Dornand, J., J.-C. Bonnafous & J.-C. Mani. 1978. Biochem. Biophys. Res. Commun. **82:** 685-692.
42. Andrée, T., W. Gutensohn & U. Kummer. 1987. Immunobiology **175:** 214-225.
43. Dosreis, G. A., A. F. Nobrega & R. P. De Carvalho. 1986. Cell. Immunol. **101:** 213-231.
44. Chaudry, I. H., M. G. Clemens & A. E. Baue. 1985. *In* Purines: Pharmacology and Physiological Role. T. W. Stone, Ed.: 115-124. Macmillan. New York, NY.
45. Gregory, S. & M. Kern. 1978. Biochem. Biophys. Res. Commun. **83:** 1111-1116.
46. Ikehara, S., R. N. Pahwa, D. G. Lunzer, R. A. Good & M. J. Modak. 1981. J. Immunol. **127:** 1834-1838.
47. Henriksson, T. 1983. Immunol. Lett. **7:** 171-176.
48. Schmidt, A., J. R. Ortaldo & R. B. Herberman. 1984. J. Immunol. **132:** 146-149.
49. Cameron, D. J. 1985. Clin. Immunol. Immunopathol. **37:** 230-235.
50. Becker. E. L. & P. M. Henson. 1975. Inflamation **1:** 71-84.
51. Sung, S.-S., J. D. E. Young, A. M. Origlio, J. M. Heiple, H. R. Kaback & S. C. Silverstein. 1985. J. Biol. Chem. **260:** 13442-13449.
52. Steinberg, T. H., A. S. Newman, J. A. Swanson & S. C. Silverstein. 1987. J. Biol. Chem. **262:** 8884-8888.
53. Lin, J., R. Krishnaraj & R. Kemp. 1985. J. Immunol. **135:** 3403-3410.
54. Chahwala, S. B. & L. C. Cantley. 1984. J. Biol. Chem. **259:** 13717-13722.
55. Wiley, J. S. & G. R. Dubyak. 1989. Blood **73:** 1316-1323.
56. El-Moatassim, C., J. Dornand & J.-C. Mani. 1987. Biochem. Biophys. Acta **927:** 437-444.
57. Heppel, L. A., G. A. Weisman & I. Friedberg. 1985. J. Membrane Biol. **86:** 189-196.
58. Barankiewicz, J. & A. Cohen. 1984. J. Biol. Chem. **259:** 15178-15181.
59. Daly, J. W. 1983. *In* Physiology and Pharmacology of Adenosine Derivatives. J. W. Daly, Y. Kuroda, J. W. Phillis, H. Simion & M. Vi, Eds.: 275-290. Raven Press, New York, NY.
60. Wohlberg, G., T. P. Zimmerman, K. Hiemstra, M. Winston & L.-C. Chu. 1975. Science **187:** 957-959.
61. Marone, G., M. Plaut & L. M. Lichtenstein. 1978. J. Immunol. **121:** 2153-2159.
62. Wohlberg, G., T. P. Zimmerman, G. S. Duncan, K. H. Singer & G. B. Elion. 1979. Biochem. Pharmacol. **27:** 1487-1495.
63. Barankiewicz, J., H.-M. Dosch, R. Cheung & A. Cohen. 1989. Adv. Exp. Med. Biol. **263B:** 475-479.
64. Thompson, L. F. 1985. Proc. Soc. Exp. Biol. Med. **179:** 432-436.
65. Thompson, L. F. 1985. J. Immunol. **134:** 3794-3797.

DISCUSSION OF THE PAPER

F. DI VIRGILIO (*Institute of General Pathology, Padua, Italy*): A brief comment in support of the presence of ecto-ATPases on the plasma membrane of differentiated T lymphocytes. We have found that unrestricted T killer cells generated by a 5-day culture with 100 U/ml of interleukin-2 have a very high ecto-ATPase activity compared to freshly isolated lymphocytes and to several lymphoma lines.

A. S. DAHMS (*San Diego State University, San Diego, CA*): 1) It was not clear whether a divalent cation was present in the T cell ecto-ATPase measurement. 2) Have you tested whether PI-specific phospholipase C releases any activity?

BARANKIEWICZ: 1) There was *no added* divalent cation above that present in the medium. 2) No, I have not tried it.

Liver Plasma Membrane Ecto-ATPase

Purification, Localization, Cloning, and Functions

SUE-HWA LIN

Department of Molecular Pathology
The University of Texas
M. D. Anderson Cancer Center
Houston, Texas 77030

INTRODUCTION

Most tissues have ecto-ATPase activity that is stimulated by Ca^{2+} or Mg^{2+}. Unlike the transport ATPases, which transport ions across membranes and use intracellular ATP as an energy source, the ecto-ATPases have their ATP-hydrolyzing site localized on the outside of the cell surface. The general characteristics of the ecto-ATPase activity are its activation to a similar extent by either Ca^{2+} or Mg^{2+}, and its ability to hydrolyze nonspecifically several different nucleotides. The ubiquitous existence of this enzyme activity in almost every tissue suggests that this enzyme serves important functions for the cells. It is suggested that the ecto-ATPase may be involved in at least two important cellular functions. One function is to hydrolyze extracellular ATP, which is a ligand for P_2-purinergic receptors, in order to terminate the response.[1] The other function is to act with 5′-nucleotidase for the formation of adenosine, which may either interact with its receptor on the cell surface to trigger cellular responses or may be transported into the cell by nucleoside transporters for the recapture of nucleosides. In order to understand the structural and functional aspects of the ecto-ATPase, we have carried out studies on the characterization, purification, localization, and molecular cloning of the liver plasma membrane ecto-ATPase.

LOCALIZATION AND CHARACTERIZATION OF HEPATOCYTE ECTO-ATPase

The properties of the liver ecto-ATPase were studied by using hepatocytes in primary culture.[2] In hepatocytes in primary culture, we can detect Ca^{2+}-ATPase and Mg^{2+}-ATPase activities by addition of ATP to the intact cells. The external localization of the active site of the ATPase was confirmed by the observation that the Ca^{2+}-ATPase and Mg^{2+}-ATPase activities were the same for intact cells, saponin-treated cells, and cell homogenates. The effect of Ca^{2+} and Mg^{2+} on the ecto-ATPase activity is not additive, indicating that both Ca^{2+}- and Mg^{2+}-ATPase activities are part of the same enzyme. This ecto-ATPase has several unique properties: 1) It has broad nucleotide-hydrolyzing activity; that is, it can hydrolyze ATP, GTP, UTP, CTP, ADP, and GDP to a similar extent. 2) The ecto-ATPase activity is not sensitive to oligomycin, vanadate, *N*-ethylmaleimide and *p*-chloromercuribenzoate, ouabain, dicyclohexylcarbodiimide, 7-chloro-4-nitrobenzo-2-oxa-1,3-diazole, F^-, *N*-ethylmaleimide, La^{3+}, and oxidized glutathione. 3) The ecto-ATPase activity is highly resistant to proteolysis. 4) The ecto-ATPase can use ATP^{4-}, Ca^{2+}-ATP, or Mg^{2+}-ATP as a substrate.

PURIFICATION OF ECTO-ATPase FROM RAT LIVER PLASMA MEMBRANES

We have further purified the ecto-ATPase from rat liver plasma membrane.[3,4] The ecto-ATPase was solubilized from liver plasma membrane using a detergent, polyoxyethylene 9-lauryl ether, and purified by column chromatography using Polybuffer Exchanger 94, concanavalin A-Sepharose 4B, and Sephadex G-200. The enzyme was purified at least 300-fold from plasma membranes. In sodium dodecyl sulfate-polyacrylamide gel electrophoresis (SDS-PAGE), one major band with molecular weight of 100,000 was observed. The purified enzyme was activated by Ca^{2+} and Mg^{2+} with apparent half-saturation constants of 0.09 and 0.16 μM, respectively. Compounds that affect Ca^{2+} pump activity (calmodulin and trifluoroperazine at concentrations of 1 μg/ml and 100 μM, respectively) had no effect on the enzymatic activity.

PREPARATION AND CHARACTERIZATION OF ANTI-ECTO-ATPase ANTIBODIES

Three different anti-ecto-ATPase antibodies have been prepared. Antibody #669, which was generated by injecting rabbits with SDS-PAGE-purified ecto-ATPase, only recognizes denatured ecto-ATPase. Antibody #708, which was generated by injecting rabbits with purified ecto-ATPase from chromatography columns, recognizes both native and denatured ecto-ATPase and is able to deplete the ecto-ATPase activity from solution when the antibody-enzyme complex is removed by protein A-Sepharose 4B. After this treatment the ecto-ATPase is found to be associated with the protein A-Sepharose 4B fraction.[3] This observation indicates that antibody #708 can bind to the native ecto-ATPase, a property that will be useful in studies using intact cells (for example, fluorescent cell sorting). Antipeptide antibody #36 was generated against a sequence obtained from trypsin digestion of the ecto-ATPase. On Western immunoblot, antipeptide antibody #36 can bind to the reduced form of ecto-ATPase but not to the nonreduced ecto-ATPase, indicating that this antibody recognizes epitopes that are protein conformation dependent. Although an ATPase activity, which has properties similar to those of the liver ecto-ATPase, could be detected in several tissues, we have found that the antibodies specific for the liver ecto-ATPase only cross-react with the ATPase of few tissues, that is, kidney and heart. This observation indicates that ecto-ATPases from various tissues are distinct, though they all have similar enzymatic properties.

LOCALIZATION OF THE ECTO-ATPase TO THE CANALICULAR REGION OF LIVER MEMBRANES

Because hepatocytes are polarized epithelial cells with distinct plasma membrane regions, unraveling the distribution of the ecto-ATPase in hepatocytes will be of importance in order to understand the functions of the ecto-ATPase. The surface distribution of the plasma membrane ecto-ATPase in rat hepatocytes was determined by several methods.[3] 1) Two polyclonal antibodies specific for the ecto-ATPase were used to examine the distribution of the enzyme in frozen sections of rat liver by immunofluorescence. Fluorescent staining was observed at the bile canalicular region of hepatocytes. 2) Plasma membranes were isolated from the canalicular and sinusoidal regions of rat liver. The specific activity of ecto-ATPase in the canalicular membranes was 22 times higher than that of sinusoidal membranes. 3) By immunoblots with polyclonal antibodies against the ecto-ATPase and the Na^+,K^+-ATPase, it was found that the ecto-ATPase protein was only detected in canalicular membranes and not in sinusoidal membranes, while the Na^+,K^+-ATPase protein was only detected in sinusoidal membranes and not in canalicular membranes. These results indicate that the liver ecto-ATPase has an unequal distribution among different domains of hepatocyte membranes and that it is more enriched in the canalicular domain. Because it was known that one of the functions of the ecto-ATPase is to regulate P_2-purinergic receptor localized in the sinusoidal domain,[5,6] the implication from this localization study is that liver may have more than one form of ecto-ATPase. We proposed that the function of the sinusoidal ecto-ATPase is to regulate the P_2-purinergic receptor/effector system while the canalicular ecto-ATPase is involved in the mechanism of retrieval of secreted nucleotides.

CLONING OF A cDNA CODING FOR A LIVER ECTO-ATPase

The amino acid sequence of the ecto-ATPase from rat liver was deduced from an analysis of cDNA clones and a genomic clone. The deduced sequence predicts a 519 amino acid protein with a calculated molecular mass of 57,388 Da. There are 16 potential asparagine-linked N-glycosylation sites in the protein. Hydropathy analysis of the deduced amino acid sequence indicates that the protein has two hydrophobic stretches. One is located at the NH_2-terminal and has properties of a membrane signal sequence and the other is near the COOH-terminal end. FIGURE 1 shows the schematic diagram of the ecto-ATPase sequence features. The structural arrangement predicts that most of the protein mass is on the outside of the cell and that the COOH-terminal of the protein is intracellular. This structural information is consistent with the fact that the protein is an ecto-ATPase with its active site localized on the outside of the cell. Sequence homology analysis shows that the ecto-ATPase has three immuno-

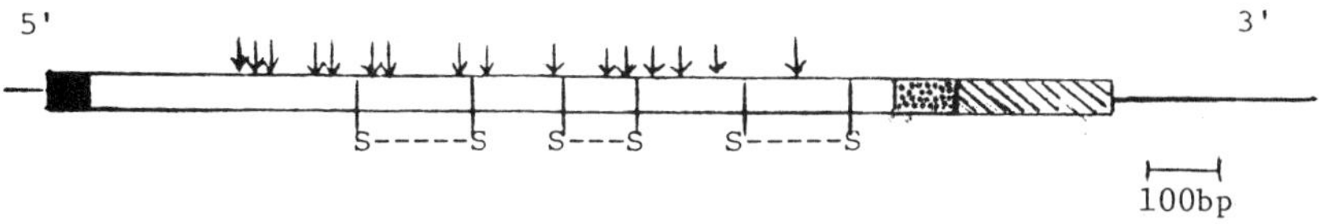

FIGURE 1. Schematic diagram of the sequence features. Untranslated regions are illustrated as horizontal line segments; the ecto-ATPase coding region is illustrated as a long rectangle, which is situated between two line segments and is divided into several regions. The signal peptide region is filled in, the extracellular region is open, the transmembrane regions are stippled, and the cytoplasmic segment is hatched. Arrows above the box indicate potential N-glycosylation sites. Vertical lines in the extracellular region indicate cysteine residues, and the proposed immunoglobulin-like disulfide bonds are indicated.

globulin-like domains that are homologous to the immunoglobulin domains of biliary glycoprotein 1 (BGP1).[7] The ecto-ATPase has a cytoplasmic domain about 71 amino acids long. This cytoplasmic domain has a phosphorylation consensus sequence for cAMP-dependent protein kinase (Lys-Arg-X-X-Ser)[8] and a phosphorylation consensus sequence for tyrosine kinase.[9] Regarding the site for nucleotide binding, there are two stretches of sequences (consisting of amino acids 92-100 and 335-348) similar to the consensus sequences for nucleotide-binding domains of other ATP-binding protein. Accordingly, the hypothetical arrangement of the ecto-ATPase in plasma membrane is shown in FIGURE 2. It should be pointed out that although in the primary sequence the two potential nucleotide-binding sites are quite far apart, they may be quite close in the folded three-dimensional structure.

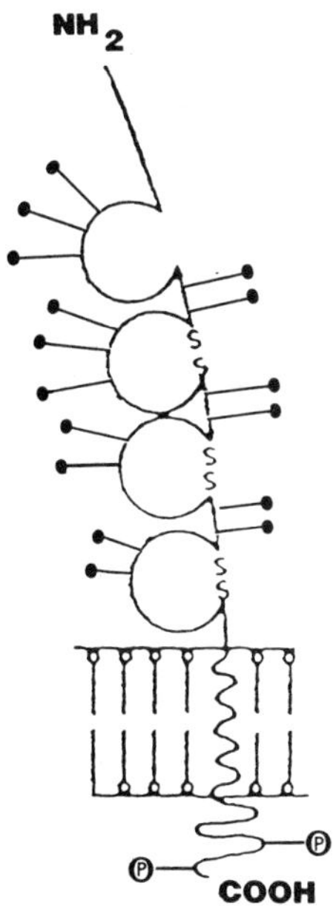

FIGURE 2. Hypothetical arrangement of the ecto-ATPase in plasma membrane. Potential N-glycosylation sites are indicated as filled circles.

EXPRESSION OF A FUNCTIONAL ECTO-ATPase IN OTHER MAMMALIAN CELLS BY TRANSFECTION

To confirm that the clone obtained is the cDNA for the ecto-ATPase, we have carried out transfection experiments. The full-length ecto-ATPase cDNA was constructed into plasmid cdm8.[10] This vector uses the cytomegalovirus promoter and SV40 origin of replication. The expression plasmid (pEXP) was transfected into mouse L cells and human HeLa cells by calcium phosphate precipitation. Immunoblot in which antipeptide antibody #36 was used showed that the ecto-ATPase can be expressed in both L cells and HeLa cells (FIG. 3). ATPase activity of the expressed protein was measured using cell lysates from cells transfected with a control plasmid and the plasmid containing the ecto-ATPase cDNA. Cells transfected with plasmid (pEXP) had higher Ca^{2+}-stimulated ATPase activity compared to that of controls (TABLE 1), indicating that the protein expressed from the plasmid is the ecto-ATPase.

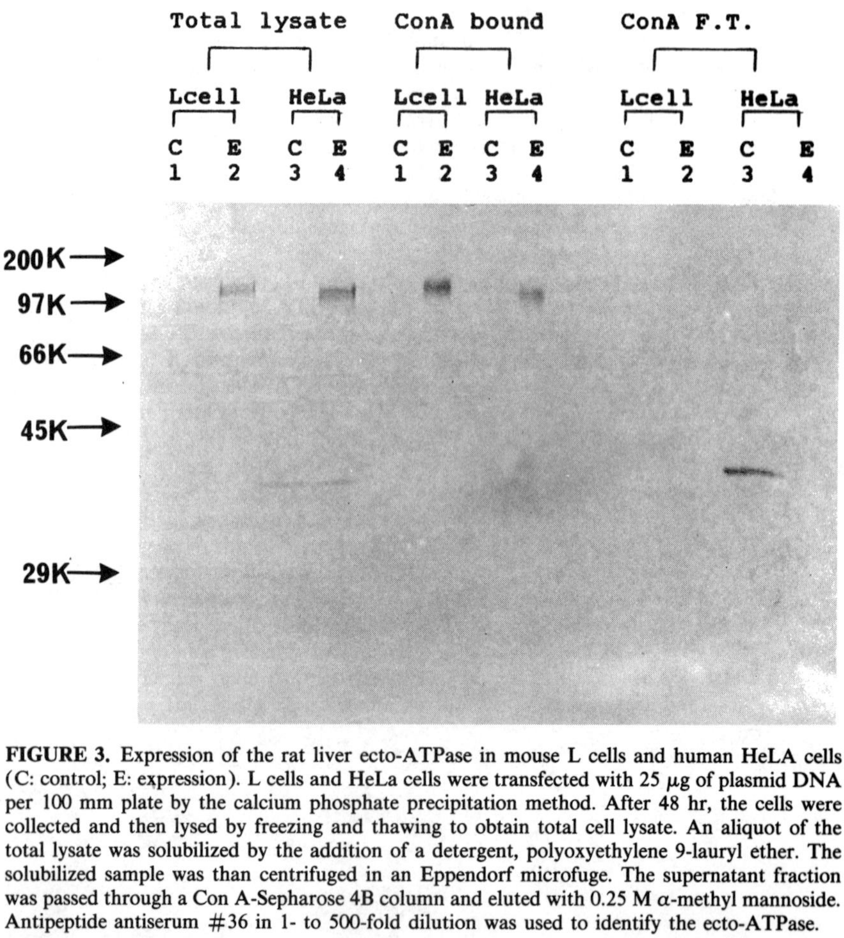

FIGURE 3. Expression of the rat liver ecto-ATPase in mouse L cells and human HeLA cells (C: control; E: expression). L cells and HeLa cells were transfected with 25 μg of plasmid DNA per 100 mm plate by the calcium phosphate precipitation method. After 48 hr, the cells were collected and then lysed by freezing and thawing to obtain total cell lysate. An aliquot of the total lysate was solubilized by the addition of a detergent, polyoxyethylene 9-lauryl ether. The solubilized sample was than centrifuged in an Eppendorf microfuge. The supernatant fraction was passed through a Con A-Sepharose 4B column and eluted with 0.25 M α-methyl mannoside. Antipeptide antiserum #36 in 1- to 500-fold dilution was used to identify the ecto-ATPase.

TABLE 1. Ca^{2+}-Stimulated ATPase Activity of L Cells and HeLa Cells Transfected with Control and Expression Plasmid-Containing Ecto-ATPase cDNA

	Ca^{2+}-ATPase Activity (nmol/mg/hr)		
Cell Type	Control	pEXP	Percentage of Control
L cells	14.6	38.0	260
HeLa cells	166.2	251.8	152

The liver plasma membrane ecto-ATPase is the first ecto-ATPase purified and cloned. The information obtained from these biochemical and molecular biological studies should be useful in the studies of this enzyme in other tissues. Moreover, the cDNA clone of the ecto-ATPase can be used to investigate the functional aspects of this interesting protein.

REFERENCES

1. GORDON, J. L. 1986. Biochem. J. **233:** 309-319.
2. LIN, S.-H. & W. E. RUSSELL. 1988. J. Biol. Chem. **263:** 12253-12258.
3. LIN, S.-H. 1989. J. Biol. Chem. **264:** 14403-14407.
4. LIN, S.-H. & J. N. FAIN. 1984. J. Biol. Chem. **259:** 3016-3020.
5. BUXTON, D. B., S. M. ROBERTSON & M. S. OLSON. 1986. Biochem. J. **237:** 773-780.
6. HAUSSINGER, D., T. STEHLE & W. GEROK. 1987. Eur. J. Biochem. **167:** 65-71.
7. HINODA, Y., M. NEUMAIER, S. A. HEFTA, Z. DRZENIEK, C. WAGENER, L. SHIVELY, L. J. F. HEFTA, J. E. SHIVELY & R. J. PAXTON. 1988. Proc. Natl. Acad. Sci. USA **85:** 6959-6963 (for a correction, see **86:** 1668, 1989).
8. KREBS, E. G. & J. A. BEAVO. 1979. Annu. Rev. Biochem. **48:** 923-959.
9. HUNTER, T. & J. A. COOPER. 1985. Annu. Rev. Biochem. **54:** 897-930.
10. SEED, B. 1987. Nature **329:** 840-842.

DISCUSSION OF THE PAPER

M. V. SITKOVSKY (*National Institutes of Health, Bethesda, MD*): Did I understand you correctly that you demonstrated phosphorylation of your ecto-ATPase in the intracellular region?

LIN: It was done by others with hepatic cells. Addition of insulin caused tyrosine phosphorylation of a protein that is most likely identical with our ecto-ATPase.

F. DI VIRGILIO (*Institute of General Pathology, Padua, Italy*): I was puzzled by the lack of homology of the ecto-ATPase you described to all known ATPases. Do you have other examples of ATPases showing this structure? Have you tested vanadate as an inhibitor of this ATPase?

LIN: As far as I know, the liver ecto-ATPase is the only ecto-ATPase whose primary structure is known. Vanadate does not inhibit the ecto-ATPase activity.

A. S. DAHMS (*San Diego State University, San Diego, CA*): A comment on the distinct nature of ecto-ATPase relative to such well-characterized pumps as the Na^+ pump, SR Ca^{2+} pump, and vacuolar H^+ pump. The ecto-ATPases are fully distinctive, and are unaffected by vanadate, oligomycin, ouabain, and other classical inhibitors. They are apparently incapable of catalyzing any other partial reactions, including P_i-ATP exchange, ADP-ATP exchange, medium ^{18}O-H_2O exchange, and cleavage of low molecular weight organophosphorous molecules such as *p*-nitrophenylphosphate and acetylphosphate. They are apparently designed for one function, that is, irreversible cleavage of ATP. There may be some structural differences between the ecto-ATPases. For example, we have amino terminal sequence information for the first 23 residues of the 85-kDa glycoprotein of the skeletal muscle transverse tubule, a putative ecto-Mg-ATPase, and there is no identity with the liver cannicular enzyme. There is another factor that bears upon the possible function of the ecto-ATPase from various sources. Dr. Lin, what is the apparent Stokes radius of the active, solubilized liver ectoenzyme? This is another important feature because it would suggest that the protein is probably not a channel former. Your data show a single bilayer helix. In the absence of oligomerization and other protein(s), this would mean it could not serve as a translocator of ions or other molecules and reinforces the conclusion that ecto-ATPases may serve to only cleave ATP.

LIN: The data are consistent with a polypeptide of M_r 85-100,000.

Ectoprotein Kinase in the Regulation of Cellular Responsiveness to Extracellular ATP[a]

YIGAL H. EHRLICH,[b] MICHAEL V. HOGAN, AND ZOFIA PAWLOWSKA

CSI/IBR Center for Developmental Neuroscience
College of Staten Island
City University of New York
and
Institute for Basic Research in Developmental Disabilities
Staten Island, New York

ULHAUS NAIK AND ELIZABETH KORNECKI

Department of Anatomy and Cell Biology
State University of New York
Health Science Center at Brooklyn
Brooklyn, New York

INTRODUCTION

The enzyme protein kinase catalyzes the transfer of the γ-phosphate from ATP to serine, threonine, or tyrosine residues within the structure of proteins, to form a covalent phosphomonoester linkage. This linkage produces a conformational change that can bring about an alteration in the functional activity of the protein. Dephosphorylation by phosphoprotein phosphatases determines the reversible nature of this regulatory process. A large body of evidence now demonstrates that the phosphorylation of intracellular proteins by several types of protein kinase plays a significant role in the regulation of multiple cellular functions.[1–3] In particular, protein kinase activity is recognized as the ubiquitous target for second messengers produced intracellularly in response to the stimulation of surface receptors by hormones, neurotransmitters, and growth factors. In addition, protein phosphorylation has been shown

[a]Studies described here from the authors' laboratories were supported in part by Grant 88-0290 from the U.S. Air Force Office of Scientific Research, Grant 669148 from the PSC-CUNY Research Award Program of the City University of New York (to Y. H. E.), and Grant RO2-HL32594 from the National Heart, Lung, and Blood Institute (to E. K.).

[b]Address for correspondence: CSI/IBR Center for Developmental Neuroscience, College of Staten Island, City University of New York, 120 Stuyvesant Place, 7-422, Staten Island, New York 10301.

to serve as a site of molecular adaptation in processes whereby hormonal, pharmacological, or behavioral inputs induce long-lasting alterations in cellular responsiveness.[4,5]

It is now well established that ATP is present in the extracellular milieu of a variety of tissues, and that extracellular ATP can elicit differing functional responses in many types of cells[6] (see also all other articles in this volume). Furthermore, several recent studies have helped to focus attention on the finding that the extracellular environment is not metabolically inert, but contains various ecto- and exoenzymes implicated in developmental processes, in intercellular communication, and in feedback regulation.[7] The utilization of extracellular ATP by protein kinases operating at cell surfaces can provide for these extracellular processes a powerful regulatory machinery analogous to that provided by protein phosphorylation systems operating intracellularly. In this article, we review in brief studies demonstrating the existence of ectoprotein kinases at cell surfaces, outline some criteria to be fulfilled in providing evidence for the presence of ectoprotein kinase activity on a cell, discuss the unique roles that extracellular protein phosphorylation may play in cells that secrete ATP by exocytosis, and describe studies from our laboratory on ectoprotein kinase activity in neuronal cells and its potential involvement in processes operating in cell development, in regulating neuronal responsiveness, and in processes underlying synaptic plasticity.

IDENTIFICATION AND CRITERIA FOR ECTOPROTEIN KINASES

Ectoenzymes constitute, by definition, a part of the structure of the plasma membrane, and are membrane-bound enzymes whose catalytic site faces the extracellular environment. One of the pioneers in the investigation of ectoenzymes, M. L. Karnovsky, has formulated a set of six criteria that should be met in providing evidence that a certain enzymatic activity indeed belongs in this category. In essence, these criteria state the following:

1. Assays using intact cells should be used to detect the activity of an ectoenzyme.
2. It should be demonstrated conclusively that the enzyme under investigation is not secreted by the assayed cells.
3. A substrate that cannot penetrate the cells should preferably be used in these assays.
4. An increase in the assayed activity may be detected when the cells are broken.
5. It should be possible to detect products of the enzymatic reaction in the extracellular environment.
6. Availability of a selective inhibitor (preferably a covalently linking reagent) that inhibits the activity in intact cells will provide a strong evidence for the ectoenzymatic nature of the activity.

In addition to these general criteria for ectoenzymes,[8] one specialized aspect of protein phosphorylation activity enables the design of tests that are unique for the detection of ectoprotein kinases. Endogenous protein phosphorylation is the process in which a kinase phosphorylates a protein that coexists with it in the same microenvironment.[9] For *in vitro* assays of endogenous protein phosphorylation activity, a radiolabeled cosubstrate of protein kinase, [γ-^{32}P]ATP, is incubated with the tissue fragments under investigation. No other additions are required. Gel electrophoresis and autoradiography of detergent-solubilized reaction products reveals the identity of the proteins that serve as specific endogenous substrates for protein kinases in the tested preparation (FIG. 1). When membrane fragments are incubated with radiolabeled ATP and the appropriate cofactors, these assays can detect the endogenous substrates of the membrane-bound protein kinases that constitute part of the structure of these membranes. When the radiolabeled ATP is added to *intact* cells incubated in a buffer with ionic composition mimicking the physiological extracellular environ-

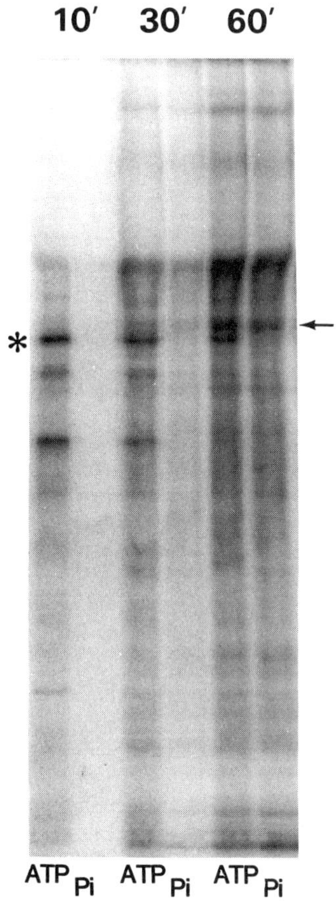

FIGURE 1. Time course of intra- and extracellular protein phosphorylation in neural cells. Cloned neural cells of the NG108-15 line were grown in a serum-free medium in 96-well cluster plates. For phosphorylation reactions, attached cells were gently rinsed and covered with 0.09 ml of Krebs-Ringer buffer, as described previously.[10] To initiate the reaction, 0.01 ml of buffer containing 10 μCi per well of inorganic ^{32}P, or 10 μCi [γ-^{32}P]ATP (final ATP concentration was 1.0 μM) was added to adjacent wells. After the incubation times indicated in the figure, the cells were solubilized in sodium dodecyl sulfate and separated by polyacrylamide gel electrophoresis. Shown is the autoradiograph of a representative gel from a study by Ehrlich *et al.*[10] The asterisk on the left side marks a protein that was seen phosphorylated when radioactive [γ-^{32}P]ATP was added to the extracellular medium, but not when intracellular ATP pools were labeled by adding inorganic ^{32}P to the cells. The arrow on the right side points at a protein that appears to be phosphorylated only intracellularly. The phosphorylation of the protein band marked by the asterisk could be inhibited by adding apyrase to the medium, but not the phosphorylation of the band marked by the arrow.

ment, the phosphoproteins detected in the assay are the cell surface proteins that serve as substrates for ectoprotein kinases.[10,11] Of course, certain requirements must be met by following appropriate controls. A main criterion, as pointed out by Karnovsky,[8] is that the assayed enzyme does not originate from intracellular sources. In the case of ectokinase assays, one can test, in addition, whether the endogenous protein substrate is of intra- or extracellular origin. Incubation of cells with radioactive inorganic phosphate ($^{32}P_i$), which penetrates cells, results in radiolabeling of intracellular ATP pools. In turn, protein substrates of intracellular protein kinases become labeled. In parallel, cells are incubated with radioactive ATP to serve as cosubstrate for ectoprotein kinase. The time course of protein phosphorylation detected by these two procedures should be different. Radiolabeling of intracellular proteins in cells incubated with $^{32}P_i$ starts after a lag period during which intracellular ATP pools are first labeled. The phosphorylation of surface proteins by radioactive ATP added directly to the medium would, therefore, be detected earlier. Furthermore, the phosphorylation of proteins in cells incubated with $^{32}P_i$, as opposed to [γ-^{32}P]ATP, would differ not only kinetically but also qualitatively. Different patterns of protein phosphorylation are revealed by the two assays outlined above because, in addition to the possibility that ectokinases phosphorylate transmembrane proteins, proteins that are unique to the cell surface can also serve as substrates for this activity. The identification of the endogenously phosphorylated protein substrates is thus a most useful tool in studies designed to establish the presence of ectoprotein kinase in a given cell population. Naturally, this identification is also a first step in investigating the potential function of this activity.

Potential complicating factors in the assays described above are the release of intracellular [γ-^{32}P]ATP to the medium from cells preincubated with $^{32}P_i$, and the rapid formation of inorganic $^{32}P_i$ from added radiolabeled ATP by ecto-ATPases. Subsequently, labeling of intracellular phosphoproteins could occur in cells incubated with [γ-^{32}P]ATP, and labeling of proteins on the surface would occur in cells incubated for a long period with $^{32}P_i$. These factors, however, can be used to our advantage. We have found that the addition of an excess (1-5 mM) of nonradioactive P_i will decrease considerably the labeling of intracellular phosphoproteins in cells incubated with $^{32}P_i$, but would not affect the early labeling of specific substrates of ectoprotein kinase when radioactive ATP is added to the medium. Conversely, addition to the medium of ATP-consuming enzymes, such as apyrase, will inhibit the labeling of substrates of ectoprotein kinase without affecting the phosphorylation of intracellular proteins. Based on these considerations, we recommend the following assays as a standard initial procedure to be used in the first phase of an investigation of ecto-protein kinase activity:

Incubate minimally disturbed cells (preferably in an attached state in 48- or 96-well plates) in a buffer with an ionic composition mimicking a physiological extracellular environment (Krebs-Ringer or Tyrode's media). The buffers should be prepared without phosphate ions. To specified rows of wells (see below) add either 1 mM cold P_i, aliquots of the enzyme apyrase (0.5 units per well), or an equal volume of buffer. To initiate the reactions, add to alternate wells equivalent amounts (in μCi) of either $^{32}P_i$ or [γ-^{32}P]ATP. In different wells, stop the reaction as a function of time at very short (10 sec-1 min), intermediate (5-10 min), and long (30-120 min) reaction periods. Thus, each of the sets is tested without additions, in the presence of 1 mM nonradioactive P_i, and with apyrase.

Proteins labeled in cells incubated with $^{32}P_i$ in the presence of apyrase are specific substrates of intracellular protein kinases. Proteins that become phosphorylated early after the addition of [γ-^{32}P]ATP to the medium, and whose labeling is insensitive to added nonradioactive P_i and is inhibited by the addition of apyrase, are the most likely candidate substrates of ectoprotein kinase activity at the cell surface.

ECTOPROTEIN KINASE ACTIVITY AT THE SURFACE OF CULTURED CELLS

Investigation of protein phosphorylation activity in extracellular domains did not lag far behind the discovery of intracellular protein kinases. In 1974, Agren and

Ronquist[12] reported that intact glia and glioma cells can catalyze the transfer of the γ-phosphate from extracellular ATP to exogenously added protein substrates, suggesting the existence of a surface ectoprotein kinase. Stimulation of a protein kinase activity at the surface of glioma cells by cyclic AMP was reported soon after.[13] Although in early studies not all of the considerations and criteria for identification outlined above for ectoprotein kinase activity were followed, these reports have provided important clues for the potential involvement of extracellular protein phosphorylation systems in a variety of cellular systems: cultured 3T3 cells,[14] Ehrlich ascites tumor cells,[15] macrophages,[16] fibroblasts,[17] fat cells,[18] hepatocytes,[19] leukocytes,[20] a macrophage-like cell line,[21] and epididymal spermatozoa.[22] In 1982, Kubler *et al.*[23] reported a most rigorous study of protein phosphorylation at the surface of cloned human HeLa cells and provided convincing evidence for the existence of ectoprotein kinase in this cell line. The enzyme phosphorylated both endogenous and exogenous protein substrates. Incubation of HeLa cells with an exogenous substrate (phosvitin or casein) effected a release of phosvitin kinase activity from the surface of intact cells.[24] This method can provide solubilized enzyme for biochemical studies of ectoprotein kinase.[25] In addition, Kubler *et al.*[26] suggest that the substrate-induced release of a surface protein kinase may have a physiological role, in that a movement of the enzyme away from the cell surface may enable novel venues of cell-environment interactions. The most recent advance in this line of investigation is the possibility that extracellular protein kinases may serve as mediators for classical intracellular second messengers in situations where the latter are released by the cells. Using the synthetic substrate Kemptide, which carries the recognition site for cyclic AMP-dependent protein kinase, Kubler *et al.*[27] demonstrated that cyclic AMP-dependent phosphorylation activity resides at the surface of intact HeLa cells. In summary, a variety of cell types has been shown to have several types of protein kinase activities at the cell surface.

A key question that has not been addressed in most of the reports cited above is that of the natural source of the extracellular ATP that serves as a cosubstrate for the ectoprotein kinases assayed in the cells under investigation. In contrast, our laboratory has focused on the extracellular protein phosphorylation activity in cells known to secrete ATP upon stimulation. These studies are described below in greater detail.

SECRETION OF ATP BY STIMULATED CELLS

Three cell types are known to store ATP within secretory vesicles and release it actively by exocytosis upon cell stimulation. These are neurons, chromaffin cells of the adrenal medulla, and platelets.[6] In these cells, depolarization or receptor activation by certain agonists induces vesicular release, which produces a pulse of extracellular ATP at the cell surface. The duration and magnitude of the pulse are determined by the extent of ATP release and by the action of ecto-ATPases present on the surface of the releasing and adjacent cells. The secreted ATP can serve as a signal that is transmitted to target cells. Unique to cells that secrete quantal amounts of vesicular ATP is the ability to provide a time- and concentration-dependent feedback control over the activity of the releasing cells. In 1954, Holton and Holton[28] were the first to report on the secretion of ATP from nerve terminals and to propose that ATP could be involved in neurotransmission. Subsequently, it was shown more directly that ATP is stored within synaptic vesicles of adrenergic and cholinergic neurons in the peripheral nervous system.[29,30] Preparations of synaptosomes from the central nervous system (CNS) were also shown to store and release ATP.[31,32] In a more recent study, we were able to demonstrate ATP release from intact CNS neurons.[33] In this study a pure (over 95%) neuronal population of embryonic cells was prepared from the neostriata dissected from the brain of 14-day mouse embryos. The neurons were maintained in culture in a chemically defined medium for a period of 15 days, during which time the neurons underwent biochemical and physiological differentiation.

Luciferin/luciferase reactions were used to measure ATP concentrations in the extracellular medium of these differentiated cells. Under basal (nonstimulated) conditions, the concentration of extracellular ATP was found to be about 0.5 nM.[33] TABLE 1 demonstrates that upon depolarization of the cells with 50 mM KCl there was over a 30-fold increase in the concentration of extracellular ATP. Over 75% of this release of ATP was dependent on the presence of calcium ions in the extracellular medium, as would be expected from a process of vesicular exocytosis. Similarly, we found in these cells a 40-fold increase in ATP release in response to stimulation by veratridine, and over 80% of this release could be blocked by tetrodotoxin.[33] Neostriatal cells such as used in the ATP release experiments were also found to incorporate $^{32}PO_4{}^{3-}$ into specific proteins when [γ-^{32}P]ATP was added to the extracellular medium. The pattern of this protein phosphorylation was different from that observed when the same cells were incubated with $^{32}P_i$.[33] These findings indicate that CNS neurons that secrete ATP upon stimulation can also use extracellular ATP in the phosphorylation of surface proteins, and suggests a role for ectoprotein kinase in brain function.

SURFACE PHOSPHOPROTEINS OF CLONED NEURAL CELLS

Great progress in the elucidation of the molecular mechanisms that regulate cellular function has been made by the investigation of homogenous populations of cloned cells grown and differentiated in culture. A model for neuronal differentiation has been provided by a hybrid clone of neuroblastoma × glioma cells called NG108-15. When differentiated in culture, these cells exhibit numerous functions of mature neurons, including the ability to form functional synapses.[34] We have used NG108-15 cells in our initial investigation of neuronal ectoprotein kinase activity. As shown in FIGURE 1, when these cells are incubated with radiolabeled ATP added to the extracellular medium, phosphorylation of specific protein components can be seen within a short incubation period. The phosphorylation of these proteins is not prominent when the intracellular ATP pools of these cells are labeled with $^{32}P_i$. Some of the evidence that these proteins are substrates of ectokinase activity is demonstrated in FIGURE 2. For these experiments NG108-15 cells were cultured in a chemically defined, serum-free medium in 96-well plates. Attached cells were washed and reacted in a Krebs-Ringer medium buffered with Hepes (pH 7.4). Lane a depicts the endogenous protein substrates phosphorylated by extracellular ATP in these cells. Phosphorylation of an exogenous substrate, α-casein, is shown in lane b. The phosphorylation of a substrate that cannot penetrate cells is one of the criteria needed to demonstrate that the kinase involved is an ectoenzyme. When the proteolytic enzyme trypsin was added to the cells at a concentration that had no detectable effect on protein composition (see the protein staining patterns in the lanes on the right in FIG. 2), it eliminated surface protein phosphorylation (lane d) and, when added after the reaction, removed the surface label (lane e). These findings indicate that not only the enzyme but also the substrates are located at the cell surface. The nonhydrolyzable analogue of ATP, AppNHp, had selective effects on the phosphorylation of specific proteins by extracellular ATP (lane h), although by itself it cannot penetrate cells. Increasing the concentration of $MgCl_2$ in the medium from 1 mM to 4 mM did not stimulate the activity of ectoprotein kinase (compare lane f to lane a). Addition of 4 mM $MnCl_2$ caused a marked increase in the phosphorylation of endogenous substrates of ectoprotein kinase (lane g), although not all were affected equally. In recent studies (Pawlowska, Hogan, and Ehrlich, in preparation), we have found that the phosphorylation of specific protein substrates, such as the one marked by an asterisk in FIGURE 2, by radiolabeled ATP can be blocked by the addition of the enzyme apyrase to the extracellular medium. This phosphorylation is not sensitive, however, to the addition of excess nonlabeled P_i. We believe that the combination of all these data provide the evidence that neural cells have an ectoprotein kinase that phosphorylates specific surface proteins by extracellular ATP.

The finding that the phosphorylation of surface proteins by ectokinase increased significantly when NG108-15 cells were induced to differentiate in culture[10] suggested

TABLE 1. Depolarization-Induced ATP Release from Neostriatal Neurons in Primary Culture[a]

Depolarization Conditions	ATP Secretion (nM)
High-potassium Krebs-Ringer buffer	17.3 ± 4.3
Calcium-free, high-potassium Krebs-Ringer buffer	4.4 ± 1.5

[a] Neostriatal cells dissected from embryonic mouse brain were maintained in culture until maturation, and assayed for ATP release by luciferin/luciferase assays, as described previously.[33] Under basal conditions, in nonstimulated cells, ATP concentration in the medium was found to be about 0.5 mM. Medium was collected every 5 min. Incubations were carried out at 37 °C. For depolarization, the medium was replaced with Krebs-Ringer buffer containing 50 mM KCl in place of NaCl. Data from Zhang *et al.*[33]

that this activity may play a role in some specialized neuronal functions of these cells. The first clue for this possibility was provided by the use of specific antibodies directed against the D2-CAM[35] class of neuronal cell adhesion molecules (N-CAMs). We have found[10] that when anti-D2-CAM antibodies are added to the medium prior to the incubation of NG108-15 cells with [γ-^{32}P]ATP, they inhibit the endogenous phosphorylation of specific protein components that migrate in polyacrylamide gels with the molecular weight expected of N-CAMs. The ability of this antibody to inhibit the phosphorylation of proteins in a specific manner without penetrating the cells provides additional evidence for the extracellular location of the inhibited kinase and its endogenous protein substrates. Moreover, this result represents the first step of an investigative approach that can be used to determine whether phosphorylation at sites located on the cell surface is involved in regulating the function of N-CAM in developing and maturing neurons.

As mentioned above, NG108-15, the cell line with which we carried out our initial studies, is a hybrid between neuroblastoma and glioma cells. The substrates of ecto-

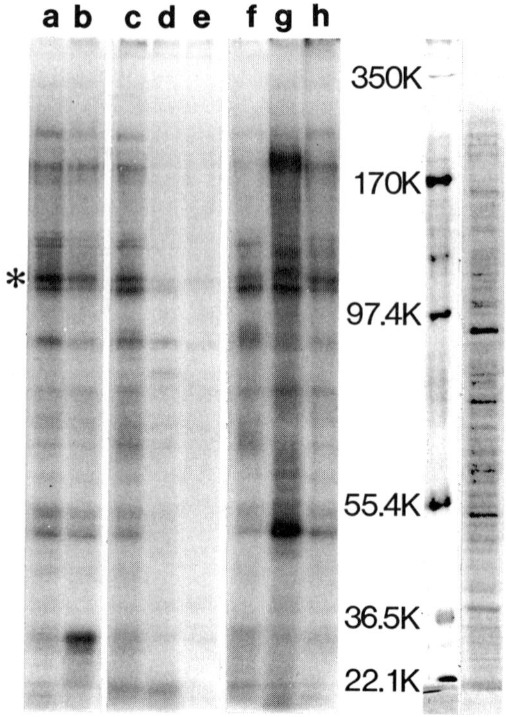

FIGURE 2. Endogenous substrates of ectoprotein kinase in neural cells. Phosphorylation reactions were carried out with radiolabeled ATP added to the extracellular medium of NG108-15 neural cells, as described in the legend to FIGURE 1. The standard reaction time was 10 min, and the standard incubation medium was Krebs-Ringer buffer (**a**), with the following modifications: (**b**) α-Casein (1 μg) was added to the medium. (**c**) Preincubation at 37 °C for 10 min preceded the addition of ATP. (**d**) Trypsin (0.01%) was added to the medium during the preincubation. (**e**) Trypsin (0.01%) was added to the medium for 10 min after the reaction. (**f**) The medium was supplemented with 4 mM $MgCl_2$. (**g**) The medium was supplemented with 4 mM $MnCl_2$. (**h**) AppNHp (10 μM) was added to the reaction medium. The reactions were stopped by adding sodium dodecyl sulfate, and solubilized proteins were separated by gel electrophoresis. The two lanes on the right show the separation of standard molecular weight markers, and the pattern of reaction products stained with Coomassie Blue, which was the same in all the lanes on the left (**a-h**). Data from Ehrlich *et al.*[10]

protein kinase identified in these cells may thus be of neuronal or of glial origin. This hybrid cell line exhibits mainly neuronal properties.[34] Comparative assays indicated that the endogenous substrates of ectoprotein kinase in these cells are, in all likelihood, of neuronal origin. We carried out identical assays with cultured neuroblastoma cells of the murine cell line N1E-115 and with NG108-15 cells. FIGURE 3 demonstrates the great similarity between the phosphorylation patterns obtained in ectoprotein kinase assays with murine neuroblastoma cells (lanes marked NE in FIG. 3) and the hybrid NG108-15 cells (lanes marked NG). The main protein substrate of ectokinase activity in these cells was found also in human neuroblastoma cells (lanes marked SK). We have also studied a cell line derived from pheochromocytoma (a tumor of the adrenal medulla) designated PC12. This cell line was developed as a model of noradrenergic neurons that respond to nerve growth factor.[36] In these cells we also detected the presence of the major substrate protein of neuronal ectokinase (lanes marked PC). Thus, several types of cloned neural cell lines that are used for investigating different aspects of neuronal function were shown to have ectoprotein kinase activity. The availability of a homogenous cell population is of particular importance in studies aimed to determine conclusively the cellular function of a specific substrate of protein kinase activity. This availability, taken together with the ability of an

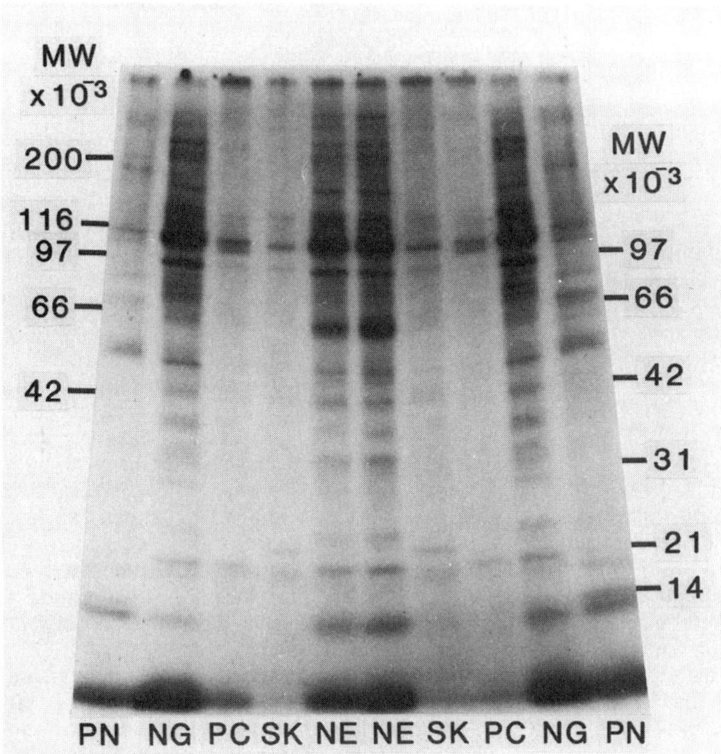

FIGURE 3. Phosphorylation of surface proteins in several types of neural cells. Phosphorylation reactions were carried out with attached cells and by adding [γ-^{32}P]ATP to the extracellular medium essentially as described in the legend to FIGURE 1, except with the following modifications: The cells were grown in 48-well cluster plates instead of 96-well plates. The final concentration of added ATP was 0.1 μM instead of 1.0 μM, and it was added at 15 μCi per well. The reaction time was 15 min. The polyacrylamide gel used to separate reaction products had a gradient of 7-15%. The products from two separate reactions of every cell type were applied to each gel. The autoradiograph of a representative gel is shown above. From the center of the gel toward the periphery, the cells are marked as follows: NE: cells of the murine neuroblastoma line N1E-115. SK: cells of the human neuroblastoma line SK-N-SH. PC: pheochromocytoma cells of the line PC12. NG: NG108-15 hybrid neuroblastoma $\times$ glioma cells. PN: primary neurons prepared from the telencephalon of 7-day chick embryos, and maintained in culture for 6 days.

antibody to inhibit phosphorylation by an ectoprotein kinase without having to penetrate the cell, makes it possible to demonstrate causal relationships between the activity of a specific protein phosphorylation system at the cell surface and the regulation of defined neuronal functions by extracellular ATP. Such studies could reveal the physiological roles of this activity.

POTENTIAL FUNCTIONS OF NEURONAL ECTOPROTEIN KINASE

The phosphorylation of cell surface proteins in neurons by extracellular ATP is most likely to be involved in the communication among cells, and between the cell surface and components of the extracellular matrix. The phosphorylation of N-CAMs, as discussed above, is one way in which ectoprotein kinase may exert functional effects on such processes. It is possible that other substrates of ectoprotein kinase are also involved in the regulation of intercellular interactions that occur during neuronal development. To investigate this possibility, we have initiated studies of ectoprotein kinase in neurons cultured from embryonic brain. We have found this activity in both mammalian and avian systems. In neurons cultured from the neostriatum and cerebral cortex of 14-day mouse embryo and maintained in culture until maturation, we identified specific proteins that became phosphorylated on addition of [γ-^{32}P]ATP to the medium. These proteins were different from those detected when labeling intracellular ATP pools.[33] In more recent studies we have used primary neurons isolated from the telencephalon of 7-day chick embryo (Hogan, Pawlowska, and Ehrlich, in preparation). The proteins phosphorylated in these neurons after the addition of radioactive ATP to the extracellular medium are shown in the lanes marked PN of FIGURE 3. Among these, we observed a duplex of phosphorylated protein bands that migrated with an apparent molecular mass just below 14,000 daltons. We have found that the phosphorylation of these proteins by ectoprotein kinase activity is maximal prior to the period of rapid neurite extension in these cells, and that the activity declines upon the maturation of these neurons. The phosphorylation of these proteins by extracellular ATP is now a priority in our investigation on the role of ectoprotein kinase in neuronal development and differentiation.

In mature neurons, a potential target for regulation by ectoprotein kinase would be those neuronal activities that are initiated at the cell surface and then propagate to intracellular locations. One such major process is the reuptake of neurotransmitter molecules by adrenergic neurons. The reuptake of released norepinephrine in adrenergic neurons is carried out by a stereoselective, high-affinity, saturable transporter located in the plasma membrane.[37] ATP is coreleased with norepinephrine from adrenergic neurons.[29,31] The unhydrolyzable analogue of ATP, AppNHp, was found to inhibit norephinephrine uptake in rat brain synaptosomes.[38] This uptake inhibition was calcium dependent, and correlated with the inhibition by AppNHp of the endogenous phosphorylation of a specific protein component with apparent molecular mass of 39,000 daltons in these synaptosomal preparations.[38] These results suggested that the high-affinity uptake of norepinephrine may be regulated by calcium-dependent phosphorylation of a neuronal surface protein. More direct studies of these phenomena, which required investigation of a homogenous population of intact cells, were carried out with a cell culture model of sympathetic neurons—PC12 cells.[39] Using the specific uptake inhibitor nisoxetine, we have determined that the high-affinity uptake of norepinephrine in PC12 cells can proceed in the presence of either 1.2 mM $MgCl_2$ *or* 1.2 mM $CaCl_2$ as well as in the presence of both divalent cations. However, while uptake measured in the presence of $MgCl_2$ (without adding $CaCl_2$) was not affected by adenine nucleotides, norepinephrine uptake measured in the presence of $CaCl_2$ was significantly stimulated by adding 0.1 μM ATP to the extracellular medium (36% increase; $p < .01$). To gain some clues on the mechanism that may mediate this effect, we have used several adenine analogues in these assays. Our results are summarized in TABLE 2. The analogue 5′-adenylylimidodiphosphate (AppNHp) had no significant effect on the uptake. Because AppNHp is an effective agonist at ATP receptors, this result suggested that the effect may involve an ATP-hydrolyzing enzyme rather than a receptor-mediated activation. Potential enzymes are ecto-ATPase and ectoprotein kinase. The analogue adenosine-5′-*O*-(3-thiotriphosphate), namely ATP-γ-S, cannot be hydrolized effectively by ATPases, but was found to serve as a substrate for neuronal protein kinase activity.[40] As can be seen in TABLE 2, ATP-γ-S was even more effective than ATP in stimulating norepinephrine uptake in PC12 cells. These results suggest

TABLE 2. Norepinephrine Uptake in PC12 Cells: Effect of Adenine Nucleotides[a]

Additions to the Extracellular Medium	High-Affinity Uptake (Percentage of Control)
None (control)	100 ± 4
ATP	136 ± 9[b]
ATP-γ-S	155 ± 18[b]
AppNHp	104 ± 5
ADP	92 ± 8
Adenosine	90 ± 9

[a] High-affinity uptake of [^{3}H]norepinephrine was measured in PC12 cells resuspended in Hepes-Krebs-Ringer buffer without magnesium ions, as described previously.[39] ATP and each of the other adenine nucleotides listed above were added at a final concentration of 0.1 micromolar. Data derived from Hardwick *et al.*[39]

[b] $p < .01$.

that an ectoprotein kinase mediates the effect of extracellular ATP on norepinephrine uptake. The products of thiophosphorylation with ATP-γ-S are known to resist phosphoprotein-phosphatase activity. Therefore, involvement of a protein kinase will render the effect of ATP-γ-S irreversible. Indeed, we found that washing the cells and incubating them for an additional 30 min prior to the uptake assay eliminated the effect of preincubation with ATP. On the other hand, extensive washing did not inhibit the stimulation of the calcium-dependent norepinephrine uptake in PC12 cells obtained by preincubation with ATP-γ-S. This result lends further support for the involvement of ectoprotein kinase in this regulation. In preliminary studies, we have detected in PC12 cells incubated with extracellular [γ-^{32}P]ATP a calcium-dependent phosphorylation of a protein with electrophoretic mobility close to that seen previously in synaptosomes. The surface phosphorylation of this protein has been implicated in uptake regulation.[38] The involvement of this surface phosphoprotein in the regulation of norepinephrine uptake by extracellular ATP can now be directly investigated in a homogenous population of PC12 cells. Based on this study we recommend the comparison of the effects of ATP, AppNHp, and ATP-γ-S as a very useful first step in investigating potential functions of ectoprotein kinases.

The involvement of protein phosphorylation activity in processes that induce long-lasting alteration in neuronal function and in mechanisms underlying synaptic plasticity has been extensively investigated.[4,41] A potential role of extracellular protein phosphorylation in these processes has been suggested based on the fate of released ATP in the synaptic cleft during repetitive, high-frequency stimulation.[11] One of the consequences of such repetitive stimulation would be the accumulation of extracellular ATP in the synaptic cleft to levels that exceed 100 μM.[42,43] Ectoprotein kinase activity with an apparent K_m for ATP in this concentration range has been found in neural cells.[44] Moreover, extracellular ATP at the very same concentration was found to cause a half-maximal increase of the intracellular level of free calcium ions in these cells.[44] Lynch *et al.*[45] have presented evidence that the increased intracellular free calcium levels induced in neurons by high-frequency, repetitive stimulation plays a key role in the chain of events leading to increased synaptic efficacy in the paradigm of long-term potentiation.[46] In a recent study, induction of long-term potentiation by extracellular ATP has been directly demonstrated.[47] It should now be investigated whether the ectoprotein kinase that can be activated uniquely by the ATP concentrations that are produced in the synaptic cleft during repetitive stimulation could help cause the induction of long-term potentiation. Involvement of ectoprotein kinase activity in neuronal differentiation has also been proposed by Tsuji *et al.*,[48] who found that the ganglioside GQ1b uniquely stimulates the phosphorylation of surface proteins in the human neuroblastoma cell line, GOTO.

ECTOPROTEIN KINASE AND ENDOGENOUS SUBSTRATES AT THE SURFACE OF HUMAN PLATELETS

Platelets store ATP within vesicles called dense granules, and release it upon stimulation of surface receptors by agonists such as thrombin and collagen. Thus, surface protein phosphorylation may be involved in the feedback regulation of platelet function by released ATP during aggregation and clot formation. We have reported on the activity of ectoprotein kinase at the surface of human platelets,[49] and have pointed out that compared to the surface protein phosphorylation observed to date in other cells, the rapid kinetics of this activity in human platelets is unique.

Multiple substrates of intracellular protein kinases can be detected in platelets incubated for 30-60 min with radioactive inorganic phosphate (see the lane marked $^{32}P_i$ in FIG. 4). Several proteins become labeled in platelets incubated with radioactive ATP for 5-10 min (results not shown). However, when washed human platelets suspended in Tyrode's buffer are incubated with [γ-^{32}P]ATP for 5-15 sec, the phosphorylation of only one protein component can be detected (see the lane marked AT^{32}P in FIG. 4). When the intracellular ATP pools of nonstimulated platelets are labeled with $^{32}P_i$, this protein is not phosphorylated (FIG. 4). Furthermore, we have found that the fast phosphorylation of this surface protein in human platelets by ectoprotein kinase is followed by very rapid dephosphorylation.[50] This unique, very fast kinetics of phosphorylation-dephosphorylation of a surface protein suggests a role for this system in the regulation of initial events occurring during platelet activation.

EXOPROTEIN KINASES

Exoenzymes are defined as activities that are present in the extracellular environment without being directly associated with cells, and that can be isolated as soluble proteins. A surface-located protein kinase, when it is released from the cell to the medium by an exogenous substrate,[24] can be referred to as an exoprotein kinase. Exoenzymes may occur naturally in the extracellular environment and act upon cell surfaces. A soluble protein kinase purified from blood plasma was shown to catalyze the phosphorylation of the collagen receptor on the platelet surface, and this phosphorylation enhanced the responsiveness of the platelets to collagen.[51] Exoenzymes

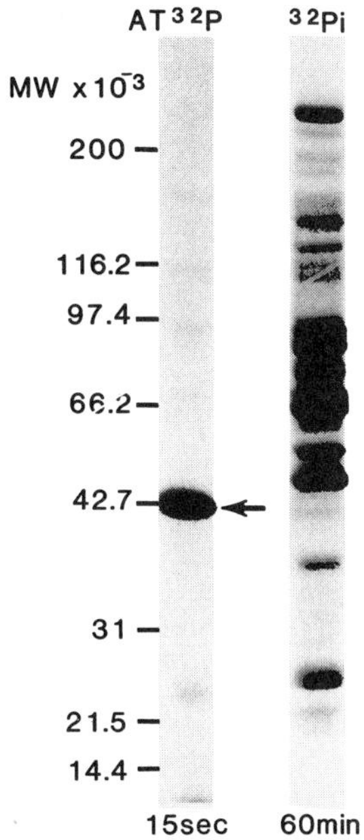

FIGURE 4. Extracellular and intracellular protein phosphorylation in human platelets. Platelets were sedimented from human platelet-rich plasma (PRP) and washed extensively as indicated previously.[49] Platelet suspension (10^9 per ml) in Tyrode's buffer were incubated at 37 °C with [γ-^{32}P]ATP (10 μM; 10 μCi per reaction) or with inorganic $^{32}P_i$ (1.0 mCi per ml in phosphate-free Tyrode's buffer). Reactions were stopped by solubilization in sodium dodecyl sulfate, and electrophoresis was carried out in a 5-15% polyacrylamide gel gradient. The autoradiograph shown here demonstrates the substrates of intracellular protein kinases detected in platelets incubated for 60 min with inorganic $^{32}P_i$ (right lane), and the major, early substrate of the platelet ectoprotein kinase, which becomes labeled when platelets are incubated with radiolabeled [^{32}P]ATP for 15 sec (left lane). Note that in nonstimulated platelets the protein indicated by the arrow is phosphorylated extracellularly but not intracellularly.

may also be stored within secretory vesicles and released to the extracellular environment only upon cell stimulation and the induction of active exocytosis. Our findings suggest that an exoprotein kinase of this nature is associated with human platelets.[52] When washed human platelets were stimulated with thrombin and the aggregated cells removed by centrifugation, a potent casein kinase activity was detected in the soluble fraction (100,000 × g supernatant) prepared from the extracellular medium. After 20 purified plasma proteins were tested as substrates for the exoprotein kinase released from platelets, it was found that the highest phosphorylation was obtained with the coagulation protein called Factor V (and its activated form, Factor V_a).[52] Another potent substrate for this enzyme was found to be the plasma protein fibrinogen, whose function is to bridge aggregating platelets. Interestingly, early in the processes of platelet aggregation and blood coagulation, both fibrinogen and Factor V bind to specific, high-affinity receptors at the platelet surface. At these binding sites they may be phosphorylated by ectoprotein kinase using ATP secreted from the stimulated platelets. The finding of ecto- and exoprotein kinases associated with platelets thus opens for investigation a novel regulatory mechanism in hemostatis.

SUMMARY AND CONCLUSIONS

The reversible phosphorylation of intracellular proteins has been established as a key regulatory mechanism in numerous cellular functions. In this process the enzyme protein kinase transfers the γ-phosphate of ATP to form a covalent bond with specific proteins. Another line of investigation has demonstrated that extracellular ATP is a potent physiological regulator in various cellular systems. Although many of the physiological effects of extracellular ATP were shown to be mediated by the action

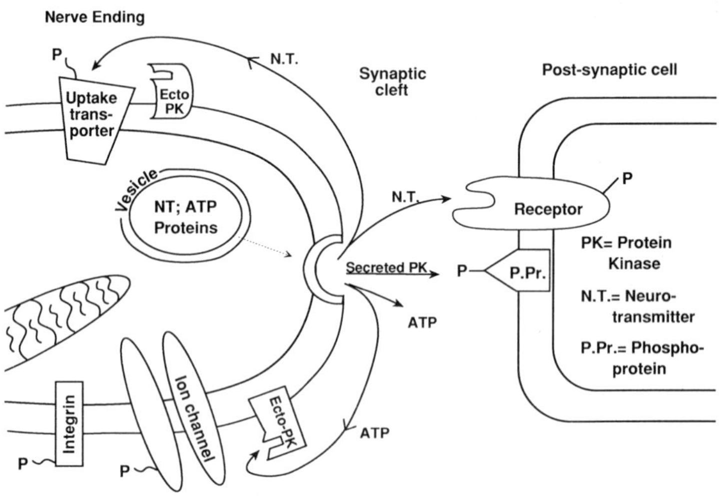

FIGURE 5. Potential roles of extracellular protein phosphorylation systems in synaptic function. Based on the findings obtained to date we have formulated the working hypothesis illustrated above. Accordingly, two types of extracellular protein kinase may operate in the synaptic cleft: a membrane-bound ectoprotein kinase and a soluble exoprotein kinase. The latter may be one of the proteins present in synaptic vesicles and coreleased with neurotransmitter molecules and ATP upon stimulation. Phosphorylation of surface sites in ion channels could provide feedback control over the activity of the releasing nerve ending. Phosphorylation of receptors and other surface proteins in target cells may play a role in the regulation of neuronal responsiveness to hormonal stimulation. The surface phosphorylation of cell-adhesion molecules and of integrins could regulate cell-cell interactions and synaptogenesis during neuronal development, and influence adaptive processes underlying synaptic plasticity. The preparation of antibodies that interact specifically with the phosphorylated sites in each of the surface phosphoproteins found in neuronal cells will provide the tools needed for direct experimental testing of these hypothetical relationships.

of purinergic receptors, it is possible that extracellular protein phosphorylation systems are also implicated in the mechanisms underlying the responsiveness of cells to extracellular ATP. The identification of ectoprotein kinases at the surface of various cells has provided evidence for the existence of such mechanisms, and revealed how the regulatory powers of protein phosphorylation systems can extend to the extracellular environment. The versatile roles that extracellular protein phosphorylation activity may play in the regulation of cellular functions is underscored by the presence of multiple protein substrates for this activity at the cell surface. Each such surface phosphoprotein may have a unique function. FIGURE 5 depicts the hypothetical relationships between the extracellular ATP secreted by exocytosis and the specific physiological function of these secreting neurons. Based on findings described in this article, we propose that extracellular ATP can be utilized by two types of extracellular protein kinase: a membrane-bound ectoprotein kinase, and a soluble exoprotein kinase. The exoprotein kinase can originate by detachment of an ectokinase from the cell surface, or be an intravesicular protein that is coreleased with ATP by exocytosis from stimulated cells. Phosphorylation of specific proteins at the surface of a secreting cell may have an important feedback control over its own presynaptic activity. The ectoprotein kinase could exert this feedback regulation by phosphorylating ion channels involved in secretion, and/or by phosphorylating transporters that carry out the reuptake of released transmitter molecules. Phosphorylation of receptors can regulate intercellular communication, and phosphorylation of integrins could regulate the interaction of the cell surface with components of the extracellular matrix. Although most of the relationships suggested in FIGURE 5 are still hypothetical, it should be possible to test them experimentally in a direct manner by raising antibodies against the phosphorylated sites of specific surface phosphoproteins. The ability of such antibodies to inhibit protein phosphorylation without penetrating the cells provides an experimental paradigm for the direct testing of potential physiological functions of ecto- and exoprotein kinase activities in a variety of cells.

REFERENCES

1. KREBS, E. G. & J. A. BEAVO. 1979. Annu. Rev. Biochem. **48:** 923-959.
2. HANKS, S. K., A. M. QUINN & T. HUNTER. 1988. Science **241:** 42-52.
3. HUNTER, T. 1987. Cell **50:** 823-829.
4. EHRLICH, Y. H., R. H. LENOX, E. KORNECKI & W. O. BERRY, EDS. 1987. Molecular Mechanisms of Neuronal Responsiveness. Plenum. New York, NY.
5. SIBLEY, D. R., J. L. BENOVIC, M. G. CARON & R. J. LEFKOWITZ. 1987. Cell **48:** 913-922.
6. GORDON, J. L. 1986. Biochem. J. **233:** 309-319.
7. KREUTZBERG, G. W., M. REDDINGTON & H. ZIMMERMANN, EDS. 1986. Cellular Biology of Ectoenzymes. Springer-Verlag. Berlin.
8. KARNOVSKY, M. L. 1986. *In* Cellular Biology of Ectoenzymes. G. W. Kreutzberg, M. Reddington & H. Zimmerman, Eds.: 3-13. Springer-Verlag. Berlin.
9. EHRLICH, Y. H. & A. ROUTTENBERG. 1974. FEBS Lett. **45:** 237-243.
10. EHRLICH, Y. H., T. B. DAVIS, E. BOCK, E. KORNECKI & R. H. LENOX. 1986. Nature **320:** 67-70.
11. EHRLICH, Y. H. 1987. Adv. Exp. Med. Biol. **221:** 187-199.
12. AGREN, G. & G. RONQUIST. 1974. Acta Physiol. Scand. **92:** 430-432.
13. SCHLAGER, E. J. & G. KOHLER. 1976. Nature (London) **2260:** 705-707.
14. MASTRO, A. M. & E. ROZENGURT. 1976. J. Biol. Chem. **251:** 7899-7906.
15. RONQUIST, G., G. AGREN, S. EKLUND & C. WERSTEDT. 1977. Upsala J. Med. Sci. **82:** 1-5.
16. REMOLD-O'DONNELL, E. 1978. J. Exp. Med. **148:** 1099-1104.

17. CHIANG, T. M., E. S. KANG & A. H. KANG. 1979. Arch. Biochem. Biophys. **195:** 518-525.
18. KANG, E. S., R. E. GATES & D. M. FARMER. 1978. Biochem. Biophys. Res. Commun. **83:** 1561-1569.
19. SOMMARIN, M., T. HENRIKSSON & B. JERGIL. 1981. FEBS Lett. **127:** 285-289.
20. EMES, C. H. & N. CRAWFORD. 1982. Biochim. Biophys. Acta **717:** 98-104.
21. AMANO, F., T. KITAGAWA & Y. AKAMATSU. 1984. Biochim. Biophys. Acta **803:** 163-173.
22. HALDAR, S. & G. C. MAJUMBER. 1986. Biochem. Biophys. Acta **887:** 291-303.
23. KUBLER, D., W. PYERIN & V. KINZEL. 1982. J. Biol. Chem. **257:** 322-329.
24. KUBLER, D., W. PYERIN, E. BUROW & V. KINZEL. 1983. Proc. Natl. Acad. Sci. USA **80:** 4021-4025.
25. PYERIN, W., E. BUROW, K. MICHAELY, D. KUBLER & V. KINZEL. 1986. Biol. Chem. Hoppe-Seyler. **368:** 215-227.
26. KUBLER, D., W. PYERIN, M. FEHST & V. KINZEL. 1986. *In* Cellular Biology of Ectoenzymes. G. W. Kreutzberg, M. Reddington & H. Zimmerman, Eds.: 191-204. Springer-Verlag. Berlin.
27. KUBLER, D., W. PYERIN, O. BILL, A. HOTZ, J. SONKA & V. KINZEL. 1989. J. Biol. Chem. **264:** 14549-14555.
28. HOLTON, F. A. & P. HOLTON. 1954. J. Physiol. **126:** 124-140.
29. GEFFEN, L. B. & B. C. LIVETT. 1971. Physiol. Rev. **51:** 98-157.
30. SILINSKY, E. M. & J. I. HUBBARD. 1973. Nature **243:** 404-405.
31. WHITE, T. D. 1977. Nature **267:** 67-68.
32. RICHARDSON, P. J. & S. J. BROWN. 1987. J. Neurochem. **48:** 622-630.
33. ZHANG, J., E. KORNECKI, J. JACKMAN & Y. H. EHRLICH. 1988. Brain Res. Bull. **21:** 459-464.
34. HAMPRECHT, B., T. GLASER, G. REISER, E. BAYER & F. PROST. 1985. Methods Enzymol. **109:** 316-341.
35. LYLES, J. M., D. LINNEMANN & E. BOCK. 1984. J. Cell Biol. **99:** 2082-2091.
36. GREENE, L. A. & A. S. TISCHLER. 1976. Proc. Natl. Acad. Sci. USA **73:** 2424-2428.
37. HENDLEY, E. D. 1984. *In* Handbook of Neurochemistry. 2nd edit. A. Lajtha, Ed. Vol. 6: 411-429. Plenum. New York, NY.
38. HENDLEY, E. D., S. R. WHITTEMORE, J. E. CHAFFEE & Y. H. EHRLICH. 1988. J. Neurochem. **50:** 263-273.
39. HARDWICK, J. C., Y. H. EHRLICH & E. D. HENDLEY. 1989. J. Neurochem. **53:** 1512-1518.
40. WHITTEMORE, S. R., S. G. GRABER, R. H. LENOX, E. D. HENDLEY & Y. H. EHRLICH. 1984. J. Neurochem. **42:** 1685-1696.
41. ROUTTENBERG, A. 1986. Prog. Brain Res. **69:** 211-244.
42. SILINSKY, E. M. 1975. J. Physiol. **247:** 145-162.
43. SILINSKY, E. M. 1984. J. Physiol. **346:** 243-256.
44. EHRLICH, Y. H., R. M. SNIDER, E. KORNECKI, M. G. GARFIELD & R. H. LENOX. 1988. J. Neurochem. **50:** 295-301.
45. LYNCH, G., J. LARSON, S. KELSO, G. BARRIONUEVO & F. SCHOTTLER. 1983. Nature **305:** 719-721.
46. TYLER, T. J. & P. DISCENNA. 1984. Brain Res. Rev. **7:** 15-28.
47. WIERASZKO, A. & T. N. SEYFRIED. 1989. Brain Res. **491:** 356-359.
48. TSUJI, S., T. YAMASHITA & Y. NAGAI. 1988. J. Biochem. (Japan) **104:** 498-503.
49. EHRLICH, Y. H. & E. KORNECKI. 1987. *In* Mechanisms of Signal Transduction by Hormones and Growth Factors. M. C. Cabot & W. L. McKeehan, Eds.: 193-204. Alan R. Liss. New York, NY.
50. NAIK, U., E. KORNECKI & Y. H. EHRLICH. 1990. Submitted for publication.
51. CHIANG, T. M., WOJCIKIEWICZ, A. H. KANG & J. N. FAIN. 1988. Throm. Res. **50:** 519-731.
52. EHRLICH, Y. H., E. KORNECKI, R. JENNY, C. S. CIERNIEWSKI & K. G. MANN. 1986. Blood **68**(Suppl. 1): 315a.

Affinity Labeling of Adenine Nucleotide Sites in Enzymes

ROBERTA F. COLMAN, JEROME M. BAILEY, DIANNE L. DECAMP, YU-CHU HUANG, AND SARA H. VOLLMER

Department of Chemistry and Biochemistry
University of Delaware
Newark, Delaware 19716

A goal of many biochemists is to identify and then to define the role of particular amino acids within the active and allosteric sites of enzymes. One approach is to chemically modify critical amino acids and to rely on the specificity of the enzyme for its substrate or normal regulatory compound to limit the extent of chemical modification to the region of the active or allosteric sites. In this approach, a reagent is designed that can react irreversibly with amino acid residues of the enzyme, but that is in addition structurally analogous to the substrate. Such a compound may mimic the substrate in forming a reversible enzyme inhibitor complex at a particular site on the enzyme; the reactive functional group of the reagent may then form a covalent bond within the active site during the existence of this complex. This is the strategy termed affinity labeling, which can potentially lead to specific but irreversible attack within purified enzymes, or even of particular enzymes or receptors when present in a mixture of proteins.

Examples of such compounds are the fluorosulfonylbenzoyl derivatives of nucleosides[1] shown in FIGURE 1. Compound a is 5′-*p*-fluorosulfonylbenzoyl adenosine (5′-FSBA), which is prepared by reaction of *p*-fluorosulfonylbenzoyl chloride with adenosine. This compound might reasonably be considered as an analogue of ADP, ATP, or NADH. In addition to the adenine and ribose moieties, it has a carbonyl group adjacent to the 5′-position, which is structurally similar to the first phosphoryl group of the naturally occurring purine nucleotides. If the molecule is arranged in an extended conformation, the sulfonyl fluoride moiety may be located in a position analogous to the terminal phosphate of ATP or to the ribose proximal to the nicotinamide ring of NADH. This sulfonyl fluoride moiety can act as an electrophilic agent in covalent reactions with several classes of amino acids, including tyrosine, lysine, histidine, serine, and cysteine.

In structure b, 5′-*p*-fluorosulfonylbenzoyl guanosine (5′-FSBG), guanine replaces the adenine moiety in the first derivative. It might be anticipated that this purine nucleotide alkylating agent would be specifically directed toward GTP sites in proteins.

Structure c is the fluorescent compound 5′-*p*-fluorosulfonylbenzoyl-1,N^6-ethenoadenosine (5′-FSBεA). This nucleotide analogue, with a fluorescence emission maximum at 412 nm, may provide a means of introducing a covalently bound fluorescent probe into nucleotide sites in proteins.

The final compound is a new bifunctional affinity label, 5′-*p*-fluorosulfonylbenzoyl-8-azidoadenosine (5′-FSBAzA), containing both an electrophilic fluorosulfonyl moiety and a photoactivatable azido group.[2] Following stoichiometric incorporation of reagent through the fluorosulfonyl at a specific site, photolysis of the tethered molecule can lead to reaction with amino acids adjacent to the residue that is initially labeled. This two-step reaction can help to elucidate the tertiary structure of the enzyme in the region of the nucleotide site.

A different class of nucleotide analogue is illustrated in FIGURE 2. These are the 2-, 6-, and 8-bromodioxonucleotide analogues of adenosine 5′-diphosphate.[3] These compounds have the adenine, ribose, and 5′-diphosphate of natural ADP. They are water soluble and negatively charged at neutral pH. The bromoketo group is potentially reactive with several nucleophiles found in proteins, including cysteine, histidine,

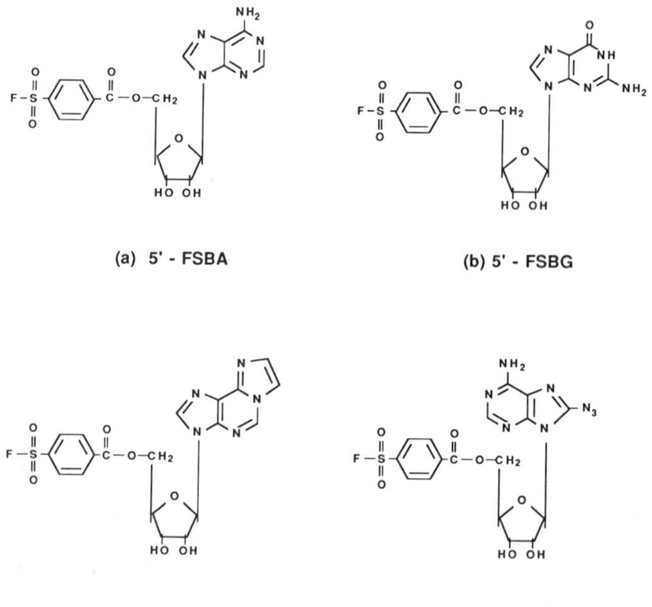

FIGURE 1. Fluorosulfonylbenzoyl nucleosides: (**a**) 5′-*p*-fluorosulfonylbenzoyl adenosine; (**b**) 5′-*p*-fluorosulfonylbenzoyl guanosine; (**c**) 5′-*p*-fluorosulfonylbenzoyl-1,N^6-ethenoadenosine; and (**d**) 5′-*p*-fluorosulfonylbenzoyl-8-azidoadenosine.

tyrosine, lysine, glutamic acid, and aspartic acid. In addition, the dioxo group lends the possibility of reaction with arginine residues. Because of the location of the functional group adjacent to the 2-, 6-, and 8-positions, these compounds might be expected to react with amino acid residues accessible from the purine region of the ADP binding site of enzymes. In this paper, bromodioxobutyl nucleotides will be used to illustrate the types of studies that can be conducted with purine nucleotide affinity labels.

Isocitrate dehydrogenase catalyzes the oxidative decarboxylation of isocitrate to yield α-ketoglutarate and carbon dioxide. The NAD-specific enzyme is regulated by ADP, which activates the enzyme by lowering the K_m for isocitrate without changing the maximum velocity.

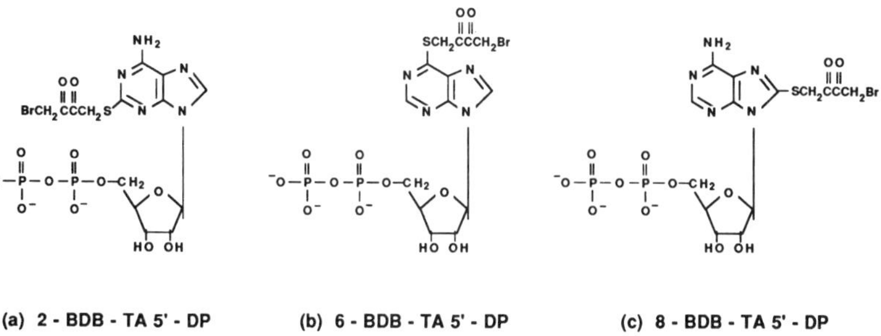

FIGURE 2. Bromodioxobutyl nucleotide affinity labels: (**a**) 2-[4-bromo-2,3-dioxobutylthio]adenosine 5′-diphosphate; (**b**) 6-[4-bromo-2,3-dioxobutylthio]adenosine 5′-diphosphate; and (**c**) 8-[4-bromo-2,3-dioxobutylthio]adenosine 5′-diphosphate.

Incubation of isocitrate dehydrogenase with 0.4 mM 2-[4-bromo-2,3-dioxobutylthio]adenosine 5′-diphosphate 2-BDB-TADP (FIG. 2a) causes a time-dependent inactivation, whereas in the absence of reagent no change is observed.[4] When 0.2-0.4 mM 2-BDB-TADP is present, a limiting value of 25% of the original activity is reached, after which no further decrease is observed. These results suggest that the reaction does not modify a group essential for catalysis. In order to calculate a pseudo first order rate constant for the decrease of enzymatic activity by 2-BDB-TADP, the limiting activity is subtracted from the activity measured at each time point, and a semilogarithmic plot is generated, allowing the calculation of a pseudo first order rate constant.[4]

When isocitrate dehydrogenase was incubated with 2-BDB-TADP over the concentration range 0.02-0.6 mM, a nonlinear dependence of the rate constant on the reagent concentration was observed. This observation is consistent with the initial formation of a reversible enzyme-reagent complex prior to the irreversible modification, as is expected for an affinity label. From a plot of $1/k_{obs}$ versus 1/[reagent], the maximum rate constant is estimated as 0.021 min^{-1}, with a K_I value of 67 μM.

TABLE 1 records the effect of substrate and nucleotides on the rate constant for inactivation of isocitrate dehydrogenase by 0.2 mM 2-BDB-TADP. Addition of the coenzyme, NAD, to the reaction mixture does not decrease the rate of inactivation; and the substrate isocitrate causes only a 2-fold reduction in the rate constant. The most striking effect is produced by the allosteric activator ADP. The addition of 1 mM ADP results in total protection against the inactivation by 2-BDB-TADP. These data indicate that the inactivation caused by 2-BDB-TADP may be attributed to modification at the ADP allosteric site of the enzyme. Conformational changes in the

TABLE 1. Effect of Substrate and Nucleotides on Rate of Inactivation of NAD-Specific Isocitrate Dehydrogenase by 0.2 mM 2-BDB-TADP

Additions to Reaction Mixture	$k \times 10^3$ (min^{-1})
None	18.6
2 mM NAD	17.9
1 mM Isocitrate	8.0
1 mM ADP	0

catalytic site upon covalent binding of reagent to the ADP site could be responsible for the large decrease in enzymatic activity.

The incorporation of 2-BDB-TADP was measured at various times up to 120 min during the incubation of isocitrate dehydrogenase with 0.4 or 0.6 mM reagent. Approximately one mol of 2-BDB-TADP is incorporated per mol of average enzyme subunit when the enzyme is maximally inactivated, suggesting that this is a limited, specific covalent modification.

The effect of the allosteric activator ADP is to lower the K_m for isocitrate. This is shown in TABLE 2, line 1, for native enzyme: the K_m for total isocitrate is 1.8 mM, and this value is decreased to 0.3 mM in the presence of 1 mM ADP. In contrast, for 2-BDB-TADP-modified enzyme, the K_m for total isocitrate is about the same in the absence or presence of 1 mM ADP. The K_m for isocitrate is not altered by reaction with 2-BDB-TADP, but the ability of ADP to strengthen the enzyme-substrate affinity is abolished. These data suggest that 2-BDB-TADP reacts covalently at the allosteric ADP site and prevents binding of free ADP, but does not itself fulfill the structural requirements for activation by nucleotide.

With the goal of isolating the critical peptide that was modified by 2-BDB-TADP, we first reacted isocitrate dehydrogenase with nonradioactive reagent and removed the excess reagent. In order to introduce a radioactive label into the 2-BDB-TADP-enzyme complex, the modified enzyme was then incubated with tritiated $NaBH_4$ to reduce the keto groups of the reagent that was covalently linked to the protein. Following carboxymethylation and digestion with trypsin, the specific modified peptide was isolated by two successive HPLC steps using a C-18 column.[5] Gas phase sequencing yielded the following peptide:

I. Leu-Gly-Asp-Gly-Leu-Phe-Leu-Gln-CmCys-CmCys-Lys

The β-carboxylate of aspartate is the only nucleophile in the peptide capable of reaction with the bromodioxobutyl functional group of the affinity label. Although the phenylthiohydantoin derivative of aspartate was detected at cycle 3, radioactivity from the reduced incorporated reagent was also measured in aliquots from that cycle. These results suggested that the reagent was released from the aspartate during the acid treatment, which is part of the Edman reaction cycle. It is most likely that the carboxylate of aspartate reacts with 2-BDB-TADP to displace the bromide and form an ester linkage to the nucleotide analogue.

Use of the Scansim program to scan sequences in the Swiss Protein Sequence Data Bank for similarity to the peptide labeled in isocitrate dehydrogenase, that is, peptide I, yielded the selection of peptides listed in TABLE 3. The amino acids at each position of the five amino acid sequences selected for comparison are either identical or reflect conservative substitutions. At position 3, where isocitrate dehydrogenase has the

TABLE 2. Effect of Reaction of 2-BDB-TADP with Isocitrate Dehydrogenase on K_m for Isocitrate

	K_m for Isocitrate	
Enzyme Sample	Without ADP (mM)	With 1 mM ADP (mM)
Native enzyme	1.8	0.3
Modified enzyme	2.0	1.8

TABLE 3. Other Proteins with Sequences Similar to Isocitrate Dehydrogenase Peptide Labeled by 2-BDB-TADP

Protein	Amino Acid Sequence
NAD-isocitrate dehydrogenase	L - G - D - G - L - F
Thymidine kinase	I - G - D - T - L - F
Transforming tyrosine kinase	L - G - E - G - A - F
Carbon catabolite derepressing protein kinase	L - G - E - G - S - F
Adenylate cyclase	L - A - E - T - L - F
ATPase, protein 6	M - N - E - N - L - F

aspartate that is the target of 2-BDB-TADP, the other sequences have an acidic residue, either aspartate or glutamate. Position 1 is a hydrophobic residue (leucine, isoleucine, or methionine). Position 2 is usually a small residue (glycine or alanine). Position 4 is a small neutral residue (usually glycine or threonine). Position 5 is usually a hydrophobic residue, and position 6 is invariably phenylalanine. These proteins all have ATP or ADP binding sites, although the aspartate or glutamate residues have not previously been implicated in these functional sites.

I want to turn now to our studies on pyruvate kinase, the key glycolytic enzyme that catalyzes the transfer of a phosphoryl group from phosphoenol pyruvate to ADP yielding pyruvate and ATP as the products. Considerable information is available now on the structure of this enzyme. Complete amino acid sequences are known for several isozymes, and the amino acid sequence of the cat muscle pyruvate kinase has been fitted to the 2.6 Å resolution electron density map. Affinity labeling with reactive nucleotide analogues offers a complementary technique for investigating the active site as it exists when the enzyme is in solution.

The first analogue that we used is 2-bromodioxobutylthio-1,N^6-ethenoadenosine 5′-diphosphate (or 2-BDB-TεA 5′-DP), shown in FIGURE 3b. This reagent has all the features of the 2-BDB-thioadenosine derivative that was used for isocitrate dehydrogenase, but it is, in addition, fluorescent.[3] When excited at 302 nm, it exhibits a fluorescence emission peak at 420 nm. This compound thus provides the possibility of inserting a covalently bound fluorescent probe at a defined site of an enzyme.

Incubation of rabbit muscle pyruvate kinase with 100-600 μM 2-BDB-TεA 5′-DP resulted in a time-dependent inactivation of the enzyme. Biphasic inactivation kinetics were observed that could be described in terms of a fast initial phase of inactivation yielding a partially active enzyme with 25% residual activity, followed by a slower phase leading to totally inactive enzyme.

Both phases exhibit a nonlinear dependence of the rate constant of inactivation on the reagent concentration, indicative of the initial formation of a reversible enzyme-reagent complex prior to irreversible modification. A double reciprocal plot of $1/k_{obs}$ versus 1/[reagent] is linear for the fast phase, and yields values of $K_1 = 133$ μM and $k_{max} = 0.193$ min^{-1} at saturating concentrations of reagent. The slow phase has a k_{max} that is 10 times lower.

Pyruvate kinase was incubated with 200 μM 2-BDB-TεA 5′-DP, and the incorporation of reagent into the enzyme was measured at various times by quantitation of the organic phosphorus or fluorescence intensity.[6] TABLE 4 shows the results of incorporation as measured by the organic phosphorus at 80 min: when the enzyme is 90% inactivated, about 1.7 mol of reagent/mol enzyme subunit is incorporated. Including either ADP + Mn^{2+} or phosphoenol pyruvate + K^+ + Mn^{2+} in the reaction mixture caused a substantial reduction in the percentage of inactivation at

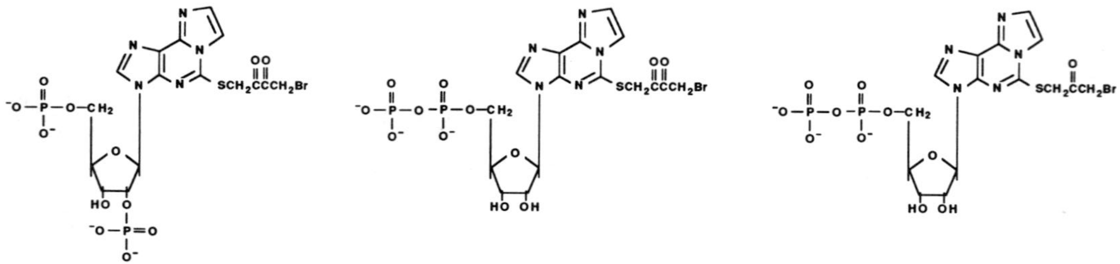

(a) 2 - BDB - T ε A 2', 5'- DP (b) 2 - BDB - T ε A 5'- DP (c) 2 - BOP - T ε A 5'- DP

FIGURE 3. Additional bromodioxobutyl nucleotide affinity labels: (**a**) 2-[4-bromo-2,3-dioxobutylthio]-1,N^6-ethenoadenosine 2′,5′-diphosphate; (**b**) 2-[4-bromo-2,3-dioxobutylthio]-1,N^6-ethenoadenosine 5′-diphosphate; and (**c**) 2-[3-bromo-oxopropylthio]-1,N^6-ethenoadenosine 5′-diphosphate.

80 min, as well as a decrease in the reagent incorporation to about 1 mol/mol enzyme subunit. These results indicate that 2-BDB-TεA 5′-DP reacts with two groups on the enzyme, one of which is at or near the active site.

In order to identify the amino acids that were reacting, modified enzyme was prepared by incubating pyruvate kinase with 2-BDB-TεA 5′-DP either in the absence or in the presence of phosphoenol pyruvate, K^+, and Mn^{2+}. The isolated enzymes, freed of reagent, were reduced with $[^3H]NaBH_4$ to introduce the radioactive tag; the enzymes were then carboxymethylated and digested with trypsin.

The tryptic digest was applied to a phenylboronate-agarose column, which forms a reversible complex with the *cis*-hydroxyl groups of the ribose of the nucleotide analogue. Nucleotidyl peptides are selectively adsorbed to the column in pH 8 buffer, while the many unmodified peptides are not retained. The diol-borate complex is dissociated by decreasing the pH or the salt concentration. The majority of the peptides was eluted with 50 mM ammonium bicarbonate buffer, pH 8.0, in the void volume. The column was subsequently washed with deionized water to elute the bound nucleotidyl peptides. This water peak, which contained 91% of the ethenoadenosine fluorescence and only a small A_{220nm} peak, was pooled and fractionated further by reverse-phase HPLC using a trifluoroacetic acid solvent system.

TABLE 4. Effect of Substrates on Inactivation of Pyruvate Kinase by 2-BDB-TεADP and on Reagent Incorporation by Enzyme

Additions to Reaction Mixture	Inactivation (%)	Incorporation (mol/mol subunit)
None	90	1.67
1.5 mM ADP + 2 mM $MnSO_4$	45	0.96
5 mM Phosphoenol pyruvate + 100 mM KCl + 2 mM $MnSO_4$	14	1.06

Tryptic digests from the inactive modified enzyme yielded two peptides with the following sequences:

II. Asn-Ile-Cys^{164}-Lys
III. Ile-Thr-Leu-Asp-Asn-Ala-Tyr^{147}-Met-Glu-Lys

Only peptide II was isolated from digests of pyruvate kinase that had been modified in the presence of phosphoenol pyruvate and had therefore retained most of its activity. Because peptide II is labeled equally whether the enzyme is active or inactive, it cannot be essential for activity. It is modification of peptide III that is associated with inactivation. The nonessential peptide II is a tetrapeptide that contains cysteine-164 as the target of the reagent. The important peptide III is a decapeptide in which tyrosine-147 is modified. Reaction of 2-BDB-TεA 5′-DP with this residue leads to inactivation.[6]

The structure of muscle pyruvate kinase in the crystalline state has been determined by Muirhead *et al.*[7] The active site is thought to lie in a pocket between Domain A and Domain B, with tyrosine-147 being located at the entrance to the active site. Specific reaction of tyrosine-147 with 2-BDB-TεA 5′-DP causes loss in activity. Cysteine-164 is an internal residue in Domain B. Its position outside the active site is consistent with its modification having little effect on catalysis.

As a second nucleotide analogue for probing the active site of pyruvate kinase, we synthesized 8-[4-bromo-2,3-dioxobutylthio]adenosine di- and triphosphates (8-BDB-TADP and 8-BDB-TATP, respectively), as illustrated in FIGURE 2c.[8] We thought that the placement of the bromoketo group adjacent to the 8-position of the adenine ring might allow the compound to react differently from 2-BDB-TεA 5′-DP. Furthermore, it is known that nucleotides featuring substituents at the 2-position of the purine ring exist predominantly in the *anti* conformation about the purine-ribose bond; whereas nucleotides with substituents in the 8-position exist predominantly in the *syn* conformation. Although the two nucleotide analogues were known to bind essentially to the same site on the enzyme, it seemed likely that different amino acid sidechains might be available to the bromoketo groups, which are tethered to either the 2-position or the 8-position of the purine ring.

Incubation of pyruvate kinase with 175 μM 8-BDB-TADP or 8-BDB-TATP at pH 7.0 and 25 °C caused a biphasic inactivation. The reactions of the two analogues are similar, but the rate constant for the triphosphate derivative is about three times greater than that of the diphosphate compound.[8]

Incorporation of reagent into the enzyme was measured at various times during the incubation by quantitation of the organic phosphorus. The triphosphate derivative is more specific, exhibiting lower incorporation for a higher degree of inactivation. At 80 min, 2.05 mol of 8-BDB-TATP per mol enzyme subunit is incorporated when the enzyme has lost 97% of its original activity.

Protection against inactivation by 8-BDB-TATP is provided by various combinations of substrates. In the presence of phosphoenol pyruvate + K^+ + Mn^{2+}, the enzyme loses only 15-20% of its activity, and the incorporation of reagent is reduced to about 1 mol/mol enzyme subunit. These results suggest that 8-BDB-TATP reacts with two groups on the enzyme, one of which is at or near the active site.

The residues modified by 8-BDB-TATP have been identified by reduction of the modified enzyme with [^{3}H]$NaBH_4$, carboxymethylation of the free cysteines, digestion with trypsin, and purification of the resultant peptides on a phenylboronate agarose column followed by HPLC.[9] The two cysteine-modified peptides that were identified in the digest of inactive enzyme modified by 8-BDB-TATP in the absence of substrates are shown below:

IV. Asn-Ile-Cys^{164}-Lys
V. Cys^{151}-Asp-Glu-Asn-Ile-Leu-Trp-Leu-Asp-Tyr-Lys

Cysteine-164 and cysteine-151 were the amino acid targets of this affinity label. Reaction in the presence of the protectants phosphoenol pyruvate, K^+, and Mn^{2+} yielded Asn-Ile-Cys-Lys as the only labeled peptide, with cysteine-164 as the labeled residue. This is the same residue modified by the 2-substituted nucleotide analogue under protecting conditions. Although cysteine-164 is clearly not required for activity and is internal, it appears to be favorably positioned to react with nucleotide analogues.

The major residue responsible for inactivation of pyruvate kinase by 8-BDB-TATP seems to be cysteine-151.[9] This result contrasts with the finding that, in the case of 2-BDB-TεA 5′-DP, it is modification of tyrosine-147 that is responsible for loss in enzyme activity. The distance between the α-carbons of cysteine-151 and tyrosine-147 in the crystal structure of pyruvate kinase is 8.9 Å. Because the target residues of the two affinity labels are approximately 9 Å apart, the question arises as to whether, if the nucleotide moieties were superimposed, the reactive -CH_2Br groups would be situated 9 Å apart.

The structure of 2-BDB-TεA 5′-DP is shown in FIGURE 4a in the *anti* conformation about the purine-ribose bond. In FIGURE 4b on the bottom, the structure of 8-BDB-TATP is pictured in the *syn* conformation about the purine-ribose bond. In the middle, the two structures are superimposed so that the ribose and phosphates coincide.

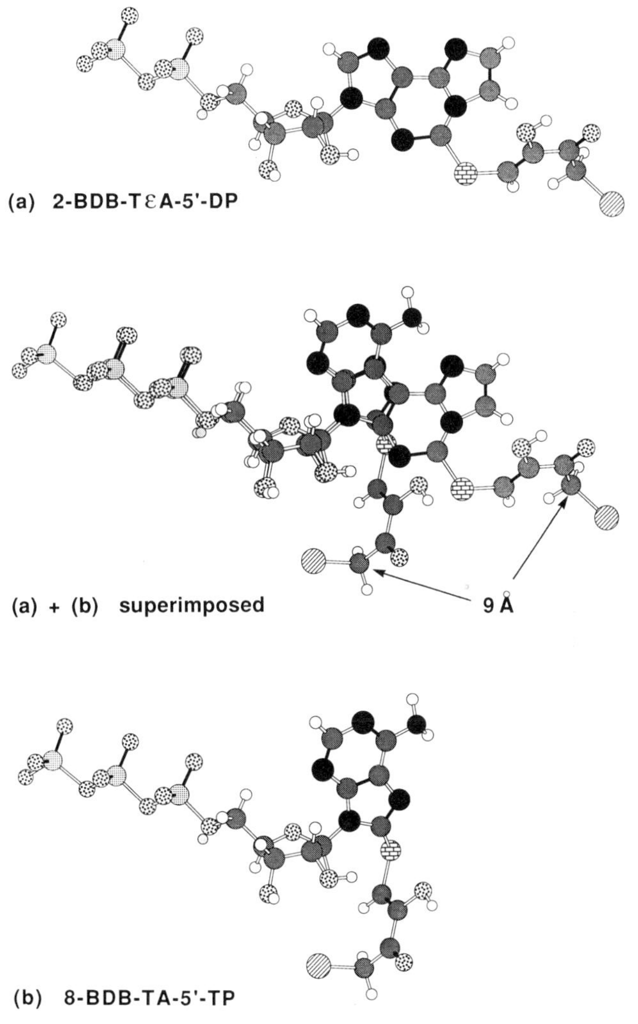

FIGURE 4. Comparison of the structures of (**a**) 2-BDB-TεA 5′-DP and (**b**) 8-BDB-TATP.

Although a range of distances between the two $-CH_2Br$ groups is possible, it *is* energetically permissible to arrange the compounds on the enzyme such that the distance between the reactive groups coincides with the 9 Å distance between cysteine-151 and tyrosine-147. If the purine, ribose, and phosphates bind similarly to the enzyme, the experimental results from affinity labeling are thus consistent with the atomic positions assumed by the enzyme in the crystalline form.

We have tried to present a representative sampling of the studies in which we use purine nucleotide affinity labels. We are hopeful that these various analogues will not only be useful for our own experiments, but that they will also be valuable to other laboratories in exploring nucleotide sites in a variety of proteins. The 5′-*p*-fluorosulfonylbenzoyl adenosine, which was the first of the reactive nucleotide analogues that we described, has already been found to yield specific labeling of NAD sites in several dehydrogenases and reductases.[10] It has also labeled ATP or ADP binding sites in many kinases and synthetases, in addition to providing an effective handle for examining an ADP receptor protein of platelet membranes. Furthermore, it has modified specific nucleotide sites in such diverse proteins as the ATPases, actin, myosin, luciferase, and oxoprolinase.[10] We anticipate that the fluorosulfonylbenzoyl derivatives of guanosine, ethenoadenosine, and azidoadenosine, as well as the new 2- and 8-bromodioxobutylthio ADP, will also have widespread applications to the elucidation of purine nucleotide sites in enzymes and receptors.

REFERENCES

1. COLMAN, R. F., 1983. Annu. Rev. Biochem. **52:** 67-91.
2. DOMBROWSKI, K. E. & R. F. COLMAN. 1989. Arch. Biochem. Biophys. **275:** 302-308.
3. COLMAN, R. F. 1989. Affinity Labelling. *In* Protein Function: A Practical Approach. T. E. Creighton, Ed. Chapter 4: 77-99. IRL Press. New York, NY.
4. HUANG, Y. C., J. M. BAILEY & R. F. COLMAN. 1986. J. Biol. Chem. **261:** 14100-14107.
5. HUANG, Y. C. & R. F. COLMAN. 1989. J. Biol. Chem. **264:** 12208-12214.
6. DECAMP, D. L. & R. F. COLMAN. 1989. J. Biol. Chem. **264:** 8430-8441.
7. MUIRHEAD, H., D. A. CLAYDEN, D. BARFORD, C. G. LORIMER, L. A. FOTHERGILL-GILMORE, E. SCHILTZ & E. SCHMITT. 1986. EMBO J. **5:** 475-481.
8. DECAMP, D. L., S. LIM & R. F. COLMAN. 1988. Biochemistry **27:** 7651-7658.
9. VOLLMER, S. H. & R. F. COLMAN. 1990. Biochemistry **29:** 2495-2501.
10. COLMAN, R. F. 1987. *In* Proteins. J. J. L'Italien, Ed.: 569-580. Plenum. New York, NY.

Extracellular ATP Causes Changes in Plasma Membrane Permeability of Mouse Lymphocytes[a]

FRANCESCO DI VIRGILIO,[b] ENZO PICELLO, VINCENZO BRONTE, PAOLA ZANOVELLO, AND DINO COLLAVO

University of Padua
Padua, Italy

Extracellular ATP (ATP_o) triggers a variety of responses in mast cells, macrophages, polymorphonuclear granulocytes, and lymphocytes: activation of secretion, inhibition of phagocytosis, generation of known intracellular second messengers, depolarization of plasma membrane potential, and permeabilization of the plasma membrane to low molecular weight aqueous solutes.[1-5] It is also documented that a prolonged exposure to ATP_o is eventually lytic to most cell types so far tested.[4,6,7] Cellular responses to ATP_o are thought to be mediated by surface receptors; the physiological role of such receptors is, however, still mysterious.

In two recent studies,[4,8] we have observed that lymphocytes with cytotoxic activity are highly insensitive to the lethal effects of ATP_o. While ATP_o caused depolarization of plasma membrane potential, a rise in $[Ca^{2+}]_i$, DNA breakdown, and eventually lysis in several lymphoid cells, none of these alterations was detectable in a cytotoxic T lymphocyte (CTL) clone (clone CHM-14), in the CTL line CTLL-L2, in lymphokine-activated killer (LAK) cells, or in peritoneal exudate-derived cytotoxic T lymphocytes. These serendipitous observations suggested to us that ATP_o had a role in cell-mediated cytotoxicity (FIG. 1).

The molecular mechanism(s) of CTL-mediated lysis is not yet understood, although several soluble cytolytic factors have been proposed as "killer" molecules, among them perforin being probably the most popular.[9] As an alternative to a fully passive role of the target in the lytic process, and in the light of the observation that one of the earliest alterations detectable during CTL-mediated lysis is disintegration of target DNA,[10] it has been proposed that the target itself collaborates in its own lysis by activating an endogenous "suicide" program.

ATP_o causes many of the cellular alterations described during cell-mediated lysis. Cells that are resistent to cytotoxic lysis are also resistent to ATP_o, and development of ATP_o resistance, when studied in LAK cells, correlates strictly with acquisition of cytotoxic activity and resistance to cell-mediated lysis.[4] Furthermore, it has also been

[a] This work was supported by the Ministry of Public Education of Italy (40% and 60%), the Consiglio Nazionale delle Ricerche, the Associazione Italiana per la Ricerca sul Cancro, and the Centro Interuniversitario per la Ricerca sul Cancro.

[b] Address for correspondence: Institute of General Pathology, University of Padua, Via Trieste 75, I-35131 Padua, Italy.

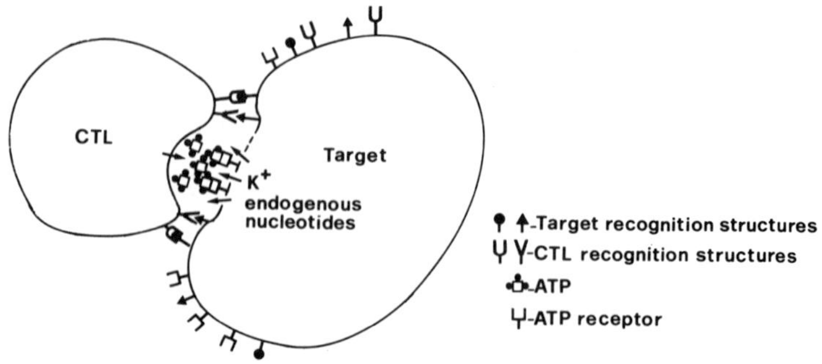

FIGURE 1. Schematic representation of a hypothetical lytic mechanism based on the release of ATP from a CTL engaged by its target. Binding of recognition structures on the CTL membrane triggers local ATP release in the pouch between the CTL and the target. Released ATP diffuses onto the target plasma membrane and activates specific ATP receptors. Ligation of these receptors causes lesions on the target plasma membrane and the efflux of endogenous monovalent cations and nucleotides. The CTL itself is immune to ATP_o because it lacks receptors for ATP_o. ATP diffusing out of the contact pouch would be quickly diluted and hydrolyzed by cell surface ecto-ATPases.

recently claimed that ATP is secreted by CTLs.[11] Finally, to underscore the danger of being exposed to a high concentration of ATP_o, it should be noted that most cells are equipped with powerful ecto-ATPases that would prevent ATP_o from accumulating in dangerous amounts in the vicinity of the surface of the plasma membrane.[7]

In view of these considerations, we suggest a role for ATP_o in cell-mediated cytotoxicity (Fig. 1).

REFERENCES

1. Cockcroft, S. & B. D. Gomperts. 1979. J. Physiol. (London) **296:** 229-243.
2. Sung, S.-S. J., J. D.-E. Young, A. M. Origlio, J. M. Heiple, H. R. Kaback & S. C. Silverstein. 1985. J. Biol. Chem. **260:** 13442-13449.
3. Dubyak, G. R. & M. B. De Young. 1985. J. Biol. Chem. **260:** 10653-10661.
4. Di Virgilio, F., V. Bronte, D. Collavo & P. Zanovello. 1989. J. Immunol. **143:** 1955-1960.
5. Steinberg, T. H., A. S. Newman, J. A. Swanson & S. C. Silverstein. 1987. J. Biol. Chem. **262:** 8884-8888.
6. Kitagawa, T., F. Amano & Y. Akamatsu. 1988. Biochim. Biophys. Acta **941:** 257-263.
7. Filippini, A., R. Taffs, T. Agui & V. Sitkovsky. 1990. J. Biol. Chem. **265:** 334-340.
8. Zanovello, P., V. Bronte, A. Rosato, P. Pizzo & F. Di Virgilio. 1990. J. Immunol.: in press.
9. Henkart, P. A. 1985. Annu. Rev. Immunol. **3:** 31-57.
10. Russel, J. H. 1983. Immunol. Rev. **72:** 77-92.
11. Filippini, A. & M. Sitkovsky. 1989. Paper presented at the International Conference on Molecular Aspects of Immune Responses. Rome.

Modulation of Vascular Tone and Hemostasis by Serotonin and ATP

GERALD SOSLAU, JANET PARKER,
ROBERT MARCUS, AND ISADORE BRODSKY

Departments of Biochemistry and Neoplastic Diseases
Hahnemann University
Philadelphia, Pennsylvania 19102

Platelet serotonin and ATP may participate significantly in the maintenance of vascular tone and hemostatic processes under normal and pathological conditions. We have previously shown that extracellular ATP may modulate platelet function by phosphorylation of surface proteins and increasing cAMP levels.[1] Bleeding times were found to correlate with platelet release of ATP and serotonin content in renal failure patients.[2] To further explore the function of serotonin and ATP within the circulatory system, we analyzed their effect on platelet aggregation and arterial contraction.

ATP and its nonhydrolyzable analogues, α,β-methylene adenosine 5′-triphosphate (α,β-ATP) and β,γ-methylene adenosine 5′-triphosphate (β,γ-ATP), were employed in these studies. ATP and its analogues (36-200 μM) inhibit collagen-induced platelet aggregations at similar levels in both whole blood and platelet-rich plasma (PRP) (TABLE 1), indicating that ATP and not a metabolite is active. Small sample size within many experimental groups accounts for several nonsignificant p values. A clear trend, however, is established. Similar results were observed with ADP-induced aggregations; however, the ATP analogues were not as effective as ATP and were virtually inactive with PRP (TABLE 1), indicating that a whole blood factor may be involved. ATP (180 μM) added to maximally ADP- or collagen-induced aggregated platelets partially reversed aggregation. Both α,β-ATP and β,γ-ATP reversed collagen-induced platelet aggregates to levels similar to those attained with ATP, but neither analogue was as effective as ATP with ADP-induced aggregates (data not shown). ATP and its analogues may act, in part, via signal transduction to increase platelet cAMP (TABLE 1).

Serotonin functions as a vasoconstrictor, vasodilator, and platelet agonist, depending upon its concentration and receptor type occupancy. At 1-30 μM, serotonin induced constriction of large (> 1 mm) and small (150-250 μM) rat pulmonary arteries (data not shown). ATP, ADP, and β,γ-ATP (16 nM-1.8 μM) relaxed whereas α,β-ATP enhanced serotonin-induced contraction of large arteries (FIG. 1A). The endothelial cells appear to participate in these responses since the effects of ADP and of ATP and its analogues were greatly diminished with the deendothelialized artery (as demonstrated by response to acetylcholine) (FIG. 1B). ATP and its analogues (20 nM-1.8 μM) enhanced serotonin-induced contraction of small arteries, with the α,β-analogue being most effective (FIG. 1C).

If an artery is injured, a cascade of events is elicited to rapidly prevent blood loss. Included in this process is the physiological activation of platelets by collagen and ADP resulting in the adhesion of the platelets to the damaged blood vessel wall and

TABLE 1. Effects of Extracellular ATP and ATP Analogues on Collagen- and ADP-Induced Platelet Aggregation and Platelet cAMP Content

	Percentage Inhibition of Agonist-Induced Aggregation[a]								
	ATP (μM)			β,γ-ATP (μM)			α,β-ATP (μM)		
	180	90	36	200	100	40	200	100	40
Collagen (42 μg/ml)									
Whole blood	59.1 ± 16.2 (7, <.001)	61.0 ± 13.4 (7, <<.001)	26.6 ± 23.8 (7, <.05)	52.3 ± 22.3 (7, <.001)	53.0 ± 30.1 (6, <.01)	44.3 ± 36.3 (6, <.05)	53.0 ± 32.9 (6, <.05)	64.2 ± 18.8 (5, <.01)	41.5 ± 27.9 (4, ~.05)
PRP	55.3 ± 22.4 (3, ~.05)	39.3 ± 39.9 (3, NS)	45.5 ± 30.4 (2, NS)	58.7 ± 22.5 (3, <.05)	41.3 ± 29.6 (3, NS)	30.3 ± 32.1 (3, NS)	50.0 ± 29.8 (3, NS)	32.3 ± 40.7 (3, NS)	38.3 ± 36.5 (3, NS)
ADP (25 μM)									
Whole blood	82.5 ± 20.1 (6, <<.001)	74.0 ± 23.0 (5, <.01)	45.4 ± 23.9 (5, <.05)	43.2 ± 31.3 (5, <.05)	20.0 ± 18.1 (4, NS)	15.8 ± 28.3 (4, NS)	29.8 ± 27.7 (4, NS)	21.3 ± 18.3 (4, NS)	18.0 ± 33.4 (4, NS)
PRP	54.8 ± 1.9 (4, <<.001)	23.0 ± 8.3 (4, ~.01)	4.0 ± 3.7 (4, NS)	13.5 ± 10.9 (4, NS)	2.3 ± 2.5 (3, NS)	0 ± 0 (2, NS)	7.5 ± 9.6 (4, NS)	1.7 ± 2.9 (3, NS)	0 ± 0 (2, NS)

Increase in Platelet cAMP Content[b]		
ATP (30 μM)	β,γ-ATP (30 μM)	α,β-ATP (30 μm)
1.31 ± 0.12 (5)	1.22 ± 0.05 (3)	1.31 ± 0.11 (3)

[a] Each value is a mean ± SEM and is accompanied by the *N* value and the *p* value in parentheses.

[b] Each increase is relative to the control value of 1.0. Each value is a mean ± SEM and is accompanied by the *N* value in parentheses.

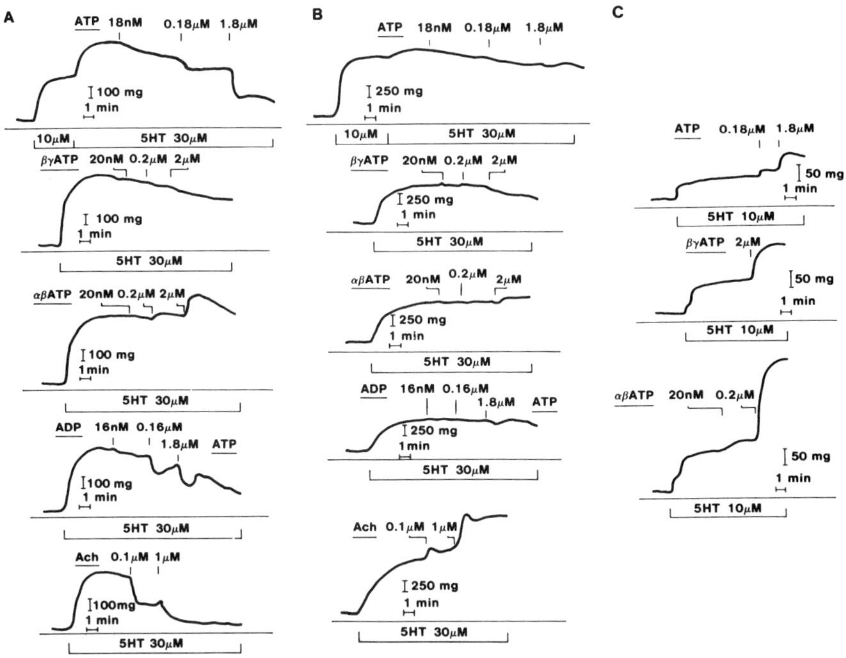

FIGURE 1. Modulation of serotonin-induced contraction of rat pulmonary artery by ADP, ATP, and ATP analogues. All responses were measured at 37 °C in K^+-free physiological salt solution in 20-ml volumes, aerated with 5% CO_2. Contractions were initiated with 10-30 μM serotonin followed by the addition of various concentrations of ADP, ATP, α,β-ATP, or β,γ-ATP. After each series of responses, the artery was relaxed back to baseline with physiological salt solution washes. (**A**) Experiments conducted with a 2-mm arterial fragment (>1 mm diameter) with an intact endothelium. (**B**) Experiments conducted with an arterial ring as in **A**, the ring in these experiments being deendothelialized by inversion and passage of air over tissue. (**C**) Experiments conducted with an arterial ring as in **A**, the ring in these experiments being <250 μM in diameter.

the release of ATP and serotonin. Our observed ATP-induced vasodilation of serotonin-contracted large arteries (>1 mm) could affect vessel architecture at the injury site and may be sufficient to cause vortexes, which have been shown to promote platelet aggregation and adhesion (clot formation).[3] In arteries less than 250 μM, ATP enhances serotonin-induced contraction. This would alter the geometry of the vessel wall and increase shear rates. This has been shown to enhance platelet adhesion to the vessel wall[3] and would promote hemostasis.

REFERENCES

1. Soslau, G. & J. Parker. 1989. Blood **74:** 984-993.
2. Soslau, G., I. Brodsky, B. Putatunda, J. Parker & A. B. Schwartz. 1990. Am. J. Hematol.: in press.
3. Turitto, V. T. 1982. *In* Progress in Hemostasis and Thrombosis. T. H. Spaet, Ed.: 139-177. Grune & Stratton. New York, NY.

Extracellular ATP Is a Mitogen for 3T3, 3T6, A431, DDT_1-MF2, BALB/MK, NIE-115, and HFF Cells[a]

L. A. HEPPEL, D. WANG, N. HUANG,
F. A. GONZALEZ, A. H. AHMED, R. G. ALFONZO,
AND M. DESHEESH

Section of Biochemistry
and
Molecular and Cell Biology
Biotechnology Building
Cornell University
Ithaca, New York 14853

Extracellular ATP, at 5-50 μM, had little or no mitogenic activity alone, but it was able to greatly enhance [^{3}H]thymidine incorporation in quiescent cultured cells treated with agents such as epidermal growth factor (EGF), phorbol 12-tetradecanoate 13-acetate (TPA), insulin, or platelet-derived growth factor (PDGF). Increases in the cell number and in the percentage of cell nuclei labeled with [^{3}H]thymidine were also observed. Similar responses were obtained with the ATP analogue, AMP-PNP, and with ADP, whereas UTP and ITP had little or no activity (TABLE 1).

Evidence that ATP and ADP were acting as nucleotides and not merely as a source of adenosine has been published.[1] Part of this evidence is based on the fact that ATP was a competence factor that needed to be present for only 45 min, during which time little or no adenosine was formed. Then it could be washed away and replaced by a progression factor such as insulin. Excellent stimulation of DNA synthesis resulted (FIG. 1).

Extracellular ATP stimulates a number of the early events observed when well-known growth factors were presented to mammalian cells (see Rozengurt[2] for a review). These early signals include the stimulation of the rate of Na^+, K^+, deoxyglucose, α-aminoisobutyric acid, and uridine entry and the enhancement of ornithine decarboxylase activity (in preparation; literature for ATP reviewed by Gonzalez *et al.*[3]).

In addition, studies from this and other laboratories[3] showed that ATP given to quiescent cells causes a transient elevation of cytosolic Ca^{2+} and stimulates Ca^{2+} efflux as well as the formation of inositol phosphates. Thus ATP resembles growth factors that activate inositol phospholipid metabolism and thereby bring about, in

[a]This work was supported by grants from the Cornell Biotechnology Program, the National Institutes of Health (DK11789), and the American Cancer Society (BC501A).

TABLE 1. Effect of Various Nucleotidies and Nucleosides on DNA Synthesis Stimulation by Growth Factors in Swiss 3T3 and 3T6 Fibroblasts[a]

Addition of Nucleotide or Nucleoside[b]	Increase in [^{3}H]Thymidine Incorporation[c]					
	Swiss 3T3 Fibroblasts			Swiss 3T6 Fibroblasts		
	No Supplement	TPA (20 ng/ml)	EGF (10 ng/ml)	No Supplement	Insulin (500 ng/ml)	PDGF (2.5 ng/ml)
None	1.0	12.0	8.1	1.0	14.1	15.2
			Low or No Synergism			
GTP	1.1	15.6	7.9	1.1	15.8	19.3
UTP	1.2	14.9	8.0	1.5	17.1	19.4
AMP	1.5	16.0	13.9	1.9	14.0	15.0
Adenosine	2.4	18.6	12.4	1.2	14.3	14.4
			Potent Synergism			
ATP	2.8	24.0	28.8	2.6	35.6	32.2
AMP-PNP	2.7	24.8	25.2	2.6	34.9	30.5
ADP	3.1	25.4	26.2	2.5	32.4	30.0

[a] The 3T3 and 3T6 cells were plated, made quiescent, and examined for [^{3}H]thymidine incorporation as described.[1]

[b] All nucleotides and nucleosides were added at 50 μM. Note good synergism with ATP, AMP-PNP, and ADP. (Synergism is here meant to indicate more than additive effects.) By contrast, adenosine was totally inactive in 3T6 cells when combined with insulin or PDGF.

[c] Each entry in the body of the table below this heading represents a fold-increase with respect to a control value. (Control values represent the incorporation of [^{3}H]thymidine into cultures incubated in serum-free growth medium alone—1235 cpm/dish for 3T3 cells; 1703 cpm/dish for 3T6 cells.) The subheadings indicate either the effect of nucleotides or adenosine with no further supplementation (No Supplement) or the effect of other growth factors (TPA, EGF, Insulin, or PDGF). The first line in the body of the table indicates stimulation achieved by other growth factors alone. Subsequent lines indicate the effects of various combinations.

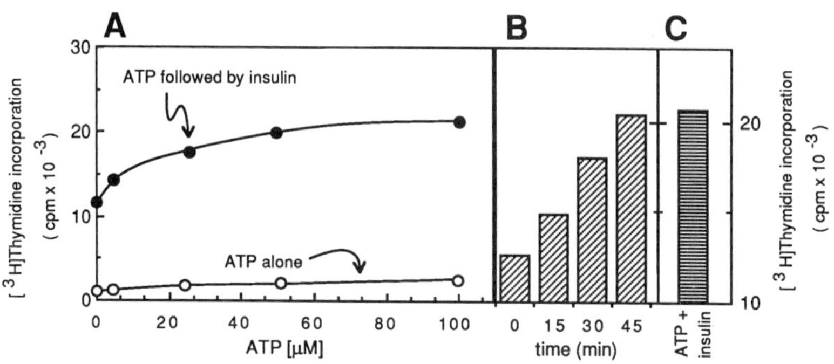

FIGURE 1. Extracellular ATP is a competence factor. In preparation for the two competence assays (**A** & **B**) and the one conventional assay (**C**) shown above, 3T3 mouse fibroblasts were plated, made quiescent, and examined for [^{3}H]thymidine incorporation as described.[1] (**A**) Cells were incubated at 37 °C in 1:1 DMEM:Waymouth serum-free medium with various levels of ATP, as shown, for 45 min. After ATP was removed by washing, the cells were again incubated, this time for 40 hr in the presence of 500 ng/ml insulin. (**B**) Similar, except that levels of ATP and insulin were fixed at 50 μM and 500 ng/ml, respectively, and the time of contact with ATP was varied as shown. (**C**) Cells were incubated with ATP (50 μM) plus insulin (500 ng/ml) for 40 hr. The stimulation of DNA synthesis compares favorably with that obtained with 500 ng/ml insulin plus 10 ng/ml EGF.

parallel fashion, activation of protein kinase C and elevation of cytosolic Ca^{2+}. However, recent results on 3T6 cells obtained in collaboration with Dr. E. Rozengurt[4] show that extracellular ATP does *not* activate protein kinase C. The fact that TPA, an activator of protein kinase C, showed powerful synergistic stimulation of DNA synthesis in the presence of ATP also suggests that ATP itself does not function by activating this kinase.[1] Furthermore, we have found that incubation of Swiss 3T6 cells with 10 ng/ml PDGF for 15 min caused an increase in the level of diacylglycerol; however, a similar incubation with 100 μM ATP did not stimulate the accumulation of diacylglycerol (in preparation). Finally, we observed that down-regulation of protein kinase C by prolonged preincubation of 3T3 or 3T6 cultures with phorbol esters did not prevent synergistic stimulation of DNA synthesis by ATP plus insulin. We conclude that ATP activates a pathway leading to DNA synthesis that remains to be elucidated. We are aware of no prior studies except for one report[5] in which thymocytes given ATP plus TPA resembled lectins in their mitogenic response.

REFERENCES

1. Huang, N., D. Wang & L. A. Heppel. 1989. Proc. Natl. Acad. Sci. USA **86:** 7904-7908.
2. Rozengurt, E. 1986. Science **234:** 161-166.
3. Gonzalez, F. A., R. G. Alfonzo, J. R. Toro & L. A. Heppel. 1989. J. Cell. Physiol. **141:** 606-617.
4. Gonzalez, F. A., E. Rozengurt & L. A. Heppel. 1989. Proc. Natl. Acad. Sci. USA **86:** 4530-4534.
5. El-Moatassim, C., J. Dornand & J. Mani. 1987. Biochim. Biophys. Acta **927:** 437-444.

ATP Effects on Secretion and Second Messenger Production in Bovine Chromaffin Cells[a]

K. T. KIM, M. DIVERSÉ-PIERLUISSI,
W. N. KOPELL, AND E. W. WESTHEAD

Program in Molecular and Cellular Biology
University of Massachusetts
Amherst, Massachusetts 01003

Cells of the adrenal medulla (chromaffin cells) cosecrete catecholamines and nucleotides, chiefly ATP. To investigate possible effects of ATP on secretion, it is important to minimize the unquantifiable effects of secreted ATP. We accomplish this by studying the effects of exogenous ATP on cultured cells in a constant stream of buffer that quickly dilutes and washes away secreted compounds. Cells are cultured on fibrinectin-coated quartz plates that become part of a 20-μl chamber through which buffer flows constantly. Stimulants are injected into the stream via an HPLC-type sample loop. The chamber is put into a fluorometer so that cytosolic Ca^+ transients can be monitored in fura-2-loaded cells. Secreted catecholamines are detected downstream electrochemically to measure secretion simultaneously with Ca^{2+} transients.

ATP alone (5-100 μM) will stimulate secretion. ATP also rapidly increases inositol triphosphate turnover and causes a rise in cAMP that plateaus in 15 min during continuous ATP perfusion. Cytosolic Ca^{2+} is elevated by extracellular ATP both in the presence and absence of extracellular Ca^{2+}. Most of the rise is from intracellular stores, but secretion requires extracellular Ca^{2+}. UTP elevates cytosolic Ca^{2+} but does not stimulate secretion in the presence or absence of extracellular Ca^{2+}.

When cells are stimulated for 5-10 sec by a nicotinic agonist, 1,1-dimethyl-4-phenylpiperazinium chloride (DMPP), or by elevated K^+, secretion is enhanced 30-50% by simultaneous addition of ATP. Enhancement of secretion by ATP is associated with an elevated cytosolic Ca^{2+} transient. Surprisingly, enhancement of secretion appears not to be simply additive secretion expected from ATP stimulation. Pretreatment of cells with cholera toxin (Ctx) has little or no effect on ATP-stimulated secretion but completely blocks the enhancing effect of ATP added with DMPP or K^+.

If cells are exposed to ATP *prior* to stimulation by DMPP or K^+, the enhancing effect gives way to an inhibitory effect that reaches a plateau at 30% inhibition within 3 min. The inhibitory, but not the enhancing effect of ATP, is mimicked by ADP, which is marginally more potent than ATP. Inhibition by ATP is seen instantly and increases over the same time course as the ATP effect.

Although Ctx pretreatment of cells does not alter the inhibitory effects of ATP and ADP, pretreatment with pertussis toxin (Ptx) eliminates the inhibitory effects

[a]This work was supported in part by the U.S. Public Health Service (1-RO1-NS26606).

while leaving the enhancement by ATP unchanged. Ptx also does not alter significantly the ability of ATP to trigger secretion.

When cells are pretreated with Ctx or Ptx, ATP or its active analogues protect specific G proteins from ADP-ribosylation. This is shown by subsequent treatment of disrupted cells or isolated membranes with [^{32}P]NAD plus fresh toxins. Western blots identify the ATP-protected Ctx substrate as G_s. Antibodies now available suggest that the ATP-protected Ptx substrate is G_o, but possibly G_i.

Because cAMP is elevated by ATP but not ADP, cAMP cannot mediate the inhibitory response. Pretreatment of cells with forskolin to raise cAMP 10-fold does not preempt the enhancing effect of ATP, so cAMP does not appear to be on the pathway of enhancement. Protein kinase A activation also seems much too slow to account for the instantaneous activation by ATP.

Hydrolysis of phosphatidylinositol phosphate stimulated by ATP or its analogues produces a complex pattern of inositol phosphate metabolites. This pattern, so far, shows no simple relationship to enhancement or inhibition.

The effects described all have $K_{0.5}$ for ATP in the range of 5-10 μM. The secretory vesicles contain ATP at about 150,000 μM, suggesting that released ATP may have feedback or cell-cell interaction effects.

ATP Alters Function in the Isolated Perfused Rabbit Heart

JOHN BIANCHI,[a] ROMUALD CICHON,
HAYMAN RAMBARAN, AND
MURALIDHARAN SEETHAPATHY

Deborah Research Institute
Brown Mills, New Jersey 08015

Adenine compounds were noted to have cardiovascular effects in 1929,[1] and since that time focus has been on the nucleosides rather than the nucleotides. The reason for this may be due to the transient effect of ATP, the rapid dephosphorylation to adenosine,[2] or the recognition of the effects of adenosine on coronary blood flow.[3] ATP is not, however, without profound effects of its own.

This study was designed to study the electropharmacology of ATP in the isolated, perfused rabbit heart. Hearts taken from rabbits weighing 1.5-2.0 kg were fixed to a perfusion apparatus. The apices were fixed to a Grass FT03 transducer with a 2.0-g preload. The coronary arteries were perfused with gassed (95% O_2/5% CO_2) normal Krebs-Henseleit buffer (pH 7.3, 35 °C) at 50 mm Hg. Bipolar, Teflon-coated stainless steel epicardial electrodes paced the hearts at 2 Hz (rectangular pulse, 2-msec duration, 2x threshold) for all but the chronotropic studies.

Addition of 10^{-6} M ATP to the perfusate produced a bradycardia, slowing the hearts to 43% of baseline (TABLE 1). As ATP increased in concentration, the force of ventricular contraction decreased from an initial value of 7.32 ± 0.28 g to 3.60 ± 0.84 g at 10^{-5} M ATP. Coronary artery flow was also altered. A low level of ATP (10^{-8} M) increased flow 25% over control value, and increases in ATP did not alter flow above this value. The increase in flow was associated with a change in oxygen extraction, but it is unknown if the flow increase was contributory. When no ATP was present, the heart removed 66% of oxygen in the perfusate. When 10^{-6} M ATP was added to the perfusate, however, the oxygen extraction was only 41%. These results were mediated directly by ATP, for similar results were obtained when the heart was blocked with atropine and propranolol, both 10^{-6} M.

The mechanism by which ATP mediates these responses is not known. When the ATP level was increased to 10^{-3} M, adenosine deaminase (0.5 U/ml) had no effect on the negative inotropic or chronotropic response, and the heart would come to standstill because of atrioventricular nodal block. Adenosine was not the mediator in this preparation. When aminophylline (10^{-5} M) was added with 10^{-3} M ATP, the contraction force decreased, but the heart rate did not alter from control. These results suggest that ATP exerts its effects directly through a nodal purinoceptor responsive

[a] Present address: Department of Pharmacology, Berlix Laboratories, Inc., 110 East Hanover Avenue, Cedar Knolls, New Jersey 07927.

TABLE 1. Effects of ATP on the Heart Rate, Contractility, Hemodynamics, and Metabolism of the Rabbit Heart *in Vitro*[a]

ATP (M)	Coronary Flow (ml/min)	Contraction Force (g)	Heart Rate (min^{-1})	Perfusate (pO_2, mm Hg)	Effluent (pO_2, mm Hg)
0	6.6 ± 0.5	7.32 ± 0.28	228 ± 6	423 ± 15	143 ± 7
	(p = .0322[b])	(p = .013[b])	(p = .005)	(p = .585)	(p = .0079[c])
10^{-8}	8.1 ± 0.6	5.48 ± 0.24			
10^{-7}	8.4 ± 0.5	4.92 ± 0.56			
10^{-6}	8.7 ± 0.5	4.68 ± 0.72	131 ± 4	407 ± 24	242 ± 22
10^{-5}	8.8 ± 0.7	3.60 ± 0.84			

[a] Each value for the effects of ATP is a mean ± SEM (N = 3).
[b] Derived from a nonparametric two-way analysis of variance.
[c] Derived from a two-sample t test.

to aminophylline that differs from the receptor mediating ventricular contractility. These results demonstrate that ATP is not dependent upon intermediate metabolism in eliciting cardiovascular responses.

REFERENCES

1. Drury, A. N. & A. Szent-Gyorgyi. 1929. J. Physiol. (London) **68:** 213.
2. Burnstock, G. 1972. Pharmacol. Rev. **24:** 509.
3. Berne, R. M. 1963. Am. J. Physiol. **204:** 317.

Extracellular ATP Opens an Amiloride-Sensitive Cation Channel in Human Lymphocytes

J. S. WILEY, W. MAYGER, E. J. CRAGOE, AND M. JOPSON

Hematology Department
Austin Hospital
Heidelberg 3084, Australia

Extracellular ATP has been reported to stimulate the activity of Na^+-H^+ countertransport in Ehrlich ascites tumor cells[1] and in bovine aortic endothelial cells.[2] In one or both of these latter cell types the pH change was half-maximally activated by 1-2 μM extracellular ATP, external Ca^{2+} was required, and UTP was equipotent to ATP in producing an alkaline intracellular pH response. These data are generally interpreted as ATP causing activation of protein kinase C and increasing intracellular $[Ca^{2+}]$—actions that activate Na^+-H^+ countertransport. Extracellular ATP increases the cation permeability of a variety of human hemopoietic cell types including fresh neutrophils, monocytes, their progenitor blast cells,[3] as well as lymphocytes from patients with chronic lymphocytic leukemia (CLL).[4] In this latter cell type ATP produces an average 12-fold increase in $^{22}Na^+$ influx, possibly by opening cation-selective channels. Such a mechanism seems likely because ATP also increases permeability to a variety of other cations by an action primarily on the plasma membrane.[4]

The kinetics of this ATP-induced response were studied by measuring unidirectional $^{22}Na^+$ influx in CLL lymphocytes incubated in media containing 10 mM NaCl without divalent cations. ATP-stimulated uptake of $^{22}Na^+$ ions was linear during a 3-min incubation, and this influx component was maximal at 0.4 mM ATP. Influx showed a sigmoid dependence on ATP concentration, and Hill analysis yielded a $K_{1/2}$ of 160 μM and an n value of 2.5. The nucleotide ATP-γ-S (1.0 mM) gave 30% of the permeability increase produced by ATP (FIG. 1), but UTP (2 mM) and deoxythymidine triphosphate (2 mM) had no effect on $^{22}Na^+$ influx.

The potent amiloride analogues ethylisopropylamiloride (40 μM) and hexamethylene amiloride (40 μM) each inhibited the ATP-stimulated $^{22}Na^+$ influx; however, both analogues had little effect on basal $^{22}Na^+$ influx. The extent of the inhibition produced by the amiloride analogues was variable, ranging from 38% to 100% decrement in the ATP-stimulated $^{22}Na^+$ influx. Sodium influx into CLL lymphocytes was stimulated 4-fold by exposure to these cells to hypertonic media, which is known to activate Na^+-H^+ countertransport. Both amiloride analogues (at 40 μM) effectively blocked the hypertonic-stimulated $^{22}Na^+$ influx. In every case the ATP-stimulated $^{22}Na^+$ influx was greater than the hypertonic-stimulated $^{22}Na^+$ influx by a mean factor of 3.7-fold. This result is consistent with ATP opening cation-selective channels in the plasma membrane rather than directly stimulating Na^+-H^+ countertransport.

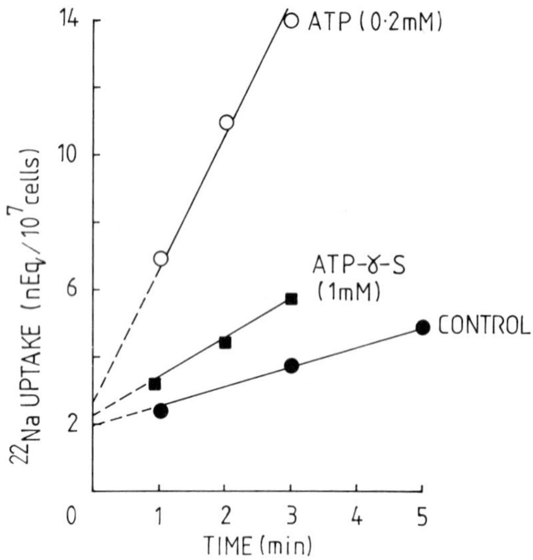

FIGURE 1. Effect of ATP and ATP-γ-S on Na^+ uptake by lymphocytes. Cells at 1×10^7/ml in low-Na^+ medium without divalent cations were preincubated for 5 min followed by addition of ^{22}NaCl with or without 0.2 mM ATP. Influx was taken as the slope of the line of best fit to the uptake over 3 min.

REFERENCES

1. WEINER, E., G. R. DUBYAK & A. SCARPA. 1986. Na^+/H^+ exchange in Ehrlich ascites tumor cells: Regulation by extracellular ATP and 12-*O*-tetradecanoylphorbol 13-acetate. J. Biol. Chem. **261:** 4529-4534.
2. KITAZONO, T., K. TAKESHIGE, E. J. CRAGOE & S. MINAKAMI. 1988. Intracellular pH changes of cultured bovine aortic endothelial cells in response to ATP addition. Biochem. Biophys. Res. Commun. **152:** 1304-1309.
3. COWAN, D. S., H. M. LAZARUS, S. B. SHURIN, S. E. STOLL & G. R. DUBYAK. 1988. Extracellular ATP activates calcium mobilization in human phagocytic leukocytes and neutrophil/monocyte progenitor cells. J. Clin. Invest. **83:** 1651-1660.
4. WILEY, J. S. & G. R. DUBYAK. 1989. Extracellular ATP increases cation permeability of chronic lymphocytic leukemic lymphocytes. Blood **73:** 1316-1323.

Evidence for ATP-Triggered Vagal Reflex in the Canine Heart *in Vivo*

AMIR PELLEG AND CARL M. HURT

Likoff Cardiovascular Institute
Department of Medicine
Hahnemann University
Philadelphia, Pennsylvania 19102

ATP and adenosine exert pronounced electrophysiologic effects in the mammalian heart. These include a negative chronotropic effect on cardiac pacemakers and a negative dromotropic effect on atrioventricular nodal conduction. Previous studies have indicated that the vagus nerve mediates, at least in part, the electrophysiologic actions of ATP in the feline and canine hearts.[1,2] This hypothesis was tested in a canine model, and the following results were obtained: 1) Both ATP and adenosine (given at equimolar doses as rapid bolus injections into the right atrium, 1-3 μmol/kg) exerted dose-dependent negative chronotropic and dromotropic effects on sinus and atrioventricular nodes, respectively. 2) These effects (that is, prolongation of sinus cycle length and the time interval between right atrial and His bundle electrograms) were transient, reaching maxima 12-15 sec following injection and subsequently recovering to baseline values within 60 to 120 sec. 3) Under baseline conditions, ATP is more potent than adenosine. 4) Propranolol (0.5 mg/kg, i.v.) did not affect the actions of either ATP or adenosine. 5) In the presence of atropine (0.2 mg/kg, i.v.), ATP and adenosine were equipotent. 6) Under conditions of total autonomic blockade, the electrophysiologic effects of both ATP and adenosine were similarly enhanced by the adenosine transport blocker, dipyridamole, and attenuated by the adenosine competitive antagonist, aminophylline.[3] It was concluded that the electrophysiologic effects of ATP are mediated by a triggered vagal reflex as well as by its degradation product, adenosine.[3]

Further studies have shown that as expected from the involvement of the vagus nerve, the electrophysiologic actions of ATP are markedly potentiated by the acetylcholinesterase inhibitor, physostigmine (50 mg/kg, i.v.).[4] Moreover, Ca^{2+} ions play an important role in neurotransmitter release. Thus, it was hypothesized that elevation of the extracellular Ca^{2+} level would enhance and that the presence of calcium channel blocker would attenuate the actions of ATP. Indeed, during $CaCl_2$ infusion (0.025 mmol/kg/min, i.v.) the negative chronotropic action of ATP, but not that of adenosine, was markedly enhanced in a time-dependent fashion.[5] This enhancement was abolished by atropine (0.2 mg/kg, i.v.). Furthermore, as shown in FIGURE 1, verapamil (0.2 mg/kg, i.v.) attenuated the negative chronotropic action of ATP but enhanced that of adenosine.[6]

Since ATP was not more potent than adenosine in suppressing sinus nodal pacemaker activity when the two compounds were given directly into the sinus nodal artery,[7] it was concluded that ATP triggers the vagal reflex by acting on chemosensitive vagal nerve terminals in either the right atrium, right ventricle, or pulmonary cir-

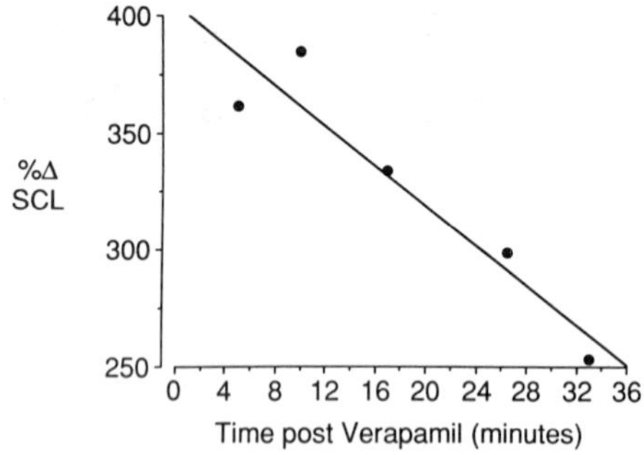

FIGURE 1. A typical example of the modulation by verapamil of the negative chronotropic action of ATP on the canine sinus node. ATP (3 μmol/kg) was given as a rapid bolus injection ($\leq$ 1 sec) into the right atrium before and repetitively after verapamil (0.2 mg/kg, i.v.). The effect of ATP was attenuated in a time-dependent fashion. In comparison, the negative chronotropic effect of adenosine under the same conditions was markedly potentiated by verapamil.[6] %Δ SCL: percentage maximal prolongation of sinus node cycle length.

culation. Further studies are being carried out to localize and characterize these receptors and to determine the physiologic role of this action of ATP.

REFERENCES

1. EMMELIN, N. & W. FELDBERG. 1948. Systemic effects of adenosine triphosphate. Br. J. Pharmacol. Chemother. **3:** 273-284.
2. CARDENAS, M., J. ACEVES & G. ALARCON. 1964. Efecto del acido adenosin-trifosforico sobre las propiedades fisiologicas del corazon. Arch. Inst. Cardiol. Mex. **34:** 485-494.
3. PELLEG, A., B. BELHASSEN, R. ILIA & S. LANIADO. 1985. Comparative electrophysiologic effects of ATP and adenosine in the canine heart: Influence of atropine, propranolol, vagotomy, dipyridamole and aminophylline. Am. J. Cardiol. **55:** 571-576.
4. PELLEG, A., H. MITAMURA & E. L. MICHELSON. 1985. Evidence for vagal involvement in the electrophysiologic actions of exogenous adenosine and adenosine triphosphate in the canine heart. J. Auton. Pharmacol. **5:** 207-212.
5. PELLEG, A. & E. L. MICHELSON. 1987. Role of the vagus in modulation by Ca^{2+} of the depressant actions of adenosine and adenosine 5′-triphosphate on the canine sinus node *in vivo.* J. Auton. Pharmacol. **7:** 127-134.
6. PELLEG, A., H. MITAMURA, T. MITSUOKA, T. MAZGALEV, E. L. MICHELSON & L. S. DREIFUS. 1987. Interactive negative chronotropic actions of adenosine and verapamil on the canine sinus node *in vivo. In* Cardiac Electrophysiology and Pharmacology of Adenosine and ATP: Basic and Clinical Aspects. A. Pelleg, E. L. Michelson & L. S. Dreifus, Eds.: 235-254. Alan R. Liss. New York, NY.
7. PELLEG, A., T. MITSUOKA & E. L. MICHELSON. 1987. Adenosine mediates the negative chronotropic action of adenosine 5′-triphosphate in the canine sinus node. J. Pharmacol. Exp. Ther. **242:** 791-795.

ATP and Adenosine

Vasoconstrictors in Human Placenta[a]

M. H. MAGUIRE, H. KITAGAWA, T. HOSOKAWA,
AND R. B. HOWARD

Department of Pharmacology, Toxicology, and Therapeutics
University of Kansas Medical Center
Kansas City, Kansas 66103

The *in vitro* preparation of a single cotyledon from human term placenta, perfused via both fetal and maternal circuits, was used to study the action of ATP, adenosine, and adenosine analogues on human fetoplacental vascular resistance. Perfusion was performed at constant flow at 37 °C using Earle's salt solution containing 4 g% dextran, gassed with 95% O_2-5% CO_2 for the maternal circuit or with 94% N_2-6% CO_2 for the fetal circuit.[1] Flow rate of maternal perfusate (Q_M) was adjusted to achieve a fetal venous perfusate pH of 7.30-7.38, a pO_2 equal to or above 170 mm Hg, and a pCO_2 of 39-47 mm Hg. The fetal perfusion pressure (PP_F) was 40-50 mm Hg, and the mean Q_M ($\pm$SD) was 1.15 $\pm$ 0.39 $ml \cdot min^{-1} \cdot g^{-1}$ (N = 21 cotyledons). PP_F was used as an index of vascular resistance. Bolus doses of adenine derivatives were administered by close arterial injection; after baseline recovery 2 min were allowed before injection of the next dose. Antagonists were infused for 10 min before injection of agonists. ATP caused reversible pressor responses (FIG. 1) that were dose dependent over the range 1-160 nmoles (N = 6 cotyledons). Adenosine and adenosine analogues elicited reversible dose-dependent increases in PP_F that were antagonized by the adenosine receptor antagonists *p*-sulfophenyltheophylline (SPT) and aminophylline (AMI). FIGURE 2 shows dose-response relationships of increases in PP_F to 5'-*N*-ethylcarboxamidoadenosine (NECA) and N^6-phenylisopropyladenosine (R-PIA), and to adenosine and 2-chloroadenosine in the presence and absence of 54 μM SPT.

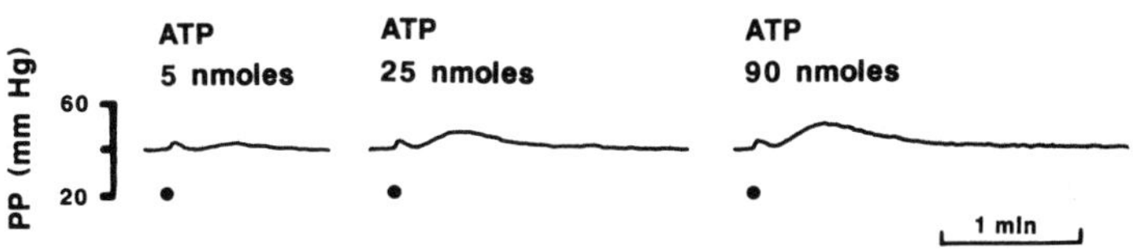

FIGURE 1. Typical pressor responses of the fetoplacental vascular bed to bolus doses of ATP.

[a] This work was supported by Grant 14888 from the National Institutes of Health.

Vasoconstriction in response to adenosine was dose-dependent over the range of 0.2-180 nmoles (N = 9 cotyledons); AMI (44 μM) caused a rightward shift of the adenosine dose-response curve (N = 2 cotyledons). In some preparations the dose-response curve to adenosine was shallow, with reduced response to the higher doses of adenosine; in these cotyledons, the increase in PP_F elicited by 10 nmoles of adenosine was 4.7 ± 0.4 mm Hg compared to 8.1 ± 1.0 mm Hg (each of these PP_F values is a mean ± SEM; N = 6 and 9 cotyledons, respectively). Reduced sensitivity to adenosine in cotyledons from some placentas may represent down-regulation of adenosine receptors as a result of exposure to adenosine during delivery. Adenosine is present in the delivered placenta,[2] and release of adenosine by the *in vitro* cotyledon is increased by ischemia.[3] ATP and adenosine did not elicit decreases in PP_F. Whether vasoconstriction of the human fetoplacental vascular bed in response to ATP and

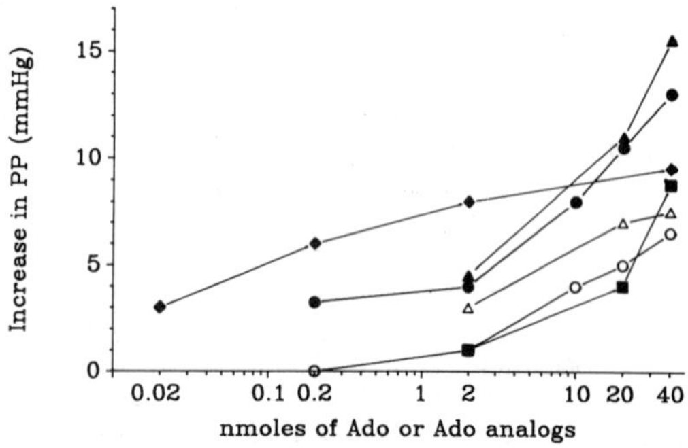

FIGURE 2. Dose-response curves in a semilogarithmic plot for increases in PP_F elicited by adenosine (●), adenosine plus 54 μM SPT (○), 2-chloroadenosine (△), 2-chloroadenosine plus 54 μM SPT (▲), NECA (◇), and R-PIA (■).

adenosine is direct or indirect is not yet established. Nevertheless, these purines may contribute to local regulation of fetoplacental vascular tone *in vivo.*

REFERENCES

1. Howard, R. B., J. Levy, T. Hosokawa & M. H. Maguire. 1987. Interrelationships of perfusion parameters in the dual-perfused human placental cotyledon. Trophoblast Res. **2:** 585-596.
2. Maguire, M. H., F. A. Westermeyer & C. R. King. 1986. Measurement of adenosine, inosine and hypoxanthine in human term placenta by reversed-phase high-performance liquid chromatography. J. Chromatogr. **380:** 55-66.
3. Slegel, P., H. Kitagawa & M. H. Maguire. 1988. Determination of adenosine in fetal perfusates of human placental cotyledons using fluorescence derivatization and reversed-phase high-performance liquid chromatography. Anal. Biochem. **171:** 124-134.

Extracellular ATP Gates a Ca^{2+}-Permeable Nonselective Cation Channel in Rat Parotid Acinar Cells

Effects of Stilbene Disulfonates and Reactive Blue on $^{45}Ca^{2+}$ Entry

S. P. SOLTOFF, M. K. McMILLIAN, B. R. TALAMO, AND L. C. CANTLEY

Department of Physiology
Tufts University
Boston, Massachusetts 02111

We investigated the effect of ATP^{4-} on $^{45}Ca^{2+}$ entry into rat parotid acinar cells. The maximum ATP-stimulated influx rate (6 nmol/mg/min, with a $K_{0.5}$ of 150 μM) was about 20 times the basal rate. 3′-*O*-(4-Benzoyl)benzoyl-ATP (Bz-ATP) was 30 times more potent and nearly twice as effective as ATP in stimulating Ca^{2+} influx. Other analogues were much less effective than ATP. The order of effectiveness of nucleotide analogues on $^{45}Ca^{2+}$ entry was Bz-ATP > ATP > ATP-γ-S > 2-methylthio-ATP > α,β-methylene ATP, ADP. The potency and effectiveness of ATP were increased 2-3-fold in the absence of extracellular Na^{2+}, which may compete with Ca^{2+} for entry through an ATP-gated Na^+- and Ca^{2+}-permeable channel. 4,4′-Diisothiocyano-2,2′-stilbene disulfonate (DIDS), which has two -NCS groups, at 100 μM completely blocked the ATP-stimulated influx (EC_{50} = 40 μM). 4-Acetamino-4′-isothiocyanostilbene-2,2′-disulfonate (SITS), which has one -NCS group, was about 50% as potent as DIDS. 4,4′-Dinitrostilbene-2,2′-disulfonate (DNDS), which lacks -NCS groups, at 1 mM was ineffective, as was K^+SCN^-. Reactive blue 2 (Cibacron Blue), a nonstilbene compound, also blocked $^{45}Ca^{2+}$ entry (EC_{50} = 50 μM). The effect of DIDS was irreversible, but the effect of reactive blue 2 was reversible by washing the cells. Reactive blue 2, added before DIDS, blocked the irreversible effect of DIDS. Our conclusions are as follows: 1) ATP activates a nonselective cation channel. 2) Stilbene compounds with -NCS groups covalently block the binding of ATP to its receptor. 3) Reactive blue 2 binds noncovalently to prevent DIDS and ATP binding.

ATP Produces Two Ca_i Responses in Rat Parotid Cells

MICHAEL McMILLIAN, STEPHEN SOLTOFF, LEWIS CANTLEY, AND BARBARA TALAMO

Department of Physiology and Neuroscience Program
Tufts University School of Medicine
Boston, Massachusetts 02111

Extracellular ATP increased intracellular calcium (Ca_i) in a biphasic manner in Fura-2-loaded rat parotid acinar cells, with a small effect apparent between 0.1 and 10 μM ATP and a much larger effect observed between 50 and 300 μM ATP (in the absence of Mg^{2+} with 1 mM Ca^{2+} present). This latter effect, which we previously described,[1] appears to be mediated by a P_{2Z}-purinergic receptor; blockade by high Mg^{2+} or Ca^{2+} suggests that ATP^{4-} is the actual agonist, and the selective P_{2Z} agonist 3′-*O*-(4-benzoyl)benzoyl-ATP (Bz-ATP) was about 30-fold more potent and 3-fold more efficacious than ATP (these differences being due, at least in part, to covalent activation by Bz-ATP). In contrast, the more sensitive ATP response was only slightly diminished by elevated Mg^{2+} or Ca^{2+} (10 mM) in the buffer (this difference probably being due to increased degradation of ATP and/or production of ADP, which blocked the ATP response). Bz-ATP was ineffective as an agonist on the more sensitive response to ATP, but other P_2-purinergic agonists showed a similar potency series for both ATP responses in parotid cells: ATP > ATP-γ-S $\geq$ 2′-MeS-ATP > > ITP and AMP-CPP. Thus, the more sensitive response, though not mediated through a P_{2Z} receptor, is more similar to this response than to P_{2X}- or P_{2Y}-purinergic responses. 4,4′-Diisothiocyano-2,2′-stilbene disulfonate (DIDS) and reactive blue 2 (Cibacron blue) blocked the large ATP^{4-} effect on Ca_i and also inhibited [^{32}P]ATP binding to parotid cells under conditions (phosphate-buffered saline with no added Mg^{2+} or Ca^{2+}) selective for ATP^{4-}. Reactive blue 2 pretreatment blocked irreversible effects of DIDS and Bz-ATP, presumably by blocking the P_{2Z} receptor and protecting it from covalent inactivation or activation, respectively. In contrast to Ca_i responses to ATP^{4-}, the more sensitive response to ATP was slightly potentiated by both DIDS and reactive blue 2 (this difference probably being due to inhibition of ATPases). Both P_2-receptors on parotid cells may open Ca^{2+}-permeable channels, since both responses are largely dependent on extracellular calcium and are quenched by extracellular Mn^{2+}, in contrast to muscarinic agonists and substance P. A weak effect of ATP on phosphatidylinositol turnover and Ca_i mobilization was due to ATP^{4-}, since this effect required high concentrations of ATP, was mimicked by Bz-ATP, and was blocked by DIDS and Mg^{2+}. The more sensitive ATP response may be physiologically important, since parasympathetic denervation produced a 3-fold increase in the maximal Ca_i response to 3 μM ATP, with no change in the ATP^{4-} response or in maximal carbachol or substance P responses.

REFERENCE

1. McMILLIAN, M. K., S. P. SOLTOFF, J. D. LECHLEITER, L. C. CANTLEY & B. R. TALAMO. 1988. Extracellular ATP increases free cytosolic calcium in rat parotid acinar cells: Differences from phospholipase C-linked receptor agonists. Biochem. J. **255:** 291-300.

Extracellular ATP-Induced Ca^{2+} Transients in Cardiac Myocytes Are Potentiated by an Increase in Cellular cAMP[a]

JING-SHENG ZHENG, MARY BETH DE YOUNG, ERIK WIENER, MATTHEW N. LEVY, AND ANTONIO SCARPA

Department of Physiology and Biophysics
Case Western Reserve University
Cleveland, Ohio 44106

It has been recently shown by this laboratory[1] that extracellular ATP in micromolar concentrations produces in isolated myocytes a transient increase in intracellular Ca^{2+} concentrations. Pretreatment with norepinephrine potentiates ATP-induced Ca^{2+} transients. Because ATP and norepinephrine are coreleased from sympathetic neurons, modulation of the ATP response by norepinephrine may have some physiological significance. We investigated the mechanism by which norepinephrine potentiates the ATP-induced Ca^{2+} transient.

We used collagenase to prepare ventricular myocytes from 250-350 g male Sprague-Dawley rats as described previously.[1,2] Intracellular Ca^{2+} was measured by the fluorescent indicator fura-2. Fluorescence was recorded with excitation and emission wavelengths of 340 nm and 510 nm, respectively. The procedure used to extract cAMP and measure it by radioimmunoassay was described previously.[3]

We have found that cAMP-modifying agents that increase the intracellular cAMP level mimic the potentiating effect of norepinephrine. FIGURE 1 shows the intracellular free Ca^{2+} concentration of fura-2-loaded myocytes exposed to 10 μM ATP without any pretreatment (**A**) or with various pretreatments (**B**-**E**). ATP response was normalized by K^+ depolarization. We have found that pretreatment with forskolin, isomethylbutylxanthine, and 8-(4-chlorophenylthio)-cAMP potentiate the ATP-induced Ca^{2+} response.

We examined the degree of intracellular cAMP changes induced by norepinephrine, forskolin, and isomethylbutylxanthine (FIG. 2B) and compared these changes with the potentiating effects of these agents (FIG. 2A). In FIGURE 2A, the data were expressed as the change in fluorescence due to the addition of ATP relative to the fluorescence changes of the cells induced by K^+ depolarization. Pretreatment with norepinephrine potentiates the ATP response from 0.36 $\pm$ 0.05 to 0.72 $\pm$ 0.03. Pretreatment with 0.5 μM forskolin and pretreatment with 50 μM isomethylbu-

[a]This work was supported by Grants HL15758 and HL18708 from the National Institutes of Health. E. W. was supported by Training Grant HL07653 from the National Institutes of Health.

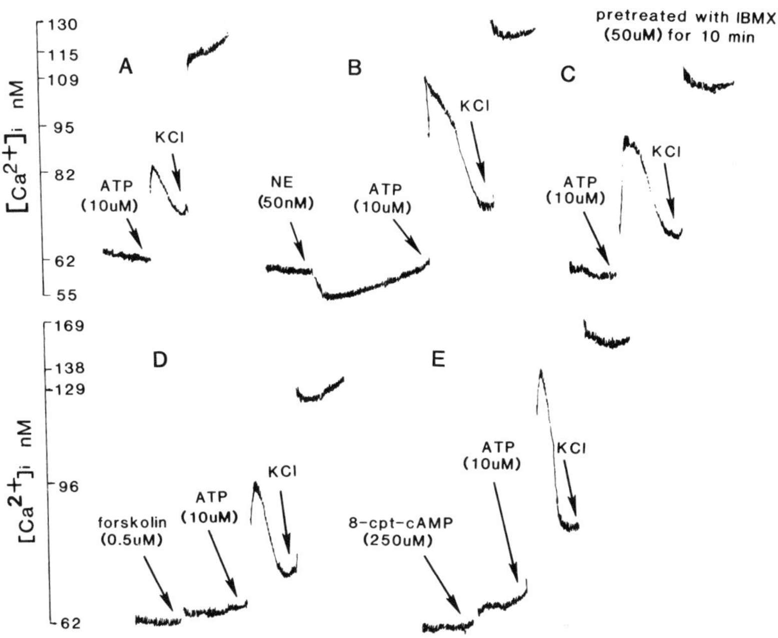

FIGURE 1. Norepinephrine, forskolin, isomethylbutylxanthine, and 8-(4-chlorophenylthio)-cAMP potentiate ATP-induced Ca^{2+} transients. Ventricular myocytes were loaded with 2 μM fura-2-AM and exposed to 10 μM ATP. **A**: no pretreatment; **B**: pretreatment with 50 nM norepinephrine (NE); **C**: pretreatment for 10 min with 50 μM isomethylbutylxanthine (IBMX), a phosphodresterase inhibiter; **D**: pretreatment with 0.5 μM forskolin, an adenylate cyclase activator; **E**: pretreatment with 250 μM 8-(4-chlorophenylthio)-cAMP (8-cpt-cAMP), a membrane-permeable cAMP analogue. The ATP response was normalized with KCl depolarization.

tylxanthine increase the ATP response to 0.58 ± 0.03 and 0.68 ± 0.07, respectively (FIG. 2A). Norepinephrine increased the basal cAMP level from 5.09 ± 0.56 to 11.46 ± 1.01 pmol/mg protein, and forskolin and isomethylbutylxanthine increased the basal cellular cAMP to 10.56 ± 1.49 and 11.14 ± 1.32, respectively (FIG. 2B). These results indicate that the noradrenergic potentiation of ATP-induced intracellular Ca^{2+} mobilization involves cAMP as a second messenger.

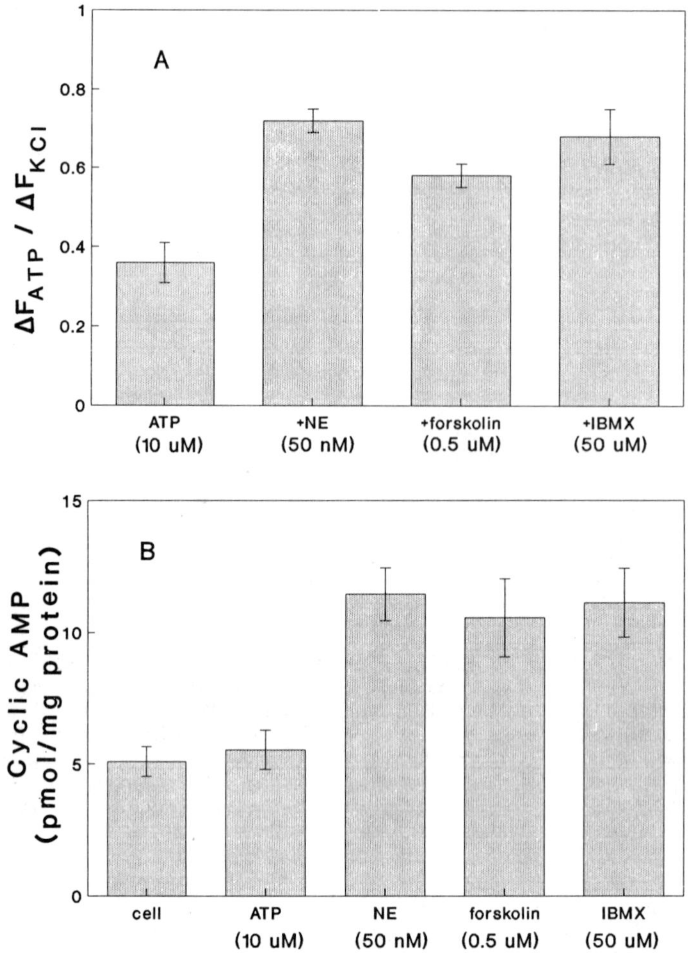

FIGURE 2. Comparison of NE, forskolin, and isomethylbutylxanthine potentiation of ATP-induced Ca^{2+} transients and increase in cellular cAMP production. (**A**) Methods are the same as in FIGURE 1. Each value is a mean ± SE ($N = 4$). (**B**) Cells were treated according to the methods indicated above. Each value is a mean ± SE of four different experiments performed in triplicate.

REFERENCES

1. DE YOUNG, M. B. & A. SCARPA. 1987. Extracellular ATP induces Ca^{2+} transients in cardiac myocytes which are potentiated by norepinephrine. FEBS Lett. **223:** 53-58.
2. DE YOUNG, M. B. & A. SCARPA. 1989. ATP-receptor-induced Ca^{2+} transients in cardiac myocytes: Sources of mobilized Ca^{2+}. Am. J. Physiol. **257:** C750-C758.
3. WIENER, E. & A. SCARPA. 1989. Activation of protein kinase C modulates the adenylate cyclase effector system of B-lymphocytes. J. Biol. Chem. **264:** 4324-4328.

P_2-Purinoceptor-Induced Inositol Phosphate Formation, Intracellular Free Calcium Increase, and Membrane Currents in DDT$_1$ MF-2 Cells

ADRIAAN NELEMANS, ARELES MOLLEMAN, BEN HOITING, MARRY DUIN, AND ADRIAAN DEN HERTOG

Department of Pharmacology/Clinical Pharmacology
University of Groningen
9713 BZ Groningen, the Netherlands

The effect of P_2-purinoceptor stimulation on inositol phosphate formation in relation to the intracellular calcium concentration and membrane currents was measured in vas deferens DDT$_1$ MF-2 smooth muscle cells. The ^{3}H-labeled inositol phosphates (7 μCi/ml; 22 hr) in the cells were analyzed by high-pressure liquid chromatography. The experiments, which were carried out at 22 °C, showed that Ins-(1,3,4,5)-P_4 formation increased rapidly upon addition of ATP to the cells to reach a maximum after about 2 min (FIG. 1). The level of the other isomers, Ins-(1,3,4,6)-P_4 and Ins-(3,4,5,6)-P_4, showed a rise with a delay of about 2 min. This pattern was also followed by Ins-(1,4,5)-P_3, which started from a relatively high level with respect to the Ins-P_4 isomers. There is considerable evidence that certain inositol phosphates are functioning as second messengers, releasing calcium from internal structures, and promoting ion fluxes across the cell membrane.[1-3] Intracellular organelles, known as the endoplasmatic reticulum and calciosomes, were recognized to possess Ins-(1,4,5)-P_3-sensitive receptors that released calcium upon stimulation. The inositol phosphate Ins-(1,3,4,5)-P_4 may function synergistically with Ins-(1,4,5)-P_3 to provide enhancement of the intracellular calcium concentration, most likely by promoting calcium influx from the extracellular space.[4] Accordingly, intracellular calcium was monitored after incubating the cells with Indo-1/AM (2 μM; 45 min). The intracellular free calcium concentration, represented by the Indo-1 fluorescence, was enhanced transiently from about 150 to 350 nM in the presence of ATP to reach a plateau of about 180 nM after 2 min (FIG. 1). Under calcium-free conditions, only the transient component was observed. This response could be evoked only once—in contrast with the response evoked in the presence of extracellular calcium. In view of the action of the inositol phosphates on cellular structures, it is tempting to assume that the transient increase in intracellular calcium observed in the presence of ATP in DDT$_1$ MF-2 cells is due to the promoting action of Ins-(1,3,4,5)-P_4 on calcium release. The sustained component of the intracellular calcium response is most likely caused by a successive replenishment of the Ins-(1,3,4,5)-P_4-sensitive calcium store and calcium release via Ins-(1,4,5)-P_3-regulated structures.

The opening of such channel-like junctions by Ins-(1,3,4,5)-P_4, which allows calcium ions to pass from the extracellular space to the cytoplasm, should be reflected by changes in membrane current. In this work, membrane current was measured in the whole-cell, patch-clamp configuration. A triphasic response, observed under voltage-clamp conditions (−50 mV), consisted of a transient inward current, followed

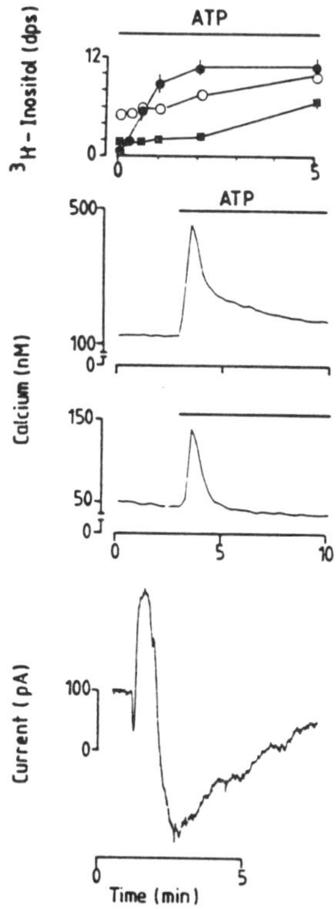

FIGURE 1. The effect of ATP (10^{-4} M) measured in DDT_1 MF-2 smooth muscle cells on inositol phosphate formation (top panel: ●, Ins-(1,3,4,5)-P_4; ○, Ins-(1,3,4,6)-P_4; ■, Ins-(3,4,5,6)-P_4), on intracellular calcium (second panel: in the presence of external calcium; third panel: under calcium-free conditions), and on membrane current (bottom panel: outward current indicated by upward deflection).

by an outward current and a sustained inward current (FIG. 1). It was found that the outward current is carried by potassium ions and is dependent on the availability of internal GTP, suggesting the involvement of a G protein.[5] The transient and sustained inward currents are at least partly carried by calcium ions. In view of the time-related events, these inward calcium currents are possibly linked with the Ins-(1,3,4,5)-P_4-induced calcium flux from the extracellular space to the cytoplasm as reflected by the enhanced internal calcium concentration.

These processes evoked by ATP in smooth muscle cells are thought to be mediated via P_2-purinoceptors.[6] Recently, it was reported that the antitrypanosomal agent suramin inhibits the response caused by stimulation of the P_2-purinoceptors.[7,8] The action of suramin on the events elicited by ATP in DDT_1 MF-2 smooth muscle cells was also investigated. This action is described here, along with the observation that the membrane currents evoked by ATP in these cells are inhibited concentration dependently (FIG. 2). It is noted that both the inward currents, which are possibly linked with the formation of Ins-(1,3,4,5)-P_4, and also the GTP-dependent outward potassium

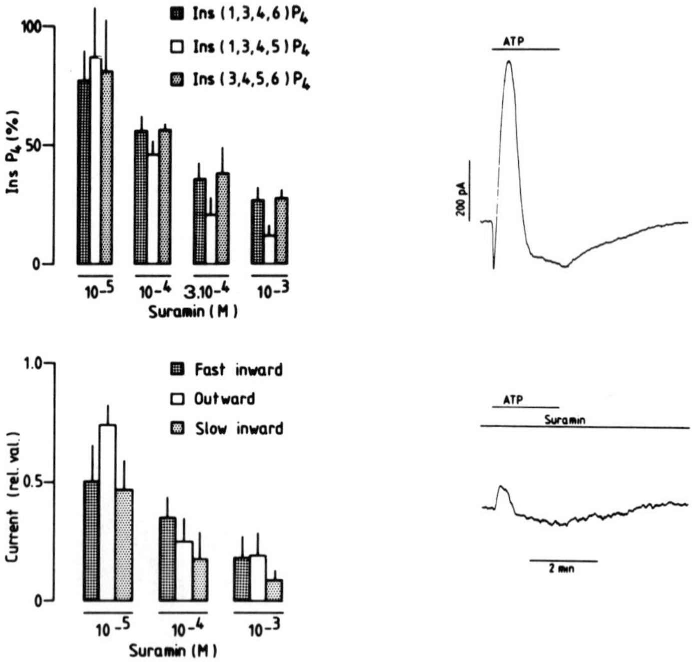

FIGURE 2. The action of suramin on the ATP-induced inositol phosphate formation and on membrane current (top left and bottom left panels: ATP at 10^{-5} M; top right panel: ATP at 10^{-4} M; bottom right panel: ATP at 10^{-4} M and suramin at 10^{-3} M). The outward current is indicated by the upward deflection.

current are inhibited by suramin. Analysis of the ATP-activated formation of the inositol phosphates (5 min) also showed inhibition of the Ins-P_4 isomers by suramin (FIG. 2). This concentration-dependent suppression of the Ins-P_4 formation by suramin is not associated with a change in Ins-(1,4,5)-P_3 formation or with the basal levels of these inositol phosphates. Thus, inhibition by suramin of both the ATP-activated inward currents and inositol phosphate formation, as well as the outward potassium current, might be explained by its antagonistic properties on the P_2-purinoceptor.

REFERENCES

1. BERRIDGE, M. 1984. Inositol phosphate and diacylglycerol as second messengers. Biochem. J. **220:** 345.
2. IRVINE, R. F. & R. M. MOOR. 1986. Microinjection of 1,3,4,5-tetrakisphosphate activates sea urchin eggs by a mechanism dependent on extracellular Ca^{2+}. Biochem. J. **240:** 301.
3. BERRIDGE, M. J. & R. F. IRVINE. 1989. Inositol phosphates and cell signalling. Nature **341:** 197.
4. IRVINE, R. F. 1989. How do inositol 1,4,5-triphosphate and inositol 1,3,4,5-tetrakisphosphate regulate intracellular Ca^{2+}? Biochem. Soc. Trans. **17:** 6.
5. MOLLEMAN, A., A. NELEMANS & A. DEN HERTOG. 1989. P_2-purinoceptor-mediated membrane currents in DDT_1 MF-2 smooth muscle cells. Eur. J. Pharmacol. **169:** 167.
6. BURNSTOCK, G. & C. KENNEDY. 1985. Is there a basis for distinguishing two types of P_2-purinoceptors? Gen. Pharmacol. **16:** 433.
7. DUNN, P. M. & A. G. H. BLAKELY. 1988. Suramin: A reversible P_2-purinoceptor agonist in the mouse vas deferens. Br. J. Pharmacol. **93:** 243.
8. DEN HERTOG, A., A. NELEMANS & J. VAN DEN AKKER. 1989. The inhibitory action of suramin on the P_2-purinoceptor response in smooth muscle cells of guinea-pig taenia caeci. Eur. J. Pharmacol. **166:** 531.

Extracellular ATP Modifies Intracellular Free Ca^{2+} Levels in Skeletal Muscle via Activation of a Purinergic Receptor-G Protein-Phospholipase Cascade

E. HEILBRONN, H. ERIKSSON, AND J. HÄGGBLAD

Unit of Neurochemistry and Neurotoxicology
University of Stockholm
S-10691 Stockholm, Sweden

This paper summarizes our recent results on the biochemical effects of extracellular ATP acting on myotubes in culture. They may play a role in events following neuromuscular transmission, especially in the regulation of the endogenous free Ca^{2+} levels needed for muscle contraction. ATP occurs in the synaptic "cholinergic" vesicle population of motor nerve endings and is released with acetylcholine by the nerve impulse. It is also released from the depolarized skeletal muscle.

The action of ATP or nonhydrolyzable derivatives on myotubes in culture is mediated by activation of a P_{2Y}-purinoceptor-G protein-phospholipase C cascade. Phosphatidylinositol (PI) turnover occurs and results in inositol trisphosphate (IP_3) and diacylglycerol (DAG) formation.[1] A biphasic rise in endogenous free Ca^{2+} levels is seen (in Fura-2- or Quin-2-loaded myotubes) and consists of two peaks: a strong, rapid fluorescence peak due to an endogenous source, and a broad, more sustained peak due to an influx of Ca^{2+} (FIGS. 1A & 1B). The second peak is abolished by 1 μM concentrations of PN200-110, suggesting that IP_3 induces the opening of a voltage-dependent Ca^{2+} channel.[2] Work with myotubes from dysgenic mice lacking L-type Ca^{2+} channels has confirmed that IP_3 indeed may be opening this type of Ca^{2+} channel. The ATP effect on PI turnover remained, but consecutive changes in cytosolic Ca^{2+} were diminished or absent (FIG. 1C).

Tonic depolarization of myotubes (>40 mM K^+) also produced a biphasic change in cytosolic Ca^{2+}, of larger amplitude than ATP and with a slowly declining second stage. Addition of ATP (25-100 μM ATP-γ-S) or of PN200-110 (FIGS. 1D & 1E) after depolarization caused a rapid decrease to baseline levels of Ca^{2+}. The ATP effect on depolarized myotubes was dependent on the degree of depolarization (that is, the K^+ concentration). It could be mimicked by the addition of phorbol esters. When ATP was added before depolarization, only the first rapid Ca^{2+} peak was seen after depolarization.[3]

Similar effects are seen after depolarization by nicotinic acetylcholine receptor (nAChR) activation with carbachol (100 μM) (FIG. 1F). PN200-110 addition before carbachol, however, gave a more complicated picture, suggesting initial Ca^{2+} influx followed by incomplete block. This suggests the involvement of both the nAChR ion channel and a voltage-dependent Ca^{2+} channel.

ATP may be one of several regulators (including certain peptides) of endogenous free Ca^{2+} levels in skeletal muscle acting via an IP_3-sensitive Ca^{2+} channel.

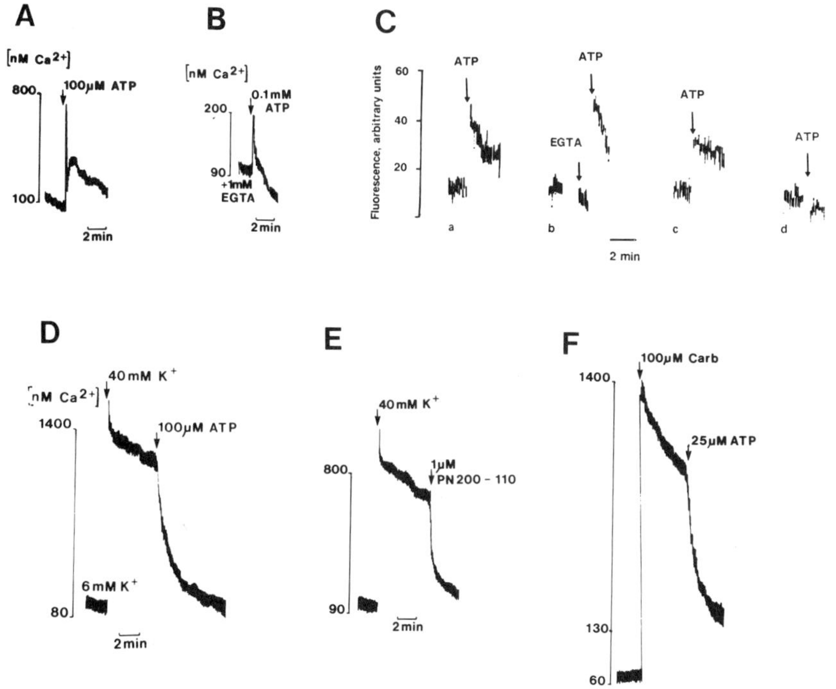

FIGURE 1. Calcium transients in Fura-2-loaded chick or rat (dysgenic) myotubes, induced by ATP, high K^+, or carbachol. Top row: nondepolarized normal (**A** & **B**) or dysgenic (**C**) myotubes. Bottom row: depolarized myotubes (**D**-**F**).

REFERENCES

1. HÄGGBLAD, J. & E. HEILBRONN. 1988. FEBS Lett. **235:** 133-136.
2. ERIKSSON, H. & E. HEILBRONN. 1989. Biochem. Biophys. Res. Commun. **159:** 878-885.
3. TASSIN, A. M., J. HÄGGBLAD & E. HEILBRONN. 1990. Muscle Nerve **13**(2): 142-145.

Bladder Purinergic Receptors[a]

MICHAEL R. RUGGIERI,[b]
KRISTENE E. WHITMORE, AND ROBERT M. LEVIN

Department of Urology
Graduate Hospital of Philadelphia
Division of Urology
University of Pennsylvania School of Medicine
and
Veterans Administration Medical Center
Philadelphia, Pennsylvania

In rabbits the contractile response of the urinary bladder is only partially via cholinergic innervation because atropine does not completely block neuronally mediated contraction. In the human bladder this atropine resistance is controversial. Results of the present investigation demonstrate that an atropine-resistant, tetrodotoxin-sensitive contraction does occur in certain human bladder strips. The importance of tetrodotoxin is that it poisons the sodium channels of the nerve cells while leaving the muscle cells able to contract. This is demonstrated by the ability of depolarization with high KCl to cause maximal contraction at the end of the experiment, indicating that the tissue is not nonspecifically damaged by the *in vitro* incubation.

The additional demonstration that the strips that are atropine sensitive do show a significant contractile response to exogenous ATP indicates that the machinery necessary for responding to the purinergic agonist (including the purinergic receptor) is present in all human bladder strips. In order for the atropine-resistant response to appear, the nerves releasing the endogenous purinergic transmitter must be present in the specimen. If the human bladder is only sparsely innervated with purinergic nerves or if these nerves become damaged, then it is quite possible that certain strips would be devoid of functional purinergic nerves and thus would not show the response. Because of the relatively small number of specimens in the present study, no correlation can be made between the presence of atropine resistance and the diagnosis of the patient. Only when we can compare the response of strips obtained from normal bladders to the response of a suitable number of strips obtained from specific dysfunctions can we begin to correlate the presence of atropine resistance to specific uropathologies. At this time, the possibility remains that the presence or absence of atropine-resistant contractions may be indicative of a pathologic state.

Evidence accumulated over the past few decades suggests that this atropine-resistant contraction may be mediated by ATP or a related purine compound. The present study is designed to develop a radioligand binding assay for this purinergic receptor. The first step in establishing a radioligand receptor binding assay is finding a suitable radiolabeled compound that will bind specifically to the receptor. We have

[a] This work was supported in part by Grants R23-NS-22087, RO1-DK26508, RO1-DK33559, and RO1-DK39086 from the National Institutes of Health and by a Veterans Administration Merit Review Grant.

[b] Address for correspondence: Temple University Department of Urology, 3400 North Broad Street, Philadelphia, Pennsylvania 19140.

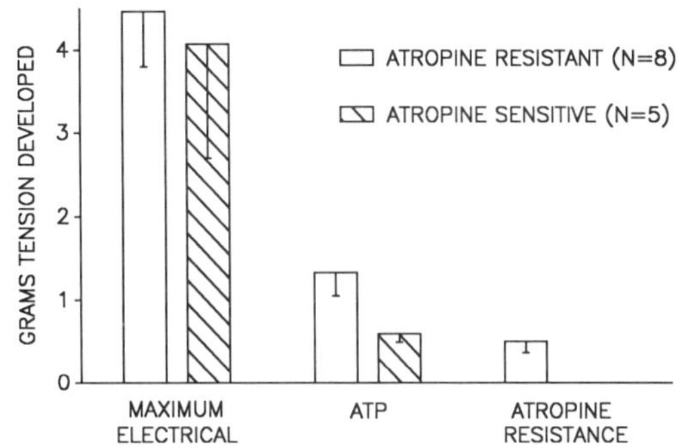

FIGURE 1. Data from atropine-resistant and atropine-sensitive specimens expressed as the actual grams tension developed. The contractile response to ATP (2 mM) was determined after three frequency response curves were generated. The atropine-resistant response is expressed as the maximal grams tension developed in the presence of atropine (10 μM) minus the maximal grams tension developed in the presence of atropine and tetrodotoxin (0.5 μM) at either 32 or 64 Hz, whichever gave maximal response. Each mean from the two to five individual strips per specimen is treated as single datum point and is expressed as a mean ± SEM.

initially considered two radioligands for the purinergic receptor: [^{3}H]ATP and [^{3}H]β,γ-methylene ATP. Several lines of evidence favors [^{3}H]β,γ-methylene ATP: 1) β,γ-Methylene ATP is 10-100-fold more potent than ATP itself in causing contractions of the urinary bladder *in vitro.*[1] 2) β,γ-Methylene ATP is very resistant to hydrolysis compared to ATP. 3) β,γ-Methylene ATP does not inhibit the hydrolysis of ATP. From these lines of evidence one would predict that β,γ-methylene ATP does not bind to ATPase but does bind to the receptor responsible for the purinergic contraction of the urinary bladder.

Although the binding affinity for β,γ-methylene ATP is rather low compared to standard radioligands for autonomic receptors, the usefulness of a receptor ligand is

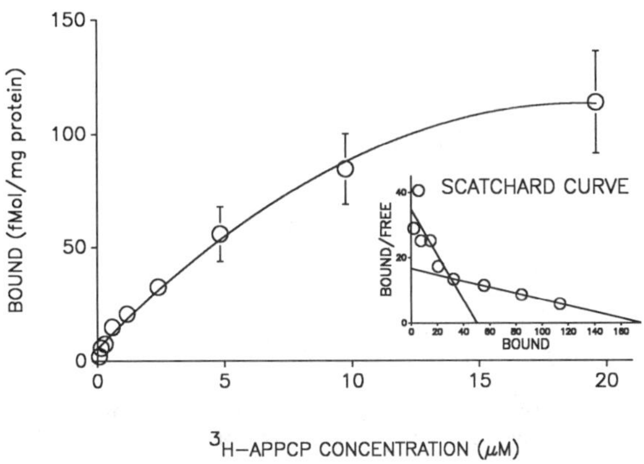

FIGURE 2. This saturation curve for [^{3}H]β,γ-methylene ATP binding is done with a 30-min incubation at 0° C, with 100 μg protein per assay tube, and with 100 μM adenosine tetraphosphate (APPCP) to define nonspecific binding. Results displayed are from a representative experiment that was repeated several times with similar results. The inset shows a Scatchard curve of this data.

primarily related to its selectivity and specificity, and not its potency. A ligand that has a high affinity for a particular receptor, but nonspecifically binds to all tissue elements, or binds to other receptors with a similar high affinity, would be a very poor ligand. Alternatively, a ligand that has a lower affinity for the same receptor, but has a low nonspecific binding, and is selective for only the one receptor type would be a very useful ligand in the study of the receptor. In addition, the affinity of a particular ligand for a receptor does not necessarily indicate anything about the affinity of the receptor for its natural transmitter. This is especially true for an agonist. Lastly, because the study of radioligand binding to purinergic receptors is in its infancy, there is no information on any cofactors that may be required for agonist binding.

The second step in establishing a radioligand receptor binding assay is finding a suitable inhibitor of specific binding. The inhibitor must also bind specifically to the receptor and thereby inhibit specific binding of the radiolabeled ligand to the receptor; however, it should not inhibit the nonspecific binding of the radiolabeled ligand. Binding assays can then be performed in the presence and absence of the inhibitor, and the binding in the presence of the inhibitor can be subtracted from the binding in the absence of the inhibitor to give the specific receptor binding of the radiolabeled ligand. Ideally, the chemical structure of the inhibitor should be completely different from the radiolabeled ligand so the inhibitor does not inhibit nonspecific binding. In addition, both the ligand and the inhibitor should be physiological competitive antagonists of the response so that the affinity of the receptor does not change as the ligand binds to the receptor (see Birdsall *et al.*[2] for a review.)

For the purinergic receptor these ideal conditions cannot be met at present. The only known antagonist of the purinergic response, arylazidoaminopropionyl adenosine triphosphate ($ANAPP_3$), is not a competitive physiological antagonist of the response. It is actually an agonist of the response until it is photolyzed and then covalently binds to the receptor, becoming a noncompetitive antagonist.[3] Its chemical structure is also quite similar to the radiolabeled ligand. Without a competitive pharmacological antagonist to use as a radiolabled ligand, the use of $[^3H]\beta,\gamma$-methylene ATP offers several advantages over $[^3H]$ATP as outlined above.

In summary: 1) Neuronally mediated, atropine-resistant contractions are present in selected human bladder strips *in vitro.* 2) A radioligand binding assay for purinergic receptors has been developed, and this assay can quantitate the presence of these receptors in tissue homogenates.

ACKNOWLEDGMENT

The authors would like to thank Ms. Debra Moore for her excellent technical assistance.

REFERENCES

1. LEVIN, R. M., R. JACOBY & A. J. WEIN. 1983. High-affinity, divalent, ion-specific binding of $[^3H]$ATP to homogenate derived from rabbit urinary bladder. Mol. Pharmacol. **23:** 1-7.
2. BIRDSALL, N. J. M., A. S. V. BURGEN & E. C. HULME. 1977. *In* Cholinergic Mechanisms and Psychopharmacology. D. J. Jenden, Ed.: 25-33. Plenum. New York, NY.
3. HOGABOOM, G. K., J. P. O'DONNELL & J. S. FEDAN. 1980. Purinergic receptors: Photoaffinity analogue of adenosine triphosphate is a specific adenosine triphosphate antagonist. Science **208:** 1273-1276.

ADP-β-F Is Not a Selective P_{2Y}-Purinoceptor Agonist

B. E. WOOD, A. SQUIRE, S. E. O'CONNOR,
AND P. LEFF

Department of Pharmacology
Pharmaceutical Division
Fisons plc
Loughborough, Leicestershire, England

INTRODUCTION

The classification of P_2-purinoceptors is limited by the absence of selective, competitive antagonists and has, therefore, relied heavily on determination of the relative order of agonist potencies. Few agonists, however, have been described that are both pharmacologically selective and resistant to degradation by ectonucleotidases. A recent report[1] identifies adenosine 5′-(2-fluorodiphosphate) (ADP-β-F) as a specific and relatively stable P_{2Y}-purinoceptor agonist in guinea pig taenia coli. As such, ADP-β-F would represent an important tool for characterizing purinoceptors. We have attempted to verify its selectivity using a vascular preparation, the rabbit isolated jugular vein. The P_{2Y}-purinoceptor agonists cause endothelial-dependent relaxation of vascular smooth muscle by release of endothelium-derived relaxing factor and/or prostacyclin.[2]

METHODS

Rings of external jugular vein from male New Zealand white rabbits were prepared for organ bath studies as described elsewhere.[3] Experiments were performed in Krebs solution containing 2.8×10^{-6} M indomethacin (37 °C, 95% O_2/5% CO_2). Some tissues were denuded of endothelium by gentle abrasion with a scored cannula. Rings were contracted with the thromboxane mimetic U46619 (10^{-8} M), and a single cumulative agonist concentration-effect curve (or E/[A] curve, where E indicates the magnitude of relaxation, and [A] indicates the concentration of agonist) was constructed for ADP-β-F (10^{-7}-10^{-3} M). Relaxant responses to ADP-β-F were determined in endothelial-intact or endothelial-denuded preparations, in the presence or absence of the selective P_1-purinoceptor antagonist 8-sulphophenyltheophylline (8-SPT) (3×10^{-4} M). Responses were expressed as a percentage of the U46619 contraction.

RESULTS

In endothelial-intact tissues ADP-β-F produced an E/[A] curve that appeared to consist of two vasorelaxant components, responses were slow and tonic over the concentration range 10^{-7}-3×10^{-5} M but fast and phasic over 10^{-4}-10^{-3} M (FIG. 1a). In endothelial-intact tissues in the presence of 8-SPT (FIG. 1b) the higher potency phase was abolished leaving the apparently monophasic lower potency component. The converse result was obtained upon endothelial denudation (FIG. 1c).

Mean data (N = 4-6) are shown in FIGURE 2. In denuded tissues relaxant responses to ADP-β-F (p[A]$_{50}$: 5.58 $\pm$ 0.13) were virtually abolished by 8-SPT, indicating that they were due to activation of P_1-purinoceptors. The lower potency endothelial-dependent component, fully described only in the presence of 8-SPT, had a p[A]$_{50}$ of 3.98 $\pm$ 0.07.

DISCUSSION

The selectivity of ADP-β-F is overtly suspect because it produces two kinetically and pharmacologically distinct relaxant responses in the rabbit jugular vein. The

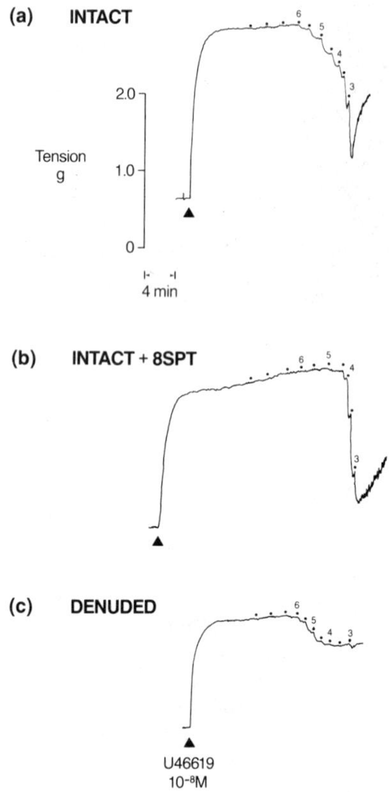

FIGURE 1. Typical original traces showing the relaxant responses of the precontracted rabbit jugular vein to ADP-β-F. (**a**) Tissue with intact endothelium. (**b**) Tissue with intact endothelium in the presence of 8-SPT (3×10^{-4} M). (**c**) Tissue denuded of endothelium.

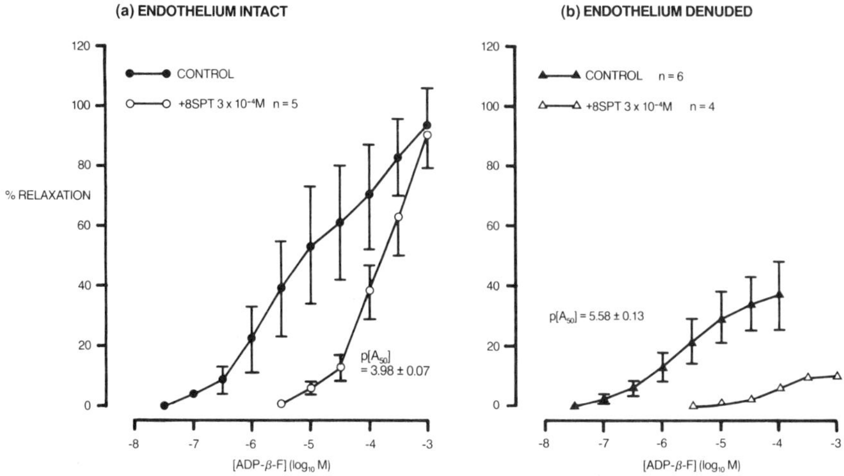

FIGURE 2. Summary of data for relaxant effects of ADP-β-F in the rabbit jugular vein (each point is a mean ± SEM). (**a**) Endothelium intact. (**b**) Endothelium denuded.

primary effect of ADP-β-F is mediated at P_1-purinoceptors located on the smooth muscle, endothelial-dependent relaxations presumably mediated at P_{2Y}-purinoceptors occur only at higher concentrations.

Hourani *et al.*[1] failed to observe P_1-agonist properties of ADP-β-F in the taenia coli. Our use of vascular rather than visceral smooth muscle allows ready separation of P_1 from P_{2Y} components by removal of the endothelium. Hourani *et al.* also report a much higher P_{2Y} potency for ADP-β-F in the taenia (2.7 μM). One possible explanation, therefore, for the lack of agreement between these two studies is a marked difference in the relative densities of the two purinoceptors in question when comparing guinea pig taenia with rabbit jugular vein. Based on the reported relative resistance of ADP-β-F to degradation,[1] it is most probable that all the effects observed in the jugular vein are directly attributable to the compound itself. We cannot, however, totally exclude the possibility that some degradation to adenosine contributed to the P_1 component of the responses.

To summarize, in the rabbit jugular vein relaxations produced by ADP-β-F were principally mediated through activation of P_1-purinoceptors. It is, therefore, unsafe to regard this agonist as a selective probe for P_{2Y}-purinoceptors. If ADP-β-F is used in the classification of P_2-purinoceptors appropriate precautions should be taken to eliminate its P_1-purinoceptor properties.

REFERENCES

1. Hourani, S. M. O., L. A. Welford, G. D. Loizou & N. J. Cusack. 1988. Eur. J. Pharmacol. **147:** 131.
2. Gordon, J. L. 1986. Biochem. J. **233:** 309.
3. Leff, P., G. R. Martin & J. M. Morse. 1987. Br. J. Pharmacol. **91:** 321.

A P_2-Purinoceptor on Cardiac Cells Mediates a Cytosolic $[Ca^{2+}]$ Response Requiring Ca^{2+} Influx, Intracellular Ca^{2+} Stores, and Extracellular Phosphate[a]

MARY BETH DE YOUNG AND ANTONIO SCARPA

Department of Physiology and Biophysics
Case Western Reserve University
Cleveland, Ohio 44106

A P_2 purinergic receptor on cardiac ventricular myocytes has been observed that has unique effects on myocardial Ca^{2+} homeostasis.[1] ATP is one of the few natural agents to induce $[Ca^{2+}]$ transients in myocytes, and affected Ca^{2+} transport mechanisms have been examined in detail.[2]

ATP-induced cytosolic $[Ca^{2+}]$ transients were measured in adult rat ventricular myocytes loaded with the Ca^{2+} dye fura-2. The $[Ca^{2+}]$ response induced by micromolar concentrations of ATP is biphasic (FIG. 1) with a transient peak phase followed by a sustained $[Ca^{2+}]$ increase. The peak phase was found to be sensitive to inhibition by both nifedipine (10 μM), which inhibits voltage-sensitive Ca^{2+} channel activity, and ryanodine (1 μM), which depletes intracellular Ca^{2+} stores. Neither agent significantly inhibited the magnitude of the sustained $[Ca^{2+}]$ increase, although ryanodine enhanced it, indicating a role for intracellular Ca^{2+} stores in regulation of this phase. Initial studies on the Ca^{2+} source of the sustained phase showed that tetrodotoxin (10 μM) was partially inhibitory. This suggested that increased intracellular $[Na^+]$ from voltage-stimulated Na^+ channels stimulated Ca^{2+} influx via Na^+/Ca^{2+} exchange. The sustained phase was inhibited more strongly, however, by removal of extracellular inorganic phosphate (P_i). Conversely, increasing extracellular P_i to 11.2

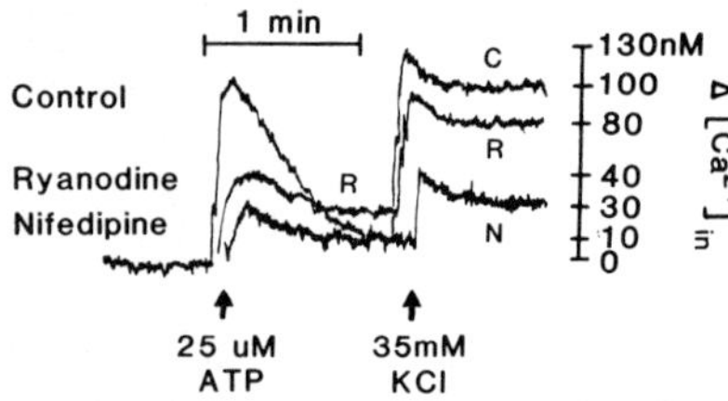

FIGURE 1. Inhibition of ATP-induced $[Ca^{2+}]$ transients by nifedipine and ryanodine.

[a] These studies were supported by Grants HL 18708 and HL 07502 from the National Institutes of Health.

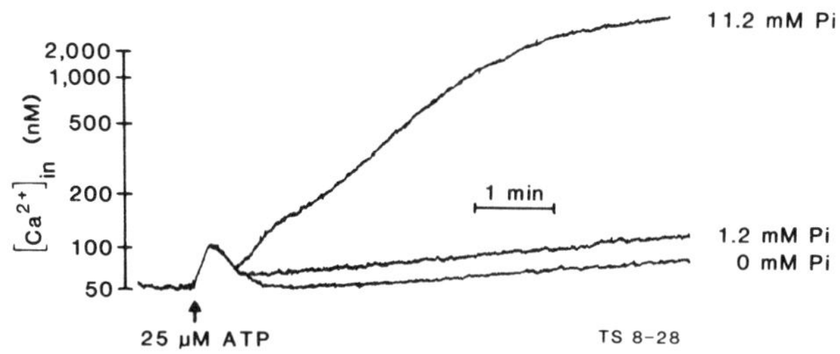

FIGURE 2. Phosphate effects on the sustained phase of the cytosolic $[Ca^{2+}]$ response to ATP.

mM dramatically increased the sustained $[Ca^{2+}]$ changes due to ATP from the nM to the μM range (FIG. 2), which resulted in cell hypercontraction or lysis. This increase was strongly Na^+ dependent, again suggesting mediation by Na^+/Ca^{2+} exchange.

The major $[Ca^{2+}]$ changes stimulated by ATP in a high P_i medium were examined in more detail. The $\Delta[Ca^{2+}]_{ATP}$ was found to increase in a dose-dependent manner with P_i concentration even though increased P_i decreases the medium Ca^{2+} concentration. The response was not due to a nonspecific effect on cell stability as the cells were stable when no agent was added, and $[Ca^{2+}]$ changes observed with cell depolarization by KCl were unaffected by high P_i. Other data indicate that the Na^+ influx involved in the $\Delta[Ca^{2+}]$ requires the presence of extracellular P_i. Ouabain has no significant effect on the magnitude of the sustained $[Ca^{2+}]$ increase in low P_i despite the expected magnification of any $\Delta[Na^+]$ and consequent $\Delta[Ca^{2+}]$. In addition, $^{22}Na^+$ uptake following addition of ATP is much greater in the presence of high phosphate. Because Na^+/P_i cotransport has been observed in these cells,[3] it appears ATP may modulate this transport mechanism. In summary, ATP receptor modulation of an Na^+- and P_i-dependent mechanism that affects Ca^{2+} transport has been observed in cardiac cells, indicating that a novel ATP receptor/effector mechanism is at work.

REFERENCES

1. DE YOUNG, M. B. & A. SCARPA. 1987. FEBS Lett. **223:** 53-58.
2. DE YOUNG, M. B. & A. SCARPA. 1989. Am. J. Physiol. **257:** C750-C758.
3. JACK, M. G., W. H. HUANG & A. ASKARI. 1989. J. Biol. Chem. **264:** 3904-3908.

P_2-Purinoceptor Subtypes in Guinea Pig Smooth Muscle

N. P. WIKLUND,[a,b] C. U. WIKLUND,[a,c] AND
L. E. GUSTAFSSON [a,c]

[a]*Department of Physiology*
[b]*Department of Urology*
and
[c]*Institute of Environmental Medicine*
Karolinska Institutet
Stockholm, Sweden

INTRODUCTION

Adenine nucleotides have diverse actions on smooth muscle, evoking relaxation in guinea pig taenia coli and contraction in guinea pig vas deferens or ileum, via activation of distinct receptors, designated P_2-purinoceptors.[1] Recent studies have indicated two subtypes of P_2-purinoceptors: P_{2X} and P_{2Y}. At P_{2X}-purinoceptors (excitatory in vas deferens), α,β-methylene ATP is more potent than ATP, and ATP is equipotent with or more potent than 2-methylthio-ATP. At P_{2Y}-purinoceptors (inhibitory in taenia coli), 2-methylthio-ATP is more potent than ATP, and ATP is more potent than α,β-methylene ATP. Furthermore, application of α,β-methylene ATP selectively desensitizes the ATP action at P_{2X}-purinoceptors, whereas no desensitization is seen at P_{2Y}-purinoceptors.[2] In contrast, reactive blue 2 selectively counteracts ATP action at P_{2Y}-purinoceptors.[3] Usually, action at P_{2X}-purinoceptors leads to contraction, whereas action at P_{2Y}-purinoceptors leads to relaxation. The aim of the present work was to further characterize the P_2-purinoceptors in smooth muscle.

MATERIALS AND METHODS

Male guinea pigs were stunned and bled. The smooth muscle preparations were suspended in organ baths containing Tyrode's solution aerated with 5% CO_2 in O_2. The ileum longitudinal muscle (3 Hz, 0.2 msec, 15 pulses at 1-min intervals), vas deferens (7 Hz, 0.3 msec, 14 pulses at 1-min intervals), and taenia coli (5 Hz, 0.3 msec, 25 pulses at 2-min intervals) were transmurally stimulated as previously described.[4] The taenia coli was stimulated in the presence of atropine (3×10^{-7} M) or during contraction with either carbachol (10^{-7} M) or histamine (3×10^{-6} M). Each contractile response to adenine nucleotides was measured as a percentage of the contractile response to the standardized transmural nerve stimulus applied to the muscle preparation.

RESULTS AND DISCUSSION

ADP and ATP derivatives dose-dependently and reversibly contracted the ileum and vas deferens. The contractile effect was unaffected by 8-*p*-sulfophenyltheophylline

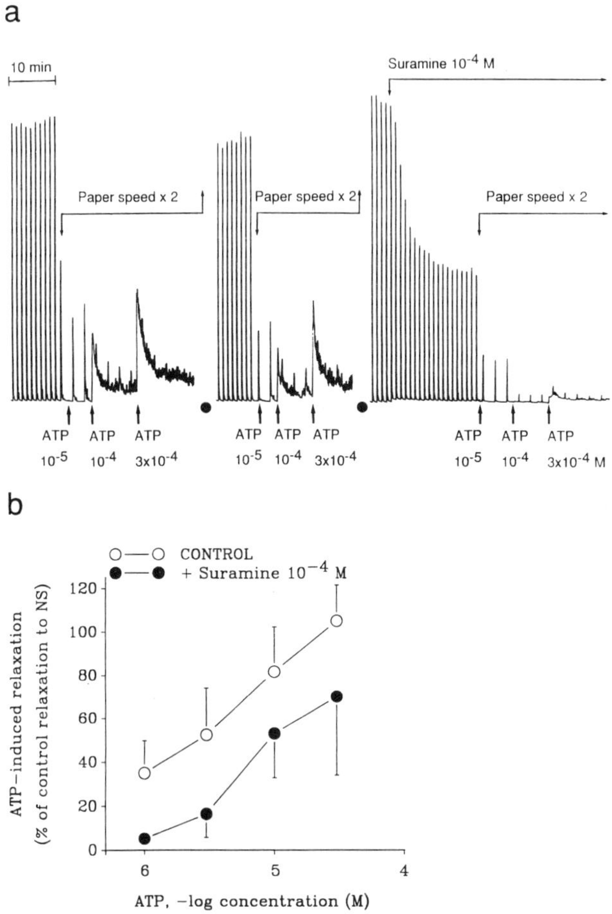

FIGURE 1. (**a**) Guinea pig ileum longitudinal muscle. Inhibition of contractile responses to transmural nerve stimulation (3 Hz, 0.3 msec, 15 pulses at 1-min intervals) by application of ATP (10^{-4}-10^{-3} M) and concomitant contractile responses evoked by the nucleotide. Antagonism of the contractile effect after application of suramine (10^{-4} M). Wash at dots. (**b**) Guinea pig taenia coli. Relaxations induced by ATP (open circles), each value expressed as a percentage of the relaxation induced by transmural stimulation (5 Hz, 0.3 msec, 25 pulses at 2-min intervals). The relaxations induced by ATP were slightly antagonized by application of suramine (10^{-4} M). The muscle preparation was precontracted by carbachol (10^{-7} M).

or tetrodotoxin. 2-Methylthio-ATP was much more potent than ATP in the ileum. 8-Bromo-ATP was without contractile activity in the ileum. In the taenia coli, ADP, ATP, and 8-bromo-ATP induced a dose-dependent and reversible relaxation that was unaffected by 8-*p*-sulfophenyltheophylline. Contractile effects of ADP and ATP were attenuated by α,β-methylene ADP or α,β-methylene ATP in vas deferens, whereas α,β-methylene derivatives enhanced the contractile effect of ATP and ADP in ileum and did not affect the relaxations induced by ADP and ATP in the taenia coli preparation. *p*-Chloromercuribenzene sulfonic acid (PCMBS) antagonized the contractile effect of ADP and ATP in ileum and vas deferens, whereas the relaxations in the taenia coli were unaltered by PCMBS treatment. Reactive blue 2 was without effect on the ATP-induced contractions in the ileum. Suramine antagonized the contractile effect of ATP in the ileum and vas deferens (FIG. 1a). Furthermore, it antagonized the inhibitory effect of ATP in the taenia coli although here it seemed less potent. In the taenia coli, suramine also inhibited relaxations evoked by nitroprusside (FIG. 1b).

The data confirm the distinction between P_2 receptors/sites in vas deferens (P_{2X}) and taenia coli (P_{2Y}). Furthermore, PCMBS is suggested to be an antagonist at excitatory P_2 receptors/sites in ileum and vas deferens. Suramine is an antagonist at both excitatory and inhibitory P_2 receptors/sites, although it was more potent at the excitatory receptors and seemed to block other agents with inhibitory actions in the taenia coli. The ileum receptor/site has agonist characteristics more like P_{2Y} receptors/sites but antagonist characteristics more like P_{2X} receptors/sites (TABLE 1), although it is not desensitized by α,β-methylene derivatives,[5] and is suggested to be of a novel type, P_{2S}.[4]

TABLE 1. Provisional Criteria for P_{2S} Receptors/Sites, as Compared to P_{2Y} and P_{2X}[a]

Agonists

Receptor Type	Agonist Potency Order
P_{2Y}	2-Methylthio-ATP[6] > ATP = ADP ≥ 8-bromo-ATP*[,7]
P_{2X}	ATP = ADP = 8-bromo-ATP[8] = 2-methylthio-ATP[9]
P_{2S}	2-Methylthio-ATP* > ATP = ADP >> 8-bromo-ATP*

Antagonists

Receptor Type	α,β-Methylene ATP	Reactive Blue 2	PCMBS	Suramine
P_{2Y}	No*	Yes[3]	No*	Yes (unspecific)*
P_{2X}	Yes*[,2]	No[3]	Yes*	Yes*
P_{2S}	No*	No*	Yes*	Yes*

NOTE: The numbers cited in superscript correspond to the numbers of the references. The asterisks indicate present results. 8-*p*-Sulfophenyltheophylline did not block the P_2 receptors.

[a] Criteria for P_{2X} and P_{2Y} receptors are discussed by Burnstock and Kennedy.[2]

REFERENCES

1. BURNSTOCK, G. 1978. A basis for distinguishing two types of purinergic receptor. *In* Cell Membrane Receptors for Drugs and Hormones: A Multidisciplinary Approach. R. W. Straub & L. Bolis, Eds.: 107. Raven Press. New York, NY.
2. BURNSTOCK, G. & C. KENNEDY. 1985. Is there a basis for distinguishing two types of P_2-purinoceptor? Gen. Pharmacol. **16:** 433.
3. BURNSTOCK, G. & J. J. I. WARLAND. 1987. P_2-Purinoceptors of two subtypes in the rabbit

mesenteric artery: Reactive Blue 2 selectively inhibits responses mediated via the P_{2Y}- but not the P_{2X}-purinoceptor. Br. J. Pharmacol. **90:** 383.
4. WIKLUND, N. P. & L. E. GUSTAFSSON. 1988. Indications for P_2-purinoceptor subtypes in guinea pig smooth muscle. Eur. J. Pharmacol. **148:** 361.
5. WIKLUND, N. P. & L. E. GUSTAFSSON. 1988. Agonist and antagonist characterization of the P_2-purinoceptors in the guinea pig ileum. Acta Physiol. Scand. **132:** 15.
6. SATCHELL, D. G. & M. H. MAGUIRE. 1975. Inhibitory effects of adenine nucleotide analogs on the isolated guinea pig taenia coli. J. Pharmacol. Exp. Ther. **195:** 540.
7. MAGUIRE, M. H. & D. G. SATCHELL. 1979. Specificity of adenine nucleotide receptor sites: Inhibition of the guinea pig taenia coli by adenine nucleotide analogues. *In* Physiological and Regulatory Functions of Adenosine and Adenine Nucleotides. H. P. Baer & G. I. Drummond, Eds.: 33. Raven Press. New York, NY.
8. SATCHELL, D. G. 1988. Differences in the structural requirements for agonist properties at P_1 and P_2 receptors in smooth muscle. *In* Adenosine and Adenine Nucleotides: Physiology and Pharmacology. D. M. Paton, Ed.: 85. Taylor & Francis. Basingstoke.
9. BURNSTOCK, G., N. J. CUSACK & L. A. MELDRUM. 1985. Studies on the stereoselectivity of the P_2-purinoceptor on the guinea pig vas deferens. Br. J. Pharmacol. **84:** 431.

Nucleotides Uncomplexed with Divalent Cations Activate a Receptor Coupled to Phosphoinositidase C in Pituitary Cells

J. S. DAVIDSON, I. WAKEFIELD,
P. A. VAN DER MERWE, U. SOHNIUS, AND
R. P. MILLAR

Department of Chemical Pathology
University of Cape Town Medical School
Cape Town, South Africa

In primary cultures of sheep pituitary cells, extracellular nucleotides stimulated rapid increases in inositol tris-, bis-, and monophosphates, accompanied by Ca^{2+} mobilization.[1,2] These effects were not due to cell permeabilization.[1] The inositol phosphate response to nucleotides was greater than any of the responses elicited by the known hypothalamic releasing peptides.[2]

The pharmacological specificity of the pituitary nucleotide receptor is different from the P_{2X}- and P_{2Y}-purinoceptors characterized in other tissues.[3] UTP, ATPγS, and ATP were the most potent agonists (their EC_{50} values for inositol phosphate production were 1.2, 2.6, and 2.7 μM, respectively, in the presence of 1 mM Ca^{2+} and 1 mM Mg^{2+}). APPCP, APCPP, and 2-methylthio-ATP were inactive.

Increasing Mg^{2+} concentrations caused a rightward shift in the ATP dose-response curves (FIGS. 1 & 2), indicating that the active species is not Mg-ATP and may be ATP^{4-}. When inositol tris- or bisphosphate was measured after brief stimulation with ATP (FIG. 1), the degree of shift due to increasing Mg^{2+} was in close agreement with the predicted shift calculated on the assumption that Mg-ATP is inactive. When total inositol phosphate accumulation after prolonged stimulation, or $^{45}Ca^{2+}$ efflux was measured (FIG. 2), however, the observed rightward shifts were somewhat less than predicted, for reasons that are not clear. Interestingly, neither Mg^{2+} nor Ca^{2+} ions are required for receptor activation, since ATP was able to stimulate inositol phosphate production with high potency (EC_{50} = 200 nM) in the presence of 1 mM EDTA (FIG. 2). This indicates that ectokinase or ecto-ATPase activity is not involved in the mechanism of receptor activation, since all phosphate transfer reactions require divalent cations. It is intriguing that a species of ATP uncomplexed with divalent cations is the active agonist. In physiological extracellular fluid these forms constitute only 6-7% of total ATP present.

Because a rise in cytosolic Ca^{2+} is a sufficient stimulus for pituitary hormone release, we considered the possibility that ATP or other nucleotides secreted by hypothalamic neurones might act as releasing factors. In experiments in which the release of luteinizing hormone, prolactin, growth hormone, corticotrophin, and thyrotrophin were measured, however, none of these pituitary hormones were secreted in response to extracellular nucleotides. This suggests that nucleotide receptors are

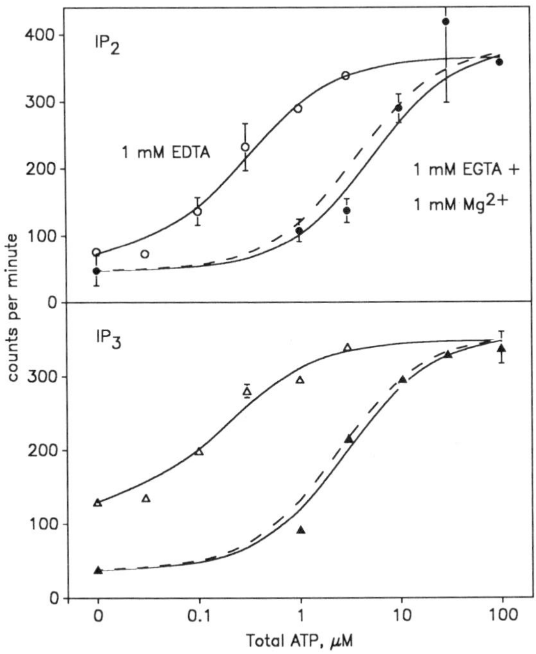

FIGURE 1. Effect of extracellular Mg^{2+} concentration on ATP-stimulated inositol bis- and trisphosphate levels. Cells were stimulated for 1 min in the absence of Li^+ and Ca^{2+}. Media contained 1 mM EDTA and 0 mM Mg^{2+} (○,△) or 1 mM EGTA plus 1.18 mM total Mg^{2+}, giving free Mg^{2+} at 1 mM (●,▲). Solid lines are the least-squares, best-fit curves for the observed data. Dashed lines show the theoretical expected position of the 1 mM free Mg^{2+} curve relative to the 0 mM Mg^{2+} curve, calculated on the assumption that Mg-ATP has no agonist activity.

located on a nonsecretory cell type. The cell type bearing these receptors is unknown, as are the functions of these receptors in the pituitary. The large magnitude of the inositol phosphate and $^{45}Ca^{2+}$ efflux responses to nucleotides, however, indicates that a substantial fraction of the cells in primary pituitary cultures bear nucleotide receptors, and suggests that extracellular nucleotides have a regulatory role in the pituitary.

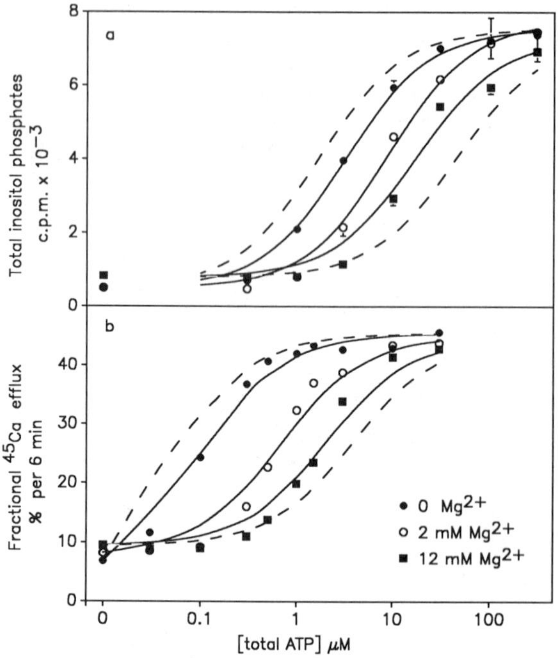

FIGURE 2. Effect of extracellular Mg^{2+} concentration on ATP stimulation of inositol phosphate production and $^{45}Ca^{2+}$ efflux. (**a**) The total inositol phosphate production: Cells prelabeled with [^{3}H]inositol were preincubated for 15 min in Mg^{2+}-free buffer containing 10 mM Li^+, then stimulated with ATP for 40 min in the presence of 10 mM Li^+; 1 mM Ca^{2+}; and 0 (●), 2 (○), or 12 mM (■) Mg^{2+}. (**b**) The $^{45}Ca^{2+}$ efflux: Cells loaded with $^{45}Ca^{2+}$ and washed extensively were stimulated with ATP for 6 min in the presence of 0.1 mM Ca^{2+} and 0 (●), 2 (○), or 12 mM (■) Mg^{2+}. Solid lines are least-squares, best-fit curves for the observed data. Dashed lines show theoretical predicted shifts of the 0 mM Mg^{2+} and 12 mM Mg^{2+} curves, relative to the central curve, calculated on the assumption that Mg-ATP is inactive.

REFERENCES

1. Van der Merwe, P. A., I. K. Wakefield, J. Fine, R. P. Millar & J. S. Davidson. 1989. FEBS Lett. **234:** 333.
2. Davidson, J. S., I. K. Wakefield, U. Sohnius, P. A. van der Merwe & R. P. Millar. 1990. Endocrinology **126:** 80.
3. Gordon, J. L. 1986. Biochem. J. **233:** 309.

ATP-Evoked Calcium Flux, Protein Phosphorylation, and Stellation in Astrocytes

J. T. NEARY, C. VAN BREEMEN, R. LASKEY,
J. BLICHARSKA, L.-O. B. NORENBERG, AND
M. D. NORENBERG

Laboratory of Neuropathology
Veterans Administration Medical Center
and
Departments of Pathology and Pharmacology
University of Miami School of Medicine
Jackson Memorial Hospital
Miami, Florida 33101

Calcium plays an important role in a number of signal transduction-regulated functions in neurons, but relatively little is known about calcium signaling in astrocytes. These cells are believed to modulate synaptic transmission and neuronal excitability by their ability to take up, release, and/or metabolize neuroactive agents.[1,2] To enhance our understanding of neuronal-glial communication, we have been investigating calcium entry in astrocytes and the mechanisms that are influenced by increases in intracellular calcium. Because one of the second messenger functions of calcium is the stimulation of protein phosphorylation systems, we have studied the effect of ATP-stimulated calcium entry on protein phosphorylation in astrocytes. In an effort to investigate possible functional roles of extracellular ATP, we have also investigated ATP-evoked morphological changes in astrocyte cultures.

Recently, it has been shown that extracellular ATP stimulates calcium influx and accumulation as well as release from internal stores in cultured astrocytes.[3-5] In our studies of ^{45}Ca flux in rat cerebral cortical astrocytes, we found that ATP (1 mM) stimulated ^{45}Ca uptake over a time course of 1 to 30 min, with a maximum increase of about 2.5-fold at 20 min.[3] A biphasic dose-response curve was obtained with EC_{50} values of 0.3 nM and 9 μM, suggesting the presence of low- and high-affinity purinergic binding sites on cultured astrocytes. Lanthanum, an inhibitor of calcium transport, was effective in blocking ATP-stimulated ^{45}Ca influx as well as a large portion of the passive calcium leak. As shown in FIGURE 1, a P_1-purinoceptor antagonist, isobutylmethylxanthine (IBMX), had no effect on ATP-stimulated ^{45}Ca influx, thereby indicating that ATP acts via a P_2-purinergic receptor in cultured astrocytes.

To further characterize the ATP-evoked calcium response, we conducted studies with fura-2. Cultures were incubated in 1 μM fura-2/AM for 45 min at 37 °C and then in physiological saline solution for 10 min to remove excess dye. In the presence of external calcium, application of extracellular ATP (100 μM) resulted in a calcium peak followed by a sustained calcium signal, whereas in the absence of external calcium only a transient peak response was observed. These findings suggest that ATP stimulates mobilization of internal calcium as well as influx of calcium across the plasma membrane.

Studies on the effect of the ATP-induced calcium signal on protein phosphorylation revealed that extracellular application of ATP increased phosphate incorporation in 55- and 52-kDa proteins (4- and 2-fold, respectively) and decreased phosphorylation by approximately 50% in 24- and 21-kDa proteins. The 52-kDa protein comigrated with glial fibrillary acidic protein (GFAP). These effects were time and dose dependent; EC_{50} was estimated to be 10 μM. An order of potency of ATP > ADP > > adenosine > GTP was observed, suggesting that the effects of ATP on phosphorylation are mediated by a P_2-purinergic receptor.

Several studies were conducted to assess the calcium dependence of the effect of extracellular ATP on protein phosphorylation. Lanthanum blocked the changes in phosphate incorporation in the 55-, 52-, 24-, and 21-kDa proteins, suggesting that the ATP-evoked effects were mediated by an increase in intracellular calcium. This was supported by the observation that the ATP-stimulated changes in phosphorylation were reduced when calcium was omitted from the bath. Moreover, the effects of

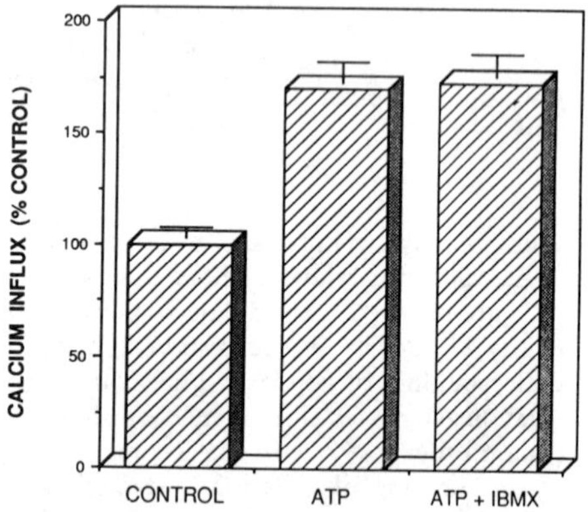

FIGURE 1. ATP-induced ^{45}Ca influx in the presence of IBMX. The influx of ^{45}Ca was conducted for 1.5 min as previously described.[3] A P_1-purinoceptor antagonist, IBMX (100 μM), did not block ^{45}Ca influx stimulated by ATP (100 μM), suggesting that ATP is acting via a P_2-purinoceptor in cultured astrocytes.

extracellular ATP on phosphate incorporation were mimicked by a calcium ionophore; application of 1 μM ionomycin for 1 min resulted in increased phosphorylation of 55- and 52-kDa proteins and decreased phosphorylation of 24- and 21-kDa proteins.

Astrocytes treated with dibutyryl cyclic AMP (Bt_2-cAMP) have enhanced calcium effector systems including an increase in the number of calcium channels[6,7] and in calcium, calmodulin-dependent protein kinase activity.[8] The Bt_2-cAMP-treated cultures also have elevated phosphate incorporation in 55- and 52-kDa proteins.[9] Application of extracellular ATP to Bt_2-cAMP-treated astrocytes decreased phosphate incorporation in the 24- and 21-kDa proteins but had little to no effect on phosphorylation in the 55- and 52-kDa proteins, probably because these proteins were already phosphorylated in the Bt_2-cAMP-treated cultures.

It is also of interest that Bt_2-cAMP treatment induces stellation in astrocytes by a mechanism that may be dependent on calcium entry.[10] In view of this, we studied the effect of extracellular ATP on process formation in untreated astrocytes and found that application of ATP (100 μM) for 1 hr resulted in marked stellation. This finding supports the idea that calcium entry may be linked to differentiation in astrocytes.

In summary, extracellular ATP evokes a calcium signal in astrocytes via mobilization of internal calcium as well as influx of calcium across the plasma membrane. Protein phosphorylation studies indicate that calcium-dependent protein kinases and phosphatases transduce the effects of the calcium signal brought about by activation of P_2-purinergic receptor-operated calcium channels in astrocytes. Because extracellular ATP induced marked stellation and resulted in phosphorylation of a protein that comigrated with GFAP, a protein known to be involved in process formation and reactive gliosis, we speculate that the purinergic-stimulated calcium signal in astrocytes may be involved in differentiation or in response to CNS injury. Furthermore, because ATP is released from purinergic nerve terminals,[11,12] a calcium signal in astrocytes stimulated by ATP may play an important role in neuronal-astrocytic communication.

REFERENCES

1. CSERR, H. F., ED. 1986. The Neuronal Microenvironment. Vol. 481. New York Academy of Sciences. New York, NY.
2. NORENBERG, M. D., L. HERTZ & A. SCHOUSBOE. 1988. The Biochemical Pathology of Astrocytes. Alan R. Liss. New York, NY.
3. NEARY, J. T., C. VAN BREEMEN, E. FORSTER & M. D. NORENBERG. 1988. Biochem. Biophys. Res. Commun. **157:** 1410-1416.
4. PEARCE, B., S. MURPHY, J. JEREMY, C. MORROW & P. DANDONA. 1989. J. Neurochem. **52:** 971-977.
5. CASTROS, H. & K. D. MCCARTHY. 1989. Trans. Am. Soc. Neurochem. **20:** 188.
6. BARRES, B. A., L. L. Y. CHUN & D. P. COREY. 1989. J. Neurosci. **9:** 3169-3175.
7. MACVICAR, B. A. & F. W. Y. TSE. 1988. Glia **1:** 359-365.
8. BABCOCK-ATKINSON, E., M. D. NORENBERG, L.-O. B. NORENBERG & J. T. NEARY. 1989. Glia **2:** 112-118.
9. NEARY, J. T., M. P. GUTIERREZ, L.-O. B. NORENBERG & M. D. NORENBERG. 1987. Brain Res. **410:** 164-168.
10. MACVICAR, B. A. 1987. Brain Res. **420:** 175-177.
11. BURNSTOCK, G. 1972. Pharmacol. Rev. **24:** 509-581.
12. RICHARDSON, P. J. & S. J. BROWN. 1987. J. Neurochem. **48:** 622-630.

Solubilization of a Guanine Nucleotide-Sensitive Form of the P_{2Y}-Purinergic Receptor from Turkey Erythrocyte Membranes

R. A. JEFFS, C. L. COOPER, J. L. BOYER, AND T. K. HARDEN

Department of Pharmacology
University of North Carolina School of Medicine
Chapel Hill, North Carolina 27599

P_{2Y}-Purinergic receptors were solubilized from turkey erythrocyte membranes with the nonionic detergent digitonin. The receptors retained the pharmacological characteristics of membrane-bound P_{2Y}-purinergic receptors, as assessed by the use of the selective radiolabeled agonist adenosine-5′-*O*-[2-thiodiphosphate] ([^{35}S]ADPβS) in an equilibrium binding assay. [^{35}S]ADPβS labeled a single population of high-affinity sites (K_d = 12.9 nM; B_{max} = 4.5 pmol/mg protein), and adenine nucleotide analogues competitively inhibited [^{35}S]ADPβS binding with a rank order of potency consistent with that for the P_{2Y}-purinergic receptor.

High-affinity binding of [^{35}S]ADPβS to solubilized P_{2Y}-purinergic receptors was sensitive to inhibition by guanine nucleotides. This effect was noncompetitive and displayed a rank order of potency that was in agreement with the potency order observed for guanine nucleotide-mediated inhibition of [^{35}S]ADPβS binding in purified turkey erythrocyte plasma membranes. Furthermore, the rate of ATPτS-induced dissociation of [^{35}S]ADPβS from solubilized receptors was increased 2.7-fold by the simultaneous addition of GTPτS.

The sensitivity of radiolabeled agonist-P_{2Y}-receptor interaction to guanine nucleotides suggests that the soluble receptor is physically associated with a guanine nucleotide regulatory protein. To test this hypothesis, the apparent molecular size of the labeled soluble receptor complex was estimated by size exclusion chromatography on a Superose-12 column. Previous studies have indicated that the turkey erythrocyte P_{2Y}-purinergic receptor can be covalently labeled with the specific photoaffinity label 3′-*O*-(4-benzoyl)benzoyl adenosine 5′-triphosphate ([α-^{32}P]BzATP). Plasma membrane P_{2Y}-purinergic receptors were prelabeled with [^{35}S]ADPβS or covalently labeled with [α-^{32}P]BzATP prior to solubilization and chromatography on Superose-12. [^{35}S]ADPβS- or [α-^{32}P]BzATP-labeled species eluted as a single peak of radioactivity of apparent M_r greater than 300,000. Addition of GTPτS to the labeled species and rechromatography resulted in a loss of protein labeling by [^{35}S]ADPβS and a shift of the covalently [α-^{32}P]BzATP-labeled species to a single peak of radioactivity eluting from the size-exclusion column as a species of much smaller apparent size (apparent M_r: 70,000).

These results suggest that a P_{2Y}-purinergic receptor-guanine nucleotide regulatory protein complex is stable to membrane solubilization with digitonin, even in the absence of prebound agonist. The development of this methodology may provide a first step toward eventual purification of the receptor, as well as eventual identification of the guanine nucleotide regulatory protein with which it interacts.

Irreversible Activation and Photoaffinity Labeling of a P_{2Y}-Purinergic Receptor Coupled to Phospholipase C Activation in Turkey Erythrocyte Membranes

JOSÉ L. BOYER, CRISTY L. COOPER, AND T. KENDALL HARDEN

Department of Pharmacology
University of North Carolina School of Medicine
Chapel Hill, North Carolina 27599

The ATP analogue, 3′-*O*-(4-benzoyl)benzoyl adenosine 5′-triphosphate (BzATP) was used as a photoaffinity probe for P_{2Y}-purinergic receptors in turkey erythrocytes. BzATP stimulated inositol phosphate formation in a concentration-dependent manner ($K_{0.5} = 172 \pm 4$ nM). This effect of BzATP on phospholipase C was strictly dependent on the presence of a guanine nucleotide (GTPγS, Gpp(NH)p, or GTP), and was not additive to the effects of other full P_{2Y}-purinergic receptor agonists. Photolysis of BzATP-preincubated [^{3}H]inositol-labeled membranes resulted in an irreversible increase in GTPγS-stimulated formation of inositol phosphates observed as a 3-fold increase in the rate of activation of phospholipase C. BzATP effects on phospholipase C activity were prevented by the presence of ATP and ATP analogues during photolysis with an order of potency, that is, ATP > ADP > App(NH)p > β,γ-methylene ATP, consistent with that for a P_{2Y}-purinergic receptor. These results indicate that BzATP is a full P_{2Y}-purinergic receptor agonist that upon photolysis irreversibly promotes P_{2Y}-purinergic receptor-mediated and guanine nucleotide-dependent activation of phospholipase C.

Based on these observations, we prepared [^{32}P]BzATP and used it as a photoaffinity probe for the labeling of the P_{2Y}-purinergic receptor in turkey erythrocyte membranes. In the absence of light, [^{32}P]BzATP bound with high affinity ($K_D \sim 5$ nM) to an apparently homogeneous population of binding sites. The binding of the photoaffinity ATP analogue was saturable and reversible. Analogues of ATP and ADP competitively inhibited the binding of [^{32}P]BzATP with a pharmacological specificity consistent with the binding of the radioligand to a P_{2Y}-purinergic receptor. Guanine nucleotides inhibited the binding of [^{32}P]BzATP in a noncompetitive manner, suggesting that the photoaffinity label interacts with a G-protein-regulated receptor protein.

Photolysis of [^{32}P]BzATP in the presence of turkey erythrocyte plasma membranes resulted in the covalent incorporation of [^{32}P]BzATP into a protein of approximately 53,000 Da, as identified by autoradiography of sodium dodecyl sulfate-polyacrylamide gels. The photoincorporation of [^{32}P]BzATP was inhibited by analogues of ATP and ADP with a rank order of potency similar to that observed for the binding of

[^{32}P]BzATP in the dark and to the pharmacological specificity observed for P_{2Y}-purinergic receptor-mediated activation of phospholipase C. The photoaffinity labeling of the 53,000-Da protein was also inhibited in a concentration-dependent manner by guanine nucleotides. A protein with identical electrophoretic mobility was also radiolabeled when [^{32}P]BzATP was photolyzed in the presence of membranes prepared from rat liver, rat brain, 1321N1 human astrocytoma cells, rat astrocytes, and bovine pulmonary aortic endothelial cells, but not in membranes from human platelets or human erythrocytes. Based on these results, [^{32}P]BzATP should prove of value for the eventual purification and characterization of P_2-purinergic receptors.

P_2-Purinergic Receptors on Vascular Endothelial Cells

Transduction Mechanisms

S. PIROTTON,[a,b] M. LECOMTE,[b] B. ROBAYE,[b]
D. DEMOLLE,[b] A. VAN COEVORDEN,[b]
A. C. NAIRN,[c] AND J. M. BOEYNAEMS [b]

[b]*Institute of Interdisciplinary Research*
Free University of Brussels
1070 Brussels, Belgium

[c]*The Rockefeller University*
New York, New York 10021

ATP and ADP, via P_{2Y} receptors, stimulate the release of nitric oxide[1] and of prostacyclin[2,3] from aortic endothelial cells and increase their rate of proliferation.[4] These actions might have a physiological importance in the interaction between platelets and the vascular endothelium. The release of prostacyclin will limit the extent of platelet aggregation following a lesion of the endothelium, while the mitogenic effect will accelerate the repair of that lesion.

Bovine aortic endothelial cells (BAECs) provided a useful model to elucidate the transduction mechanisms associated with P_{2Y} receptors. The occupancy of these receptors induces the hydrolysis of several phospholipids by various phospholipases, generating different second messengers. These responses have different time courses and are triggered by distinct mechanisms (FIG. 1). ATP and ADP induce a rapid and transient accumulation of inositol trisphosphate (IP_3), reflecting the hydrolysis of phosphatidylinositol bisphosphate (PIP_2) by phospholipase C (FIG. 1A).[5] The rise of cytoplasmic Ca^{2+} that results[6] is responsible for a burst of prostacyclin release, probably via the activation of a Ca^{2+}-dependent phospholipase A_2.[7] The coupling of the P_{2Y} receptors and the phospholipase C seems to involve a GTP-binding protein (FIG. 1C).[8] ATP and ADP also induce a sustained increase of the choline level inside endothelial cells (FIG. 1B), an increase likely to result from the activation of phospholipase D, which hydrolyzes phosphatidylcholine into phosphatidic acid and choline.[9] Phosphatidic acid might play a role in the mitogenic effect of ATP.[10] Depletion of protein kinase C by a prolonged exposure to phorbol 12-myristate 13-acetate abolished the ATP-stimulated release of choline metabolites (FIG. 1D). This indicates that the activation of protein kinase C by diacylglycerol released from PIP_2 plays a crucial role in the stimulation of phospholipase D.[9]

[a] Address for correspondence: Institute of Interdisciplinary Research, Free University of Brussels, 808 Route de Lennik, Building C, 1070 Brussels, Belgium.

In many systems, the effects of second messengers are mediated by the activation of specific protein kinases that phosphorylate a variety of substrates. In BAECs, the transient rise of cytoplasmic free Ca^{2+} induced by ATP leads to the phosphorylation of several substrates, in particular 95-kDa and 28-kDa proteins.[11] The 95-kDa protein has been immunologically identified as the elongation factor-2 (EF-2) (FIG. 2). Phosphorylation of the EF-2 decreases the rate of polypeptide elongation: the significance of this event in the action of ATP on BAECs remains unclear. The phosphorylation of the 28-kDa proteins is also stimulated by tumor necrosis factor (TNF) (FIG. 3),[13]

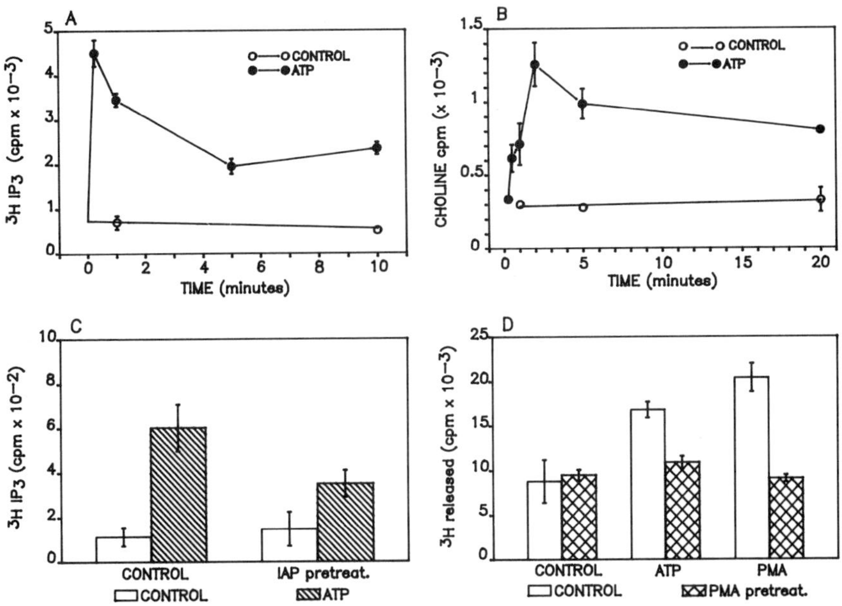

FIGURE 1. Effect of ATP on PIP_2 and PC metabolism in BAECs. (**A**) ATP (100 μM) induces a rapid and transient accumulation of IP_3 in BAECs labeled with [^{3}H]inositol. (**B**) ATP (100 μM) also produces a sustained accumulation of choline inside BAECs prelabeled with [^{3}H]choline. (**C**) The production of IP_3 induced by ATP is partially inhibited when the cells are pretreated for 15 hr with pertussis toxin (200 ng/ml). Such a treatment has no effect on the formation of choline in response to ATP. (**D**) Down-regulation of kinase C (24-hr exposure to 500 nM phorbol 12-myristate 13-acetate) abolishes the stimulatory effect of ATP on choline and phosphorylcholine release from prelabeled BAECs.

via another signaling pathway than that of ATP, because TNF does not increase the concentration of IP_3 and cytoplasmic Ca^{2+} in endothelial cells.[13] Furthermore, the same proteins are phosphorylated following exposure of these cells to arsenite, indicating that they are identical to the well-known 27-kDa stress proteins.[13] The fact that the increased phosphorylation of these proteins is an event common to distinct signaling mechanisms suggests that these proteins might play an important, but still unknown, role in endothelial cell regulation.

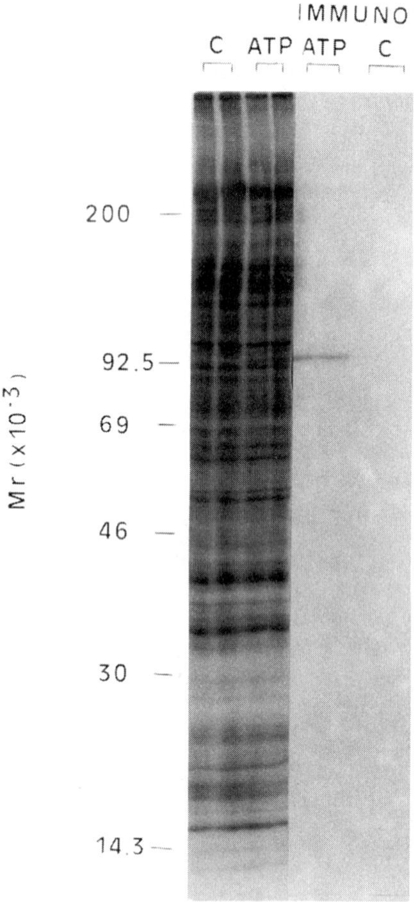

FIGURE 2. Identification of the 95-kDa protein phosphorylated in the BAECs in response to ATP. Immunological studies using an anti-EF-2 antibody[12] show that the 95-kDa substrate phosphorylated in BAECs in response to ATP 50 (μM) is identical to this protein.

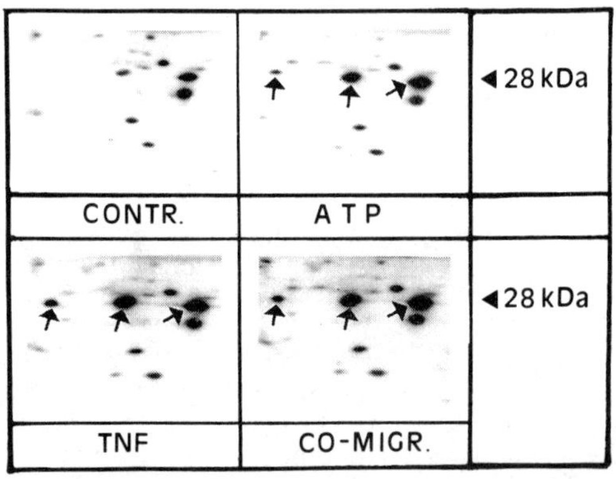

FIGURE 3. Identification of the 28-kDa protein phosphorylated in BAECs in response to ATP. ATP (100 μM) and TNF (2000 U/ml) stimulate the phosphorylation of three proteins of 28-kD on two-dimensional gel electrophoresis. The comigration of samples of ATP-treated cells and of TNF-treated cells shows that these three proteins phosphorylated in response to these agents are identical.

REFERENCES

1. KELM, M. *et al.* 1988. Biochem. Biophys. Res. Commun. **154:** 236-244.
2. PEARSON, J. D. *et al.* 1983. Biochem. J. **214:** 273-276.
3. VAN COEVORDEN, A. & J. M. BOEYNAEMS. 1984. Prostaglandins **27:** 615-626.
4. VAN COEVORDEN, A. *et al.* 1989. Thromb. Haemostas. **62:** 190.
5. PIROTTON, S. *et al.* 1987. J. Biol. Chem. **262:** 17461-17466.
6. HALLAM, T. J. & J. D. PEARSON. 1986. FEBS Lett. **207:** 95-99.
7. CARTER, T. D. *et al.* 1988. Br. J. Pharmacol. **95:** 1181-1190.
8. PIROTTON, S. *et al.* 1987. Biochem. Biophys. Res. Commun. **147:** 1113-1120.
9. PIROTTON, S. *et al.* 1990. J. Cell. Physiol. **142:** 449-457.
10. VAN CORVEN, E. J. *et al.* 1989. Cell **59:** 45-54.
11. DEMOLLE, D. *et al.* 1988. J. Biol. Chem. **263:** 18459-18465.
12. NAIRN, A. C. & H. C. PALFREY. 1987. J. Biol. Chem. **262:** 17299.
13. ROBAYE, B. *et al.* 1989. Biochem. Biophys. Res. Commun. **161:** 301-308.

Excitatory Action of Extracellular ATP on Chick Skeletal Muscle

STEVEN A. THOMAS AND RICHARD I. HUME[a]

Department of Biology
Natural Science Building
University of Michigan
Ann Arbor, Michigan 48109

Micromolar concentrations of extracellular ATP elicit a rapid excitatory response in developing chick skeletal muscle.[1] These experiments were designed to determine the ionic basis of the excitatory response. Experiments were performed on myoballs using the whole-cell patch-clamp technique. The reversal potential of the excitatory response was consistent with an increase in conductance to either cations or anions. To distinguish between these possibilities, myoballs were bathed with a solution that would produce a negative reversal potential for cations, and a positive reversal potential for anions. The ATP reversal potential was between the two extremes, indicating that both cations and anions were permeant. In contrast, acetylcholine responses always reversed close to the predicted cation reversal potential. We estimated the relative permeability of ions by measuring the shift in reversal potential when one ion was

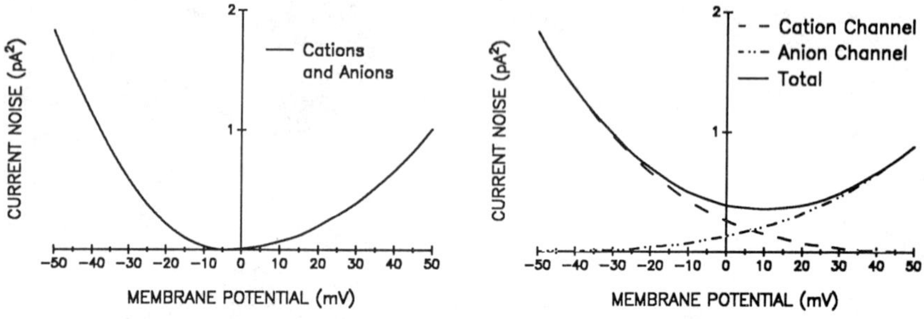

FIGURE 1. Theoretical basis for deciding whether ATP activates a single class of nonselective channels or separate cation and anion channels during the excitatory response. As first pointed out by Dionne and Ruff,[2] if the reversal potentials of two types of permeant ions are set far apart, the predicted noise is very different for the case of one or two channels. For any single channel type, the noise is the product of a term characteristic of the behavior of the single channels times the driving force squared ($Var_i = K(V_m - E_{rev})^2$). If a single class of nonselective channels is involved, then the noise should go to zero at the macroscopic reversal potential (left panel). In contrast, if two different channel types are involved, then the noise at any potential will be the sum of the noise through the two sets of channels. Because neither type of channel is at its reversal potential at the macroscopic reversal potential, noise should not go to zero (right panel).

[a]To whom correspondence should be sent.

TABLE 1. Measurements of ATP-Activated Noise in Myoballs[a]

	Potential (mV)		
	−50	−5	+40
Current (pA)	−223 ± 31	−1 ± 6	+200 ± 86
Variance (pA^2)	1.83 ± 0.61	0.00 ± 0.04	0.66 ± 0.12
Numbers of cells	10	11	4

[a]The reversal potential for cations was set near +50 mV, and the reversal potential for anions near −50 mV, by making whole-cell recordings with appropriate solutions (external solution: 150 mM $CsNO_3$, 3 mM TEA·OH, 1 mM $BaCl_2$, 10 mM HEPES, and 40 mM glucose; internal solution: 21 mM $CsNO_3$, 43 mM TEA·OH, 10 mM BAPTA, 1 mM $CaCl_2$, 10 mM HEPES, and 200 mM sucrose). With these solutions the macroscopic reversal potential of the excitatory ATP response was −5 mV. Although there was a clear increase in ATP-activated noise at −50 and +40, there was no excess noise at the reversal potential. We conclude that there is a single class of nonselective channels activated by ATP. Based on the ratio of the variance to the mean, we estimate that the unitary conductance of these channels is about 0.3 pS.

substituted for another. We found that small monovalent cations, small divalent cations, and small anions all permeate the membrane during the excitatory ATP response, and that there was only moderate selectivity between many of these ions. However, large organic cations and anions (such as tetraethylammonium and glucuronate) were not permeant.

To determine whether ATP activates a single class of channels that conduct both cations and anions, or whether ATP activates separate cation and anion channels, we analyzed the fluctuations about the mean current induced by ATP. As first pointed out by Dionne and Ruff,[2] if the reversal potentials of two types of permeant ions are set far apart, the predicted noise is very different for the cases of one and two channels. For any single channel, the noise is the product of a term characteristic of the behavior of the single channels (K) times the driving force ($V_m - E_{rev}$) squared. If a single class of nonselective channels is activated by ATP, then the noise should go to zero at the macroscopic reversal potential (FIG. 1, left panel). In contrast, if two different classes of channels are involved, then the noise at all potentials will be the sum of the noise through the two sets of channels. Because neither channel is at its reversal potential at the macroscopic reversal potential, noise should not go to zero (FIG. 1, right panel). In our experiments ionic conditions were arranged so that the reversal potential for cations was +50 mV and the reversal potential for anions was −50 mV. At both +40 mV and −50 mV ATP elicited a clear increase in noise, but at the ATP reversal potential of −5 mV no increase in noise above background was seen (TABLE 1). These results indicate that there is only a single excitatory ATP-activated channel type, which does not select by charge. Analysis of the current variance indicates that the unitary conductance of the channel is approximately 0.3 pS.

REFERENCES

1. HUME, R. I. & M. G. HONIG. 1986. Excitatory action of ATP on embryonic chick muscle. J. Neurosci. **6:** 681-690.
2. DIONNE, V. E. & R. L. RUFF. 1977. Endplate current fluctuations reveal only one type of channel type at frog neuromuscular junction. Nature **266:** 263-265.

Activation of Potassium Channels in Chick Skeletal Muscle by Extracellular ATP

RICHARD I. HUME[a] AND STEVEN A. THOMAS

Department of Biology
Natural Science Building
University of Michigan
Ann Arbor, Michigan 48109

In developing chick skeletal muscle, micromolar concentrations of extracellular ATP elicit an early excitatory response followed by a late potassium conductance increase.[1] The potassium conductance activates with a delay of approximately one second and is greatly reduced at low temperature, suggesting that a second messenger may be involved. To examine the mechanism of activation of the potassium channels, we recorded from myoballs using the whole-cell and single-channel patch-clamp configurations.

When whole-cell recordings were made from myoballs held at the reversal potential for the early excitatory current, ATP elicited an outward potassium current. Noise analysis yielded spectral estimates for the single-channel conductance of approximately 20 pS. To determine whether these channels are activated by a second messenger, recordings were made from cell-attached patches while ATP was applied outside the pipette. It seemed unlikely that a highly charged molecule like ATP could cross the membrane-pipette seal, so we expected that channels within the patch should only be activated if this system used a second messenger. We observed the opening of potassium channels in response to ATP application outside of the pipette (FIG. 1) in about 60% of the cell-attached patches. To be sure that ATP was not gaining direct access to the pipette interior, some of these experiments were performed in the presence of the ATP receptor antagonist DIDS. When both DIDS and ATP were added to the solution outside of the pipette there was no activation of channels in the patch. However, when DIDS was present inside of the patch pipette, but not in the bathing solution (so as to selectively block the response to any ATP that managed to get across the seal), ATP applied outside of the pipette was still able to evoke potassium channel opening. These experiments indicated that the potassium channels must be activated by a second messenger.

The opening of potassium channels in response to ATP could also be observed in outside-out patches (FIG. 2). This result strongly suggests that the second messenger is restricted to the plasma membrane. In excised patch recordings the cytoplasmic face of the membrane is bathed by a huge volume of solution, and it seems unlikely that the concentration of a cytoplasmic second messenger could rise enough to elicit a response. A membranous second messenger would not be diluted by the solution inside the patch pipette.

[a] To whom correspondence should be addressed.

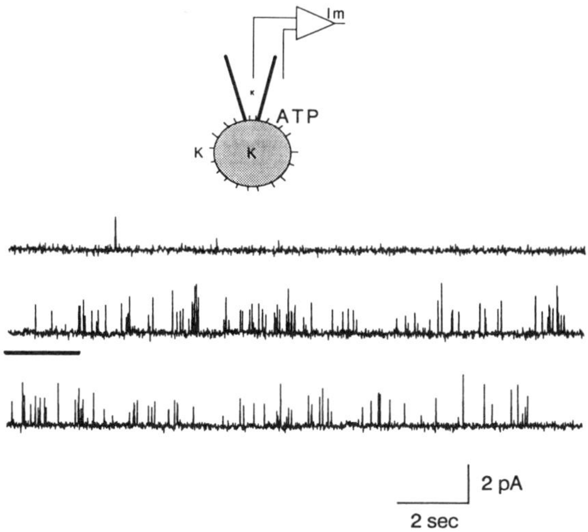

FIGURE 1. ATP activates potassium channels in chick skeletal muscle via a diffusible second messenger. Recordings were made from cell-attached patches in response to ATP applied outside of the pipette. The top trace was taken before ATP was applied. Very few potassium channels were active. ATP was applied during the middle trace (bar under trace). Many more potassium channels were active. The bottom trace was taken immediately after the middle trace. Once ATP was removed the potassium channel activity decreased.

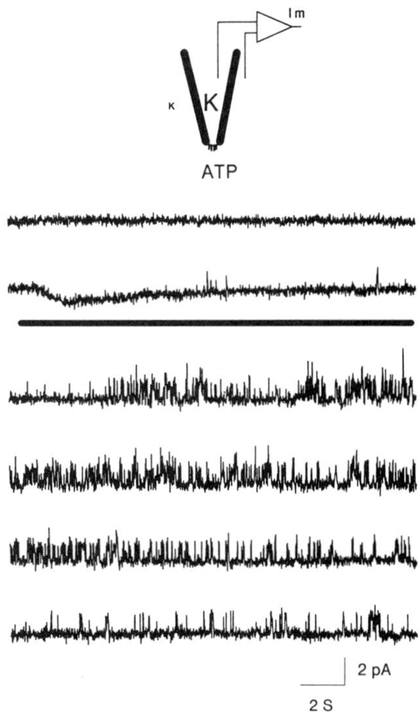

FIGURE 2. The second messenger is probably restricted to the membrane. These traces show recordings from an outside-out patch in response to ATP. The top trace is the current across an outside-out patch before ATP was applied. ATP (10 μM) was applied during the second sweep (bar under trace). The next four sweeps followed immediately after the application of ATP. Potassium channel openings in response to ATP are clearly seen. Because the volume of the solution bathing the inside of the patch is huge, it would have been expected that a cytoplasmic messenger would have been diluted too much to be active. The most likely explanation is that the second messenger remains within the membrane, and therefore is not diluted.

Pharmacological manipulation of myoballs indicated that activation of the potassium current did *not* depend on: 1) a rise in internal free calcium, 2) the activation of a G protein, or 3) protein phosphorylation. Taken together, these experiments indicate that the system that activates potassium currents in chick muscle is unlike any second messenger system that has yet been described.

REFERENCE

1. HUME, R. I. & S. A. THOMAS. 1988. Multiple actions of adenosine 5′-triphosphate on chick skeletal muscle. J. Physiol. **406:** 503-524.

Characteristics of ATP-Induced Plasma Membrane Lesions in Mast Cells

P. E. R. TATHAM

Department of Physiology
University College London
London WC1E 6JJ, England

M. LINDAU

Physics Department
Free University of Berlin
D-1000 Berlin 33
Federal Republic of Germany

Among the actions of extracellular ATP is its ability to bring about the reversible permeabilization of the plasma membrane of certain cells including mast cells.[1,2] The cell surface receptor involved exhibits a pattern of agonist specificities that indicate that it is distinct from all other known ATP receptors.[3] Because the nature of the pores and the mechanism of their formation is poorly understood, we have attempted to characterize them by three complementary methods, namely by observation of the uptake of fluorescent dyes, by investigation of their electrical properties under patch-clamp conditions, and by detection of cell activation in response to externally applied Ca^{2+} and the GTP analogue GTP-γ-S.

The normally impermeant dyes ethidium bromide and trimethylammonium diphenylhexatriene (TMA-DPH) detect permeabilization by entering the cell and binding at sites that enhance their fluorescence efficiency. The rate of increase in fluorescence is then a measure of the degree of permeabilization. In media containing millimolar concentrations of Ca^{2+} and Mg^{2+}, the permeabilizing species ATP^{4-} is present as a minor equilibrium component, and its concentration is calculated by computer program. The dependence of rate of increase of fluorescence upon ATP^{4-} concentration is sigmoid for each dye with EC_{50} values of 11.1 μM and 6.2 μM, respectively, and Hill coefficients close to 2. These data are in agreement with the observed EC_{50} of 19 μM for ATP-induced conductance changes in patch-clamped cells where the Hill coefficient is also close to 2.

Cells treated with ATP in media containing divalent cations are activated to secrete by entry of Ca^{2+} ions. At high levels of ATP, secretion is inhibited as endogenous nucleotides are lost. This inactivation is removed when GTP-γ-S is provided, showing that the cells are still capable of exocytosis. We conclude that GTP-γ-S has access only at the higher levels of ATP, indicating that there is a progressive increase in the size of the pores with increasing ATP concentration.

The patch-clamp data reveal that the lesions are formed very rapidly (within 65 msec), and removal of ATP brings about very fast resealing of the pores (within seconds). The whole-cell conductance of the permeabilized cells is 35-70 nS, but in excised outside-out patches, discrete channel openings are not observed (at a resolution of 40 pS). Noise analysis indicates a unitary conductance of 2-10 pS, which is far below the conductance expected for a pore large enough to allow the passage of substances of molecular mass 300-900 daltons (that is, $\approx$300 pS).

Our data support a model in which ATP forms lesions by binding to specific membrane sites and indicate that at least two molecules of the nucleotide are required to form a pore. As the concentration of ATP^{4-} is increased the pores increase in size by small steps as further units are recruited.

REFERENCES

1. Cockcroft, S. & B. D. Gomperts. 1979. Nature **279:** 541-542.
2. Bennett, J. P., S. Cockcroft & B. D. Gomperts. 1981. J. Physiol. (London) **317:** 335-345.
3. Tatham, P. E. R., N. J. Cusack & B. D. Gomperts. 1988. Eur. J. Pharmacol. **147:** 13-21.

Electrophysiologic Effects of ATP on Rat Ventricular Myocytes

ALLEN CHRISTIE[a] AND SHEY-SHING SHEU

Department of Pharmacology
University of Rochester
Rochester, New York 14642

Since the original description of the effect of micromolar concentrations of extracellular ATP on the cytosolic calcium concentration ($[Ca^{2+}]_i$) of ventricular cells,[1] other groups have reported similar findings.[2,3] A common finding was that ATP stimulates calcium entry into cells from the extracellular solution primarily through dihydropyridine-sensitive calcium channels. It was not known, however, whether this was a direct or indirect action of ATP on these calcium channels.

The whole-cell configuration of the patch-clamp technique was used to investigate the electrophysiologic effects of extracellular ATP on rat ventricular myocytes. The L-type calcium channel was isolated by depolarizing the cell from a holding potential (V_H) of -45 or -40 mV to various testing potentials (V_T). Outward potassium currents were inhibited by substituting cesium for potassium in the pipette filling solution and the extracellular solution. ATP (50-500 μM) had no effect on the current-voltage (I-V) relationship ($N = 10$) of the L-type current or on the peak current elicited by stepping to $V_T = 0$ mV ($N = 7$). These results suggest that ATP does not directly effect the L-type calcium current.

During the recording of the L-type calcium current, there was an increase in the inward holding current when ATP was applied to the cell. This phenomenon was similar to that described in smooth muscle.[4] To record the ATP-activated current, cells were held at $V_H = -70$ mV and the current was continuously monitored. FIGURE 1A is a representative experiment demonstrating the ability of 100 μM to produce an inward current. Soon after the application of ATP, a slowly activating, long-lasting inward current was elicited. The I-V relationship of the current shown in FIGURE 1B was generated by computer subtraction of the ramp (-100 to $+30$ mV) current before ATP application from the ramp current after ATP application. The current of each cell, at every 10-mV interval, was normalized to the current at -100 mV. The I-V relationship was linear and had a reversal potential of -1.9 ± 6.5 mV ($N = 12$).

Associated with the ATP-activated current was membrane depolarization. When cells were bathed in an extracellular solution devoid of voltage-sensitive channel blockers, ATP depolarized cells to the threshold for the firing of action potentials. (FIG. 2A). The first action potential was typical of a fast sodium-dependent action

[a] Present address: Department of Physiology and Biophysics, Case Western Reserve University, 2119 Abington Road, Cleveland, Ohio 44106.

potential whereas the subsequent action potentials were typical of slow calcium-dependent action potentials. In the presence of channel blockers, ATP depolarized cells 13 ± 5 mV (N = 6), but the cells did not fire action potentials (FIG. 2B).

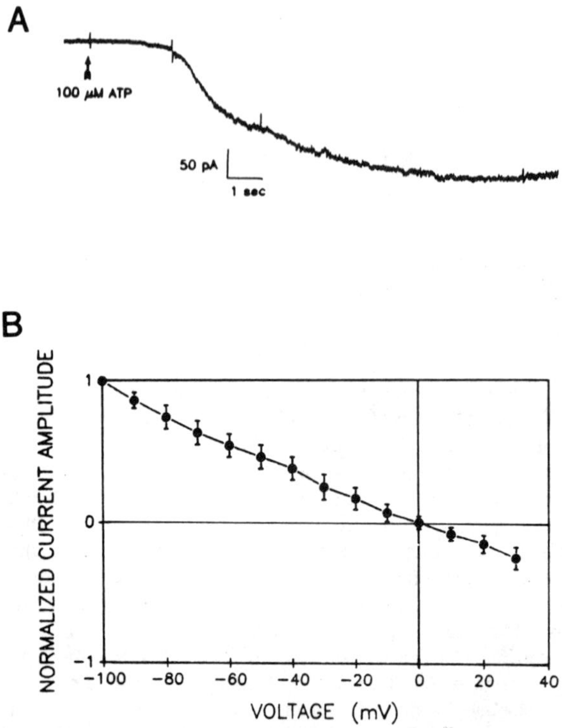

FIGURE 1. ATP activates an ionic current. (**A**) Typical effect of 100 μM ATP on the holding current of a cell held continuously at −70 mV. Upon addition of ATP, a slowly activating, long-lasting current developed that had a peak amplitude of −226 pA. (**B**) Current-voltage relationship for the ATP-activated current. The reversal potential was −1.9 ± 6.5 mV (N = 9). Pipette filling solution contained (in mM) 136 CsCl, 1 $MgCl_2$, 2.5 $CaCl_2$, 10 EGTA, 10 HEPES. Extracellular solution contained 136 mM NaCl, 1 mM $MgCl_2$, 1 mM $CaCl_2$, 10 mM CsCl, 10 mM glucose, 10 mM HEPES, 5 μM tetrodotoxin, 1 μM nitrendipine, and 2 mM 4-aminopyridine.

In summary, extracellular ATP has no direct effect on the L-type, dihydropyridine-sensitive calcium current of rat ventricular myocytes. At negative membrane potentials, ATP activates an inward current similar to the nonspecific cation conducting current described in vascular smooth muscle. It is concluded that the ATP-activated current and depolarization are the underlying mechanisms by which ATP increases $[Ca^{2+}]_i$.

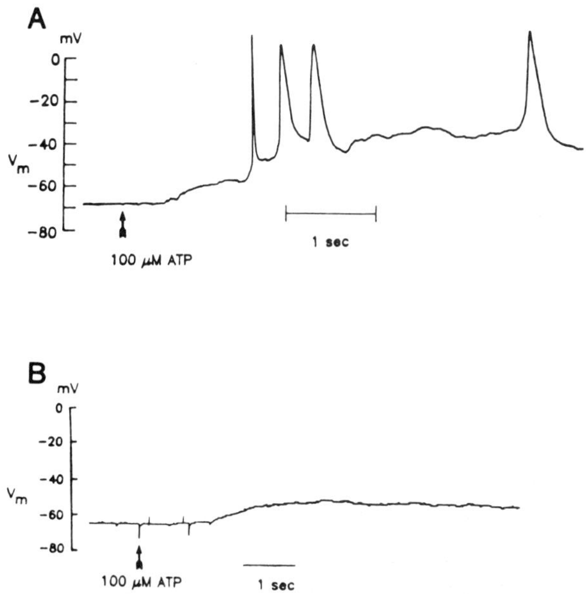

FIGURE 2. ATP depolarizes cells. (**A**) A current-clamped cell depolarized from −68 mV to −49 mV upon application of 100 μM ATP, at which time the cell fired multiple action potentials. Internal and external solutions contained KCl in place of CsCl, and the external solution was devoid of voltage-activated channel blockers. (**B**) In the presence of voltage-activated channel blockers (tetrodotoxin, nitrendipine, 4-aminopyridine, and CsCl), ATP still depolarized the cell but no action potentials were elicited. Upon addition of 100 μM ATP, the cell depolarized from −65 mV to −51 mV and then repolarized.

REFERENCES

1. SHARMA, V. K. & S.-S. SHEU. 1986. Biophys. J. **49:** 351a.
2. DE YOUNG, M. B. & A. SCARPA. 1987. FEBS Lett. **223:** 53-58.
3. DANZINGER, R. S., S. RAFFAELI, R. MORENO-SANCHEZ, M. SAKAI, M. C. CAPOGROSSI, H. A. SPURGEON & E. G. LAKATTA. 1988. Cell Calcium **9:** 193-199.
4. BENHAM, C. D. & R. W. TSIEN. 1988. Soc. Gen. Physiol. Ser. **42:** 45-64.

Involvement of ATP as a Neurotransmitter in the Hippocampus[a]

A. WIERASZKO[b] AND T. N. SEYFRIED

Department of Biology
Boston College
Chestnut Hill, Massachusetts 02176

Increasing evidence suggests the involvement of extracellular ATP in the regulation of neuronal excitability.[1,2] We have studied the neurotransmitter role of ATP in the mouse hippocampus.

Two major criteria should be fulfilled for an endogenous compound to be recognized as a neurotransmitter. One is the stimulation-dependent release of ATP, and the other is the influence of exogenously applied ATP on excitability of neurons.

We used the hippocampal slice preparation described previously[3] as a model for our studies. In all experiments, the stimulating electrode was placed on Schaffer collaterals, which use glutamate as a primary neurotransmitter. The extracellular records (population spikes) were taken from CA1 pyramidal neurons. To follow ATP release, we used a specially designed apparatus, which allows monitoring of ATP release during stimulation of the slices.[4] The determination of ATP release was performed using a luciferin-luciferase system.[4] As shown in FIGURE 1, high-frequency stimulation of Schaffer collaterals (300 Hz for 50 msec at 2-sec intervals for a total of 30 sec) evokes ATP release. This pattern of stimulation also induces a permanent increase in the size of the population spike (upper part of FIG. 1). This well-known phenomenon is called long-term potentiation (LTP).[5] Omitting calcium from the incubation medium eliminated ATP release. Blocking postsynaptic glutamate receptors with 2 mM kynurenic acid had no influence on ATP release. These data demonstrate that ATP can be released from stimulated nerve terminals (Schaffer collaterals) in a calcium-dependent way. It is interesting to note that ATP release was not observed following low-frequency stimulation, which evokes postsynaptic response but is not able to induce LTP. It appears that ATP is coreleased with a primary neurotransmitter (glutamate in this case) only with a certain pattern of stimulation. We suggest that ATP, while not involved in normal neurotransmission, may participate in the potentiation effect after high-frequency stimulation. Further support for this suggestion comes from our experiments showing an influence of exogenous ATP on the size of the population spike.[6] As shown in FIGURE 2, 400 nM ATP gradually increased the size of the population spike. The maximal effect was observed after about 20 min

[a]This work was supported by grants from the National Science Foundation (BNS 8644955) and the National Institutes of Health (24826, 27866, and 23355).

[b]Present address: Department of Biology, College of Staten Island, City University of New York, CSI/IBR Center for Developmental Neuroscience, Staten Island, New York 10301.

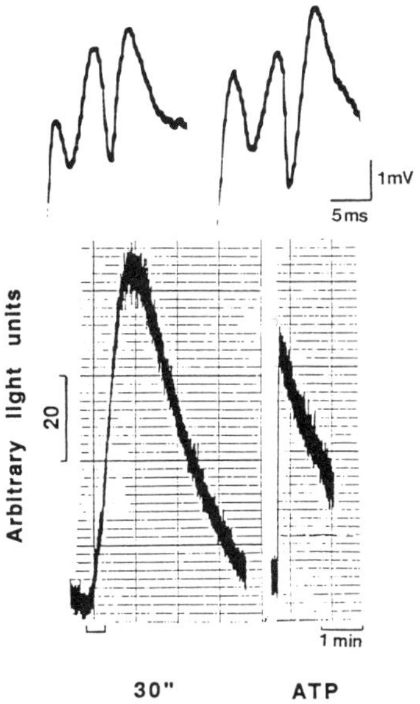

FIGURE 1. The release of ATP from a slice of mouse hippocampus following high-frequency stimulation of Schaffer collaterals (left bottom). The standard solution of ATP (5.5×10^{-10} M) applied later to the same preparation (right bottom) allowed calculation of the concentration of released ATP, which was 0.61 ± 0.06 pmoles/slice (N = 10). The upper part of the figure shows the size of the population spike before (left) and after (right) high-frequency stimulation. Note the increase in the size of the potential (LTP).

(FIG. 2b). The amplified potential was slightly reduced by removing ATP through perfusion (FIG. 2c). The size of the potential, however, still remained elevated and did not return to the control level (FIG. 2d).

On the other hand 71 μM of ATP temporarily reduced the size of the potential, which recovered by itself (no perfusion), reaching a value higher than that observed before ATP application. It is interesting to note that induction of LTP is often accompanied by transient reduction in the size of the potential immediately following high-frequency stimulation. This may be due to excessive ATP release, which could depress synaptic responses, but which simultaneously triggers the processes responsible for later synaptic facilitation. The synaptic potential was depressed by higher ATP concentrations (250 μM) and did not recover by itself. The cells were not killed, however, as the potential could be restored by 0.71 μM 3,4-diaminopyridine. Because ATP is coreleased with the primary neurotransmitter and can potentiate synaptic responses when applied exogenously, new insights can be obtained on the mechanism

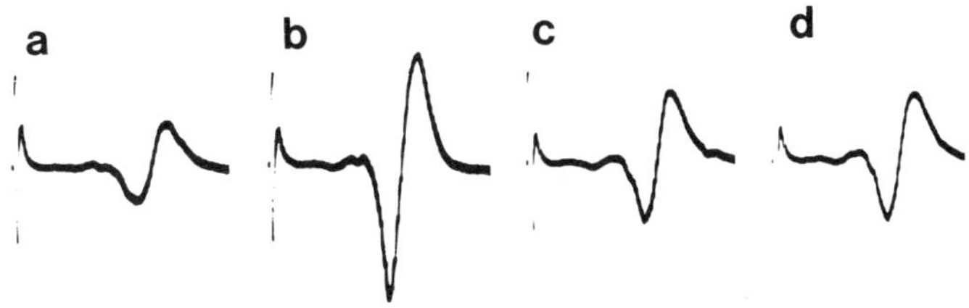

FIGURE 2. The effect of exogenously applied ATP (400 nM) on the population spike recorded from CA1 pyramidal neurons. The control response (**a**) is followed by the response recorded 20 min after ATP application (**b**). Ten minutes of perfusion reduces the size of the potential (**c**), but even 60 min after perfusion the potential remains elevated (**d**). Calibration: 5 msec, 1 mV.

of LTP. In our opinion, the action of ATP is exerted not through activation of P_2 receptors,[7] but rather through the regulation of an extracellular enzymatic process. A strong candidate is protein phosphorylation by exogenous kinases,[8] especially in the view of recently published data showing the influence of nM concentrations of ATP on extracellular protein phosphorylation.[9] We also do not exclude the possibility of an action of ATP on a second messenger system. It has been demonstrated in synaptosomal preparation that ATP can influence inositol trisphosphate production,[10] which in turn elevates intracellular calcium levels necessary for LTP induction.[11] The diversity of ATP action on neuron excitability in the central nervous system suggests an important role of ATP in neurotransmission.

REFERENCES

1. PHILLIS, J. W. & P. WU. 1981. Progr. Neurobiol. **16:** 187-239.
2. STONE, T. W. 1981. Neuroscience **16:** 523-555.
3. WIERASZKO, A. 1982. Brain Res. **237:** 449-457.
4. WIERASZKO, A., G. GOLDSMITH & T. N. SEYFRIED. 1988. Brain Res. **485:** 244-250.
5. BLISS, T. V. P. & T. LOMO. 1973. J. Physiol. (London) **232:** 310-356.
6. WIERASZKO, A. & T. N. SEYFRIED. 1989. Brain Res. **491:** 356-359.
7. BURNSTOCK, G. 1981. J. Physiol. **313:** 1-35.
8. EHRLICH, Y. 1987. Adv. Exp. Med. Biol. **221:** 187-199.
9. EHRLICH, Y. 1989. J. Neurochem. **53:** 1512-1518.
10. HUANG, H. M., & G. Y. SUN. 1988. J. Neurochem. **50:** 366-374.
11. ECCLES, J. C. 1983. Neuroscience **10:** 1071-1081.

Sources of Adenosine Released from Hippocampal Slices following Electrical and Hypoxic/ Hypoglycemic Stimulation

BERTIL B. FREDHOLM AND HILARY G. LLOYD[a]

Department of Pharmacology
Karolinska Institutet
Stockholm, Sweden

INTRODUCTION

Adenosine is known to be released from central and peripheral tissues following nerve stimulation.[1] The released adenosine could derive 1) from intracellularly formed adenosine (cytosolic 5′-nucleotidase or *S*-adenosylhomocysteine (SAH) hydrolase[2]) or 2) from nucleotides degraded extracellularly by ecto-5′-nucleotidase. There is excellent evidence that adenine nucleotides, particularly ATP, can be released by nerve stimulation,[2] but it is not known whether such released nucleotides actually constitute a *quantitatively* important source of adenosine. We have tested this using several different pharmacological tools.

METHODS

Rat hippocampal slices (400-μm thick) were incubated with [^{3}H]adenine to label the intracellular adenine nucleotides, placed in perfusion chambers, and superfused with artificial cerebrospinal fluid. The efflux of endogenous and radiolabeled purines was determined by HPLC. The efflux of purines from the slices was stimulated either by electrical field stimulation (10 Hz, 20 V, 5-15 min) or by combined hypoglycemia and hypoxia (25-35 min).

RESULTS AND DISCUSSION

Under basal conditions adenosine constituted less than 10% of the total radioactive purines, and the proportion was not substantially increased by an inhibitor of adenosine

[a]Present address: Department of Physiology, University of Sydney, Australia.

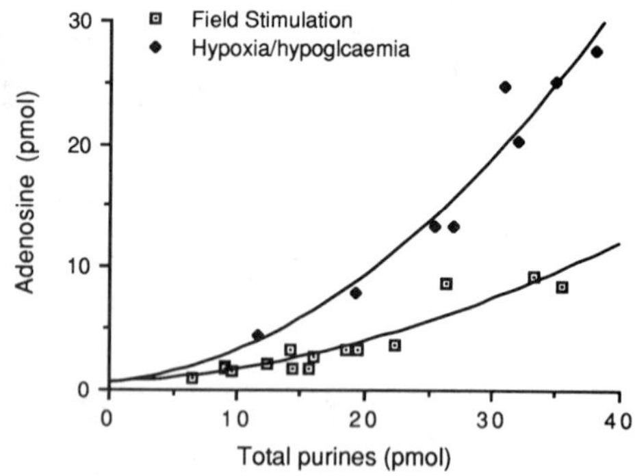

FIGURE 1. Relationship between release of adenosine and total release of purines following stimulation with electrical pulses or with combined hypoxia/hypoglycemia. The experiments were carried out in the presence of EHNA to block adenosine deaminase. Note that adenosine constitutes a much larger proportion of released purines in the latter case.

deaminase, erythro-9-hydroxy-3-nonyl adenine (EHNA, 3 μM). Similarly, nerve stimulation caused a frequency-dependent release of adenosine, which was essentially unaffected by EHNA. Hypoxia/hypoglycemia, by contrast, caused a marked increase in adenosine release (FIG. 1), and this was further increased by EHNA. Inhibition of adenosine kinase by 5′-iodotubericidin caused a great increase in adenosine release. These results indicate that only a minor part of the total purines released from [^{3}H]ATP stores under basal conditions and following nerve activity ever was adenosine. This implies a major intracellular formation of the purines, since the AMP-IMP-inosine pathway is intracellular.

Following the addition of the ecto-5′-nucleotidase inhibitor α,β-methylene ADP (AOPCP), there was no reduction of adenosine release under any of the conditions studied (FIG. 2). This indicates that a minor part of the purines were formed by the action of ecto-5′-nucleotidase. Addition of L-homocysteine thiolactone, which shifts the SAH hydrolase toward synthesis,[2] caused a net accumulation of SAH in the slices and a marked decrease in the efflux of adenosine, indicating that much of the adenosine is formed intracellularly.

These data indicate that adenosine release from hippocampal slices is independent of ATP release. Therefore the roles of ATP and adenosine can be defined independently of each other. ATP could fulfill specialized functions in acting as a signal between cells,[3,4] whereas adenosine may be more important as a local hormone involved in energy homeostasis.

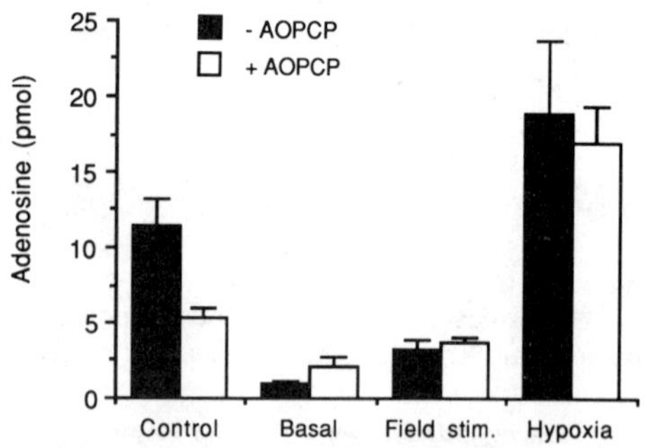

FIGURE 2. Effect of inhibition of ecto-5-nucleotidase by AOPCP on the release of endogenous adenosine under basal conditions, following field stimulation and hypoxia/hypoglycemia. The columns labeled Control represent the ability of AOPCP to block the conversion of AMP to adenosine.

REFERENCES

1. FREDHOLM, B. B. & P. HEDQVIST. 1980. Biochem. Pharmacol. **29:** 1635-1643.
2. LLOYD, H. G. E., A. DEUSSEN, H. WUPPERMANN & J. SCHRADER. 1988. Biochem. J. **252:** 489-494.
3. WHITE, T. D. 1988. Pharmacol. Ther. **38:** 129-168.
4. WESTFALL, D. P., G. K. HOGABOOM, J. COLBY, J. P. O'DONNELL & J. S. FEDAN. 1982. Proc. Natl. Acad. Sci. USA **79:** 7041-7045.

Evidence for a Presynaptic P_{2X}-Purinoceptor Involved in Facilitation of Acetylcholine Release

E. S. VIZI,[a,b] B. SPERLAGH,[a] AND A. LAJTHA[b,c]

[a]*Institute of Experimental Medicine*
Hungarian Academy of Sciences
H-1450 Budapest
Hungary

[b]*Center for Neurochemistry*
Nathan S. Kline Institute for Psychiatric Research
Ward's Island, New York 10035

The first evidence that purine nucleotides and adenosine have presynaptic inhibitory effects on acetylcholine (ACh)[1] and norepinephrine[2] release was obtained in 1976. In these experiments the presynaptic inhibitory receptors proved to be theophylline sensitive. Now it is generally accepted that the actions of purines are mediated via two different types of receptors. According to Burnstock,[3] the P_1-purinoceptors are activated by adenosine and the P_2-purinoceptors are activated mainly by ATP, ADP, and their nondegradable analogues, but not by adenosine. Previous work has suggested that P_2-purinoceptors are mainly located postjunctionally and the P_1-purinoceptors prejunctionally.[3] The presynaptic receptor that mediates these inhibitory actions has been called the $P_1(A_1)$ receptor since it has been noted that the inhibitory actions of both ATP and adenosine are prevented by methylxanthines.

In this study an attempt was made to study the effect of ATP and its stable analogue α,β-methylene ATP (α,β-Me-ATP) on ACh release. Guinea pig ileal longitudinal muscle strip with Auerbach's plexus attached was used as described by Paton and Vizi.[4] The tissue was incubated in Krebs solution containing [*methyl*-^{3}H]choline (4 μCi/ml).[5] Two stimulations (2 Hz, 240 shocks, S_1 and S_2) were applied, and drugs were added 15 min before the second (S_2) stimulation. Changes in the S_2/S_1 ratio represent the effect of drugs on ACh release.

Whereas ATP inhibited it, the nonhydrolyzable derivative α,β-Me-ATP (1-50 μM) enhanced the stimulation-evoked release of ACh (TABLE 1). The inhibitory effect of ATP, when added 15 min before the S_2, was partly antagonized by 8-phenyltheophylline (8-PT), whereas the facilitatory effect of α,β-Me-ATP was not. These findings indicate that the effect of ATP is due to its breakdown product, adenosine, and is mediated via the presynaptic $P_1(A_1)$ receptors (FIG. 1). In contrast, when ATP was added 3 sec prior to stimulation, it reduced the release in an 8-PT-sensitive way. 8-PT itself enhanced the release of ACh, as indicated by the increase of the S_2/S_1 ratio from 0.79 $\pm$ 0.04 to 1.74 $\pm$ 0.24 (TABLE 1). Adenosine (50-100 μM) inhibited release, even if release was previously enhanced by α,β-Me-ATP.

[c]To whom correspondence should be addressed.

TABLE 1. Effect of ATP and its Stable Analogue on [^{3}H]ACh Release Evoked by Field Stimulation[a]

Drugs[b]	S_2/S_1	8-PT (10 μM)
Control	0.79 ± 0.04	1.74 ± 0.24[c]
Adenosine (100 μM)	0.38 ± 0.10[c]	0.70 ± 0.09[c]
ATP (100 μM)	0.38 ± 0.08[c]	0.67 ± 0.06[c]
ATP (120 μM)	0.76 ± 0.07	0.35 ± 0.02[c]
α,β-Me-ATP (30 μM)	1.40 ± 0.15[c]	2.35 ± 0.56[c]

[a] Longitudinal muscle strip with Auerbach's plexus attached was used throughout.
[b] All drugs were added 15 min before the S_2, except for 120 μM ATP, which was added 3 sec before the S_2.
[c] Significant difference from control ($p < .05$). Averages of four experiments are given.

Because 2-methylthio-ATP (30 μM) did not enhance the release of ACH (data not shown) and α,β-Me-ATP did, it seems likely that receptors involved in facilitation of transmitter release are of the P_{2X} type. The facilitatory effect of α,β-Me-ATP on the release of ACh evoked by electrical stimulation may be related to an ATP receptor-mediated effect on the calcium channel and its depolarizing effect, similar to those produced by nicotinic receptor stimulation. When ATP is released,[6] either from neurons or from other sources, its presynaptic facilitatory action may be overshadowed by its inhibitory action on release, mediated through its breakdown product, adenosine.

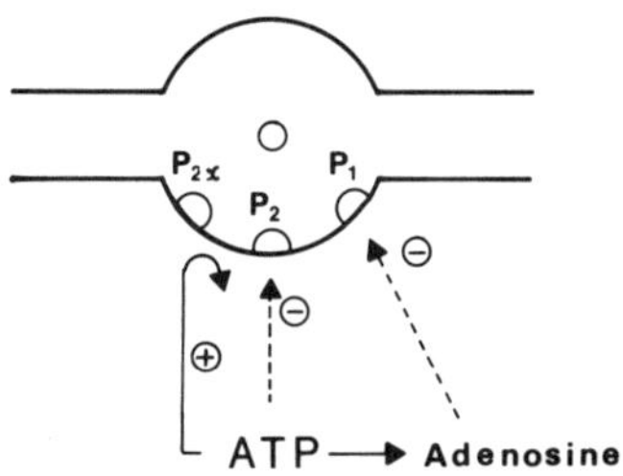

FIGURE 1. Scheme of presynaptic purinoceptors sensitive to adenosine (P_1) and ATP (P_2) involved in modulation of ACh release. The presynaptic P_2-purinoceptors are not sensitive to 8-PT, even though they are operative under conditions in which the P_1-purinoceptors are inhibited. ATP released either from the axon terminals[6] or from postsynaptic sites[6] acts on both pre- and postsynaptic P_2-purinoceptors. After hydrolysis by ectonucleotidase, the presynaptic inhibitory effect of ATP on ACh release is exerted through its breakdown product. It seems likely that there are ATP-sensitive receptors on the cholinergic neurons as well. 2-Methylthio-ATP had no effect on the release of ACh; therefore it is suggested that the P_{2X}-purinoceptor is responsible for the facilitation of ACh release, and its agonist is ATP and α,β-Me-ATP. In addition, when the tissue was exposed to ATP for a short period of time, the release was reduced even in the presence of 8-PT.

REFERENCES

1. Vizi, E. S. & J. Knoll. 1976. Neuroscience **1:** 391-398.
2. Hedqvist, P. & B. B. Fredholm. 1976. Naunyn-Schmiedeberg's Arch. **293:** 217-233.
3. Burnstock, G. 1981. J. Physiol. **313:** 1-35.
4. Paton, W. D. M. & E. S. Vizi. 1969. Br. J. Pharmacol. **35:** 10-28.
5. Somogyi, G. T. & E. S. Vizi. 1987. Brain Res. Bull. **21:** 575-579.
6. Vizi, E. S. & G. Burnstock. 1988. Eur. J. Pharmacol. **158:** 69-77.

α-Adrenergic Receptor-Stimulated Release of ATP from Cardiac Endothelial Cells in Primary Culture[a]

IAIN L. O. BUXTON, JAMES WALTHER, AND DAVID P. WESTFALL

Department of Pharmacology
University of Nevada School of Medicine
Reno, Nevada 89557

INTRODUCTION

Release of ATP from endothelial cells was first described in 1979,[1] and a role for adenyl purines in the regulation of vascular tone under conditions such as hypoxia has been well established, as noted in a recent review.[2] Despite demonstration of the actions of ATP and its breakdown product adenosine in blood vessels, it is not clear how, if at all, ATP release from endothelium would be regulated in nonpathological states. We have developed an improved HPLC fluorescence method for the detection of picogram quantities of adenyl purines in physiological solutions and have employed this method to detect the release of ATP from cardiac endothelial cells in primary tissue culture. When norepinephrine (NE) is added to the bathing medium, cells rapidly release large quantities of ATP that can be detected following fluorescence derivation of the sample.[3] The effect of NE is blocked by the addition of prazosin, suggesting that the response follows agonist occupation of an α_1 receptor on the endothelial cell. We suggest that endothelial cell ATP, released following sympathetic nerve activity, may act at several sites. ATP acting via P_2-purinoceptors on smooth muscle and/or endothelium may regulate vascular tone directly. In addition, ATP can act at a P_3-purinoceptor[4] on sympathetic nerve to regulate neurotransmitter release. Furthermore, the rapid breakdown of ATP to adenosine acting downstream from the site of release at P_1-purinoceptors to relax blood vessels is also likely. These actions of ATP derived from endothelial cells stimulated by neurotransmitters may constitute a paracrine mechanism for the regulation of regional blood flow in the heart.

[a]This work was supported by grants from the National Institutes of Health (HL35416 and HL38126) and a grant from the American Heart Association.

MATERIALS AND METHODS

Preparation of Endothelial Cells

Cells are prepared by collagenase (1.5 mg/ml) perfusion of hearts from young adult male guinea pigs (300 g) submerged in 20% sucrose solution after the method of Nees *et al.*[5] Cells released in the effluent are collected from the buoyant layer on top of the sucrose and purified by centrifugation (4,000 × g × 15 min) at room temperature on a preformed Percoll gradient (20,000 × g × 15 min). The endothelial cell layer is removed from the Percoll, washed in growth medium (Dulbecco's modified Eagle's medium) containing fetal bovine serum (10% v/v), and seeded (1 ml) in multiwell tissue culture plates. Cultures are maintained until confluent (5-7 days), and experiments are initiated by removing the growth medium and washing cells with Krebs buffer followed by addition of Krebs buffer containing drug.

Assay for ATP Release

Adenyl purines are assayed by HPLC using a modification of the method of Levitt *et al.*[3] based on the formation of etheno-purine. Briefly, medium bathing the cells is quickly withdrawn, frozen in liquid nitrogen, and lyophilized to dryness. Dried samples are placed in a Hewlett-Packard (1090M) autoinjector where they are sequentially resuspended in 35 μl of water and 4 μl of 0.8 M chloracetylaldehyde, and the resulting solution is then heated with mixing to 80 °C for 10 min and injected on a C18 reverse-phase column. Adenyl purines are separated using a linear gradient from 100% KPO_4 buffer (pH 6.0) to 30% methanol over 20 min at 0.75 ml/min. The etheno-purine derivatives are detected using a fluorescence detector (HP 1046) capable of excitation at 216 nm and emission at 420 nm. Derivation of standards yields linear standard curves from which sample purine content is deduced. Calibration using *bona fide* etheno-ATP reveals that sample derivation proceeds to better than 95% completion. The detection limit for ATP is 10.0 pg.

RESULTS AND DISCUSSION

Radioligand binding studies employing the α_1-adrenergic receptor radioligand [^{3}H]prazosin and isolated intact endothelial cells harvested from primary cultures, reveal that the cells we study possess a significant number (8,000 sites/cell) of α_1-adrenergic receptors with high affinity for the antagonist radioligand (K_D = 40 pM). Competition studies employing subtype-specific α_1 antagonists confirm that the receptor is of the α_1 subtype (TABLE 1). Stimulation of α_1 receptors by the addition of NE (10 μM) results in the immediate release of ATP (FIG. 1) that can be blocked by addition of the α_1-adrenergic antagonist prazosin (1 μM) but is unaffected by addition of the α_2 antagonist yohimbine (TABLE 1). The dose-response curve for NE (FIG. 1, inset) yields an EC_{50} for NE of 50 nM and is consistent with action of the agonist at an α_1 receptor. The time course of release suggests that the cells possess

TABLE 1. Characterization of α_1-Adrenergic Receptors on Cardiac Endothelial Cells[a]

	Antagonists			
	Affinity (K_D)	Density (B_{max})	Antagonistic Affinities (K_I)	Inhibition of ATP Release at 1 μM
[^{3}H]Prazosin	40 ± 6 pM	8,000 ± 900 cells	—	—
Phentolamine	—	—	1 μM	20%
Yohimbine	—	—	1 μM	0%
Prazosin	—	—	50 pM	92%
		Agonists		
		Affinity[b] (K_D)		
		High-Affinity State	Low-Affinity State	Enhancement of ATP Release
NE (1 μM)		10 nM (30%)	800 nM (70%)	515%[c]
NE + GTP-γ-S (10 μM)		—	500 nM (100%)	—

[a] Data for the affinity of the α_1-adrenergic receptor for radioligand and receptor number were determined in saturation binding experiments employing intact cells ($N = 3$). Antagonist affinities were determined from competition experiments performed in triplicate. Inhibition of ATP release was determined as a percentage of the release produced by 1 μM NE in a single experiment performed in duplicate. Affinity of the receptor for agonist was assessed in competition experiments employing [^{3}H]prazosin binding in endothelial cell membranes.

[b] Each of the values in parentheses below indicates the percentage of receptors assumed to be in either a high- or low-affinity state. Data were best fit by least-squares analysis assuming the presence of both high- and low-affinity states of the receptor. In the presence of GTP-γ-S (a hydrolysis-resistant form of GTP), data were best fit assuming a single low-affinity class of binding sites for the agonist.

[c] An increase from 20 to 123 pg ATP/10^6 cells.

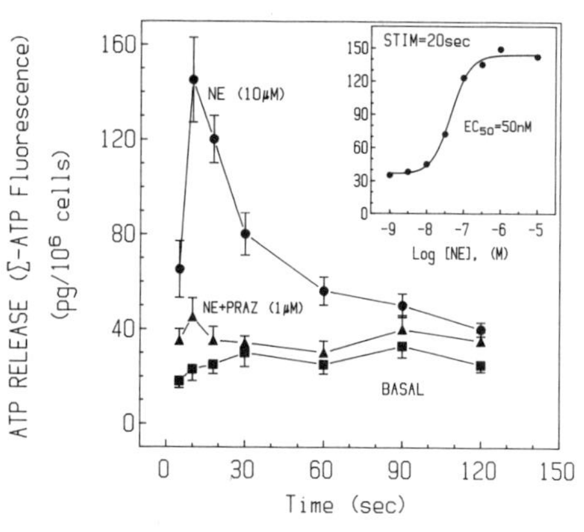

FIGURE 1. α_1-Adrenergic regulation of ATP release from cardiac endothelial cells. Cultured guinea pig endothelial cells were stimulated and ATP release was determined as described in Methods. Each data point represents a mean ± a range of values for etheno-ATP measured in triplicate from replicate culture wells in a single experiment. Cells were stimulated with buffer alone, ■; NE at 10 μM, ●; or NE in the presence of prazosin (PRAZ) at 1 μM, ▲. The dose dependence of NE-stimulated ATP release is shown in the inset. Replicate culture wells were stimulated for 20 sec with increasing concentrations of NE. Data points, fit with a sigmoid curve (Hill = 1.4), are mean values obtained in a single experiment.

the expected nucleotidase activity since ATP levels fall within seconds following release and the levels of AMP and adenosine rise over the same time course (data not shown). Control experiments reveal that release of ATP is not the result of cell damage since addition of NE to cultures returned to growth conditions for 90 min following the first stimulation results in the continued ability of the cells to release ATP.

Our data suggest a role for neurotransmitter-stimulated ATP release in blood vessels in the heart in a fashion consistent with the actions of ATP on nerves, as well as both other endothelial cells and smooth muscle cells in the same blood vessel. ATP is thought to act at a unique purinergic site, P_3,[4] to limit sympathetic neurotransmitter release from nerves. In addition, ATP is likely to act at the immediate site of release at a P_2-purinoceptor on either smooth muscle, endothelium, or both. The rapid action of endothelial ectonucleotidases will then convert ATP to adenosine, which is likely to act at a P_1-purinoceptor on smooth muscle distant from the site of release. Thus, ATP may be released from the endothelial compartment of the blood vessel during periods of increased sympathetic nerve activity. This ATP might be expected to serve to inhibit further transmitter release, as well as dilate blood vessels to accommodate increased blood flow. We suggest that endothelial cell ATP release acting directly and as an origin for adenosine constitutes a paracrine mechanism for the regulation of regional blood flow in the heart and may be an important general phenomenon in many mammalian blood vessels during increased sympathetic nerve activity.

REFERENCES

1. Pearson, J. D. & J. L. Gordon. 1979. Vascular endothelial and smooth muscle cells in culture selectively release adenine nucleotides. Nature **281:** 384-386.
2. Burnstock, G. & C. Kennedy. 1986. A dual function for adenosine 5′-triphosphate in the regulation of vascular tone. Circ. Res. **58:** 319-330.
3. Levitt, B., R. Head & D. P. Westfall. 1984. High-pressure liquid chromatographic-fluorometric detection of adenosine and adenine nucleotides: Application to endogenous content and electrically induced release of adenyl purines in guinea pig vas deferens. Anal. Biochem. **137:** 93-100.
4. Shinozuka, K., R. A. Bjur & D. P. Westfall. 1988. Characterization of prejunctional purinoceptors on adrenergic nerves of the rat caudal artery. Naunyn-Schmiedeberg's Arch. Pharmacol. **338:** 221-227.
5. Nees, S., A. L. Gerbes & E. Gerlach. 1981. Isolation, identification, and continuous culture of coronary endothelial cells from guinea pig heart. Eur. J. Cell. Biol. **24:** 289-297.

Purine Modulation of Norepinephrine Release in the Rat Vas Deferens[a]

K. M. FORSYTH, K. SHINOZUKA, R. A. BJUR, AND D. P. WESTFALL

Department of Pharmacology
University of Nevada School of Medicine
Reno, Nevada 89557

INTRODUCTION

There are several mechanisms that regulate release of norepinephrine (NE) from adrenergic nerves. One mechanism may involve the prejunctional action of adenine nucleotides and nucleosides, and both adenosine (ADO) and adenosine triphosphate (ATP) have been shown to reduce the release of NE in a variety of sympathetically innervated tissues.[1-3] The generally accepted view is that this inhibitory effect is mediated by prejunctional P_1-purinoceptors and that the effect of ATP results from its metabolism to ADO, with ADO then acting via prejunctional P_1-purinoceptors, rather than with ATP acting directly on P_2-purinoceptors.[4,5] Work carried out in this laboratory with the rat caudal artery, however, indicates that prejunctional purinoceptors cannot simply be defined in terms of the existing P_1- and P_2-purinoceptor classification.[6] This study sought to determine the nature of prejunctional purinoceptors in the rat vas deferens and to determine whether or not endogenously released purines contribute to the physiological modulation of NE release.

METHODS

Vasa deferentia from Fisher rats were incubated in organ baths containing 3.5 ml of a modified Krebs solution, gassed with 95% O_2/5% CO_2, and electrically field stimulated with 0.5 msec pulses for 3 min at a frequency of 2 Hz. Following the period of stimulation, the organ baths were drained and the NE released into the surrounding Krebs solution quantified by HPLC electrochemical detection.

[a]This work was supported by a grant from the National Institutes of Health (HL38126).

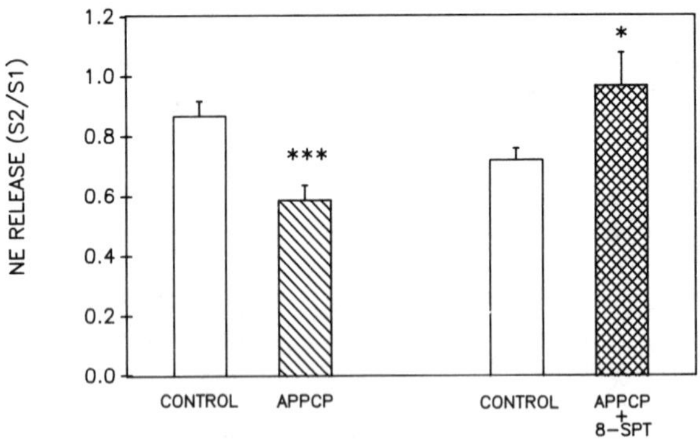

FIGURE 1. Inhibition by 3 μM APPCP of electrically evoked NE release from the rat vas deferens stimulated at 2 Hz for 3 min, and reversal of this effect by 30 μM 8-SPT. Pretreatment with APPCP was for 3 min; pretreatment with 8-SPT, for 15 min. Asterisks indicate a significant difference from control.

RESULTS AND DISCUSSION

The stable analogue of ATP, β,γ-methylene ATP (APPCP), reduced the release of NE from the rat vas deferens (FIG. 1). A concentration of 3 μM produced an inhibition of 30%. Despite the evidence that APPCP interacts with P_2-purinoceptors, the P_1-purinoceptor antagonist 8-(*p*-sulfophenyl)-theophylline (8-SPT) significantly reduced the inhibition produced by APPCP (FIG. 1). Furthermore, 8-SPT, when added alone, potentiated NE release (FIG. 2). These results support the idea that the receptor that mediates purine-induced inhibition of NE release exhibits characteristics different from known P_1- or P_2-purinoceptors. Furthermore, it appears that this putative P_3-purinoceptor is involved in a physiological modulation of NE release, since antagonism of the receptor with 8-SPT produced an enhancement of NE release.

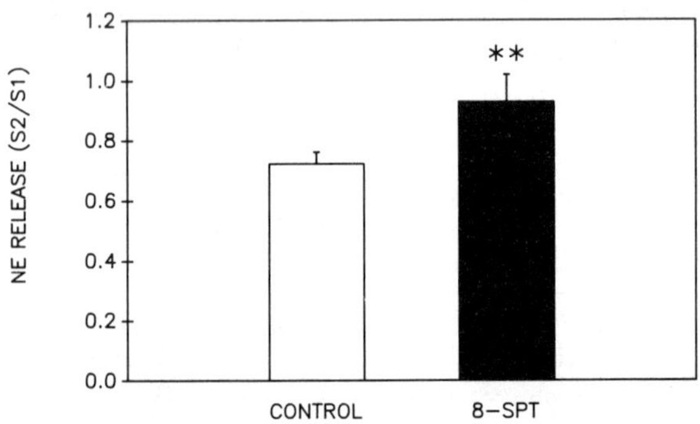

FIGURE 2. Potentiation by 30 μM 8-SPT of electrically evoked NE release from the rat vas deferens stimulated at 2 Hz for 3 min. The asterisks indicate a significant difference from the control.

REFERENCES

1. Clanachan, A. S., A. Johns & D. M. Paton. 1977. Presynaptic actions of adenine nucleotides and adenosine on neurotransmission in the rat vas deferens. Neuroscience **2:** 597-602.
2. Moylan, R. D. & T. C. Westfall. 1979. Effect of adenosine on adrenergic neurotransmission in the superfused rat portal vein. Blood Vessels **16:** 302-310.
3. Paton, D. M. 1981. Presynaptic neuromodulation mediated by purinergic receptors. *In* Purinergic Receptors. G. Burnstock, Ed.: 199-219. Chapman & Hall. London.
4. Burnstock, G. 1981. Neurotransmitters and trophic factors in the autonomic nervous system. J. Physiol. **313:** 1-35.
5. Burnstock, G. 1983. A comparison of receptors for adenosine and adenine nucleotides. *In* Regulary Function of Adenosine. R. M. Berne, T. W. Hall & R. Rubio, Eds.: 49-62. Martinus Nijhoff. Boston, MA.
6. Shinozuka, K., A. Bjur & D. P. Westfall. 1988. Characterization of prejunctional purinoceptors on adrenergic nerves of the rat caudal artery. Naunyn-Schmiedeberg's Arch. Pharmacol. **338:** 221-227.

Norepinephrine and ATP as Cotransmitters in the Guinea Pig Portal Vein and Rabbit Saphenous Artery[a]

W. L. CHAU, E. M. DE ANGELIS, AND
D. P. WESTFALL

Department of Pharmacology
University of Nevada School of Medicine
Reno, Nevada 89557

INTRODUCTION

Norepinephrine (NE) and ATP have been demonstrated to function as cotransmitters in numerous smooth muscle tissues such as the vas deferens[1] and blood vessels.[2,3] In the portal veins of the guinea pig and rat, exogenously applied ATP elicited a contraction that was blocked by arylazido aminopropionyl ATP ($ANAPP_3$), a selective P_2-purinoceptor antagonist.[4] Transmural stimulation of the guinea pig portal vein preincubated with [^{3}H]adenosine resulted in a 6-fold increase in the release of ^{3}H-labeled purines.[5] α,β-Methylene ATP, an agent known to desensitize the P_2-purinoceptor and thus antagonize the effect of ATP, abolished the first phase of the neurogenic biphasic contraction of the rabbit saphenous artery.[6] Stimulation of the perivascular nerve of the guinea pig saphenous artery elicited excitatory junction potentials (ejp's) in the smooth muscle cells, and these ejp's were inhibited by $ANAPP_3$.[7] These observations suggest the existence of purinoceptors for ATP on the vascular smooth muscle cells, and that ATP may play a role in neurotransmission of these two vessels. In the present study, we investigated the possibility that NE and ATP are cotransmitters in both the guinea pig portal vein and the rabbit saphenous artery.

METHODS

Tissues were isolated and mounted in an organ bath containing a modified Krebs solution that was kept at 37 °C and bubbled with 95% O_2/5% CO_2. Transmural

[a]This work was supported by grants from the National Institutes of Health (AM07478 and HL38126).

electrical stimulation of the portal vein (0.5 msec pulse duration, supramaximal voltage) was applied across a pair of platinum ring electrodes (7 mm apart) surrounding the tissue. Tension generated was measured along the longitudinal axis of the tissue by a force transducer (Grass FT03) and recorded on a Grass polygraph. Electrical stimulation of the saphenous artery (0.1 msec pulse duration, supramaximal voltage) was delivered by a pair of parallel platinum wire electrodes: one located outside and the other located inside the lumen of the vessel. Tension generated was measured perpendicular to the longitudinal axis of the vessel by a force transducer and recorded on a polygraph.

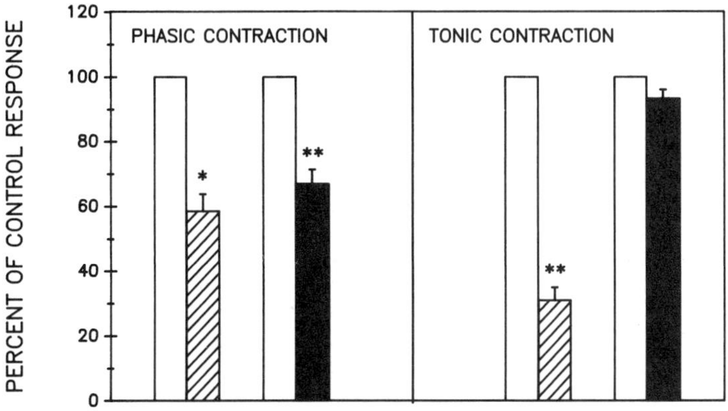

FIGURE 1. Effect of prazosin and α,β-methylene ATP on the neurogenic contraction of the guinea pig portal vein: □: control; ▨: prazosin (0.3 μM) ($N = 5$); ■: α,β-methylene ATP (10 μM) ($N = 7$). The tissue was stimulated at 16 Hz for 1 min. Antagonist was introduced into the organ bath 20 min before the next stimulation. Contractile response was expressed as a percentage of the control response. The biphasic neurogenic contraction was separated into phasic and tonic components depicted in left and right panels, respectively. Left panel: Both prazosin and α,β-methylene ATP attenuated the first phase or phasic component of the biphasic contraction. Right panel: Only prazosin depressed the second phase or tonic component of the biphasic contraction. *: $p < .01$; * *: $p < .001$.

RESULTS AND CONCLUSIONS

Transmural stimulation of the guinea pig portal vein and rabbit saphenous artery elicited a biphasic contraction consisting of a rapid initial phase followed by a slower sustained phase. In the portal vein, prazosin (0.3 μM), an α_1-adrenoceptor antagonist, depressed both phases of the neurogenic contraction; however, the second phase was much more reduced than the first phase. α,β-Methylene ATP (10 μM), which desensitizes the P_2-purinoceptor, attenuated only the first phase of the contraction (FIG. 1). In the rabbit saphenous artery, prazosin (0.3 μM) reduced both phases of the neurogenic contraction almost equally (70-80%). Further addition of α,β-methylene ATP (10 μM) to a tissue in which the response was already attenuated by prazosin abolished the first and significantly depressed the second phase of the contraction (FIG. 2). These results suggest that NE and ATP are excitatory cotransmitters in both blood vessels. In the guinea pig portal vein, both NE and ATP appear to contribute

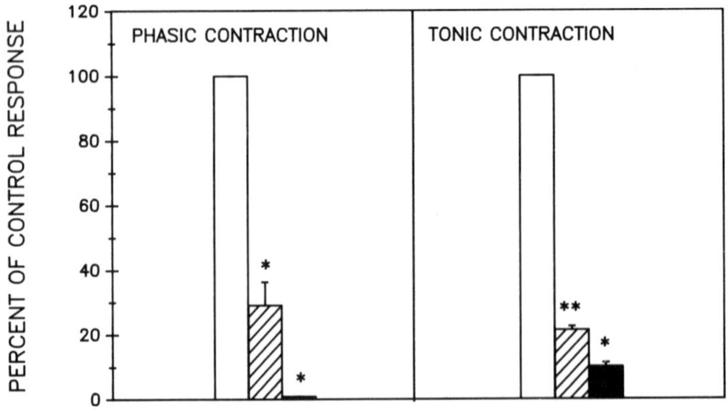

FIGURE 2. Effect of prazosin and α,β-methylene ATP on the neurogenic contraction of the rabbit saphenous artery: □: control; ▨: prazosin (0.3 μM); ■: prazosin (0.3 μM) plus α,β-methylene ATP (10 μM). The tissue was stimulated at 8 Hz for 1 min. Antagonist was introduced into the organ bath 15 min before the next stimulation. Contractile response was expressed as a percentage of the control response. The biphasic neurogenic contraction was separated into phasic and tonic components, depicted in left and right panels, respectively. Note that prazosin attenuated both phases of the biphasic neurogenic contraction. Addition of α,β-methylene ATP to the tissue in the presence of prazosin further depressed both phases of the contraction. The response in the presence of both prazosin and α,β-methylene ATP was statistically compared to the response in the presence of prazosin alone. *: $p < .05$; * *: $p < .001$ ($N = 4$).

to the initial phase of the neurogenic contraction, whereas NE alone mediates the slower sustained phase. In the rabbit saphenous artery, both NE and ATP contribute to the first and second phases of the neurogenic contraction.

REFERENCES

1. SNEDDON, P. & D. P. WESTFALL. 1984. J. Physiol. **347:** 561-580.
2. VON KÜGELGEN, I. & K. STARKE. 1985. J. Physiol. **367:** 435-455.
3. SNEDDON, P. & G. BURNSTOCK. 1984. Eur. J. Pharmacol. **106:** 149-152.
4. CHAU, W. L., T. HARBACH & D. P. WESTFALL. 1988. FASEB J. **2:** A1814.
5. BURNSTOCK, G., R. CROWE & H. K. WONG. 1979. Br. J. Pharmacol. **65:** 377-388.
6. BURNSTOCK, G. & J. J. I. WARLAND. 1987. Br. J. Pharmacol. **90:** 111-120.
7. CHEUNG, D. W. & M. FUJIOKA. 1986. Br. J. Pharmacol. **89:** 3-5.

The Role of Surface Protein Kinase in the ATP-Induced Growth Inhibition in Transformed Mouse Fibroblasts[a]

ILAN FRIEDBERG

Department of Microbiology
G. S. Wise Faculty of Sciences
Tel-Aviv University
69978 Tel-Aviv, Israel

DIETER KÜBLER

Institute of Experimental Pathology
German Cancer Research Center
D-6900 Heidelberg, Federal Republic of Germany

Addition of ATP to cultures of transformed mouse fibroblasts (Swiss mouse 3T6 cells) results in cell growth inhibition, whereas the growth of the nontransformed counterparts of 3T6 cells, namely 3T3 cells, is only slightly affected.[1,2] The mechanisms underlying the growth inhibition are only partly understood. The uptake of adenine nucleotides[3] and the alteration of ion fluxes[4] were suggested to mediate the inhibition, in adenocarcinoma and erythroleukemia cells, respectively. These effects, however, were not detected in mouse fibroblasts. We have shown that hydrolysis of ATP, and uptake of the adenosine generated, has a role in the inhibition process.[5] Because ATP and its products are metabolized within one day,[5] the continuation of the inhibition is probably mediated by additional mechanisms. We found that conditioned medium from ATP-treated cells inhibits cell proliferation, indicating that extracellular factors are involved in the inhibition.[2] The selectivity for transformed cells and the specificity for ATP (and to a lesser extent to other adenine nucleotides) indicate that the effects exerted by ATP are mediated by cell surface enzymes or receptors.

In this study we examined the possible role of cell surface protein kinase (SPK, casein type II protein kinase[6–8]) in the ATP-induced growth inhibition, using mouse fibroblasts, Balb/c 3T3 cells, and their transformed derivatives, Balb/c SV40-3T3 cells.

The SPK activity found in these cells could be removed by washings with an assay

[a]This work was supported by the National Council for Research and Development, Jerusalem, Israel, and the Deutsches Krebsforschungszentrum, Heidelberg, Federal Republic of Germany. Partial support was provided by a short-term fellowship from the European Molecular Biology Organization.

TABLE 1. Effect of Cell Washing on ATP-Induced Growth Inhibition[a]

Treatment	Growth Inhibition by ATP (%)	
	SV40-3T3	3T3
None	53	0
Washing with assay mixture	67	—
Washing with assay mixture + phosvitin	8	—

[a] Balb/c 3T3 cells and their transformed derivatives, SV40/3T3, were grown on Dulbecco's modified Eagle's medium (DMEM), containing 10% fetal calf serum (FCS), in 33-mm dishes. One day after innoculation, some of the cultures were washed three times with assay mixture, consisting of 70 mM NaCl, 30 mM Tris(OHAc), 5 mM $Mg(OAc)_2$, 5 mM potassium phosphate, 0.5 mM EDTA, and 75 mM glucose (pH 7.2, osmolarity 290 ± 10 mOsm[7]). Other dishes were washed with assay mixture containing 1 mg/ml phosvitin. Then DMEM + 10% FCS was added to the dishes either with or without 0.3 mM ATP. After an additional day the cells were detached by trypsin and counted in a coulter counter, and the percentage of inhibition was calculated in comparison to untreated control cultures. Similar inhibition was obtained after two days of growth. Washings with casein, a substrate of SPK, gave results similar to phosvitin, whereas washings with albumin, a nonsubstrate protein, brought about results comparable to those obtained with the assay mixture by itself.

mixture that contained a substrate protein. In this study (see TABLE 1), the substrate protein was phosvitin (also used in a study with HeLa cells[7]).

TABLE 1 shows that exogenous ATP does not inhibit the growth of 3T3 cells, whereas the growth of SV40-3T3 cells is markedly inhibited. Washings with the assay mixture did not prevent the inhibition by ATP. When phosvitin (1 mg/ml) was added to the washing solution, however, the ATP-induced inhibition was markedly reduced. The washings with phosvitin-containing assay mixture resulted in a decrease in the SPK activity, up to 80%, whereas washings with the assay mixture by itself did not affect this activity. Thus, there is a correlation between the activity of SPK and the ability of external ATP to inhibit the growth of the cells.

This correlation was further supported by experiments in which the SPK activity was altered by changing the density of the cultures (as shown in HeLa cells[6]). FIGURE 1 shows that the activity of SPK and, in parallel, the ability of ATP to inhibit cell growth are decreased with the increase of cell density. The curve in FIGURE 1, which

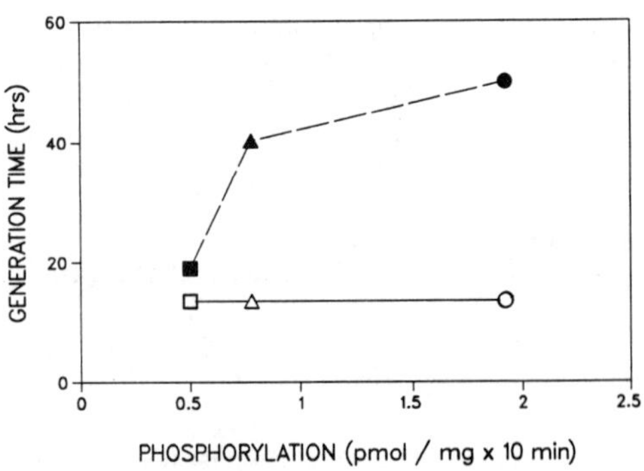

FIGURE 1. Effect of cell density on phosphorylation and growth rate in Balb/c SV40-3T3 cells. Cells were inoculated at various densities and were grown as described in TABLE 1. Cell growth was determined by cell counting, every 24 hr, over three days. Phosphorylation was performed in the presence of reaction mixture (see TABLE 1), with [γ-^{32}P]ATP (10 min, 37 °C), using phosvitin as a substrate.[6] The SPK activity determinations were performed in cultures with a cell density of 80×10^3 (○), 240×10^3 (△), and 800×10^3 (□) cells per dish. ATP was added to parallel dishes, with the same cell densities (●, ▲, ■), to a final concentration of 0.3 mM.

indicates that the ratio between phosphorylation and growth rate is not linear, also indicates a threshold value for SPK activity, above which the ATP-induced inhibition is more effective.

The results of this study indicate that SPK might be involved in the process of ATP-induced cell growth inhibition, by catalyzing protein phosphorylation, which could activate growth inhibitor, or its receptor, or inhibit growth-stimulating factor or its receptor.

REFERENCES

1. BELZER, I. & I. FRIEDBERG. 1984. Isr. J. Med. Sci. **20:** 483.
2. BELZER, I. & I. FRIEDBERG. 1989. J. Cell Physiol. **140:** 524-529.
3. RAPPAPORT, E. 1983. J. Cell Physiol. **114:** 279-283.
4. CHAHWALA, S. B. & L. C. CANTLEY. 1984. J. Biol. Chem. **259:** 13717-13722.
5. WEISMAN, G. A., K. D. LUSTIG, E. LANE, N. HUANG, I. BELZER & I. FRIEDBERG. 1988. J. Biol. Chem. **263:** 12367-12372.
6. KÜBLER, D., W. PYERIN & V. KINZEL. 1982. J. Biol. Chem. **257:** 322-329.
7. KÜBLER, D., W. PYERIN, E. BUROW & V. KINZEL. 1983. Proc. Natl. Acad. Sci. USA **80:** 4021-4025.
8. PYERIN, W., E. BUROW, K. MICHAELY, D. KÜBLER & V. KINZEL. 1987. Biol. Chem. Hoppe-Seyler **368:** 215-227.

Cell Surface Protein Kinase-Mediated Protein Phosphorylation[a]

DIETER KÜBLER AND WALTER PYERIN

Institute of Experimental Pathology
German Cancer Research Center
D-6900 Heidelberg, Federal Republic of Germany

Protein phosphorylation catalyzed by a number of protein kinase (PK) activities is known to occupy a key role in diverse metabolic pathways as well as in transmembranal signal transduction. We have found that a wide range of mammalian cells carry two types of cell surface-located (ecto)-PK activities with characteristics of phosvitin/casein type II PK[1,2] and cAMP-dependent PK.[3] These ecto-PK's, which use exogenous nucleotide triphosphates at low micromolar concentrations as cosubstrate, are discernable provided intact cells are employed (a prerequisite) and provided the plasma membrane barrier is carefully controlled (even a small degree of cell membrane perturbation and leakage is detectable).[4] Under such conditions, intact cells exert phosphorylation of a specific set of integral cell surface proteins (sensitive against exogenous trypsin) via their phosvitin/casein type II ecto-PK, as exemplified for HeLa cells in FIGURE 1, as well as extrinsic proteins such as phosvitin, casein, and fibrinogen.[1,5] Both the exogenous substrate phosphorylation reaction and phosphorylation of surface proteins used ATP and/or GTP as cosubstrates, and enzyme activity was inhibited by heparin ($ID_{50} < 1$ μg/ml). Although the phosvitin/casein type II ecto-PK proved to be stably associated with the cell surface even during extensive cell washes with the isotonic incubation fluid (for composition, see FIG. 1), it can be detached to a considerable degree from intact cells by interaction with exogenous substrates by a yet unknown mechanism.[6] Because the amount of released ecto-PK does not allow the study of molecular aspects of the enzyme, as an intermediate step we are analyzing its intracellular counterpart. The highly purified enzyme was found to be composed of three different subunits: α, α', and β (α and α' are the catalytic subunits; β is probably the regulatory subunit[2]). Recently we succeeded in the molecular cloning and sequencing of the β subunit, which turned out to be a highly conserved protein.[7]

Unambiguous demonstration of cAMP-dependent ecto-PK was obtained recently[3] through the use of the synthetic peptide Kemptide (Leu-Arg-Arg-Ala-Ser-Leu-Gly), with the consensus sequence -R-R-X-S- being specific for cAMP-PK. Ecto-cAMP-PK activity that can utilize only ATP as cosubstrate appears to be directed exclusively toward exogenous substrates because phosphorylation of cell surface proteins (as in FIG. 1) was influenced by neither added cAMP nor added heat- and acid-stable Walsh-Krebs inhibitory protein specific for cAMP-PK.

To understand the physiological role of ecto-PK's, it is necessary to search for

[a]This work was supported by the National Council for Research and Development, Jerusalem, Israel, and the Deutsches Krebsforschungszentrum (grant Ca37), Heidelberg, Federal Republic of Germany.

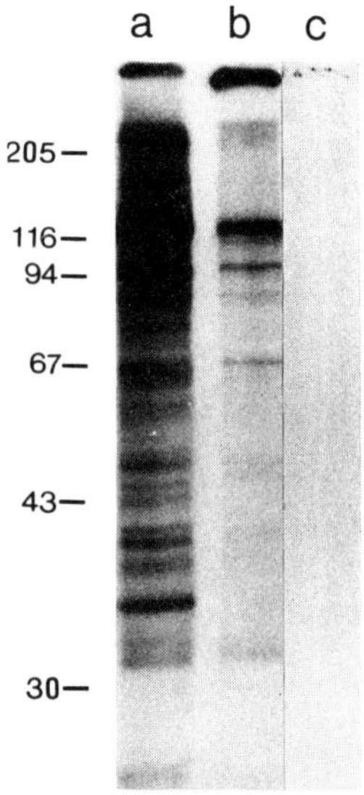

FIGURE 1. Phosphorylation patterns of intact HeLa cells as generated after exposure to radiolabeled ATP. Culture plates 5 cm across containing 10^6 cells were rinsed and incubated with assay mixture consisting of 70 mM NaCl, 30 mM Tris (OHAc), 5 mM $Mg(OAc)_2$, 5 mM potassium phosphate, 0.5 mM EDTA, and 75 mM glucose (pH 7.2, osmolarity 290 mOsm). The phosphorylation reaction, which was started by addition of 0.5 μM [γ-^{32}P]ATP to the assay mixture, proceeded for 15 min. Labeled cells were rinsed with the mixture containing 1 mM unlabeled ATP and then lysed with SDS-sample buffer. The total cell lysates were subjected to SDS-PAGE (8.75% acrylamide). Shown are the autoradiographic patterns observed (**a**) with reducing agent (5% v/v β-mercaptoethanol), (**b**) in the absence of reducing agent, and (**c**) after labeled cells were treated with 0.05% trypsin for 3 min prior to lysis. Molecular weight marker proteins are indicated in kDa.

and analyze natural substrates and the significance of their phosphorylation. Certain extracellularly occurring natural peptides, such as the atrial natriuretic peptide (ANP), also carry the sequence -R-R-X-S-. We were able to phosphorylate the ANP with intact cells (FIG. 2), suggesting a new model of posttranslational processing of this peptide. Controls in the absence of cells or with cell incubation fluids generated only negligible phospho-ANP (as was also observed with phospho-Kemptide), suggesting that ecto-cAMP-PK remained membrane bound, a property in contrast to the ecto-phosvitin/casein type II-PK.

Ecto-PK-catalyzed protein phosphorylation might also be considered as a mechanism controlling cell growth inhibition caused by extracellular ATP on transformed cells.[8]

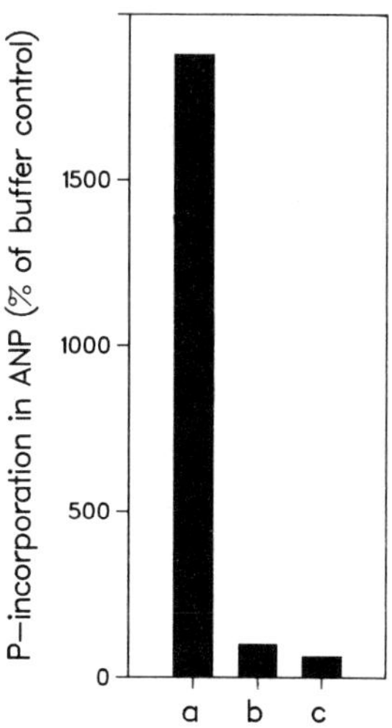

FIGURE 2. Phosphorylation of the ANP by cAMP-dependent PK of intact cells. The phosphorylation reaction, which proceeded under the conditions described in the legend of FIGURE 1, was carried out in the presence of 5 μM cAMP and 1 μM ANP for 10 min with intact HeLa cells or with cell supernatants. To obtain the latter, cells were preincubated for 10 min in the absence of radiolabeled ATP with assay buffer (with or without cAMP/ANP), and after completion to 5 μM cAMP and 1 μM ANP, the cell-free phosphorylation reaction was carried out for another 10 min. Analysis of ANP labeling was carried out by SDS-PAGE (18% acrylamide) and autoradiography. Quantitation of phospho-ANP was performed by cutting out gel bands and determining radioactivity by scintillation counting. Given are the amounts of phospho-ANP as generated through (**a**) reaction in the presence of cells, (**b**) reaction with preincubation buffer (control set to 100%), and (**c**) reaction with preincubation buffer containing cAMP/ANP.

REFERENCES

1. KÜBLER, D., W. PYERIN & V. KINZEL. 1982. J. Biol. Chem. **257:** 322-329.
2. PYERIN, W., E. BUROW, K. MICHAELY, D. KÜBLER & V. KINZEL. 1987. Biol. Chem. Hoppe-Seyler **368:** 215-227.
3. KÜBLER, D., W. PYERIN, O. BILL, A. HOTZ, J. ŠONKA & V. KINZEL. 1989. J. Biol. Chem. **264:** 14549-14555.
4. KÜBLER, D., W. PYERIN & V. KINZEL. 1982. Eur. J. Cell Biol. **26:** 306-309.
5. ŠONKA, J., D. KÜBLER & V. KINZEL 1989. Biochim. Biophys. Acta **997:** 268-277.
6. KÜBLER, D., W. PYERIN, E. BUROW & V. KINZEL. 1983. Proc. Natl. Acad. Sci. USA **80:** 4021-4025.
7. JACOBI, R., H. VOSS & W. PYERIN. 1989. Eur. J. Biochem. **183:** 227-233.
8. FRIEDBERG, I. & D. KÜBLER. 1990. Ann. N.Y. Acad. Sci. This volume.

Effect of Inhibition of Protein Synthesis, RNA Synthesis, and Tyrosine Kinase on the Induction of Ecto-ATPases of Human Hepatoma (Li-7A) Cells

AILEEN F. KNOWLES AND SANDRA L. MURRAY

Department of Biology
Northeastern University
Boston, Massachusetts 02115

Human hepatoma Li-7A cells have an ecto-ATPase activity that is activated by either Mg^{2+} or Ca^{2+}, and is sensitive to inhibition by *p*-chloromercuriphenyl sulfonate (pCMPS). When Li-7A cells are cultured in the presence of epidermal growth factor (EGF) and cholera toxin, or forskolin, or cAMP analogues, the ecto-ATPase activity is enhanced 5-10-fold. The induced ecto-Ca^{2+}-ATPase, as it is more active with Ca-ATP, is not inhibited by pCMPS.[1-3] Moreover, this activity is significantly different from the activity on the untreated cells (referred to as ecto-Mg^{2+}-ATPase) in its enzymatic properties.[1] Ecto-Mg^{2+}-ATPase, which is more active with Mg-ATP than with Ca-ATP, is enhanced 5-10-fold when Li-7A cells are treated with 2-5 mM butyrate,[4] a short-chain fatty acid that affects cell growth and gene expression in many cells.[5] Induction of the two ecto-ATPase activities by different modulators further supports the conclusion that they are manifested by different enzymes.

Maximal induction of the ecto-Ca^{2+}-ATPase by the synergistic action of EGF and a cAMP-elevating agent requires exposure of the cells to the modulators for 4-5 days. The primary action of EGF is considered to be the activation of the tyrosine kinase of the EGF receptor. Cyclic AMP activates protein kinase A, a serine protein kinase. Increase of the ecto-Ca^{2+}-ATPase activity could be the result of 1) enzyme activation due to simultaneous phosphorylation of the enzyme protein by EGF-stimulated tyrosine kinase and protein kinase A, or 2) increased gene expression regulated by phosphorylated proteins, these proteins being the targets of the two protein kinases. Inhibitors of tyrosine kinase will prevent increases of enzyme activity by either mechanism. Inhibitors of protein synthesis and RNA synthesis should not affect the former, but will have an effect on the latter.

Genistein, a tyrosine kinase-specific inhibitor,[6] inhibited the autophosphorylation of the EGF receptor/kinase by [γ-^{32}P]ATP in Li-7A cell membrane preparations (FIG. 1A). When added together with EGF/cholera toxin, genistein inhibited the induction of ecto-Ca^{2+}-ATPase but had no effect on ecto-Mg^{2+}-ATPase (FIG. 1B). Induction of ecto-Ca^{2+}-ATPase was also inhibited by cycloheximide or actinomycin D (experiment 1 in TABLE 1). We conclude from these results that induction of ecto-Ca^{2+}-ATPase requires increased synthesis of mRNA and protein, the most interesting

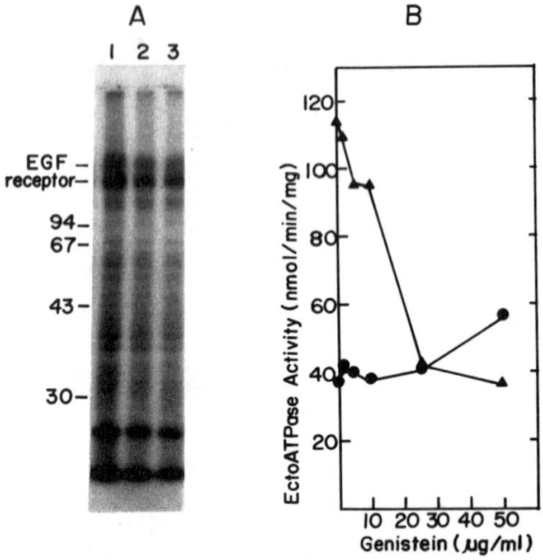

FIGURE 1. Effect of genistein on phosphorylation of Li-7A membranes and induction of ecto-Ca^{2+}-ATPase by EGF/cholera toxin. (**A**) Membranes from Li-7A cells (20 μg) were preincubated for 10 min at 0 °C in 45 μl of reaction mixture containing 10 nmol HEPES (pH 7.4), 1 nmol $MnCl_2$, 0.05 nmol Na_3VO_4, and 200 ng EGF in the absence and presence of genistein. Phosphorylation was initiated by the addition of 10 nmol [γ-^{32}P]ATP in 5 μl (900,000 cpm/nmol). After 10 min at 0 °C, the reaction was terminated by the addition of 25 μl of a 3×-concentrated solution of Laemmli's sample buffer. The samples were heated, and an aliquot (50 μl) was applied to a 7.5% acrylamide gel containing SDS. Autoradiography was obtained from dried gel after staining by Commassie blue and destaining. The amounts of genistein added for each sample were 0 (lane 1), 0.5 μg/ml (lane 2), and 1 μg/ml (lane 3). (**B**) Li-7A cells (1×10^5 cells/well) were plated on day 1. EGF (100 ng/ml), cholera toxin (5 ng/ml), hydrocortisone (25 nM), and the indicated concentrations of genistein (0-50 μg/ml) were added on day 2. Cells were allowed to grow for two days. Ecto-Mg^{2+}-ATPase activity (●) and ecto-Ca^{2+}-ATPase activity (▲) were determined on day 4.

implication being that the ecto-Ca^{2+}-ATPase gene contains elements responsive to both EGF and cAMP. Induction of ecto-Mg^{2+}-ATPase by butyrate was also inhibited by actinomycin D and cycloheximide (experiments 2 and 3 in TABLE 1), adding ecto-Mg^{2+}-ATPase to the list of enzymes whose gene expression is regulated by butyrate.

Definitive proof that expression of the genes of the two ecto-ATPases is activated by their respective modulators is yet to be obtained. Preliminary results indicated the ecto-Mg^{2+}-ATPase of Li-7A cells is similar to the rat liver ecto-ATPase,[7] whereas the ecto-Ca^{2+}-ATPase is different. The Li-7A cells whose two ecto-ATPases are enhanced differentially will be useful in addressing questions of isozymes, physiological function, and gene regulation of the ecto-Mg^{2+}-ATPase and the ecto-Ca^{2+}-ATPase.

TABLE 1. Effect of Cycloheximide and Actinomycin D on Induction of Ecto-Ca^{2+}-ATPase by EGF and Cholera Toxin and Induction of Ecto-Mg^{2+}-ATPase by Sodium Butyrate[a]

			Ecto-Ca^{2+}-ATPase		Ecto-Mg^{2+}-ATPase	
Expt	Modulator	Inhibitor	nmol/min/well	nmol/min/mg	nmol/min/well	nmol/min/mg
1	None	None	9.2	22.0		
	EGF + CT	None	18.2	46.7		
	EGF + CT	CHX (0.2 μg/ml)	3.3	16.0		
	EGF + CT	Act D (10 nM)	2.7	30.0		
2	None	None			9.0	36.0
	Bu	None			50.7	242.0
	Bu	CHX (1 μg/ml)			7.3	81.3
3	None	None			12.9	29.3
	Bu	None			60.2	144.7
	Bu	Act D (10 nM)			1.4	20.0

ABBREVIATIONS: EGF, epidermal growth factor; CT, cholera toxin; Bu, sodium butyrate; CHX, cycloheximide; Act D, actinomycin D.

[a] In experiment 1, Li-7A cells were plated in Nunclon 6-well plates at 1×10^5 cells/well. After 48 hr, EGF (100 ng/ml), cholera toxin (5 ng/ml), hydrocortisone (25 nM), and the indicated concentrations of inhibitors were added. Ecto-Ca^{2+}-ATPase activity was determined after 30 hr. In experiments 2 and 3, Li-7A cells were plated at 1.7×10^5 cells/well and 2.5×10^5 cells/well, respectively. Cells were allowed to attach overnight, and then treated with or without butyrate (2.5 mM) and the indicated concentrations of inhibitors. Ecto-Mg^{2+}-ATPase activity was determined after 24 hr. Activity was reported for per well of cells and per mg of cell protein.

REFERENCES

1. Knowles, A. F. 1988. Arch. Biochem. Biophys. **263:** 264-271.
2. Knowles, A. F. 1988. J. Cell. Physiol. **134:** 109-116.
3. Knowles, A. F. & S. L. Murray. 1988. FASEB J. **2:** A357.
4. Murray, S. L. & A. F. Knowles. 1988. J. Cell Biol. **107:** 124a.
5. Kruh, J. 1982. Mol. Cell. Biochem. **42:** 65-82.
6. Akiyama, T., J. Ishida, S. Nakagawa, H. Ogawara, S.-I. Watanabe, N. Itoh, M. Shibuya & Y. Fukami. 1987. J. Biol. Chem. **262:** 5592-5595.
7. Lin, S.-H. & G. Guidotti. 1989. J. Biol. Chem. **264:** 14408-14414.

Characterization of Ectonucleotidases in Chromaffin Cells

M. T. MIRAS-PORTUGAL, J. PINTOR, P. ROTLLÁN, AND M. TORRES

Departamento de Bioquímica
Facultad de Veterinaria
Universidad Complutense de Madrid
28040 Madrid, Spain

Secretory vesicles of chromaffin cells store catecholamines and ATP, in high concentrations.[1] We recently found that diadenosine polyphosphates (Ap_4A and Ap_5A) are also stored in these vesicles.[2] The granular content is released to the extracellular medium during the secretory response.[3] It is known that ATP and its main product, adenosine, influence many biological processes,[4] including the secretory response in chromaffin cells, through specific plasma membrane receptors.[5] Also, it is known that the diadenosine polyphosphates, both Ap_4A and Ap_5A, inhibit the nicotinic and K^+-evoked release of catecholamines on bovine chromaffin cells,[6] although their receptors have not been characterized yet.

The extracellular nucleotide actions are ended by an enzymatic degradative mechanism. The enzymes responsible for nucleotide hydrolysis have been characterized in some tissues, like endothelial cells,[7] cholinergic synapses from mammalian brain,[8] and glial cells.[9] In the present work we report on the presence of ectonucleotidases in chromaffin cells,[10] as well as on an ectophosphodiesterase able to hydrolyze the diadenosine polyphosphate -Ap_4A-.

The extracellular ATP and ADP were rapidly degraded by cultured chromaffin cells. The kinetic parameters for ecto-ATPase and ecto-ADPase are shown in TABLE 1. The affinity constants were similar to those described for endothelial cells, and the V_{max} values were higher than those found in a variety of cells.[7–9]

The ecto-5′-nucleotidase activity was more specific for purine nucleotides, AMP and IMP, than for pyrimidine nucleotides, CMP and TMP. The V_{max} value for AMP was higher than that reported in other tissues. Nevertheless, the affinity constants were very similar (TABLE 1). All of these studies were carried out in the presence of 20 mM β-glycerol phosphate to avoid nonspecific phosphatase activities. The nonhydrolyzable ADP and ATP analogues were competitive inhibitors of this enzyme. The α,β-methylene adenosine 5′-diphosphate presented a K_i value of 73 $\pm$ 3.5 nM, and the adenylyl(β,γ-methylene)disphosphonate presented a K_i value of 193 $\pm$ 29 nM.

The phosphatidylinositol-specific phospholipase C (PI-PLC) released the ecto-5′-nucleotidase from chromaffin cells in culture, thus suggesting an anchorage through phosphatidylinositol to plasma membranes. Nevertheless, no ATPase or ADPase activity can be released in our experimental conditions (TABLE 1).

TABLE 1. Kinetic Parameters of Ectonucleotidase Activities in Cultured Chromaffin Cells[a]

Enzyme	K_m (μM)	V_{max} nmol/10^6 cells/min	V_{max} μmol/mg prot/min	Activity Released by PI-PLC[d]	*N*
ATPase[b]	250 ± 18	167 ± 25	1.67 ± 0.23	Nonreleased	4
ADPase[b]	375 ± 40	125 ± 20	1.25 ± 0.20	Nonreleased	4
5′-Nucleotidase[c]	55 ± 5	4.3 ± 0.8	0.043 ± 0.007	50%	4

[a] Cultured chromaffin cells were obtained according to the procedure of Miras-Portugal *et al.*[11] Cells were plated in 24-well Costar cluster dishes at a density of 250,000 cells/well for ectonucleotidase activity determinations. The experiments with PI-PLC were done in petri dishes containing 3×10^6 cells.

[b] ATPase and ADPase activities were measured by an HPLC method, according to the procedure of Rodriguez del Castillo *et al.*[2]

[c] 5′-Nucleotidase activity was measured by HPLC[2] and a radiometric technique, the latter being a modification of adenosine kinase measurement, according to Rotllán and Miras-Portugal.[13] The kinetic parameters obtained were identical with both techniques.

[d] Experiments with PI-PLC to release the ectonucleotidase activities from plasma membranes were carried out with 0.5 I.U. of enzyme activity in a volume of 1 ml, with 3×10^6 cells/petri dish, for 1 hr of incubation.

Cultured chromaffin cells can also hydrolyze the extracellular Ap_4A. The hydrolysis of this compound is nonsymmetrical, ATP and AMP being formed as products of the reaction (FIG. 1A). In FIGURE 1B the Michaelis-Menten and Lineweaver-Burk kinetic representations are shown—a K_m value of 1.58 ± 0.5 μM and a V_{max} value of 248 ± 64 pmol/10^6 cells/min were obtained. Both AMP and ATP were inhibitors of this enzyme. The apparent K_i values were 3.6 ± 0.4 μM and 6.5 ± 0.3 μM for ATP and AMP, respectively, when 1 μM of Ap_4A was employed.

From these results, it seems that chromaffin cells are able to recycle the granular

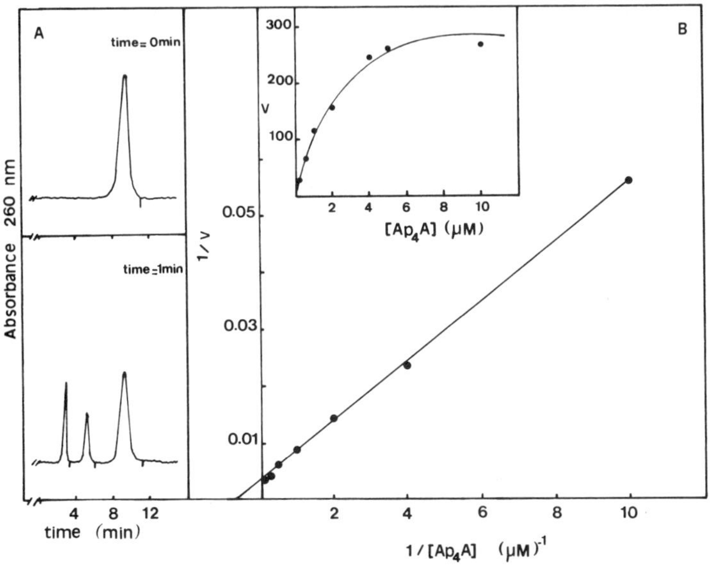

FIGURE 1. Hydrolysis of extracellular Ap_4A by cultured chromaffin cells. (**A**) Elution profile of 5 μM Ap_4A hydrolysis by HPLC at 0 and 1 min after incubation with cultured chromaffin cells. The retention times were 3 min for AMP, 5.2 min for ATP, and 9.4 min for Ap_4A, obtained in isocratic conditions with 10 mM KPO_4H_2 buffer (pH = 7.4) containing 2 mM tetrabutylammonium and 15% acetonytrile (v/v), with a C_{18} reversed-phase column. (**B**) Kinetics of Ap_4A extracellular hydrolysis by cultured chromaffin cells. Experiments were done in petri dishes containing 3 × 10^6 cells each. All velocity values were from the lineal period of enzyme activity, which is always expressed as pmol/10^6 cells/min. The main plot is the Lineweaver-Burk representation; the Michaelis representation is shown as an insert.

ATP after exocytosis. Three enzymatic ectonucleotidase activities are present, together with efficient adenosine transport.[11] On the other hand, adenosine can regulate its own transport by acting on the A_2 receptors[12] present in chromaffin cells.

The extracellular hydrolysis of Ap_4A to ATP and AMP allows the complete recycling of the most representative nucleotides in neural storage granules.

REFERENCES

1. WINKLER, H. 1976. Neuroscience **1:** 65-80.
2. RODRIGUEZ DEL CASTILLO, A., M. TORRES, E. G. DELICADO & M. T. MIRAS-PORTUGAL. 1988. J. Neurochem. **51:** 1969-1703.

3. Burgoyne, R. D. 1984. Biochim. Biophys. Acta **779:** 201-216.
4. Gordon, J. L. 1986. Biochem. J. **233:** 309-319.
5. Williams, M. 1987. Annu. Rev. Pharmacol. Toxicol. **27:** 315-345.
6. Castro, E., M. Torres, M. P. González & M. T. Miras-Portugal. 1990. Br. J. Pharmacol. **100:** 360-364.
7. Gordon, E. L., J. D. Pearson & L. L. Slakey. 1986. J. Biol. Chem. **261:** 15496-15504.
8. Richardson, P. J., S. J. Brown, E. M. Bailyes & J. P. Luzio. 1987. Nature **327:** 232-234.
9. Grondal, E. J. M., A. Janetzko & H. Zimmermann. 1988. Neuroscience **24:** 351-363.
10. Torres, M., F. Luzini & M. T. Miras-Portugal. 1989. J. Neurochem. **52**(Suppl.): 89.
11. Miras-Portugal, M. T., M. Torres, P. Rotllán & D. Aunis. 1986. J. Biol. Chem. **261:** 1712-1719.
12. Delicado, E. G., A. Rodrigues, R. P. Sen, A. M. Sebastiao, A. Ribeiro & M. T. Miras-Portugal. 1990. J. Neurochem. **54:** 1941-1946.
13. Rotllán, P. & M. T. Miras-Portugal. 1985. Eur. J. Biochem. **151:** 365-371.

Ecto-Phosphoryltransfer-Transduced Contraction to ATP in Isolated Vas Deferens of the Guinea Pig[a]

SHEILA J. LAMPORT[b] AND JEFFREY S. FEDAN[b,c]

[b]*Department of Pharmacology and Toxicology*
West Virginia University
Morgantown, West Virginia 26506

[c]*Physiology Section*
Division of Respiratory Disease Studies
National Institute for Occupational Safety and Health
Morgantown, West Virginia 26505

ATP contracts the guinea pig vas deferens by stimulating cell surface P_{2X}-purinoceptors. The response is biphasic at concentrations $\geq 3 \times 10^{-5}$ M, indicating that ATP may act at multiple sites. The first phase is inhibited by the P_2-purinoceptor antagonist arylazido aminopropionyl ATP ($ANAPP_3$), while the second phase is inhibited by periodate-oxidized ATP (P-ATP; 10^{-2} M, 5 min).[1,2] We suggested[1-3] that the first contractile phase is mediated by $ANAPP_3$-sensitive receptors, while the second contractile phase is transduced by a P-ATP-sensitive phosphate chain cleavage. Therefore, biochemical evidence was sought for the incorporation of ^{32}P from [γ-^{32}P]ATP and ^{35}S from [^{35}S]ATPγS into vas deferens proteins.

Using protocols employed in contraction studies, one intact vas deferens of a pair was incubated with [γ-^{32}P]ATP (3 mM) or [^{35}S]ATPγS (3 mM) for 5 sec and served as the reference for the contralateral vas deferens, which was incubated for 10, 20, 30, or 60 sec. After 5, 10, 20, 30, or 60 sec, sodium dodecyl sulfate-polyacrylamide gel electrophoresis (SDS-PAGE) dissociation buffer (100 °C) was added, and the tissue was homogenized and subjected to SDS-PAGE.

Autoradiograms (FIG. 1) indicated incorporation of ^{32}P and ^{35}S into a single band (19-21 kD). The time courses of incorporation followed the time course of the transient contraction; that is, labeling occurred within 5 sec and returned to basal level by 60 sec. These findings suggest that ecto-phosphoryltransfer is involved in the transduction of the second phase of contraction to ATP.

The incorporation of ^{35}S at 60 sec decreased following $ANAPP_3$ and P-ATP treatment. In protection experiments, the incorporation was reduced following incubation of the preparation with ATP (10^{-2} M), while histamine (10^{-3} M) and norepinephrine (10^{-3} M) did not reduce incorporation.

In homogenized vas deferens, ^{35}S incorporating into the 19-21-kD protein *increased* at 60 sec, which is in contrast to what was observed in intact vas deferens. P-ATP

[a]This work was supported, in part, by a grant from the National Institutes of Health (PTR 5 T32 GM07039-13).

treatment of intact tissues decreased incorporation of ^{35}S at 5 sec in homogenate, but not at 60 sec.

These results indicate that the response of the guinea pig vas deferens to ATP and ATPγS may be associated with the phosphorylation and thiophosphorylation, respectively, of a 19-21-kD protein associated with the cell surface of the vas deferens. It had been proposed[3] that the contractile response of the guinea pig vas deferens to adenine nucleotides resulted, in part, from a transduction arising from phosphate chain cleavage. The results shown here indicate that there is a reversible incorporation of

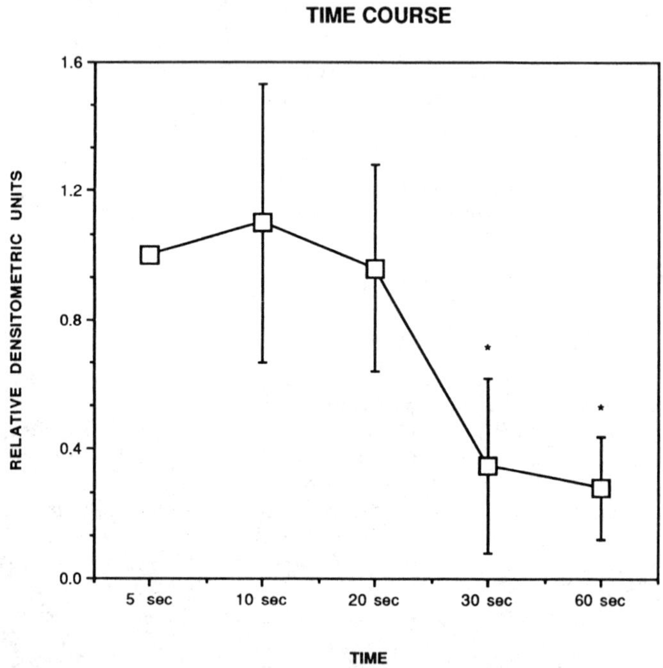

FIGURE 1. Time course of ^{35}S incorporation from [^{35}S]ATPγS into proteins of the intact vas deferens. The gels were scanned, and the incorporation seen at 5, 10, 20, 30, and 60 sec in the test vas deferens was normalized (as relative densitometric units) in relation to that observed in the reference vas deferens (5 sec). The incorporation of radiolabel followed the time course of the contractile response to ATPγS in the isolated vas deferens. The incorporation of ^{35}S into vas deferens proteins was significantly decreased (*: $p < .05$) at 30 and 60 sec. This figure depicts results (mean ± SEM) from five experiments.

γ-PO_3 or γ-SPO_2 into a protein in the vas deferens, and that this process may account, in part, for the production of the second contractile phase induced by adenine nucleotides.

REFERENCES

1. FEDAN, J. S. 1986. Periodate-oxidized ATP (P-ATP) discriminates two phases of the response of the guinea-pig isolated vas deferens to ATP. Pharmacologist **28:** 238a.

2. FEDAN, J. S. 1987. Comparison of the effects of periodate-oxidized ATP (P-ATP) and $ANAPP_3$ on responses of the smooth muscle of the guinea-pig isolated vas deferens to ATP. Fed. Proc. **46:** 699a.
3. FEDAN, J. S., G. K. HOGABOOM, D. P. WESTFALL & J. P. O'DONNELL. 1982. Comparison of the effects of arylazido aminopropionyl ATP ($ANAPP_3$), an ATP antagonist, on responses of the smooth muscle of the guinea-pig vas deferens to ATP and related nucleotides. Eur. J. Pharmacol. **85:** 277-290.

Structural, Enzymatic, and Regulatory Properties of the Skeletal Muscle Transverse Tubule Mg-ATPase

Its Role as a Receptor for Diacylglycerol[a]

JAW-JOU KANG, H. BRADLEY CUNNINGHAM, CHRIS JACHEC-SCHMIDT, KURT C. NORTON, ANNE M. PRIEST, ROGER A. SABBADINI, AND A. STEPHEN DAHMS[b]

Departments of Chemistry and Biology
Molecular Biology Institute
and
Rees-Stealy Research Foundation
San Diego State University
San Diego, California 92182

In striated muscle cells, the contraction-relaxation cycle is controlled by a sarcotubular membrane system consisting of the transverse tubule (TT) system and the sarcoplasmic reticulum. The TT membranes are continuous invaginations of the plasma membrane or sarcolemma, and constitute over 80% of the surface membrane of striated muscle. The topic of TT membranes has been reviewed recently.[1] The TT system is the locus of a number of metabolically significant proteins, including receptors for the Ca^{2+} channel blockers (the richest source studied), insulin receptors (80% of the insulin receptors in striated muscle and perhaps 90% of organismal insulin receptors are in the TT system), adrenergic receptors, dystrophin, and glucose transporters (50% of cytochalasin-sensitive glucose transporters are in the TT system); a panoply of TT proteins have been found to serve as substrates for endogenous and exogenous protein kinases.[1] This cholesterol-rich membrane possesses a characteristic complement of ion pumps, channels, and enzymes.

One of the most prominent enzymatic activities found in highly purified TT membranes is a Ca^{2+} or Mg^{2+}-dependent ATPase (Mg-ATPase, EC 3.6.1.3). This concanavalin A-modulated enzyme is exclusively localized at junctional or perijunctional regions where the TT system makes periodic contact with the sarcoplasmic

[a]This work was supported by grants from the National Science Foundation (DCB 8613881 and INT 8515846) and by the California Metabolic Research Foundation.

[b]To whom correspondence should be sent.

reticulum. It possesses a number of physical, enzymatic, kinetic, and thermodynamic properties that easily distinguish it from other striated muscle ATPases dependent on divalent cations and present in the sarcolemma, in the sarcoplasmic reticulum, or in intracellular organelles. This Mg-ATPase appears to be designed to catalyze the hydrolytic turnover of ATP under specifically controlled conditions and mediates unidirectional catalysis (hydrolysis) without any "partial reaction" characteristic of ATPases of other types.[1]

Ecto-ATPases have been detected in a variety of tissues[1–6] and have been defined by criteria proposed by Pearson.[3] At this time, the sidedness of the TT Mg-ATPase has not been fully resolved. As seen in TABLE 1, the enzyme shares many properties with ecto-ATPases, but these individual properties are not absolutely unique to the latter. Histochemical studies show deposition of hydrolytic products intracellularly and support an *endo* function of the TT Mg-ATPase,[1] in agreement with our preliminary data using side-specific radioligand (saxitoxin and ouabain) binding. The effects of concanavalin A upon the ATPase could be used to support an *ecto* role, but these could be occurring through mediation by intracellular glycosylated conjugates. The intact single-cell approach used to characterize ecto-ATPases in a variety of types[1,3] is virtually impossible to conduct with differentiated striated muscle. This approach

TABLE 1. Similar Properties of the Skeletal Muscle TT Mg-ATPase and Ecto-ATPases

Lack of covalent modification by fluorescein isothiocyanate
Lack of inhibition by ouabain, oligomycin, azide, and *N*-ethylmaleimide
Lack of inhibition by vanadate and thereby a probable lack of a phosphoenzyme
Prominent inhibition by detergents
A broad nucleoside triphosphate substrate specificity
Activation by Mg^{2+} or Ca^{2+}
Glycosylation
Modulation by lectins
Absence of "partial reactions"

could be used in undifferentiated striated muscle, but TT formation occurs well after myoblast fusion. Minimally, the TT Mg-ATPase possesses a glycosylated protein, with an M_r of 85 kDa, that can be reduced to a protein core, with an M_r of about 65 kDa (similar to a rat liver ecto-ATP/ADPase[4]), but comparative N-terminal sequence analyses show no commonality. On the other hand, isoforms appear to exist among the ecto-ATPases that appear to differ in their ability to cleave ADP and in their sensitivity to site-labeling reagents. Studies employing triad-derived TT membrane vesicles, *in situ* labeling reagents, and molecular cloning are underway to absolutely define the sidedness of the TT Mg-ATPase.

MODULATION BY PHOSPHOLIPID-DERIVED SECOND MESSENGERS AND LIPOPHILIC AGENTS

Because of the prominent Triton X-100 inhibition of the TT Mg-ATPase at subcritical micelle concentrations (mimicking ATP at a low-affinity nucleotide site[1,2]), evaluation of the effects of known lipophilic and amphipathic regulatory agents was initiated. The enzyme was found to be responsive to free fatty acids (inhibition) and fatty acyl coenzyme A's (activation) at subcritical micelle concentrations, suggestive of a highly sophisticated regulatory mechanism for the control of ATP hydrolysis.[1]

The high affinity of the putative regulatory site(s) for the above agents has led to the evaluation of lipid second messenger effects upon the TT Mg-ATPase. As seen in TABLE 2, the chicken TT Mg-ATPase is inhibited by submicromolar concentrations of phorbol esters (phorbol 12-myristate-13-acetate and 4-α-phorbol 12, 13-didecanoate) and diacylglycerols (*rac*- and *sn*-oleoyl-2-acetylglycerol). It is significant that the enzyme is inhibited by both phorbol esters noted above: The regulatory (inhibitory) site on the TT Mg-ATPase, because it is sensitive to both phorbol 12-myristate-13-acetate and 4-α-phorbol didecanoate, is distinguishable from the activation site on protein kinase C (PKC), which, like the TT Mg-ATPase site, is sensitive to diacylglycerols, but which, unlike the TT Mg-ATPase site, is only sensitive to the former phorbol ester. Other unpublished data from our laboratories show that several PKC inhibitors are without effect upon either the TT Mg-ATPase or its responses to phorbol esters or diacylglycerols. Of further significance, the known inhibition of PKC by sphingosine is directly complemented by its activation of TT Mg-ATPase. Related to lectin modulation of the TT Mg-ATPase,[1,2] the above phorbol ester and diacylglycerol inhibitions are protected by concanavalin A.

TABLE 2. Effects of Diacylglycerols, Phorbol Esters, and other PKC Modifiers upon the TT Mg-ATPase

	Effect Upon Mg-ATPase ($K_{0.5}{}^{a}$)	Effect Upon PKC
Phorbol 12-myristate-13-acetate	Inhibition (0.2 ± 0.02 μM)	Activation
4-α-Phorbol 12,13-didecanoate	Inhibition (0.3 ± 0.01 μM)	No effect
rac-Oleoyl-2-acetylglycerol	Inhibition (1.7 ± 0.4 μM)	Activation
sn-Oleoyl-2-acetylglycerol	Inhibition (0.5 ± 0.1 μM)	Activation
sn-Stearoyl-2-arachidonylglycerol	No effect	Activation
Sphingosine	Activation (1 ± 0.1 μM)	Inhibition
H-7	No effect at 24 μM	Inhibition

[a] Concentration for half-maximal effect.

This is the first report of diacylglycerol regulation of an enzymatic system other than PKC. Importantly, although the concentration of phorbol 12-myristate-13-acetate necessary for inhibition of the TT Mg-ATPase is 50-fold higher than that for phorbol 12-myristate-13-acetate activation of PKC, the ATPase is 2-5 times more sensitive to the natural regulatory diacylglycerols than is PKC. It is probable that the TT Mg-ATPase is stoichiometrically a more potent diacylglycerol receptor than PKC in striated muscle. As with the fatty acyl coenzyme A's and free fatty acids, there is a chain length dependence for diacylglycerols because *sn*-stearoyl-2-arachidonylglycerol was without effect; the absence of *sn*-stearoyl-2-arachidonylglycerol effects may be indicative of regulation by diacylglycerols generated from phospholipase C- and phospholipose D-mediated turnover of phospholipids other than the phosphoinositides.

In summary, the TT Mg-ATPase responds to traditional PKC activators and inhibitors but in a completely opposite sense to PKC. The chain length specificities and saturability under low subcritical micelle concentrations are indicative of specific binding and regulation rather than a general lipid effect. The effects of these lipophilic and amphipathic agents upon the TT Mg-ATPase are independent of PKC. Under conditions where the TT Mg-ATPase is maximally inhibited, PKC would be expected to be fully activated. From an endo-ATPase perspective, regulation of intra- or perijunctional ATP hydrolysis would occur under conditions that would maximize phospholipase/PKC and inositol trisphosphate modulation of Ca^{2+} release and place TT Mg-ATPase activity in phase with the sarcoplasmic reticulum Ca-ATPase during muscle relaxation. From an ecto-ATPase perspective, a regulation of extracellular ATP hydrolysis would occur that could maximize PKC effects mediated by purinergic P_2 receptors. The level of striated muscle P_2 receptors and the activity of PKC in TT membranes and triads are being evaluated.

REFERENCES

1. Sabbadini, R. A. & A. S. Dahms. 1989. J. Bioenerg. Biomembr. **21:** 163-213.
2. Moulton, M. M., R. A. Sabbadini, K. C. Norton & A. S. Dahms. 1986. J. Biol. Chem. **261:** 12244-12251.
3. Pearson, J. D. 1986. Methods Phamacol. **6:** 83-107.
4. Lin, S. H. & G. Guidotti. 1989. J. Biol. Chem. **264:** 14408-14414.
5. Gordon, E. L., J. D. Pearson & L. L. Slakey. 1986. J. Biol. Chem. **261:** 15496-15504.
6. Filipini, A., R. E. Taffs, T. Agui & M. V. Sitkovsky. 1990. J. Biol. Chem. **265:** 334-340.

Desensitization of Cardiac Ventricular Myocytes to Extracellular ATP-Induced Increases in $[Ca^{2+}]_i$

Studies of the $[Ca^{2+}]_i$ Response in Fura-2-Loaded Single Myocytes and Bulk Suspensions of Myocytes from the Rat[a]

I. R. SIEMENS, A. P. BOULET,[b] J. R. MONCK,
J. R. WILLIAMSON, AND Ó. G. BJÖRNSSON[c]

Department of Biochemistry and Biophysics
University of Pennsylvania
Philadelphia, Pennsylvania 19104

INTRODUCTION

Previously we have reported[1] that cardiac ventricular myocytes of the rat contain P_{2Y}-purinoceptors[2] on the external surface of the sarcolemma, which upon activation lead to transient depolarization and activation of cation channels in the plasma membrane. This results in influx of extracellular Ca^{2+} into the cytosol and a further flux of Ca^{2+} into the cytosol from intracellular stores.[1] Previous data also suggested that hydrolysis of high-energy phosphate bonds was essential for activity of extracellular adenine nucleotides.[1] The present work on cardiac ventricular myocytes using either single cells or cells in bulk suspension further investigates the properties of $[Ca^{2+}]_i$ transients induced by extracellular ATP and the mechanism of extracellular ATP action. In particular, the desensitization (which rapidly develops in ventricular myocytes upon repeated application of extracellular ATP and to subsequent electrical stimulation) is examined.

[a]This work was supported by grants from the National Heart, Lung, and Blood Institute (HL-14461), the American Heart Association (Southeast Pennsylvania Chapter), and the Icelandic Science Foundation (Vísindasjódur Íslands).

[b]A postdoctoral fellow supported by Le Fonds de la Recherche Santé du Québec, Canada.

[c]To whom correspondence should be addressed. Address for correspondence: Metabolic Research Laboratory, Radcliffe Infirmary, Woodstock Road, Oxford OX2 6HE, England.

METHODS

The isolation of rat cardiac ventricular myocytes and the measurement of $[Ca^{2+}]_i$ in bulk suspensions of these cells by fura-2 have been described previously.[1,3] In single cell experiments, 100 μl of cell suspension (5.0 mg dry wt · ml^{-1} buffer) were mixed with 900 μl of oxygen-saturated Hepes-Krebs-Henseleit buffer ($CaCl_2$ = 1.25 mM) that contained 15-20 μM of fura-2/AM (Calbiochem).[4] The whole aliquot was transferred to a fibronectin-coated glass coverslip (4 mg % w/v). The fura-2 loading period was 20-30 min, and during this time the cells became attached to the coverslip. The coverslip with the loaded cells was washed three times with buffer and transferred to a cell chamber containing 3 ml of the above-named buffer, which was kept at 35 °C. The cell chamber was mounted on the stage plate of an upright Nikon Inverted Diaphot TMD epifluorescence microscope (100X Nikon fluor objective, Nippon Kogaku) and illuminated with an alternating 340-nm and 380-nm light by a spinning wheel (100 Hz) fluorometer.[5] The emitted light was filtered with a broad bandpass filter (470-600 nm). Changes in cell fluorescence were measured with a photomultiplier coupled to an IBM-AT clone computer with a CLAB data acquisition system (Indec System, Sunnyvale, CA). Data were collected at 10 points · sec^{-1} with averaging (64 values per point). The Ca^{2+} measurements were performed on single cells of average size and healthy appearance (rod shaped with no blebs) that were isolated within the field of view and that had an average fluorescence intensity. The fura-2 signal was recorded at excitation wavelengths of 340 nm and 380 nm (indicating the Ca^{2+}-bound fura-2 and free fura-2, respectively) to assess changes in fluorescence reflecting changes in $[Ca^{2+}]_i$. Results were expressed as $[Ca^{2+}]_i$ in nM after direct calibration in individual cells. Calibration was carried out by increasing $[Ca^{2+}]_o$ to 6.25×10^{-3} M, which allowed rapid saturation of intracellular fura-2 with Ca^{2+} upon addition of ionomycin (final concentration: 2×10^{-5} M) (R_{max}).[4] The R_{min} value[4] was obtained by adding EGTA (final concentration: 2×10^{-2} M) to the system after alkalinization to pH 8.3 with Tris base.

Compounds to be tested, that is, isoproterenol, forskolin, BAY K 8644, KCl, and ATP (ATP either alone or mixed with tetrodotoxin), were injected into the cell chamber at low pressure (<3 psi) through a micropipette (outlet diameter: 1.5 μm), positioned about 20 μm from the surface of the cell to be studied (injection system, Multi-Channel Picospritzer II, General Valve, CA). Field electrical stimulation was carried out using platinum wires (thickness: 25 μm) placed in the cell chamber and a Grass SD9 electrical stimulator (100 V, 10 msec, stimulation rate variable, controlled manually).

In bulk cell suspension studies, the compounds being studied, that is, ATP (sodium, Tris, or magnesium salts), KCl, caffeine, and tetrodotoxin, were injected in appropriate amounts into the cell cuvette to give the final concentration required. Changes in fluorescence were also continuously monitored in single myocytes or bulk suspensions of myocytes incubated in buffer containing bisoxonol[1] (DiSBa-C_2(3), Molecular Probes, Eugene, OR) (1×10^{-6} M-1×10^{-5} M), using the fluorometer described above equipped with an excitation interference filter centered on 540 nm and an emission filter centered on 580 nm. Bisoxonol fluorescence in suspensions of gramicidin (200 nM)-treated ventricular myocytes incubated in buffer containing varying Na^+ concentrations (Na^+ being replaced to a varying degree by *N*-methyl-*D*-glucamine) was recorded and plotted, as shown in FIGURE 1, as a function of calculated transmembrane potential values (E_M) according to the following equation:

$$E_M = -59 \log [K^+ + Na^+]_i / [K^+ + Na^+]_o$$

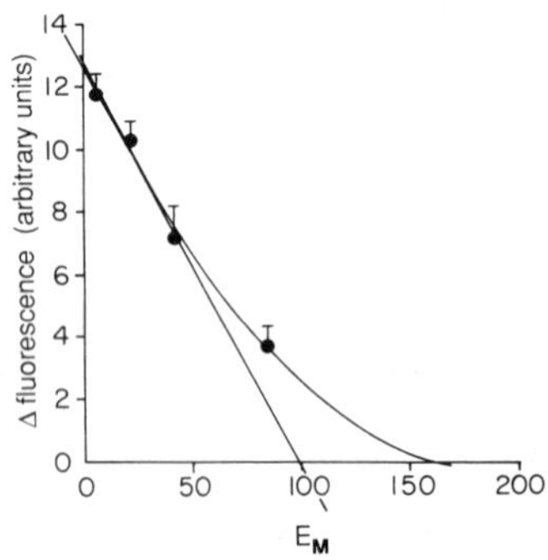

FIGURE 1. Bisoxonol fluorescence from gramicidin (200 nM)-treated cardiac ventricular myocytes in suspensions plotted as a function of transmembrane potential (E_M) (calculated values). Suspended cardiac ventricular myocytes (2.5 mg dry weight · mg^{-1}) were equilibrated in buffer containing bisoxonol (1×10^{-5} M). The Na^+ of the incubation buffer was varied by replacing it with *N*-methyl-*D*-glucamine of equimolar concentrations. Bisoxonol fluorescence was recorded in arbitrary units (Δf). The E_M was calculated as described in Methods. The straight line is the first-order polynomial (least-squares fit regression line) for the first three plotted data points ($\Delta f = -0.10579t + 12.233$; $r^2 = 0.984$); the curve is a second-order polynomial ($\Delta f = 12.684 - 0.14115t + 3.8845 \cdot 10^{-4} x^2$; $r^2 = 0.991$) for all four plotted data points. The plotted data points represent $\bar{x}$ + SEM (N = 5-9). According to this calibration method, the E_M of cardiac ventricular myocytes in suspension in the presence of extracellular ATP (1×10^{-4} M) (magnesium salt) was estimated at -58 ± 11 mV (SEM, N = 4) (E_M being assumed to be between -85 mV and -95 mV in these cells at rest).

This procedure, which has been described by Breuer *et al.*,[6] allowed calibration of the transmembrane potential and quantitative assessment of ATP-induced depolarization.

RESULTS AND DISCUSSION

In the present study, we have monitored $[Ca^{2+}]_i$ in fura-2-loaded cardiac ventricular myocytes at rest, during electrical stimulation, and during and after extracellular exposure to ATP. As shown in FIGURES 2 & 3, a considerable heterogeneity of $[Ca^{2+}]_i$ response to extracellular ATP (1×10^{-4} M) was observed in experiments using bulk suspensions of cells as well as in those using single cells. Extracellular ATP-induced increases in $[Ca^{2+}]_i$ were usually transient. In some single cells (FIG. 4), however, the $[Ca^{2+}]_i$ sometimes remained high despite the fact that ATP was applied for a limited time only (usually 30 or 60 sec), and was subsequently diluted away after cessation of agonist puffing (as could be confirmed in dummy experiments with trypan blue in the pipette). In both cell preparations, the initial rate of $[Ca^{2+}]_i$ rise after ATP stimulation was usually fast and was followed by a further increase with a slower rate of rise (FIGS. 2A-2D). Sometimes, however, the opposite could happen (FIGS. 2E, 2F & 3B). In 36 bulk cell suspension experiments, the "time-to-peak" of the $[Ca^{2+}]_i$ response to extracellular ATP stimulation was estimated at 14.2 ± 1.2 sec (SEM; range: 5.9-23.0 sec). The rate of decay of the extracellular ATP-induced transient also varied greatly, the $t_{0.5\text{-off}}$ being 33.2 ± 3.9 sec (SEM; N = 24; range: 7.5-99.0 sec). Variations in the $[Ca^{2+}]_i$ response can be explained, in part, by the heterogeneity of the cell population.[7] Some cells, for example, can be intolerant to extracellular Ca^{2+} at the concentration of Ca^{2+} used in the incubation buffer (1.25 mM). This may result in elevated $[Ca^{2+}]_i$ and various states of contractility, as can easily be confirmed by light microscopy. These cells tended to respond to extracellular

ATP in a sluggish way. In single cell experiments such cells could be avoided by selecting for long, rod-shaped, noncontracted cells. According to our observations, however, these morphological criteria do not guarantee a homogeneity of cell response since apparently normal looking rod-shaped myocytes can respond to external stimuli like ATP in a very different manner. Sometimes the $[Ca^{2+}]_i$ response is multiphasic, and the main effect on $[Ca^{2+}]_i$ may occur late or after the cessation of injection of the agonist (FIG. 2F). Of course, variation in experimental conditions cannot be fully excluded, such as variation in injection pressure, diameter of the micropipette, or position of the cell relative to the injection pipette. Another factor that might have caused variability in cell response relates to the time of use of the freshly prepared myocytes. We put great effort into maintaining our fresh cells by frequently changing the oxygenated suspension buffer (every 5 min) and by keeping the cells well oxy-

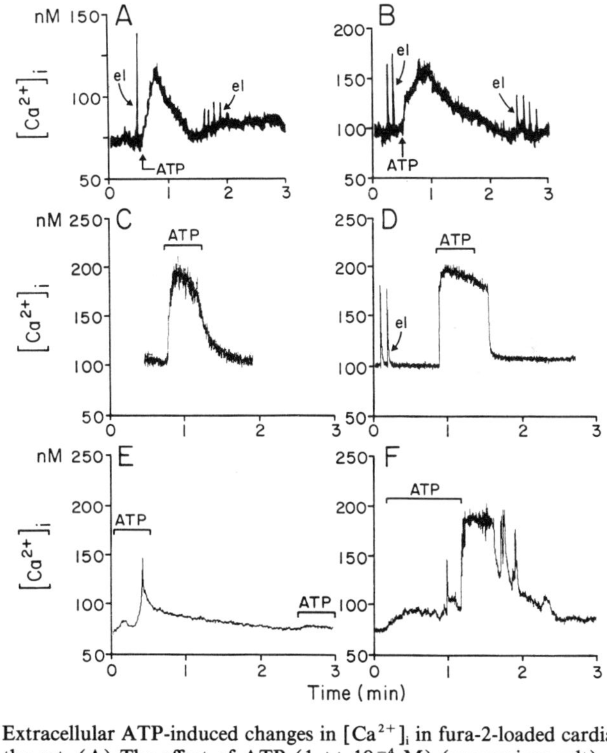

FIGURE 2. Extracellular ATP-induced changes in $[Ca^{2+}]_i$ in fura-2-loaded cardiac ventricular myocytes of the rat. (**A**) The effect of ATP (1×10^{-4} M) (magnesium salt) and electrical stimulations in bulk suspension of myocytes incubated in normal buffer. (**B**) The effect of ATP (1×10^{-4} M) (Tris salt) and electrical stimulations in bulk suspensions of myocytes incubated in buffer where Na^+ had been replaced by an equimolar concentration of Li^+. (**C-F**) The effect of ATP (magnesium salt) on $[Ca^{2+}]_i$ in single cardiac ventricular myocytes. In the single cell experiments ATP (1×10^{-4} M) was injected into the buffer close ($\sim$20 μm) to the cell being studied (injection time 30 sec or 60 sec, as marked on the figures). Note the desensitization to a second addition of ATP in experiment E, and the suppressed response to electrical stimulations after addition of ATP in experiments **A** and **B**. Note also that replacing Na^+ (**A**) by Li^+ (**B**) in the incubation buffer had little effect on the extracellular ATP-induced increase in $[Ca^{2+}]_i$.

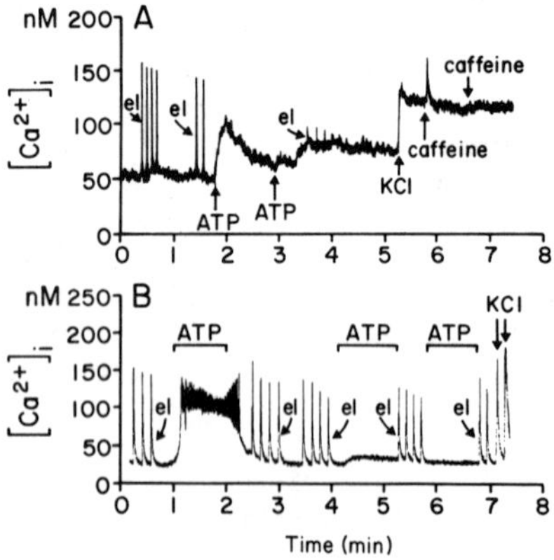

FIGURE 3. Effects of repeated additions of ATP and electrical stimulations on $[Ca^{2+}]_i$ in cardiac ventricular myocytes. (**A**) The effect of ATP (1×10^{-4} M) (sodium salt) (applied twice) on suspension of fura-2-loaded ventricular myocytes; also, the effect of a subsequent addition of KCl (4×10^{-2} M) and caffeine (5×10^{-2} M) (twice) (cytosolic concentration of fura-2: 46 μM). (**B**) The effect of repeated injections (60 sec) of ATP on a single ventricular myocyte (sodium salt, applied at 1×10^{-4} M); also, the effect of subsequent injections of KCl (2×10^{-2} M) (2-sec injections, twice). Note the lack of suppression of electrically induced $[Ca^{2+}]_i$ transients following injection of ATP in the single cell (**B**). Unfortunately, electrical stimulation was not carried out during the actual injection of ATP in the present study, but this might have revealed ATP-induced suppression of the electrical transients.

genated. Despite these efforts, we observed a progressive decline in response and an increased variability in response pattern of the myocytes to ATP or electrical stimulation 3-4 hr after the cells were prepared (data not shown). We have therefore limited our experiments to that time frame.

Resting levels of $[Ca^{2+}]_i$ were usually estimated at between 50 nM and 100 nM in both single cells and in bulk suspensions of cells. Upon electrical stimulation, or stimulation by extracellular ATP, $[Ca^{2+}]_i$ usually did not rise above 200-300 nM, which is in agreement with recently published studies.[8,9] In spite of these low $[Ca^{2+}]_i$ values,[7] electrical stimulation, or stimulation by extracellular ATP, caused cell contraction as confirmed by light microscopy. In bulk suspensions of cells desensitization toward repeated extracellular addition of ATP rapidly developed. Responses to subsequent electrical stimulation were also attenuated (FIGS. 2A & 2B). Similar results were usually observed in single cells. However, no suppression of electrical stimulation was observed in some of the single cells following addition of ATP, in spite of full suppression of the response to a second addition of extracellular ATP (FIG. 3B). Repeated additions of ATP also resulted in an inability of extracellular ATP to

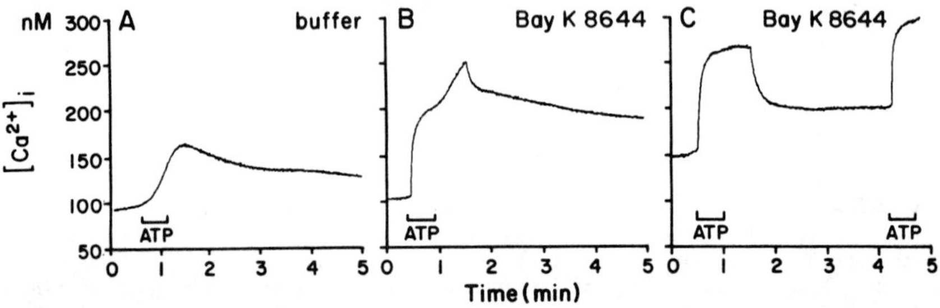

FIGURE 4. The effect of the 1,4-dihydropyridine calcium agonist BAY K 8644 on the extracellular ATP-induced $[Ca^{2+}]_i$ increase in a fura-2-loaded single cardiac ventricular myocyte. (**A**) The cell incubated in buffer alone. (**B & C**) The same cell as in **A** incubated in buffer containing BAY K 8644 (5×10^{-6} M). ATP (1×10^{-4} M) (magnesium salt) was injected for 30 sec each time, in **B**, about 5 min after the addition of BAY K 8644, in **C**, about 12 min and about 16 min after the addition of BAY K 8644.

depolarize the myocytes (increase bisoxonol fluorescence) (FIG. 5A).[1] The observation that tetrodotoxin[10,11] blocks the effect of extracellular ATP, not only on the $[Ca^{2+}]_i$,[1] but also on the transmembrane potential (FIGS. 5B & 5D-5F), suggests that fast Na^+ channels of the sarcolemma are involved in the effect of extracellular ATP on $[Ca^{2+}]_i$ in ventricular myocytes, and that transmembrane depolarization is a necessary causal

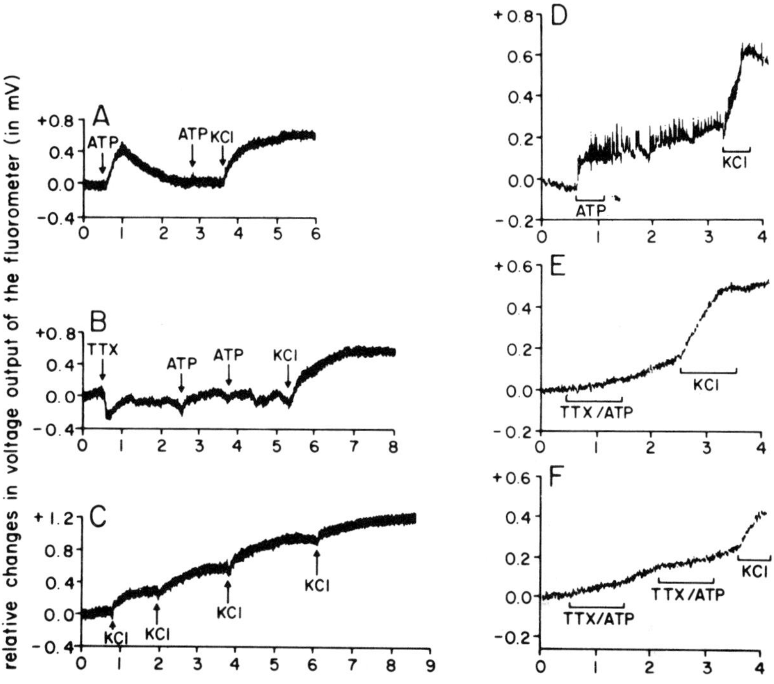

FIGURE 5. Effects of extracellular ATP (magnesium salt) and KCl on bisoxonol fluorescence in cardiac ventricular myocytes of the rat. (**A**) The effect of ATP (1×10^{-4} M) (applied twice) on suspension of myocytes, followed by addition of KCl (2×10^{-2} M). (**B**) The effect of ATP (1×10^{-4} M) (applied twice) on suspension of myocytes pretreated with tetrodotoxin (TTX) (5×10^{-5} M), followed by addition of KCl (2×10^{-2} M). (**C**) The effect of repeated additions of KCl (5×10^{-3} M) on cell suspension of myocytes (cumulative concentration response study). (**D**) The effect of ATP (1×10^{-4} M) on a single ventricular myocyte (applied for 30 sec), followed by addition of KCl (2×10^{-2} M) for 30 sec. (**E**) The effect of a simultaneous addition of TTX (2×10^{-4} M) and ATP (1×10^{-4} M) on a single ventricular myocyte (applied for 60 sec), followed by addition of KCl (2×10^{-2} M for 60 sec). (**F**) The effect of a simultaneous addition of TTX (2×10^{-4} M) and ATP (1×10^{-4} M) on a single cardiac ventricular myocyte (applied twice, for 60 sec each time), followed by addition of KCl (2×10^{-2} M for 30 sec). TTX and ATP were added together as a mixture.

factor in the mechanism of this effect. In the presence of 1,4-dihydropyridine Ca^{2+} agonist BAY K 8644 (5×10^{-6} M),[12–14] suppression of the extracellular ATP-induced $[Ca^{2+}]_i$ increase was not observed in single cells upon repeated additions of ATP (FIG. 4). In fact, a second addition of ATP resulted in a further increase in the $[Ca^{2+}]_i$ (FIG. 4A-4C). Also, pretreatment of single myocytes with isoproterenol or forskolin, compounds with strong positive inotropic effects on the heart,[3,15] prevented ATP-induced suppression of electrically induced $[Ca^{2+}]_i$ transients (data not shown). This indicates that the ATP-induced increase in $[Ca^{2+}]_i$ results from Ca^{2+} influx through dihydropyridine-sensitive Ca^{2+} channels,[13] and that the prolonged open time of these channels in the presence of BAY K 8644[12] can compensate for the suppressive effect of repeated additions of ATP. Phosphorylation of these channels, induced by isoproterenol or forskolin, may also increase the Ca^{2+} influx through these channels during depolarization of the sarcolemma[11,13] induced by extracellular ATP.

REFERENCES

1. BJÖRNSSON, Ó. G., J. R. MONCK & J. R. WILLIAMSON. 1989. Identification of P_{2Y}-purinoceptors associated with voltage-activated cation channels in cardiac ventricular myocytes of the rat. Eur. J. Biochem. **186:** 395-404.
2. BURNSTOCK, G. & C. KENNEDY. 1985. Is there a basis for distinguishing two types of P_2-purinoceptors? Gen. Pharmacol. **16:** 433-440.
3. BJÖRNSSON, Ó. G. & J. R. WILLIAMSON. 1988. Prolonged suppression of coronary flow rate and cardiac work induced by leukotriene D_4, and the reversal of this effect by activators of adenylate cyclase. Ann. N.Y. Acad. Sci. **524:** 75-90.
4. GRYNKIEWICZ, G., M. POENIE & R. Y. TSIEN. 1985. A new generation of Ca^{2+} indicators with greatly improved fluorescence properties. J. Biol. Chem. **260:** 3440-3450.
5. CHANCE, B., V. LEGALLAIS, J. SORGE & N. GRAHAM. 1975. A versatile time-sharing multichannel spectrophotometer, reflectometer, and fluorometer. Anal. Biochem. **66:** 498-514.
6. BREUER, W. V., E. MACK & A. ROTHSTEIN. 1988. Activation of K^+ and Cl^- channels by Ca^{2+} and cyclic AMP in dissociated kidney epithelial (MDCK) cells. Pflügers Arch. (Eur. J. Physiol.) **411:** 450-455.
7. WIER, W. G., M. B. CANNELL, J. R. BERLIN, E. MARBAN & W. J. LEDERER. 1987. Cellular and subcellular heterogeneity of $[Ca^{2+}]_i$ in single heart cells revealed by fura-2. Science **235:** 325-328.
8. MONCK, J. R., K. KOBAYASHI, Ó. G. BJÖRNSSON & J. R. WILLIAMSON. 1988. Measurement of spontaneous and electrically stimulated Ca^{2+} transients in single rat ventricular myocytes. *In* Biology of Isolated Adult Cardiac Myocytes. W. A. Clark, R. S. Decker & T. K. Borg, Eds.: 328-331. Elsevier. New York, NY.
9. CALLEWAERT, G., L. CLEEMANN & M. MORAD. 1988. Epinephrine enhances Ca^{2+} current-regulated Ca^{2+} release and Ca^{2+} reuptake in rat ventricular myocytes. Proc. Natl. Acad. Sci. USA **85:** 2009-2013.
10. BAER, M., P. M. BEST & H. REUTER. 1976. Voltage-dependent action of tetrodotoxin in mammalian cardiac muscle. Nature **263:** 344-345.
11. CATTERALL, W. A. 1988. Structure and function of voltage-sensitive ion channels. Science **242:** 50-61.
12. SCHRAMM, M., G. THOMAS, R. TOWART & G. FRANCKOWIAK. 1983. Novel dihydropyridines with positive inotropic action through activation of Ca^{2+} channels. Nature **303:** 535-537.
13. SEAGAR, M. J., M. TAKAHASHI & W. A. CATTERALL. 1988. Molecular properties of dihydropyridine-sensitive calcium channels. Ann. N.Y. Acad. Sci. **522:** 162-175.
14. BJÖRNSSON, Ó. G., K. KOBAYASHI & J. R. WILLIAMSON. 1988. Modulation of coronary flow rate and cardiac contractility by the divalent cation ionophore A23187 and inhibitors of the cyclooxygenase and 5-lipoxygenase pathways: Development of heterogeneous patterns of myocardial ischemia. Naunyn-Schmiedeberg's Arch. Pharmacol. **337:** 191-202.
15. BJÖRNSSON, Ó. G., K. KOBAYASHI & J. R. WILLIAMSON. 1987. Inducers of adenylate cyclase reverse the effect of leukotriene D_4 in isolated working guinea pig heart. Am. J. Physiol. **252:** H1235-H1242.

Index of Contributors

(Italic page numbers refer to comments made in discussion)